Multidisciplinary Approach of Musculoskeletal Metastasis

Juan Pretell-Mazzini
Santiago A. Lozano Calderón
Editors

Multidisciplinary Approach of Musculoskeletal Metastasis

Current Trends and Future Directions

Springer

Editors
Juan Pretell-Mazzini
Division of Musculoskeletal Oncology -
Miami Cancer Institute
Baptist Health of South Florida / Herbert
Wertheim College of Medicine - Florida
International University
Miami, FL, USA

Santiago A. Lozano Calderón
Division of Musculoskeletal Oncology
Department of Orthopedic Surgery /
Massachusetts General Hospital - Harvard
Medical School
Boston, MA, USA

ISBN 978-3-032-09497-1 ISBN 978-3-032-09498-8 (eBook)
https://doi.org/10.1007/978-3-032-09498-8

This Springer imprint is published by the registered company Springer Nature Switzerland AG
The registered company address is: Gewerbestrasse 11, 6330 Cham, Switzerland

This book is dedicated first and foremost to my family. To my wife, Juliana (Juju), whose love, energy, and unwavering support keep me moving forward—without her understanding and encouragement, none of my dreams would have been possible. To my children—George, Julia, Giovanna, Gianluca, and Emma—whose patience, love, and joy have sustained me throughout every stage of my personal and professional journey.

I also hold close in my heart my parents, Mariella and Augusto, whose values and guidance remain with me always; my sister, Mariella, who showed me that you can always make things better; my aunt Gina, a constant source of support; and in loving memory of my Nonna, whose strength and example have been the inspiration behind so much of what I have accomplished and whose memory continues to guide me every day.

I dedicate this work as well to the memory of my uncle Gianfranco, whose brilliant mind and extraordinary intellect left a profound impression on me. He taught me that the value of a person is not measured solely by outward success but by the depth of their thinking, their curiosity, and their integrity. His example continues to remind me of the importance of intellectual honesty and rigor in everything I do.

To the memory of my uncle David, whose life was defined by loyalty, kindness, and devotion to family. Married to my aunt Gina, he embraced my sister and me as if we were his own children. He remained a unifying force who kept his family together through love, generosity, and quiet strength. His legacy lives on in the family he cherished so deeply.

Finally, to my cousins—Marianna, David, and Juan Francisco (Pancho)—who have always been more like siblings than cousins. I have learned just as much from them: about resilience, shared history, and the enduring bonds of family. Their affection and trust have been a constant source of motivation and joy throughout my life.

Beyond my family, I extend this dedication to my mentors, whose wisdom and guidance have profoundly shaped my career; to my colleagues and friends from medical school and residency, who enriched my formation through their collaboration and camaraderie; and most importantly, to my patients, whose courage and perseverance in the face of illness continue to inspire me and give meaning to the work reflected in these pages.

—Juan Pretell-Mazzini, MD

This book is not an exception to the common denominator of all the meaningful milestones in my career, my wife, Kate, whose love, unconditional support, understanding, motivation, and generosity make everything possible. This is not my work but our work. I dedicate this book to her, to my children—Santiago, Juan Sebastián, Mateo, Nicolás y Ana Lucia; my parents, Santiago and Marta; my aunt and second mother, Coco; my sisters, Adriana, Catalina, and Maria José; and my niece and nephews, Juliana, Gabriel, Alejandro y Andrés. You all always have believed in me and my dreams, your unlimited love always gives me the inspiration, the energy, and the motivation to get and keep going.

I dedicate this book to my teachers and mentors and to all of those that have shaped my professional career with wisdom, good example, and their passion. Also, to my patients and their families, especially those that have died. Their resilience, love for life, and tenacity are the main driver and motivation to write this book.

Lastly and more importantly, I humbly offer this work to the good Lord. Without him nothing starts, happens, or ends. "Pray as though everything depended on God. Work as though everything depended on you" (St Augustin). Ad Majorem Dei Gloriam.

—Santiago A. Lozano Calderón, MD, PhD

Foreword

Skeletal metastatology is a scientific and medical discipline that provides foundation for comprehensive state-of-the-art oncological, pharmacological, biological, orthopedic, and interventional care through rigorous scientific investigation and technological advances.

Medical oncological care has been rapidly evolving from traditional cytotoxic drug therapies to more targeted therapies in an aim to decrease cancer burden if at all possible. It is a common practice to switch oncologic drugs to kill heterogeneous populations of cancer cells unlike the era of using a few choices of cytotoxic drugs. Radiation therapy has been commonly employed for skeletal metastases. Anti-resorptive agents such as bisphosphonates or denosumab have been given for skeletal metastases. Despite all these available therapies, patients with advanced cancers still develop skeletal metastases, pain, and pathological fractures.

Most orthopedic surgeons use surgical stabilization techniques and implants that are designed for trauma surgeries or arthroplasties. Reduction or killing of cancer cells is not always considered in intramedullary nailing or reconstruction of peri-articular pathological fractures. Intramedullary reaming or insertion of an antegrade long intramedullary nail may mechanically transport radiation-resistant or drug-resistant cancer cells distally. Likewise, retrograde nails can spread cancers in the joint and proximal bone.

In an era of molecular medicine and precision cancer therapies, it becomes important for orthopedic surgeons to understand basic bone biology, cancer-induced bone loss, and biomechanics to provide most scientific evidence-based optimal care. Likewise, the main goal of orthopedic care should facilitate life-saving systemic therapies without causing prolonged recovery or complications that may delay systemic cancer therapies. It would be ideal to consider local cancer killing by ablation or modifying local bone environment to make up for cancer-induced bone loss.

This book will build a common ground for many scientists and care providers for modern oncologic care in 2020s and beyond. I hope this first edition is updated by a second edition in few years with more updated knowledge and newer technologies to enhance care in patients with musculoskeletal metastases.

Francis Y. Lee

Wayne O. Southwick Professor,
Department of Orthopaedics
and Rehabilitation
Yale University School
of Medicine
New Haven, CT, USA

President of the Musculoskeletal
Tumor Society (2025–2026)
Rosemont, IL, USA

Board Member, International Society
of Limb Salvage (ISOLS)
Vienna, Austria

Preface

Musculoskeletal metastasis remains one of the most challenging conditions encountered in modern oncology and orthopedics. Its management requires not only technical expertise but also the integration of multiple disciplines—medical oncology, radiation oncology, orthopedic surgery, pathology, radiology, palliative care, and ethics—working collaboratively to provide patients with the most effective and compassionate care. As advances in systemic therapies continue to prolong survival, the demand for improved quality of life and preservation of function in patients with skeletal metastases has never been greater.

This book, *Multidisciplinary Approach of Musculoskeletal Metastasis: Current Trends and Future Directions*, was conceived to address this pressing need for a comprehensive and integrative perspective. The volume is structured into sections that explore the medical and surgical aspects of care, while also dedicating important attention to ethical considerations that often accompany complex decision-making in this patient population. Each chapter has been authored by experts in the field, bringing forth both evidence-based knowledge and clinical experience to guide readers through the latest advances and ongoing debates in practice.

Our goal is to provide a resource that is not only scientifically rigorous but also clinically practical. We aim to assist physicians, surgeons, allied health professionals, and trainees in understanding the diverse elements involved in the care of patients with musculoskeletal metastases. The multidisciplinary lens through which this book has been developed reflects the very reality of patient care today—where collaboration and communication across specialties define the standard of excellence.

We would like to express our gratitude to all contributing authors for their dedication and expertise. Their willingness to share knowledge and insight ensures that this work will serve as a reference and guide for years to come. We also extend our appreciation to the patients and families whose resilience inspires our efforts and reminds us of the profound human dimension behind every medical decision.

It is our hope that this book not only provides a foundation for current practice but also sparks dialogue, research, and innovation to shape the future directions of musculoskeletal metastasis management.

Miami, FL, USA — Juan Pretell-Mazzini
Boston, MA, USA — Santiago A. Lozano Calderón

Acknowledgments

We would like to express our deepest gratitude to the many individuals whose support made this book possible. We are indebted to our contributing authors, who generously shared their time, expertise, and insights to create a comprehensive and multidisciplinary perspective on the management of musculoskeletal metastasis. We thank our mentors and colleagues, whose guidance and collaboration have continually enriched our professional development. We are also grateful to the editorial and publishing teams for their invaluable assistance throughout the preparation of this volume. Finally, we acknowledge our patients and their families, whose resilience and trust inspire our ongoing commitment to advancing knowledge and improving care in this challenging field.

Contents

Part I General Concepts

1 Overview of Musculoskeletal Metastases ... 3
Lucy L. Hederick, Joseph J. Connolly, Marcos R. Gonzalez, and Juan Pretell-Mazzini

2 Understanding Musculoskeletal Metastases ... 13
Rayna S. Kuthiala, Marcos R. Gonzalez, Carolina Mendez-Guerra, and Juan Pretell-Mazzini

3 Multidisciplinary Treatment of Musculoskeletal Metastases ... 29
Marcos R. Gonzalez, Carolina Mendez-Guerra, and Juan Pretell-Mazzini

Part II Patient Assessment and Diagnostic Approach

4 Patient Assessment and Selection ... 45
Jacob Kopeld, Jorge Valverde, and Juan Pretell-Mazzini

5 Nutritional Evaluation of Patients with Metastatic Disease ... 69
Berenice De La Mota, Karla Pfaeffle-Palomo, and Daniel Otero-Pfaeffle

6 Radiological Assessment: From Plain Radiographs to Advanced Cross-Sectional Studies ... 87
James A. Strahan and Zachary E. Stewart

7 Obtaining Tissue Samples: What Is the Best Approach? ... 103
Felipe Souza, Erik J. Geiger, and Ty K. Subhawong

8 Pathological Assessment of Metastatic Disease to the Musculoskeletal System ... 117
Julio A. Diaz-Perez and Andrew E. Rosenberg

9 Has My Cancer Spread? – Staging Strategies ... 131
Lukas M. Nystrom

10 I Am Not Crazy: Why Do I Need to See a Mental Health Specialist? 139
Elana Monchar, J Wittig, and J Valverde

Part III Principles of Surgical Intervention

11 Overview of Surgical Indications and Outcomes 149
Varna R. Kanapuram and Jonathan A. Forsberg

12 "Oops...This Is About to Break!": Impending Pathologic Fractures of Long Bones 163
Joseph J. Connolly, Neha Jejurikar, Sarah Ballatori, Megan H. Goh, and Santiago A. Lozano Calderón

13 Management of Spinal Metastatic Disease 181
Spencer M. Richardson, Zayd Almaya, and Motasem Al Maaieh

14 Management of Pelvic Metastasis – Minimally Invasive Pelvic Surgery 201
Barlas Goker, Joseph J. Connolly, Jichuan Wang, Bang H. Hoang, David S. Geller, and Rui Yang

15 Management of Pelvic Metastasis: Open Surgical Techniques 217
Herrick J. Siegel

16 Management of Shoulder Gridle Metastatic Disease 235
Alexandra M. Arguello, Leilani Garayua Cruz, Jonathan D. Barlow, and Matthew T. Houdek

17 Management of Diaphyseal Humeral Metastases 253
Sarah Ballatori, Lucy Hederick, Megan Goh, Neha Jejurikar, Joseph Benevenia, and Santiago A. Lozano Calderón

18 Management of Metastatic Disease Around the Elbow 269
Niveditta Ramkumar and Dipak B. Ramkumar

19 Management of Metastatic Disease Around the Forearm, Wrist, and Hand 281
Melinda J. Choi and Seth D. Dodds

20 Management of Metastatic Disease to the Proximal Femur 295
Costantino Errani

21 Management of Metastatic Disease to the Femoral Diaphysis 307
Wesley M. Durand and Adam S. Levin

22 Management of Metastatic Disease Around the Knee 323
Marcos Galli Serra, Bruno Terrarrossa Retegui, and Walter Parizzia

23 Management of Metastatic Disease to the Tibial Shaft 349
Jestin Williams, Irfan A. Khan, Christopher Bloise, and Matthew G. Cable

24 Management of Metastatic Disease to the Foot and Ankle 369
Taha Aksoy and H. Thomas Temple

Part IV Principles of Medical Treatment

25 Medical Management of Musculoskeletal Metastasis from Breast Cancer 389
Naomi Dempsey, Yolcar Chamorro, Carlos Hernandez, Priya Bhatt, and Reshma Mahtani

26 Medical Management of Musculoskeletal Metastasis from Prostate Cancer 403
Chinmay T. Jani and Janaki Neela Sharma

27 Medical Management of Musculoskeletal Metastasis from Lung Cancer 427
Gavin Wayne Stoel, Erik Geiger, and Estelamari Rodriguez

28 Medical Management of Musculoskeletal Metastasis from Renal Cell Carcinoma 443
Palash Asawa and Jaime Rafael Merchan

29 Medical Management of Musculoskeletal Metastasis from Thyroid Carcinoma 457
Cesar Ochoa

30 Medical Management of Multiple Myeloma: An Orthopedic Blood Cancer 467
Yonatan Najman-Licht and James E. Hoffman

31 Radiation Oncology Strategies for the Management of Musculoskeletal Metastasis 479
Matthew D. Hall, Keitrina Mair, and Adam L. Holtzman

Part V Rehabilitation and Post-operative Care

32 Physical Therapy Strategies 501
Kathryn Quaglia, Greer Gilbert, and David Crandell

33 Pain Management Protocols 513
Rafael Vazquez, Brinda Kamdar, and Ana Gutierrez-Hasbach

34 Multidisciplinary Care Clinics for Patients with Metastatic Bone Disease 539
Adrian Torres, Carl Quinion, and John Alexander

Part VI Oligometastatic Disease

35 Metastasectomy for Musculoskeletal Oligometastases—Surgical Point of View 555
Yvette Ho

36 Principles of Surgical Management: Ablative Techniques for Skeletal Metastases 563
Andrew D. Kleven and John C. Neilson

37 Radiation Strategies in Oligometastatic Disease 589
Armando Vera, Lauren Nagel, and Raphael Yechieli

Part VII Especial Scenarios

38 Skeletal Metastasis of Unknown Primary: Now What Should I Do? 605
Mohamad Cherry and Michael P. Rogers

39 Isolated Lytic Lesions: Where Is This Coming from? 621
Alysia K. Kemp and Michelle Ghert

40 Soft Tissue Metastases from Carcinomas and Sarcomas: What Is This About? 631
Javier Delgado Obando

41 Post-Radiation Pathological Fractures - this Bone is not Healing !!! 643
Nicole Butler and Sean P. Kelly

42 Ethical Considerations in Musculoskeletal Metastasis Management 655
Kyle E. Walker, J. B. Smith, and R. Lor Randall

43 Comprehensive Percutaneous Intervention for Skeletal Metastases: Ablation-Osteoplasty-Reinforcement-Internal Fixation (AORIF) 667
Francis Y. Lee

Part VIII What Comes After Treatment and Future Directions?

44 Survivorship and Quality of Life 683
Tanios Dagher, Matthew Hess, and Tessa Balach

45 Future Directions in Multidisciplinary Musculoskeletal Oncology . . . 699
Timothy A. Damron

Index 713

Part I

General Concepts

Overview of Musculoskeletal Metastases

1

Lucy L. Hederick, Joseph J. Connolly, Marcos R. Gonzalez, and Juan Pretell-Mazzini

Introduction

As medical technology and cancer treatments improve, oncologic patients are living longer with a subsequent increase in the prevalence of metastatic disease. Although treatment is increasingly curative, metastatic disease remains the leading cause of cancer-related death [1]. Musculoskeletal metastases, or cancer that has spread from its original location to bones, muscles, cartilage, or other soft tissues of the musculoskeletal system, are no exception to this trend. Once cancer spreads to bone, it is largely incurable and management is traditionally palliative rather than curative in nature. Therefore, clinicians and scientists have a significant interest in improving the treatment and prevention of musculoskeletal metastatic disease [2].

Metastatic cancer cells tend to settle in tissues with high vascularity and consistent blood flow. Consequently, the most common sites of metastasis are the lungs, liver, and bones. Within the musculoskeletal system, skeletal metastases are far more likely than soft tissue metastases [2]. Bone requires constant blood flow and undergoes perpetual remodeling through the activity of osteoclasts and osteoblasts; as such, it is a welcoming environment for the establishment, growth, and survival of malignant cells [3]. On the other hand, blood flow through soft tissues, especially muscles, fluctuates in response to tissue pressure. This variable vascularity combined with soft tissues' comparatively unique microenvironment in terms of pH, metabolites, and temperature helps explain why metastases to soft tissues are rare [4].

L. L. Hederick · J. J. Connolly · M. R. Gonzalez
Division of Orthopaedic Oncology, Department of Orthopaedic Surgery, Massachusetts General Hospital, Harvard Medical School, Boston, MA, USA

J. Pretell-Mazzini (✉)
Division of Musculoskeletal Oncology - Miami Cancer Institute, Baptist Health of South Florida / Herbert Wertheim College of Medicine - Florida International University, Miami, FL, USA
e-mail: juan.pretell@baptisthealth.net

J. Pretell-Mazzini, S. A. Lozano Calderón (eds.), *Multidisciplinary Approach of Musculoskeletal Metastasis*, https://doi.org/10.1007/978-3-032-09498-8_1

Metastases disrupt the normal process of remodeling and homeostasis in bone. They therefore increase patients' risk of experiencing skeletal-related events (SREs), which include hypercalcemia, pathologic fractures, spinal cord compression, radiation to bone, or bone surgery. Just one SRE greatly decreases patients' physical, emotional, and functional well-being [5]. The increasing prevalence of musculoskeletal metastases and the destructive potential of SREs underscore the importance of their prevention and early treatment [6].

Epidemiology

In 2022, roughly 20 million new cases of cancer were diagnosed and 9.7 million cancer deaths occurred worldwide. In 2050, 35 million new cancer diagnoses are predicted worldwide [7]. In the United States, more than two million new cancer diagnosis and 600, 000 cancer deaths were projected to occur in 2024 [8]. An estimated 90% of cancer deaths overall are from metastatic disease [8]. Worldwide, lung cancer accounted for the highest incidence of new cases (2.48 million or 12.4% of new diagnoses), followed by breast, colorectal, and prostate cancers [7]. American men are most likely to be diagnosed with prostate cancer (29% of new diagnoses), followed by lung (11%) and colorectal (8%) cancers, while American women are most likely to be diagnosed with breast (32% of new diagnoses), followed by lung (12%) and colorectal (7%) cancers [8]. Notably, three of the four most common cancers in both men and women worldwide and in the United States have a high propensity to metastasize to bone [2, 8].

In 2012, an estimated 1.5 million patients had bone metastases worldwide [1]. Bone metastasis occurs in at least 100, 000 patients in the United States every year [9]. The projected increase in cancer incidence, combined with an increasing proportion of cancer survivors, indicates a concurrent rise in the incidence of metastatic bone disease [6]. Indeed, between 2010 and 2015, the diagnosis rate of bone metastases increased by 4.6% each year for patients with prostate cancer [10].

Primary bone and soft tissue sarcomas are considered rare cancers. Of newly diagnosed cancer cases in 2024, only 3970 (0.20%) are projected to be bone sarcomas and 13, 590 (0.67%) soft tissue sarcomas [8]. Primary bone and soft tissue sarcomas spread hematogenously, with the most common sites of metastasis being the lungs, bones, and lymph nodes [11, 12].

Musculoskeletal metastases are thus far more common than primary musculoskeletal tumors. Patients with carcinomas, namely, prostate, breast, lung, and renal cancers, most commonly develop bone metastases, with breast and prostate cancer patients accounting for up to 70% of bone metastases [2]. According to a retrospective study analyzing the incidence of bone metastases for 82, 000 patients followed up for up to 10 years after a primary solid tumor diagnosis, 6.9% of the patients were diagnosed with bone metastases over the course of their follow-up [5]. In all, 70% of breast and prostate cancer patients and 40% of lung cancer patients showed signs of metastatic bone disease upon postmortem examination [13]. Bone is the second-most common site of metastasis for primary bone and soft tissue sarcomas.

However, this phenomenon occurs at a comparatively low rate: 4.8% of patients with primary bone sarcomas present with musculoskeletal metastases, and up to 10% of patients with primary soft tissue sarcomas will develop musculoskeletal metastases over the course of their disease [11, 12].

Broadly speaking, the incidence of bone metastases fluctuates markedly based on the type of primary tumor. However, no matter the primary tumor type, patients with more advanced disease are more likely to be diagnosed with bone metastases. The incidence of bone metastases in patients with stage IV cancer increases exponentially in the months following diagnosis. While the risk of bone metastases was decreased in patients with stage I–III disease, the risk of skeletal metastases still increases in an approximately linear manner over time after the primary tumor diagnosis [5].

Despite their rarity, soft tissue metastases should still be acknowledged since they are easily misdiagnosed as soft tissue sarcomas. Treatment and prognosis for the two diagnoses are highly dissimilar, so it is important to make this distinction. In a study from the Ohio State University Medical Center evaluating 118 cases of soft tissue metastases, the most frequent locations of metastasis were the abdominal wall, back, thigh, chest wall, and shoulder. The most common primary tumors (70%) were carcinomas, most of which derived from the lung, breast, and kidney. Malignant melanomas accounted for the next 17% of primary cancers, followed by primary soft tissue sarcomas, which accounted for 9% of soft tissue metastases. In some cases, soft tissue metastases were identified before the primary tumor or misdiagnosed as a primary soft tissue sarcoma [4, 12].

The propensity of some cancers to spread to soft tissues correlates with their overall incidence in the population. Lung carcinoma, for example, is the most prevalent malignancy overall and is the second-most likely cancer to spread to soft tissues [4]. For other primary tumor types, like renal cell carcinomas, the propensity to metastasize to soft tissues is unrelated to the population incidence.

Musculoskeletal metastases carry a significant economic burden as well. Medical costs are more than doubled in patients with bone metastases (approximately $75, 000 per month) compared to counterparts without metastatic bone disease (approximately $31, 000 per month). Aggregated on the national scale, metastatic bone disease accounts for 17%, or $12.6 billion, of the $74 billion of cancer-associated costs in the United States [14].

Pathogenesis

In 1889, Steven Paget published the "seed and soil" hypothesis about cancer metastasis. Paget postulated that cancer cells (the "seeds") have an affinity for friendly organs (the "soil"), eventually taking root and forming metastases [15]. Different primary tumors exhibit different metastatic patterns; for example, breast cancer frequently spreads to the bones, lungs, and brain, while colon cancer metastasizes more commonly to the liver [15]. Interactions between metastatic cancer cells and host organs dictate whether a metastatic lesion will form in a given location.

Although the molecular mechanisms underlying metastasis formation in bone are not well-understood, there must be a "reciprocal interaction" between the bone microenvironment and traveling cancer cells for metastatic lesions to form [16, 17].

Due to their high vascularity, molecular composition, degree of metabolic activity, and constant cellular remodeling, bones are a well-suited soil for traveling cancer cells from multiple different types of primary tumors [18]. Metastasis itself is a complex, stepwise process involving cancer cell detachment from the primary tumor, migration, intravasation, extravasation, and, finally, cell growth. Metastatic lesions begin as micrometastases—small collections of cancer cells that have successfully colonized another part of the body but are too small to be detected by imaging. The transition from micrometastases to full-fledged macroscopic metastases relies on the nature of the bone microenvironment [16, 17].

Metastases colonize red bone marrow before exhibiting effects on a bone's architecture. Thus, bones with a greater supply of red marrow—the vertebrae, the pelvis, and the ends of long bones (Fig. 1.1)—more commonly harbor metastases than bones in the hands and feet, which have comparatively less marrow [18]. There are three types of metastatic bone lesions: osteolytic, osteoblastic, and mixed. Osteolytic lesions, the most common type, cause bone destruction. Multiple myeloma, melanoma, and certain carcinomas like renal, breast, thyroid, and nonsmall cell lung cancers characteristically produce osteolytic bone metastases [2]. In osteolytic lesions, cancer cells overexpress parathyroid hormone-related peptide (PTHrP), either inherently or spurred by the bone microenvironment. PTHrP upregulates osteoclast formation and thus bone resorption and is often associated with hypercalcemia [2, 19]. Additionally, the receptor activator of nuclear factor-κB (RANK) ligand stimulates differentiation of osteoclast precursors into mature osteoclasts, resulting in further bone destruction [2, 20, 21] (Fig. 1.2).

Osteoblastic lesions lead to increased bone formation and are common in prostate cancer, Hodgkin's lymphoma, carcinoid tumors, small cell lung cancer, and medulloblastoma. Bone resorption may or may not occur before deposition of a new

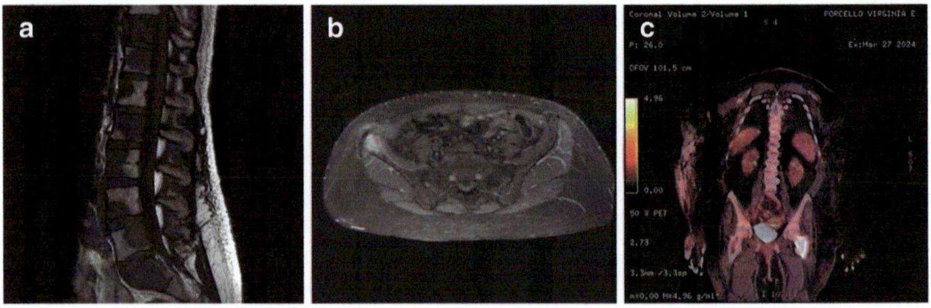

Fig. 1.1 Most common locations of musculoskeletal metastasis. (**a**) A 46-year-old female with diffuse breast metastatic disease to the lumbar and thoracic spine: Sagittal T1 lumbar spine MRI. (**b**) A 75-year-old female with lung metastatic disease to the right iliac bone: Axial T2 pelvic MRI. (**c**) A 74-year-old female with oligometastatic breast carcinoma with left proximal femur metastasis: Coronal PET-CT. PET positron emission tomography

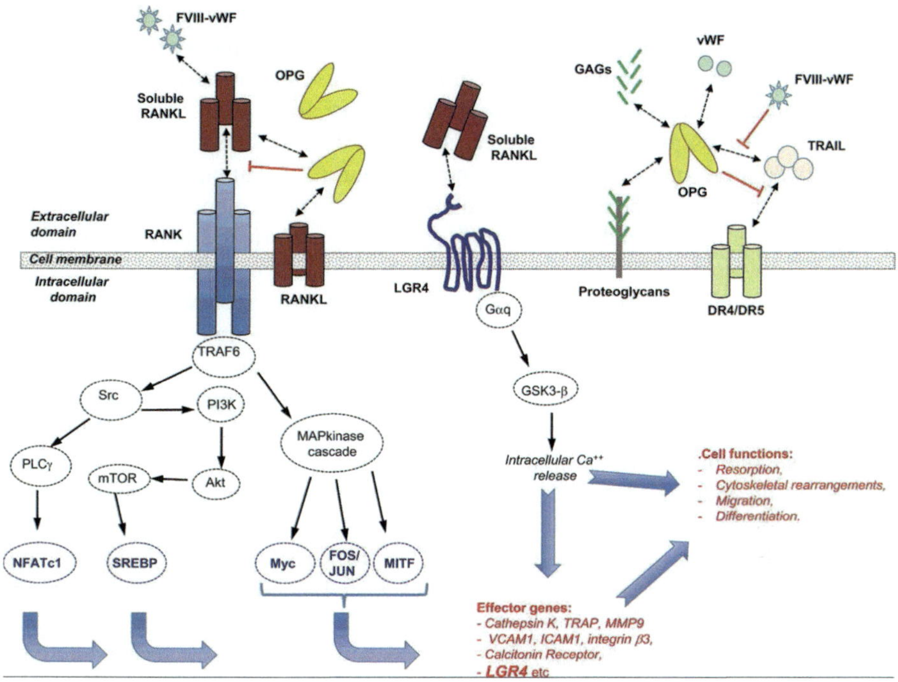

Fig. 1.2 RANK/RANKL signaling in cancer cells. **Notes**: Reprinted from Renema et al. [21] (Copyright © 2016 The Authors. This is an open access journal published by Portland Press Limited on behalf of the Biochemical Society and distributed under the Creative Commons Attribution License 4.0 (CC BY) [21]. RANKL receptor activator of nuclear factor-κB ligand)

bone. Osteoblast activation may be upregulated by the actions of bone morphogenic proteins (BMPs), endothelin-1, and transforming growth factor. Additionally, PTHrP destruction caused by elevated prostate-specific antigen (PSA) may decrease bone resorption [2].

Mixed lesions refer to those with both osteolytic and osteoblastic characteristics and result from three types of carcinomas: breast, gastrointestinal, and squamous cell.

Symptoms and Skeletal-Related Events (SREs)

Bone pain is a major indicator of metastases in cancer patients. Tumors may lead to inflammation through periosteal aggravation, cytokine release, and activation of nerves within the vertebrae. The physical presence of the tumor mass can lead to bone weakening, which causes weight-bearing-associated pain [2]. Pain tends to be constant and can increase in severity over time. Functional limitation may accompany bone metastases, characterized by difficulty in lying down comfortably, sitting, and/or walking [22].

Bone pain that increases with movement and/or weight-bearing can help predict a patient's risk of pathologic fracture. Pathologic fractures in the long bones, pelvis, ribs, and vertebrae are common in oncologic patients, and the risk of a pathologic fracture increases with time after bone metastasis occurs [2].

Spinal cord compression is indicated by the sudden onset of back pain, numbness or weakness in the legs, and/or incontinence [22]. Patients presenting with these symptoms should be treated immediately to avoid permanent nerve damage [2].

Hypercalcemia, a metabolic disorder, results from disrupted bone homeostasis and renal dysfunction [2]. On a basic metabolic panel, hypercalcemia is indicated by a serum calcium >2.6 mmol/L. [22] It is paramount to measure calcium levels with the appropriate correction of serum protein levels. Symptoms may be digestive, renal, or neurologic in nature, ranging from constipation to polyuria to fatigue and confusion. Hypercalcemia may also weaken bones by triggering bone resorption, making them prone to bone pain and pathologic fracture. Untreated hypercalcemia has a median survival of 10–12 weeks [2].

Metastatic bone disease often disrupts bone marrow. Bone marrow aplasia commonly presents with nonspecific symptoms like fatigue, weakness, and shortness of breath [22]. Patients may also exhibit pallor, headaches, bruising, mucosal bleeding, menorrhagia, and fever as a result of low levels of red blood cells, white blood cells, and platelets [23].

Metastases to soft tissue may be confused with primary soft tissue sarcomas since both can present as a mass that may be painful or increasing in size. However, soft tissue metastasis from a primary carcinoma, namely, lung or breast cancer, commonly presents as a painful soft tissue mass [4].

Imaging

Early detection of musculoskeletal metastases is critical to reduce patients' risk of experiencing SREs. Imaging is particularly important to evaluate the extent of the disease and permit accurate oncologic assessment and surgical planning. Plain X-rays are useful as the primary diagnostic technique if a patient begins presenting with symptoms of a metastasis since they are quick and inexpensive. Although plain films have high specificity, their lack of sensitivity means that smaller, early-stage lesions may not be detectable. Additionally, poor contrast in trabecular bone makes medullary lesions harder to spot compared to cortical ones. In those instances where metastases do appear on plain films, osteolytic lesions are darker than healthy bone while osteoblastic lesions are whiter [2].

CT scans are more sensitive than plain radiographs, generating images with high resolution and contrast. Osteolytic and osteoblastic metastases are visible along with bone metastases that have invaded neighboring soft tissues. CT scans should be performed in preparation for a biopsy [2].

Bone scans identify areas of high vascularity and osteoblast activity, reflective of lesions causing inflammation, trauma, or abnormal growth. This method is highly sensitive but has low specificity, leading to a high degree of false positives, so it should be clinically correlated and used in combination with other imaging techniques [2, 22].

MRI is needed to visualize the extent and effect of metastases on soft tissues and bone marrow. This technique has a similar specificity but higher sensitivity than CT. For cancer patients with back pain, MRI should be pursued promptly to diagnose spinal cord compression [22].

Positron emission tomography (PET) scans allow for early detection of bone metastases along with visualization of metastatic cancer through the whole body. Fast-growing lesions from lung and breast cancers are more accurately visualized with PET than slow-growing ones from prostate and renal cancers [2].

Imaging for soft tissue metastases relies more heavily on MRI, CT, and PET imaging, as X-rays and bone scans are ineffective in detecting these lesions. Importantly, a thorough histologic analysis is necessary to rule out primary soft tissue sarcomas when diagnosing metastases to soft tissues; imaging is not sufficient for diagnosis.

For poorly differentiated neoplasms, immunohistochemical stains should be used in addition to a full histopathologic workup. Epithelial, neural, melanocytic, and other markers should be used to help distinguish between a soft tissue sarcoma and a soft tissue metastasis. Soft tissue metastases of malignant melanomas are particularly susceptible to misdiagnosis as soft tissue sarcomas, and the use of immunohistochemical stain aids in differentiating between the two diagnoses [4].

Ultimately, a combination of imaging techniques should be pursued for accurate and timely diagnosis of musculoskeletal metastases. Each technique has strengths and limitations; diagnostic testing should begin with plain radiographs, while bone scans, CT, MRI, and/or PET scans should be ordered as needed to gain a full picture of the extent of each patient's metastatic disease [2, 4, 22].

An Overview of Treatment

Musculoskeletal metastases may be treated operatively or nonoperatively based largely on the patient's pain, extent of functional limitation, and risk of pathologic fracture. Treatment is palliative rather than curative, though it can slow the progression of metastatic disease within the musculoskeletal system. In the case of oligometastatic disease, in which metastases are limited in number and location rather than widely disseminated throughout the body, lesions are best treated with surgical resection with or without reconstruction as well as radiation treatment to improve prognosis [6, 24].

Pharmacologic treatment with bone antiresorptive agents, namely, denosumab and bisphosphonates, is needed to address SREs. These drugs target osteoclasts to slow bone destruction, thus helping alleviate bone pain, prevent and treat hypercalcemia, and reduce the risk of pathologic fractures [18, 22].

Radiation is useful in treating localized bone pain, neurologic symptoms from spinal cord compression, and impending fractures with or without prophylactic fixation. However, patients with diffuse bone pain or pain in previously radiated sites may benefit more from bisphosphonates. Radio-frequency ablation and cryoablation can be used to shrink or eradicate lesions, thus alleviating pain [2]. Ablation procedures are becoming more commonly used, given their effectiveness and lack of cumulative effect when compared to radiation.

Surgery can be used for patients with impending or displaced pathologic fractures or spinal cord compression. Fracture fixation can be achieved using intramedullary rods, plates, and screws [2, 17]. Prompt spinal cord decompression is required to avoid permanent neurologic damage for patients with spine metastases causing compression. Decompression can be achieved surgically, often in tandem with spinal stabilization or tumor resection and reconstruction [25].

Overall, decision-making should take into consideration a patient's potential overall survival, weighing the short- and long-term consequences of each possible treatment on pain level, functionality, and quality of life. Orthopedic oncologists must discern patients who, despite the high risk of complications, will benefit from surgical intervention [26]. Given the nuance of treating patients with bone and soft tissue metastases, a multidisciplinary team of specialists including a radiation oncologist, a medical oncologist, a pathologist, a radiologist, an orthopedic surgeon, a psychiatrist or psychologist, and an occupational or physical therapist should work closely together [27].

Why Is the Management of Musculoskeletal Metastasis Important?

Cancer that has spread to the bones and soft tissues of the musculoskeletal system is, for the most part, incurable. Thus, it must be managed appropriately to minimize pain and maximize functionality and quality of life.

Prognosis takes a sharp turn for the worse after bone metastases take root. In breast cancer, just 20–30% of patients with bone metastases are expected to survive 5 years beyond their diagnosis, while the overall 5-year survival rate for breast cancer is estimated to be 85% [28]. Men with bone metastases from prostate cancer have a median survival duration ranging from 30 to 53 months, depending on their performance status [2]. One study found that <1% of patients with prostate cancer, bone metastases, and SREs survived 5 years beyond their diagnosis [29]. Management of musculoskeletal metastases is critical to prevention and early treatment of SREs. Bone metastases put patients at risk for SREs. Just one SRE is correlated with increased mortality, and the risk of subsequent SREs increases after the first one takes place [5]. The correlation between SREs and decreased survival may be explained by the fact that SREs indicate more aggressive disease. Alternately, the effects of and treatments for SREs (diminished mobility and independence and surgery, respectively) may worsen prognosis after bone metastases occur. Bone surgery, for example, increases mortality as the risk of postoperative complications is significant [5].

Patients' quality of life decreases significantly after their cancer spreads to bone or soft tissues. Since treatment for bone and soft tissue metastases focuses largely on addressing patients' symptoms to improve their quality of life, these outcomes are important to consider when managing musculoskeletal metastases. Several studies have demonstrated a negative correlation between quality-of-life outcomes and SREs, Eastern Cooperative Oncology Group (ECOG) score, and pain level. Patients with SREs, pain, and a low functional score are more likely to suffer from psychological disorders and systemic symptoms like fatigue, insomnia, and appetite loss [27].

Ultimately, musculoskeletal metastases pose a significant challenge to the successful treatment of cancer patients. Lesions in the bones and soft tissues cause pain, pathologic fractures, hypercalcemia, bone marrow aplasia, and spinal cord compression. Such events worsen oncologic patients' prognoses, decrease their functionality, and diminish their overall quality of life [2, 22]. The following chapters will further elucidate the nuances of musculoskeletal metastases, including their pathophysiology, diagnosis, and management.

References

1. Zhang J, Cai D, Hong S. Prevalence and prognosis of bone metastases in common solid cancers at initial diagnosis: a population-based study. BMJ Open. 2023;13(10):e069908. https://doi.org/10.1136/bmjopen-2022-069908.
2. Macedo F, Ladeira K, Pinho F, et al. Bone metastases: an overview. Oncol Rev. 2017;11(1):321. https://doi.org/10.4081/oncol.2017.321.
3. Shupp AB, Kolb AD, Mukhopadhyay D, Bussard KM. Cancer metastases to bone: concepts, mechanisms, and interactions with bone osteoblasts. Cancer. 2018;10(6):182. https://doi.org/10.3390/cancers10060182.
4. Metastases to soft tissue - Plaza - 2008 - Cancer - Wiley Online Library. September 12, 2024. https://acsjournals.onlinelibrary.wiley.com/doi/10.1002/cncr.23151. Accessed 12 Sept 2024.
5. Hernandez RK, Wade SW, Reich A, Pirolli M, Liede A, Lyman GH. Incidence of bone metastases in patients with solid tumors: analysis of oncology electronic medical records in the United States. BMC Cancer. 2018;18 https://doi.org/10.1186/s12885-017-3922-0.
6. Cirstoiu C, Cretu B, Iordache S, Popa M, Serban B, Cursaru A. Surgical management options for long-bone metastasis. EFORT Open Rev. 2022;7(3):206–13. https://doi.org/10.1530/EOR-21-0119.
7. Bray F, Laversanne M, Sung H, et al. Global cancer statistics 2022: GLOBOCAN estimates of incidence and mortality worldwide for 36 cancers in 185 countries. CA Cancer J Clin. 2024;74(3):229–63. https://doi.org/10.3322/caac.21834.
8. Siegel RL, Giaquinto AN, Jemal A. Cancer statistics, 2024. CA Cancer J Clin. 2024;74(1):12–49. https://doi.org/10.3322/caac.21820.
9. Chin H, Kim J. Bone metastasis: concise overview. Fed Pract. 2015;32(2):24–30.
10. Ryan C, Stoltzfus KC, Horn S, et al. Epidemiology of bone metastases. Bone. 2022;158:115783. https://doi.org/10.1016/j.bone.2020.115783.
11. Xiao ZW, Guo H ling, Chen H chao, et al. Prevalence, risk factors, and prognostic factors of primary malignant bone neoplasms with bone metastasis at initial diagnosis: a population-based study. J Oncol 2022;2022:9935439. doi:https://doi.org/10.1155/2022/9935439.
12. Pretell-Mazzini J, Seldon CS, D'Amato G, Subhawong TK. Musculoskeletal metastasis from soft-tissue sarcomas: a review of the literature. JAAOS. 2022;30(11):493. https://doi.org/10.5435/JAAOS-D-21-00944.
13. Lipton A, Balakumaran A. Denosumab for the treatment of cancer therapy-induced bone loss and prevention of skeletal-related events in patients with solid tumors. Expert Rev Clin Pharmacol. 2012;5(4):359–71. https://doi.org/10.1586/ecp.12.35.

14. Schulman KL, Kohles J. Economic burden of metastatic bone disease in the U.S. Cancer. 2007;109(11):2334–42. https://doi.org/10.1002/cncr.22678.
15. Akhtar M, Haider A, Rashid S, Al-Nabet ADMH. Paget's "seed and soil" theory of cancer metastasis: an idea whose time has come. Adv Anat Pathol. 2019;26(1):69–74. https://doi.org/10.1097/PAP.0000000000000219.
16. Clézardin P, Coleman R, Puppo M, et al. Bone metastasis: mechanisms, therapies, and biomarkers. Physiol Rev. 2021;101(3):797–855. https://doi.org/10.1152/physrev.00012.2019.
17. Cecchini MG, Wetterwald A, van der Pluijm G, Thalmann GN. Molecular and biological mechanisms of bone metastasis. EAU Updat Ser. 2005;3(4):214–26. https://doi.org/10.1016/j.euus.2005.09.006.
18. Fornetti J, Welm AL, Stewart SA. Understanding the bone in cancer metastasis. J Bone Miner Res. 2018;33(12):2099–113. https://doi.org/10.1002/jbmr.3618.
19. Lai NB, Martinez D. Physiological roles of parathyroid hormone-related protein. Acta Biomed. 2019;90(4):510–6. https://doi.org/10.23750/abm.v90i4.7715.
20. Wc D, M G, K C, et al. RANK is essential for osteoclast and lymph node development. Genes Dev. 1999;13(18) https://doi.org/10.1101/gad.13.18.2412.
21. Renema N, Navet B, Heymann MF, Lezot F, Heymann D. RANK–RANKL signalling in cancer. Biosci Rep. 2016;36(4):e00366. https://doi.org/10.1042/BSR20160150.
22. Farrell C. Bone metastases: assessment, management and treatment options. Br J Nurs. 2013;22(Sup7):S4–S11. https://doi.org/10.12968/bjon.2013.22.Sup7.S4.
23. Furlong E, Carter T. Aplastic anaemia: current concepts in diagnosis and management. J Paediatr Child Health. 2020;56(7):1023–8. https://doi.org/10.1111/jpc.14996.
24. Hellman S, Weichselbaum RR. Oligometastases. JCO. 1995;13(1):8–10. https://doi.org/10.1200/JCO.1995.13.1.8.
25. Bakar D, Tanenbaum JE, Phan K, et al. Decompression surgery for spinal metastases: a systematic review. Published online August 1, 2016. https://doi.org/10.3171/2016.6.FOCUS16166.
26. Nooh A, Goulding K, Isler MH, et al. Early improvement in pain and functional outcome but not quality of life after surgery for metastatic long bone disease. Clin Orthop Relat Res. 2018;476(3):535–45. https://doi.org/10.1007/s11999.0000000000000065.
27. Shinoda Y, Sawada R, Yoshikawa F, et al. Factors related to the quality of life in patients with bone metastases. Clin Exp Metastasis. 2019;36(5):441–8. https://doi.org/10.1007/s10585-019-09983-0.
28. Harries M, Taylor A, Holmberg L, et al. Incidence of bone metastases and survival after a diagnosis of bone metastases in breast cancer patients. Cancer Epidemiol. 2014;38(4):427–34. https://doi.org/10.1016/j.canep.2014.05.005.
29. Nørgaard M, Jensen AØ, Jacobsen JB, Cetin K, Fryzek JP, Sørensen HT. Skeletal related events, bone metastasis and survival of prostate cancer: a population based cohort study in Denmark (1999 to 2007). J Urol. 2010;184(1):162–7. https://doi.org/10.1016/j.juro.2010.03.034.

Understanding Musculoskeletal Metastases

2

Rayna S. Kuthiala, Marcos R. Gonzalez,
Carolina Mendez-Guerra, and Juan Pretell-Mazzini

Introduction

Metastasis—the process by which cancer cells spread from the primary tumor to distant organs—is the leading cause of cancer-related death worldwide. The ability of cancer cells to invade, disseminate, and colonize distant tissues involves complex molecular and cellular changes. These processes are influenced by genetic mutations, altered signaling pathways, and the tumor microenvironment (TME). This chapter examines the mechanisms of metastasis, with a focus on bone and soft tissue metastases. It explores key genetic alterations, molecular pathways, and the contributions of the TME to metastatic spread.

A General Overview

The Process of Metastasis

Metastasis is a multistep process that includes invasion, dissemination, and colonization. Each step requires a specific set of cellular and molecular changes that allow the tumor cells to survive and thrive in distant organs.

R. S. Kuthiala · M. R. Gonzalez
Division of Orthopaedic Oncology, Department of Orthopaedic Surgery, Massachusetts General Hospital, Harvard Medical School, Boston, MA, USA
e-mail: rkuthiala@mgh.harvard.edu

C. Mendez-Guerra
Facultad de Ciencias de la Salud, Universidad Peruana de Ciencias Aplicadas, Lima, Peru

J. Pretell-Mazzini (✉)
Division of Musculoskeletal Oncology - Miami Cancer Institute, Baptist Health of South Florida / Herbert Wertheim College of Medicine - Florida International University, Miami, FL, USA
e-mail: juan.pretell@baptisthealth.net

J. Pretell-Mazzini, S. A. Lozano Calderón (eds.), *Multidisciplinary Approach of Musculoskeletal Metastasis*, https://doi.org/10.1007/978-3-032-09498-8_2

Invasion

The first step in metastasis is invasion, where cancer cells breach the surrounding tissue barriers. This step is driven by epithelial-to-mesenchymal transition (EMT), a process in which epithelial cells lose their cell–cell adhesion properties and acquire a more migratory phenotype. EMT is critical for facilitating the invasive capacity of tumor cells [1]. During this process, the expression of epithelial markers such as E-cadherin is reduced, while those of mesenchymal markers like N-cadherin and vimentin are upregulated. These changes promote cell motility and invasiveness, allowing the tumor to infiltrate surrounding tissues [2].

Furthermore, tumor cells secrete matrix metalloproteinases (MMPs) that degrade the extracellular matrix (ECM), enabling the cells to move through tissue barriers [3].

Dissemination

Once tumor cells invade the surrounding tissue, they can enter the bloodstream or lymphatic system, allowing them to travel to distant organs. Tumor cells interact with components of the vascular endothelium through adhesion molecules like integrins, facilitating their entry into blood vessels [4]. The bloodstream offers a route for dissemination, while the lymphatic system, which is involved in regional spread, often serves as a secondary metastatic site [5].

One of the critical features of disseminated tumor cells is their ability to survive the harsh conditions of the bloodstream. Many tumor cells form emboli, which are clusters of cells that are more likely to survive circulatory shear forces [6].

Colonization

The final step of metastasis is colonization, where disseminated tumor cells adapt to a new tissue environment and begin to proliferate. Tumor cells must overcome numerous challenges, including immune surveillance and the need for a new blood supply. This step is facilitated by the pre-metastatic niche, where the primary tumor secretes factors such as growth factors, cytokines, and extracellular vesicles that prepare distant organs to support tumor growth [7].

Molecular and Cellular Changes in Metastatic Cells

For cancer cells to metastasize, they must acquire specific genetic alterations and signaling changes. These alterations help tumor cells survive in distant organs and support their ability to invade and proliferate.

Key Genes and Mutations

The first major alteration involves the *p53* tumor suppressor gene, which is frequently mutated in metastatic cancers. Mutant *p53* enhances cell migration and promotes epithelial-to-mesenchymal transition (EMT), both critical for the invasion and dissemination of tumor cells [8]. Additionally, mutant *p53* increases the production of growth factors such as vascular endothelial growth factor (VEGF), which supports angiogenesis, a process essential for tumor growth and metastasis [9].

Other genes associated with an increased risk of metastasis are *BRCA1* and *BRCA2*, which are involved in DNA repair. The loss of BRCA function leads to genomic instability, promoting a more aggressive tumor phenotype and resistance to therapy [10]. Tumors with *BRCA* mutations are more likely to acquire additional mutations that drive metastasis [10].

Lastly, human epithelial growth factor receptor 2 (HER2) is prone to mutation. When overexpressed or mutated, HER2 drives uncontrolled cell proliferation and enhances the metastatic potential of cancer cells. HER2 signaling also promotes angiogenesis and increases the survival of disseminated tumor cells, making it a key factor in the metastatic spread of breast cancer and other cancers [11].

Signaling Pathways

Several key signaling pathways are implicated in metastasis, including the Wnt/β-catenin and transforming growth factor-β (TGF-β) pathways, both of which play critical roles in regulating cell behavior and promoting metastasis.

The Wnt signaling pathway regulates cellular processes such as adhesion, migration, and invasion. In bone metastasis, Wnt/β-catenin signaling promotes osteoclastogenesis, leading to bone resorption and the release of growth factors that support metastatic growth [12]. This pathway is particularly crucial for metastasis to the bone and other tissues.

On the other hand, TGF-β is a potent inducer of EMT and contributes to the conversion of epithelial tumor cells into mesenchymal-like cells, which are more capable of invasion and metastasis. Additionally, TGF-β plays a role in immune evasion by suppressing immune responses within the tumor microenvironment [13].

Role of the Tumor Microenvironment

The tumor microenvironment (TME) is an essential factor that supports metastasis. It is composed of various cellular components, including fibroblasts, immune cells, endothelial cells, and ECM. Tumor cells communicate with these components to enhance their survival, migration, and invasion.

Tumor–Stroma Interaction

Stromal cells within TME play a critical role in promoting metastasis. For example, fibroblasts in the stroma secrete cytokines and growth factors like TGF-β, which promote EMT and the invasive phenotype of cancer cells [14]. Additionally, immune cells such as tumor-associated macrophages (TAMs) can promote metastasis by secreting pro-inflammatory cytokines and growth factors that enhance angiogenesis and tumor cell survival [15].

Blood Supply and Lymphatic Drainage

Angiogenesis plays a critical role in tumor growth and metastasis. Tumor cells secrete VEGF, which stimulates the formation of new blood vessels, providing oxygen and nutrients to the growing tumor [16]. This newly formed vascular network also offers a route for cancer cells to disseminate and reach distant organs.

Similarly, the lymphatic system serves as a common pathway for cancer cell spread, especially in cancers like breast cancer and melanoma. Cancer cells often spread to regional lymph nodes, which can act as secondary metastatic sites [17]. The interaction between tumor cells and lymphatic endothelial cells is facilitated by adhesion molecules such as integrins and chemokines like CXCL12 [18].

Bone Versus Soft Tissue Metastasis

Metastasis to bone and soft tissue presents distinct challenges and pathways.

Bone Metastasis

Tumor cells that metastasize to bone interact with osteoblasts and osteoclasts. Osteoblasts release growth factors like TGF-β and platelet-derived growth factor (PDGF), which support tumor growth in bone [19]. Concurrently, tumor cells secrete parathyroid hormone-related protein (PTHrP), which activates osteoclasts and drives bone resorption. This process results in the release of additional growth factors, further promoting tumor growth [20]. When bone metastases are associated with osteolytic lesions, fractures, and hypercalcemia, it reflects a disruption in the tightly regulated balance between osteoblast and osteoclast activity, resulting in progressive bone destruction [19].

In contrast, some metastases—particularly those originating from prostate cancer—manifest as sclerotic rather than lytic lesions. These lesions are marked by abnormal and excessive bone formation driven by a different pathological imbalance in the bone remodeling process [19, 20]. A key contributor to this osteogenic shift is dysregulated Wnt signaling, which plays a pivotal role in osteoblast differentiation and activity. The canonical Wnt pathway, through β-catenin stabilization and nuclear translocation, upregulates osteogenic genes such as *Runx2* and *Osterix*, enhancing osteoblast function and matrix mineralization [19]. In sclerotic lesions, this pathway is often hyperactivated, either by tumor-derived Wnt ligands or through the influence of TGF-β superfamily members like bone morphogenetic proteins (BMPs), which interact with Wnt signaling to amplify osteogenic effects. The noncanonical Wnt pathways (e.g., Wnt/planar cell polarity and Wnt/Ca^{2+}) also contribute by modulating cytoskeletal dynamics and osteoblast recruitment, though their effects are more indirect [19]. Together, the interplay between Wnt and TGF-β signaling—particularly via Smad1/5/8 activation—leads to excessive osteoblast proliferation and matrix deposition. While this results in increased bone density, the newly formed bone is often disorganized and mechanically weak, increasing the risk of skeletal complications despite its seemingly hardened appearance [19].

Soft Tissue Metastasis

Soft tissue metastasis occurs when cancer cells spread from their primary site to distant soft tissues, such as muscles, adipose tissue, or connective tissues. This process is a significant challenge in cancer treatment, as soft tissues are less commonly targeted by conventional treatments compared to more typical metastatic sites like the lungs or liver. Once cancer cells enter the bloodstream or lymphatic system, they can travel to distant tissues, where they may survive and form secondary tumors. Vascularity plays a crucial role in soft tissue metastasis. Tumors with rich blood vessel networks can more easily spread by entering the bloodstream or lymphatic system. Increased vascularization facilitates cancer cell dissemination, making high-grade and larger tumors more likely to metastasize to soft tissues [21]. Additionally, metastatic cells must adapt to the new soft tissue microenvironment. Their ability to survive in these tissues depends on the tumor cells' ability to evade immune detection, often through altered cell surface proteins or secretion of immune-suppressing molecules [22].

There should also be a stronger emphasis on the adverse environment of muscle tissue in significantly decreasing the chances of metastasis. The muscle tissue environment is inherently less hospitable to metastatic cells due to several key factors. The muscle's blood flow is inherently turbulent, which limits colonization by metastatic cells [22]. For instance, the muscle's lower vascularity restricts the supply of nutrients and oxygen, limiting the growth potential of metastasizing cells [22]. The presence of specific immune cells within the muscle tissue can actively target and eliminate foreign cells, including those of a metastatic origin. Additionally, the extracellular matrix in muscle is less conducive to the survival and proliferation of metastatic cells, as its dense, rigid structure obstructs their ability to invade and colonize [21]. These factors collectively create a hostile environment for metastatic cells. This highlights the dual nature of soft tissue metastasis—on one hand, the promoting factors, such as increased vascularization and immune evasion in some tissues, facilitate metastatic spread. On the other hand, the counteracting environment in muscle tissue, marked by its unfavorable characteristics stated above, actively reduces the ability of metastatic cells to establish secondary tumors [22].

Extracellular matrix (ECM) remodeling is also a key factor in soft tissue metastasis. Cancer cells can produce enzymes that break down ECM, allowing them to invade soft tissues more easily [23]. Furthermore, immune evasion mechanisms, including immune checkpoint inhibition, help metastatic cells persist in distant tissues, contributing to tumor growth [24]. Soft tissue metastasis remains a significant challenge for early detection and treatment, requiring innovative strategies to target both the vascular and immune aspects of metastasis.

Patterns of Metastasis Based on Primary Tumor Location (Axial Versus Appendicular)

The pattern of metastasis in various types of cancers varies significantly between axial and appendicular locations. Axial tumors, which are found in the trunk, pelvis, and head/neck, are more likely to metastasize early and to a greater extent compared to appendicular tumors, which occur in the limbs. This difference is primarily due to the axial region's direct access to major blood vessels and lymphatic systems, providing a more accessible pathway for cancer cells to spread to vital organs such as the lungs, liver, and lymph nodes [21].

Axial tumors tend to have a higher risk of distant metastasis, often spreading to organs like the lungs, liver, and bones at an earlier stage. This is particularly true for cancers located in regions with rich vascular networks and proximity to large blood vessels, which facilitates the entry of cancer cells into the bloodstream or lymphatic system. In contrast, appendicular tumors, found in the limbs, are less likely to spread early. The vascular anatomy and lymphatic drainage of the limbs are more restrictive, limiting the immediate dissemination of cancer cells compared to the more centrally located axial region [22].

As a result, cancers in the axial region are often diagnosed with a higher metastatic burden and require more aggressive treatment strategies. Metastasis from axial tumors commonly occurs through the bloodstream, with the lungs and liver being the most frequent sites of spread. On the other hand, tumors in the appendicular region typically metastasize later, most often to the lungs [25]. These differences in metastatic patterns are influenced by factors such as vascular accessibility, tumor biology, and immune response profiles [24].

Cancers in the axial region are associated with worse overall survival and higher rates of metastasis, underscoring the critical role of tumor location in determining metastatic potential and guiding treatment decisions [22].

Patterns of Metastatic Spread in Carcinomas and Sarcomas

Metastatic Spread of Carcinomas

Carcinomas, which account for most malignancies, often exhibit aggressive metastatic behavior. In this regard, a large autopsy study found that the most common metastatic targets in carcinomas were the local and regional lymph nodes, followed in frequency by the liver, lungs, and bones [25]. Pertaining to the latter, metastatic bone disease (MBD) is pivotal in carcinoma progression, as it significantly compromises the patient's health-related quality of life and overall survival [26, 27]. Additionally, although skeletal muscle metastasis constitutes an uncommon phenomenon, its presence is indicative of an aggressive primary malignancy with adverse survival outcomes [28].

Given the information outlined in previous lines, in the following section, we will discuss the metastatic spread of carcinoma to the bones and, to a lesser extent, to the muscles.

Metastasis to the Bone

Bone metastasis is frequently observed in several carcinomas, with most cases arising from breast, prostate, lung, thyroid, and kidney malignancies [29]. Notably, postmortem examination studies reported that 70% of patients with breast and prostate cancers had evidence of MBD at their time of death, whereas an incidence of

40% was reported in patients with thyroid, kidney, and lung malignancies [30]. An in-depth understanding of the genetic and molecular basis of bone metastasis is crucial for developing effective treatment strategies. Accordingly, in the following lines, we will discuss the role of bone microenvironment in metastatic tropism. For a more detailed discussion regarding the key genes and signaling pathways involved in bone metastasis, refer to the previous sections of this chapter (Fig. 2.1).

The "Vicious Cycle": Why Do Certain Malignancies Preferentially Target Bone?

Metastasis is a complex, dynamic process in which tumor outgrowth and distant dissemination are not solely driven by the intrinsic migratory and invasive properties of tumor cells but also by the fitness advantages that may be conferred by certain microenvironments [31]. The bone microenvironment, which is primarily composed of bone marrow and a mineralized extracellular matrix [32], constitutes an extremely fertile soil for primary and secondary malignancies to seed and thrive. The term "vicious cycle" was introduced in the early 90 s to describe the tighter intercellular interactions between tumor cells and bone cells [33]. This theory portrays that tumor cells promote the expression of receptor activator of nuclear factor kappa-B ligand (RANKL) in osteoblasts, which ultimately leads to osteolysis, by secreting parathyroid hormone-related protein (PTHrP) [34]. This osteoclast-mediated bone destruction prompts the release of bone matrix-embedded growth factors that promote metastatic niche growth [34]. Additionally, transforming growth factor-beta (TGF-β), released during dysregulated bone remodeling, has been shown to enhance PTHrP secretion [35]. Furthermore, evidence suggests that

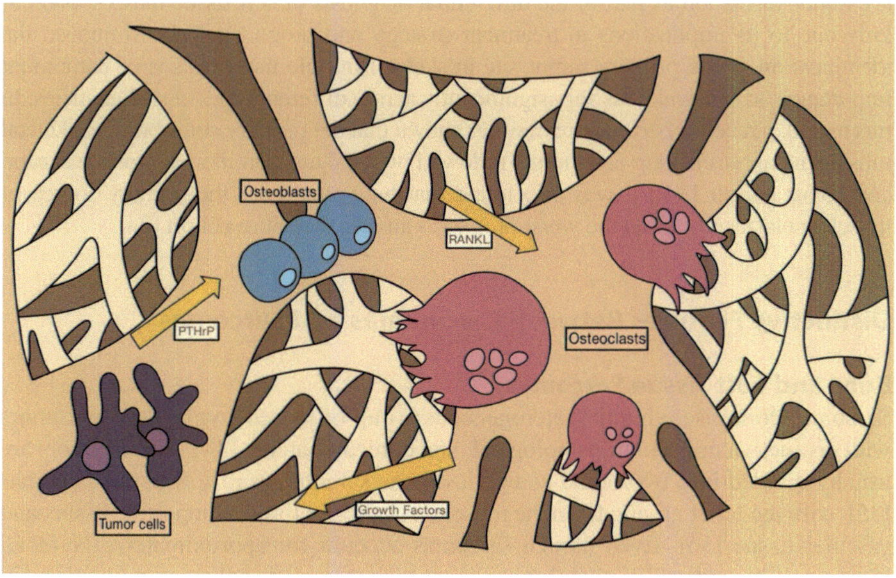

Fig. 2.1 Mechanisms of bone metastasis

matrix metalloproteinases (MMPs) function as key regulators of the bioavailability and bioactivity of RANKL and TGF-β, implicating these molecules as potential master mediators of the vicious cycle [36].

While osteolytic lesions are common in many cancers, sclerotic (osteoblastic) metastases represent a distinct pathological manifestation characterized by abnormal new bone formation. In these cases, tumor-secreted factors such as endothelin-1, bone morphogenetic proteins (BMPs), and Wnt proteins stimulate osteoblast activity, leading to excessive and disorganized bone deposition [19]. Despite appearing denser radiologically, these sclerotic lesions are structurally unsound and contribute to significant skeletal-related events (SREs). Importantly, the RANKL–receptor activator of nuclear factor-κB (RANK) pathway remains relevant in this context, as osteoblast–osteoclast coupling continues to play a role in lesion development. Accordingly, the activation of the RANKL–RANK pathway has been found to contribute to the metastatic phenotype of both breast and prostate cancer cells [37]. Further investigation into this topic is warranted, as it may serve as a crucial target for developing novel therapies in the management of MBD.

Metastasis to the Skeletal Muscle

Although bones constitute a frequent site for secondary tumor formation, skeletal muscle metastasis remains relatively uncommon. Indeed, it was estimated that skeletal muscle metastasis accounted for only 3% of all soft tissue malignancies diagnosed at an orthopedic oncology center over a 14-year period [31]. Despite its relatively low prevalence, recent advancements in imaging modalities, coupled with the adoption of a multidisciplinary approach, have enabled an earlier and broader recognition of skeletal muscle metastasis [32]. This change in landscape prompts the consideration of secondary tumor formation as part of the differential diagnosis of soft tissue masses, particularly due to its implications in treatment strategy and prognosis [33]. Moreover, the identification of the primary tumor site in skeletal muscle metastasis is of paramount importance, as survival rates vary significantly across different types of malignancies. In this regard, a recent systematic review has shown that the primary contributor to skeletal muscle metastasis is lung carcinoma, followed by renal cell carcinoma, prostate cancer, and breast cancer [34]. It was also found that lung cancer, as the primary source of muscle metastasis, carried the worst prognosis among the entire cohort [34].

Distinctive Features Between Carcinomas and Sarcomas

Bone and Soft Tissue Sarcomas

Sarcomas comprise a highly heterogeneous group of mesenchymal-derived tumors with distinctive molecular, pathological, and clinical features [44]. These tumors are usually divided into two broad categories: bone sarcomas and soft tissue sarcomas [35], with the latter arising from the muscle, connective tissue, supportive tissue, and vascular tissue [36]. Even though sarcomas account for approximately 1% of all human malignancies [37], their highly aggressive metastatic behavior, particularly in high-grade sarcomas, depicts them as an important group of secondary malignancies

[38]. In this regard, the most common sites for metastasis in osteosarcoma are the lungs and bones [39]; whereas, in soft tissue sarcomas, the most frequent metastatic targets include the lungs, bones, liver, brain, and subcutaneous tissue [40].

A Comparative Overview of the Metastatic Patterns of Carcinomas and Sarcomas (Table 2.1)

Distinctive patterns of metastasis are exhibited by carcinomas and sarcomas, owing to their differences in tumor cell origin, molecular mechanism of metastasis, tumor microenvironment, and routes of dissemination. Given the information outlined above, a comparative overview of the metastatic patterns of carcinomas and sarcomas will be presented in the following lines.

Phenotypic Plasticity

Phenotypic plasticity constitutes a dynamic process in which tumor cells undergo reversible phenotypical changes critical for the metastatic cascade [41]. It is well-established that carcinomas, tumors of an epithelial origin, undergo a process known as epithelial-to-mesenchymal transition (EMT), in which these epithelial tumor cells acquire invasive and migratory properties by switching to a mesenchymal phenotype [42]. In contrast, even though sarcomas are, by definition, derived from mesenchymal cells, it has been proposed in the current literature that these cells may also undergo EMT-related changes [43]. As described in previous sections, TGF-β signaling plays an essential role in EMT. It has been shown that TGF-β induces transdifferentiation of breast epithelial cells to a mesenchymal phenotype [44]. Additionally, the role of TGF-β in sarcomas has also been described in the literature. High levels of serum TGF-β have been found in patients with osteosarcoma compared to healthy patients [45]. The same study reported augmented levels of serum TGF-β in those with metastasis [45]. Furthermore, it has also been

Table 2.1 Comparison of metastasis in sarcomas and carcinomas

Feature	Sarcoma metastasis	Carcinoma metastasis
Origin	Derived from mesenchymal tissues, including bone, muscle, fat, cartilage, and connective tissue	Derived from epithelial tissue lining organs, glands, and mucosal surfaces
Histological organization	Disorganized growth pattern More aggressive and infiltrative with poorly defined margins Often invades surrounding tissues	Structured growth pattern Follows an orderly pattern; remains localized longer before dissemination
Molecular mechanisms	May also undergo EMT-related processes MMPs facilitate angiogenesis and tumor dissemination	EMT driven by the TGF-β signaling pathway
Route of dissemination	Primarily hematogenous (via the bloodstream)	Primarily through lymphatic channels
Common sites of dissemination	Lungs, bones, liver, soft tissues	Lymph nodes, lungs, liver, brain, bones

EMT epithelial-to-mesenchymal transition; *MMPs* matrix metalloproteinases; *TGF-β* transforming growth factor-beta hypothesized that osteosarcoma tumor-induced changes in bone microenvironment, partially driven by TGF-β, play an important role in chemoresistance [46].

Tumor Microenvironment

Distinctive features in the tumor microenvironments of carcinomas and sarcomas have been described in the current literature. On one side, the tumor microenvironment in carcinomas is mainly composed of two compartments: an epithelial compartment of malignant cells and a stromal compartment of fibroblasts and myofibroblasts, with the latter accounting for 30–90% of the specimens [2]. Furthermore, shared cellular and noncellular elements have been identified in the tumor microenvironments of colon, lung, and breast cancers, owing to their common epithelial lineage [47]. On the other side, it was once believed that the tumor microenvironment in sarcomas was mostly composed of only one compartment of malignant mesenchymal cells [2]. Nevertheless, current knowledge illustrates that stromal cell heterogeneity is also present in the sarcoma tumor microenvironment [48]. Moreover, it is worth pointing out that, on account of their mesenchymal origin, distinctive features between malignant tumor cells and benign stromal cells are particularly challenging to evaluate [48]. The role of tumor microenvironment in tumor progression is currently under study, as it is proposed that its multifaceted nature may play a crucial role in acquiring angiogenic, invasive, and migratory properties for metastasis [47]. Indeed, further investigation into the tumor microenvironment is warranted, as new therapeutic strategies may be developed.

Route of Dissemination

The route of dissemination differs between carcinomas and sarcomas. Extensive evidence indicates that tumor outgrowth and dissemination in carcinomas follow a structured, sequential process, in which malignant cells initially spread from the primary tumor site to the regional or sentinel lymph nodes with subsequent dissemination to distant sites [6]. In contrast, the main route of dissemination in sarcomas is the vascular system, except for epithelial sarcoma, angiosarcoma, and embryonal rhabdomyosarcoma [6]. Indeed, the current literature states that lymph node involvement in sarcomas is relatively uncommon and that lymph node metastasis surveillance is mostly warranted in the aforementioned histological subtypes [18]. Furthermore, it is important to emphasize the role of MMPs in sarcoma dissemination. It has been found that the activation of endothelial MMP-2, linked to MMP-14 expression, facilitates angiogenesis by modulating the extracellular matrix of sarcomas [49].

Clinical Impact of Metastasis on Bone and Soft Tissues

Bone Metastasis

Signs and symptoms of metastatic bone disease arise from disruptions in the normal functions of the axial and appendicular skeleton, including spinal cord protection, calcium homeostasis regulation, and structural support of the body. When these functions are compromised by a neoplastic process, complications such as pain, spinal cord compression, hypercalcemia, and pathological fractures can occur. Collectively, these complications are referred to as skeletal-related events (SREs).

(a) *Pain*

Bone pain is the most common type of pain among cancer patients, affecting 60–84% of patients with bone metastases [50]. Although the pathophysiology of bone cancer pain remains complex and poorly understood, it involves both inflammatory and neuropathic mechanisms at the peripheral and central nervous system levels [51]. Peripherally, pain may result from tumor-induced distortion of the bony trabeculae through microfractures, periosteal stretching due to tumor expansion, and mechanical stress on the weakened bone [52]. Additionally, reactive muscle spasms can exacerbate pain intensity [53]. Clinically, patients often report severe pain that is only partially responsive to analgesics. While pain may have multiple contributing factors beyond the neoplastic process, physicians should remain vigilant for red flags such as night pain, unintentional weight loss, pain from weight-bearing, or an enlarging mass [54]. Pain management strategies include pharmacological therapy, radiation therapy, minimally invasive procedures, and surgical interventions.

(b) *Spinal cord compression*

Malignant spinal cord compression (MSCC) is a devastating cancer complication and a medical emergency due to the risk of permanent neurological damage and death [55]. While its exact incidence remains unknown, Loblaw et al. reported a 5-year cumulative incidence of 2.5%, ranging from 0.2% in pancreatic cancer to 7.9% in multiple myeloma [56]. The most common mechanism of MSCC is hematogenous spread, leading to vertebral infiltration and compression of the dural sac, root sleeves, and their contents through tumor expansion [57]. Symptom reversibility varies widely; a slow onset allows for spinal cord adaptation and is associated with a better prognosis, whereas sudden, rapidly progressive symptoms carry a worse prognosis with limited functional recovery [55]. Overall, survival after MSCC remains poor, with fewer than 10% of patients living beyond 18 months after symptom onset [58]. Given the urgency of this condition, early diagnosis and intervention are critical. The National Institute for Health and Care Excellence (NICE) recommends an urgent whole-spine MRI within 24 hours, immediate administration of dexamethasone upon suspicion, and definitive treatment with surgery and/or radiation therapy within 24 hours [59].

(c) *Hypercalcemia*

Hypercalcemia of malignancy is the most common metabolic disorder in patients with advanced stage cancer, occurring in 10–30% of all cancer patients [60]. Although most frequently observed in hematological malignancies, it also occurs in solid tumors, particularly breast, prostate, and squamous cell carcinomas [61]. Hypercalcemia of malignancy primarily develops through three mechanisms: overproduction of parathyroid hormone-related protein (PTHrP), bone metastases that release osteoclast-activating factors, and increased synthesis of 1,25-dihydroxy vitamin D (calcitriol) [62]. While mild hypercalcemia (serum calcium: 10.5–11.9 mg/dL) may be asymptomatic, moderate-to-severe hypercalcemia is typically associated with a combination of gastrointestinal (e.g., nausea, vomiting, constipation), neurological (e.g., muscle weakness, fatigue), renal (e.g., polyuria, polydipsia), and cardiovascular (e.g., shortened ST segment and QT interval) signs and symptoms [61]. Management typically involves intravenous (IV) hydration with normal saline and bisphosphonates, as fluids alone may not be sufficient to correct calcium levels [63]. Notably, cancer-related hypercalcemia is associated with a poor prognosis, often indicating widespread disease. As a result, its diagnosis should prompt discussions regarding goals of care and advance directives between physicians and patients.

(d) *Pathological fractures*

Pathological fractures due to bone metastases occur in 5–29% of patients and are associated with an extremely poor prognosis [64]. These fractures result from compromised bone integrity, with the mechanism depending on lesion type: osteolytic lesions arise from tumor-induced osteoclast activation via RANKL upregulation, while osteoblastic lesions are caused by tumor-secreted endothelin-1 [65]. The impact of pathological fractures on survival is significant, with studies by Saad et al. showing a 32% increased risk of death in patients with fractures [66], and Groot et al. reporting lower 1-year survival rates for complete pathological fractures compared to impending fractures (38% vs. 48%; $p = 0.03$) [67]. Consequently, in cases of imminent fracture (impending pathological fracture), prophylactic stabilization is recommended. Given that most patients with impending or complete pathological fractures have advanced metastatic disease, surgery is typically palliative, aiming to restore ambulation and control pain. While fixation of complete fractures is standard, surgical intervention for impending fractures requires accurate risk assessment. Although a Mirels score ≥ 9 has traditionally guided the decision for prophylactic fixation, its low specificity has led to concerns about overtreatment [68]. Recent studies have suggested using CT scans for risk assessment, given their superior sensitivity and specificity compared to radiographic tools like the Mirels score [69, 70]. While a full review of surgical criteria and fixation strategies is beyond this chapter's scope, physicians should consider (1) patient characteristics, (2) tumor type, and (3) fracture location and bone stock in their decision-making process [71].

Soft Tissue Metastases

Although soft tissue accounts for approximately 55% of body mass, soft tissue metastases are extremely rare, and the literature on the subject is limited. Plaza et al. reviewed the pathology files of 7237 soft tissue tumors over 30 years and found that only 118 (1.6%) were soft tissue metastases [72]. Factors such as changes in pH, metabolite accumulation, and local temperature at soft tissue sites may explain this low prevalence [72]. In their cohort, Plaza et al. found that 70% of soft tissue metastases originated from carcinomas, with lung carcinoma being the most common histology [72]. Conversely, soft tissue metastases from a primary soft tissue sarcoma (STS) were extremely rare, accounting for <10% of cases [72]. Signs and symptoms of soft tissue metastases, including painless or painful masses with or without compressive effects, edema, and inflammation, are similar to those of other malignant soft tissue neoplasms. However, Tuoheti et al. reported that painful masses were more common in patients with soft tissue metastasis than in those with primary sarcomas [73]. Based on their findings, the authors suggested that a painful mass with extensive peritumoral enhancement in patients with a known carcinoma is highly indicative of soft tissue metastasis [73]. When soft tissue metastasis is suspected, biopsy is essential to differentiate from a primary soft tissue sarcoma and guide management.

Conclusions

Metastasis is a highly complex process that involves genetic mutations, altered signaling pathways, and interactions between cancer cells and the tumor microenvironment. Understanding the molecular mechanisms underlying metastasis provides crucial insights into the development of more effective therapies. Given the challenges of bone and soft tissue metastasis, continued research into these processes is essential for improving patient outcomes.

References

1. Lamouille S, Xu J, Derynck R. Molecular mechanisms of epithelial–mesenchymal transition. Nat Rev Mol Cell Biol. 2014;15(3):178–96.
2. Tomlinson J, Barsky SH, Nelson S, Singer S, Pezeshki B, Lee MC, et al. Different patterns of angiogenesis in sarcomas and carcinomas. Clin Cancer Res. 1999;5(11):3516–22.
3. Kessenbrock K, Plaks V, Werb Z. Matrix metalloproteinases: regulators of the tumor microenvironment. Cell. 2010;141(1):52–67.
4. Friedl P, Gilmour D. Collective cell migration in morphogenesis, regeneration and cancer. Nat Rev Mol Cell Biol. 2009;10(7):445–57.
5. Achen MG, Stacker SA. Tumor lymphangiogenesis and metastatic spread—New players begin to emerge. Int J Cancer. 2006;119(8):1755–60.
6. Leong SPL, Cady B, Jablons DM, Garcia-Aguilar J, Reintgen D, Jakub J, et al. Clinical patterns of metastasis. Cancer Metastasis Rev. 2006;25(2):221–32.
7. Peinado H, Zhang H, Matei IR, Costa-Silva B, Hoshino A, Rodrigues G, et al. Pre-metastatic niches: organ-specific homes for metastases. Nat Rev Cancer. 2017;17(5):302–17.
8. Muller PAJ, Vousden KH. p53 mutations in cancer. Nat Cell Biol. 2013;15(1):2–8.
9. Ahn JH, Kim TJ, Lee JH, Choi JH. Mutant p53 stimulates cell invasion through an interaction with Rad21 in human ovarian cancer cells. Sci Rep. 2017;7(1).
10. Rosen E, Pishvaian M. Targeting the BRCA1/2 tumor suppressors. Curr Drug Targets. 2014;15(1):17–31.

11. Kreutzfeldt J, Rozeboom B, Dey N, De P. The trastuzumab era: current and upcoming targeted HER2+ breast cancer therapies [Internet]. Vol. 10. Am J Cancer Res. 2020. Available from: www.ajcr.us/
12. Fang F, VanCleave A, Helmuth R, Torres H, Rickel K, Wollenzien H, et al. Targeting the Wnt/β-catenin pathway in human osteosarcoma cells. Oncotarget. 2018;9(95):36780–92.
13. Padua D, Massagué J. Roles of TGFβ in metastasis. Vol. 19, Cell Research. 2009. pp. 89–102.
14. Zhao M, Mishra L, Deng CX. The role of TGF-β/SMAD4 signaling in cancer. Int J Biol Sci. 2018;14(2):111–23.
15. Qian BZ, Pollard JW. Macrophage diversity enhances tumor progression and metastasis. Cell. 2010;141(1):39–51.
16. D'Alessio A, Moccia F, Li JH, Micera A, Kyriakides TR. Angiogenesis and vasculogenesis in health and disease. Biomed Res Int. 2015;2015:126582.
17. Park M, Kim D, Ko S, Kim A, Mo K, Yoon H. Breast cancer metastasis: mechanisms and therapeutic implications. Int J Mol Sci. 2022;23(12).
18. Fong Y, Coit DG, Woodruff JM, Brennan MF. Lymph node metastasis from soft tissue sarcoma in adults. Analysis of data from a prospective database of 1772 sarcoma patients. Ann Surg [Internet]. 1993 [cited 2025 Mar 31];217(1):72–7. Available from: https://pubmed.ncbi.nlm.nih.gov/8424704/
19. Jann J, Gascon S, Roux S, Faucheux N. Influence of the TGF-β superfamily on osteoclasts/osteoblasts balance in physiological and pathological bone conditions. Int J Mol Sci [Internet]. 2020 [cited 2025 Apr 20];21(20):1–58. Available from: https://pubmed.ncbi.nlm.nih.gov/33066607/
20. Cox TR, Erler JT. Remodeling and homeostasis of the extracellular matrix: implications for fibrotic diseases and cancer. Dis Model Mech [Internet]. 2011 [cited 2025 Apr 6];4(2):165–78. Available from: https://pubmed.ncbi.nlm.nih.gov/21324931/
21. Carmignani CP, Sugarbaker TA, Bromley CM, Sugarbaker PH. Intraperitoneal cancer dissemination: mechanisms of the patterns of spread. Cancer Metastasis Rev [Internet]. 2003 [cited 2025 Apr 1];22(4):465–72. Available from: https://pubmed.ncbi.nlm.nih.gov/12884919/
22. Ahmed AA, Strong MJ, Zhou X, Robinson T, Rocco S, Siegel GW, et al. Differential immune landscapes in appendicular versus axial skeleton. PLoS One [Internet]. 2022 [cited 2025 Apr 1];17(4):e0267642. Available from: https://pmc.ncbi.nlm.nih.gov/articles/PMC9045623/
23. Leong SPL, Cady B, Jablons DM, Garcia-Aguilar J, Reintgen D, Jakub J, et al. Clinical patterns of metastasis. Cancer Metastasis Rev. [Internet]. 2006 [cited 2025 Apr 1];25(2):221–32. Available from: https://pubmed.ncbi.nlm.nih.gov/16770534/
24. Yuan Z, Li Y, Zhang S, Wang X, Dou H, Yu X, et al. Extracellular matrix remodeling in tumor progression and immune escape: from mechanisms to treatments. Mol Cancer [Internet]. 2023 [cited 2025 Apr 1];22(1):48. Available from: https://pmc.ncbi.nlm.nih.gov/articles/PMC10007858/
25. Disibio G, French SW. Metastatic patterns of cancers: results from a large autopsy study. Arch Pathol Lab Med. 2008;132(6):931–9.
26. Heidary Z, Ghaemi M, Hossein Rashidi B, Kohandel Gargari O, Montazeri A. Quality of life in breast cancer patients: a systematic review of the qualitative studies. Cancer Control. 2023;30:10732748231168318.
27. Zacharia B, Joy J, Subramaniam D, Pai PK. Factors affecting life expectancy after bone metastasis in adults—results of a 5-year prospective study. Indian J Surg Oncol. 2021;12(4):759–69.
28. Pop D, Nadeemy AS, Venissac N, Guiraudet P, Otto J, Poudenx M, et al. Skeletal muscle metastasis from non-small cell lung cancer. J Thorac Oncol. 2009;4(10):1236–41.
29. Kreps LM, Addison CL. Targeting intercellular communication in the bone microenvironment to prevent disseminated tumor cell escape from dormancy and bone metastatic tumor growth. Int J Mol Sci. 2021;22(6).
30. Coleman RE. Clinical features of metastatic bone disease and risk of skeletal morbidity. Clin Cancer Res. 2006;12(20 Pt 2):6243s–9s.

31. Sammon J, Jain A, Bleakney R, Mohankumar R. Magnetic resonance imaging appearance of soft-tissue metastases: our experience at an orthopedic oncology center. Skeletal Radiol. 2017;46(4):513–21.
32. Pretell-Mazzini J, Younis MH, Subhawong T. Skeletal muscle metastases from carcinomas: a review of the literature. JBJS Rev. 2020;8(7):e1900114–8.
33. Molina-Garrido MJ, Guillén-Ponce C. Muscle metastasis of carcinoma. Clin Transl Oncol. 2011;13(2):98–101.
34. Pretell-Mazzini J, de Neyra JZS, Luengo-Alonso G, Shemesh S. Skeletal muscle metastasis from the most common carcinomas orthopedic surgeons deal with. A systematic review of the literature. Arch Orthop Trauma Surg. 2017;137(11):1477–89.
35. Demetri GD, Baker LH, Beech D, Benjamin R, Casper ES, Conrad EU, et al. Soft tissue sarcoma clinical practice guidelines in oncology. J Natl Compr Canc Netw. 2005 Mar;3(2):158–94.
36. Lanzkowsky P. Rhabdomyosarcoma and other soft-tissue sarcomas. In: Manual of pediatric hematology and oncology. 2011. pp. 715–38.
37. Hogendoorn PCW. WHO classification of tumors of soft tissue and bone [Internet]. 2013. Available from: https://www.researchgate.net/publication/302956200
38. Grünewald TG, Alonso M, Avnet S, Banito A, Burdach S, Cidre-Aranaz F, et al. Sarcoma treatment in the era of molecular medicine. EMBO Mol Med. 2020;12(11):e11131.
39. Sheng G, Gao Y, Yang Y, Wu H. Osteosarcoma and metastasis. Front Oncol. 2021;11:780264.
40. Pretell-Mazzini J, Seldon CS, D'Amato G, Subhawong TK. Musculoskeletal metastasis from soft-tissue sarcomas: a review of the literature. J Am Acad Orthop Surg. 2022;30(11):493–503.
41. Jehanno C, Vulin M, Richina V, Richina F, Bentires-Alj M. Phenotypic plasticity during metastatic colonization. Trends Cell Biol. 2022;32(10):854–67.
42. Tsai JH, Yang J. Epithelial-mesenchymal plasticity in carcinoma metastasis. Genes Dev. 2013;27(20):2192–206.
43. Sannino G, Marchetto A, Kirchner T, Grünewald TGP. Epithelial-to-mesenchymal and mesenchymal-to-epithelial transition in mesenchymal tumors: a paradox in sarcomas? Cancer Res. 2017;77(17):4556–61.
44. Miettinen PJ, Ebner R, Lopez AR, Derynck R. TGF-beta induced transdifferentiation of mammary epithelial cells to mesenchymal cells: involvement of type I receptors. J Cell Biol. 1994;127(6 Pt 2):2021–36.
45. Xu S, Yang S, Sun G, Huang W, Zhang Y. Transforming growth factor-beta polymorphisms and serum level in the development of osteosarcoma. DNA Cell Biol [Internet]. 2014 [cited 2025 Mar 31];33(11):802–6. Available from: https://pubmed.ncbi.nlm.nih.gov/25098449/
46. Mintz MB, Sowers R, Brown KM, Hilmer SC, Mazza B, Huvos AG, et al. An expression signature classifies chemotherapy-resistant pediatric osteosarcoma. Cancer Res. 2005;65(5):1748–54.
47. Saggioro M, D'Angelo E, Bisogno G, Agostini M, Pozzobon M. Carcinoma and sarcoma microenvironment at a glance: where we are. Front Oncol. 2020;10:76.
48. Ehnman M, Chaabane W, Haglund F, Tsagkozis P. The tumor microenvironment of pediatric sarcoma: mesenchymal mechanisms regulating cell migration and metastasis. Curr Oncol Rep. 2019;21(10):90.
49. Roebuck MM, Helliwell TR, Chaudhry IH, Kalogrianitis S, Carter S, Kemp GJ, et al. Matrix metalloproteinase expression is related to angiogenesis and histologic grade in spindle cell soft tissue neoplasms of the extremities. Am J Clin Pathol. 2005;123(3):405–14.
50. Mercadante S. Malignant bone pain: pathophysiology and treatment. Pain. 1997;69(1–2):1–18.
51. Zajączkowska R, Kocot-Kępska M, Leppert W, Wordliczek J. Bone pain in cancer patients: mechanisms and current treatment. Int J Mol Sci. 2019;20(23).
52. Jimenez-Andrade JM, Bloom AP, Stake JI, Mantyh WG, Taylor RN, Freeman KT, et al. Pathological sprouting of adult nociceptors in chronic prostate cancer-induced bone pain. J Neurosci. 2010;30(44):14649–56.
53. Twycross RG. Management of pain in skeletal metastases. Clin Orthop Relat Res. 1995;312:187–96.
54. Jayarangaiah A, Kemp AK, Theetha Kariyanna P. Bone metastasis. 2025.

55. Boussios S, Cooke D, Hayward C, Kanellos FS, Tsiouris AK, Chatziantoniou AA, et al. Metastatic spinal cord compression: unraveling the diagnostic and therapeutic challenges. Anticancer Res. 2018;38(9):4987–97.
56. Loblaw DA, Laperriere NJ, Mackillop WJ. A population-based study of malignant spinal cord compression in Ontario. Clin Oncol (R Coll Radiol). 2003;15(4):211–7.
57. Perrin RG, Laxton AW. Metastatic spine disease: epidemiology, pathophysiology, and evaluation of patients. Neurosurg Clin N Am. 2004;15(4):365–73.
58. Quraishi NA, Esler C. Metastatic spinal cord compression. BMJ. 2011;342:d2402.
59. Metastatic spinal cord compression: diagnosis and management of patients at risk of or with metastatic spinal cord compression - PubMed [Internet]. [cited 2025 Mar 31]. Available from: https://pubmed.ncbi.nlm.nih.gov/22171401/
60. Turner JJO. Hypercalcaemia - presentation and management. Clin Med (Lond). 2017;17(3):270–3.
61. Asonitis N, Angelousi A, Zafeiris C, Lambrou GI, Dontas I, Kassi E. Diagnosis, pathophysiology and management of hypercalcemia in malignancy: a review of the literature. Horm Metab Res. 2019;51:770–8.
62. Vakiti A, Anastasopoulou C, Mewawalla P. Malignancy-related hypercalcemia. 2025.
63. Stewart AF. Clinical practice. Hypercalcemia associated with cancer. N Engl J Med. 2005;352(4):373–9.
64. Buggay D, Jaffe K. Metastatic bone tumors of the pelvis and lower extremity. J Surg Orthop Adv. 2003;12(4):192–9.
65. Roodman GD. Biology of osteoclast activation in cancer. J Clin Oncol. 2001;19(15):3562–71.
66. Saad F, Lipton A, Cook R, Chen YM, Smith M, Coleman R. Pathologic fractures correlate with reduced survival in patients with malignant bone disease. Cancer. 2007;110(8):1860–7.
67. Groot OQ, Lans A, Twining PK, Bongers MER, Kapoor ND, Verlaan JJ, et al. Clinical outcome differences in the treatment of impending versus completed pathological long-bone fractures. J Bone Joint Surg Am. 2022;104(4):307–15.
68. Jawad MU, Scully SP. In brief: classifications in brief: Mirels' classification: metastatic disease in long bones and impending pathologic fracture. Clin Orthop Relat Res. 2010;468(10):2825–7.
69. Tatar Z, Soubrier M, Dillies AF, Verrelle P, Boisgard S, Lapeyre M. Assessment of the risk factors for impending fractures following radiotherapy for long bone metastases using CT scan-based virtual simulation: a retrospective study. Radiat Oncol. 2014;9:227.
70. Oh E, Seo SW, Yoon YC, Kim DW, Kwon S, Yoon S. Prediction of pathologic femoral fractures in patients with lung cancer using machine learning algorithms: Comparison of computed tomography-based radiological features with clinical features versus without clinical features. J Orthop Surg (Hong Kong). 2017;25(2):2309499017716243.
71. Gonzalez MR, Bryce-Alberti M, Pretell-Mazzini J. Management of long bones metastatic disease: concepts that we all know but not always remember. Orthop Res Rev. 2022;14:393–406.
72. Plaza JA, Perez-Montiel D, Mayerson J, Morrison C, Suster S. Metastases to soft tissue: a review of 118 cases over a 30-year period. Cancer. 2008;112(1):193–203.
73. Tuoheti Y, Okada K, Osanai T, Nishida J, Ehara S, Hashimoto M, et al. Skeletal muscle metastases of carcinoma: a clinicopathological study of 12 cases. Jpn J Clin Oncol. 2004;34(4):210–4.

Multidisciplinary Treatment of Musculoskeletal Metastases

3

Marcos R. Gonzalez, Carolina Mendez-Guerra, and Juan Pretell-Mazzini

Introduction

Musculoskeletal metastases affect an estimated 300,000–600,000 patients annually in the United States, a number projected to grow in the coming decades [1, 2]. With the rise in cancer survivors, the economic burden of metastatic bone disease (MBD) is substantial, accounting for nearly one-fifth of total oncology costs [2]. Shulman et al. estimated that the mean direct medical cost for all cancers combined was \$75,329 for patients with MBD and \$31,382 for controls (patients with cancer without MBD) [2]. Skeletal-related events (SREs) such as pathologic fractures, spinal cord compression, hypercalcemia of malignancy, and severe bone pain requiring palliative radiotherapy or surgery are a major cost driver, often requiring prolonged hospital stays, multiple procedures, and numerous outpatients visits [3].

Musculoskeletal metastases and SREs are also associated with worsening quality of life (QoL) and dire prognosis. SREs, which are detected simultaneously with MBD in 22.4% of patients with breast cancer, 22.4% of those with lung cancer, and 10% of those with prostate cancer, further diminish the survival rate of patients [4]. Notably, the Scandinavian Sarcoma Group Skeletal Metastasis Registry report of 1195 patients with surgically treated skeletal metastases found a survival of 41% and 2% at 1 and 5 years' posttreatment, respectively [5]. With regard to QoL, studies on metastatic prostate cancer have reported clinically meaningful reduction in

M. R. Gonzalez
Division of Orthopaedic Oncology, Department of Orthopaedic Surgery, Massachusetts General Hospital, Harvard Medical School, Boston, MA, USA

C. Mendez-Guerra
Facultad de Ciencias de la Salud, Universidad Peruana de Ciencias Aplicadas, Lima, Perú

J. Pretell-Mazzini (✉)
Division of Musculoskeletal Oncology - Miami Cancer Institute, Baptist Health of South Florida / Herbert Wertheim College of Medicine - Florida International University, Miami, FL, USA
e-mail: juan.pretell@baptisthealth.net

J. Pretell-Mazzini, S. A. Lozano Calderón (eds.), *Multidisciplinary Approach of Musculoskeletal Metastasis*, https://doi.org/10.1007/978-3-032-09498-8_3

health-related QoL instruments, such as FACT-P (Functional Assessment of Cancer Therapy-Prostate) and ED-5D (EuroQol 5 Dimensions), after presenting SREs [6, 7].

Historically, the diagnosis and management of musculoskeletal metastases were not prioritized due to their association with poor prognosis and imminent death. However, advancements in medical treatments, alongside a focus on pain management and functional improvement, highlight the importance of a multidisciplinary approach. This chapter will explore the key components of multidisciplinary care for patients with musculoskeletal metastases and discuss the tools available for prompt diagnosis and treatment.

Importance of Multidisciplinary Care

With greater understanding of cancer's foundations, the landscape of cancer care is continuously evolving. These advances in the field of cancer have greatly improved the life expectancy of advanced-stage cancer patients [8]. Conversely, cancer patients with protracted clinical courses are more likely to face complications resulting from cancer, such as bone metastases [9].

The management of metastatic bone disease has evolved in recent times, prompting the refinement of current treatment objectives and strategies. Therefore, preventing skeletal-related events (SREs) as well as restoring function in patients with bone metastases has become the main goal of current therapies [10].

Treatment strategies for bone metastases must be comprehensive to allow for appropriate and individualized decision-making, tailored to each patient's characteristics. Several factors must be taken into consideration when assessing a cancer patient with a bone lesion, such as the origin and nature of the lesion itself, the effect of the lesion on the bone, the degree of disease dissemination, and the patient's overall heath and prognosis [11]. A comprehensive assessment of these factors requires a multidisciplinary team, comprised of, but not limited to, a medical oncologist, a radiation oncologist, a radiologist, a pathologist, an orthopedic oncologic surgeon, an anesthesiologist, a palliative care specialist, a rehabilitation specialist, a cancer nurse specialist, an occupational therapist, and a medical social worker [10, 11]. Each specialist plays a critical role in determining the most suitable treatment option as well as the best environment possible so that these interventions are adequate for each patient. The main goals are to reduce the morbidity of cancer patients, including limiting psychosocial burden, as well as to improve their quality of life [12]. The usefulness of a multidisciplinary center for cancer patients has been supported, especially in terms of reducing psychophysical morbidity [12].

Life Expectancy and Prognosis Estimation

Adequate estimation of life expectancy and prognosis in a patient with metastatic bone disease is of paramount importance, as it frames the most appropriate care for each patient's condition [11]. A stratification system for the prediction of expected survival in patients with metastatic bone disease has been designed based on the primary tumor type, the Karnofsky performance score (KPS), and the presence of visceral and/or cerebral metastases [13]. This stratification model classifies patients into four categories, each one with different survival estimates [13]. The median survival estimate for category A is 31.2 months, 15.4 months for category B, 4.8 months for category C, and 1.6 months for category D [13] (Fig. 3.1).

KPS is a rating scale widely used for the assessment of functional impairment in cancer patients. It has proven to be reliable and valid when gauging the functional status of oncologic patients, and, nowadays, it is routinely used as a standard assessment tool by oncologists (Table 3.1) [14, 15].

The Eastern Cooperative Oncology Group (ECOG) Performance Status is also a valid assessment tool frequently employed by clinicians when assessing the functional status of oncologic patients (Table 3.2) [16].

For both performance scales, assessment is based on the clinician's observation of the patient's activity level, the ability to care for oneself, and the ability to work [16]. As a prognostic tool, KPS has shown to be more reliable for predicting survival when assessing patients with lower scores [17]. Additionally, both performance scales have been shown to yield lower sensitivities when assessing elderly cancer patients [18]. Moreover, it is also important to highlight that both scales are performed by observing the patient's functional status rather than asking questions, and, therefore, due to its nature, more reliable scores are often rated by experienced clinicians [19].

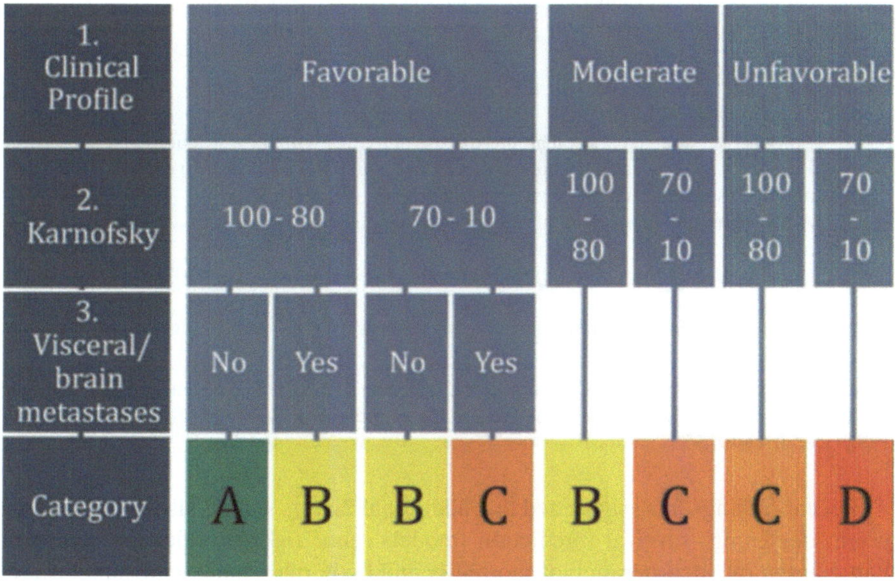

Fig. 3.1 Stratification system for patients with symptomatic spinal metastases. (Displayed with permission from Bollen et al. [13])

Table 3.1 Karnofsky Performance Status scale definitions rating (%) criteria

Health status	Score %	Description
Able to carry on normal activity and work; no special care needed	100	Normal, no complaints; no evidence of disease
	90	Able to carry on normal activity; minor signs or symptoms of disease
	80	Normal activity with effort; signs or symptoms of disease
Unable to work; able to live at home and care for most personal needs; varying amounts of assistance needed.	70	Cares for self; unable to carry on normal activity or do active work
	60	Requires occasional assistance but is able to care for most personal needs
	50	Requires considerable assistance but is able to care for most personal needs
Unable to care for self; requires equivalent of institutional or hospital care; disease may be progressing rapidly	40	Disabled; requires special care and assistance
	30	Severely disabled; hospital admission is indicated, although death is not imminent
	20	Very sick; hospital admission necessary; active supportive treatment necessary
	10	Moribund; fatal processes progressing rapidly
	0	Dead

Reference: Karnofsky DA, Abelmann WH, Craver LF, Burchenal JH. The use of the nitrogen mustards in the palliative treatment of carcinoma. Cancer. 1948 Nov;1(4):634-56

Table 3.2 Eastern Clinical Oncology Group (ECOG) Performance Status

Grade	Description
0	Fully active, able to carry on all pre-disease activities without restriction
1	Restricted in physically strenuous activity but ambulatory and able to carry out work of a light or sedentary nature, e.g., light housework, office work
2	Ambulatory and capable of all self-care but unable to carry out any work activities; up and about >50% of waking hours
3	Capable of only limited self-care; confined to the bed or chair for >50% of waking hours
4	Completely disabled; cannot carry on any self-care; totally confined to the bed or chair
5	Death

Reference: Oken MM, Creech RH, Tormey DC, Horton J, Davis TE, McFadden ET, Carbone PP. Toxicity and response criteria of the Eastern Cooperative Oncology Group. Am J Clin Oncol. 1982 Dec;5(6):649–55

Recent findings have suggested that additional factors should also be considered when designing survival prediction models and risk stratification systems. Particularly, the incorporation of morbidity and body mass index into survival prediction models has been recommended by the current literature, as they have been associated with poorer survival rates [20]. Interestingly, the incorporation of patient's specific data into calculators, using machine learning (ML) algorithms, for predicting survival in metastatic bone disease, has been described [21]. While these findings are innovative, advancing in the field of ML algorithms poses significant challenges, such as improving model interpretability and diversifying data [21–24].

Prevention of Skeletal-Related Events

Preventing SREs and preserving limb function and stability in patients with musculoskeletal metastasis are currently framed as the primary objectives in the management of metastatic bone disease [10]. SREs comprise a series of complications resulting from bone metastasis, such as cancer-related pain, hypercalcemia, pathologic bone fractures, and spinal cord compression [9]. SREs have been found to cause significant impairment in patient's activities of daily living and quality of life; therefore, they are strongly associated with increased morbidity, mortality, and cancer-related costs [10]. As outlined in previous lines, the prevention of SREs constitutes a major priority in the management of metastatic bone disease. For the past decade, bisphosphonates have been established as first-line pharmacotherapy for the prevention and treatment of SREs in patients with bone metastasis [25]. Zoledronic acid was the first bisphosphonate found to be effective in decreasing skeletal complications in patients with bone metastasis from solid tumors other than breast and prostate cancers [26]. Notably, in recent times, alternative treatment options have also proven to be effective in preventing SREs. Indeed, denosumab, a monoclonal antibody directed against receptor activator of nuclear kappa-B ligand (RANK-L), has shown to moderately reduce the likelihood of pathologic fractures compared to zoledronic acid in patients with bone metastasis [27]. Further discussion regarding current treatment strategies aimed at preventing SREs in oncologic patients will be presented in the following sections.

Pathologic fractures are among the most disabling skeletal complications resulting from bone metastasis. Early diagnosis and prompt treatment of pathologic fractures, as well as their prevention, are fundamental to prevent disability and death in cancer patients with metastatic bone disease. In this regard, patients undergoing prophylactic stabilization of impending pathologic fractures display a survival benefit in comparison to patients treated for displaced pathologic fractures [28, 29]. Moreover, estimating the risk of fracture in patients with bone metastasis is of paramount importance. A more detailed discussion regarding the assessment and management of impending fractures will be presented later in this book.

Surgical treatment is tailored on the basis of the overall prognosis and risk of fracture in patients with metastatic bone disease [30]. Several factors such as the site of bone metastasis, extent of bone involvement, and the number of lesions should also be taken into consideration when choosing the appropriate surgical procedure [31]. Conventionally, intramedullary nailing has been extensively used for the management of impending and complete pathologic fractures in the setting of bone metastasis, as it preserves mobility and increases quality of life [32]. Nevertheless, in recent years, selected patients have been treated with limb reconstruction surgeries, such as wide resection and endoprosthetic reconstruction, as it enables for pain relief, potential avoidance of adjuvant radiation therapy when Ro resections are achieved as well as early weight-bearing capacity, with low complication and failure rates [33]. The current literature describes the use of intramedullary nailing for older patients with multiple metastases and poor prognosis, whereas resection of bone lesions and endoprosthetic reconstruction is recommended for younger patients with single or oligometastatic disease and an overall good prognosis [34]. Regardless of the surgical procedure selected, the method of fixation utilized should be durable for the expected survival and the rehabilitation time should not surpass the life expectancy of the patient [30]. Further discussion regarding the current surgical options for the management of metastatic bone disease will be provided in the following sections.

The Extent of the Disease and Clinical Decision-Making

The extent of the disease of the primary tumor is an important factor when assessing a patient with bone metastasis, as it determines treatment strategy and the patient's overall prognosis [30]. In the current literature, dissemination status is categorized into three groups: solitary lesions, oligometastases (defined as two to four bone lesions), and diffuse lesions [30]. Current evidence suggests that patients with oligometastases of the bone should be distinguished from patients with diffuse disease, as different prognostic values are encountered in each group. Indeed, the current literature reveals that patients with oligometastases, who are treated with adequate local control procedures such as radiation treatment and/or surgical intervention, exhibit better survival and disease prognosis [11]. Moreover, with the advent of new local treatments, such as minimally invasive surgical techniques, stereotactic radiotherapy, and radio-frequency ablation, new strategies in the management of metastatic bone disease are emerging [35]. Adequate assessment of the dissemination status is of paramount importance, since, as mentioned in previous lines, it greatly influences the decision-making process in the management of a patient with metastatic bone disease. The preferred test for assessing dissemination status is positron emission tomography/computer tomography (PET/CT) scanning, due to its superior sensitivity and specificity compared to bone scintigraphy, which has constituted the test of choice in past years [36].

The complexity of managing metastatic bone disease underscores the critical need for multidisciplinary care in cancer treatment. As the landscape of cancer therapies evolves, so must our approach to patient care, integrating insights from a diverse array of specialists. This collaborative framework not only enhances individualized treatment plans but also addresses the multiple challenges patients face, from preventing skeletal-related events to optimizing quality of life. By performing a comprehensive evaluation of each patient's unique circumstances—considering factors such as disease progression, functional status, and psychosocial needs—morbidity can be significantly reduced and outcomes improved.

Case Example: What Can Go Wrong When Multidisciplinary Treatment Is Not Performed?

We look at the case of a 75-year-old female with a history of 2 months of pain in the right thigh when weight-bearing. After twisting of the right lower extremity while going to the bathroom, she suffered a fall from standing height with subsequent pain and an inability to stand up. The patient was transferred to the ER of a community hospital where she was diagnosed with a right femoral shaft fracture (Fig. 3.2a). After this, she underwent closed reduction and fixation with intramedullary nail (Fig. 3.2b). The patient was discharged and followed up by an orthopedic surgeon for 6 weeks. After this, the patient continued to be in a rehabilitation center and had increasing pain in the right thigh with difficulty on weight-bearing. After 4 months, the patient continued to have increasing pain in the right thigh with increasing volume and was wheelchair-bound. The patient was taken to the ER of a local hospital and was diagnosed with a "destructive lesion of the right femur" (Fig. 3.2c). After this, the patient was transferred to our institution, and, after undergoing staging studies and biopsy of the soft tissue component of thigh mass seen on MRI (Fig. 3.2d, e), she was diagnosed with stage IV lung carcinoma. Unfortunately, the patient developed massive pulmonary embolism (PE) and died.

Discussion In this case, the orthopedic surgeon who initially performed the surgery was not aware of the lytic lesion at the fracture site. So, there was no adequate initial workup, with no staging studies or biopsy. Due to the unawareness of this pathologic fracture, the patient was not followed up properly and was lost to follow-up in a rehabilitation center. Unfortunately, due to progression of disease, the patient was transferred to a cancer center very late and died.

Even though, technically, a "femoral shaft fracture" can be fixed in a community hospital, patients with a suspicious mechanism for a fracture like in this case should be assessed with precaution. Not identifying a "bone lesion" and missing the

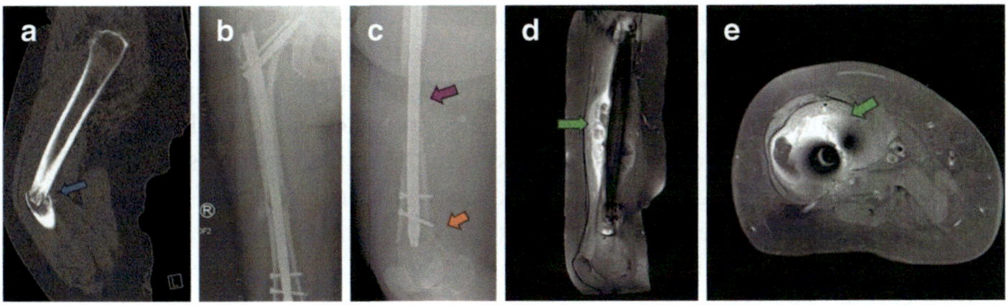

Fig. 3.2 A 75-year-old female with stage IV lung carcinoma. (**a**) Right femur CT sagittal view showing a lytic lesion (blue arrow) and a pathologic fracture. (**b**) AP view plain radiograph of the right femur after closed reduction and internal fixation with intramedullary nail. (**c**) Lateral view plain radiograph of the right femur showing extensive femoral shaft destruction (purple arrow) with calcification of soft tissue, soft tissue mass shadow, and distal peri-implant fracture. (**d, e**) T2-weighted signal MRI sagittal and axial views showing a large soft tissue component in the femoral shaft (green arrows)

opportunity to give the patient adequate care had devastating consequences. If there is any doubt when assessing a patient, it is better to refer them to a specialized center where the patient can receive multidisciplinary care.

Multidisciplinary Care

Systemic Treatment

Systemic treatment of musculoskeletal metastases refers to both antineoplastic and bone-modifying agents (BMAs). Antineoplastic agents directly or indirectly target and destroy cancerous cells, stopping the spread of the tumor. While there is some overlap among classes, antineoplastic agents are often classified into one of the following categories: chemotherapy, immunotherapy, endocrine/hormonal therapy, and targeted therapy. The treatment choice relies on the biology of the primary tumor, but the ultimate goal is to either palliate pain in patients with multifocal metastases or prolong survival in patients with solitary bone metastases. A full discussion of available systemic treatments for each cancer type is beyond the scope of this chapter.

Bone-modifying agents (BMAs) are typically classified into two classes: bisphosphonates and RANK-L inhibitors (denosumab). Bisphosphonates act by inhibiting osteoclast bone resorption, thus stabilizing the bone architecture. Denosumab is a monoclonal antibody that binds avidly to RANK-L, preventing its interaction with its receptor RANK and therefore suppressing bone resorption by inhibition of osteoclast activity [37, 38]. In patients with MBD, there are three main indications for BMAs: (1) hypercalcemia, (2) prevention of SREs, and (3) nonmechanical metastatic bone pain. While radiation therapy (RT) remains the first-line treatment for localized bone pain, BMAs have demonstrated significant pain relief across different tumor types [39].

No randomized controlled trials (RCTs) have yet determined the optimal timing to initiate BMA therapy in patients with bone metastases. Moreover, available tools to estimate the risk of SREs have not been widely validated and are usually not incorporated in treatment guidelines. However, a pivotal RCT comparing zoledronic acid to denosumab found that 37% of patients presented SREs at the time of study inclusion after a median of only 2 months from the initial diagnosis of bone metastases [40]. Due to the high prevalence and associated morbidity of SREs, the 2020 European Society for Medical Oncology (ESMO) Clinical Practice Guidelines recommend starting BMA as soon as bone metastases are diagnosed, regardless of patient symptoms [38].

While current guidelines do not recommend one BMA over another, RCTs have demonstrated different efficacy profiles between agents. The ZICE (Zoledronate versus Ibandronate Comparative Evaluation) trial demonstrated that zoledronate was superior to ibandronate in preventing SREs in patients with bone metastases from breast cancer [41]. Likewise, a randomized, double-blind RCT showed that zoledronate reduced the overall risk of SREs by an additional 16% compared to pamidronate

[42]. When comparing bisphosphonates to denosumab, studies have reported a higher efficacy with the latter [40, 43, 44]. In patients with advanced breast cancer, Stopeck et al. found that denosumab was superior to zoledronate in delaying time to the first on-study SRE (hazard ratio (HR) = 0.82) [40]. In addition, denosumab has a more favorable safety profile with less incidence of renal toxicity and acute-phase reactions. Despite its potential advantages, limitations to a wider use of denosumab include short post-marketing surveillance time, increased costs, and potential increase in infection rates in patients with osteoporosis and breast cancer [45].

Despite their multiple benefits, there are serious adverse effects associated with the use of BMAs. Osteonecrosis of the jaw (ONJ) is the most important adverse effect and is associated with prolonged BMA administration. The incidence of ONJ is about 1% per year on treatment for both zoledronate and denosumab [46]. Atypical femoral fractures (AFFs) are another severe complication associated with the long-term use of bisphosphonates or denosumab. Notably, long-term use of BMAs has been associated with an absolute risk of 100 per 100,000 person-years [47]; the risk, however, may decline when treatment is stopped. Since nephrotoxicity is another well-known adverse effect associated with BMAs, evaluation of renal function is mandatory before starting bisphosphonates. Recent guidelines have recommended lower doses of clodronate and zoledronate and longer infusion times for pamidronate in patients with mild-to-moderate renal impairment (creatinine clearance 30–60 mL) [48]. Unlike osteoporosis, for which "drug holidays" are recommended, BMAs are typically continued indefinitely due to the increased risk of SREs associated with treatment interruption [38]. However, treatment may be interrupted in patients with good prognostic features such as oligometastatic disease, a perceived low risk of bone complications, and durable response to systemic treatment.

Local Treatment

Radiation Therapy

The main indications of RT in patients with bone metastases are pain unresponsive to oral analgesia, malignant spinal cord compression, and treatment of symptomatic bone lesions. External beam RT (EBRT) is highly effective in alleviating pain, with 60% of patients reporting pain relief and 20% complete response [49]. Available regimens include a single 8-Gray (Gy) fraction and multiple-fraction regimens, such as 20 Gy in 5 fractions, 24 Gy in 6 fractions, and 30 Gy in 10 fractions. Multiple RCTs performed in the last few decades have demonstrated pain relief equivalency between single- and multiple-fraction regimens [50, 51]. Moreover, a recent systematic review on RT fractionation for bone metastases has reported similar overall response rates but higher retreatment rates in the single-fraction arm (20%) compared to the multiple-fraction one (8%) [49]. In light of these findings, a single-fraction RT schedule is recommended for the treatment of symptomatic and uncomplicated bone metastases or in patients with a poor prognosis [52] However, due to lack of experience with large fraction sizes, participation in RCTs, and reimbursement-related factors, a wider use of single-fraction regimens remains restricted [53].

Despite its critical role in managing painful bone metastases, toxicity and radiation-associated fractures (RAFs) are important concerns and a reason for multidisciplinary management. RAFs occur mostly after high doses of RT (>50 Gy) and within 1 year of irradiation [54]. Treatment of RAFs is difficult, and <50% achieve union after initial fixation [55]. Due to the high morbidity associated with these, prophylactic internal fixation before RT may be considered.

Surgery

The goal of surgery in patients with musculoskeletal metastases is rarely curative; rather, it seeks to provide pain relief, restore patient function, and restore skeletal stability. Ideally, patients should be able to bear weight immediately after surgery.

The decision to prophylactically stabilize an impending pathologic fracture hinges on an accurate estimation of fracture risk, which will be discussed later in this book. Available fixation strategies for impending and displaced pathologic fractures may be classified into three categories: intramedullary nailing (IMN), open reduction and internal fixation (ORIF), and prosthetic reconstruction. IMN is the most widely used technique, as it provides whole-bone fixation, is minimally invasive, and allows for early mobilization. However, it is associated with a higher revision rate compared to endoprostheses, especially in patients with poor bone stock [56]. ORIF offers improved visualization and allows for tumor debulking, especially in radioresistant tumors; however, it is associated with higher revision rates than IMN [45, 56]. Prosthetic reconstruction offers a valuable alternative in tumors with extensive bone destruction and/or for those nonresponsive to medical treatment as they allow for early mobilization and do not rely on fracture healing. Janssen et al. compared these three surgical strategies and found that fixation failure was most common after ORIF (13% vs. 3% in IMN and 0% in endoprosthetic reconstruction) [56]. However, the authors also found that deep infection occurred more frequently with endoprostheses (8.6%) than other strategies. Ultimately, the choice of surgical treatment will depend on the patient's comorbidity profile and life expectancy.

Conclusions

Multidisciplinary care is critical to the current management of musculoskeletal malignancies. Adequate treatment selection relies on accurate estimation of patient prognosis, risk of fracture, and extent of disease. Current treatment strategies include a combination of local and systemic therapies offered by different subspecialties.

References

1. Li S, Peng Y, Weinhandl ED, et al. Estimated number of prevalent cases of metastatic bone disease in the US adult population. Clin Epidemiol. 2012;4:87–93. https://doi.org/10.2147/CLEP.S28339.

2. Schulman KL, Kohles J. Economic burden of metastatic bone disease in the U.S. Cancer. 2007;109(11):2334–42. https://doi.org/10.1002/cncr.22678.
3. Durán I, Garzón C, Sánchez A, et al. Cost analysis of skeletal-related events in Spanish patients with bone metastases from solid tumours. Clin Transl Oncol. 2014;16(3):322–9. https://doi.org/10.1007/s12094-013-1077-2.
4. Oster G, Lamerato L, Glass AG, et al. Natural history of skeletal-related events in patients with breast, lung, or prostate cancer and metastases to bone: a 15-year study in two large US health systems. Support Care Cancer. 2013;21(12):3279–86. https://doi.org/10.1007/s00520-013-1887-3.
5. Ratasvuori M, Wedin R, Keller J, et al. Insight opinion to surgically treated metastatic bone disease: Scandinavian Sarcoma Group Skeletal Metastasis Registry report of 1195 operated skeletal metastasis. Surg Oncol. 2013;22(2):132–8. https://doi.org/10.1016/j.suronc.2013.02.008.
6. Saad F, Ivanescu C, Phung D, et al. Skeletal-related events significantly impact health-related quality of life in metastatic castration-resistant prostate cancer: data from PREVAIL and AFFIRM trials. Prostate Cancer Prostatic Dis. 2017;20(1):110–6. https://doi.org/10.1038/pcan.2016.62.
7. Ivanescu C, Phung D, Loriot Y, et al. Skeletal-related events (SRES) impact significantly the health-related quality of life (HRQOL) of chemo-naive men with metastatic castration resistant prostate cancer (MCRPC). Value Health. 2014;17(7):A650–1. https://doi.org/10.1016/j.jval.2014.08.2365.
8. Lichtenberg FR. The impact of pharmaceutical innovation on premature mortality, hospital separations, and cancer survival in Australia. Econ Rec. 2017;93(302):353–78. https://doi.org/10.1111/1475-4932.12332.
9. Tsuzuki S, Park SH, Eber MR, Peters CM, Shiozawa Y. Skeletal complications in cancer patients with bone metastases. Int J Urol. 2016;23(10):825–32. https://doi.org/10.1111/iju.13170.
10. Kimura T. Multidisciplinary approach for bone metastasis: a review. Cancers (Basel). 2018;10(6) https://doi.org/10.3390/cancers10060156.
11. Ardakani AHG, Faimali M, Nystrom L, et al. Metastatic bone disease: early referral for multidisciplinary care. Cleve Clin J Med. 2022;89(7):393–9. https://doi.org/10.3949/ccjm.89a.21062.
12. Ibrahim T, Flamini E, Fabbri L, et al. Multidisciplinary approach to the treatment of bone metastases: Osteo-oncology center, a new organizational model. Tumori J. 2009;95(3):291–7. https://doi.org/10.1177/030089160909500304.
13. Bollen L, van der Linden YM, Pondaag W, et al. Prognostic factors associated with survival in patients with symptomatic spinal bone metastases: a retrospective cohort study of 1,043 patients. Neuro-Oncology. 2014;16(7):991–8. https://doi.org/10.1093/neuonc/not318.
14. Schag CC, Heinrich RL, Ganz PA. Karnofsky performance status revisited: reliability, validity, and guidelines. J Clin Oncol. 1984;2(3):187–93. https://doi.org/10.1200/JCO.1984.2.3.187.
15. Perlow HK, Cassidy V, Farnia B, et al. Impact of performance status and comorbidity on palliative radiation treatment tolerance and end-of-life decision-making. Adv Radiat Oncol. 2019;4(1):127–33. https://doi.org/10.1016/j.adro.2018.09.002.
16. Garman KS, Cohen HJ. Functional status and the elderly cancer patient. Crit Rev Oncol Hematol. 2002;43(3):191–208. https://doi.org/10.1016/s1040-8428(02)00062-8.
17. Yates JW, Chalmer B, McKegney FP. Evaluation of patients with advanced cancer using the Karnofsky performance status. Cancer. 1980;45(8):2220–4. https://doi.org/10.1002/1097-0142(19800415)45:8<2220::aid-cncr2820450835>3.0.co;2-q.
18. Extermann M, Aapro M. Assessment of the older cancer patient. Hematol Oncol Clin North Am. 2000;14(1):63–77, viii–ix. https://doi.org/10.1016/s0889-8588(05)70278-1
19. Maltoni M, Nanni O, Derni S, et al. Clinical prediction of survival is more accurate than the Karnofsky performance status in estimating life span of terminally ill cancer patients. Eur J Cancer. 1994;30A(6):764–6. https://doi.org/10.1016/0959-8049(94)90289-5.
20. Janssen SJ, van der Heijden AS, van Dijke M, et al. 2015 Marshall Urist young investigator award: prognostication in patients with long bone metastases: does a boosting algorithm

improve survival estimates? Clin Orthop Relat Res. 2015;473(10):3112–21. https://doi.org/10.1007/s11999-015-4446-z.

21. Rizk PA, Gonzalez MR, Galoaa BM, et al. Machine learning-assisted decision making in orthopaedic oncology. JBJS Rev. 2024;12(7) https://doi.org/10.2106/JBJS.RVW.24.00057.
22. Willeumier JJ, van der Linden YM, van der Wal CWPG, et al. An easy-to-use prognostic model for survival estimation for patients with symptomatic long bone metastases. J Bone Joint Surg Am. 2018;100(3):196–204. https://doi.org/10.2106/JBJS.16.01514.
23. de Groot TM, Ramsey D, Groot OQ, et al. Does the SORG machine-learning algorithm for extremity metastases generalize to a contemporary cohort of patients? Temporal validation from 2016 to 2020. Clin Orthop Relat Res. 2023;481(12):2419–30. https://doi.org/10.1097/CORR.0000000000002698.
24. Tseng TE, Lee CC, Yen HK, et al. International validation of the SORG machine-learning algorithm for predicting the survival of patients with extremity metastases undergoing surgical treatment. Clin Orthop Relat Res. 2022;480(2):367–78. https://doi.org/10.1097/CORR.0000000000001969.
25. Iranikhah M, Wilborn TW, Wensel TM, Ferrell JB. Denosumab for the prevention of skeletal-related events in patients with bone metastasis from solid tumor. Pharmacotherapy. 2012;32(3):274–84. https://doi.org/10.1002/j.1875-9114.2011.01092.x.
26. Rosen LS, Gordon D, Tchekmedyian S, et al. Zoledronic acid versus placebo in the treatment of skeletal metastases in patients with lung cancer and other solid tumors: a phase III, double-blind, randomized trial--the Zoledronic Acid Lung Cancer and Other Solid Tumors Study Group. J Clin Oncol. 2003;21(16):3150–7. https://doi.org/10.1200/JCO.2003.04.105.
27. Al Farii H, Frazer A, Farahdel L, AlFayyadh F, Turcotte R. Bisphosphonates versus denosumab for prevention of pathological fracture in advanced cancers with bone metastasis: a meta-analysis of randomized controlled trials. J Am Acad Orthop Surg Glob Res Rev. 2020;4(8):e20.00045. https://doi.org/10.5435/JAAOSGlobal-D-20-00045.
28. Ristevski B, Jenkinson RJ, Stephen DJG, et al. Mortality and complications following stabilization of femoral metastatic lesions: a population-based study of regional variation and outcome. Can J Surg. 2009;52(4):302–8. http://www.ncbi.nlm.nih.gov/pubmed/19680515
29. Philipp TC, Mikula JD, Doung YC, Gundle KR. Is there an association between prophylactic femur stabilization and survival in patients with metastatic bone disease? Clin Orthop Relat Res. 2020;478(3):540–6. https://doi.org/10.1097/CORR.0000000000000803.
30. Willeumier JJ, van der Linden YM, van de Sande MAJ, Dijkstra PDS. Treatment of pathological fractures of the long bones. EFORT Open Rev. 2016;1(5):136–45. https://doi.org/10.1302/2058-5241.1.000008.
31. Johnson SK, Knobf MT. Surgical interventions for cancer patients with impending or actual pathologic fractures. Orthop Nurs. 2008;27(3):160–71; quiz 172–173. https://doi.org/10.1097/01.NOR.0000320543.90115.d5
32. Arpornsuksant P, Morris CD, Forsberg JA, Levin AS. What factors are associated with local metastatic lesion progression after intramedullary nail stabilization? Clin Orthop Relat Res. 2022;480(5):932–45. https://doi.org/10.1097/CORR.0000000000002104.
33. Park DH, Jaiswal PK, Al-Hakim W, et al. The use of massive endoprostheses for the treatment of bone metastases. Sarcoma. 2007;2007:62151. https://doi.org/10.1155/2007/62151.
34. Pala E, Procura A, Trovarelli G, Berizzi A, Ruggieri P. Intramedullary nailing for impending or pathologic fracture of the long bone: titanium vs carbon fiber peek nailing. EFORT Open Rev. 2022;7(8):611–7. https://doi.org/10.1530/EOR-22-0001.
35. Haussmann J, Matuschek C, Bölke E, Orth K, Ghadjar P, Budach W. The role of local treatment in oligometastatic and oligoprogressive cancer. Dtsch Arztebl Int. 2019;116(50):849–56. https://doi.org/10.3238/arztebl.2019.0849.
36. Pires AO, Borges US, Lopes-Costa PV, Gebrim LH, da Silva BB. Evaluation of bone metastases from breast cancer by bone scintigraphy and positron emission tomography/computed tomography imaging. Eur J Obstet Gynecol Reprod Biol. 2014;180:138–41. https://doi.org/10.1016/j.ejogrb.2014.06.021.

37. de Groot AF, Appelman-Dijkstra NM, van der Burg SH, Kroep JR. The anti-tumor effect of RANKL inhibition in malignant solid tumors - a systematic review. Cancer Treat Rev. 2018;62:18–28. https://doi.org/10.1016/j.ctrv.2017.10.010.
38. Coleman R, Hadji P, Body JJ, et al. Bone health in cancer: ESMO Clinical Practice Guidelines. Ann Oncol. 2020;31(12):1650–63. https://doi.org/10.1016/j.annonc.2020.07.019.
39. D'Oronzo S, Coleman R, Brown J, Silvestris F. Metastatic bone disease: pathogenesis and therapeutic options: up-date on bone metastasis management. J Bone Oncol. 2019;15:4. https://doi.org/10.1016/j.jbo.2018.10.004.
40. Stopeck AT, Lipton A, Body JJ, et al. Denosumab compared with zoledronic acid for the treatment of bone metastases in patients with advanced breast cancer: a randomized, double-blind study. J Clin Oncol. 2010;28(35):5132–9. https://doi.org/10.1200/JCO.2010.29.7101.
41. Barrett-Lee P, Casbard A, Abraham J, et al. Oral ibandronic acid versus intravenous zoledronic acid in treatment of bone metastases from breast cancer: a randomised, open label, non-inferiority phase 3 trial. Lancet Oncol. 2014;15(1):114–22. https://doi.org/10.1016/S1470-2045(13)70539-4.
42. Rosen LS, Gordon D, Kaminski M, et al. Long-term efficacy and safety of zoledronic acid compared with pamidronate disodium in the treatment of skeletal complications in patients with advanced multiple myeloma or breast carcinoma: a randomized, double-blind, multicenter, comparative trial. Cancer. 2003;98(8):1735–44. https://doi.org/10.1002/cncr.11701.
43. Lipton A, Steger GG, Figueroa J, et al. Extended efficacy and safety of denosumab in breast cancer patients with bone metastases not receiving prior bisphosphonate therapy. Clin Cancer Res. 2008;14(20):6690–6. https://doi.org/10.1158/1078-0432.CCR-07-5234.
44. Fizazi K, Lipton A, Mariette X, et al. Randomized phase II trial of denosumab in patients with bone metastases from prostate cancer, breast cancer, or other neoplasms after intravenous bisphosphonates. J Clin Oncol. 2009;27(10):1564–71. https://doi.org/10.1200/JCO.2008.19.2146.
45. Gonzalez MR, Bryce-Alberti M, Pretell-Mazzini J. Management of long bones metastatic disease: concepts that we all know but not always remember. Orthop Res Rev. 2022;14:393–406. https://doi.org/10.2147/ORR.S379603.
46. Saad F, Brown JE, Van Poznak C, et al. Incidence, risk factors, and outcomes of osteonecrosis of the jaw: integrated analysis from three blinded active-controlled phase III trials in cancer patients with bone metastases. Ann Oncol. 2012;23(5):1341–7. https://doi.org/10.1093/annonc/mdr435.
47. Black DM, Geiger EJ, Eastell R, et al. Atypical femur fracture risk versus fragility fracture prevention with bisphosphonates. N Engl J Med. 2020;383(8):743–53. https://doi.org/10.1056/nejmoa1916525.
48. Aapro M, Abrahamsson PA, Body JJ, et al. Guidance on the use of bisphosphonates in solid tumours: recommendations of an international expert panel. Ann Oncol. 2008;19(3):420–32. https://doi.org/10.1093/annonc/mdm442.
49. Rich SE, Chow R, Raman S, et al. Update of the systematic review of palliative radiation therapy fractionation for bone metastases. Radiother Oncol. 2018;126(3):547–57. https://doi.org/10.1016/j.radonc.2018.01.003.
50. 8 Gy single fraction radiotherapy for the treatment of metastatic skeletal pain: randomised comparison with a multifraction schedule over 12 months of patient follow-up. Bone Pain Trial Working Party. Radiother Oncol. 1999;52(2):111–121. http://www.ncbi.nlm.nih.gov/pubmed/10577696
51. Hartsell WF, Scott CB, Bruner DW, et al. Randomized trial of short- versus long-course radiotherapy for palliation of painful bone metastases. J Natl Cancer Inst. 2005;97(11):798–804. https://doi.org/10.1093/jnci/dji139.
52. Wu JSY, Wong RKS, Lloyd NS, et al. Radiotherapy fractionation for the palliation of uncomplicated painful bone metastases - an evidence-based practice guideline. BMC Cancer. 2004;4:71. https://doi.org/10.1186/1471-2407-4-71.

53. Fairchild A, Barnes E, Ghosh S, et al. International patterns of practice in palliative radiotherapy for painful bone metastases: evidence-based practice? Int J Radiat Oncol Biol Phys. 2009;75(5):1501–10. https://doi.org/10.1016/j.ijrobp.2008.12.084.
54. Cannon CP, Lin PP, Lewis VO, Yasko AW. Management of radiation-associated fractures. J Am Acad Orthop Surg. 2008;16(9):541–9. http://www.ncbi.nlm.nih.gov/pubmed/18768711
55. Lin PP, Schupak KD, Boland PJ, Brennan MF, Healey JH. Pathologic femoral fracture after periosteal excision and radiation for the treatment of soft tissue sarcoma. Cancer. 1998;82(12):2356–65. http://www.ncbi.nlm.nih.gov/pubmed/9635528
56. Janssen SJ, Kortlever JTP, Ready JE, et al. Complications after surgical Management of Proximal Femoral Metastasis: a retrospective study of 417 patients. J Am Acad Orthop Surg. 2016;24(7):483–94. https://doi.org/10.5435/JAAOS-D-16-00043.

Part II

Patient Assessment and Diagnostic Approach

Patient Assessment and Selection

4

Jacob Kopeld, Jorge Valverde, and Juan Pretell-Mazzini

Introduction

Metastatic bone disease (MBD) affects more than 4.5 million Americans every year [1]. The prevalence of affected individuals will continue to gradually increase given the advancements in medical oncology. Therefore, appropriate clinical evaluation and functional assessment of these patients is essential for treatment, management, and improving outcomes [1].

The anatomical region affected by metastatic disease depends on the site of the primary cancer. The skeleton is the third-most common site of metastatic disease due to the bone marrow and bone micronutrient environment in which cancer cells can thrive and develop. In metastatic bone disease (MBD), breast, lung, kidney, prostate, and thyroid are the most common carcinomas, accounting for 80% of bone metastases [2, 3]. The most common sites are the spine, pelvis, ribs, skull, proximal humerus, and the long bones of the lower extremity [4]. There are multiple mechanisms by which cancer can damage bone, such as direct invasion by cancer cells or activation of pathways via lytic or blastic mechanisms, the toxic effects of chemotherapy or radiation, and a local increase of inflammatory cytokines. Some of the most vulnerable individuals suffering from bone metastases are elderly patients because of the compounded effect of their preexisting conditions such as low mineral bone density combined with skeletal metastases that accelerate reduction in bone quality. This leads to the more common presentation of chronic bone pain and higher risks of pathologic fractures [5].

J. Kopeld · J. Valverde
Baptist Health Cancer Care, Baptist Health of South Florida, Miami, FL, USA

J. Pretell-Mazzini (✉)
Division of Musculoskeletal Oncology - Miami Cancer Institute, Baptist Health of South Florida / Herbert Wertheim College of Medicine - Florida International University, Miami, FL, USA
e-mail: juan.pretell@baptisthealth.net

J. Pretell-Mazzini, S. A. Lozano Calderón (eds.), *Multidisciplinary Approach of Musculoskeletal Metastasis*, https://doi.org/10.1007/978-3-032-09498-8_4

Given the complexity in the diagnosis and rising number of patients with MBD, it is imperative that the orthopedic oncologists going into practice today grasp the fundamentals of assessing and managing an individual with MBD. Assessment should be thorough, excluding other diseases that may present like metastases, including multiple myeloma (MM), primary lymphoma of bone (PLB), and any type of primary malignancy, as the treatment options between these conditions are different. Appropriate treatment options should be tailored toward creating the most stable and lasting construct and potentially decreasing patient morbidity and mortality [1].

In this chapter, we dive into the step-by-step workup and assessment of patients with or without an oncologic history suffering from a bone lesion likely representing musculoskeletal metastasis with a specific focus on clinical assessment, functional assessment, and quality of life (QoL).

Clinical Assessment

Malignant solitary bone lesions visualized on plain radiographs are 500 times more likely to be a metastatic carcinoma than a primary bone sarcoma, thus creating such a disparity that can lead to a hastened diagnosis and a possible oversight by physicians [1]. Although a patient's clinical scenario may point toward bone metastasis, a staging workup is fundamental to improve a patient's diagnosis, establish a prognosis, and rule out all possibilities of the lesion being a primary tumor. Utilizing a systematic approach to staging may aid the treating surgeon in avoiding unintentional errors such as placing a rod through a sarcoma and subjecting the patient to an amputation that may have been unnecessary [6].

Moreover, the osteolytic or osteoblastic pattern of bone destruction by skeletal metastasis may significantly impact a patient's prognosis [7]. Patients with breast or prostate cancer have the potential to survive for years after the diagnosis because their pathologies are associated with either blastic or mixed lytic/blastic lesions. However, the survival times of those with osteolytic lesions originating from lung, thyroid, or kidney primaries may best be measured in months rather than years due to the higher mortality and higher risk of pathologic fractures [8, 9]. The extent and evolution of the disease's progression along with its ability to spread to the soft tissue also help predict the patient's outcome more precisely. This relationship is supported by the strong correlation between the patient's morbidity and their rate of bone resorption. Thus, along with objective studies such as complete imaging, laboratory values, present comorbidities, and histologic analysis, the clinician must assess each patient on a case-by-case basis by factoring in their functional status and quality of life [8].

History and Physical Examination

Ideally, management of a patient with MBD should include an oncologist, whether orthopedic, radiation, or medical, if not a combination of all the three every 2–4 months [10]. When evaluating a patient, certain clues may be indicative of metastases and cannot be ignored, such as recent weight loss and symptoms that may be linked to the primary tumor such as smoking history, alcoholism, cirrhosis, hematuria, or abdominal pain. Upon physical exam, the breasts, thyroid gland, skin, lymph nodes, and rectum should be examined along with a urinalysis [11]. Close to 75% of patients with skeletal metastasis will present with generalized or localized pain of the involved region and may require further workup [8].

The first step in the initial evaluation is to take a proper medical history, which entails identifying the onset, duration, and nature of the patient's symptoms with specific details about the patient's experienced pain. Determining the characteristics of a patient's pain is essential, whether it is constant or intermittent, worse at night, and responds to analgesics [12]. In addition, a history of trauma, even a questionable one, does not rule out the chances of a pathologic fracture. Hu et al. reported that >75% of pathologic fractures are caused by metastasis occurring spontaneously or during everyday activities without the occurrence of a significant traumatic event [11, 13]. Therefore, a clinician must be aware of any associated prodromal symptoms, which can be highly indicative of an impending pathologic fracture. A patient's morbidity or decreased quality of life is associated with the skeletal-related events (SREs) they may suffer. These SREs not only include pathologic fractures but also other factors that affect the bone such as radiotherapy, bone surgery, spinal cord compression, hypercalcemia, and/or bone pain [9]. Once the diagnosis of a bone lesion and/or an SRE has been established, the next important step involves confirming a true bone metastasis rather than a primary malignancy, which would require a completely different treatment plan [14]. Patients should be subjected to a chest imaging study including radiographs or computed tomography and a series of laboratory tests including serum calcium, blood cell counts, C-reactive protein (CRP), erythrocyte sedimentation rate (ESR), serum electrolytes and creatinine, and serum protein electrophoresis [11]. Further investigation includes testing for tumor markers. Serum tumor markers improve the ability to make the distinction between metastatic carcinoma, a primary bone lesion, and a hematologic malignancy [15]. Finally, advanced imaging will provide key information for further staging and to truly comprehend the extent of MBD.

Complete Blood Count (CBC)

A CBC provides essential information about the patient's hematologic status. Anemia can be a sign of chronic disease or bone marrow involvement, leukocytosis may indicate an inflammatory or infectious process, and thrombocytopenia can suggest marrow infiltration by metastatic disease or multiple myeloma [16, 17].

A Biochemical Profile

A biochemical profile includes tests such as calcium levels, liver function tests, and renal function tests. Elevated calcium levels can be indicative of malignant hypercalcemia, a common complication in metastatic bone disease. Liver and kidney function tests help assess the overall metabolic and clearance functions, which are crucial for planning systemic therapies [8, 18]. Overall, additional laboratory testing for tumor markers such as PSA (prostate-specific antigen), CA 15-3, CA 19-9, and CEA (carcinoembryonic antigen) can provide valuable clues about the primary tumor from which the metastasis developed, unless of course the primary is of an unknown origin (discussed in Chap. 38). Elevated levels of these markers can confirm the presence of specific cancers and guide the choice of systemic treatment [19].

Pathophysiology

Once tumor cells infiltrate bone, they will initiate bone resorption and release transforming growth factor-β (TGF-β) from the bone's stores [20]. Subsequently, this causes a downstream effect of suppressing T-cell proliferation and inhibiting natural killer cells [20]. Elevated serum levels of parathyroid hormone-related protein (PTHrP) can be an indicator of breast cancer bone metastases. PTHrP stimulates bone resorption and osteoclasts through the cytokine receptor activator of nuclear factor-kappa B ligand (RANKL), which plays a key role in osteolysis [21]. Guise et al. (1993) and Edwards et al. (2024) found that the role of PTHrP in malignancy-associated hypercalcemia points toward the potential of PTHrP as a therapeutic target for the possible prevention of bone metastasis formation in patients with breast cancer [22, 23]. In addition to PTHrP, interleukin-6 (IL-6), prostaglandin E2 (PGE2), tumor necrosis factor (TNF), and macrophage colony-stimulating factor (M-CSF) all increase RANK ligand (RANKL) production and are produced in the bone marrow by disseminated tumor cells [24]. Santini et al. reported in a human clinical study that the level of RANK expression, the receptor for RANKL, in primary breast, colorectal, renal, lung, and prostate tumors positively correlates with the frequency of bone metastasis. Higher RANK gene expression correlated with a worse prognosis [25].

Plain Radiographs

The initial imaging modality for evaluating bone lesions is a plain radiograph. Radiographs are quick, accessible, cost-effective, and provide a good overview of the bone architecture. They can reveal critical information about the lesion's size, location, and characteristics (i.e., whether it is lytic, blastic, or mixed). These details help in forming a preliminary diagnosis and guide further imaging studies [26].

Computed Tomography

Computed tomography (CT) has a much higher sensitivity than plain radiography to both osteolytic and osteosclerotic metastases. Unlike plain radiographs, CT images have a higher resolution and no superimposition of structures and, as such, can highlight anatomical structures with precision and in greater detail. This allows CT to be unique in its ability to reveal and evaluate the full extent of osseous destruction and fracture risk, especially in the ribs that have a high cortex-to-marrow ratio [30]. A large advantage of CT imaging is that it can be performed in tandem with the staging or restaging of other organs, thus helping reduce the patient's radiation exposure by limiting the number of separate imaging studies that must be performed. Additionally, although it has a limited soft tissue resolution compared to MRI, CT will still reveal the presence of bone metastasis in the marrow much earlier than a plain radiograph [28]. A lesion involving bone marrow would appear on CT as an increase in attenuation as compared to that of the normal fatty bone [31].

MRI

MRI is invaluable for providing detailed information about a lesion. It offers superior soft tissue contrast compared to other imaging modalities, making it ideal for determining the extent of the lesion and its involvement in the surrounding soft tissues and bone marrow. MRI is particularly useful in surgical planning and determining the exact margins of a tumor. An MRI has better contrast resolution in comparison to a CT scan and allows for distinguishing subtle bone marrow changes, which has led to its primary and widespread use in imaging bone and soft tissues in musculoskeletal oncology. Additionally, the lack of ionizing radiation, ability to image in multiple plains without loss of image resolution, and the tissue characteristics provided by different pulse sequences are all advantages of MRI [26]. The ability of MRI to distinguish variations in the bone marrow makes it an important imaging modality for catching metastatic spread prior to cortical destruction seen on radiographs and the pathologic osteoblastic process that registers on a bone scan. Hence, MRI's specificity for detecting bone metastasis can be as high as 90–95%. Where MRI is inferior to CT is in the detection of cortical instability or destruction because the cortical bone appears dark on both T1 and T2 modalities [28, 32].

Bone Scintigraphy

Bone scintigraphy, or bone scans, are essential for detecting multiple metastatic lesions throughout the skeleton. This whole-body imaging technique highlights areas of increased bone turnover, indicative of metastatic disease, and helps determine the extent of the disease. In bone scintigraphy, the sites of bony metastasis will show "hot" or brighter. These scans remain the primary method of evaluating osseous metastases due to their widespread availability [27, 33].

Unlike positron emission tomography (PET) scans that demonstrate metabolic activity, radionuclide bone scans detect areas of bony repair and can be negative in purely lytic-destructive processes such as myeloma and in metastatic lesions arising from the kidney or thyroid [34]. Highly aggressive metastases may even show up as photopenic areas on a bone scan, which can lead to false-negative results. If the metastasis is diffused enough throughout the bone, it may also present a uniform pattern that can be a false-negative [35, 36]. False positives arise on bone scintigraphy due to its nonspecificity and usually present as multiple benign osseous lesions like eosinophilic granuloma, fibrous dysplasia, and enchondroma [37].

PET/CT

The advantage of PET scans includes superior spatial resolution when compared to conventional bone scans. Two types of radiotracers used in skeletal metastasis include ^{18}F-sodium fluoride (^{18}F-NaF) and ^{18}F-fluorodeoxyglucose (^{18}F-FDG) [32]. These two radiopharmaceuticals each assess different parts of tumor biology. NaF has higher specificity and sensitivity compared to bone scintigraphy for detecting osteoblastic activity, while ^{18}F-FDG reveals sites of high metabolic activity [38]. This type of imaging is useful for detecting metastases that may not be visible in other imaging studies and can reveal the aggressiveness of the tumor based on its metabolic activity. The combination of PET with ^{18}F-FDG and CT is a major advance in tumor identification [39, 40]. The sensitivity of ^{18}F-FDG/PET has proven to be beneficial in the diagnosis, staging, and treatment of bone tumors [41]. Compared to bone scintigraphy, FDG-PET separates itself not by diagnosing the presence of skeletal metastasis but instead confirming its absence in a much larger proportion of patients [41]. In a recent meta-analysis, Liu et al. have reported that the best modality to detect bone metastasis in lung cancer patients is ^{18}F-FDG-PET on a per-patient and per-lesion basis [42]. Prostate-specific membrane antigen (PSMA) can be labeled with fluorine-18 or ^{68}Ga-based radiotracers to determine the extent of prostate bone metastasis. PSMA PET/CT imaging continues to become more popular in diagnostic and therapeutic capacities, improving patient outcomes [82]. Estrogen receptor (ER) positivity affects about 80% of breast cancers. One of the most common radiotracers used to image ER-positive breast cancer is 16α-[^{18}F]fluoro-17β-estradiol (FES). Estradiol binds to ER as an agonist with a high binding affinity, giving clear imaging of metastatic breast cancer [83].

In conclusion, a combination of MRI and PET would significantly increase the chances of an early diagnosis of skeletal metastasis. In their 1993 study, Rougraff et al. reported the effectiveness of each diagnostic modality including patient history and physical examination, laboratory studies, and imaging studies. In total, they were able to identify the primary tumor in 85% of patients. Just history and physical exam alone were able to reveal the primary tumor in 10% of the patients [43]. Even with innovations in imaging technology, this number has crept up by only 4% (to 89%) over the past 30 years [44].

Functional Assessment

Given the palliative treatment of many patients with MBD, it is critical to assess the patient's reported outcomes with regard to the patient's quality of life (QoL) and activities of daily living (ADL) to ensure a more holistic case-by-case approach to the patient's unique needs. In addition to general assessments, more specific classifications may be referenced to further guide treatment and diagnosis.

Patient-Reported Outcome Measurement Information System (PROMIS)

PROMIS was developed to measure outcomes across all health domains. It allowed for further evaluation while avoiding disease-specific measures. The patient's quality of life can be measured separately from their underlying medical condition. PROMIS was developed in 2004 and is patient-derived instead of physician-derived [45].

PROMIS measures the patient's score as a T-score, normalized to the average within the US population. This allows for comparison of patients to the general population and provides orthopedic oncologists with a better measurement of the relative and clinical outcomes of different treatment options. Ho et al. found that in foot and ankle patients, preoperative PROMIS scores were highly predictive of postoperative success. As it continues to evolve, this method of assessment may continue to play a larger role in cross-specialist care and demonstrate a larger predictive value in surgical implications.

Quality of Life

In 2023, Rajeswaran et al. completed a systematic review of 588 studies to identify the quality-of-life (QoL) issues faced by patients with bone metastasis and to comprehensively evaluate the accuracy of the various patient-reported outcome measures (PROMs) used to assess their QoL [46]. Commonly observed issues included interference with mobility and strenuous physical exercise. Patients had difficulty standing, sitting, and even lying down. Studies have found that almost 80% of patients pending surgical stabilization for BM lost their ability to walk unaided, with 13% remaining bedridden [47]. Akakura et al. reported that 90% of patients found negative effects to their daily lives and/or physical activities, with 73% recording a direct negative impact on their families and 41% with their sleep affected [48]. Significant factors pointing to the patients' reduced function included other disabilities, tumor type, and female gender [49].

Generally, the clinical presentation of bone pain in cancer patients, particularly its propensity for flaring when weight is placed on the affected limb, has direct implications on the patient's daily activities and their treatment. For example, patients may not tolerate certain radiotherapies like stereotactic body radiation therapy (SBRT) because they cannot remain in the same position for the allotted or required time. New or growing bone pain may also alter the way in which patients view their own condition, inducing fear of the cancer's progression [50–52]. Pain intensity was also higher in patients with multiple bone metastases, and it varied based on the tumor type [49].

The three bone metastasis-specific QoL assessment tools found in Rajeswaran's review were the European Organisation for Research and Treatment of Cancer Quality of Life Questionnaire for patients with Bone Metastases (EORTC QLQ-BM22), Functional Assessment of Cancer Therapy-Bone Pain (FACT-BP), and the Bone Metastases Quality-of-Life-10 (BOMET-QoL-10). The most frequently used tool was the EORTC QLQ-BM22, which was developed by the EORTC Quality of Life Group in 2009 [53].

EORTC QLQ-BM22 was developed to supplement the European Organization for Research and Treatment of Cancer Core Questionnaire (EORTC QLQ-C30). Phase one of development included interviewing patients and healthcare providers to generate a list of health-related quality of life (HRQoL) items specific to bone metastases. A total of 61 relevant symptoms were found and divided into 6 categories: symptom (13 issues), function (15 issues), side effects from treatment (3 issues), psychosocial (27 issues), treatment expectation (2 issues), and financial (1 issue). These 61 items were given to 413 patients in a questionnaire format where they ranked each one from 1 (not relevant) to 4 (very relevant). Further studies and testing resulted in the development of a 22-item module with symptom scales including 5 painful sites and 3 pain characteristics along with functional scales that include 8 functional interference and 6 psychosocial aspects. The patients would rank each item from 1 (not at all) to 4 (very much), which would then be converted into a standardized score out of 100. High symptom scores indicate increased distress, while high functional scores indicate better functional ability. Chow et al. concluded that EORTC QLQ-C30 supplemented by EORTC QLQ-BM22 would provide the best comprehensive HRQoL for the longitudinal analysis of symptoms and the effects of treatment on patient lives [53]. Later in 2016, Raman et al. completed a large prospective study of patients undergoing palliative radiotherapy that found that there were meaningful changes in their EORTC QLQ-BM22 scores (Fig. 4.1) [54].

FACT-BP PROM is a more focused tool that can be quickly completed by patients in <10 minutes. The questionnaire uses a Likert-type scale (0–4) to evaluate a patient's bone pain. It can be used in conjunction with the Bone Treatment Convenience and Satisfaction Questionnaire (FACIT-TS-BTCSQ), which evaluates a patient's response to the treatment they receive for their MBD. The FACT-BP scale has meaningful implications for use on a broader scale, while FACIT-TS-BTCSQ still requires further investigation to be implemented on a wider scale [51].

BOMET-QoL was originally a 25-item questionnaire that was then validated and reduced to 10 items and renamed BOMET-QoL-10. Validation of BOMET-QoL-10 revealed that MBD's impact on HRQoL varied significantly by the presence, number, and duration of irruptive pain crisis, pain management index, and Eastern Cooperative Oncology Group Performance Status (ECOG PS) [55].

In its final version, BOMET-QoL-10 is unidimensional with Likert-scale-based questions (0–4) and a set time frame of the week leading up to the patient's visit. It has a global score of 40, which is then converted into a standardized score out of 100. BOMET-QoL-10 was developed to serve as a short and easy method to quickly assess a patient's quality of life. Nowadays, it is difficult to convince clinicians to use HRQoL questionnaires due to difficulty in interpreting results and the time

Timeframe: Past Week	Not at all	A Little	Quite a Bit	Very Much
	Painful Sites			
1) Back	1	2	3	4
2) Leg(s) or hip(s)	1	2	3	4
3) Arm(s) or shoulder(s)	1	2	3	4
4) Chest or ribs	1	2	3	4
5) Buttocks	1	2	3	4
	Pain Characteristics			
6) Constant Pain	1	2	3	4
7) Intermittent Pain	1	2	3	4
8) Pain not relieved by meds	1	2	3	4
	Functional Interference			
9) Pain while lying down	1	2	3	4
10) Pain While sitting	1	2	3	4
11) Pain when trying to stand up	1	2	3	4
12) Pain while walking	1	2	3	4
13) Pain with bending or stairs	1	2	3	4
14) Pain with strenuous activity	1	2	3	4
15) Pain that interferes with sleep	1	2	3	4
16) Modfication of daily activities	1	2	3	4
	Psychosocial Aspects			
17) Feelings of isolation	1	2	3	4
18) Worried about mobility loss	1	2	3	4
19) Worried about dependency on others	1	2	3	4
20) Worried about health in the future	1	2	3	4
21) Felt hopeful that pain will get better	1	2	3	4
22) Felt positive about health	1	2	3	4

Fig. 4.1 The EORTC QLQ-BM22 symptom scale. EORTC QLQ-BM22 European Organisation for Research and Treatment of Cancer Quality of Life Questionnaire for patients with Bone Metastases

commitment. This questionnaire is effective due to how easy it is for patients to complete it and how user-friendly it is for the physician as an assessment tool while the measurement properties still provide meaningful indications (Fig. 4.2) [56].

Activities of Daily Living

Two of the oldest and most used ADL assessment scales for the elderly patient population are the Barthel Index and the Katz Index of Independence in ADL (Katz ADL). ADLs can be divided into two types: instrumental activities of daily living (IADL) and basic ADLs. Basic ADLs refer to the personal activities that a person completes in their home such as eating, bathing, or dressing. IADLs refer to the complex activities a person completes to serve as an active member of a greater community such as driving, financial management, shopping, medication management, and meal preparation [57]. In these patients, it is not so much the direct

Past Week's Symptoms	Not at all	A little	Quite a Bit	Very Much
1) I feel tired	1	2	3	4
2) I find it difficult to get out of bed	1	2	3	4
3) I have a feeling of general discomfort	1	2	3	4
4) I feel depressed and feel like crying	1	2	3	4
5) I have little interest in going out of the house	1	2	3	4
6) I avoid doing activities with my family	1	2	3	4
7) I feel pain in certain parts of my body which affects my life	1	2	3	4
8) I'm in constant pain which affects my life	1	2	3	4
9) I have an intense pain that constantly bothers me	1	2	3	4
10) The pain prevents me from enjoying life like before	1	2	3	4

Fig. 4.2 BOMET-QoL-10. BOMET-QoL-10 Bone Metastases Quality-of-Life-10

disabilities such as problems with communication, partial or total blindness, hearing, or cognitive functions that cause the patient's dependency but rather that they tend to impede the patient's basic and instrumental ADLs [57, 58].

The Barthel Index was designed to accurately reflect a patient's dependency status and more specifically to measure the severity of disability in patients whose disease interfered with the independent movement of their limbs. Developed in 1964 and proven reliable by Collins et al. in 1988, it has shown success whether administered as a self-report, a Registered Nurse's clinical observation, testing by a nurse, or testing by a physiotherapist [57, 59, 60].

The score measures a patient's success when completing 10 common basic ADLs. The activities include feeding, grooming, bathing, dressing, bowel control, bladder control, toilet use, chair/bed transfers, ambulation, and stair climbing. The patient can be assigned a score of 0 or 5 for each activity, with more important activities including higher scores such as a 10 or 15. The highest total score is 100, which signifies a full functional status [60–62].

Based on this original scoring system that was convenient for statistical analysis, a simpler adapted version was formed and is more commonly used in practical clinical scenarios. The 10 items are scored as 0, 1, or 2 with a total score of 20. A total score < 9 signifies dependency and that <5 signifies severe dependency. A maximum score of 20, however, still does not mean the patient can definitively live alone [57, 60, 62]. Clinicians commonly use the Barthel Index to assess patients with chronic debilitating conditions such as stroke recovery or metastatic disease. It has proven to be a useful tool for determining the patient's functional status and need for further invasive treatment (Fig. 4.3) [61].

Similarly, the Katz ADL also measures a patient's dependency status as one cumulative score through a Guttman-type scale. Rather than 10 separate measures that are each scored individually like the Barthel Index, the Katz ADL uses 6 hierarchically scaled tasks under the assumption that the patient will return to independence in a predictable manner. The six ADL functions are bathing, dressing, toileting, transferring, continence, and feeding. The scoring is binary (independent/dependent) for each function, with A representing total independence, C indicating independence in bathing, dressing, and one other function, and G meaning that the patient is dependent in all categories. Independence is defined as the ability to complete the function without supervision, direction, or active personal assistance,

Fig. 4.3 The Barthel Index: A general assessment of the functional dependency of a patient [57]

Activity	Score
Feeding	
Unable	0
Needs help cutting, spreading butter etc. requires modified diet	5
Independent	10
Bathing	
Dependent	0
Independent	5
Grooming	
Needs help with personal care	0
Independent face/hair/teeth/shaving	5
Dressing	
Dependent	0
Needs help but can do about half unaided	5
Independent	10
Bowels	
Incontinent	0
Occassional accident	5
Continent	10
Bladder	
Incontinent, or catheterized	0
Occasional Accident	5
Continent	10
Toilet Use	
Dependent	0
Needs some help, but can do something alone	5
Independent	10
Transfers (bed to chair and back)	
Unable, no sitting balance	0
Major help, one or two people, physical), can sit	5
Minor help (verbal or physical)	10
Independent	15
Mobility	
Immobile or < 50 yards	0
Wheelchair independent, including corners, > 50 yards	5
Walks with help of one person > 50 yards	10
Independent (may use aid) > 50 yards	15
Stairs	
Unable	0
Needs help	5
Independent	10
TOTAL SCORE	

except when specified by the original Katz ADL [63]. A patient who refuses to perform the action is assumed incapable of performing it. The philosophy behind the Katz ADL stems from the belief that recovery from a disabling illness mirrors the development of primary functions in a child. By this logic, if development has order, then one's regression must also follow a similar pattern (Fig. 4.4) [57, 63–65].

Functional Scores and Prognostic Tools

Having the ability to consistently and accurately predict patient survival is invaluable to oncologists. Various tools have been created to guide clinicians' expertise, some focusing on one complex factor like function and others combining those

Katz Index of Independence in ADL	
Score	Dependency
A	independent in feeding, continence, transferring going to toilet, dressing, and bathing
B	Independent in all but one of these functions
C	Independent in all but bathing and one additional function
D	Independent in all but bathing, dressing, and one additional function
E	Independent in all but bathing, dressing, going to toilet, and one additional function
F	Independent in all but bathing, dressing, going to toilet, transferring, and one additional function
G	Dependent in at all 6 functions

Fig. 4.4 The Katz ADL: Functional assessment of dependency [63]. ADL activities of daily living

various factors into an overall prognostic tool. The most common functional tools include Karnofsky Performance Status (KPS), the Eastern Cooperative Oncology Group (ECOG) Performance Status, the Palliative Performance Scale (PPS), and the Enneking Staging System. The Palliative Prognostic Score (PaP) and the Palliative Prognostic Index (PIP) are broader prognostic tools that integrate the functional tools into their scoring [66].

KPS includes 11 functional levels at increasing intervals of 10% ranging from 0% (dead) to 100% (normal functioning). Its development stemmed from a need to document physical function and need for assistance in cancer patients, with its use becoming widespread. In their 1984 study, Schag et al. found strong relationships between KPS and energy level, driving ability, weight loss, grooming, and work. He found that an accurate interpretation of KPS combined with consistent use could allow for earlier discovery of specific areas of difficulty for the patient, leading to earlier referrals. This, in turn, improved quality of life and care received by patients with advanced cancer (Fig. 4.5) [66, 67].

Since its development, multiple versions of KPS have been developed including Australia-modified KPS (AKPS), ECOG Performance Status, and PPS versions 1 and 2. A simplified version of KPS and direct adaptation of the Zubrod performance score is the ECOG Performance Status. This tool is specifically used for the functional assessment of cancer patients incorporating ambulatory status and need for assisted care. It has five levels of performance ranging from 0 (normal functioning and performance) to 5 (dead). A 0–1 is comparable to a KPS of 80–100%, a 2 is comparable to a KPS of 60–70%, and a 3–4 is comparable to a KPS of 10–50% (Fig. 4.6) [66, 68].

ECOG Performance Status may serve as a proper initial assessment score that is quick and easy; however, due to its simplicity, it demonstrates significant variability amongst clinicians. Datta et al. reported poor agreement between raters for most cases ranging over multiple oncologic diseases. With training agreement, it slightly improved but never reached above 60%. This can be attributed to the difficulties of combining quantitative and qualitative data to form a prognosis [69]. Further training and education of upcoming clinicians with an emphasis on a holistic view of the patient will improve the use of this prognostic tool along with how clinicians provide reasonable cancer care [66, 69].

Fig. 4.5 Karnofsky Performance Status criteria [79]

Definition	% Score	Criteria
Able to carry on normal activity and to work. No special care is needed	100%	*Normal; no complaints; no evidence of disease.*
	90%	*Able to carry on normal activity; minor signs or symptoms of disease.*
	80%	*Normal activity with effort; some signs or symptoms of disease.*
Unable to work. Able to live at home, care for most personal needs. A varying amount of assistance is needed.	70%	*Cares for self. Unable to carry on normal activity or to do active work.*
	60%	*Requires occasional assistance, but is able to care for most of his needs.*
	50%	*Requires considerable assistance and frequent medical care.*
Unable to care for self. Requires equivalent of institutional or hospital care. Disease may be progressing rapidly.	40%	*Disabled; requires special care and assistance.*
	30%	*Severely disabled; hospitalisation is indicated although death not imminent.*
	20%	*Very sick; hospitalisation necessary; active supportive treatment necessary.*
	10%	*Moribund; fatal processes progressing rapidly.*
	0%	*Dead.*

Fig. 4.6 ECOG Performance Status provides quick information on the patient's ambulatory status [69]. ECOG Eastern Cooperative Oncology Group

Grade	ECOG
0	Fully active, able to carry on all pre-disease performance without restriction
1	Restricted in physically strenuous activity but ambulatory and able to carry out work of a light or sedentary nature, e.g. light housework, office work
2	Ambulatory and capable of all selfcare but unable to carry out any work activities. Up and about more than 50% of waking hours
3	Capable of only limited selfcare, confined to bed or chair more than 50% of waking hours
4	Completely disabled. Cannot carry on any selfcare. Totally confined to bed or chair
5	Dead

PPS, which has two versions, was developed specifically to measure the changes in the physical status of patients undergoing palliative care. It factors in the patient's ambulation, activity level, evidence of disease, self-care, intake, and level of consciousness. It has been proven to have a reliable, valid, and strong association with survival for both cancer and noncancer patient populations in various countries [66, 70, 71].

Like the Karnofsky score, the PPS scoring system ranges from 0% to 100% in 10% increments, with 0% meaning deceased. The score is determined by the palliative care physician completing the note. Its strongest use is in the inpatient setting with short-term mortality such as a 1-month projected lifespan [72].

Both PaP and PPI are newer, more complex assessment tools created for the terminally ill cancer patient that integrate symptoms, laboratory data, and functional assessments to create a prognostic score. PaP incorporates KSP, while PIP uses PPS [66].

While palliative care contributes mostly to patient comfort rather than survival, knowledge of patients' projected lifespan and use of prediction tools can aid the clinician's treatment plan. For example, in most cases, a depressed cancer patient would be prescribed tricyclic antidepressants, but, if the patient was only projected to live for another 3 weeks, the antidepressants would rarely be prescribed since they would only result in an improved emotional status in a minimum of 2–4 weeks [73]. If the same patient had a better prognosis with a predicted survival of over a month, the antidepressants would be encouraged [73].

PaP divides terminally ill cancer patients into three different risk groups based on their 30-day survival probability. Their total score signifies which category they fall into. The group with a > 70% chance of a 30-day survival (Group A) corresponds to a PaP score of 0–5.5, a score of 5.6–11 points to 30–70% (Group B), and a score of 11.1–17.5 predicts a < 30% chance of a 30-day survival (Group C). The assessment includes six factors including dyspnea, anorexia, KPS, the provider's clinical prediction of survival (CPS), total white blood count, and lymphocyte rate (Fig. 4.7) [74].

Unlike PaP, PPI is purely based on clinical variables, so it is easily implemented and does not require laboratory results. A total score is formed factoring in PPS, oral intake, edema, dyspnea at rest, and delirium [75]. These factors were found to have the most useful and consistent prognostic indications. In their development of the PPI system in 1999, Morita et al. divided their patient population into three groups: PPI ≤ 2 (Group A), $2 \leq$ PPI ≤ 4 (Group B), and PPI ≥ 4 (Group C). Their mean survival times were 155 ± 20 days (Group A), 89 ± 7.7 days (Group B), and 18 ± 2.0 days (Group C) [73]. Thus, the higher the PPI score, the more dismal the prognosis. Morita et al. deemed this scoring system sufficiently accurate to predict survival [73].

Later, in 2012, Maltoni et al. carried out a prospective comparison of prognostic palliative scores and found that both PaP and PPI scores showed statistically significant predictive capacity, with PaP demonstrating slightly better overall accuracy [76].

Although both PaP and PPI are tools that primarily evaluate survival, they can indirectly suggest treatment responses. For example, if a patient's PaP score improves after receiving treatment, it may suggest effective palliative care and an improved prognosis [73, 74]. In terms of PPI, a decreased score at the end of treatment may suggest an appropriate treatment response with an improved prognosis [75, 76]. However, it is essential to point out that these tools do not directly evaluate treatment response, but, if used in conjunction with other clinical assessments, it may provide some further clinical insight. In conclusion, both PPI and PaP are prognostic tools that can help a clinician determine whether aggressive treatment regimens are likely to benefit a patient or if the treatment plan should shift its focus primarily to around comfort-oriented care (Fig. 4.8).

Fig. 4.7 The Palliative Prognostic Score with a legend indicating the estimate survival based on the total score found [80]

The Palliative Prognostic Score (PaP)		
Criterion	**Assessment**	**Partial Score**
Dyspnea		
	Yes	0
	No	1
Anorexia		
	Yes	0
	No	1.5
KPS		
	≥50	0
	30-40	0
	10-20	2.5
CPS		
	>12	0
	11-12	2
	9-10	2.5
	7-8	2.5
	5-6	4.5
	3-4	6
	1-2	8.5
Total WBC		
	Normal (4800-8500 cell/mm^3)	0
	High (8501-11,000 cell/mm^3)	0.5
	Very high (>11,000 cell/mm^3)	1.5
Lympocyte Percentage		
	Normal (20-40%)	0
	Low (12-19.9%)	1
	Very low (0-11.9%)	2.5
	TOTAL SCORE	

Legend		
Risk Groups	**Explanation**	**Total Score**
A	*30-day Survival Probability > 70%*	**0-0.5**
B	*30-day Survival Probability 30-70%*	**5.6-11**
C	*30-day survival probability <30%*	**11.1-17.5**

Fig. 4.8 Palliative Prognostic Index with the legend indicating survival times [73, 81]

Palliative Prognostic Index (PPI)		
Criterion	**Assessment**	**Partial Score**
Palliative Performance Scale (PPS)		
	10-20	4
	30-50	2.5
	>60	0
Oral Intake		
	Mouthfuls or less	2.5
	Reduced but more than mouthfuls	1
	Normal	0
Edema		
	Present	1
	Absent	0
Dyspnea at rest		
	Present	3.5
	Absent	0
Delirium		
	Present	4
	Absent	0
	TOTAL SCORE	

Legend	
PPI Score:	**Survival Time**
>6	**Less than 3 weeks**
>4	**Less than 6 weeks**
≤4	**More than 6 weeks**

Treatment

Patients with MBD experience skeletal-related events every 3–6 months on average, with the events occurring irregularly and more frequently as the disease continues to spread and worsen [8]. When evaluating patients with MBD, it is not only important to evaluate activity levels but also their life expectancy and short-term benefits from the treatment [1]. Therefore, patients and their families should be reminded that in many cases the procedure is palliative rather than curative and that their expectations should be realistic and in-line with the surgeon's expectations and with knowledge of the extensive list of comorbidities with which a patient may present [1].

Patients with skeletal metastasis who exercise regularly can still maintain or improve their physical function and quality of life. In their study, Guinan et al. found that patients who adhered to physical activity guidelines reported lower pain interference scores, higher EORTC QLQ-30 scores, lower insomnia scores, and higher scores in functional interference (higher scores indicating higher function). The benefits of exercising, however, should always be carefully weighed against the risks of a resulting SRE. This involves a comprehensive analysis of all the conditions present that are specific to each patient [77].

Nooh et al. published a study in 2018, which retrospectively evaluated 184 patients who reported functional outcomes and pain scores improving as early as

2 weeks after surgical treatment of MBD, while, surprisingly, their quality of life did not improve at all [78]. Surgical intervention for bone metastasis is only indicated in response to severe symptoms of pain or neurologic deficit [29].

Conclusions

With a diligent, systematic, and comprehensive approach to patients with musculoskeletal metastasis, their quality of life and functional status may be significantly improved. The physician must approach the patient using both clinical and functional assessment tools to formulate a detailed diagnosis and treatment plan. Specific tools and classification systems have been developed to guide decision-making skills and help standardize treatment. The clinician should begin with a detailed history and physical examination, followed by the appropriate laboratory test and imaging studies. Once the patient's diagnosis becomes clearer and staging has been completed, their ADLs and QoL should be measured using reliable, objective classification systems like the Barthel Index and EORTC QLQ-BM22. These tools, combined with objective functional assessments like KPS, ECOG, or PPS, will allow clinicians to form complex prognostic scores such as PaP or PIP that reliably predict the patient's expected lifespan. Moreover, new treatment options and assessment tools continue to surface on a yearly basis, as the prevalence of musculoskeletal metastasis rises. It is the duty of orthopedic oncologists and other providers to stay on top of new trends in this field so that patients receive the best options possible to improve their quality of life.

References

1. Biermann JS, Holt GE, Lewis VO, Schwartz HS, Yaszemski MJ. Metastatic bone disease: diagnosis, evaluation, and treatment. J Bone Joint Surg Am. 2009;91(6):1518–30.
2. Fares J, Fares MY, Khachfe HH, Salhab HA, Fares Y. Molecular principles of metastasis: a hallmark of cancer revisited. Signal Transduct Target Ther. 2020;5(1):28.
3. Mundy GR. Metastasis to bone: causes, consequences and therapeutic opportunities. Nat Rev Cancer. 2002;2:584–93.
4. Coleman RE. Metastatic bone disease: clinical features, pathophysiology and treatment strategies. Cancer Treat Rev. 2001;27:165–76.
5. Kommalapati A, Tella SH, Esquivel MA, Correa R. Evaluation and management of skeletal disease in cancer care. Crit Rev Oncol Hematol. 2017;120:217–26. Epub 2017 Sep 8. Erratum in: Crit Rev Oncol Hematol. 2018 Apr;124:73.
6. Peabody T. The rodded metastasis is a sarcoma: strategies to prevent inadvertent surgical procedures on primary bone malignancies. Instr Course Lect. 2004;53:657–61. PMID: 15116656.
7. Okamoto K. Role of RANKL in cancer development and metastasis. J Bone Miner Metab. 2021;39(1):71–81. https://doi.org/10.1007/s00774-020-01182-2. Epub 2021 Jan 2. PMID: 33387063.
8. Coleman RE. Clinical features of metastatic bone disease and risk of skeletal morbidity. Clin Cancer Res. 2006;12(20 Pt 2):6243s–9s. https://doi.org/10.1158/1078-0432.CCR-06-0931. PMID: 17062708.

9. Damron TA, Mann KA. Fracture risk assessment and clinical decision making for patients with metastatic bone disease. J Orthop Res. 2020;38(6):1175–90. https://doi.org/10.1002/jor.24660. Epub 2020 Mar 23. PMID: 32162711; PMCID: PMC7225068.
10. Rougraff BT, Frassica FJ. Presentation and staging of metastatic bone disease. Clin Orthop Relat Res. 2003;415:S129–31. https://doi.org/10.1097/01.blo0000092980.12414.32.
11. Anract P, Biau D, Boudou-Rouquette P. Metastatic fractures of long limb bones. Orthop Traumatol Surg Res. 2017;103(1S):S41–51. https://doi.org/10.1016/j.otsr.2016.11.001. Epub 2017 Jan 11. PMID: 28089230.
12. Macedo F, Ladeira K, Pinho F, Saraiva N, Bonito N, Pinto L, Goncalves F. Bone metastases: an overview. Oncol Rev. 2017;11(1):321. https://doi.org/10.4081/oncol.2017.321. PMID: 28584570; PMCID: PMC5444408.
13. Hu YC, Lun DX, Wang H. Clinical features of neoplastic pathological fracture in long bones. Chin Med J. 2012;125:3127–32.
14. Soldatos T, Chalian M, Attar S, McCarthy EF, Carrino JA, Fayad LM. Imaging differentiation of pathologic fractures caused by primary and secondary bone tumors. Eur J Radiol. 2013;82:e36–42. [Epub Sep 25].
15. Tsukushi S, Katagiri H, Kataoka T, Nishida Y, Ishiguro N. Serum Tumor Markers in Skeletal Metastasis. Jpn J Clin Oncol. 2006;36(7):439–44.
16. Betsch DM, Gray S, Zed SE. A case of metastatic prostate cancer and immune thrombocytopenia. Curr Oncol. 2017;24(5):e434–6. https://doi.org/10.3747/co.24.3592. Epub 2017 Oct 25. PMID: 29089814; PMCID: PMC5659168.
17. Dixon LR. The complete blood count: physiologic basis and clinical usage. J Perinat Neonatal Nurs. 1997;11(3):1–18.
18. Amin MB, Greene FL, Edge SB, Compton CC, Gershenwald JE, Brookland RK, Meyer L, Gress DM, Byrd DR, Winchester DP. The Eighth Edition AJCC Cancer Staging Manual: Continuing to build a bridge from a population-based to a more "personalized" approach to cancer staging. CA Cancer J Clin. 2017;67(2):93–9. https://doi.org/10.3322/caac.21388. Epub 2017 Jan 17. PMID: 28094848.
19. Byun BH, Kong C-B, Lim I, et al. Comparison of (18)F-FDG PET/CT and (99 m)Tc-MDP bone scintigraphy for detection of bone metastasis in osteosarcoma. Skeletal Radiol. 2013;42:1673–81.
20. Suva LJ, Washam C, Nicholas RW, Griffin RJ. Bone metastasis: mechanisms and therapeutic opportunities. Nat Rev Endocrinol. 2011;7(4):208–18. https://doi.org/10.1038/nrendo.2010.227. Epub 2011 Jan 4. PMID: 21200394; PMCID: PMC3134309.
21. Yasuda H, et al. Osteoclast differentiation factor is a ligand for osteoprotegerin/osteoclastogenesis-inhibitory factor and is identical to TRANCe/RANKL. Proc Natl Acad Sci USA. 1998;95:3597–602.
22. Guise TA, Yin JJ, Taylor SD, Kumagai Y, Dallas M, Boyce BF, Yoneda T, Mundy GR. Evidence for a causal role of parathyroid hormone-related protein in the pathogenesis of human breast cancer-mediated osteolysis. J Clin Invest. 1996;98(7):1544–9. https://doi.org/10.1172/JCI118947. PMID: 8833902; PMCID: PMC507586.
23. Edwards CM, Kane JF, Smith JA, Grant DM, Johnson JA, Diaz MAH, Vecchi LA 3rd, Bracey KM, Omokehinde TN, Fontana JR, Karno BA, Scott HT, Vogel CJ, Lowery JW, Martin TJ, Johnson RW. PTHrP intracrine actions divergently influence breast cancer growth through p27 and LIFR. Breast Cancer Res. 2024;26(1):34. https://doi.org/10.1186/s13058-024-01791-z. PMID: 38409028; PMCID: PMC10897994.
24. Kommalapati A, Tella SH, Esquivel MA, Correa R. Evaluation and management of skeletal disease in cancer care. Crit Rev Oncol Hematol. 2017;120:217–26. https://doi.org/10.1016/j.critrevonc.2017.09.003. Epub 2017 Sep 8. Erratum in: Crit Rev Oncol Hematol. 2018 Apr;124:73. doi: 10.1016/j.critrevonc.2018.02.007. PMID: 29032892; PMCID: PMC5773131.

25. Santini D, Schiavon G, Vincenzi B, Gaeta L, Pantano F, Russo A, Ortega C, Porta C, Galluzzo S, Armento G, La Verde N, Caroti C, Treilleux I, Ruggiero A, Perrone G, Addeo R, Clezardin P, Muda AO, Tonini G. Receptor activator of NF-kB (RANK) expression in primary tumors associates with bone metastasis occurrence in breast cancer patients. PLoS One. 2011;6(4):e19234. https://doi.org/10.1371/journal.pone.0019234. PMID: 21559440; PMCID: PMC3084800.
26. Mintz DN, Hwang S. Bone tumor imaging, then and now: review article. HSS J. 2014;10(3):230–9. https://doi.org/10.1007/s11420-014-9403-y. Epub 2014 Jul 16. PMID: 25264439; PMCID: PMC4171450.
27. Rosenthal DI. Radiologic diagnosis of bone metastases. Cancer. 1997;80(8 Suppl):1595–607. https://doi.org/10.1002/(sici)1097-0142(19971015)80:8+<1595::aid-cncr10>3.3.co;2-z. PMID: 9362427.
28. O'Sullivan GJ, Carty FL, Cronin CG. Imaging of bone metastasis: an update. World J Radiol. 2015;7(8):202–11. https://doi.org/10.4329/wjr.v7.i8.202. PMID: 26339464; PMCID: PMC4553252.
29. Bailey S, Hackney D, Vashishth D, Alkalay RN. The effects of metastatic lesion on the structural determinants of bone: current clinical and experimental approaches. Bone. 2020;138:115159. https://doi.org/10.1016/j.bone.2019.115159. Epub 2019 Nov 21. PMID: 31759204; PMCID: PMC7531290.
30. Bäuerle T, Semmler W. Imaging response to systemic therapy for bone metastases. Eur Radiol. 2009;19:2495–507.
31. Helms CA, Cann CE, Brunelle FO, Gilula LA, Chafetz N, Genant HK. Detection of bone-marrow metastases using quantitative computed tomography. Radiology. 1981;140:745–50.
32. Choi J, Raghavan M. Diagnostic imaging and image-guided therapy of skeletal metastases. Cancer Control. 2012;19(2):102–12. https://doi.org/10.1177/107327481201900204. PMID: 22487972.
33. Cuccurullo V, Cascini GL, Tamburrini O, Rotondo A, Mansi L. Bone metastases radiopharmaceuticals: an overview. Curr Radiopharm. 2013;6:41–7.
34. Ilaslan H, Sundaram M. Advances in musculoskeletal tumor imaging. Orthop Clin North Am. 2006;37(3):375–91, vii. https://doi.org/10.1016/j.ocl.2006.05.003. PMID: 16846768
35. Parker BR, Pinckney L, Etcubanas E. Relative efficacy of radiographic and radionuclide bone surveys in the detection of the skeletal lesions of histiocytosis X. Radiology. 1980;134(2):377–80. https://doi.org/10.1148/radiology.134.2.6965429. PMID: 6965429.
36. Roberts CC, Daffner RH, Weissman BN, Bancroft L, Bennett DL, Blebea JS, Bruno MA, Fries IB, Germano IM, Holly L, Jacobson JA, Luchs JS, Morrison WB, Olson JJ, Payne WK, Resnik CS, Schweitzer ME, Seeger LL, Taljanovic M, Wise JN, Lutz ST. ACR appropriateness criteria on metastatic bone disease. J Am Coll Radiol. 2010;7(6):400–9. https://doi.org/10.1016/j.jacr.2010.02.015. Erratum in: J Am Coll Radiol. 2010 Sep;7(9):e1. PMID: 20522392.
37. Zhang Y, Zhao C, Liu H, Hou H, Zhang H. Multiple metastasis-like bone lesions in scintigraphic imaging. J Biomed Biotechnol. 2012;2012:957364.
38. Cook GJ. PET and PET/CT imaging of skeletal metastases. Cancer Imaging. 2010;10(1):1–8. https://doi.org/10.1102/1470-7330.2010.0022. PMID: 20663736; PMCID: PMC2922743.
39. Gallagher BM, Ansari A, Atkins H, Casella V, Christman DR, Fowler JS, Ido T, MacGregor RR, Som P, Wan CN, Wolf AP, Kuhl DE, Reivich M. Radiopharmaceuticals XXVII. 18F-labeled 2-deoxy-2-fluoro-d-glucose as a radiopharmaceutical for measuring regional myocardial glucose metabolism in vivo: tissue distribution and imaging studies in animals. J Nucl Med. 1977;18(10):990–6. PMID: 903484.
40. Beyer T, Townsend DW, Brun T, Kinahan PE, Charron M, Roddy R, Jerin J, Young J, Byars L, Nutt R. A combined PET/CT scanner for clinical oncology. J Nucl Med. 2000;41(8):1369–79. PMID: 10945530.

41. Franzius C, Sciuk J, Daldrup-Link HE, Jürgens H, Schober O. FDG-PET for detection of osseous metastases from malignant primary bone tumours: comparison with bone scintigraphy. Eur J Nucl Med. 2000;27(9):1305–11. https://doi.org/10.1007/s002590000301. PMID: 11007511.
42. Liu T, Xu JY, Xu W, Bai YR, Yan WL, Yang HL. Fluorine-18 deoxyglucose positron emission tomography, magnetic resonance imaging and bone scintigraphy for the diagnosis of bone metastases in patients with lung cancer: which one is the best?--a meta-analysis. Clin Oncol (R Coll Radiol). 2011;23:350–8.
43. Rougraff BT, Kneisl JS, Simon MA. Skeletal metastases of unknown origin. A prospective study of a diagnostic strategy. J Bone Joint Surg Am. 1993;75(9):1276–81. https://doi.org/10.2106/00004623-199309000-00003. PMID: 8408149.
44. Argentiero A, Solimando AG, Brunetti O, Calabrese A, Pantano F, Iuliani M, Santini D, Silvestris N, Vacca A. Skeletal metastases of unknown primary: biological landscape and clinical overview. Cancers (Basel). 2019;11(9):1270. https://doi.org/10.3390/cancers11091270. PMID: 31470608; PMCID: PMC6770264.
45. Le D, Martin K, Clark SC, Ruso D, Hoyen A, Hunter K, Kim TWB. Trends in functional outcome measures in orthopedic oncology. J Orthop Res. 2024;42(9):2088–94. https://doi.org/10.1002/jor.25846. Epub 2024 Apr 5. PMID: 38578623.
46. Rajeswaran T, Wong HCY, Zhang E, Kennedy SKF, Gojsevic M, Soliman H, Vassiliou V, Rades D, Bonomo P, Lee SF, Chan AW, Rembielak A, Oldenburger E, Maranzano E, Pergolizzi S, Finkelstein JA, Larouche J, Zhang N, Zhang X, Marta GN, Yee AJM, Yu S, van der Velden JM, van der Linden YM, Chow E; EORTC Quality of Life Group. Quality of life issues in patients with bone metastases: a systematic review. Support Care Cancer. 2023;32(1):18. https://doi.org/10.1007/s00520-023-08241-0. PMID: 38091116.
47. Kim Y, Krishnan CK, Kim HS, Cho HS, Han I. Ambulation recovery after surgery for metastases to the femur. Oncologist. 2020;25(1):e178–85.
48. Akakura K, Uemura H, Miyazaki K, et al. A qualitative research study in Japan investigating patients' experience with metastatic castration-resistant prostate cancer: from diagnosis to decision for Ra-223 treatment. Future Oncol. 2021;17(36):5103–18.
49. Janssen SJ, Pereira NRP, Thio QCBS, Raskin KA, Bramer JAM, Lozano-Calderon SA, Schwab JH. Physical function and pain intensity in patients with metastatic bone disease. J Surg Oncol. 2019;120(3):376–81. https://doi.org/10.1002/jso.25510. Epub 2019 May 29. PMID: 31140605.
50. Clamp A, Danson S, Nguyen H, Cole D, Clemons M. Assessment of therapeutic response in patients with metastatic bone disease. Lancet Oncol. 2004;5(10):607–16. https://doi.org/10.1016/S1470-2045(04)01596-7. PMID: 15465464.
51. Broom R, Du H, Clemons M, Eton D, Dranitsaris G, Simmons C, Ooi W, Cella D. Switching breast cancer patients with progressive bone metastases to third-generation bisphosphonates: measuring impact using the Functional Assessment of Cancer Therapy-Bone Pain. J Pain Symptom Manage. 2009;38(2):244–57. https://doi.org/10.1016/j.jpainsymman.2008.08.005. Epub 2009 Apr 11. PMID: 19364633.
52. Pielkenrood BJ, van der Velden JM, van der Linden YM, et al. Pain response after stereotactic body radiation therapy versus conventional radiation therapy in patients with bone metastases - A phase 2 randomized controlled trial within a prospective cohort. Int J Radiat Oncol Biol Phys. 2021;110(2):358–67.
53. Chow E, Hird A, Velikova G, Johnson C, Dewolf L, Bezjak A, Wu J, Shafiq J, Sezer O, Kardamakis D, van der Linden Y, Ma B, Castro M, Arnalot PF, Ahmedzai S, Clemons M, Hoskin P, Yee A, Brundage M, Bottomley A; EORTC Quality of Life Group. The European Organisation for Research and Treatment of Cancer Quality of Life Questionnaire for patients with bone metastases: the EORTC QLQ-BM22. Eur J Cancer. 2009;45(7):1146–52.

54. Raman S, Ding K, Chow E, Meyer RM, Nabid A, Chabot P, Coulombe G, Ahmed S, Kuk J, Dar AR, Mahmud A, Fairchild A, Wilson CF, Wu JSY, Dennis K, DeAngelis C, Wong RKS, Zhu L, Brundage M. Minimal clinically important differences in the EORTC QLQ-BM22 and EORTC QLQ-C15-PAL modules in patients with bone metastases undergoing palliative radiotherapy. Qual Life Res. 2016;25(10):2535–41. https://doi.org/10.1007/s11136-016-1308-4. Epub 2016 May 2. PMID: 27138964.
55. Barnadas A, Muñoz M, Margelí M, Chacón JI, Cassinello J, Antolin S, Adrover E, Ramos M, Carrasco E, Jimeno MA, Ojeda B, González X, González S, Constenla M, Florián J, Miguel A, Llombart A, Lluch A, Ruiz-Borrego M, Colomer R, Del Barco S; GEICAM, Spanish Breast Cancer Group. BOMET-QoL-10 questionnaire for breast cancer patients with bone metastasis: the prospective MABOMET GEICAM study. J Patient Rep Outcomes. 2019;3(1):72. https://doi.org/10.1186/s41687-019-0161-y. PMID: 31865481; PMCID: PMC6925605.
56. Sureda A, Isla D, Cózar J, Ruiz M, Domine M, Margelí M, et al. Final development and validation of the BOMET-QoL questionnaire for assessing quality of life in patients with malignant bone disease due to neoplasia. J Med Econ. 2007;10(1):27–39. https://doi.org/10.3111/200710027039.
57. Hartigan I. A comparative review of the Katz ADL and the Barthel Index in assessing the activities of daily living of older people. Int J Older People Nursing. 2007;2(3):204–212. https://doi.org/10.1111/j.1748-3743.2007.00074.x. PMID: 20925877.
58. Wellwood DMS. A comparison of the Barthel Index and the OPCS disability instrument used to measure outcome after acute stroke. Age Ageing. 1995;24:54–7.
59. Wylie CM, White BK. A measure of disability. Arch Environ Health. 1964;8:834–9.
60. Collins C, Wade D, Davies S, Horne V. The Barthel ADL index: a reliability study. Int Disabil Stud. 1988;10:61–3.
61. Morishima T, Sato A, Nakata K, Matsumoto Y, Koeda N, Shimada H, Maruhama T, Matsuki D, Miyashiro I. Barthel index-based functional status as a prognostic factor in young and middle-aged adults with newly diagnosed gastric, colorectal and lung cancer: a multicentre retrospective cohort study. BMJ Open. 2021;11(4):e046681. https://doi.org/10.1136/bmjopen-2020-046681. PMID: 33853804; PMCID: PMC8054075.
62. Gresham GE, Philips TF, Labi MLC. ADL status in stroke: relative merits of three standard indexes. Arch Phys Med Rehabil. 1980;61:355–8.
63. Katz S, Ford AB, Moskowitz RW, Jackson BA, Jaffee MW. Studies of illness in the aged: the index of ADL: a standardized measure of biological and psychosocial function. JAMA. 1963;185:914–9.
64. Cohen ME, Marino RJ. The tools of disability outcomes research functional status measures. Arch Phys Med Rehabil. 2000;81:s21–9.
65. Katz S, Akpom CA. A measure of primary sociobiological functions. Int J Health Serv. 1976;6:493–507.
66. de Kock I, Mirhosseini M, Lau F, Thai V, Downing M, Quan H, Lesperance M, Yang J. Conversion of Karnofsky Performance Status (KPS) and Eastern Cooperative Oncology Group Performance Status (ECOG) to Palliative Performance Scale (PPS), and the interchangeability of PPS and KPS in prognostic tools. J Palliat Care. 2013;29(3):163–9. PMID: 24380215.
67. Schag CC, Heinrich RL, Ganz PA. Karnofsky performance status revisited: reliability, validity, and guidelines. J Clin Oncol. 1984;2(3):187–93. https://doi.org/10.1200/JCO.1984.2.3.187. PMID: 6699671.
68. Sok M, Zavrl M, Greif B, et al. Objective assessment of WHO/ECOG performance status. Support Care Cancer. 2019;27:3793–8.

69. Datta SS, Ghosal N, Daruvala R, Chakraborty S, Shrimali RK, van Zanten C, Parry J, Agrawal S, Atreya S, Sinha S, Chatterjee S, Gollins S. How do clinicians rate patient's performance status using the ECOG performance scale? A mixed-methods exploration of variability in decision-making in oncology. Ecancermedicalscience. 2019;13:913.
70. Ho F, Lau F, Downing MG, et al. A reliability and validity study of the palliative performance scale. BMC Palliat Care. 2008;7:10. https://doi.org/10.1186/1472-684X-7-10.
71. Weng LC, Huang HL, Wilkie DJ, Hoenig NA, Suarez ML, Marschke M, Durham J. Predicting survival with the Palliative Performance Scale in a minority-serving hospice and palliative care program. J Pain Symptom Manage. 2009;37(4):642–8. https://doi.org/10.1016/j.jpainsymman.2008.03.023. Epub 2008 Sep 26. PMID: 18823751; PMCID: PMC2699378.
72. Bischoff KE, Patel K, Boscardin WJ, O'Riordan DL, Pantilat SZ, Smith AK. Prognoses associated with palliative performance scale scores in modern palliative care practice. JAMA Netw Open. 2024;7(7):e2420472.
73. Morita T, Tsunoda J, Inoue S, et al. The Palliative Prognostic Index: a scoring system for survival prediction of terminally ill cancer patients. Support Care Cancer. 1999;7:128–33. https://doi.org/10.1007/s005200050242.
74. Scarpi E, Nanni O, Maltoni M. Development and validation of the PaP score nomogram for terminally ill cancer patients. Cancer. 2022;14(10):2510. https://doi.org/10.3390/cancers14102510.
75. Fernandes M, Branco TP, Fernandez MCN, Paparelli C, Braz MS, Kishimoto CS, de Freitas Medeiros HM, Ebina K, Cabral LRB, Nagashima S, de Avó Cortizo SA, Borges F, Monteiro MR, Abrahao ABK, Moreira RB, Dos Santos Tavares AP, Aguiar PN Jr. Palliative Prognostic Index accuracy of survival prediction in an inpatient palliative care service at a Brazilian tertiary hospital. Ecancermedicalscience. 2021;15:1228. https://doi.org/10.3332/ecancer.2021.1228. PMID: 34158832; PMCID: PMC8183653.
76. Maltoni M, Scarpi E, Pittureri C, Martini F, Montanari L, Amaducci E, Derni S, Fabbri L, Rosati M, Amadori D, Nanni O. Prospective comparison of prognostic scores in palliative care cancer populations. Oncologist. 2012;17(3):446–54.
77. Campbell KL, Cormie P, Weller S, Alibhai SMH, Bolam KA, Campbell A, Cheville AL, Dalzell MA, Hart NH, Higano CS, Lane K, Mansfield S, McNeely ML, Newton RU, Quist M, Rauw J, Rosenberger F, Santa Mina D, Schmitz KH, Winters-Stone KM, Wiskemann J, Goulart J. Exercise recommendation for people with bone metastases: expert consensus for health care providers and exercise professionals. JCO Oncol Pract. 2022;18(5):e697–709. https://doi.org/10.1200/OP.21.00454. Epub 2022 Jan 6. PMID: 34990293; PMCID: PMC9810134.
78. Nooh A, Goulding K, Isler MH, Mottard S, Arteau A, Dion N, Turcotte R. Early improvement in pain and functional outcome but not quality of life after surgery for metastatic long bone disease. Clin Orthop Relat Res®. 2018;476(3):535–45. https://doi.org/10.1007/s11999.0000000000000065.
79. Timmermann C. 'Just give me the best quality of life questionnaire': the Karnofsky scale and the history of quality of life measurements in cancer trials. Chronic Illn. 2013;9(3):179–90. https://doi.org/10.1177/1742395312466903. Epub 2012 Dec 13. PMID: 23239756; PMCID: PMC3837542.
80. Pirovano M, Maltoni M, Nanni O, Marinari M, Indelli M, Zaninetta G, Petrella V, Barni S, Zecca E, Scarpi E, Labianca R, Amadori D, Luporini G. A new palliative prognostic score: a first step for the staging of terminally ill cancer patients. Italian Multicenter and Study Group on Palliative Care. J Pain Symptom Manag. 1999;17(4):231–9. https://doi.org/10.1016/s0885-3924(98)00145-6. PMID: 10203875.
81. Stone CA, Tiernan E, Dooley BA. Prospective validation of the palliative prognostic index in patients with cancer. J Pain Symptom Manage. 2008;35(6):617–622. https://doi.org/10.1016/j.jpainsymman.2007.07.006. Epub 2008 Feb 8. PMID: 18261876.

82. Huang S, Ong S, McKenzie D, et al. Comparison of 18F-based PSMA radiotracers with [68Ga]Ga-PSMA-11 in PET/CT imaging of prostate cancer—a systematic review and meta-analysis. Prostate Cancer Prostatic Dis. 2024;27:654–64. https://doi.org/10.1038/s41391-023-00755-2.
83. Heesch A, Maurer J, Stickeler E, Beheshti M, Mottaghy FM, Morgenroth A. Development of radiotracers for breast cancer—the tumor microenvironment as an emerging target. Cells. 2020;9(10):2334. https://doi.org/10.3390/cells9102334.

Nutritional Evaluation of Patients with Metastatic Disease

5

Berenice De La Mota, Karla Pfaeffle-Palomo, and Daniel Otero-Pfaeffle

Introduction

Cancer is a life-threatening illness affecting millions globally, with approximately 18–19 million new cases diagnosed annually, over half of which progress to metastasis. Metastatic spread is responsible for >90% of cancer-related deaths [1]. The World Cancer Research Fund emphasizes nutrition and lifestyle modifications for cancer prevention and survivorship, as survivors face increased risks of second primary cancers and chronic conditions [2]. While research primarily targets prevention, the role of nutrition therapy in metastatic cancer warrants further exploration. A healthy lifestyle, including specific dietary choices and physical activity, is foundational for cancer survivors [2–4].

Nutrition guidelines from organizations like the American Cancer Society and the National Comprehensive Cancer Network typically recommend maintaining a healthy weight, limiting alcohol, emphasizing plant-based diets, increasing fiber, and reducing processed foods and red meat [2, 4]. In some cancer survivors, obesity is linked to decreased effectiveness of cancer treatment [5]. In advanced cancer cases, improved treatment outcomes have led to not only higher survival rates but also a higher likelihood of developing bone metastases [5]. The Mediterranean diet, rich in whole grains, fruits, vegetables, fish, nuts, and olive oil, may plausibly alleviate skeletal-related events and enhance the quality of life for metastatic patients [1].

Proper nutrition assessment and intervention are vital in palliative care for patients with metastatic cancer, aiding in the management of treatment side effects and enhancing overall well-being [3, 4]. Malnutrition significantly impacts physical and mental health, with common issues like appetite loss and gastrointestinal problems affecting nutritional status. Evaluating nutritional status is crucial for improving physical function, weight, and quality of life [3, 4,

B. De La Mota · K. Pfaeffle-Palomo (✉) · D. Otero-Pfaeffle
Miami Cancer Institute, Miami, FL, USA
e-mail: karlao@baptisthealth.net

J. Pretell-Mazzini, S. A. Lozano Calderón (eds.), *Multidisciplinary Approach of Musculoskeletal Metastasis*, https://doi.org/10.1007/978-3-032-09498-8_5

6–8]. This chapter will delve into the nutritional evaluation of patients with metastatic disease and strategies for optimizing their nutritional intake to support treatment and well-being.

Nutritional Requirements

Energy

Resting energy expenditure (REE) changes significantly in cancer patients due to disease-related factors. REE accounts for 55–75% of total energy expenditure (TEE) [9, 10] and is influenced by tumor burden, inflammation, and cachexia. Some patients experience elevated REE from the tumor's metabolic demands, increasing glycolysis and lactate production [11]. Tumor cells often show high lactate levels due to anaerobic glycolysis [12]. Heightened glucose turnover contributes to increased REE and muscle breakdown [13–15]. Larger tumors or liver metastases particularly elevate energy demands [11]. However, REE varies by cancer type; aggressive cancers like pancreatic and lung cancers raise REE, while breast and prostate cancers are linked to normal or decreased REE [11].

In advanced cancer, energy expenditure is altered due to the tumor's metabolic demands, systemic inflammation, and treatments like chemotherapy and radiotherapy. Hypermetabolism is driven by the tumor's increased energy demands, with large or metastatic tumors exacerbating this increase in energy expenditure. Individualized nutritional interventions are crucial to prevent malnutrition and cachexia [11, 16].

Predictive equations are widely used to estimate REE in cancer patients because they are easy and cost-effective. However, these formulas, including Harris–Benedict and Mifflin–St Jeor, often lack accuracy compared to indirect calorimetry (IC), the gold standard for measuring energy expenditure. Predictive equations are based on population averages and do not account for metabolic changes in cancer patients, such as inflammation or tumor burden, leading to unreliable estimates [9]. Studies indicate that these equations frequently miscalculate REE, underscoring the importance of IC for precise assessments [17].

When indirect calorimetry is unavailable, several "rules of thumb" guide energy estimation for cancer patients. The European Society of Parenteral and Enteral Nutrition (ESPEN) guidelines recommend individualized nutritional therapy, estimating energy needs at 25–30 kcal/kg/day for weight maintenance [16, 17]. For weight gain, 30–35 kcal/kg/day is suggested, while for weight loss, intake should be reduced to 20–25 kcal/kg/day [16].

Estimation of Energy Needs: ESPEN guidelines [16]

For weight maintenance	25–30 Kcal/kg
To promote weight gain	30–35 Kcal/kg
To promote weight loss	20–25 Kcal/kg

ESPEN European Society of Parenteral and Enteral Nutrition

Protein needs are elevated in cancer patients due to increased metabolic demands and the risk of muscle wasting. ESPEN guidelines recommend 1.0–1.5 g of protein per kilogram of body weight (BW) per day to support lean mass and protein synthesis. For advanced cancer or cachexia, intake may increase to 1.5–2.0 g/kg/day [17].

Estimation of Protein Needs: ESPEN guidelines [16]

Adults with cancer	1.0–1.5 g/kg
Advanced cancer or cachexia	1.5–2.0 g/kg

ESPEN European Society of Parenteral and Enteral Nutrition

Carbohydrates and fats are vital for cancer patients' energy needs. Carbohydrates should make up 45–55% of energy intake and fats 25–35% [17]. Carbohydrates help preserve muscle by sparing protein, while fats provide dense energy. Both macronutrients must be tailored to the patient's metabolic state, treatment, and nutritional needs to prevent malnutrition.

Fluid management is crucial for cancer patients, especially in advanced stages. Rehydration is essential for malnourished or dehydrated patients, but excessive hydration can cause fluid retention, particularly with renal or cardiac impairment. In palliative care, artificial hydration is recommended only for clear short-term goals, such as preserving cognitive function [16].

Micronutrients, including vitamins and minerals, are essential for health, growth, and development, though they do not serve as energy sources but aid metabolic processes. Micronutrient deficiencies are common in cancer patients, especially malnourished ones, with water-soluble vitamins being particularly affected [18]. Minerals like zinc and selenium are also frequently deficient due to reduced intake or malabsorption. Maintaining adequate levels supports immune function and tissue repair and prevents complications [16]. For patients with low oral intake, a multivitamin–multimineral supplement near the recommended daily allowance (RDA) may be safe, even during chemo or radiation therapy [19]. However, high-dose supplements are discouraged unless deficiencies are confirmed.

Nutritional Screening

Nutritional screening is essential for metastatic cancer patients and should occur at diagnosis and throughout treatment to improve outcomes with timely interventions. It helps establish baseline nutritional status and identify risks that may impact the treatment plan. Tools like the Nutritional Risk Screening (NRS), Malnutrition Screening Tool (MST), and Patient-Generated Subjective Global Assessment (PG-SGA) are effective for this purpose. These tools evaluate weight loss, dietary intake, and anorexia, common in metastatic cancer patients. While MST and NRS are quick screening tools, PG-SGA offers a more detailed assessment by including body composition, biochemical markers, and functional status [20].

Nutritional screening must continue beyond diagnosis throughout cancer treatment, as treatments like chemotherapy and radiotherapy can worsen malnutrition through side effects such as nausea, anorexia, and mucositis, leading to weight loss and nutrient deficiencies. Regular screening ensures timely detection of nutritional decline. For instance, at our institution, chemotherapy patients are screened at each cycle, while radiation therapy patients are screened weekly during treatment. This approach enables early interventions to enhance treatment tolerance, reduce complications, and improve outcomes [20].

For metastatic cancer patients, proper nutrition is vital since malnutrition increases the risk of pulmonary complications, weakens the immune response, and lowers survival rates. Even moderate weight loss (<5% of body weight) worsens prognosis, and losses exceeding 10% significantly impair chemotherapy response and survival [20].

Malnutrition Screening Tool

The Malnutrition Screening Tool (MST) is a quick and effective way to identify nutritional risk, particularly in cancer patients. Validated for use in both inpatient and outpatient oncology settings, MST focuses on two key indicators: recent weight loss and appetite loss, the most sensible indicators of malnutrition. In metastatic cancer patients, weight loss strongly predicts malnutrition and correlates with poor treatment response and lower survival rates [20].

MST has demonstrated high sensitivity and specificity (93%) compared to more detailed assessments like the Patient-Generated Subjective Global Assessment (PG-SGA) [21]. This accuracy ensures at-risk patients are identified promptly for nutritional interventions. Its simplicity also makes it ideal for outpatient settings with limited time and resources [20].

MST is particularly useful for metastatic cancer patients, offering a quick initial screening that can lead to more comprehensive assessments. If a patient is identified as at-risk for malnutrition, a registered dietitian should conduct a detailed assessment, followed by individualized nutritional interventions.

Nutrition-Focused Physical Exam (NFPE)

Nutrition-Focused Physical Exam (NFPE) is a systematic head-to-toe assessment used to evaluate nutritional status by detecting malnutrition, nutrient deficiencies, and body composition changes. It examines muscle mass, subcutaneous fat, fluid accumulation, and functional capacity—key areas affected by malnutrition [22].

Physical assessments have evolved from traditional measures like triceps skinfolds and mid-arm circumference to more comprehensive evaluations of body composition [22]. This shift is essential since malnutrition often manifests as muscle wasting, fat loss, and reduced strength, particularly in chronic conditions such as metastatic cancer.

In clinical practice, NFPE plays a key role in assessing metastatic cancer patients, who are at a high risk of muscle wasting and nutrient deficiencies. Its head-to-toe approach ensures a thorough evaluation of nutritional status, enabling accurate and timely malnutrition diagnoses. Additionally, NFPE helps differentiate between malnutrition-related changes and those caused by inflammation or disease progression [22].

Registered Dietitian Nutritionists (RDNs) can be trained to conduct NFPEs, ensuring they follow a standardized approach for diagnosing malnutrition.

Key features of NFPE	Description	Examples of signs to look for
Observation	Visual assessment of the patient's overall appearance	Hollow look, dark circles, and loose skin in the orbital *area* is indicative of subcutaneous fat loss. Hollow, sunken, and narrow *cheeks* are indicative of subcutaneous fat loss. Protruding, prominent bone in the *clavicle bone region* is indicative of muscle loss
Palpitation	Physical touch used to evaluate the degree of muscle wasting or fat loss. Palpation helps detect subtle changes that may not be visible through observation alone	Fingers touch with minimal fat in the upper arm region (*triceps*). Deep depression with minimal-to-no muscle tone or resistance in the temple *region*. Minimal-to-no bounce back of fat pads in the *cheek region*. Minimal-to-no fat can be pinched in the *thoracic and lumbar area*
Functional capacity assessment	One of the unique aspects of NFPE is its inclusion of a handgrip strength test, which provides an insight into the patient's functional status and strength	Reduced grip strength

NFPE Nutrition-Focused Physical Exam
NFPE Pocket Guide. 3rd edition. Publisher: Academy of Nutrition and Dietetics, 2022.

Criteria for Diagnosis of Malnutrition

In 2012, the Academy of Nutrition and Dietetics (AND) and the American Society for Parenteral and Enteral Nutrition (ASPEN) established standardized criteria for diagnosing malnutrition to ensure consistency in identification and documentation, especially for coding and reimbursement. NFPE is a key component of the AND/ASPEN consensus for malnutrition diagnosis.

An AND/ASPEN diagnosis requires at least two of the following six clinical characteristics [23]:

1. Inadequate energy intake (<75% of estimated energy requirement)
2. Unintended weight loss (≥5% in 1 month)
3. Loss of muscle mass
4. Loss of subcutaneous fat stores
5. Fluid retention masking weight loss
6. Diminished functional status measured by handgrip strength

These criteria offer a comprehensive framework for diagnosing malnutrition and ensuring timely interventions in metastatic cancer patients. Identifying two or more criteria confirms malnutrition, and early management is critical for improving outcomes, survival, and quality of life in cancer care [22].

Nutritional Challenges in Advanced Cancer

Nutritional challenges in advanced cancer, including malnutrition, cachexia, and sarcopenia, are severe and interconnected, contributing to morbidity and mortality.

Malnutrition affects nearly 40% of cancer patients, increasing with disease progression and treatment intensity, with 60% of advanced cancer patients impacted [24–26]. Many patients already present with metastatic disease at their first oncology visit [26].

Malnutrition is driven not only by reduced intake but also by the tumor's metabolic demands, treatment side effects (e.g., anorexia, nausea, dysphagia), and inflammation. It leads to poorer outcomes, including reduced therapy tolerance, higher toxicities, and lower survival [24, 25].

Cachexia affects about 50% of cancer patients, especially in advanced stages [25]. Unlike simple starvation, it involves involuntary loss of lean body mass, with or without fat loss, driven by systemic inflammation, anorexia, insulin resistance, and increased muscle catabolism [25].

Cachexia predicts mortality, regardless of tumor size, and results from reduced intake and abnormal metabolism caused by tumor factors and host cytokines, creating a hypercatabolic state that cannot be reversed by standard nutrition [24].

Cachexia progresses through three stages: pre-cachexia, cachexia, and refractory cachexia, with the latter often occurring in the final 3 months of life [27]. See Fig. 5.1 for details.

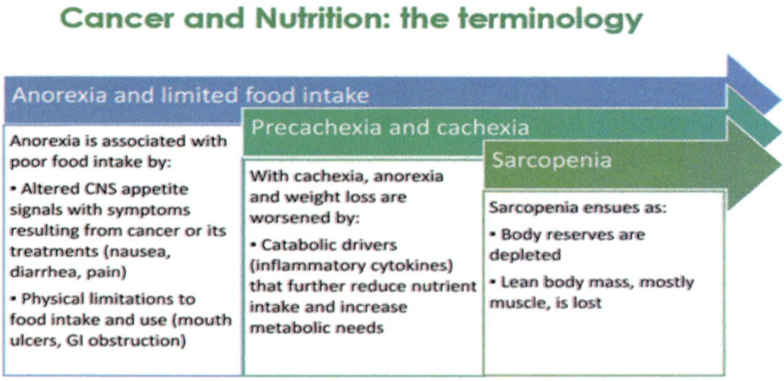

Fig. 5.1 Universally accepted definitions for malnutrition, cachexia, and sarcopenia. These definitions aim to guide clinicians in detecting and managing the metabolic and nutritional problems linked to chronic and acute illnesses, such as cancer. (Arends, Jann, et al. "ESPEN expert group recommendations for action against cancer-related malnutrition." *Clinical Nutrition* 36.5 (2017): 1187–1196)

Cachexia-associated inflammation accelerates muscle wasting and disrupts metabolism, increasing energy expenditure and reducing nutrient use. As a result, patients lose weight and muscle mass even with adequate nutrition. Early intervention with targeted therapies and nutrition is essential to mitigate its impact and improve outcomes. Monitoring inflammatory markers like C-reactive protein can help assess cachexia severity and guide nutritional interventions [24, 25, 27].

Sarcopenia, the loss of skeletal mass and function, often accompanies cachexia and plays a key role in cancer-related malnutrition. It directly affects physical function, treatment outcomes, and quality of life. CT scans, particularly at the lumbar vertebrae, are commonly used to assess muscle mass, with grip strength serving as a functional measure linked to sarcopenia, inflammation, and survival [28, 29]. Sarcopenia predicts poor outcomes, including higher treatment toxicity, more postoperative complications, and reduced survival [25].

The interplay between malnutrition, cachexia, and sarcopenia creates a cycle in which each condition worsens the others, driving a decline in health. Early, individualized nutritional intervention is crucial, though challenging in advanced cancer, to stabilize weight and address the metabolic and inflammatory disturbances underlying these conditions [25].

In conclusion, managing nutrition in advanced cancer requires a multifaceted approach, recognizing that these conditions are integral to the disease and significantly impact survival and quality of life. Integrating nutritional care into oncology treatment protocols is essential for enhancing therapeutic outcomes and improving patient well-being.

Nutritional Interventions

Nutrition Impact Symptoms (NISs)

Oncologic therapies can lead to malnutrition and metabolic changes in cancer patients, resulting in complex symptoms known as nutrition impact symptoms (NISs) [7]. These symptoms adversely affect oral intake, nutritional status, health, and even survival rates. NISs can influence treatment outcomes, acting as dose-limiting side effects and affecting patients' willingness to continue therapy [7]. Effective symptom management is linked to improved quality of life for both patients and their families [7, 8, 30].

Patients undergoing treatment commonly report symptoms like anorexia, early satiety, constipation, diarrhea, mucositis, nausea, and fatigue. Early identification and management through education, behavior modification, and medication can help mitigate these symptoms, sustaining nutritional status and enhancing treatment tolerance [4].

Systematic assessment of NISs is essential for quality cancer care, with validated tools aiding practitioners in evaluating the presence and severity of these symptoms [31]. Patient-reported outcomes provide critical information for managing treatment side effects and ensuring consistency in care. Many oncology centers use the National Cancer Institute's Common Terminology Criteria for Adverse Events (CTCAE) grading system to document and report these symptoms [32].

The CTCAE displays grades 1 through 5 with unique clinical descriptions of severity for each AE based on this general guideline [3, 32]

Grade 1	Mild; asymptomatic or mild symptoms; clinical or diagnostic observations only; intervention not indicated
Grade 2	Moderate; minimal, local or noninvasive intervention indicated; limiting age-appropriate instrumental activities of daily living (ADL)[a]
Grade 3	Severe or medically significant but not immediately life-threatening; hospitalization or prolongation of hospitalization indicated; disabling; limiting self-care activities of daily living (ADL)[b]
Grade 4	Life-threatening consequences: urgent intervention indicated
Grade 5	Death related to AE

CTCAE common terminology criteria for adverse events; *AE*, adverse event; *ADL* activities of daily living

[a]Instrumental ADL refers to preparing meals, shopping for groceries or clothes, using the telephone, managing money, etc

[b]Self-care ADL refers to bathing, dressing and undressing, feeding self, using the toilet, taking medications, and not bedridden

Implementing medical nutrition therapy (MNT) is crucial, as dietary interventions can help manage NISs, enhance food intake, maintain body weight, improve quality of life, decrease treatment toxicity, and potentially prolong survival. While ongoing research continues to explore nutrition interventions, registered dietitian nutritionists (RDNs) lack evidence-based guidelines for all oncology-related nutrition interventions and rely on various evidence-based guidelines (including reviews, case studies, expert onions, etc.) and reputable organizations (i.e., Oncology Nursing Society, American Society for Clinical Oncology, American Society for Parenteral and Enteral Nutrition, etc.).

Studies on patients with advanced cancer receiving palliative treatment or medical oncology treatment for 1 year have reported prevalent symptoms like appetite loss, constipation, early satiety, weight loss, and xerostomia [3]. Two-thirds of patients experienced at least one severe symptom or an extremely low quality of life, within the first 2 months after diagnosis of advanced cancer [33]. Between 31% and 53% of patients with advanced cancer experience distressing symptoms like nausea, constipation, xerostomia, weight loss, and anorexia, while 11–23% experience constipation [4]. Fatigue, which affects 32–90% of patients with advanced cancer [8], is associated with weight loss, anorexia, and poor performance status [4]. Pain in advanced cancer impacts 35–96% of patients [7]. This can affect a patient's desire to eat, and analgesics may also contribute to constipation and nausea [4].

The following Table 5.1 outlines some common nutrition impact symptoms throughout oncologic treatment in advanced cancer, additional considerations, and interventions involving medical nutrition therapy and pharmacotherapy.

Table 5.1 The CTCAE displays Grades 1 through 5 with unique clinical descriptions of severity for each AE based on this general guideline

Nutrition problems	Nutrition recommendations [4]
Anorexia	Consume frequent small meals and snacks throughout the day, which are rich in nutrients and include fluids Establish a schedule for eating five to eight times daily Optimize food intake during that period of the day when your appetite is strongest If eating becomes challenging, consider incorporating oral nutrition supplements. Explore methods to boost calorie and protein content in shakes, such as adding protein powder or dried milk Consider alternative high-calorie shake options Drink fluids between meals rather than concurrently Increase the nutrient density of food choices as tolerated Select easily prepared and served foods to conserve energy Have convenient options like frozen meals, nutrition bars, and nutrient-dense beverages readily available View eating as an essential component of your overall treatment plan Engage in light physical activity to aid in the digestion process and promote movement of food through the gastrointestinal tract
Constipation	Educate individuals about the significance of maintaining adequate hydration, fiber intake, and overall food consumption for bowel regularity Promote a gradual rise in dietary fiber intake, incorporating foods rich in insoluble fiber such as legumes, fruits with skin, whole grains, and wheat bran If using medicinal fibers, consider increasing fluid intake by an additional 32 ounces daily Incorporate light activity or stretching exercises to enhance bowel regularity, if medically feasible When opioids are introduced, adjust the bowel regimen by combining a stimulant laxative with a stool softener, as this combination is typically well-tolerated Discuss the impact of medications on bowel function and encourage bowel regimens tailored to individual experiences, such as gradually increasing interventions in response to insufficient bowel movements
Diarrhea	Consume small, frequent meals if improvement is noticed Maintain a consistent meal schedule and ensure adequate hydration by consuming 8–10 cups of fluids daily, adding an extra cup (8 oz) for each loose bowel movement Opt for a low-fat, low-insoluble-fiber, or low-lactose diet or a combination of these based on individual needs Increase consumption of soluble-fiber foods like pectin, applesauce, bananas, oatmeal, potatoes, and rice If experiencing gas or bloating, restrict intake of gas-forming foods such as cruciferous vegetables and legumes as well as carbonated beverages Avoid caffeine, alcohol, and highly spiced foods, along with products containing sugar alcohols like sorbitol found in sugar-free gum and candy Consume electrolyte-rich foods and fluids as necessary. Consider using oral rehydration salts (ORS) products or recipes if suitable and lactase enzyme products and substitutes if recommended

(continued)

Table 5.1 (continued)

Nutrition problems	Nutrition recommendations [4]
Dysgeusia	Accommodating changes in taste is a highly personalized experience. Some general suggestions include the following. To counteract a bitter/acidic/metallic taste: Replace metal silverware with plastic or bamboo utensils Include sweet foods in meals such as melons, juices, and marinades Experiment with adding acidic elements like lemon, orange, or balsamic vinegars to dishes if mucositis is not present Try varying protein sources like eggs, beans, tofu, dairy, chicken, fish, or oral nutrition supplements
Fatigue	Eat snacks and small meals, if difficulty with regular meals Try oral nutrition supplements high in protein and/or calories Opt for foods and drinks high in calories and protein Consume easy-to-chew foods, if eating is challenging Keep nonperishable snacks at the bedside Plan a larger meal for when your appetite is best (i.e., breakfast) Stay hydrated by consuming at least eight cups of fluids daily Eat every 3–5 hours Continue to perform ADLs and light activities Consider complementary therapies: Consult with a physical therapist for strength training, yoga, massage therapy, and psychosocial interventions for fatigue management
Nausea and vomiting	Consume small and frequent amounts of food Opt for mild, starchy foods and clear liquids that are served at room temperature Stay away from oily, high-fat, and heavily seasoned dishes Drink liquids in between meals rather than during a meal Minimize exposure to cooking smells by avoiding the kitchen and using exhaust fans or opening windows Avoid or reduce the use of strong-smelling lotions, soaps, perfumes, and air fresheners Rest with the head elevated for half an hour after eating Schedule meals around the time when anti-nausea medications are most effective Take pain relievers with crackers or light snacks
Parosmia (distorted smell)	Opt for foods served at cold or room temperature like smoothies, sandwiches, cottage cheese, yogurt, puddings, custards, nut butters, and fruits Avoid foods with strong odors such as fish, onions, and cabbage Refrain from being in cooking areas during meal preparation Minimize lengthy cooking processes like using a crockpot Ensure cooking areas are ventilated by utilizing exhaust fans or opening a window
Xerostomia	Alternate bites and sips at mealtime Add sauces and/or broths to meals and moisten dry foods Sip liquids throughout the day and maintain hydration Suck on hard candy or frozen grapes Avoid alcoholic beverages and mouthwash that has alcohol Practice good oral hygiene Use of baking soda/salt oral rinses might be helpful Use homemade baking soda/salt oral rinses

Nutrition Support

Nutrition support encompasses oral nutrition supplements (ONSs), enteral nutrition (EN), and parenteral nutrition (PN) [34, 35].

Oral Nutrition Supplements

The European Society for Clinical Nutrition and Metabolism (ESPEN) recommends a minimum protein intake of 1.0 g/kg body weight per day to maintain muscle mass in cancer patients [36, 37]. ONSs should be considered when oral intake is <75% of the estimated needs. [34] ONSs are commercially manufactured products in liquid, semisolid, or powder form that are designed to provide a variety of nutrients to enhance oral nutrition consumption. These supplements are used for patients who are unable to meet their nutritional needs through regular food. They can be milk-, soy-, or juice-based and may or may not contain fiber. They are typically lactose-free [12]. ONSs must comply with specific guidelines, which include requirements for labeling, safety, and manufacturing practices [34]. Disease-specific ONSs are tailored to meet the unique nutritional requirements of individuals with specific illnesses, digestive issues, or metabolic disorders. For example, oncology-specific feeds are often energy- and protein-dense. [34]

ONSs are typically recommended to be consumed between meals to avoid affecting appetite. This approach helps maximize their benefits without affecting appetite for regular meals. To ensure effectiveness, it is suggested to prescribe at least two servings of ONSs daily for up to 3 months. Patients receiving ONSs should be regularly monitored to track adherence, assess progress toward treatment objectives, and determine whether discontinuation is necessary due to factors such as disease progression, end-of-life considerations, or poor adherence [34]. The use of ONSs in palliative care should be evaluated based on individual needs, focusing on quality of life [34]. While ONSs can enhance nutritional intake, their impact on clinical outcomes, like survival, remains inconclusive [36–39].

The utilization of ONSs is more successful when under the guidance of a registered dietitian. Results were observed in a meta-analysis involving cancer patients, where it was noted that nutritional counseling, accompanied by ONS when required, consistently led to improvements in various aspects of nutritional status. In contrast, using ONSs alone resulted in only minor and statistically insignificant results [34]. There was no statistical significance in body weight (BW), between study groups receiving ONSs and standard care [36]. However, data on BW and ONSs have yielded mixed findings. A review on studies examining malnourished cancer patients prescribed nutrition formulas enriched with immunomodulating nutrients (i.e., omega 3, glutamine, and/or arginine) demonstrated nearly 30% improvement in BW, compared to standard formulas, isocaloric control products, or a standard diet [12]. While these studies demonstrated positive effects of ONSs on body weight, none was observed for treatment toxicity or survival [34].

Enteral and Parenteral Nutrition

Patients with advanced cancer may transition through many disease phases that present unique and sometimes difficult challenges to meet nutritional needs [40–44]. The use of EN and PN is selected depending on cancer type, its extent, complications, oncologic therapy, and prognosis [40–45]. Nutrition support may be essential for patients unable to meet their nutritional needs [2]. EN is preferred due to its lower complication rates, while PN is reserved for cases where EN and/or oral nutrition is not feasible [40].

The use of nutrition support as a standard practice for all cancer patients undergoing antineoplastic therapy has not been shown to improve survival rates [2, 39]. Cancer patients who are malnourished, losing weight, undergoing anticancer treatment, and are expected to be unable to consume or absorb enough nutrients for over 1–2 weeks may benefit from artificial nutrition, preferably through the enteral route. Research comparing enteral nutrition (EN) to parenteral nutrition (PN) suggests that EN may have fewer instances of reduced tumor response compared to PN [2, 40]. In cases of severe mucositis or obstructive tumors in the head, neck, or thorax, enteral feeding through nasogastric or gastrostomy tubes is recommended [2]. For oncology patients without gastrointestinal issues (i.e., impaired absorption, mechanical bowel obstruction, motility disorders, etc.) leading to intestinal failure, PN is not effective and may even be harmful [2, 35, 41].

Patients with Advanced Cancer Receiving No Antineoplastic Therapy

Patients with advanced cancer who are not undergoing anticancer treatment should have their nutritional status evaluated, as deficiencies are linked to decreased quality of life and performance levels [2, 44]. The decision to provide nutritional support should be carefully considered, taking into account factors such as the patient's expected survival, current nutritional status, potential benefits, and the preferences of both the patient and their family [2, 45]. Evidence supporting the benefits of nutritional support in advanced cancer patients, especially in their final weeks of life, is limited [2].

It is recommended that regular nutritional status assessments be conducted for all patients with advanced cancer [2]. However, the decision to implement nutritional support should only be made after thorough consideration of potential benefits [2]. In the advanced stages of terminal illness, artificial nutrition is unlikely to offer significant benefits for most patients [2].

Nutrients for Supporting Patients with Advanced Cancer

Addressing nutrient deficiencies is essential, given the catabolic nature of cancer and its treatments on nutritional status.

Supplements, such as vitamin D and calcium, enhance bone health and reduce fracture risk, particularly important for patients with limited sun exposure or malabsorption [46]. ESPEN guidelines recommend 600–800 IU/day of vitamin D, with higher doses for deficiencies [4]. Due to its longer half-life, 25-hydroxyvitamin D is used to assess vitamin D status, with D3 (cholecalciferol) having better bioavailability than D2 (ergocalciferol). In 2010, the tolerable upper intake of vitamin D was raised to 4000 IU/day. The Endocrine Society suggests maintaining 25-hydroxyvitamin D levels between 40 and 60 ng/mL, with up to 100 ng/mL considered safe [4]. Calcium intake should reach 1200 mg/day from food or supplements if necessary [4].

Monitoring for vitamin D toxicity, though rare, is essential, as it can cause hypercalcemia, leading to neurologic, psychiatric issues, or renal failure. Toxicity usually results from iatrogenic causes or labeling errors [47]. Hypercalcemia can also occur independently of diet, requiring medical management [4]. The most common cause, accounting for 80% of cases, is excessive secretion of parathyroid hormone-related Peptide (PTHrP), typically seen in solid tumors and some non-Hodgkin's lymphomas [48].

Iron deficiency is prevalent, particularly in cancer-related anemia, which is multifactorial, involving chronic inflammation, poor absorption, and reduced red blood cell production. Chemotherapy exacerbates iron deficiency, lowering hemoglobin and increasing fatigue. Iron supplementation can alleviate fatigue and functional decline, but monitoring is essential, as excessive iron intake carries risks [30, 46].

Omega-3 fatty acids, particularly eicosapentaenoic acid (EPA) and docosahexaenoic acid (DHA), benefit metastatic cancer patients due to their anti-inflammatory effects, which help counteract the systemic inflammation that worsens cachexia. Studies suggest omega-3 supplementation may improve appetite, reduce inflammation, and enhance quality of life with minimal risk of adverse effects [30, 46]. EPA and DHA may lower pro-inflammatory cytokines like tumor necrosis factor-α (TNF-α) and interleukin (IL)-6 [49]. While omega-3 s have shown mixed results on body weight and lean mass, they consistently improve quality of life and median survival [49, 50]. Omega-3 s may also enhance muscle function when combined with protein supplementation and exercise [51]. Despite variable effects on weight, their anti-inflammatory and well-being benefits make omega-3 polyunsaturated fatty acids (PUFAs) a valuable tool for managing cancer cachexia [49, 51]. Omega-3 fatty acids, such as EPA and DHA, improve muscle function in cancer patients by reducing inflammation and promoting muscle protein synthesis. They activate the mechanistic target of rapamycin complex 1 (mTORC1) pathway, supporting muscle health, and enhance mitochondrial function, reducing oxidative stress and improving muscle metabolism [50]. In cancer care, fish oil or EPA-enriched supplements are recommended, with typical doses of 1.5–3 grams per day of combined EPA and DHA [51].

β-Hydroxy β-methylbutyrate (HMB), a metabolite of leucine found in muscle cells, avocados, and citrus fruits, helps counteract muscle loss. It inhibits the ubiquitin–proteasome pathway and promotes protein synthesis via the mTOR pathway [52].A systematic review found HMB supplementation improved muscle mass in all high-quality studies, which used bioelectrical impedance analysis (BIA), CT scans, and air-displacement plethysmography for assessment. It enhanced muscle function in two studies, reduced hospitalizations in three, and improved survival in five out of seven. However, evidence on quality of life and body weight was mixed.

No serious adverse effects were reported, and HMB was safe at 3.0 g/day. While promising, further trials are needed to confirm its benefits in cancer care [53].

Conclusions

Cancer is a complex disease affecting millions, with metastatic spread presenting significant treatment challenges. Nutrition is vital for survivorship and managing advanced cancer, yet research often focuses on prevention rather than the role of nutrition therapy in metastatic cases. A balanced diet rich in proteins, healthy fats, carbohydrates, and calories—preferably plant-inclusive—is essential. Dietary recommendations should consider patients' motivations and personal needs, empowering them to make choices aligned with their lifestyles and health goals.

Malnutrition is a frequent barrier in cancer survivorship, making nutritional screening a crucial aspect of care, conducted at diagnosis and throughout treatment. Timely interventions are necessary to support patients at risk of malnutrition and to preserve their health. Enhancing nutritional status can be achieved through key nutrients and nutrition support, though clinical judgment is essential, especially in advanced cases. Overall, optimizing nutrition and addressing related symptoms are critical for comprehensive cancer care, enhancing quality of life and well-being. Personalized nutrition counseling, alongside a holistic support framework, can improve treatment outcomes, necessitating a collaborative approach among healthcare professionals to manage treatment-related side effects.

References

1. Maroni P, Bendinelli P, Fulgenzi A, Ferraretto A. Mediterranean diet food components as possible adjuvant therapies to counteract breast and prostate cancer progression to bone metastasis. Biomolecules. 2021;11(9) https://doi.org/10.3390/biom11091336.
2. de las Peñas R, Majem M, Perez-Altozano J, et al. SEOM clinical guidelines on nutrition in cancer patients (2018). Clin Transl Oncol. 2019;21(1):87. https://doi.org/10.1007/S12094-018-02009-3.
3. Oostra DL, Burse NR, Wolf LJ, et al. Understanding nutritional problems of metastatic breast cancer patients: opportunities for supportive care through eHealth. Cancer Nurs. 2021;44(2):154–62. https://doi.org/10.1097/NCC.0000000000000788.
4. Oncology Nutrition for Clinical Practice. Oncology Nutrition Practice Group. Voss AC, Williams V, editors. Second Edition. Academy of Nutrition and Dietetics. 2021. https://www.eatrightstore.org/product-type/books/oncology-nutrition-for-clinical-practice-2nd-edition. Accessed 13 Aug 2024.
5. Lee K, Kruper L, Dieli-Conwright CM, Mortimer JE. The impact of obesity on breast cancer diagnosis and treatment. Curr Oncol Rep. 2019;21(5) https://doi.org/10.1007/S11912-019-0787-1.
6. Rock CL, Thomson CA, Sullivan KR, et al. American Cancer Society nutrition and physical activity guideline for cancer survivors. CA Cancer J Clin. 2022;72(3):230–62. https://doi.org/10.3322/CAAC.21719.
7. Henson LA, Maddocks M, Evans C, Davidson M, Hicks S, Higginson IJ. Palliative care and the management of common distressing symptoms in advanced cancer: pain, breathlessness,

nausea and vomiting, and fatigue. J Clin Oncol. 2020;38(9):905–14. https://doi.org/10.1200/JCO.19.00470.
8. Kang YE, Yoon JH, Park N, Ahn YC, Lee EJ, Son CG. Prevalence of cancer-related fatigue based on severity: a systematic review and meta-analysis. Sci Rep. 2023;13(1):12815. https://doi.org/10.1038/S41598-023-39046-0.
9. Soares MJ, Müller MJ. Resting energy expenditure and body composition: critical aspects for clinical nutrition. Eur J Clin Nutr. 2018;72(9):1208–14.
10. McMurray RG, et al. Examining variations of resting metabolic rate of adults: a public health perspective. Med Sci Sports Exerc. 2014;46(7):1352.
11. Tüccar TB, Tek NA. Determining the factors affecting energy metabolism and energy requirement in cancer patients. J Res Med Sci. 2021;26(1):124.
12. Gupta G. Lactate: metabolic hallmark of cancer in the 21st century. Clin Oncol. 2017;2(1375):18.
13. Purcell SA, et al. Key determinants of energy expenditure in cancer and implications for clinical practice. Eur J Clin Nutr. 2016;70(11):1230–8.
14. Friesen DE, Baracos VE, Tuszynski JA. Modeling the energetic cost of cancer as a result of altered energy metabolism: implications for cachexia. Theor Biol Med Model. 2015;12:1–18.
15. Shyh-Chang N. Metabolic changes during cancer cachexia pathogenesis. In: Translational research in breast cancer: biomarker diagnosis, targeted therapies and approaches to precision medicine; 2017. p. 233–49.
16. Arends J, et al. ESPEN guidelines on nutrition in cancer patients. Clin Nutr. 2017;36(1):11–48.
17. Mazzo R, Ribeiro FB, Vasques ACJ. Accuracy of predictive equations versus indirect calorimetry for the evaluation of energy expenditure in cancer patients with solid tumors–an integrative systematic review study. Clin Nutr ESPEN. 2020;35:12–9.
18. Mamede AC, et al. The role of vitamins in cancer: a review. Nutr Cancer. 2011;63(4):479–94.
19. Norman HA, et al. The role of dietary supplements during cancer therapy. J Nutr. 2003;133(11):3794S–9S.
20. Castillo-Martínez L, et al. Nutritional assessment tools for the identification of malnutrition and nutritional risk associated with cancer treatment. Rev Investig Clin. 2018;70(3):121–5.
21. Ferguson M, et al. Development of a valid and reliable malnutrition screening tool for adult acute hospital patients. Nutrition. 1999;15(6):458–64.
22. Hummell AC, Cummings M. Role of the nutrition-focused physical examination in identifying malnutrition and its effectiveness. Nutr Clin Pract. 2022;37(1):41–9.
23. White JV, et al. Consensus statement of the Academy of Nutrition and Dietetics/American Society for Parenteral and Enteral Nutrition: characteristics recommended for the identification and documentation of adult malnutrition (undernutrition). J Acad Nutr Diet. 2012;112(5):730–8.
24. Laviano A, Di Lazzaro L, Koverech A. Nutrition support and clinical outcome in advanced cancer patients. Proc Nutr Soc. 2018;77(4):388–93.
25. Hébuterne X, et al. Prevalence of malnutrition and current use of nutrition support in patients with cancer. J Parenter Enter Nutr. 2014;38(2):196–204.
26. Muscaritoli M, et al. Prevalence of malnutrition in patients at first medical oncology visit: the PreMiO study. Oncotarget. 2017;8(45):79884.
27. Fearon K, et al. Definition and classification of cancer cachexia: an international consensus. Lancet Oncol. 2011;12(5):489–95.
28. Kilgour RD, et al. Handgrip strength predicts survival and is associated with markers of clinical and functional outcomes in advanced cancer patients. Support Care Cancer. 2013;21:3261–70.
29. Gaafer OU, Zimmers TA. Nutrition challenges of cancer cachexia. J Parenter Enteral Nutr. 2021;45(S2):S16–25.
30. Kim DH. Nutritional issues in patients with cancer. Intest Res. 2019;17(4):455–62. https://doi.org/10.5217/ir.2019.00076.
31. Simão D, Barata PC, Alves M, Papoila AL, Oliveira S, Lawlor P. Symptom clusters in patients with advanced cancer: A prospective longitudinal cohort study to examine their stability and prognostic significance. Oncologist 2024;29(1):E152-E163. doi:https://doi.org/10.1093/ONCOLO/OYAD211

32. Cancer Institute N. Common terminology criteria for adverse events (CTCAE) version 4.0. Published online 2009. http://www.meddramsso.com. Accessed 16 Sept 2024.
33. Siemens W, Schönsteiner SS, Orellana-Rios CL, et al. Severe symptoms and very low quality-of-life among outpatients newly diagnosed with advanced cancer: data from a multicenter cohort study. Support Care Cancer. 2020;28(11):5547. https://doi.org/10.1007/S00520-020-05388-Y.
34. de van der Schueren MAE. Use and effects of oral nutritional supplements in patients with cancer. Nutrition. 2019:67–8. https://doi.org/10.1016/J.NUT.2019.07.002.
35. Druml C, Ballmer PE, Druml W, et al. ESPEN guideline on ethical aspects of artificial nutrition and hydration. Clin Nutr. 2016;35(3):545–56. https://doi.org/10.1016/J.CLNU.2016.02.006.
36. Dingemans AM, van Walree N, Schramel F, et al. High protein Oral nutritional supplements enable the majority of cancer patients to meet protein intake recommendations during systemic anti-cancer treatment: a randomised controlled parallel-group study. Nutrients. 2023;15(24) https://doi.org/10.3390/NU15245030.
37. Baldwin C, Spiro A, Ahern R, Emery PW. Oral nutritional interventions in malnourished patients with cancer: a systematic review and meta-analysis. JNCI J Natl Cancer Inst 2012;104(5):371–385. doi:https://doi.org/10.1093/JNCI/DJR556
38. Lee JLC, Leong LP, Lim SL. Nutrition intervention approaches to reduce malnutrition in oncology patients: a systematic review. Support Care Cancer 2016;24(1):469–480. doi:https://doi.org/10.1007/S00520-015-2958-4
39. Elia M, Van Bokhorst-De Van Der Schueren MAE, Garvey J, et al. Enteral (oral or tube administration) nutritional support and eicosapentaenoic acid in patients with cancer: a systematic review. Int J Oncol. 2006;28(1):5–23. https://doi.org/10.3892/IJO.28.1.5.
40. Arends J, Bachmann P, Baracos V, et al. ESPEN guidelines on nutrition in cancer patients. Clin Nutr. 2017;36(1):11–48. https://doi.org/10.1016/J.CLNU.2016.07.015.
41. Bozzetti F, Arends J, Lundholm K, Micklewright A, Zurcher G, Muscaritoli M. ESPEN guidelines on parenteral nutrition: non-surgical oncology. Clin Nutr 2009;28(4):445–454. doi:https://doi.org/10.1016/J.CLNU.2009.04.011
42. Gavazzi C, Bhoori S, LoVullo S, Cozzi G, Mariani L. Role of home parenteral nutrition in chronic radiation enteritis. Am J Gastroenterol 2006;101(2):374–379. doi:https://doi.org/10.1111/J.1572-0241.2006.00438.X
43. Tong H, Isenring E, Yates P. The prevalence of nutrition impact symptoms and their relationship to quality of life and clinical outcomes in medical oncology patients. Support Care Cancer 2009;17(1):83–90. doi:https://doi.org/10.1007/S00520-008-0472-7/METRICS
44. Laird BJ, Kaasa S, McMillan DC, et al. Prognostic factors in patients with advanced cancer: a comparison of clinicopathological factors and the development of an inflammation-based prognostic system. Clin Cancer Res. 2013;19(19):5456–64. https://doi.org/10.1158/1078-0432.CCR-13-1066.
45. Tong H, Isenring E, Yates P. The prevalence of nutrition impact symptoms and their relationship to quality of life and clinical outcomes in medical oncology patients. Support Care Cancer. 2009;17(1):83–90. https://doi.org/10.1007/S00520-008-0472-7/METRIC.
46. Johal J, Han CY, Joseph R, et al. Dietary supplements in people with metastatic cancer who are experiencing malnutrition, cachexia, sarcopenia, and frailty: A scoping review. Nutrients. 2022;14(13):2642. https://doi.org/10.3390/nu14132642.
47. Galior K, Grebe S, Singh R. Development of vitamin D toxicity from overcorrection of vitamin D deficiency: a review of case reports. Nutrients. 2018;10(8) https://doi.org/10.3390/nu10080953.
48. Vakiti A, Anastasopoulou C, Mewawalla P. Malignancy-related hypercalcemia. StatPearls. Published online April 16, 2023. Accessed 26 Sept 2024.
49. Jin X, Xu XT, Tian MX, Dai Z. Omega-3 polyunsaterated fatty acids improve quality of life and survival, but not body weight in cancer cachexia: A systematic review and meta-analysis of controlled trials. Nutr Res. 2022;107:165–78. https://doi.org/10.1016/j.nutres.2022.09.009.
50. Therdyothin A, Phiphopthatsanee N, Isanejad M. The effect of omega-3 fatty acids on sarcopenia: mechanism of action and potential efficacy. Mar Drugs. 2023;21(7):399. https://doi.org/10.3390/md21070399.

51. Freitas R, Campos MM. Protective effects of omega-3 fatty acids in cancer-related complications. Nutrients. 2019;11(5):945. https://doi.org/10.3390/nu11050945.
52. J Cruz-Jentoft A. Beta-hydroxy-beta-methyl butyrate (HMB): from experimental data to clinical evidence in sarcopenia. Curr Protein Pept Sci. 2018;19(7):668–72. https://doi.org/10.2174/1389203718666170529105026.
53. Prado CM, Orsso CE, Pereira SL, Atherton PJ, Deutz NEP. Effects of β-hydroxy β-methylbutyrate (HMB) supplementation on muscle mass, function, and other outcomes in patients with cancer: a systematic review. J Cachexia Sarcopenia Muscle. 2022;13(3):1623–41. https://doi.org/10.1002/jcsm.12952.

6 Radiological Assessment: From Plain Radiographs to Advanced Cross-Sectional Studies

James A. Strahan and Zachary E. Stewart

Introduction

The presence or absence of metastases greatly influences treatment and prognostication for most, if not all, hematologic and solid cancers. Metastasis location can be stratified as osseous, soft tissue (non-osseous), or mixed. Non-osseous metastases are less common than osseous metastases and most commonly affect the lymph nodes, lung, or liver. Diagnosis is most often confirmed via open or percutaneous biopsy. The following discussion will focus on osseous metastasis related to both hematologic and solid malignancies.

Bone is among the most common sites of metastatic disease along with the lung and liver, and spread is most often hematogenous [1]. The most common sources of osseous metastases in adults over age 25 include cancers of the lung, prostate, and breast. Of patients over age 18 diagnosed with any solid cancer, 8.4% will develop osseous metastasis within 10 years of the initial diagnosis; however, rates are much higher for certain primary malignancies such as prostate cancer. The most common sites of osseous metastasis are the spine and pelvis, including the proximal femurs [2].

A variety of imaging modalities are utilized to evaluate metastatic lesions of bone, including plain film radiography (X-rays), computed tomography (CT), magnetic resonance imaging (MRI), and a variety of nuclear medicine examinations such as technetium 99 m (99mTc)-labeled bone scintigraphy and positron emission tomography (PET), which is often fused with CT (PET–CT) or MRI (PET–MRI) images to aid in lesion localization and morphologic characterization.

J. A. Strahan · Z. E. Stewart (✉)
Department of Radiology, Massachusetts General Hospital, Boston, MA, USA

Harvard Medical School, MA, Boston, USA
e-mail: ZSTEWART@mgh.harvard.edu

J. Pretell-Mazzini, S. A. Lozano Calderón (eds.), *Multidisciplinary Approach of Musculoskeletal Metastasis*, https://doi.org/10.1007/978-3-032-09498-8_6

Basic Principles of Bone Metastasis Imaging

Metastases are often first detected or evaluated by either X-rays or CT, which both use ionizing radiation to construct two- and three-dimensional (3D) images, respectively. Osseous lesions are broadly stratified as lytic, blastic (also referred to as sclerotic), or mixed based on X-ray and CT findings. MRI and nuclear medicine studies have demonstrated complementary findings, which are described using different terminology specific to the imaging modality.

Lytic Metastases

Lytic lesions destroy the involved bone via osteoclast activation and replace it with soft tissue, resulting in lucency of the involved bone (Fig. 6.1) [3]. The most common lytic metastases include breast cancer, non-small cell lung cancer, and renal cell carcinoma [4]. Other common sources include melanoma, thyroid cancer, and non-Hodgkin's lymphoma.

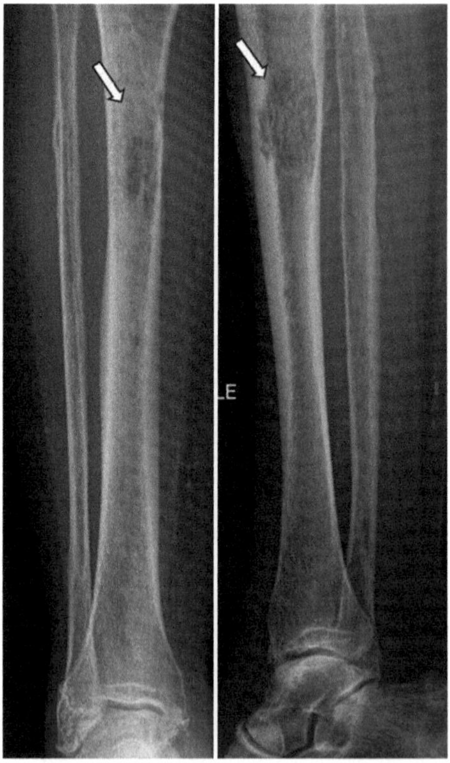

Fig. 6.1 AP and lateral X-ray images of the right tibia–fibula show a lytic lesion (arrows) in the proximal tibia caused by multiple myeloma. AP anteroposterior

Blastic Metastases

Blastic lesions result from the stimulation of osteoblast activity yielding increased formation of bone and a sclerotic appearance on X-rays and CT (Fig. 6.2) [3]. The two most common blastic metastases are breast and prostate cancers [5]. Other common sources include carcinoid, Hodgkin's lymphoma, and medulloblastoma.

Mixed-Feature Metastases

Mixed lytic and blastic lesions exhibit intermixed components of osseous lucency and sclerosis (Fig. 6.2). Common mixed metastases include cancers of the digestive tract and squamous cell carcinoma [6]. Breast cancer may also show a mixed appearance.

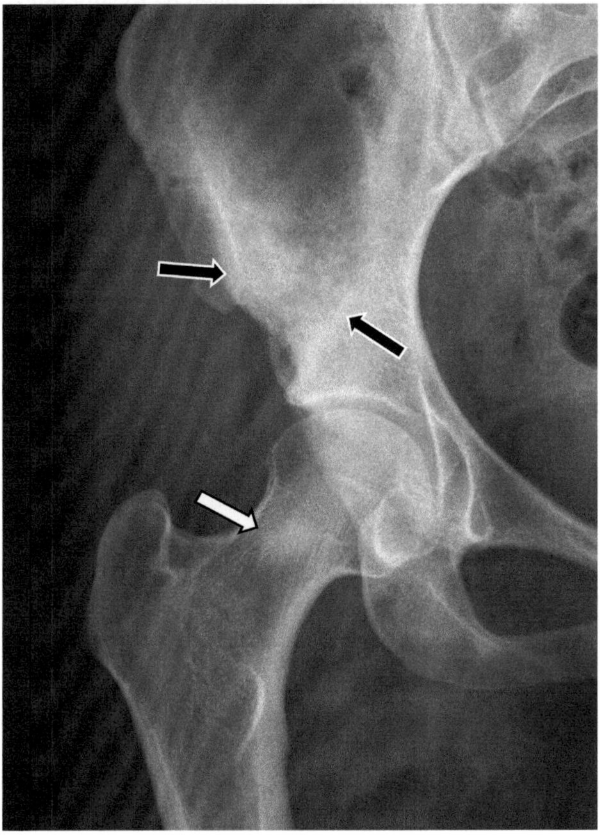

Fig. 6.2 An AP X-ray image of the right hip shows a sclerotic lesion in the right femoral neck (white arrow) and a mixed lytic and sclerotic lesion in the right iliac bone (black arrows) caused by invasive ductal carcinoma of the breast. AP anteroposterior

Plain Film Radiography (X-Rays)

An X-ray utilizes ionizing radiation to make use of the differences in tissue density, allowing reconstruction of greyscale, two-dimensional images of the imaged structures. On an X-ray, normal bone exhibits a continuous peripheral rim of dense (white) cortex and a network of reticular trabeculations filling the medullary cavity. Given the relative similarity of the density of different soft tissue structures, an X-ray lacks sensitivity for evaluation of soft tissues. Therefore, the utility of X-rays in cancer imaging remains predominantly limited to evaluation of bones and readily offers characterization of osseous lesions as lytic, blastic, or mixed.

Lytic lesions demonstrate lucency or discontinuity of the trabeculae with or without the presence of cortical destruction (Fig. 6.3). However, lesions are not visible on X-rays until at least 30% of the trabecular bone has been destroyed [7].

Blastic lesions appear more dense or sclerotic (whiter) on an X-ray relative to adjacent, uninvolved bone (darker). The cortex usually remains intact in the absence of a pathologic fracture and may even be focally thickened.

Mixed lesions can exhibit a variety of characteristics with differing contributions of bony lysis or sclerosis.

Computed Tomography

Like X-rays, CT utilizes ionizing radiation to reconstruct images but with the added advantage of 3D evaluation due to the acquisition of hundreds of sequential X-rays. Lesions can likewise be characterized by CT as lytic, blastic, or mixed.

Unlike X-rays, CT is also useful for assessing extraosseous soft tissue extension of an osseous lesion, a finding that often helps guide the approach for biopsy. CT also offers the possible addition of intravenous iodinated contrast media (ICM) for

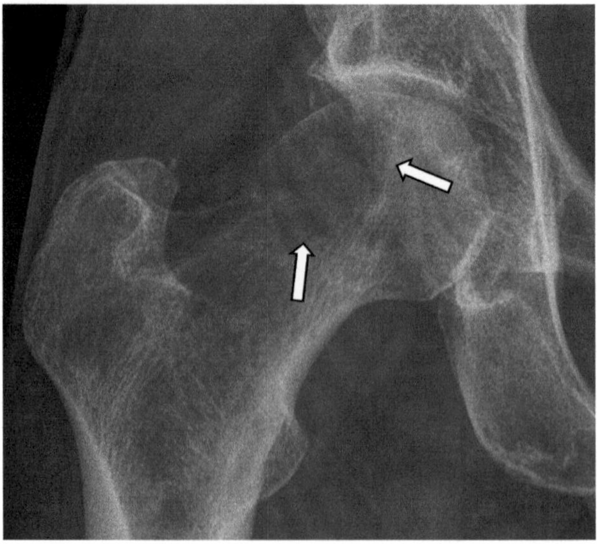

Fig. 6.3 An AP X-ray image of the right hip shows a lytic lesion in the right femoral head caused by metastatic gastric carcinoma. AP anteroposterior

improved evaluation of soft tissue lesions, but ICM offers little further information of osseous metastases. Emerging CT techniques assessing dynamic contrast enhancement for a characterized lesion may show increased utility in the future; however, this is not routinely used at this time and is beyond the scope of this text [8].

Other advantages of CT include better assessment of the presence of a pathologic fracture or the risk of fracture development. Although several classification systems exist, which stratify the risk of a pathologic fracture based on imaging characteristics, the general CT findings that portend a higher risk of fracture include the extent of cortical destruction and the percentage of the bone cross section affected. A well-validated, comprehensive assessment of pathologic fracture risk has not been established, but efforts are underway, which aim to provide a more clinically relevant assessment of fracture risk [9]. CT also offers better assessment of periosteal reaction, a finding frequently associated with more aggressive lesions.

The 3D evaluation offered by CT helps discern between nonmalignant bone changes such as spinal endplate degenerative change and true lytic, sclerotic, or mixed lesions. CT is also helpful for planning surgical interventions. Although CT exhibits higher sensitivity for lesion detection compared to X-rays, its sensitivity remains low relative to those of more advanced modalities such as MRI and those utilizing radiotracers [7].

Lytic lesions can be better characterized with CT, relative to X-rays, regarding the extent of trabecular and cortical destruction and the presence or absence of adjacent extraosseous soft tissue components. CT offers significantly increased sensitivity over X-rays for detection of osteolysis (Fig. 6.4) [10].

Blastic lesions demonstrate circumscribed regions of sclerosis within the medullary cavity with or without extension to the cortex. These are unlikely to exhibit extraosseous extension or a corresponding soft tissue component. CT allows for measurement of the mean and maximum density within a lesion, measured in standardized Hounsfield units (HU), which has been shown to help distinguish benign entities such as enostoses from untreated or treated blastic metastasis [11]. Untreated blastic metastases will generally show a mean CT attenuation value of <885 HU and a maximum value of <1060 HU [12]. Enostoses, by comparison, usually demonstrate a mean attenuation value of >885 HU and a maximum attenuation value of >1060 HU. However, this threshold HU cutoff is less reliable after treatment of blastic lesions [11].

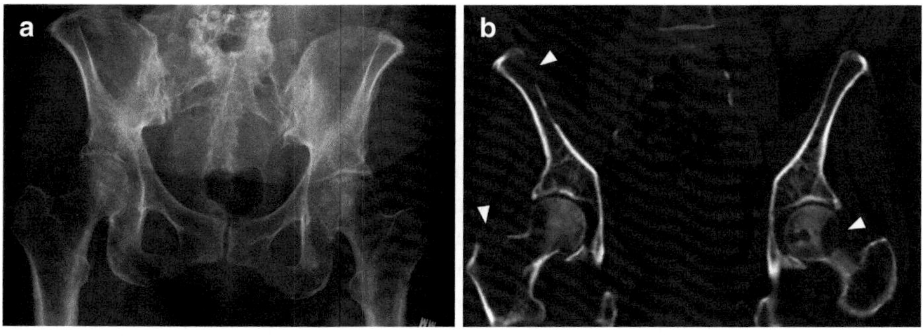

Fig. 6.4 An AP X-ray (**a**) and coronal CT slice (**b**) of the pelvis. The CT image reveals lytic lesions (arrowheads), which are difficult to visualize by an X-ray. AP anteroposterior

As with X-rays, mixed lesions can exhibit a wide variety of appearances ranging from predominantly lytic to predominantly sclerotic. Extraosseous extension of a soft tissue component may also be present.

Magnetic Resonance Imaging

MRI constructs images with high spatial and soft tissue contrast resolution by utilizing magnetic fields and radio-frequency pulses to leverage the differences in the molecular behavior of different tissues when exposed to these fields. This method of image acquisition does not involve the use of ionizing radiation. Varying sequences can therefore be obtained by manipulating the MRI parameters, resulting in different tissues exhibiting predictably different brightness, also referred to as signal or intensity, for a given sequence. "Brighter" structures on any sequence are referred to as "hyperintense," whereas "darker" structures are referred to as "hypointense." A tissue's intensity is mostly referred to as hyperintense or hypointense by comparing it to normal skeletal muscle, which is therefore designated "isointense" [13].

T1-weighted (T1w) sequences are considered "anatomical" sequences, in which normal tissues such as fat and bone marrow are hyperintense and cerebrospinal fluid (CSF) is hypointense. Gadolinium is also hyperintense on T1-weighted imaging, and gadolinium-containing contrast material (GCM) can be injected intravenously to assess a lesion's enhancement. T2-weighted (T2w) and short tau inversion recovery (STIR) sequences are considered "fluid-sensitive" with tissues being hyperintense in direct correlation with their water content. Employing some features of both T1 and T2 weighting, proton density (PD) images are often used in musculoskeletal MRI protocols for the evaluation of joints but play little role in characterizing a malignancy or metastasis. Another practical sequence often used for assessment of tumors and metastasis is diffusion-weighted imaging (DWI) in which more densely packed, water-containing tissues such as neoplasms appear hyperintense to normal soft tissues. Multiparametric MRI (mpMRI) refers to a single exam incorporating multiple MRI sequences (T1w, T2w, STIR, DWI, post-contrast, etc.) of the same target tissue or body region, thus offering detailed assessment of the lesion in question [13].

An important overarching principle of all bone metastases is that of marrow replacement, a process in which the normal medullary yellow or red marrow tissue is replaced by tumor cells. This is best discerned on T1w images by replacement of the high-signal yellow (fatty) marrow by a low-signal tissue. In general, if any trace of fatty marrow (high T1 signal) is preserved within a lesion, it is more likely to be benign. In rare instances, a high T1 signal can be seen within bone metastases. This may occur due to the paramagnetic effects of intralesional hemorrhage or due to melanin (in the case of melanoma metastases) [14]. In these cases, use of a fat-suppressed sequence can help differentiate fat from either blood or melanin.

Lytic, blastic, and mixed lesions, like on X-rays and CT, exhibit distinct characteristics on MRI. Table 6.1 displays the varied appearances of lytic and blastic lesions compared to the normal marrow on different standard MRI sequences.

MRI offers far superior sensitivity for detection of lytic lesions compared to CT and is often the preferred modality for screening as well as workup of skeletal lesions (Fig. 6.5) [15]. MRI also allows detailed evaluation of the periosteum and the adjacent soft tissues.

Table 6.1 Normal marrow, blastic lesions, and lytic lesions depicted on multiple MRI sequences. Created with permission on Biorender.com

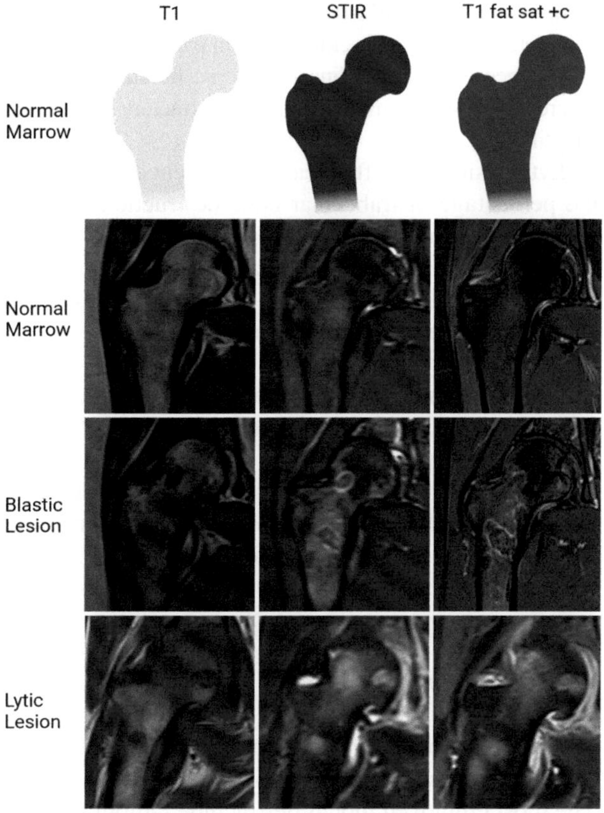

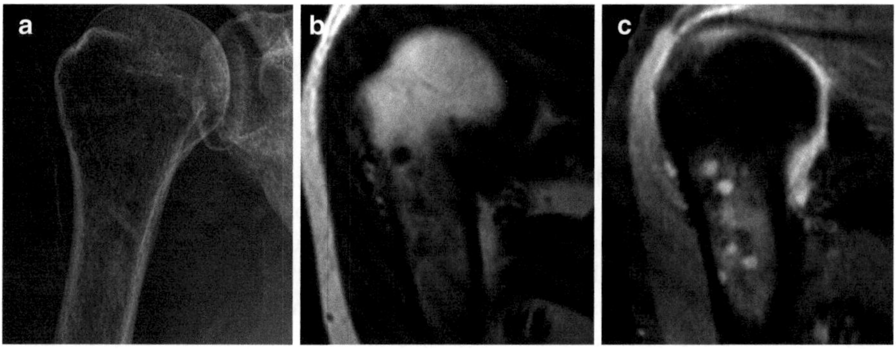

Fig. 6.5 AP X-ray and coronal T1w and STIR MR images of the right shoulder. MRI demonstrates osseous metastases that are not readily apparent on an X-ray. AP anteroposterior, STIR short tau inversion recovery, MR magnetic resonance

Importantly, extraosseous extension of spinal metastases or displacement of bone fragments in the setting of pathologic fracture are best assessed by MRI, which offers a detailed evaluation of the spinal canal and neural foramina to identify neural encroachment or compression. Standard MRI protocols offer high sensitivity for detection of impending or completed pathologic fractures [16]. In particular, the presence of periosteal edema should raise concern for impending pathologic fracture, although as discussed separately with regard to CT, shortcomings with current fracture risk stratification systems limit their clinical utility [9].

Lytic lesions are often readily apparent on fluid-sensitive sequences regardless of the percentage of trabecular bone destruction, thus offering higher sensitivity for lesion detection compared to CT. A lytic lesion typically appears as increased signal intensity on fluid-sensitive sequences and decreased signal intensity on T1w images. The addition of fat saturation to T1w and T2w images further improves lesion detection as pathologic tissues may be readily distinguished from the adjacent hypointense fatty marrow.

On MRI, blastic (sclerotic) lesions appear markedly hypointense on both T1- and T2-weighted images and fat saturation can further decrease the signal difference between sclerotic lesions and the surrounding normal marrow. For this reason, sequences without fat saturation are often more helpful for lesion identification. Blastic lesions often exhibit little or no enhancement after the administration of GCM.

Mixed lesions can be identified and characterized by MRI based predominantly on their lytic component for the reasons described previously.

Whole-Body Imaging

A variety of whole-body imaging modalities may be utilized to screen for the presence of osseous metastasis or to assess the treatment response of known metastasis. The most employed modalities include radiographic surveys, whole-body CT (wb-CT), whole-body MRI (wb-MRI), and numerous nuclear medicine studies including bone scintigraphy, PET–CT, and PET–MRI.

Radiographic Surveys

Radiographic surveys usually consist of X-rays of the calvarium, ribs, and the axial and proximal appendicular skeleton. Once a mainstay of disease evaluation for systemic processes such as multiple myeloma, skeletal survey has been largely supplanted by other whole-body imaging modalities due its far inferior sensitivity and specificity compared to most other options (Figs. 6.4 and 6.5). Sensitivity is further degraded by the coexistence of low bone mineral density [7]. Many current consensus recommendations discourage radiographic surveys for skeletal screening when more sensitive modalities such as whole-body CT or MRI are available [17].

Whole-Body CT and MRI

Whole-body CT (wb-CT) has increased in popularity due to widespread accessibility, short scan acquisition time, relative low cost, and lack of patient preparation needed compared to other more advanced whole-body imaging modalities such as

MRI or nuclear medicine scans. Compared to radiographic survey, wb-CT offers superior lesion localization and increased sensitivity for lesion detection [17].

Compared to wb-CT, however, whole-body MRI (wb-MRI) offers further increased sensitivity for detection of bone metastasis [8]. Limitations, however, include increased cost relative to X-rays and CT, decreased accessibility and long wait times, increased scan acquisition time, which can lead to motion artifacts that degrade image quality, and patient discomfort associated. Nonetheless, the benefits of MRI over radiography and CT often outweigh the drawbacks of higher cost and longer wait and scan times. Over the past decade into the present day, wb-MRI has, and is currently, replacing other modalities as the preferred method of whole-body bone screening, particularly for certain disease processes such as multiple myeloma. On the contrary, for other disease processes such as prostate cancer, melanoma, or breast cancer, nuclear medicine scans remain the dominant screening and staging modality [10, 18].

Multiple myeloma is a common indication for whole-body imaging, and recommendations for the optimal screening modality have evolved over time as technologies have advanced and become more widely deployed and cost-effective. Historically, radiographic surveys were used as the mainstay for diagnosing and monitoring patients with multiple myeloma. However, given their increased sensitivity, specificity, and diagnostic accuracy, the most recent 2019 International Myeloma Working Group (IMWG) consensus recommendations have affirmed the preference for utilizing wb-CT or wb-MRI rather than radiographic surveys [17]. The recommendations also acknowledge the increasing utility of nuclear medicine studies as acceptable modalities for disease diagnosis and monitoring.

Nuclear Medicine

Nuclear medicine is used for metastatic disease imaging by complexing a radioisotope, or radiotracer, to a compound that is preferentially targeted to areas of bone turnover (osteotropic), areas of increased metabolism (nonspecific oncotropic radioisotopes), or areas of specific cell type/receptor (specific oncotropic) [19]. The avidity of radiotracer activity corresponding to a lesion can indicate the severity of disease activity; however, the absolute value of the radiotracer activity is more useful for follow-up of lesions after treatment. The radiotracer activity of a lesion is expected to decline with favorable response to treatment, though imaging findings and changes in radiotracer activity may lag the clinical or biologic response to treatment [20].

Osteotropic Radioisotopes

The most utilized osteotropic radiotracer is Tc99m methyl diphosphonate (Tc99m MDP). A Tc99m MDP scan is often referred to as "bone scintigraphy" or "bone scan," and, generally, the entire skeleton is imaged using a gamma camera. ^{18}F-sodium fluoride (^{18}F-NaF) is another osteotropic radioisotope that is used in positron emission tomography (PET) imaging [21].

Tc99m MDP is formed by complexing technetium-99 m (Tc99m) to a phosphate analogue, namely, methylene diphosphonate (MDP) [22]. MDP acts as a substrate to produce crystalline hydroxyapatite by osteoblasts, and, thus, the regional deposition of Tc99m MDP seen on a bone scan is proportional to the osteoblast activity [23]. Due to its mechanisms of actions, bone scintigraphy is most useful for osteoblastic metastases, like prostate cancer, and is less reliable for detecting osteolytic lesions like renal cell carcinoma and multiple myeloma [24].

On a normal bone scan, faint uptake is seen throughout the skeleton and within the kidneys. Metastases will generally appear as intense focal areas of radiotracer uptake. When there are diffuse osteoblastic metastases, it may not be possible to see focal uptake of Tc99m MDP on these scans, but the normal faint activity that should be seen in the kidneys on a normal bone scan will no longer be evident, and this is commonly referred to as a "super scan" [19]. Focal areas of radiotracer uptake are also seen in benign conditions, most commonly with osteoarthritis—seen on almost every bone scan—but also with fracture and benign osseous lesions [25].

Oncotropic Radioisotopes

The most utilized nonspecific oncotropic radioisotope is ^{18}F-fluorodeoxyglucose (^{18}F-FDG). This radioisotope is used in routine positron emission tomography (PET) often in conjunction with CT or MRI. ^{18}F-FDG acts by functioning as a glucose analogue, which is trapped intracellularly after uptake [26]. This imaging relies on the hypermetabolism/glycolysis that is generally present in malignancies. ^{18}F-FDG, given its nonspecific mechanism of action, is versatile, but there are certain metastases to which it is known to be poorly sensitive. This includes blastic lesions with low metabolic activity (i.e., prostate metastases) and indolent or well-differentiated tumors (i.e., neuroendocrine and bronchial tumors) [27, 28]. Additionally, increased ^{18}F-FDG uptake in bone can be seen with conditions leading to regional hypermetabolism, such as trauma or infection.

There are a growing number of disease-specific radiotracers that utilize specific tumor cell-type biology to localize metastatic lesions. For example, gadolinium-68 complexed to prostate-specific membrane antigen (Ga68–PSMA) is commonly employed for assessing metastatic disease in prostate cancer. Molecular imaging targets have also long been used for imaging neuroendocrine tumors and thyroid cancers. There is ongoing research to identify more disease-specific molecular targets [29].

Mimics of Metastatic Disease

A number of clinical entities can mimic metastatic disease on imaging. This can be seen with trauma, infection, autoimmune conditions, degenerative conditions, or benign bone lesions. Often, a mimic of metastatic disease on one imaging modality can be differentiated with subsequent use of a second modality. For instance, a lipid-poor hemangioma of bone may appear as a round, low-T1-signal-intensity lesion by MRI that may be suspicious for metastatic disease. However, if a CT is performed, it may show the characteristic trabeculated appearance of an intraosseous hemangioma. In some circumstances, tissue sampling may be necessary to confidently exclude metastases.

Trauma

Marrow edema on MRI, as can be seen with trauma, can mimic the appearance of a marrow replacement by a neoplasm. This can be particularly difficult to distinguish in compression fractures of the spine. Generally, benign fractures show less homogeneous low-T1 marrow signal intensity. Benign fractures should also not show convex or well-defined margins of T1 hypointense marrow signal—features that are commonly seen with pathologic fractures [30]. Trauma can also show abnormal uptake on radioisotope studies (Fig. 6.6).

Infection

Metastatic disease and infection/osteomyelitis share many overlapping characteristics on imaging. They can both appear aggressive and cause cortical destruction and are characterized by marrow replacement. Differentiating these entities is often based upon clinical, historical, and laboratory correlations.

In adults, most infection of the long bones is not multifocal and is usually adjacent to a soft tissue infection or skin defect/ulcer. Osteomyelitis–discitis most often occurs by hematogenous spread, and there should generally be disc/endplate involvement, which is not a prerequisite for metastases. The presence of epidural or adjacent soft tissue abscesses can also help in distinguishing infection from metastatic disease (Fig. 6.7).

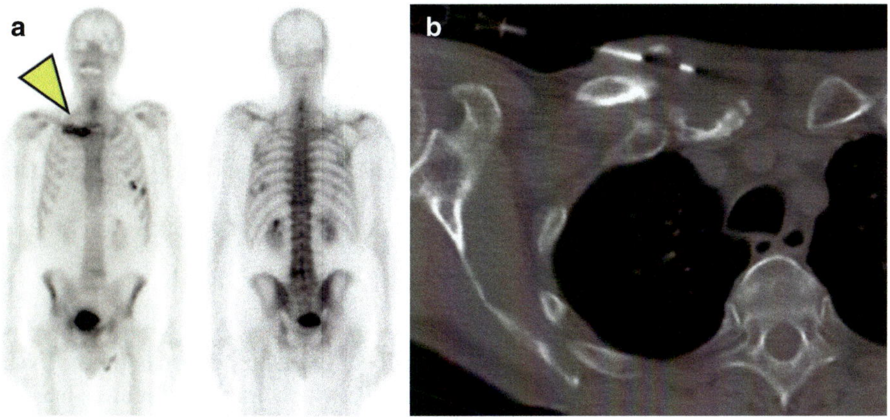

Fig. 6.6 Bone scintigraphy (**a**) showing radiotracer uptake of the right sternoclavicular joint that is concerning for metastasis. CT-guided biopsy of a lytic lesion (**b**) in the corresponding region revealed nonunion of fracture without a malignancy

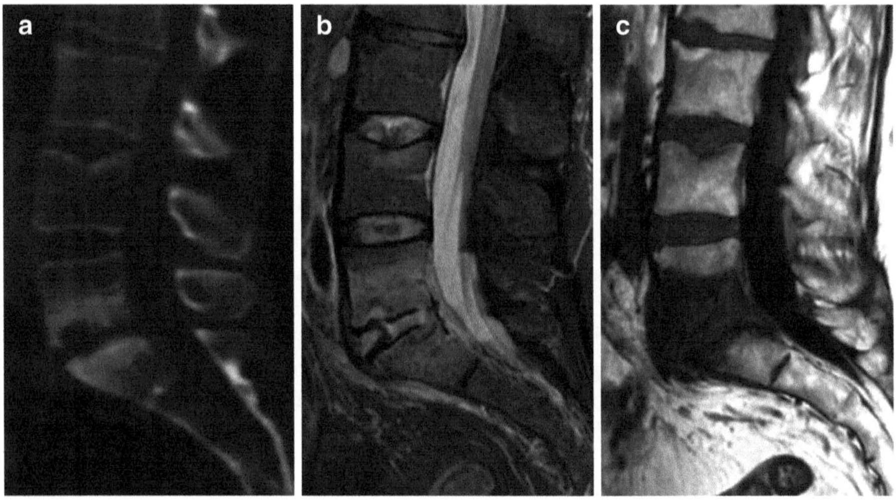

Fig. 6.7 Sagittal CT (**a**), T2w MRI with fat saturation (**b**), and T1w MRI (**c**) images depicting discitis of the L5–S1 disc and osteomyelitis of the adjacent L5 and S1 vertebral bodies. The MRI images show ventral epidural extension of infection. T1w T1-weighted, T2w T2-weighted

Autoimmune and Congenital Conditions

Certain autoimmune and congenital conditions can affect the bones. In these circumstances, multifocal bone lesions can be easily mistaken for metastatic disease.

A relatively commonly encountered mimic of metastatic disease is sarcoidosis. Like metastatic disease, sarcoidosis of the bones may present as multifocal marrow-replacing lesions. Generally, these lesions will have more angular and less rounded margins than metastases. They are typically less hypointense on T1w MRI and are often occult or less conspicuous on CT examinations (Fig. 6.8). Of course, other clinical features of sarcoidosis are generally present, and, if this diagnosis is suspected, lung imaging should be performed to assess for pulmonary involvement and the typical mediastinal lymphadenopathy.

Congenital conditions can also be a source of confusion when assessing the bones for metastatic disease. For instance, multifocal sclerotic lesions can be seen in the bones of patients with tuberous sclerosis complex and may mimic blastic metastases [31].

Degenerative Conditions

Less commonly, degenerative conditions, particularly in the spine, can mimic metastatic disease. Features of degenerative change, which may be seen as marrow edema or even gas, can appear as abnormal signal on MRI. For example, a

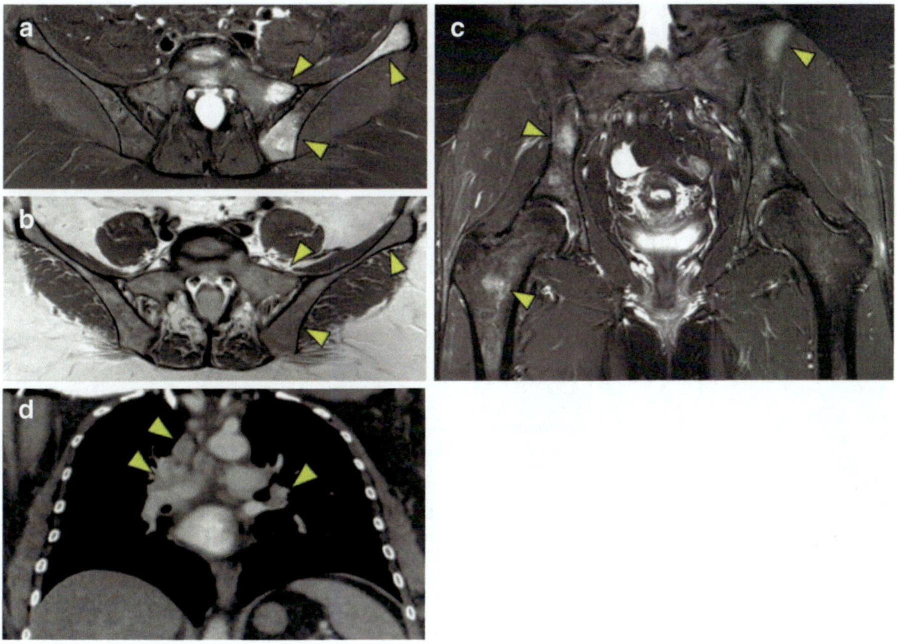

Fig. 6.8 Axial STIR (**a**), axial proton density (**b**), and coronal STIR (**c**) MRI images showing multiple sacral and left iliac bone marrow-replacing lesions (arrowheads) proven by biopsy to reflect osseous involvement of sarcoidosis. Notice the angular margins of the lesion in the right femur (**c**). Coronal CT chest (**d**) of the same patient shows mediastinal and hilar lymphadenopathy of sarcoidosis. STIR short tau inversion recovery

gas-containing a Schmorl's node may appear on MRI as a low-signal-intensity rounded lesion on all sequences, which could mimic a lytic or blastic metastatic lesion. As in this case, CT can often help distinguish degenerative change from metastases (Fig. 6.9).

Benign Bone Lesions

Occasionally, benign bone lesions can be difficult to distinguish from metastatic disease. Common entities that may cause confusion on MRI include lipid-poor intraosseous hemangiomas and enostoses (bone islands). Often, CT can be used to help distinguish these entities. Hemangiomas will usually have a characteristic woven/trabeculated appearance on CT. Bone islands may mimic blastic metastases but will generally show a mean CT attenuation value of >885 HU and a maximum value of at least 1060 HU [12] (Fig. 6.10).

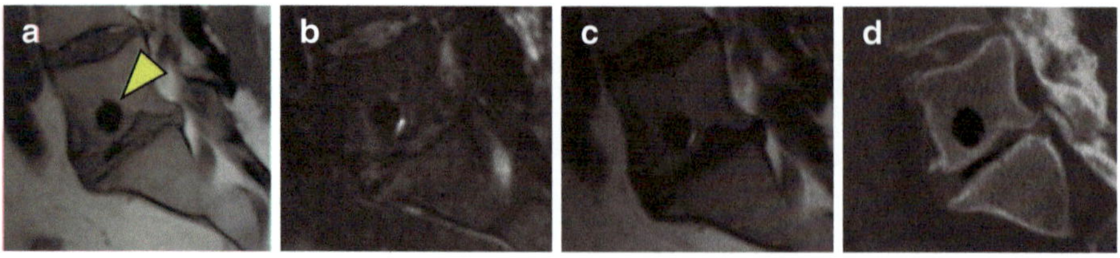

Fig. 6.9 Sagittal T2w (**a**), STIR (**b**), T1w (**c**), and CT (**d**) images of a Schmorl's node in the L5 vertebral body. T2w T2-weighted, STIR short tau inversion recovery, T1w T1-weighted

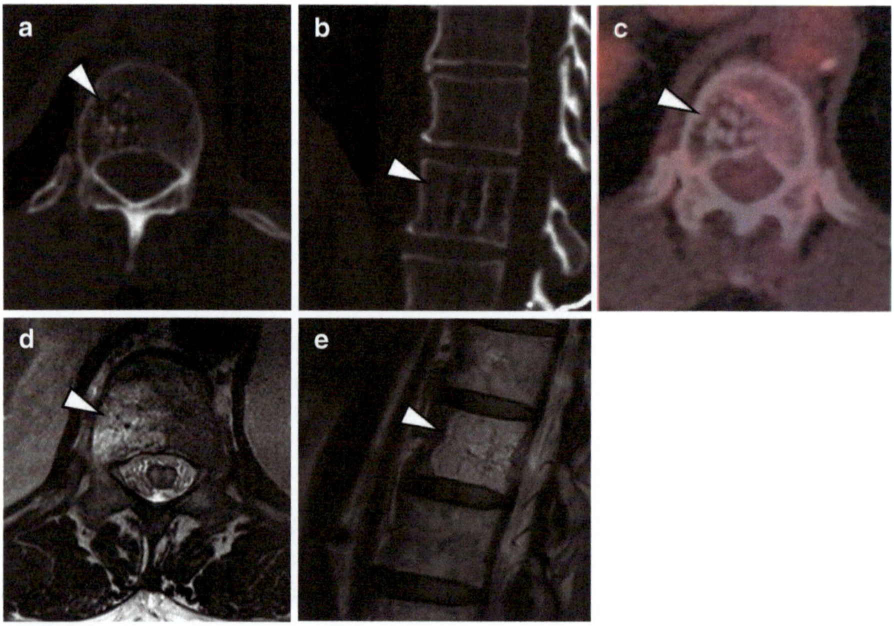

Fig. 6.10 Axial (**a**) and sagittal (**b**) CT, axial PET–CT (**c**), and axial (**d**) and sagittal (**e**) T2w MRI images of a benign T12 vertebral body hemangioma. Note the lesion's typical trabeculated appearance on CT and the lack of hypermetabolic activity on PET–CT. PET positron emission tomography, T2w T2-weighted

Conclusions and Future Directions

Imaging plays a crucial role in the management of osseous metastatic disease. Lytic and blastic lesions have characteristic imaging features. Metastatic lesions of bone, whether lytic or blastic, share the common feature of marrow replacement on T1w imaging. MRI remains the gold-standard imaging modality due to its sensitivity for detecting marrow replacement. The contrast resolution of MRI also allows for assessment of soft tissue extension and impending pathologic fractures. Other imaging modalities, primarily CT and radioisotope scans, play a complementary role in the management of metastatic disease.

There is a plethora of research into imaging methods, which may improve diagnostic imaging in metastatic disease. Routinely, new disease-specific oncotropic radioisotopes are developed and investigated [32]. Photon-counting CT is a new technology in the early stage of clinical implementation. This technology has higher spatial resolution and less noise than conventional CT and may ultimately improve the sensitivity of CT for lesion detection. Furthermore, and particularly relevant for orthopedic management, it produces less artifacts when imaging tissues around metal implants [33]. Advancements in MRI technology may lead to faster and higher-resolution exams that could ultimately allow for wider implementation of whole-body imaging.

References

1. Riihimäki M, Thomsen H, Sundquist K, Sundquist J, Hemminki K. Clinical landscape of cancer metastases. Cancer Med. 2018;7(11):5534.
2. Hernandez RK, Wade SW, Reich A, Pirolli M, Liede A, Lyman GH. Incidence of bone metastases in patients with solid tumors: analysis of oncology electronic medical records in the United States. BMC Cancer. 2018;18(1):44.
3. Wang M, Xia F, Wei Y, Wei X. Molecular mechanisms and clinical management of cancer bone metastasis. Bone Res. 2020;8:30.
4. Clayer MT. Lytic bone lesions in an Australian population: the results of 100 consecutive biopsies. ANZ J Surg. 2006;76(8):732.
5. Body JJ. Metastatic bone disease: clinical and therapeutic aspects. Bone. 1992;13:S57.
6. Macedo F, Ladeira K, Pinho F, Saraiva N, Bonito N, Pinto L, et al. Bone metastases: an overview. Oncol Rev. 2017;11(1):321.
7. Mikhael J, Ismaila N, Cheung MC, Costello C, Dhodapkar MV, Kumar S, et al. Treatment of multiple myeloma: ASCO and CCO joint clinical practice guideline. J Clin Oncol. 2019;37:1228.
8. Ellmann S, Beck M, Kuwert T, Uder M, Bäuerle T. Multimodal imaging of bone metastases: from preclinical to clinical applications. J Orthop Translat. 2015;3(4):166–77.
9. Damron TA. Update on predicting fracture risk: what's new, what's old, what's accurate. Tech Orthop. 2018;33(3):191–200.
10. Wolf MB, Murray F, Kilk K, Hillengass J, Delorme S, Heiss C, et al. Sensitivity of whole-body CT and MRI versus projection radiography in the detection of osteolyses in patients with monoclonal plasma cell disease. Eur J Radiol. 2014;83(7):1222.
11. Elangovan SM, Sebro R. Accuracy of CT attenuation measurement for differentiating treated osteoblastic metastases from enostoses. Am J Roentgenol. 2018;210(3):615.
12. Ulano A, Bredella MA, Burke P, Chebib I, Simeone FJ, Huang AJ, et al. Distinguishing untreated osteoblastic metastases from enostoses using CT attenuation measurements. Am J Roentgenol. 2016;207:362.
13. Major N, Anderson M. Musculoskeletal MRI. 3rd ed. Philadelphia: Elsevier; 2019. p. 1–22.
14. Wong VK, Lubner MG, Menias CO, Mellnick VM, Kennedy TA, Bhalla S, et al. Clinical and imaging features of noncutaneous melanoma. Am J Roentgenol. 2017;208:942.
15. Costelloe CM, Rohren EM, Madewell JE, Hamaoka T, Theriault RL, Yu TK, et al. Imaging bone metastases in breast cancer: techniques and recommendations for diagnosis. Lancet Oncol. 2009;10:606.
16. Isaac A, Dalili D, Dalili D, Weber MA. State-of-the-art imaging for diagnosis of metastatic bone disease. Radiologe. 2020;60:1.
17. Hillengass J, Usmani S, Rajkumar SV, Durie BGM, Mateos MV, Lonial S, et al. International myeloma working group consensus recommendations on imaging in monoclonal plasma cell disorders. Lancet Oncol. 2019;20:e302.

18. Liu F, Dong J, Shen Y, Yun C, Wang R, Wang G, et al. Comparison of PET/CT and MRI in the diagnosis of bone metastasis in prostate cancer patients: a network analysis of diagnostic studies. Front Oncol [Internet]. 2021;11. Available from: https://www.frontiersin.org/journals/oncology/articles/10.3389/fonc.2021.736654.
19. O'Sullivan GJ. Imaging of bone metastasis: an update. World J Radiol. 2015;7(8):202–11.
20. Wald C, Scheirey CD, Tran TM, Erbay N. An update on imaging of colorectal cancer. Surg Clin N Am. 2006;86:819.
21. Cuccurullo V, Lucio Cascini G, Tamburrini O, Rotondo A, Mansi L. Bone metastases radiopharmaceuticals: an overview. Curr Radiopharm. 2013;6(1):41.
22. Adams C, Banks K. StatPearls. 2023. Bone Scan.
23. Subramanian G, McAfee JG, Blair RJ, Kallfelz FA, Thomas FD. Technetium 99m methylene diphosphonate: a superior agent for skeletal imaging: comparison with other technetium complexes. J Nucl Med. 1975;16(8):744.
24. Sadeghi MV, Sedaghat S. Is 99m Tc-methylene diphosphonate bone scintigraphy a sensitive method for detecting bone lesions in multiple myeloma? Caspian J Intern Med. 2018;9(2):140–3.
25. Zhang Y, Zhao C, Liu H, Hou H, Zhang H. Multiple metastasis-like bone lesions in scintigraphic imaging. J Biomed Biotechnol. 2012;2012:1.
26. Khan N, Islam MM, Mahmood S, Hossain GA, Chakraborty RK. 18F-fluorodeoxyglucose uptake in tumor. Mymensingh Med J. 2011;20:332–42.
27. Bäuerle T, Semmler W. Imaging response to systemic therapy for bone metastases. Eur Radiol. 2009;19:2495.
28. HHM Y, Tsai YY, Hoffe SE. Overview of diagnosis and management of metastatic disease to bone. Cancer Control. 2012;19:84–91.
29. Altena R, Tzortzakakis A, Af Burén S, Tran TA, Frejd FY, Bergh J, et al. Current status of contemporary diagnostic radiotracers in the management of breast cancer: first steps toward theranostic applications. EJNMMI Res. 2023;13:43.
30. Marshall RA, Mandell JC, Weaver MJ, Ferrone M, Sodickson A, Khurana B. Imaging features and management of stress, atypical, and pathologic fractures. Radiographics. 2018;38(7):2173.
31. Boronat S, Barber I, Thiele EA. Sclerotic bone lesions in tuberous sclerosis complex: a genotype–phenotype study. Am J Med Genet A. 2017;173(7):1891.
32. Crişan G, Moldovean-cioroianu NS, Timaru DG, Andrieş G, Căinap C, Chiş V. Radiopharmaceuticals for PET and SPECT imaging: a literature review over the last decade. Int J Mol Sci. 2022;23:5023.
33. Esquivel A, Ferrero A, Mileto A, Baffour F, Horst K, Rajiah PS, et al. Photon-counting detector CT: key points radiologists should know. Korean J Radiol. 2022;23:854.

Obtaining Tissue Samples: What Is the Best Approach?

7

Felipe Souza, Erik J. Geiger, and Ty K. Subhawong

Introduction

Musculoskeletal metastases present a multitude of problems in the care of oncology patients, with adverse influence on their quality of life through significant physical limitations and negative psychologic impacts [1]. Skeletal metastases, in particular, are a common problem, being the third-most common site of distant spread after the lung and liver [2], with an estimated 300,000 cases annually [3]. Bone metastases lead to considerable suffering due to pain, reduced mobility, pathologic fractures, neurologic compromise, and disruptions in hematopoiesis and hormonal balance (e.g., hypercalcemia) [4]. A majority of patients receiving palliative care will develop bone metastases, and up to 17% of healthcare spending in oncology is driven by skeletal metastases [5, 6].

As patient survival times lengthen with emerging therapies, a multidisciplinary approach to managing the challenges posed by musculoskeletal metastases will remain essential. Foundational to that approach is obtaining a lesional tissue sample, frequently via a minimally invasive technique but occasionally via an open

F. Souza
Department of Radiology, Sylvester Comprehensive Cancer Center and the University of Miami Miller School of Medicine/Jackson Memorial Hospital, Miami, FL, USA

Department of Interventional Radiology, Sylvester Comprehensive Cancer Center and the University of Miami Miller School of Medicine, Miami, FL, USA

E. J. Geiger
Department of Orthopaedic Surgery, Sylvester Comprehensive Cancer Center and the University of Miami Miller School of Medicine/Jackson Memorial Hospital, Miami, FL, USA

T. K. Subhawong (✉)
Department of Radiology, Sylvester Comprehensive Cancer Center and the University of Miami Miller School of Medicine/Jackson Memorial Hospital, Miami, FL, USA
e-mail: tsubhawong@miami.edu

J. Pretell-Mazzini, S. A. Lozano Calderón (eds.), *Multidisciplinary Approach of Musculoskeletal Metastasis*, https://doi.org/10.1007/978-3-032-09498-8_7

biopsy when circumstances require. This chapter discusses the factors that radiologists (both diagnostic and interventional) and orthopedic surgeons consider when determining indications for percutaneous needle biopsy, along with the principles and techniques employed to achieve diagnostic accuracy and reduce complications.

Indications

Although skeletal metastases affect many oncology patients at some point in their clinical course, they are the initial presenting finding in a small minority of patients. This has been estimated at 5% based on the Surveillance, Epidemiology, and End Results (SEER) database, most commonly from a lung primary [7]. Even in patients with synchronous multiorgan disease that includes the liver and lung, bone lesions may be preferable for biopsy risk mitigation, given the low bleeding risks of a bone biopsy [8] and the risk of pneumothorax from a lung biopsy or that of a hemorrhage during biopsies of visceral sites [9]. Initial staging studies should be scrutinized for bone lesions amenable to biopsy, particularly in the iliac bones since the axial skeleton is frequently involved when osseous metastases are present [10] and this location is particularly conducive to safe percutaneous biopsy.

Incidental bone lesions should be considered primary bone tumors until proven otherwise, as inadvertent nailing of a primary bone tumor may jeopardize limb salvage surgery, resulting in amputation and lower functional outcomes [11]. However, in the setting of a known primary malignancy, the question of whether to biopsy a suspected bone lesion is situation-dependent. If there are no other bone lesions or known sites of metastasis, percutaneous biopsy is prudent. Toomayan and Major found that in the scenario of an indeterminate bone lesion with a known primary malignancy, up to 12% of patients were either ultimately diagnosed with a second primary tumor or confirmed negative for bone metastasis [12]. However, other studies have only found a 2% rate of discordant diagnoses in patients with a known primary tumor and an incident new bone lesion [13]. Histologic confirmation of metastasis may be required for other practical reasons, such as insurance coverage, or enrollment in a clinical trial. Finally, a bone biopsy may be requested to obtain tissue for additional molecular testing, particularly in the case of progressive osseous disease that has failed first- or second-line therapies, in the hopes of identifying targetable mutations.

Given their hematogenous spread, bone metastases predominate in the axial skeleton and long bone metaphysis, echoing hematopoietic marrow distribution. This has important clinical ramifications in prioritizing detection and treatment of lesions, most likely leading to patient morbidity. Symptoms often arise from bone pathology or from tumors pressing on nearby structures like nerves. The risk of a pathologic fracture is greatest with lytic lesions involving the proximal femur, but use of pathologic fracture risk stratification scoring, such as that based on the Mirels criteria, can aid in decision-making regarding the need for prophylactic fixation [14].

Biopsy Considerations

In 1996, Mankin et al. reported an approximately 20% rate of poorly performed biopsies that either precluded the optimal surgical approach or necessitated additional adjunctive treatments (chemotherapy or irradiation). Of their cohort, 10% experienced a worsened outcome attributable to a poorly performed biopsy [15]. Biopsy technical errors and adverse outcomes were 2–12 times greater ($p < 0.001$) when the biopsy was performed outside of a treatment center. These older data skewed toward an open surgical biopsy, with only 85/597 (14%) performed percutaneously with a needle; of these, only 29 were performed by a radiologist under CT guidance. Additionally, the diagnostic yield in this series (60%) is much lower than the more recent estimates of CNB accuracy (85–95%) [16, 17]. Poorly performed biopsies, whether by unnecessary compartment violation or contamination of the peritumoral environment after incisional biopsy, continue to confound optimal oncologic management, emphasizing the need for careful planning and coordination in a multidisciplinary environment and in dedicated cancer centers where definitive treatment is offered to patients with musculoskeletal tumors.

Planning

When a definitive diagnosis of a musculoskeletal lesion is required, the benefits of percutaneous needle biopsy typically outweigh their risks when compared to an open surgical biopsy, although certain scenarios demanding the latter will be subsequently discussed. Both core needle biopsies and fine-needle aspirations (FNA) are minimally invasive means of obtaining a tissue sample, often sufficiently with local anesthesia alone for soft tissues, and as such are routine in-office procedures. For bone lesions, image guidance and use of moderate sedation is warranted.

During pre-biopsy planning, one must consider patient- and tumor-specific factors and characteristics that influence patient safety. Because bleeding risk is generally low for these types of procedures, thrombocytopenia and anticoagulation are not contraindications to percutaneous biopsy in the extremities, particularly as injectable procoagulants can expedite hemostasis [8, 18]. A caveat is that for spine biopsies, anticoagulants should be held with a target interquartile range (INR) of ≤ 1.5–1.8 and platelets $>50 \times 10^9$/L, since spinal or epidural hematomas can be difficult to manage and carry high morbidity risks [19].

Bone biopsies are typically conducted under moderate sedation, using both opioids and anxiolytics, supplemented with local anesthetic infiltration into the soft tissues and periosteum [20]. Monitored anesthesia care (MAC) is prudent for patients with pulmonary or other medical morbidities, severe systemic disease, or American Society of Anesthesiologists (ASA) score ≥ 3 as well as for patients with sleep apnea, Pickwickian syndrome, and high Mallampati scores so that a qualified anesthesia provider can convert to general anesthesia, if needed [21]. Routine periprocedural antibiotic prophylaxis for percutaneous biopsies is not required for musculoskeletal lesions and may increase the risk of colitis [22, 23].

Image Guidance Modality

The different modalities used in image-guided biopsy include ultrasound, fluoroscopy, CT, and, rarely, MRI. A thorough examination of preprocedural imaging is essential to identify the most suitable image guidance modality: CT and ultrasound most commonly, though fluoroscopy may suffice instead of CT in many circumstances. MRI is only rarely used as the vast majority of musculoskeletal lesions can be targeted with CT or ultrasound [24]; those marrow lesions that are revealed on MRI but occult on CT may be sampled by a judicious choice of anatomic landmarks to match the biopsy target with the area of involvement. Appropriate positioning of the patient during the procedure should be anticipated—for instance, in large patients with extremity lesions, positioning in the lateral decubitus may better center the target within the CT gantry.

Advantages of ultrasound include real-time image guidance, which helps direct the needle into the target while avoiding damage to nearby structures such as nerves, vessels, tendons, and ligaments. Ultrasound also offers superior visibility of superficial soft tissues, is easier to site within facilities, requires low capital expense, and does not require radiation shielding. However, its image quality is highly dependent on the operator and can be influenced by patient factors—deep tumors in obese patients may be difficult or impossible to visualize adequately through sonography.

MR guidance may be helpful as a problem-solving adjunct in certain situations, such as a deep-seated soft tissue tumor that is not amenable to ultrasound guidance and where low soft tissue contrast precludes confident targeting by CT. Augmented reality has even been used to facilitate needle guidance in the MRI suite [25]. However, MR-compatible needles still introduce susceptibility artifacts that may obscure small lesions, and patient discomfort/claustrophobia limits routine use of MRI guidance for musculoskeletal needle biopsy. Usually, adequate anatomic landmarks can be identified that allow confident targeting of subtle or occult bone lesions on CT by careful comparison with MRI and/or PET–CT (Fig. 7.1). Whether diagnostic yield is consistent regardless of lesion visibility on CT remains controversial, with some authors contending equivalence [26] and others arguing that the yield drops significantly for occult (37%) vs. visible lesions (77%) [27].

Percutaneous Biopsy Execution

While most musculoskeletal malignancies are metastases or myelomas, it is crucial to adopt an agnostic biopsy approach to an indeterminate lesion so as not to preclude or impair curative resection in the event of a primary musculoskeletal sarcoma. The needle track should avoid multicompartment transgression, neurovascular injury, and be in line with the planned surgical approach. Perhaps counterintuitively to the radiologist, directing the needle through muscle is preferred to an intermuscular path, since muscle fibers act as a barrier to tumor seeding and surgeons can more easily resect an intramuscular tract [15, 28]. There are several excellent resources that outline preferred compartmental approaches for percutaneous needle

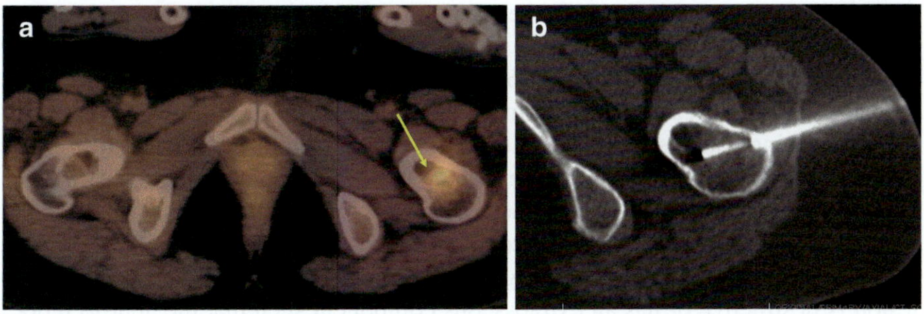

Fig. 7.1 A 29-year-old woman with a history of Ewing sarcoma. (**a**) An axial ^{18}F-FDG PET–CT image shows increased radiotracer uptake in the left intertrochanteric femur (arrow), without visible trabecular bone destruction or a soft tissue component. (**b**) The patient underwent a CT-guided biopsy, by meticulous attention to anatomic landmarks to target the area of increased metabolic activity. The final pathology confirmed Ewing sarcoma metastasis. ^{18}F-FDG PET–CT ^{18}F-fluorodeoxyglucose positron emission tomography–CT

biopsies that avoid compromising muscle function or crossing soft tissue structures essential for future flap or soft tissue reconstruction [20, 29, 30]. Lesions in the proximal humerus should be biopsied through the anterior third of the deltoid muscle (avoiding the deltopectoral groove) because axillary nerve innervation runs from posterior to anterior and en bloc resection of a posterior needle tract could render functionless the more anterior muscle fibers. Biopsies of lesions in the femur should not involve the rectus femoris because of its significance in knee extension; typically, a lateral or medial approach through the vastus muscles is favored depending on the surgical plan. In the lower leg, an approach through the anteromedial tibial cortex minimizes soft tissue and muscle contamination and is preferable to an approach through the posterior or lateral calf.

The standard practice of biopsy tract resection after open biopsy procedures does not seem to apply to the definitive surgery after needle biopsies. A growing body of empirical evidence has failed to show an adverse influence on outcome and disease recurrence risk to the extent once believed [31], especially with the coaxial needle systems, thus reducing the chance for tract seeding. In a series of 53 cases in which the needle tract was not excised, Binitie et al. observed only a 9% local recurrence rate, equivalent to historical controls where the needle tract was excised [32]. Importantly, 97% of the patients in this cohort received radiation therapy to augment local control and 83% received chemotherapy. These results were subsequently corroborated in a study of 72 patients that failed to show a difference in local recurrence-free survival rates between otherwise risk-matched groups [33]. Nevertheless, we believe it is prudent to coordinate with the orthopedic surgeon prior to the biopsy to ensure that definitive surgical treatment is not compromised by the planned biopsy approach.

We typically use a coaxial technique with a 14-gauge cutting needle for soft tissue biopsies and a 12-gauge trephine needle in conjunction with a battery-powered driver for bone biopsies. Battery-powered drivers shorten procedure times and reduce radiation exposure, with similar or better diagnostic yields compared to manual systems [34, 35], although others have found that a manual biopsy produces longer cores and less histologic artifacts [36]. A small volume of saline can be injected to displace nerves or vessels away from the needle track. Additionally, use of gravity and positioning patients in the decubitus position can facilitate moving critical tissue and mobile organs away from the needle path [31]. For cystic or friable intraosseous tumors, core biopsy using the trephined needle may fail to capture an adequate core, while use of a soft tissue cutting needle allows tissue capture in a chamber via a spring-loaded cutting mechanism (Fig. 7.2). It should be emphasized that if the bone lesion is destructive or has an extraosseous component such that the approach does not require passage through cortical bone, the lesion may be amenable to ultrasound guidance instead of CT (Fig. 7.3). Sampling of the viable soft tissue component is sufficient and obviates the need to drill into bone; a minimum of three cores should be obtained, aiming for a total specimen length of at least 2 cm [16].

Because of the difficulty in differentiating aggressive tumors from infection, tissue should be sent for microbiology in addition to histology and genetics. Samples intended for genetic testing should be clearly marked as such so that strong acid decalcification is avoided, as this may render the sample unsuitable for PCR-based molecular testing [37]. If lymphoma is in the differential, placing at least one or two cores of tissue in Roswell Park Memorial Institute (RPMI) solution permits analysis by flow cytometry. Occasionally, and particularly with densely sclerotic bone lesions, the specimen may become lodged in the biopsy needle; this stuck core may then itself need to be biopsied by a smaller-gauge needle. Alternatively, we have also found it useful to apply a portable electrocautery device to the biopsy needle to widen the core by thermal expansion and facilitate core specimen ejection [38], though prolonged heating may char the tissue.

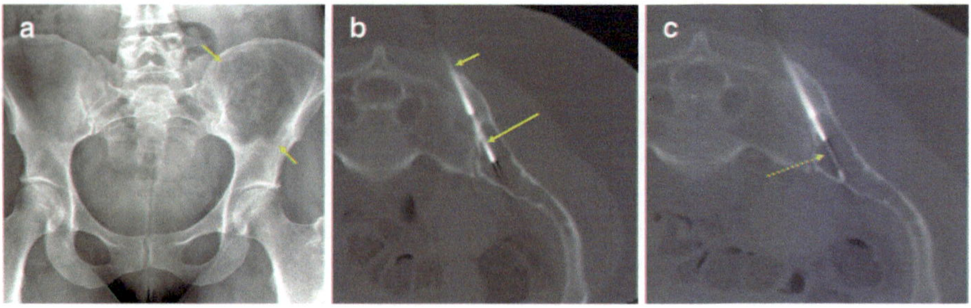

Fig. 7.2 A 35-year-old woman with left hip pain. (**a**) An AP pelvis radiograph reveals a large lytic lesion in the left iliac wing (arrows) with sclerotic margins. (**b**) A CT-guided biopsy performed first with a 12-gauge trephine needle (long arrow) coaxially through a 10-gauge access cannula (short arrow); note deliberate throw through the lesion margin to attempt retrieval of specimen that captures a rim of uninvolved bone and the interface with lesional tissue. (**c**) Additional sampling with a 14-gauge side-cutting semiautomated biopsy gun with a notched trocar (dotted arrow) that captures the cut tissue after releasing a spring-loaded firing mechanism; it was placed coaxially through the same 10-gauge access cannula. The final pathology confirmed cystic fibrous dysplasia. AP anteroposterior

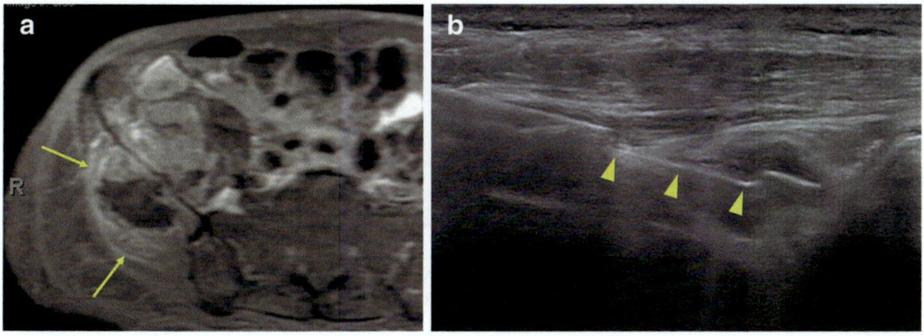

Fig. 7.3 A 71-year-old woman with an iliac mass. (**a**) A post-contrast, fat-suppressed, T1-weighted MRI image shows a large destructive mass arising from the iliac wing with a necrotic extraosseous component posteriorly (arrow). (**b**) The extraosseous component was successfully biopsied under ultrasound guidance; note the needle trough (arrowheads) that functions as a specimen chamber after spring-loaded firing of the cutting component. Ultrasound allows real-time guidance of needle positioning to ensure that peripheral viable components of the tumor are sampled

Accuracy and Yield

The diagnostic accuracy for core needle biopsies is generally higher than that for fine-needle aspiration [39]; although FNA carries a lower procedural risk, for molecular studies, core biopsies are typically needed. A higher diagnostic yield is associated with a larger lesion size, lytic rather than sclerotic lesions in bone, more cores submitted, and perhaps, most critically, longer core specimens [16, 40–42]. We typically aim to obtain a minimum of three cores to balance the increased diagnostic yield from specimen number against risks to adjacent structures with multiple passes and those incurred by longer procedures. Sclerotic bone lesions may benefit from powered drilling to increase the diagnostic yield [35]. In a large retrospective study of 800 bone biopsies, 31% were indeterminate; of those initially indeterminate, 39% were later found to be malignant, stressing the importance of resampling suspicious lesions or carefully following those lesions where an initial attempt at definitive diagnosis has failed [43]. Initial nondiagnostic biopsies of radiologically suspicious osseous lesions frequently turn out to be lymphoma (Fig. 7.4) [44]. Conversely, when osseous lesions appear nonaggressive with a high pretest probability of benignity, a nondiagnostic biopsy may still usefully guide management in up to 60% of cases, offering reassurance and averting unnecessary surgery [45].

Importantly, the extraosseous or soft tissue component of a destructive bone lesion will yield better diagnostic tissue than samples obtained from the lesion's center in the destroyed bone. In centrally necrotic tumors, it is imperative to target the solid peripheral components to retrieve viable tumors (Fig. 7.5). On-site cytology or rapid frozen sections performed on core specimens, confirming retrieval of diagnostic lesional tissue, eliminates delays in care associated with nondiagnostic biopsy and expedites preoperative treatment planning [44]. For soft tissue tumors,

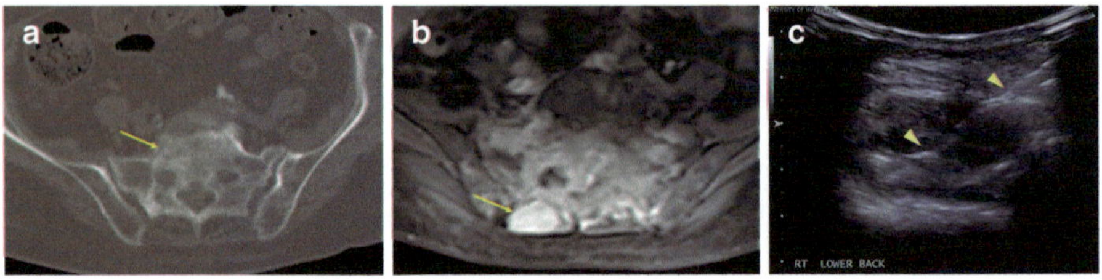

Fig. 7.4 A 73-year-old woman with 15-pound weight loss, lower back pain, and difficulty in walking. (**a**) An axial CT image shows a mixed lytic and sclerotic lesion in the sacrum (arrow) with ill-defined margins; a biopsy at an outside hospital was inconclusive at this time. (**b**) An axial, fat-suppressed, contrast-enhanced T1-weighted MRI image 4 months later shows an infiltrative enhancing mass in the sacrum with an ovoid extraosseous component in the lower paravertebral soft tissues (arrow), amenable to percutaneous biopsy. (**c**) An ultrasound-guided biopsy of the extraosseous component with a semiautomated biopsy gun (arrowheads) confirmed diffuse large B-cell lymphoma. Aggressive or marrow-infiltrating bone lesions with initial inconclusive biopsies are often ultimately diagnosed as lymphoma

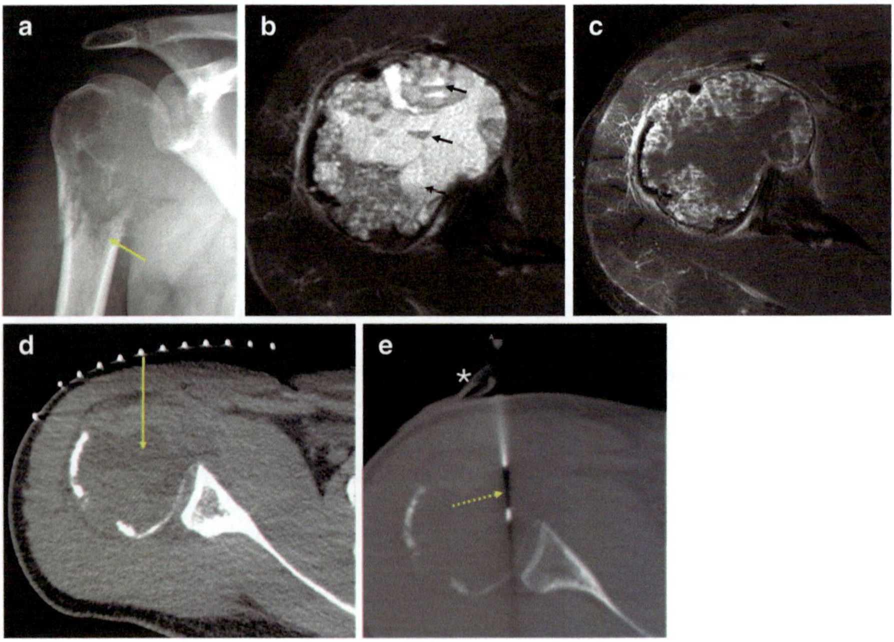

Fig. 7.5 A 21-year-old man with right shoulder pain. (**a**) An AP radiograph shows a destructive lytic lesion in the proximal humerus with a wide transition zone at the inferior margin (arrow) and cortical breakthrough medially. (**b**) An axial, fat-suppressed, T2-weighted MRI image shows a large cystic lesion with innumerable fluid–fluid levels (small arrows), a nonspecific finding that indicates the presence of blood products settling in a gravity-dependent manner. (**c**) An axial, contrast-enhanced, fat-suppressed, T1-weighted MRI image showing microcystic enhancement at the tumor periphery but no large solid nodular components; the radiologic differential includes an aneurysmal bone cyst, a secondary aneurysmal bone cyst formation in a giant cell tumor of bone, and a telangiectatic osteosarcoma. (**d**) An axial CT shows grid lines placed on the skin to mark the biopsy needle entry site (sixth from lateral, arrow). (**e**) A semiautomated biopsy gun is advanced through the anterior third deltoid and into the lesion, with the notch spanning the peripheral viable tumor components (dotted arrow). Note towels (*) are being used to buttress the needle handle in place while the table is moved into the CT gantry for imaging confirmation of desired intralesional positioning. The final pathology confirmed telangiectatic osteosarcoma

frozen section accuracy has been reported to be as high as 95%, though it is lower for adipocytic (84%) and fibroblastic/myofibroblastic (87%) tumors [46]. Ultimately, multidisciplinary consultation with the surgeon and pathologist allows for appropriate management of clinic–radiologic discordancy [47].

Surgical Biopsy

When percutaneous biopsy results are inconclusive, an open incisional biopsy may be required. A careful review of imaging facilitates planning of the incision and positioning of the patient. A longitudinal incision is made over the lesion, carefully avoiding neurovascular structures. Meticulous hemostasis is maintained throughout the approach. For a bone lesion, the periosteum is incised in line with the skin incision and an osteotome or drill is used to create a cortical window. A representative sample of viable tissue is obtained, typically at least 0.5 cm^2 in size [28], ensuring an adequate specimen for histopathology, microbiology, and molecular studies. To further increase the chance that a definitive diagnosis is obtained from the biopsy, a frozen section can be performed to confirm adequate lesional tissue has been retrieved. Wallace et al. showed that frozen section analysis of 485 open bone biopsies was accurate in 54%, partially diagnostic in 21% (a definitive diagnosis was not possible but discriminated nonneoplastic or benign disease from a malignancy or sarcoma from another malignancy), nondiagnostic in 21%, and discordant with the final pathology in 3.5% [48]. Excisional biopsies may be appropriate for small subcutaneous nodules highly likely to be benign, particularly those with characteristic imaging features (e.g., lipoma, schwannoma); but indeterminate tumors, even when <2 cm, should be approached with caution [49]. Otherwise, tumor excision should aim to achieve the definitive margin necessary for local control, with a wide margin required for malignant musculoskeletal tumors. An intralesional or marginal margin is acceptable if there is certainty of a benign inactive lesion or if the goal is palliation. Hemostasis should be meticulously maintained to minimize the risk of tumor contamination in the surgical bed. Postoperatively, patients are monitored for complications such as infection and limited weight-bearing may be prudent to reduce the risk of fracture through the stress riser after bone biopsies [50].

Conclusions

In summary, a biopsy of a musculoskeletal lesion must fulfill the twin aims of obtaining sufficient material for a correct diagnosis while not compromising the surgical management or clinical outcome of the patient. Given the consequences of improperly performed biopsies, a multidisciplinary approach and execution by experienced practitioners is essential. A percutaneous needle biopsy is often the preferred method due to its minimally invasive nature, though an open biopsy may be necessary in certain scenarios. Proper planning, including preprocedural imaging and consideration of anatomical factors, helps maximize diagnostic yield while minimizing complications.

Different imaging modalities, including CT, ultrasound, and MRI, play important roles in guiding biopsies, each with specific advantages. The choice of biopsy technique, including needle type and approach, must be carefully tailored to the lesion's characteristics, with particular attention to avoiding contamination of surrounding tissues. Diagnostic yield is improved through multiple core samples, targeted sampling of viable tumor components, and the use of rapid cytology or frozen section assessment.

While percutaneous biopsies are generally safe, proper coordination with orthopedic oncologists ensures biopsy tract management in cases requiring surgical intervention. In instances of inconclusive biopsy results, open biopsy remains an option. Ultimately, a well-executed biopsy strategy significantly enhances definitive treatment planning and patient outcomes in musculoskeletal oncology.

Conflict of Interest The authors declare no conflicts of interest.

Funding/Disclosures None.

References

1. Rajeswaran T, Wong HCY, Zhang E, Kennedy SKF, Gojsevic M, Soliman H, et al. Quality of life issues in patients with bone metastases: a systematic review. Support Care Cancer. 2023;32(1):18.
2. Cuccurullo V, Cascini GL, Tamburrini O, Rotondo A, Mansi L. Bone metastases radiopharmaceuticals: an overview. Curr Radiopharm. 2013;6(1):41–7.
3. Li S, Peng Y, Weinhandl ED, Blaes AH, Cetin K, Chia VM, et al. Estimated number of prevalent cases of metastatic bone disease in the US adult population. Clin Epidemiol. 2012;4:87–93.
4. Macedo F, Ladeira K, Pinho F, Saraiva N, Bonito N, Pinto L, et al. Bone metastases: an overview. Oncol Rev [Internet]. 2017 [cited 2021 Mar 2];11(1). Available from: https://www.ncbi.nlm.nih.gov/pmc/articles/PMC5444408/.
5. Oster G, Lamerato L, Glass AG, Richert-Boe KE, Lopez A, Chung K, et al. Natural history of skeletal-related events in patients with breast, lung, or prostate cancer and metastases to bone: a 15-year study in two large US health systems. Support Care Cancer. 2013;21(12):3279–86.
6. Schulman KL, Kohles J. Economic burden of metastatic bone disease in the U.S. Cancer. 2007;109(11):2334–42.
7. Younis MH, Fuentes-Rivera L, Summers S, Pretell-Mazzini J. Survival in patients with carcinomas presenting with bone metastasis at diagnosis: a SEER population-based cohort study. Arch Orthop Trauma Surg. 2021;141(3):367–73.
8. Foremny GB, Pretell-Mazzini J, Jose J, Subhawong TK. Risk of bleeding associated with interventional musculoskeletal radiology procedures. A comprehensive review of the literature. Skeletal Radiol. 2015;44(5):619–27.
9. Hoffmann P, Balik M, Hoffmannova M, Kopecky J, Ryska P, Draganovicova J, et al. Computed tomography-guided percutaneous biopsy in diagnosis of suspected metastatic renal cell carcinoma: which location is the most suitable? Wideochir Inne Tech Maloinwazyjne. 2024;19(3):361–9.
10. Hage WD, Aboulafia AJ, Aboulafia DM. Incidence, location, and diagnostic evaluation of metastatic bone disease. Orthop Clin North Am. 2000;31(4):515–28, vii.
11. Adams SC, Potter BK, Mahmood Z, Pitcher JD, Temple HT. Consequences and prevention of inadvertent internal fixation of primary osseous sarcomas. Clin Orthop. 2009;467(2):519–25.

12. Toomayan GA, Major NM. Utility of CT-guided biopsy of suspicious skeletal lesions in patients with known primary malignancies. AJR Am J Roentgenol. 2011;196(2):416–23.
13. Cronin CG, Cashell T, Mhuircheartaigh JN, Swords R, Murray M, O'Sullivan GJ, et al. Bone biopsy of new suspicious bone lesions in patients with primary carcinoma: prevalence and probability of an alternative diagnosis. AJR Am J Roentgenol. 2009;193(5):W407–10.
14. Mirels H. Metastatic disease in long bones. A proposed scoring system for diagnosing impending pathologic fractures. Clin Orthop. 1989;249:256–64.
15. Mankin HJ, Mankin CJ, Simon MA. The hazards of the biopsy, revisited. Members of the Musculoskeletal Tumor Society. J Bone Joint Surg Am. 1996;78(5):656–63.
16. Wong H, Tarr GP, Anand R, Atkinson N, Flint M, Clarke A, et al. Diagnostic yield and concordance of image-guided biopsy in musculoskeletal lesions. Skeletal Radiol. 2024;53(1):75–84.
17. Winkler WL, Baker JC, Tomasian A, Vander Velde TL, Hillen TJ, Luo C, et al. Diagnostic efficacy of image-guided core needle biopsy of suspected malignant osseous lesions: a retrospective cohort study from a single academic institution. Eur Radiol. 2024;34(9):5760–72.
18. Liu B, Limback J, Kendall M, Valente M, Armaly J, Grekoski V, et al. Safety of CT-guided bone marrow biopsy in thrombocytopenic patients: a retrospective review. J Vasc Interv Radiol. 2017;28(12):1727–31.
19. Patel IJ, Rahim S, Davidson JC, Hanks SE, Tam AL, Walker TG, et al. Society of Interventional Radiology Consensus Guidelines for the Periprocedural Management of Thrombotic and Bleeding Risk in patients undergoing percutaneous image-guided interventions-part II: recommendations: endorsed by the Canadian Association for Interventional Radiology and the Cardiovascular and Interventional Radiological Society of Europe. J Vasc Interv Radiol. 2019;30(8):1168–1184.e1.
20. Espinosa LA, Jamadar DA, Jacobson JA, DeMaeseneer MO, Ebrahim FS, Sabb BJ, et al. CT-guided biopsy of bone: a radiologist's perspective. AJR Am J Roentgenol. 2008;190(5):W283–9.
21. Moran TC, Kaye AD, Mai AH, Bok LR. Sedation, analgesia, and local anesthesia: a review for general and interventional radiologists. Radiogr Rev Publ Radiol Soc N Am Inc. 2013;33(2):E47–60.
22. Filippiadis DK, Charalampopoulos G, Mazioti A, Keramida K, Kelekis A. Bone and soft-tissue biopsies: what you need to know. Semin Interv Radiol. 2018;35(4):215–20.
23. Venkatesan AM, Kundu S, Sacks D, Wallace MJ, Wojak JC, Rose SC, et al. Practice guidelines for adult antibiotic prophylaxis during vascular and interventional radiology procedures. Written by the Standards of Practice Committee for the Society of Interventional Radiology and Endorsed by the Cardiovascular Interventional Radiological Society of Europe and Canadian Interventional Radiology Association [corrected]. J Vasc Interv Radiol. 2010;21(11):1611–30; quiz 1631.
24. Tomasian A, Hillen TJ, Jennings JW. Bone biopsies: what radiologists need to know. AJR Am J Roentgenol. 2020;215(3):523–33.
25. Fritz J, U-Thainual P, Ungi T, Flammang AJ, McCarthy EF, Fichtinger G, et al. Augmented reality visualization using image overlay technology for MR-guided interventions: cadaveric bone biopsy at 1.5 T. Invest Radiol. 2013;48(6):464–70.
26. Hillen TJ, Talbert RJ, Friedman MV, Long JR, Jennings JW, Wessell DE, et al. Biopsy of CT-occult bone lesions using anatomic landmarks for CT guidance. AJR Am J Roentgenol. 2017;209(1):214–21.
27. Kim W, Sun K, Kung JW, Wu JS. CT-guided Core needle biopsy of nonspinal bone lesions: comparison of occult and visible bone lesions. AJR Am J Roentgenol. 2023;220(1):104–14.
28. Springfield DS, Rosenberg A. Biopsy: complicated and risky. J Bone Joint Surg Am. 1996;78(5):639–43.
29. Liu PT, Valadez SD, Chivers FS, Roberts CC, Beauchamp CP. Anatomically based guidelines for core needle biopsy of bone tumors: implications for limb-sparing surgery. Radiogr Rev Publ Radiol Soc N Am Inc. 2007;27(1):189–205; discussion 206.

30. Meek RD, Mills MK, Hanrahan CJ, Beckett BR, Leake RL, Allen H, et al. Pearls and pitfalls for soft-tissue and bone biopsies: a cross-institutional review. Radiographics. 2020;40(1):266–90.
31. Lai C, Long JR, Larsen BT, Iturregui JM, Wilke BK, Goulding KA. Percutaneous biopsy of musculoskeletal tumors and the potential for needle tract seeding: technical considerations, current controversies, and outcomes. Skeletal Radiol. 2023;52(3):505–16.
32. Binitie O, Tejiram S, Conway S, Cheong D, Temple HT, Letson GD. Adult soft tissue sarcoma local recurrence after adjuvant treatment without resection of core needle biopsy tract. Clin Orthop. 2013;471(3):891–8.
33. Siddiqi MA, Kim HS, Jede F, Han I. Association of core needle biopsy tract resection with local recurrence in extremity soft tissue sarcoma. Skeletal Radiol. 2017;46(4):507–12.
34. Kihira S, Koo C, Lee A, Aggarwal A, Pawha P, Doshi A. Reduction of radiation dose and scanning time while preserving diagnostic yield: a comparison of battery-powered and manual bone biopsy systems. AJNR Am J Neuroradiol. 2020;41(3):387–92.
35. Cohen MG, McMahon CJ, Kung JW, Wu JS. Comparison of battery-powered and manual bone biopsy systems for core needle biopsy of sclerotic bone lesions. AJR Am J Roentgenol. 2016;206(5):W83–6.
36. Ho CK, Gimarc D, Carroll HF, Clay M, Schowinsky J, Jesse MK, et al. Evaluating bone biopsy quality by technique in an animal model. Res Diagn Interv Imaging. 2022 Jun;2:100008.
37. Chen H, Luthra R, Goswami RS, Singh RR, Roy-Chowdhuri S. Analysis of pre-analytic factors affecting the success of clinical next-generation sequencing of solid organ malignancies. Cancer. 2015;7(3):1699–715.
38. Crawford DA, Bauer B, Jennings JW. Portable cautery device as an aid to eject a lodged sclerotic bone biopsy specimen. J Vasc Interv Radiol. 2023;34(9):1631–3.
39. Dupuy DE, Rosenberg AE, Punyaratabandhu T, Tan MH, Mankin HJ. Accuracy of CT-guided needle biopsy of musculoskeletal neoplasms. AJR Am J Roentgenol. 1998;171(3):759–62.
40. Roberts CC, Morrison WB, Leslie KO, Carrino JA, Lozevski JL, Liu PT. Assessment of bone biopsy needles for sample size, specimen quality and ease of use. Skeletal Radiol. 2005;34(6):329–35.
41. Wu JS, Goldsmith JD, Horwich PJ, Shetty SK, Hochman MG. Bone and soft-tissue lesions: what factors affect diagnostic yield of image-guided core-needle biopsy? Radiology. 2008;248(3):962–70.
42. Monfardini L, Preda L, Aurilio G, Rizzo S, Bagnardi V, Renne G, et al. CT-guided bone biopsy in cancer patients with suspected bone metastases: retrospective review of 308 procedures. Radiol Med (Torino). 2014;119(11):852–60.
43. Hwang S, Lefkowitz RA, Landa J, Zheng J, Moskowitz CS, Maybody M, et al. Percutaneous CT-guided bone biopsy: diagnosis of malignancy in lesions with initially indeterminate biopsy results and CT features associated with diagnostic or indeterminate results. AJR Am J Roentgenol. 2011;197(6):1417–25.
44. Chang CY, Huang AJ, Bredella MA, Torriani M, Halpern EF, Rosenthal DI, et al. Percutaneous CT-guided needle biopsies of musculoskeletal tumors: a 5-year analysis of non-diagnostic biopsies. Skeletal Radiol. 2015;44(12):1795–803.
45. Didolkar MM, Anderson ME, Hochman MG, Rissmiller JG, Goldsmith JD, Gebhardt MG, et al. Image guided core needle biopsy of musculoskeletal lesions: are nondiagnostic results clinically useful? Clin Orthop. 2013;471(11):3601–9.
46. Miwa S, Yamamoto N, Hayashi K, Takeuchi A, Igarashi K, Tada K, et al. Accuracy of histological grades from intraoperative frozen-section diagnoses of soft-tissue tumors. Int J Clin Oncol. 2020;25(12):2158–65.
47. Banks JS, Garner HW, Chow AZ, Peterson JJ, Bestic JM, Wessell DE. Radiology-pathology correlation for bone and soft tissue tumors or tumor-like masses: single institutional experience after implementation of a weekly conference. Skeletal Radiol. 2021;50(4):731–8.

48. Wallace MT, Lin PP, Bird JE, Moon BS, Satcher RL, Lewis VO. The accuracy and clinical utility of intraoperative frozen section analysis in open biopsy of bone. J Am Acad Orthop Surg. 2019;27(11):410–7.
49. Pham K, Ezuddin NS, Pretell-Mazzini J, Subhawong TK. Small soft tissue masses indeterminate at imaging: histological diagnoses at a tertiary orthopedic oncology clinic. Skeletal Radiol. 2019;48(10):1555–63.
50. Ho KWK, Gilbody J, Jameson T, Miles AW. The effect of 4 mm bicortical drill hole defect on bone strength in a pig femur model. Arch Orthop Trauma Surg. 2010;130(6):797–802.

Pathological Assessment of Metastatic Disease to the Musculoskeletal System

8

Julio A. Diaz-Perez and Andrew E. Rosenberg

Introduction

Metastases are the most common malignant tumors of bone, and the skeleton ranks third after the lungs and liver as the most frequent organ system affected by metastatic disease [1]. Metastatic bone disease is a frequent cause of cancer-related morbidity and death [2].

The prevalence of metastatic disease from solid tumors to bone in individuals older than 18 years in the US population was 0.001% in 2008 [3]. The exact incidence of bone metastasis is unknown, but it is estimated that 210,000–400,000 new cases of metastatic bone disease are diagnosed annually in the United States [4]. The most common primary sites are the lung (24.8%), prostate (19.4%), breast (19.3%), gastrointestinal tract (9.4%), and urological tract (6.5%) [3, 4].

Bone metastases tend to occur in areas that possess a rich vasculature, including the vertebrae and large, long bones such as the femur, humerus, and tibia. These bones contain hematopoietic marrow rich in growth factors and cytokines that facilitate the implantation and growth of metastatic cells [5, 6]. Dissemination also

The authors have no conflicts of interest to disclose. All authors have critically reviewed and approved the final version of this chapter.

This chapter is a unique submission and is not being considered for publication by any other source. Further, it has not been published in any form, in part or whole.

J. A. Diaz-Perez
Departments of Dermatology and Pathology, Virginia Commonwealth University, Richmond, VA, USA
e-mail: julio.diazperez@vcuhealth.org

A. E. Rosenberg (✉)
Department of Pathology and Laboratory Medicine, University of Miami, Miami, FL, USA
e-mail: ARosenberg@med.miami.edu

J. Pretell-Mazzini, S. A. Lozano Calderón (eds.), *Multidisciplinary Approach of Musculoskeletal Metastasis*, https://doi.org/10.1007/978-3-032-09498-8_8

occurs through the lymphatics affecting regional lymph nodes or via direct extension to the neighboring skeletal structures [7].

The focus of the pathological examination of skeletal metastasis is to identify the tumor type, the anatomical site of origin, and the expression of relevant biomarkers that aid in offering appropriate adjuvant therapy and skeletal stabilization [8, 9]. This process begins with proper selection of the best site to sample and determining whether a targeted needle or open biopsy is required. Intraoperative frozen section examination helps assure the quality of the sample. Careful consideration should be given to ensuring retrieval of an appropriate quantity of viable tumor, adequate fixation, and decalcification in a manner that preserves morphology, antigens, RNA, and DNA.

Handling Tissue Specimens and Intraoperative Examination

Image-guided biopsy frequently improves tumor sampling, facilitating a more complete pathological evaluation. The sample must represent the viable part of the tumor; therefore, the biopsy is often taken from the periphery of the lesion, avoiding central necrosis [10]. Tissue samples should undergo careful gross inspection. Some cases require examination of the entire specimen and additional deeper tissue levels to identify rare tumor cells in limited or necrotic samples. In our experience, needle core biopsies should include a minimum of three separate needle cores of viable tumor so that all necessary pathology testing can be completed. Each core should be processed individually in the laboratory to maximize the tissue available for ancillary testing. Touch preparations, frozen section analysis of one core, or rapid on-site evaluation using cytology can be performed to help confirm the presence of diagnostic tissue and provide immediate diagnostic and sampling information.

Intraoperative examination of samples suspected of metastasis improves the detection of small metastatic deposits and reduces sampling errors [11]. In this manner, discrepancies between intraoperative examination and the final diagnosis can be minimized. We recommend combining the analysis of histological slides via frozen section with intraoperative touch imprint cytology as an adjunct technique to assist in this process. Cytology is also valuable in providing rapid information in cases involving the evaluation of mineralized tissue.

Surgical resection specimens should be serially sectioned, searching for viable lesional tissue. The percentage of viable, necrotic, and hemorrhagic tumor should be recorded on the gross description. This information is especially relevant to evaluate neoadjuvant therapy response [12].

Recently, the use of digital pathology and artificial intelligence (AI) for the histological evaluation of skeletal metastases has attracted considerable interest. Although the technological capabilities are promising, the practical usefulness of AI algorithms in clinical settings still needs to be established [13].

Histological Examination of Common Types of Skeletal Metastases

Bone metastases are often the first sign of occult cancer and indicate advanced disease. Pathological examination involves a careful, detailed assessment of cells, tissue, and their architectural arrangements to confirm the presence of metastatic disease and differentiate it from primary bone neoplasms. The metastatic tissue should be quantified by the greatest diameter of the tumor deposit, and the percentage of viable versus necrotic tissue should be recorded. Micrometastases or small metastatic deposits are only observed under a microscope and are generally <2 mm in diameter [14]. Additional hematoxylin and eosin-stained sections from deeper tissue levels provide relevant information in cases with limited numbers of tumor cells.

Immunohistochemistry significantly enhances the ability to subtype metastatic malignancies and identify their site of origin. It also aids in detecting micrometastases and potential therapeutic targets. However, relying solely on a single immunohistochemical stain can be a source of error because individual immunohistochemical markers are not specific and may be present in different types of neoplasms and metastases often exhibit phenotypic plasticity (Table 8.1). Using a combination of immunostains is more helpful, especially if they are interpreted in conjunction with clinical, radiological, and histological findings.

Table 8.1 Anatomical site typing of metastatic carcinoma to the skeleton with immunohistochemistry

IHC	GATA3	CDX2	TTF-1	PAX8
CK7+/ CK20+	Breast [TRPS1] Urothelial [p63]	Pulmonary (mucinous AC) Ovarian (mucinous AC) [CA125] Gastric (AC) Pancreatic [CK19] Biliary [CK19]	Pulmonary (AC) Pulmonary (Small cell) [INSM1]	Endometrial [ER]
CK7+/CK20–	Breast [TRPS1] Urothelial [p63]	GI (Upper) [PDX1] Pancreatic [CK19] Biliary [CK19]	Pulmonary (AC papillary) Thyroid [Thyroglobulin] Thymic [CD5]	Ovarian (serous) [WT1] Pancreatic Endometrial Cervix [p16] Mesothelioma Thyroid medullar [calcitonin] Salivary gland Kidney (papillary)
CK7–/ CK20+	Merkel cell [INSM1]	Colorectal [SATB2] Gastric (AC) Small bowel Salivary gland Bladder (AC)	Neuroendocrine [INSM1]	Bladder (AC) Kidney (papillary)

(continued)

Table 8.1 (continued)

IHC	GATA3	CDX2	TTF-1	PAX8
CK7−/CK20−	Prostate [NKX3.1] SCC [EMA] BCC [BerEP4] Sebaceous Ca. [AR] Hepatocellular Adrenal cortical [SF1]	Bladder (AC) Colon (medullary) [SATB2] NUT [NUT]	Pulmonary (small cell) [Syn] Neuroendocrine [Syn]	Kidney (clear cell)

IHC immunohistochemistry, *AC* adenocarcinoma, *SCC* squamous cell carcinoma, *BCC* basal cell carcinoma

Carcinomas

Carcinomas are the most common source of metastatic disease to bone. Based on histological features, metastatic carcinomas can be classified into various types, including adenocarcinomas, squamous cell carcinomas (SCCs), and poorly differentiated carcinomas. Adenocarcinomas are the most common metastasis type, followed by SCCs and neuroendocrine carcinomas. Their identification is usually straightforward but can be difficult in cases with limited samples, posttreatment, or those that have undergone high-grade transformation. Posttreatment metastatic deposits often consist of single cells and small clusters of cells scattered within fibrotic and inflamed tissue, which also has hemosiderin deposits and dystrophic calcifications. High-grade transformation may present as sarcomatoid differentiation, marked by spindle cell morphology with significant nuclear pleomorphism, increased mitotic activity, and tumor necrosis. In such cases, determining the site and type of metastasis can be challenging.

Cytokeratins help confirm the epithelial origin in poorly differentiated malignancies, although other tumor types, including mesothelioma and some sarcomas, also express cytokeratins. A cocktail of AE1/AE3 and CAM 5.2 is an effective panel for identifying the epithelial lineage. This combination detects up to 40% of occult metastases [14]. Therefore, the use of cytokeratin immunohistochemistry can facilitate the identification of small numbers of carcinoma cells and is recommended before a biopsy is considered negative. Cytokeratins are also often helpful in highlighting tumor cells within treatment-related changes and in confirming the true extent of residual disease [15].

Adenocarcinomas exhibit diverse cell morphology and tend to grow diffusely in cell aggregates separated by delicate septa. Tumor cells are polyhedral and generally exhibit significant cellular atypia, characterized by an increased nuclear-to-cytoplasmic ratio and frequent mitoses. Tubular, cribriform, and papillary formations are observed in well and moderately differentiated adenocarcinomas. In comparison, tumor cells grow in cords, trabeculae, or single cells in poorly differentiated variants. Metastases of lobular breast carcinoma and signet ring cell adenocarcinoma often manifest as noncohesive tumor cells. Metastatic malignancies with abundant tumor giant cells include metastatic choriocarcinomas, pheochromocytomas, large cell carcinomas, and poorly differentiated SCCs.

Among the cytokeratins, CK7 and CK20 are the most valuable to predict the anatomical site of origin, and their expression pattern is summarized in Table 8.1. Adenocarcinomas are generally positive for low-molecular-weight keratins, such as CK8 (including clone CAM5.2) and CK18 [16]. Some adenocarcinomas, especially those from the breast, can be associated with numerous non-neoplastic osteoclast-type giant cells, which can mimic giant cell tumor of bone [17]. Immunohistochemistry for keratin and H3.3 G34W can be extremely helpful in rendering an accurate diagnosis (Fig. 8.1). In other situations, an adenocarcinoma can induce the development of osteoblasts and abundant new bone formation that may manifest as a sclerotic tumor on radiographs.

SCCs are characterized by keratinization of tumor cells. The primary site of this kind of tumor is, in most cases, determined clinically. High-molecular-weight keratins, such as CK5 and CK14 (including clone 34bE12, which recognizes CK1, CK5, CK10, and CK1410), react predominantly with the squamous epithelium and urothelium. p63 and p40 are also extremely helpful markers of the squamous phenotype. p16 may be useful to identify cervical, anal, penile, and other carcinomas associated with human papillomavirus (HPV). In these cases, molecular studies (such as polymerase chain reaction (PCR) or in situ hybridization) to detect HPV are also helpful [18].

In men, metastatic cancers originating from the lung, followed by the prostate gland and kidney, commonly spread to the bones. Prostate origin can be confirmed through the expression of prostate-specific membrane antigen (PSMA), prostein (also called P501S), and NKX3.1 [19] (Fig. 8.3). In women, the most common primary tumor site is the breast, followed by the lung and female genital organs. Squamous differentiation is common in endometrioid and breast carcinomas. Endometrioid adenocarcinomas are typically diffusely positive for CK7, CA125,

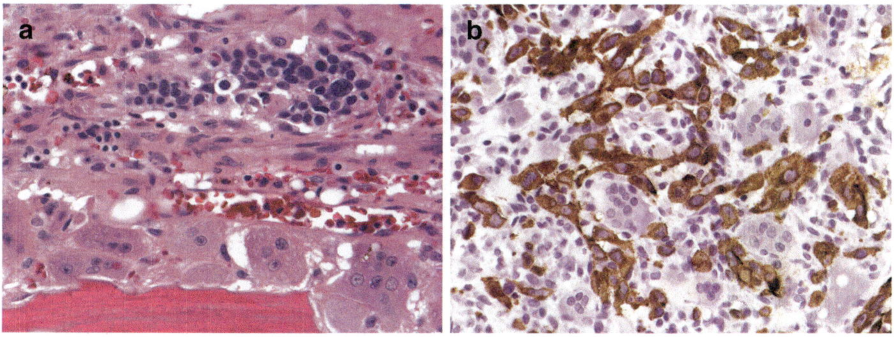

Fig. 8.1 Metastatic adenocarcinoma can display diverse morphology, as seen in this case of metastatic breast carcinoma rich in non-neoplastic osteoclast-like giant cells (**a**, HE 40x) that mimicked a giant cell tumor of bone. The immunohistochemical positivity of cytokeratin (**b**, 40x) and negativity of H3.3 G34W confirmed the diagnosis. HE hematoxylin and eosin

and ER and negative for CK20, CEA, and CDX2. Metastatic breast carcinoma often loses luminal differentiation. The immunohistochemical expression of GATA binding protein 3 (GATA3), gross cystic disease fluid protein (GCDFP), and trichorhinophalangeal syndrome type 1 (TRPS1) suggests a breast origin (Fig. 8.2) [20, 21].

Thyroid transcription factor-1 (TTF-1) is valuable for identifying metastatic carcinomas originating from the lung and thyroid. Histologically, metastatic thyroid carcinomas can appear quite bland and sometimes lack well-defined nuclear features. Among thyroid neoplasms, TTF-1 expression has been observed in nearly 100% of papillary and follicular carcinomas as well as poorly differentiated (insular) carcinomas [22, 23]. Even rare thyroid carcinomas, such as mucoepidermoid carcinoma of the thyroid, test positive for TTF-1. PAX-8 is also typically positive in both thyroid and thymic carcinomas. In the lung, TTF-1 stains tumors with acinar, tubular, bronchioloalveolar, mucinous (colloid), and signet ring-like morphology. TTF-1 is also positive in 7–10% of lung SCCs and neuroendocrine carcinomas [24, 25].

In metastatic gastrointestinal adenocarcinomas, the glands are typically lined by less well-differentiated cells with increased nuclear hyperchromatism and loss of polarity compared to other adenocarcinomas. In the upper gastrointestinal tract, a combination of PDX1 and CDX2 may help identify these lesions. Dirty necrosis is more commonly observed in metastatic intestinal carcinomas, and CDX2 and SATB2 are useful markers for their detection. Metastatic pancreatic or biliary tract carcinomas show a similar immunophenotype, being positive for CK7 and CA19–9, and they may be negative or display diffuse or focal positivity for CK20, CEA, and CDX2 (Fig. 8.3). The immunophenotype of gastric carcinomas generally resembles that of pancreatic and biliary tract adenocarcinomas.

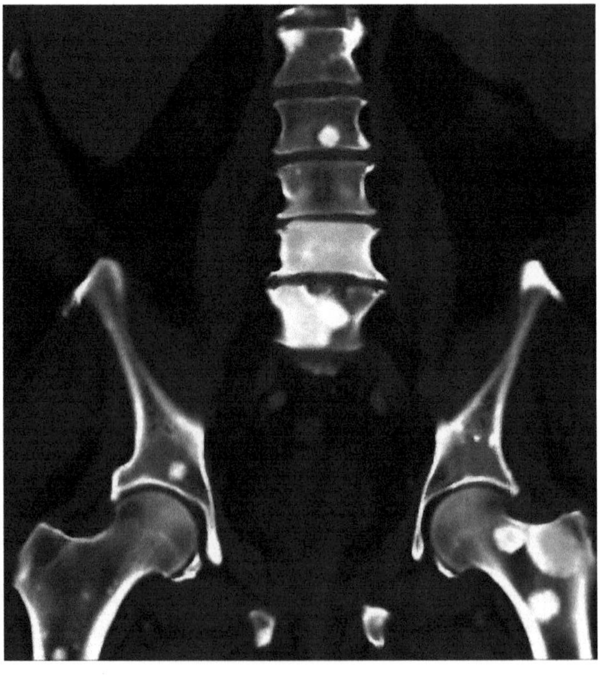

Fig. 8.2 Radiological correlation is useful to establish a diagnosis. A metastatic carcinoma of the prostate usually manifests with osteoblastic or sclerotic lesions that are hyperdense on CT

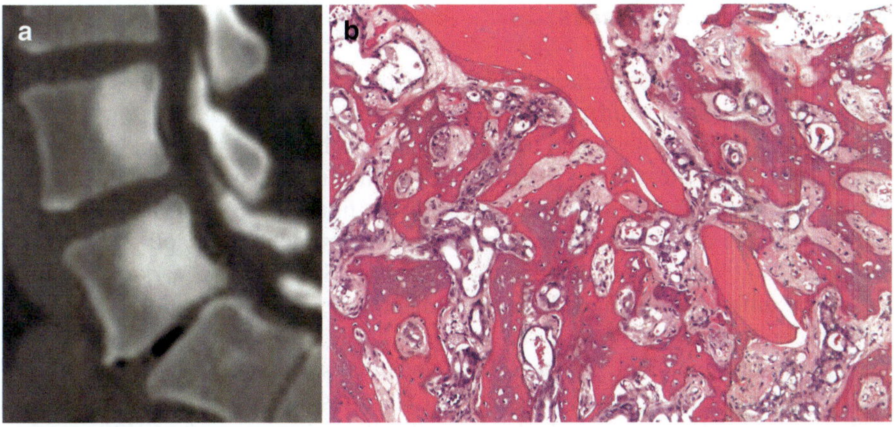

Fig. 8.3 Other metastatic adenocarcinomas can also present as osteoblastic or sclerotic lesions, as seen in this metastatic pancreatic adenocarcinoma involving the spine (**a**). Malignant glands within the sclerotic bone are present (**b**, HE 20x). HE hematoxylin and eosin

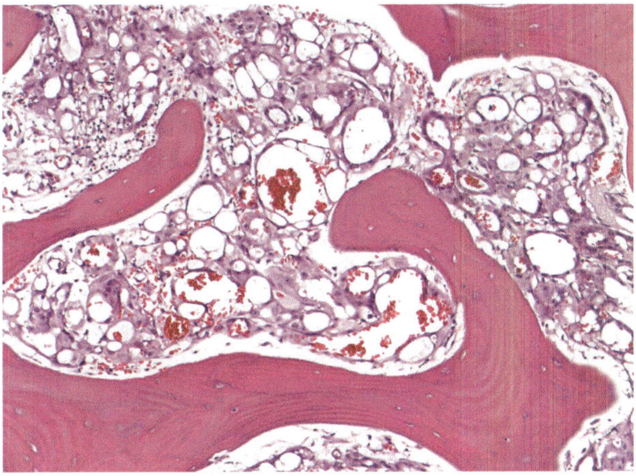

Fig. 8.4 Some metastatic carcinomas can display diverse morphology, including sarcomatoid features, as in this case of metastatic renal cell carcinoma with an angioformative pattern that mimicked an angiosarcoma (HE, 10x). HE hematoxylin and eosin

PAX8 and PAX2 are commonly expressed in various renal and uterine carcinomas, including clear cell renal cell carcinomas (RCCs), papillary RCCs, chromophobe RCCs, acquired cystic kidney-related RCCs, mucinous tubular and spindle cell carcinomas, mixed epithelial and stromal tumors, cystic nephroma, and oncocytoma (Fig. 8.4) [26]. Tumor types such as collecting duct RCCs or sarcomatoid RCCs, which are almost always negative for traditional RCC markers like RCCM, CD10, or kidney-specific cadherin, do express PAX2 [27].

Metastatic mucinous carcinomas to the skeleton can originate from several primary sites, including the colorectum, pancreas, biliary tract, appendix, stomach, cervix, and breast. Salivary gland carcinomas often involve the craniofacial skeleton. The most common type is mucoepidermoid carcinoma [28].

Neuroendocrine Tumors

Identifying neuroendocrine morphological features should lead to the use of neuroendocrine markers to confirm the diagnosis. Neuroendocrine tumors are made up of densely cellular, solid nests and trabeculae of cells that range from spindle to plasmacytoid to large clear cells; these tumors exhibit nuclear molding and salt-and-pepper chromatin and are separated by delicate fibrovascular stroma. They often have noticeable areas of necrosis. Common markers include synaptophysin, chromogranin, and insulinoma-associated protein 1 (INSM1). Other markers of neuroendocrine differentiation include neurofilament protein and CD56 [29].

Neuroendocrine carcinomas can originate from various organs, including the skin, gastrointestinal tract, pancreas, lung, and other sites. Merkel cell carcinoma, or neuroendocrine carcinoma of the skin, is CK20-positive in a dot-like pattern and TTF-1-negative. However, rarely, lung small cell carcinoma can be CK20-positive, but most of these cases are TTF-1-positive. Merkel cell polyomavirus immunohistochemistry is expressed in up to 80% of cases [30]. Fli-1 and CD99 are also positive in Merkel cell carcinoma, which can potentially create confusion with Ewing sarcoma [31]. The presence of dual positivity for PDX-1 and NESP-55 staining, along with negativity for TTF-1 and CDX-2, is 97% specific for pancreatic neuroendocrine carcinoma [32].

Pheochromocytomas stain positive for GATA3, neuroendocrine markers, and INSM1 and are almost always cytokeratin-negative [33, 34]. Neuroblastoma can be identified using a combination of PHOX2B or beta-catenin with PGP9.5, synaptophysin, or tyrosine hydroxylase [35].

Melanomas

Melanomas frequently metastasize to the skeleton. They exhibit great phenotypic diversity, and accurately determining whether a metastatic melanoma is present provides crucial information for a patient's prognosis, management, and therapy planning. While most cases are straightforward, some can be challenging, especially when there are limited numbers of tumor cells, the tumor has undergone dedifferentiation, or when the neoplastic cells are small or spindled (Fig. 8.5). In metastatic melanomas, melanin pigment is often inconspicuous or absent in approximately 50% of the cases.

Immunohistochemistry and molecular pathology provide phenotypic and genotypic markers that are valuable for confirming a metastatic melanoma diagnosis. SOX10, S100, Melan A, and MART-1 are positive in most melanomas. Some melanomas might lose the expression of one or more melanocytic markers, such as spindle cell, desmoplastic, and rhabdoid melanomas. Rare melanoma cases may show loss of all the diagnostic immunohistochemical markers used in routine practice, indicating an undifferentiated/dedifferentiated state [36]. Melanoma metastases can mimic various nonmelanocytic neoplasms such as carcinoma and sarcoma, showing heterologous transdifferentiation. Between 2% and 5% of melanomas can

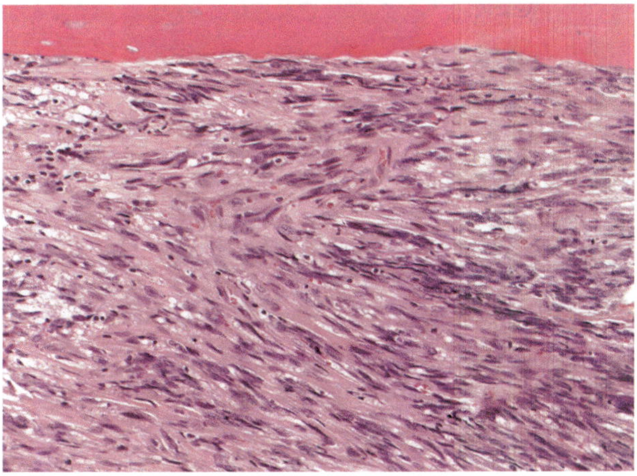

Fig. 8.5 Metastatic spindle cell and round cell melanoma can show loss of expression of one or more melanocytic immunohistochemical markers (HE, 20x). HE hematoxylin and eosin

aberrantly express keratins (CAM 5.2 and AE1/AE3), 2% are positive for epithelial membrane antigen, and some cases may express myogenic markers. Molecular analysis for *BRAF/NRAS* mutations and identification of an ultraviolet mutation signature can be extremely useful in rendering an accurate diagnosis [37].

Sarcomas

Sarcomas more frequently metastasize to bone than any other cancers and can present significant diagnostic challenges. The pathologist's interpretation has important therapeutic implications related to distinguishing primary from metastatic tumors, classifying the aggressiveness, and determining their line of differentiation.

Pathological examination generally reveals a high-grade sarcoma composed of fascicles of spindle cells or sheets of epithelioid cells with nuclear atypia and many mitoses. Most cases lack a classic epithelial component, except in metastases of epithelioid hemangioendothelioma, epithelioid angiosarcoma, synovial sarcoma, epithelioid sarcoma, desmoplastic small round cell tumor, chordoma, and adamantinoma. In these cases, cytokeratins are usually positive. Many sarcomas often exhibit focal expression of cytokeratin, such as leiomyosarcoma, pleomorphic fibroblastic sarcoma, rhabdomyosarcoma, malignant rhabdoid tumor, malignant peripheral nerve sheath tumor, and clear cell sarcoma. Leiomyosarcoma often expresses SMA, caldesmon, and desmin, rhabdomyosarcoma expresses desmin, myogenin, and MYOD1, and malignant peripheral nerve sheath often loses expression of H3k27me3. These tumors can undergo molecular testing for the identification of diagnostic molecular abnormalities.

Metastatic osteosarcoma is identified through clinical and radiological correlation after documenting a malignant tumor with bone matrix formation—the tumor cells are usually positive for SATB2. In many cases, metastatic chondrosarcoma corresponds to metastasis of dedifferentiated chondrosarcoma. Leiomyosarcoma and malignant peripheral nerve sheath tumors can metastasize to bone, and it is

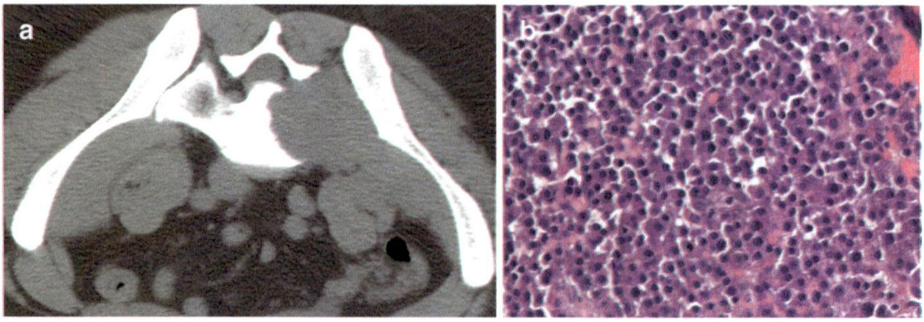

Fig. 8.6 Plasma cell neoplasms commonly involve the bone, showing well-defined, lytic lesions on images (**a**, CT). The histology displays sheets of plasma cells (**b**, HE, 20x). HE hematoxylin and eosin

important to differentiate metastases from skeletal primaries. Both leiomyosarcoma and peripheral nerve sheath tumors can exhibit heterologous differentiation, including rhabdomyosarcomatous, osteosarcomatous, chondrosarcomatous, and liposarcomatous differentiation.

Round cell sarcomas commonly cause skeletal metastases. Immunohistochemistry for phenotypic markers and gene surrogates are usually required [38]. Additionally, fluorescence in situ hybridization (FISH) and next-generation sequencing (NGS) may be necessary in cases in which a previous diagnosis is unknown [39].

Hematolymphoid Tumors

Many types of malignant hematolymphoid tumors affect the skeleton. CD45 alone is not sufficient to exclude a potential diagnosis of a hematopoietic neoplasm. Some diffuse, large B-cell lymphomas, plasmablastic lymphomas, myeloid sarcomas, and anaplastic lymphomas may be negative for CD45. We recommend combining CD45 with CD43 and CD138 to better detect myeloid and plasma cell neoplasms (Fig. 8.6). It is important to note that CD138 can stain many types of carcinomas and is a potential diagnostic pitfall.

Selected Immunohistochemical Markers for Targeted Therapy

Molecularly tailored therapy significantly enhances clinical outcomes in metastatic disease involving the skeleton. Immunohistochemistry and genetic analysis help identify targeted therapy options. Estrogen receptor (ER) protein and progesterone receptor (PR) protein are important markers in breast cancer. Human epidermal growth factor receptor 2 (HER-2/neu) protein expression is relevant for treating various carcinomas. HER2 amplification detected by in situ hybridization is also valuable. PD-L1 expression serves as a biomarker for using immune checkpoint inhibitors. High tumor mutational burden (TMB-H), defined as $\geq$10 mutations per megabase (mut/Mb), is another useful biomarker to help initiate immunotherapy.

Genetic Analysis of Skeletal Metastasis

Genomic profiling technology has been widely used in the supplementary examination and targeted treatment of metastatic cancers. Somatic mutations identified through targeted genomic or whole exome/genome sequencing of tumors can be categorized into driver mutations, which include a relatively small set of functional oncogenic mutations, and passenger mutations, which consist of a larger number of nonfunctional alterations [40]. Driver mutations are also valuable biomarkers for targeted therapies. Molecular genetic analysis can also provide definitive information in cases in which there is a high mutational burden or a ultraviolet (UV) mutational signature (MS).

NGS offers insights into the suitability for clinical trials for patients with molecular alterations that lack approved targeted therapies. However, some mutations found in metastases can differ considerably from those in the primary tumor. Driver mutations, including *BRAF*, *NRAS*, and *NF1*, detected in established melanomas via next-generation sequencing (NGS), are useful in diagnosing de/transdifferentiated melanomas since the conventional and de/transdifferentiated components share similarly mutated genomes [37]. *EGFR* mutations are common in metastatic lung carcinoma and have the potential for targeted therapy. Detecting mutational signatures (MSs) or distinct patterns resulting from accumulated somatic mutations of both exogenous and endogenous origins—such as ultraviolet (UV)-induced DNA damage signatures, *MDM2* amplification, and other genetic profiles—also proves beneficial. Finally, genome-wide methylation profiling combined with machine learning classifiers is being developed as a valuable method for determining tissue lineage or cell-of-origin, which can aid in tumor classification [41].

Conclusions

Skeletal metastases present a significant clinical and diagnostic challenge and remain a major contributor to cancer-related morbidity and mortality. An accurate diagnosis of metastatic disease should be based on a clinicopathological correlation to exclude differential diagnoses and is achieved through proper handling of the tissue sample, careful morphological analysis, and integration of ancillary testing such as immunohistochemistry, cytogenetics, and molecular testing.

References

1. Selvaggi G, Scagliotti GV. Management of bone metastases in cancer: a review. Crit Rev Oncol Hematol. 2005;56(3):365–78. https://doi.org/10.1016/j.critrevonc.2005.03.011. Epub 2005 Jun 23. PMID: 15978828.
2. Coleman RE. Clinical features of metastatic bone disease and risk of skeletal morbidity. Clin Cancer Res. 2006;12(20 Pt 2):6243s–9s. https://doi.org/10.1158/1078-0432.CCR-06-0931. PMID: 17062708.
3. Li S, Peng Y, Weinhandl ED, Blaes AH, Cetin K, Chia VM, Stryker S, Pinzone JJ, Acquavella JF, Arneson TJ. Estimated number of prevalent cases of metastatic bone disease in the US adult

population. Clin Epidemiol. 2012;4:87–93. https://doi.org/10.2147/CLEP.S28339. Epub 2012 Apr 10. PMID: 22570568; PMCID: PMC3345874.
4. Christ AB, Piple AS, Gettleman BS, Duong A, Chen M, Wang JC, Heckmann ND, Menendez L. Prevalence of primary malignant tumours, rates of pathological fracture, and mortality in the setting of metastatic bone disease. Bone Jt Open. 2023;4(6):424–31. https://doi.org/10.1302/2633-1462.46.BJO-2023-0042.R1. PMID: 37272304; PMCID: PMC10240245.
5. Cocchi C, Dapporto M, Guerrieri AN, Liverani C, Tavoni M, Bellotti C, Spadazzi C, Tampieri A, Gambarotti M, Miserocchi G, Sprio S, Lucarelli E, Iafisco M, Ibrahim T, De Vita A, Mercatali L. A modular biomimetic preclinical platform to elucidate the interaction between cancer cells and the bone metastatic niche. J Funct Biomater. 2025;16(6):220. https://doi.org/10.3390/jfb16060220. PMID: 40558906; PMCID: PMC12194721.
6. Coghlin C, Murray GI. Current and emerging concepts in tumour metastasis. J Pathol. 2010;222(1):1–15. https://doi.org/10.1002/path.2727. PMID: 20681009.
7. Akhtar M, Haider A, Rashid S, Al-Nabet ADMH. Paget's "seed and soil" theory of cancer metastasis: an idea whose time has come. Adv Anat Pathol. 2019;26(1):69–74. https://doi.org/10.1097/PAP.0000000000000219. PMID: 30339548.
8. Ehne J, Tsagozis P. Current concepts in the surgical treatment of skeletal metastases. World J Orthop. 2020;11(7):319–27. https://doi.org/10.5312/wjo.v11.i7.319. PMID: 32908816; PMCID: PMC7441493.
9. Kitagawa T, Kitagawa Y, Aoyagi Y, Majima T. Risk of pathological fracture and paralysis during rehabilitation for patients with bone metastases: a questionnaire-based study. J Nippon Med Sch. 2023;90(4):326–32. https://doi.org/10.1272/jnms.JNMS.2023_90-304. Epub 2023 Jun 2. PMID: 37271550.
10. O'Sullivan GJ, Carty FL, Cronin CG. Imaging of bone metastasis: an update. World J Radiol. 2015;7(8):202–11. https://doi.org/10.4329/wjr.v7.i8.202. PMID: 26339464; PMCID: PMC4553252.
11. Alani A, Diaz-Perez JA. Intraoperative examination in bone and soft tissue pathology. Semin Diagn Pathol. 2025;42(3):150904. https://doi.org/10.1016/j.semdp.2025.150904. Epub 2025 Apr 17. PMID: 40250309.
12. Bonnefoi H, Litière S, Piccart M, MacGrogan G, Fumoleau P, Brain E, Petit T, Rouanet P, Jassem J, Moldovan C, Bodmer A, Zaman K, Cufer T, Campone M, Luporsi E, Malmström P, Werutsky G, Bogaerts J, Bergh J, Cameron DA. EORTC 10994/BIG 1-00 Study investigators. Pathological complete response after neoadjuvant chemotherapy is an independent predictive factor irrespective of simplified breast cancer intrinsic subtypes: a landmark and two-step approach analyses from the EORTC 10994/BIG 1-00 phase III trial. Ann Oncol. 2014;25(6):1128–36. https://doi.org/10.1093/annonc/mdu118. Epub 2014 Mar 11. PMID: 24618153; PMCID: PMC4037859.
13. Lorkowski SW, Dermawan JK, Rubin BP. The practical utility of AI-assisted molecular profiling in the diagnosis and management of cancer of unknown primary: an updated review. Virchows Arch. 2024;484(2):369–75. https://doi.org/10.1007/s00428-023-03708-1. Epub 2023 Nov 24. Erratum in: Virchows Arch. 2024 Feb;484(2):377. https://doi.org/10.1007/s00428-023-03728-x. PMID: 37999736.
14. Qiao Y, Wang M, Hui K, Jiang X. Diagnosis progress of carcinoma of unknown primary. Front Oncol. 2024;14:1510443. https://doi.org/10.3389/fonc.2024.1510443. PMID: 39659790; PMCID: PMC11628522.
15. Behzatoglu K. Osteoclasts in tumor biology: metastasis and epithelial-mesenchymal-myeloid transition. Pathol Oncol Res. 2021;27:609472. https://doi.org/10.3389/pore.2021.609472. PMID: 34257573; PMCID: PMC8262221.
16. Alessandrini L, Zanoletti E, Cazzador D, Sbaraglia M, Franz L, Tealdo G, Frigo AC, Blandamura S, Nicolai P, Mazzoni A, Marioni G. Tumor budding to investigate local invasion, metastasis and prognosis in temporal bone squamous cell carcinoma. Pathol Res Pract. 2022;229:153719. https://doi.org/10.1016/j.prp.2021.153719. Epub 2021 Nov 28. PMID: 34953406.

17. Patton A, Ilaslan H, Armstrong SM, Bakhshwin A, Cheng YW, Minhas F, Fritchie KJ. Keratin-positive Giant cell-rich tumor of bone harboring an *HMGA2::NCOR2* fusion: two cases, including a patient with metastatic disease, and review of the literature. Int J Surg Pathol. 2024;32(3):556–64. https://doi.org/10.1177/10668969231185076. Epub 2023 Jul 17. PMID: 37461232.
18. Montero-Macías R, Coronado PJ, Robillard N, Veyer D, Villefranque V, Le Frére-Belda MA, Auberger E, Bitolog P, Stankovic I, Bélec L, Bats AS, Lécuru F, Péré H. HPV molecular genotyping as a differential diagnosis tool in cervical cancer metastasis. J Pers Med. 2023;13(2):177. https://doi.org/10.3390/jpm13020177. PMID: 36836411; PMCID: PMC9961249.
19. Gurel B, Ali TZ, Montgomery EA, Begum S, Hicks J, Goggins M, Eberhart CG, Clark DP, Bieberich CJ, Epstein JI, De Marzo AM. NKX3.1 as a marker of prostatic origin in metastatic tumors. Am J Surg Pathol. 2010;34(8):1097–105. https://doi.org/10.1097/PAS.0b013e3181e6cbf3. PMID: 20588175; PMCID: PMC3072223.
20. Westbrook JA, Wood SL, Cairns DA, McMahon K, Gahlaut R, Thygesen H, Shires M, Roberts S, Marshall H, Oliva MR, Dunning MJ, Hanby AM, Selby PJ, Speirs V, Mavria G, Coleman RE, Brown JE. Identification and validation of DOCK4 as a potential biomarker for risk of bone metastasis development in patients with early breast cancer. J Pathol. 2019;247(3):381–91. https://doi.org/10.1002/path.5197. Epub 2019 Jan 25. PMID: 30426503; PMCID: PMC6618075.
21. Lennartz M, Löhr N, Höflmayer D, Dwertmann Rico S, von Bargen C, Kind S, Reiswich V, Viehweger F, Lutz F, Bertram V, Fraune C, Gorbokon N, Weidemann S, Blessin NC, Hube-Magg C, Menz A, Schlichter R, Krech T, Hinsch A, Burandt E, Sauter G, Simon R, Kluth M, Marx AH, Lebok P, Dum D, Minner S, Jacobsen F, Clauditz TS, Bernreuther C, Steurer S. TRPS1 is a highly sensitive marker for breast cancer: a tissue microarray study evaluating more than 19,000 tumors from 152 different tumor entities. Am J Surg Pathol. 2024;48(6):637–51. https://doi.org/10.1097/PAS.0000000000002213. Epub 2024 Apr 18. PMID: 38647255; PMCID: PMC11093513.
22. Conner JR, Hornick JL. Metastatic carcinoma of unknown primary: diagnostic approach using immunohistochemistry. Adv Anat Pathol. 2015;22(3):149–67. https://doi.org/10.1097/PAP.0000000000000069. PMID: 25844674.
23. Chernock RD, Lewis JS. Approach to metastatic carcinoma of unknown primary in the head and neck: squamous cell carcinoma and beyond. Head Neck Pathol. 2015;9(1):6–15. https://doi.org/10.1007/s12105-015-0616-2. Epub 2015 Mar 25. PMID: 25804376; PMCID: PMC4382479.
24. Ordóñez NG. Value of thyroid transcription factor-1 immunostaining in tumor diagnosis: a review and update. Appl Immunohistochem Mol Morphol. 2012;20(5):429–44. https://doi.org/10.1097/PAI.0b013e31825439bc. PMID: 22531688.
25. Vidarsdottir H, Tran L, Nodin B, Jirström K, Planck M, Mattsson JSM, Botling J, Micke P, Jönsson P, Brunnström H. Comparison of three different TTF-1 clones in resected primary lung cancer and epithelial pulmonary metastases. Am J Clin Pathol. 2018;150(6):533–44. https://doi.org/10.1093/ajcp/aqy083. PMID: 30169783.
26. Ozcan A, de la Roza G, Ro JY, Shen SS, Truong LD. PAX2 and PAX8 expression in primary and metastatic renal tumors: a comprehensive comparison. Arch Pathol Lab Med. 2012;136(12):1541–51. https://doi.org/10.5858/arpa.2012-0072-OA. PMID: 23194047.
27. Yang C, Cimera RS, Aryeequaye R, Jayakumaran G, Sarungbam J, Al-Ahmadie HA, Gopalan A, Sirintrapun SJ, Fine SW, Tickoo SK, Epstein JI, Reuter VE, Zhang Y, Chen YB. Adverse histology, homozygous loss of CDKN2A/B, and complex genomic alterations in locally advanced/metastatic renal mucinous tubular and spindle cell carcinoma. Mod Pathol. 2021;34(2):445–56. https://doi.org/10.1038/s41379-020-00667-9. Epub 2020 Sep 2. PMID: 32879414; PMCID: PMC7855055.
28. Benchetrit L, Mehra S, Mahajan A, Rahmati RW, Judson BL, Edwards HA. Major salivary gland cancer with distant metastasis upon presentation: patterns, outcomes, and imaging implications. Otolaryngol Head Neck Surg. 2022;167(2):305–15. https://doi.org/10.1177/01945998211058354. Epub 2021 Nov 16. PMID: 34784258.

29. Berner AM, Pipinikas C, Ryan A, Dibra H, Moghul I, Webster A, Luong TV, Thirlwell C. Diagnostic approaches to neuroendocrine neoplasms of unknown primary site. Neuroendocrinology. 2020;110(7–8):563–73. https://doi.org/10.1159/000504370. Epub 2019 Oct 28. PMID: 31658461.
30. Matsushita M, Nonaka D, Iwasaki T, Kuwamoto S, Murakami I, Kato M, Nagata K, Kitamura Y, Hayashi K. A new in situ hybridization and immunohistochemistry with a novel antibody to detect small T-antigen expressions of Merkel cell polyomavirus (MCPyV). Diagn Pathol. 2014;9:65. https://doi.org/10.1186/1746-1596-9-65. PMID: 24649978; PMCID: PMC3994555.
31. Juhlin CC, Zedenius J, Höög A. Metastatic neuroendocrine neoplasms of unknown primary: clues from pathology workup. Cancers (Basel). 2022;14(9):2210. https://doi.org/10.3390/cancers14092210. PMID: 35565339; PMCID: PMC9100271.
32. Srivastava A, Hornick JL. Immunohistochemical staining for CDX-2, PDX-1, NESP-55, and TTF-1 can help distinguish gastrointestinal carcinoid tumors from pancreatic endocrine and pulmonary carcinoid tumors. Am J Surg Pathol. 2009;33(4):626–32. https://doi.org/10.1097/PAS.0b013e31818d7d8b. PMID: 19065104.
33. Perrino CM, Ho A, Dall CP, Zynger DL. Utility of GATA3 in the differential diagnosis of pheochromocytoma. Histopathology. 2017;71(3):475–9. https://doi.org/10.1111/his.13229. Epub 2017 Jun 5. PMID: 28374498.
34. Wang XQ, Fu J, Zhang YT, Xu Y. INSM1 expression in primary and metastatic neuroendocrine neoplasms at distinct locations. Pathol Res Pract. 2024;253:155067. https://doi.org/10.1016/j.prp.2023.155067. Epub 2023 Dec 28. PMID: 38171081.
35. Hata JL, Correa H, Krishnan C, Esbenshade AJ, Black JO, Chung DH, Mobley BC. Diagnostic utility of PHOX2B in primary and treated neuroblastoma and in neuroblastoma metastatic to the bone marrow. Arch Pathol Lab Med. 2015;139(4):543–6. https://doi.org/10.5858/arpa.2014-0255-OA. PMID: 25822764.
36. Lezcano C, Pulitzer M, Moy AP, Hollmann TJ, Jungbluth AA, Busam KJ. Immunohistochemistry for PRAME in the distinction of nodal nevi from metastatic melanoma. Am J Surg Pathol. 2020;44(4):503–8. https://doi.org/10.1097/PAS.0000000000001393. PMID: 31633488; PMCID: PMC7071962.
37. Agaimy A, Specht K, Stoehr R, Lorey T, Märkl B, Niedobitek G, Straub M, Hager T, Reis AC, Schilling B, Schneider-Stock R, Hartmann A, Mentzel T. Metastatic malignant melanoma with complete loss of differentiation markers (undifferentiated/dedifferentiated melanoma): analysis of 14 patients emphasizing phenotypic plasticity and the value of molecular testing as surrogate diagnostic marker. Am J Surg Pathol. 2016;40(2):181–91. https://doi.org/10.1097/PAS.0000000000000527. PMID: 26448190.
38. Meredith DM, Charville GW, Fletcher CDM, Hornick JL. Distantly metastatic retinoblastoma to soft tissue and bone: a challenging diagnosis highlighting the utility of CRX. Am J Surg Pathol. 2021;45(6):820–4. https://doi.org/10.1097/PAS.0000000000001620. PMID: 33165094.
39. Negri GL, Grande BM, Delaidelli A, El-Naggar A, Cochrane D, Lau CC, Triche TJ, Moore RA, Jones SJ, Montpetit A, Marra MA, Malkin D, Morin RD, Sorensen PH. Integrative genomic analysis of matched primary and metastatic pediatric osteosarcoma. J Pathol. 2019;249(3):319–31. https://doi.org/10.1002/path.5319. Epub 2019 Aug 28. PMID: 31236944; PMCID: PMC9263032.
40. Akın D, Kahraman Çeti NN, Erdoğdu İH, Öztürk H, Meteoğlu İ. Clinicopathological significance of mutation profile detected by next generation sequencing in different metastatic organs of non-small cell lung cancers. Pathol Res Pract. 2024;260:155463. https://doi.org/10.1016/j.prp.2024.155463. Epub 2024 Jul 14. PMID: 39013258.
41. Walker A, Fang CS, Schroff C, Serrano J, Vasudevaraja V, Yang Y, Belakhoua S, Faustin A, William CM, Zagzag D, Chiang S, Acosta AM, Movahed-Ezazi M, Park K, Moreira AL, Darvishian F, Galbraith K, Snuderl M. Deep learning-based classifier for carcinoma of unknown primary using methylation quantitative trait loci. J Neuropathol Exp Neurol. 2025;84(2):147–54. https://doi.org/10.1093/jnen/nlae123. PMID: 39607989; PMCID: PMC11747144.

Has My Cancer Spread? – Staging Strategies

9

Lukas M. Nystrom

Introduction

Staging is the process of prognosticating the behavior of a cancer by assessing its local extent, the presence of metastases, and its histologic grade as a surrogate for its biologic activity. This chapter will focus on how to determine the presence of metastases when staging sarcomas and how to detect new metastases during the period of surveillance. Metastases can be both regional and distant and have the most significant impact on the patient's prognosis. Thus, accurate determination of metastatic disease is crucial to inform *treatment planning* and in counseling the patient on their disease.

When compared to staging, although highly similar techniques are employed, surveillance is the process of monitoring a patient with a malignant diagnosis after a definitive treatment has been rendered. A sarcoma diagnosis can still present with local or distant recurrence due to microscopic spread that is undetectable at the time of initial treatment. This microscopic disease can then blossom into clinically and radiographically apparent disease. Surveillance is important because although carrying a much poorer prognosis, metastatic disease can still be "cured" when presenting in an oligometastatic manner or treated with palliative intent when more widespread [10].

Initial Staging of a Soft Tissue Sarcoma

Standard initial staging of a soft tissue sarcoma involves a computed tomography (CT) scan of the chest and a magnetic resonance imaging (MRI) of the primary site. More than 95% of metastatic disease in soft tissue sarcomas involves the lungs

L. M. Nystrom (✉)
Department of Orthopaedic Surgery, Cleveland Clinic, Cleveland, OH, USA
e-mail: NYSTROL@ccf.org

J. Pretell-Mazzini, S. A. Lozano Calderón (eds.), *Multidisciplinary Approach of Musculoskeletal Metastasis*, https://doi.org/10.1007/978-3-032-09498-8_9

(Fig. 9.1). This is therefore the highest priority when staging this malignancy. Chest radiographs are an option for staging but are far less sensitive in comparison to CT scans. The most common practice is to perform at least the initial staging with a CT scan for this reason.

Local staging of the malignancy involves cross-sectional imaging, with MRI being favored due to its ability to better discern different tissue types and pathologic tissue. An additional benefit of MRI is it can quite readily identify pathologically enlarged lymph nodes in the immediate area (regional metastases). Pathologically enlarged lymph nodes warrant further investigation, likely including biopsy of the concerning node. If MRI is contraindicated, most commonly for implanted medical devices that are not compatible or for claustrophobia, CT evaluation is the next best choice.

While most soft tissue sarcomas metastasize exclusively to the lungs, certain histologic subtype staging considerations can be far more nuanced. For example, tumors that have a predilection for nodal metastases require evaluation in a more thorough manner. Then, there are also others such as myxoid liposarcoma and angiosarcoma that behave in a unique manner. Classically, these subtypes are rhabdomyosarcoma, angiosarcoma, clear cell sarcoma, epithelioid sarcoma, and synovial sarcoma. Notably, synovial sarcoma has been called into question given evidence demonstrating an extremely low rate of nodal metastases [15]. Because of the propensity for nodal metastases in these subtypes, there is evidence to suggest initial staging of these tumors with not only a CT scan of the chest but also of the abdomen and pelvis [17] or with a whole-body positron emission tomography (PET)/CT [2].

It is important to note and understand that nodal spread can occur with any malignancy and any pathologically enlarged lymph node should be investigated, as mentioned previously. When evaluating any one of the above histologic diagnoses, or if a pathologically enlarged node is identified (Fig. 9.2), further staging should be considered with a whole-body PET/CT [4, 9, 21, 30] (Fig. 9.3) and/or sentinel lymph node biopsy.

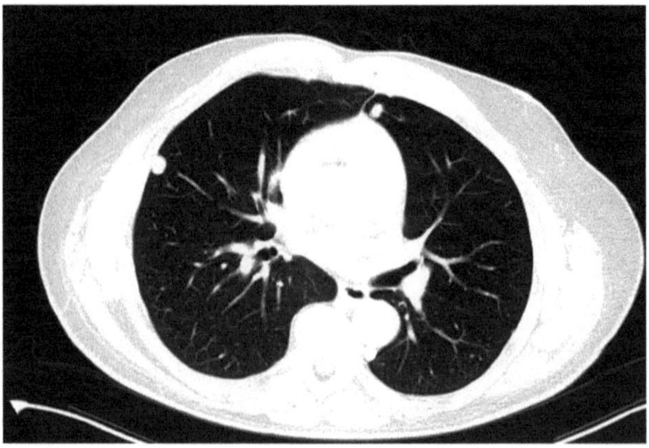

Fig. 9.1 The axial section of CT chest demonstrates large pulmonary nodules in the lateral right upper lobe and the anterior left upper lobe. The rounded and sharply demarcated appearance of these nodules are typical for sarcoma pulmonary metastases

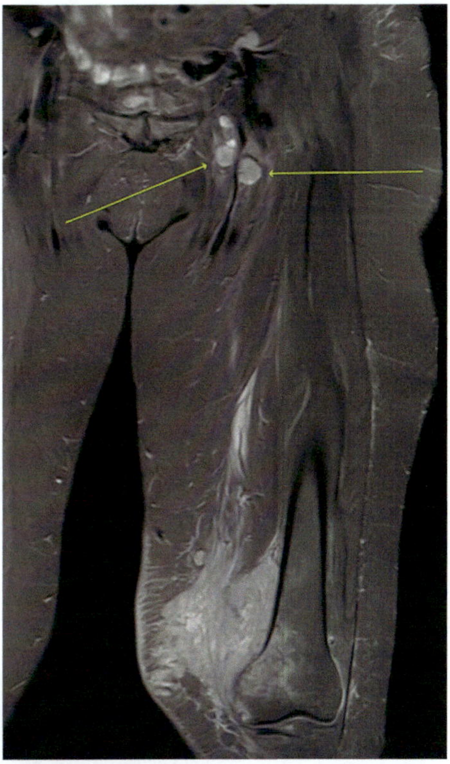

Fig. 9.2 Coronal STIR MRI demonstrating a clear cell sarcoma in the left distal medial thigh. By imaging the entire limb, it is apparent that there are pathologically enlarged lymph nodes in the left inguinal region

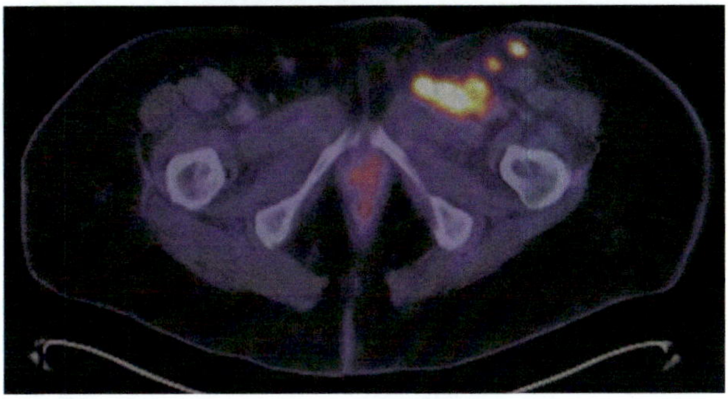

Fig. 9.3 Axial PET/CT in the same patient from Fig. 9.2 demonstrating the high Standardized Uptake Value (SUV) uptake in the left inguinal nodal region. This demonstrates how most sarcoma nodal metastases can be ascertained on PET/CT. PET positron emission tomography

Myxoid liposarcoma is a specific histologic subtype that deserves a special mention in the staging evaluation. This malignancy has a propensity to metastasize to unusual locations including the spinal column, retroperitoneal space, other organs, and pleural space in addition to the standard pulmonary metastasis [13, 19, 25, 26]. It is therefore recommended by the National Comprehensive Cancer Network (NCCN) that staging evaluation include a CT scan of the chest/abdomen/pelvis as well as an MRI of the entire spinal column [20]. Because this staging evaluation requires significant time, radiation exposure, and cost, there has been great interest in evaluating alternative whole-body staging regimens including whole-body PET/MRI and whole-body MRI alone [1, 11, 13, 23].

Staging of a Bone Sarcoma

While there is significant overlap in the staging of bone and soft tissue sarcomas, there are some unique considerations when considering bone staging. Like soft tissue sarcoma patients, all bone sarcoma patients require staging with chest CT given that the dominant pattern of metastatic spread is to the lungs. In addition to plain radiographs, local staging is ideally completed with an MRI of the entire bone harboring the sarcoma with and without gadolinium enhancement. MRI is helpful for understanding not only the extent of the soft tissue mass but also the marrow involvement, which is defined in detail by the MRI (Fig. 9.4a, b). For surgical planning, it is critical to have a coronal T1 sequence reconstruction, which will identify not only the proximal extent of the marrow involvement of the primary tumor but also any skip metastases. Axial T1 sequence reconstructions are also critical to define the anatomy at the site of the primary tumor, most notably neurovascular structures.

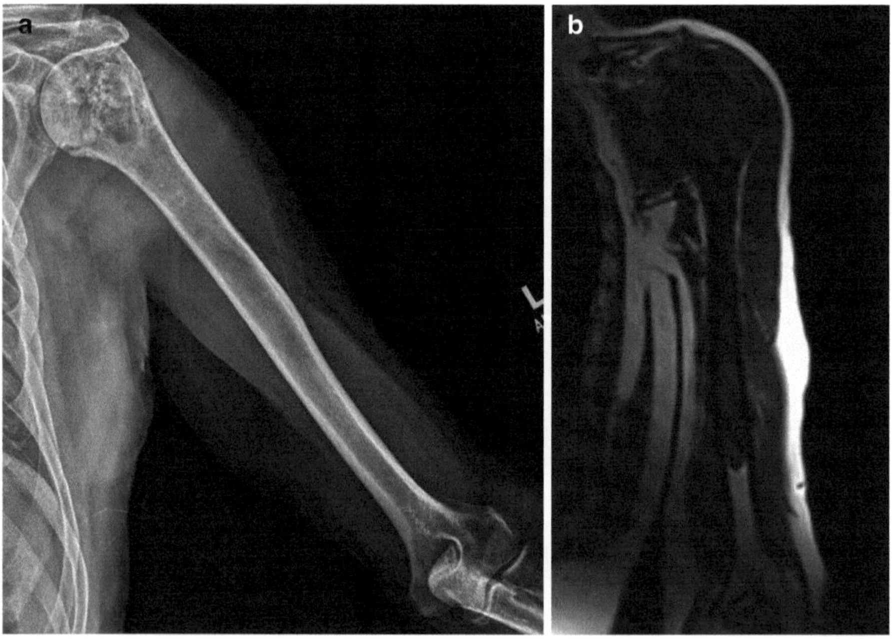

Fig. 9.4 (**a**) A plain AP radiograph of the left humerus in a patient with chondrosarcoma. There is clear evidence of a chondroid matrix in the left humeral head. (**b**) T1 coronal MRI of the left humerus in the same patient demonstrating the large extent of bone marrow replacement. This extensive replacement is not evident on the plain radiograph

In addition to a CT scan of the chest and an MRI of the entire bone, a bone sarcoma is also staged with a nuclear medicine whole-body bone scan to aid in the identification of skip metastases and any other sites of bone metastases.

An Ewing sarcoma is considered a unique bone sarcoma in terms of staging protocol given its potential for bone marrow metastases. Bone scans have been shown to have a high false-negative rate in detecting marrow metastases in Ewing sarcoma, with PET/CT being a superior test in place of a bone scan. While it is still considered standard to complete the staging of an Ewing sarcoma with a bone marrow biopsy, PET/CT has been utilized routinely in many centers to supplant bone marrow biopsy in the setting of a normal scan [7, 16]. In the setting of abnormal imaging findings, a bilateral posterior iliac crest bone marrow biopsy and aspirate is recommended [12].

Staging of a Carcinoma

A patient with a carcinoma arising from a solid organ will typically come to the attention of an orthopedic surgeon when there is a lytic lesion in the bone (Fig. 9.5) or a completed pathologic fracture, either of which may cause pain and disability. If the patient has no known cancer diagnosis, or a known diagnosis but no previous

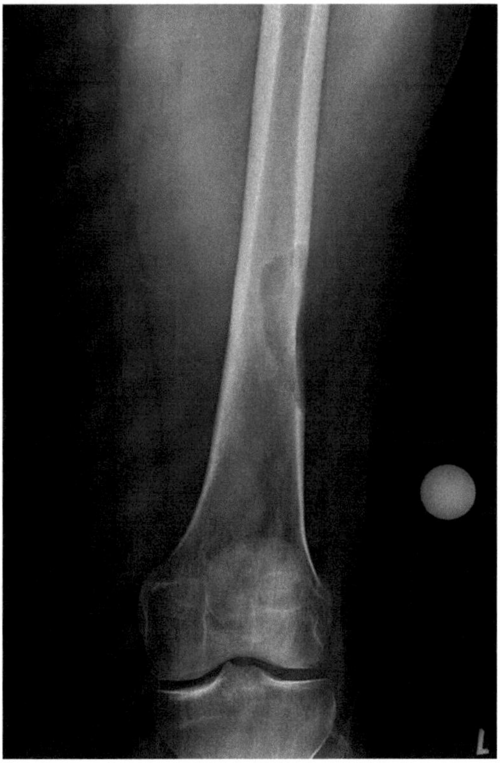

Fig. 9.5 An AP radiograph of the left distal femur demonstrating an osteolytic lesion with an impending pathologic fracture

histologic or radiographically confirmed metastatic bone disease, a complete staging evaluation is required. This is particularly important in the setting of a patient who has no diagnosis, a previous low-grade cancer diagnosis, or a long disease-free interval of any diagnosis and was presumed in remission. The most common sites of cancer to metastasize to bone are breast, prostate, lung, renal, thyroid, and colon [6]. However, it is important to understand that any solid tumor is capable of bone metastasis. There is a highly systematic process toward achieving the underlying diagnosis. In addition to a thorough history and physical examination, the diagnostic process involves: a CT scan of the chest/abdomen/pelvis with oral and IV contrast, a nuclear medicine whole-body bone scan, and a laboratory evaluation that includes a complete blood count with differential, complete metabolic panel, serum and urine protein electrophoresis (to rule out multiple myeloma), and prostate-specific antigen (PSA) in men. The diagnosis is then confirmed with a biopsy of the most accessible lesion. Using this approach, an accurate diagnosis is achieved approximately 85% of the time [24]. Once a diagnosis of previous bone metastases is established, either by biopsy or, in some instances, radiographic evidence of widespread disease, further biopsy confirmation of new lesions is not required unless the clinical presentation is not characteristic of the known metastasis.

In the event that a diagnosis is not achieved with the above process, the most common scenario is that it involves a lung cancer that is not readily visible on chest CT, or from an area not well-diagnosed with these modalities. Additional considerations are malignancies of the gastrointestinal (GI) tract, which warrant evaluation with upper and lower GI endoscopy. Patients with a smoking or smokeless tobacco history should also be considered strongly for a thorough head and neck exam, possibly including panendoscopy [28]. While a PET/CT is often considered in these clinical scenarios, it is often poor in identifying a malignancy that was not detected by the abovementioned algorithm [18]. Additionally, MRI of the abdomen has gained traction as a diagnostic tool in cases in which the primary tumor is not easily identifiable or the CT scans are equivocal [29, 31].

In specific histologies, such as lung cancer, or in patients with carcinomas with known nodal metastases, there has been a trend to use PET/CT for surveillance purposes as well as for assessment of systemic treatment response [8, 22, 27]. With the advent of new capabilities in combining PET/CT and specific antigen markers such as prostate-specific antigen (PSA), there is also new evidence demonstrating higher performance, specificity, and sensitivity when following patients with metastatic and nonmetastatic prostate cancer with this technology (PET/CT PSMA) and its systemic treatment response [3, 14]. This tool is of limited use in anaplastic prostate cancer where the expression of PSA is nonexistent. With development of new markers, this technology may be expanded to other carcinomas.

Lastly, special attention and suspicion is needed to diagnose unconventional sites of metastasis. In lung carcinoma, brain metastasis is more common than in other carcinomas despite their intrinsic scarcity. Some advocate for central nervous screening with either a brain MRI or a head CT [5]. Such evaluation should certainly be pursued if patients present symptoms such as headaches, blurred vision, muscular weakness, tremors, or focalization.

The Role of Biopsy

In the event of a new imaging finding suspicious for metastatic disease, no matter how definitive the imaging may appear, it is worthwhile to pursue tissue confirmation. This tissue confirmation can direct medical therapy and has the logistical benefits of providing diagnostic assurance for participation in clinical trials and for payers who may require it to approve certain treatment modalities. In general, a biopsy is not required for patients with prior confirmed metastatic disease unless it has an unusual presentation (i.e., a new/unusual site, atypical appearance).

References

1. Ahlawat S, Debs P, Amini B, Lecouvet FE, Omoumi P, Wessell DE. Clinical applications and controversies of whole-body MRI: AJR expert panel narrative review. AJR Am J Roentgenol. 2023;220:463–75.
2. Annovazzi A, Rea S, Zoccali C, et al. Diagnostic and clinical impact of 18F-FDG PET/CT in staging and restaging soft-tissue sarcomas of the extremities and trunk: mono-institutional retrospective study of a sarcoma referral center. J Clin Med. 2020;9:2549.
3. Bagguley D, Ong S, Buteau JP, et al. Role of PSMA PET/CT imaging in the diagnosis, staging and restaging of prostate cancer. Future Oncol Lond Engl. 2021;17:2225–41.
4. Burkhard-Meier A, Jurinovic V, Berclaz LM, et al. Differentiation of benign and metastatic lymph nodes in soft tissue sarcoma. Clin Exp Metastasis. 2024;41:131–41.
5. Diaz ME, Debowski M, Hukins C, Fielding D, Fong KM, Bettington CS. Non-small cell lung cancer brain metastasis screening in the era of positron emission tomography-CT staging: current practice and outcomes. J Med Imaging Radiat Oncol. 2018;62:383–8.
6. D'Oronzo S, Coleman R, Brown J, Silvestris F. Metastatic bone disease: pathogenesis and therapeutic options. J Bone Oncol. 2018;15:004.
7. Du Y, Zhao Z, Yang C. Is bone marrow biopsy and aspiration still mandatory when 18F-FDG PET/CT is available for the initial assessment of bone marrow metastasis in pediatric Ewing sarcoma? Front Oncol. 2024;14:1372481.
8. Evangelista L, Sepulcri M, Pasello G. PET/CT and the response to immunotherapy in lung cancer. Curr Radiopharm. 2020;13:177–84.
9. Federico SM, Spunt SL, Krasin MJ, et al. Comparison of PET-CT and conventional imaging in staging pediatric rhabdomyosarcoma. Pediatr Blood Cancer. 2013;60:1128–34.
10. Ferguson PC, Deheshi BM, Chung P, et al. Soft tissue sarcoma presenting with metastatic disease: outcome with primary surgical resection. Cancer. 2011;117:372–9.
11. Gorelik N, Reddy SMV, Turcotte RE, et al. Early detection of metastases using whole-body MRI for initial staging and routine follow-up of myxoid liposarcoma. Skeletal Radiol. 2018;47:369–79.
12. Gupta A, Riedel RF, Shah C, et al. Consensus recommendations in the management of Ewing sarcoma from the National Ewing Sarcoma Tumor Board. Cancer. 2023;129:3363–71.
13. Hanna SA, Qureshi YA, Bayliss L, et al. Late widespread skeletal metastases from myxoid liposarcoma detected by MRI only. World J Surg Oncol. 2008;6:62.
14. Henríquez I, Malave B, Campos FL, et al. PSMA PET/CT SUVmax as a prognostic biomarker in patients with metachronous metastatic hormone-sensitive prostate cancer (mHSPC). Clin Transl Oncol. 2025;27:706–15.
15. Jacobs AJ, Morris CD, Levin AS. Synovial sarcoma is not associated with a higher risk of lymph node metastasis compared with other soft tissue sarcomas. Clin Orthop. 2018;476:589–98.

16. Kasalak Ö, Glaudemans AWJM, Overbosch J, Jutte PC, Kwee TC. Can FDG-PET/CT replace blind bone marrow biopsy of the posterior iliac crest in Ewing sarcoma? Skeletal Radiol. 2018;47:363–7.
17. King DM, Hackbarth DA, Kilian CM, Carrera GF. Soft-tissue sarcoma metastases identified on abdomen and pelvis CT imaging. Clin Orthop. 2009;467:2838–44.
18. Lawrenz JM, Gordon J, George J, et al. Does PET/CT aid in detecting primary carcinoma in patients with skeletal metastases of unknown primary? Clin Orthop. 2020;478:2451–7.
19. Lin S, Gan Z, Han K, Yao Y, Min D. Metastasis of myxoid liposarcoma to fat-bearing areas: a case report of unusual metastatic sites and a hypothesis. Oncol Lett. 2015;10:2543–6.
20. von Mehren M, Kane JM, Agulnik M, et al. Soft tissue sarcoma, version 2.2022, NCCN clinical practice guidelines in oncology. J Natl Compr Cancer Netw. 2022;20:815–33.
21. Metser U, Kulanthaivelu R, Salawu A, et al. [18F]FDG PET/CT in the initial staging and restaging of soft-tissue or bone sarcoma in patients with negative or equivocal findings for metastases or limited recurrence on conventional work-up: results of a prospective multicenter registry. J Nucl Med. 2023;64:1371–7.
22. Owens C, Hindocha S, Lee R, Millard T, Sharma B. The lung cancers: staging and response, CT, 18F-FDG PET/CT, MRI, DWI: review and new perspectives. Br J Radiol. 2023;96:20220339.
23. Paruthikunnan S, Gorelik N, Turcotte RE, Jung S, Alcindor T, Powell TI. Clinical impact of whole-body MRI in staging and surveillance of patients with myxoid liposarcoma: a 14-year single-centre retrospective study. Eur Radiol. 2024;34:6688–700.
24. Rougraff BT, Kneisl JS, Simon MA. Skeletal metastases of unknown origin. A prospective study of a diagnostic strategy. J Bone Joint Surg Am. 1993;75:1276–81.
25. Schwab JH, Boland P, Guo T, et al. Skeletal metastases in myxoid liposarcoma: an unusual pattern of distant spread. Ann Surg Oncol. 2007;14:1507–14.
26. Schwab JH, Boland PJ, Antonescu C, Bilsky MH, Healey JH. Spinal metastases from myxoid liposarcoma warrant screening with magnetic resonance imaging. Cancer. 2007;110:1815–22.
27. Sheikhbahaei S, Mena E, Yanamadala A, et al. The value of FDG PET/CT in treatment response assessment, follow-up, and surveillance of lung cancer. AJR Am J Roentgenol. 2017;208:420–33.
28. Sokoya M, Chowdhury F, Kadakia S, Ducic Y. Combination of panendoscopy and positron emission tomography/computed tomography increases detection of unknown primary head and neck carcinoma. Laryngoscope. 2018;128:2573–5.
29. van der Strate I, Kazemzadeh F, Nagtegaal ID, et al. International consensus on the initial diagnostic workup of cancer of unknown primary. Crit Rev Oncol Hematol. 2023;181:103868.
30. Vaarwerk B, Breunis WB, Haveman LM, et al. Fluorine-18-fluorodeoxyglucose (FDG) positron emission tomography (PET) computed tomography (CT) for the detection of bone, lung, and lymph node metastases in rhabdomyosarcoma. Cochrane Database Syst Rev. 2021;11:CD012325.
31. JRJ W, Lahaye MJ, Goedegebuure EP, et al. Added value of body MRI to detect primary abdominal malignancies in the diagnostic work-up of patients with adenocarcinoma of unknown primary. Eur Radiol. 2025;35(5):2702–11.

I Am Not Crazy: Why Do I Need to See a Mental Health Specialist?

10

Elana Monchar, J Wittig, and J Valverde

Introduction

There is increasing awareness that cancer is not just a physical disease but one that profoundly impacts a person's sense of emotional well-being and their mental health. Understanding this is imperative to provide comprehensive care to cancer patients. Psycho-oncology is an interdisciplinary field focusing on the social, emotional, and behavioral dimensions of cancer patients. A cancer diagnosis is a life-altering event, one that can trigger a wide range of emotional responses, varying throughout the cancer journey. Oftentimes, these emotional responses are appropriate and manageable by patients without intervention. At times, the emotional distress can lead to diagnosable mental illness requiring further assessment and intervention. It is of utmost importance to identify these mood disorders, due to the strong association between depression and increased morbidity and mortality in patients with cancer [1].

Major depressive disorder (MDD) in the United States is recognized as one of the most common mental health disorders [1, 2], affecting 1 in 10 adult Americans and almost 1 in 5 adolescents and young adults [2]. From the years 2015 to 2019, the prevalence of depression significantly increased, even prior to the COVID pandemic[2]. Although depression is a public health issue, oncology patients are at a higher risk of developing mood disorders secondary to the psychosocial stressors of their diagnoses when compared to the general population [3]. Studies have demonstrated that MDD is present in approximately 22–29% of patients with newly diagnosed cancer [3, 4].

E. Monchar (✉)
Department of Psychiatry, Morristown Medical Center, Morristown, NJ, USA
e-mail: Elana.Monchar@atlantichealth.org

J. Wittig · J. Valverde
Morristown, NJ, USA

J. Pretell-Mazzini, S. A. Lozano Calderón (eds.), *Multidisciplinary Approach of Musculoskeletal Metastasis*, https://doi.org/10.1007/978-3-032-09498-8_10

It is also important to recognize the prevalence and impact of anxiety in cancer patients. While anxiety can exist independently of other mental illnesses, it is often comorbid with major depression in this patient population. With multiple recent studies showing a strong association between depression/anxiety and increased cancer mortality, decreased treatment compliance and lower end-of-life quality, the impact of these disorders cannot be underscored [5, 6]. Studies have shown that anxiety is present in about 17% of this patient population. Unfortunately, due to an overlap between cancer and anxiety symptoms, including fatigue, restlessness, and impaired sleep, it is often underrecognized and underdiagnosed [5–7].

Screening

There are various times at which an oncology patient is most vulnerable, i.e., the time of diagnosis, the days leading up to the onset of the proposed treatment, the days leading up to restaging imaging, and at the completion of treatment. It is at these pivotal stages where a provider must be cognizant of assessing a patient for emotional distress so that they can provide emotional support and/or refer patients for appropriate services and support. Doing so can have significant positive effects on a patient, ranging from decreased isolation to strengthening resilience and coping skills, all of which can favorably impact the patient's outcome. Early screening and identification of mood disorders, with referral to appropriate resources, is a key element in improving the quality of life of cancer patients.

Multiple screening tools are available for assessing a patient's psychological state. The Patient Health Questionnaire (PHQ)-9 is a commonly used tool that can assess for the various symptoms of depression [8]. Similarly, the Generalized Anxiety Disorder (GAD-7) Anxiety Scale is used to assess for anxiety symptoms [9]. While these are commonly used in various medical and mental health settings, the Emotional Distress Thermometer (EDT) (Fig. 10.1), shown below, was developed as an easy-to-use visual scale to quickly assess a patient's level of emotional distress [10]. It can be used prior to a patient's visit with their healthcare specialist. This serves to not only promote early screening of a patient but also help identify when further psychological evaluation, including through the more comprehensive PHQ-9 and GAD-7 screening tools, and management may be warranted.

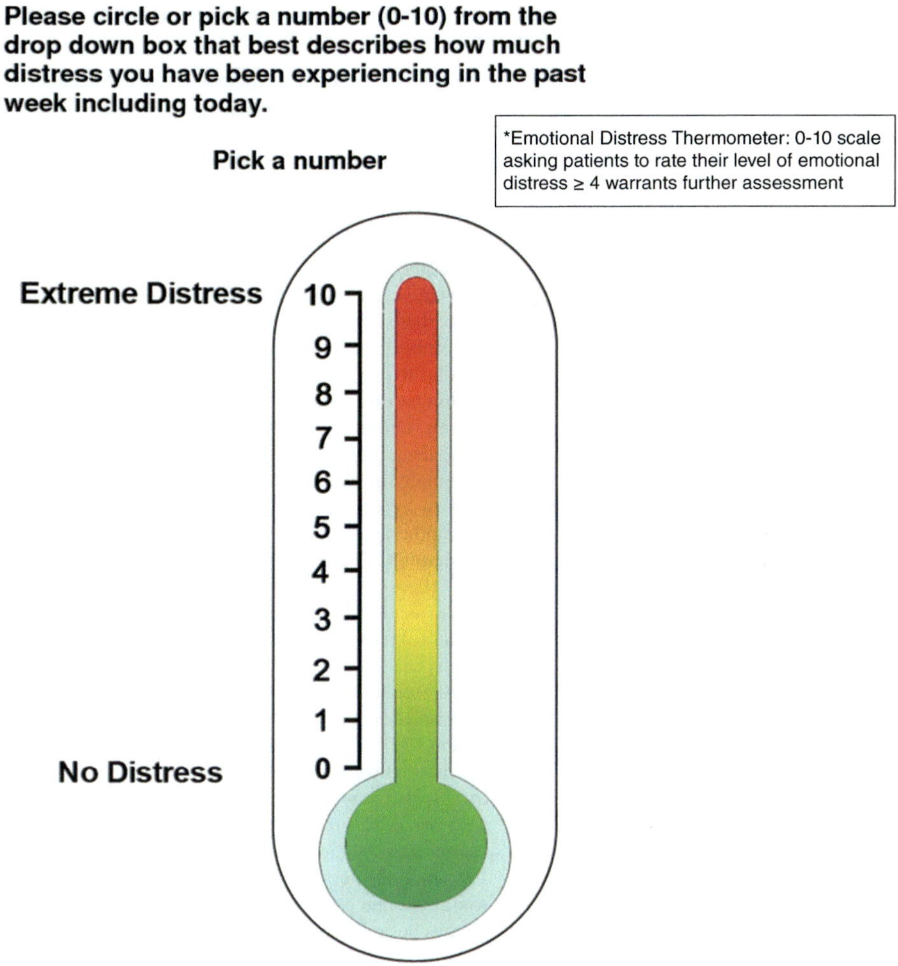

Fig. 10.1 Emotional distress thermometer. (*Ownby [10])

Pharmacologic Interventions

One unique challenge with this specific patient population is determining which antidepressant is the most appropriate given the many potential interactions and side effects when used in combination with the specific chemotherapy regimen a patient may receive [11]. In general, selective serotonin reuptake inhibitors (SSRIs) are considered first-line pharmacotherapy for MDD. However, many of these medications can interact with cytochrome P450 and, as such, the provider should be mindful of the possible interactions with chemotherapy regimens a patient is receiving [12]. The following table (Table 10.1) describes the different pharmacologic agents and the interactions with different chemotherapeutic agents [11–14]:

Table 10.1 Pharmacologic agents used in the treatment of mood disorders and their potential interactions with chemotherapeutic agents

Class	Medication	Potential concerns with chemotherapeutic agents
SSRIs	Fluoxetine	May inhibit CYP2D6, increasing the toxicity of chemotherapeutic agents metabolized by this enzyme (e.g., tamoxifen). Possible QT prolongation risk
	Sertraline	Less inhibition of CYP2D6 than fluoxetine but still possible. QT prolongation is a risk with certain chemotherapies
	Citalopram	Can cause QT prolongation, which may be exacerbated by some chemotherapies (e.g., doxorubicin, arsenic trioxide)
	Escitalopram	Similar to citalopram but with a lower risk of QT prolongation
	Paroxetine	Strong inhibitor of CYP2D6, increasing the risk of drug–drug interactions (e.g., tamoxifen). QT prolongation risk
SNRIs (Serotonin–norepinephrine reuptake inhibitors)	Venlafaxine	Risk of hypertension, which may worsen with chemotherapies that increase blood pressure. Moderate CYP2D6 inhibition
	Duloxetine	Can affect liver enzymes, increasing the risk of hepatotoxicity when combined with chemotherapy agents
	Desvenlafaxine	Similar concerns to venlafaxine, including potential hypertension and interaction with CYP enzymes
Atypical antidepressants	Bupropion	Potent CYP2D6 inhibitor, increasing the potential for interaction with chemotherapeutics metabolized by this pathway
	Mirtazapine	Minimal CYP enzyme interaction but may exacerbate sedation from chemotherapeutics
	Buspirone	Low risk of interactions but possible serotonergic toxicity when combined with serotonergic chemotherapeutics
Mood stabilizers	Lithium	Risk of nephrotoxicity and dehydration, which can be worsened by certain chemotherapies (e.g., cisplatin)
	Lamotrigine	Rare risk of Stevens–Johnson syndrome may be increased when used with certain chemotherapeutics (e.g., methotrexate)
Antipsychotics	Risperidone	Risk of QT prolongation, which can be worsened with chemotherapeutic agents. Risk of hyperprolactinemia
	Olanzapine	Risk of metabolic syndrome (weight gain, hyperlipidemia), which may be worsened by chemotherapy agents
	Quetiapine	Can cause sedation and metabolic side effects; may interact with chemotherapeutics that have similar side effects

(continued)

Table 10.1 (continued)

Class	Medication	Potential concerns with chemotherapeutic agents
Benzodiazepines	Lorazepam	Increased sedation risk when combined with chemotherapy that has sedative effects. No significant CYP interaction
	Clonazepam	Similar to lorazepam but may also carry a risk of increased sedation or respiratory depression with chemotherapeutics

MDD major depressive disorder, *PHQ-9* Patient Health Questionnaire-9, *GAD-7* generalized anxiety disorder, *EDT* emotional distress thermometer, *SSRIs* selective serotonin reuptake inhibitors, *SNRIs* serotonin–norepinephrine reuptake inhibitors, *CYP* cytochrome P, *CBT* cognitive behavioral therapy, *NICE* National Institute for Health and Clinical Excellence, *APOS* American Psychosocial Oncology Society

Remarkably, some SSRIs, such as sertraline and fluoxetine, show antitumor effects with studies demonstrating their role in increasing mitochondrial matrix calcium levels, which, in turn, resulted in apoptosis of cancerous cells visualized in vivo and in vitro in different types of cancers such as oral, prostate, and bladder cancers as well as glioblastoma [15–18]. This field is still evolving, and there are many clinical trials currently underway to best define the role of antidepressants and cancer cells [15–18].

Cognitive Behavioral Therapy (CBT)

Aside from pharmacologic management, there are many other therapies that are tailored toward improving a patient's emotional distress. Cognitive behavioral therapy (CBT) is an effective, evidence-based treatment modality based on the idea that thoughts, feelings, and behaviors are all interrelated and that a modification in any one of these dictates a change in the others [19]. CBT helps individuals identify and change their negative thought patterns and behaviors that contribute to emotional distress. CBT aims to help patients recognize irrational thoughts so that they can challenge and change their maladaptive automatic thoughts while aiming to help them adjust their behaviors and engage in more positive activities [19, 20]. With oncology patients, one goal of CBT can be to enhance a person's sense of control and self-efficacy in the face of illness and navigating through the unknown [20].

Mindfulness

There has also been increasing interest in the field of mindfulness-based practices in oncology patients. Mindfulness is a practice derived from Buddhist traditions and entails nonjudgmental awareness of the present, with attention focused on the body and breathing. This, in turn, allows the mind to rest from intrusive inner thoughts such as worry and contemplation [21, 22]. One of the most unique and well-known protocols was established in the 1980s by Jon Kabat-Zinn, who developed the mindfulness-based stress reduction protocol [21, 23]. This protocol consists of weekly 2-hour group sessions that teach nonreligious meditation including yoga, breathing exercises, sitting meditations, and body scans. Participants are also given tasks to complete at home, which consist of 45-minute daily meditations 6 days a week as well as audio recordings and awareness exercises. After the sixth week of the protocol, the participants attend an 8-hour long retreat. A study conducted to determine patient compliance to completing this protocol, as well as the effects of mindfulness practices, found a 76% compliance rate with women having higher compliance and, during meditation and yoga, patient's reported feelings of relaxation, acceptance, and well-being, demonstrating that this protocol may be useful for many patients and can be applied to their daily lives.

Exercise

Exercise is another fundamental cost-effective adjunct to pharmacotherapy and CBT. Studies have shown that exercise not only provides mental health benefits but also improves physical and cognitive outcomes [24, 25]. In the United Kingdom, the National Institute for Health and Clinical Excellence (NICE) established an exercise protocol, for the general population, which consists of three exercise sessions per week of moderate duration, anywhere from 45 min to 1 h, over a 10–14-week span [25]. The authors also stated that the easiest and safest modality to achieve these goals is by walking, targeting at least 10,000 steps per day or 6000 steps per day in addition to other daily activities [25]. There are many ongoing trials that are seeking to standardize an exercise program that can be applied to a broad patient population. Although the authors did not specifically address this, it would be most beneficial for a patient to have a customized exercise regimen of their choosing, as that is most likely to result in enhanced compliance. As physicians, we need to promote the various beneficial effects of exercise and physical activity to all our patients, oncology patients notwithstanding, who should be encouraged to exercise as tolerated.

Conclusions

Involving a multidisciplinary team equipped with various tools affords our oncology patients the very best possible outcomes, even in the face of difficult and, at times traumatic, diagnoses. We must remain sensitive to the hardships oncology patients experience throughout their treatment journey. We must always remember that we are the gate keepers of health and that we should be treating patients and not only their diseases.

As a provider, regardless of our specialty, we must have the ability to recognize potential mental health issues and be armed with information and resources necessary for a patient. Simply acknowledging the emotional impact of cancer on well-being and offering support and appropriate referrals can be of tremendous value. The following resources are widely available and can be offered to a patient (Fig. 10.2).

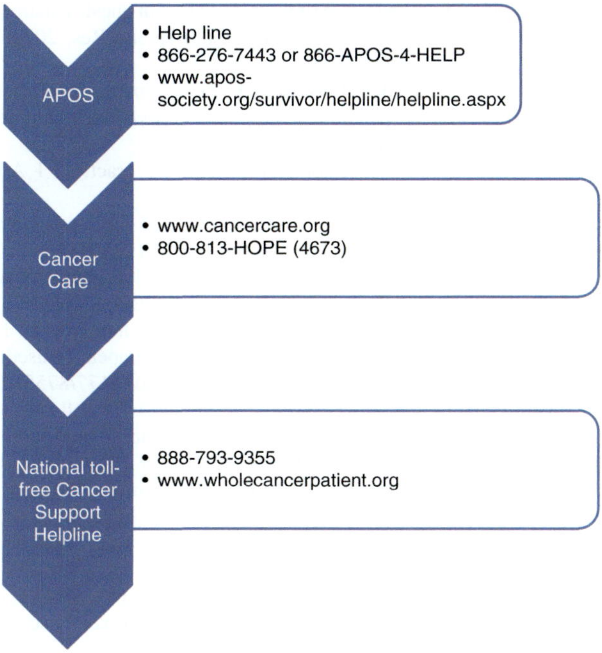

Fig. 10.2 Different resources that providers can use to refer cancer patients for support

References

1. Polfer EM, Alici Y, Baser RE, Healey JH, Bartelstein MK. What proportion of patients with musculoskeletal sarcomas demonstrate symptoms of depression or anxiety? Clin Orthop Relat Res. 2022;480(11):2148–60. https://doi.org/10.1097/CORR.0000000000002295. Epub 2022 Jun 30. PMID: 35901433; PMCID: PMC9556120.
2. Goodwin RD, Dierker LC, Wu M, Galea S, Hoven CW, Weinberger AH. Trends in U.S. depression prevalence from 2015 to 2020: the widening treatment gap. Am J Prev Med. 2022;63(5):726–33. https://doi.org/10.1016/j.amepre.2022.05.014. Epub 2022 Sep 19.
3. Naser AY, Hameed AN, Mustafa N, Alwafi H, Dahmash EZ, Alyami HS, Khalil H. Depression and anxiety in patients with cancer: a cross-sectional study. Front Psychol. 2021;12:585534. https://doi.org/10.3389/fpsyg.2021.585534.
4. Brothers BM, Yang HC, Strunk DR, Andersen BL. Cancer patients with major depressive disorder: testing a biobehavioral/cognitive behavior intervention. J Consult Clin Psychol. 2011;79(2):253–60. https://doi.org/10.1037/a0022566.
5. Cooper K, Campbell F, Harnan S, Sutton A. Association between stress, depression or anxiety and cancer: rapid review of reviews. Compr Psychoneuroendocrinol. 2023;16:100215. https://doi.org/10.1016/j.cpnec.2023.100215.
6. Papadopoulou A, Govina O, Tsatsou I, Mantzorou M, Mantoudi A, Tsiou C, Adamakidou T. Quality of life, distress, anxiety and depression of ambulatory cancer patients receiving chemotherapy. Med Pharm Rep. 2022;95(4):418–29. https://doi.org/10.15386/mpr-2458. Epub 2022 Oct 27.
7. Alwhaibi M, AlRuthia Y, Sales I. The impact of depression and anxiety on adult cancer patients' health-related quality of life. J Clin Med. 2023;12(6):2196. https://doi.org/10.3390/jcm12062196.

8. Ford J, Thomas F, Byng R, McCabe R. Use of the patient health questionnaire (PHQ-9) in practice: interactions between patients and physicians. Qual Health Res. 2020;30(13):2146–59. https://doi.org/10.1177/1049732320924625.
9. Johnson SU, Ulvenes PG, Øktedalen T, Hoffart A. Psychometric properties of the general anxiety disorder 7-item (GAD-7) scale in a heterogeneous psychiatric sample. Front Psychol. 2019;10:1713. https://doi.org/10.3389/fpsyg.2019.01713.
10. Ownby KK. Use of the distress thermometer in clinical practice. J Adv Pract Oncol. 2019;10(2):175–9. Epub 2019 Mar 1.
11. Grassi L, Nanni MG, Rodin G, Li M, Caruso R. The use of antidepressants in oncology: a review and practical tips for oncologists. Ann Oncol. 2018;29(1):101–11.
12. Rodriguez-Antona C, Ingelman-Sundberg M. Cytochrome P450 pharmacogenetics and cancer. Oncogene. 2006;25(11):1679–91. https://doi.org/10.1038/sj.onc.
13. Turossi-Amorim ED, Camargo B, do Nascimento DZ, Schuelter-Trevisol F. Potential drug interactions between Psychotropics and intravenous chemotherapeutics used by patients with cancer. J Pharm Technol. 2022;38(3):159–68. https://doi.org/10.1177/87551225211073942.
14. Williams JW Jr, Mulrow CD, Chiquette E, Noël PH, Aguilar C, Cornell J. A systematic review of newer pharmacotherapies for depression in adults: evidence report summary. Ann Intern Med. 2000;132(9):743–56. https://doi.org/10.7326/0003-4819-132-9-200005020-00011.
15. Liu KH, Yang ST, Lin YK, Lin JW, Lee YH, Wang JY, Hu CJ, Lin EY, Chen SM, Then CK, Shen SC. Fluoxetine, an antidepressant, suppresses glioblastoma by evoking AMPAR-mediated calcium-dependent apoptosis. Oncotarget. 2015;6(7):5088–101. https://doi.org/10.18632/oncotarget.3243.
16. Chien JM, Chou CT, Pan CC, Kuo CC, Tsai JY, Liao WC, Kuo DH, Shieh P, Ho CM, Chu ST, Su HH, Chi CC, Jan CR. The mechanism of sertraline-induced $[Ca2+]_i$ rise in human OC2 oral cancer cells. Hum Exp Toxicol. 2011;30:1635–43.
17. Huang JK, Chang HT, Chou CT, Shu SS, Kuo CC, Tsai JY, Liao WC, Wang JL, Lin KL, Lu YC, Chen IS, Liu SI, Ho CM, Jan CR. The mechanism of sertraline-induced [Ca(2+)](i) rise in human PC3 prostate cancer cells. Basic Clin Pharmacol Toxicol. 2011;109:103–10.
18. Tang KY, Cheng JS, Lee KC, Chou KJ, Huang JK, Chen WC, Jan CR. Fluoxetine-induced Ca2+ signals in Madin-Darby canine kidney cells. Naunyn Schmiedeberg's Arch Pharmacol. 2001;363:16–20.
19. Curtiss JE, Levine DS, Ander I, Baker AW. Cognitive-behavioral treatments for anxiety and stress-related disorders. Focus (Am Psychiatr Publ). 2021;19(2):184–9. https://doi.org/10.1176/appi.focus.20200045.
20. Kaczkurkin AN, Foa EB. Cognitive-behavioral therapy for anxiety disorders: an update on the empirical evidence. Dialogues Clin Neurosci. 2015;17(3):337–46. https://doi.org/10.31887/DCNS.2015.17.3/akaczkurkin.
21. Kabat-Zinn J. Full catastrophe living: using the wisdom of your body and mind to face stress, pain, and illness. Bantam Books; 2013.
22. Felsted KF. Mindfulness, stress, and aging. Clin Geriatr Med. 2020;36(4):685–96. https://doi.org/10.1016/j.cger.2020.06.010.
23. Kabat-Zinn J, Chapman-Waldrop A. Compliance with an outpatient stress reduction program: rates and predictors of program completion. J Behav Med. 1988;11(4):333–52.
24. Chen C, Beaunoyer E, Guitton MJ, Wang J. Physical activity as a clinical tool against depression: opportunities and challenges. J Integr Neurosci. 2022;21(5):132. https://doi.org/10.31083/j.jin2105132.
25. Noetel M, Sanders T, Gallardo-Gómez D, Taylor P, Del Pozo Cruz B, van den Hoek D, Smith JJ, Mahoney J, Spathis J, Moresi M, Pagano R, Pagano L, Vasconcellos R, Arnott H, Varley B, Parker P, Biddle S, Lonsdale C. Effect of exercise for depression: systematic review and network meta-analysis of randomised controlled trials. BMJ. 2024;384:e075847. https://doi.org/10.1136/bmj-2023-075847.

Part III

Principles of Surgical Intervention

Overview of Surgical Indications and Outcomes

11

Varna R. Kanapuram and Jonathan A. Forsberg

Introduction

Most, if not all, metastatic tumors demonstrate organotropism, a nonrandom process in which distant metastases are distributed to specific organs. This is regulated by the molecular features of the cancer cells, the cancer subtype, the host immune microenvironment, and cross talk interactions with local cells [1]. Tumor cells can disseminate from the primary tumor and be attracted to the vascular, osteoblastic, and hematopoietic components of bones, which can eventually progress to metastatic bone disease (MBD). The most common primary cancers that metastasize to bone are lung and bronchus, prostate, breast, and thyroid cancers, with up to 30% of patients presenting with bone metastases from unknown primary cancers [2]. However, different cancer types and subtypes exhibit different organotropisms, such as prostate cancer, often metastasizing to bone, while uveal melanoma usually spreads to the liver [3].

Bone is one of the most common organs targeted by metastasis, following the lungs and liver. Although most malignancies can metastasize to bone, they most commonly result from breast, prostate, renal, thyroid, and lung cancers [4–6]. In the United States, these primary cancers constitute around 70% of metastatic bone disease (MBD) cases, with a total estimated prevalence of 600,000 cases each year and with breast, prostate, and lung cancers predominating [4]. The most common sites of bone metastases include the spine, pelvis, femur, humerus, and tibia [4, 6, 7].

Metastatic disease can develop after a latent period following surgical resection of the primary tumor. This can be attributed to low circulating tumor cells, disseminated tumor cells, and micrometastases remaining in the body after resecting the primary tumor [8]. Disseminated tumor cells may be present and dormant within

V. R. Kanapuram · J. A. Forsberg (✉)
Orthopaedic Service, Department of Surgery, Memorial Sloan Kettering Cancer Center, New York, NY, USA
e-mail: kanapuv1@mkscc.org; forsbej@mskcc.org

J. Pretell-Mazzini, S. A. Lozano Calderón (eds.), *Multidisciplinary Approach of Musculoskeletal Metastasis*, https://doi.org/10.1007/978-3-032-09498-8_11

bones for extended periods, anywhere from several years to even decades, without causing symptoms [9]. The duration of this metastatic latency period is highly dependent on the type of cancer: lung cancer usually has a short latency of a few weeks, breast cancer typically has a medium or long latency, and prostate cancer has a very long latent period [8]. With breast cancer, the latent period duration is dependent on the primary tumor's volume, stage, molecular subtype, and estrogen receptor (ER) status. R-tumor subtypes are more aggressive with short medium-length or medium-length latent periods with recurrence peaking around 2 years after diagnosis. R+ tumor subtypes have more extended latent periods, with more than half of metastases occurring 5 years or longer after diagnosis and primary tumor resection [10]. Meanwhile, prostate cancer has a very long latent period where nearly 100% of patients survive 5 years post-diagnosis and 82% survive 15 years, according to a statistics review from 1975 to 2012 through the National Cancer Institute [8].

MBD is the loss of bone homeostasis and dysregulation of bone metabolism through the uncoupling of osteoblasts and osteoclasts. This process is dependent on the pathophysiology of the dysregulation, resulting is osteolysis or osteogenesis.

Diagnostic Workup

Evaluation should include a thorough history and physical examination of information about possible primary cancers that may cause bone metastases. Patients should undergo appropriate laboratory workup as well as comprehensive examinations of the breast, thyroid gland, skin, and lymph nodes for a new diagnosis with routine laboratory analysis, plain radiographs of the area of interest, MRI of the area of interest, and CT chest/abdomen/pelvis [11]. Nuclear medicine studies, such as whole-body bone scan and positron emission tomography (PET)/CT scans, are also useful in vigilant monitoring, accurate detection, and assessment of therapeutic response of bone involvement. A whole-body bone scan allows for whole-body exploration in a short amount of time with low irradiation and high specificity to osteoblastic activity. PET/CT scans provide superior sensitivity and specificity in detecting bone metastases [12, 13].

Skeletal-Related Events (SREs)

SREs are commonly seen in patients with MBD and can be dominant factors in maintaining functional mobility and quality of life [14]. In a 15-year study across 2 large health systems, the rate of skeletal complications was documented in 1819 patients with bone metastases from breast, lung, and prostate cancers [15]. In this population-based study, 22% of patients with breast or lung cancer and 10% of patients with prostate cancer had an SRE at diagnosis. Notably, the rate of SREs increased to 62%, 58%, and 51% in patients with breast, lung, or prostate cancer, respectively, indicating the need for continued skeletal monitoring during their lifetimes [15].

Treatment Indications

Treatment options for metastatic bone disease vary widely based on the patient's prognosis, ranging from nonsurgical palliative radiotherapy to surgical endoprosthesis reconstructions. Recent advances in prognostic models have improved our ability to estimate patient survival, even leveraging machine learning to help select the appropriate treatment for MBD patients. This next section details preoperative planning, prognostic models, and indications for the wide array of approaches to treat MBD.

Preoperative Planning

Survival estimates are essential in setting patient, family, and physician expectations when treating metastatic bone disease [16]. By identifying which patients will likely benefit from surgery or require more durable implants, surgeons can carefully avoid under- or overtreatment of disease [17]. Survival estimates are particularly important for patients with short life expectancies (<1 month) to avoid unsuccessful and expensive procedures at the end of life [18]. Historically, physician estimates alone have been notoriously inaccurate and overly optimistic, requiring more objective methods of estimating life expectancy [16, 19].

Background on Objective Survival Estimates

Preoperative prognostic models play a crucial role in objectively assessing the survival time of patients with metastatic bone disease. Various prognostic models, such as the Katagiri model, PATHFx model, Scandinavian Sarcoma Group (SSG) score, Janssen nomogram, OPTIModel, and SPRING 13 nomogram, have been developed. However, each model accounts for factors impacting survival in different ways, highlighting the importance of comparing the accuracy and reliability of each model's survival estimates [20].

A retrospective study was conducted for estimating survival of patients that underwent surgical fixation for femoral MBD at a single institution between 2003 and 2014. Univariate and multivariate Cox regression proportional hazard models revealed that the OPTIModel had the highest accuracy for predicting survival at 12 and 24 months after surgery, while PATHFx was most accurate at predicting survival at 3 and 6 months after surgery [20]. PTIModel and PATHFx are both validated clinical decision-support tools to estimate the likelihood of survival after surgery for patients with metastatic bone disease. The PTIModel was specifically developed for patients with symptomatic long bone metastases, leveraging Cox regression analyses and Kaplan–Meier curves to estimate survival [21]. Meanwhile, PATHFx uses modern machine learning technology that can accommodate uncertainty from incomplete information, which is common in clinical settings [16]. These models use readily available clinical, laboratory, disease-related, patient performance, and surgeon input [17, 21].

Importantly, PATHFx has been updated twice using contemporary patient information, and a third iteration is underway. Models such as this must be updated from time to time to ensure that the estimations reflect changes in patient survival resulting from new therapeutics and updates in treatment philosophies. As with any decision-support tool, its effectiveness depends on several factors, including external validation statistics and decision analysis [22].

Prognostic models can be stratified into classical or regression-driven scores and machine learning-driven scores. Between 1990 and 2016, Cox proportional hazards or multivariable logistic regressions were popular for estimating the survival of patients with spinal metastatic disease [19]. These methods were used to identify prognostic factors and predict short- or long-term survival of MBD patients. It was particularly successful in determining the association between prognostic factors and patient outcomes but less so for predicting patient survival. However, machine learning methods have grown in popularity due to their unique ability to identify complex, nonlinear relationships and interactions between different prognostic factors [19]. A study was conducted on a training set of 186 patients undergoing surgery for skeletal metastases at the Memorial Sloan Kettering Cancer Center, comparing 3- and 12-month survival predictions by various prognostic model types. Results showed that the artificial neural network (ANN) outperformed the Bayesian belief network (BBN) and logistic regression-based models, recommending the use of ANN-based models in surgical decision-making for MBD [23]. Machine learning algorithms have also shown to successfully predict clinical outcomes, often outperforming traditional regressions, in numerous clinical disciplines [24–26]. Furthermore, their ability to assess model discrimination, model calibration, and decision curve analysis are all crucial elements of understanding a model's utility [19]. Ultimately, estimating a personalized survival trajectory is useful for surgical and medical decision-making. These clinical decision-support tools are intended to guide, rather than replace, good clinical judgment and are also useful as a pedagogical tool when training residents and fellows on how to approach patients with metastatic bone disease.

Clinical Application of Objective Survival Estimates

To derive individualized prognostic information, one must be familiar with several variables. These include: age at the time of surgery, sex, indication for surgery (impending or completed pathologic fracture), number of bone metastases (solitary or multiple), surgeon's estimate of survival (postoperatively, in months), the presence or absence of visceral metastases, the presence or absence of lymph node metastases, preoperative hemoglobin concentration (g/dL, on admission to the hospital, prior to transfusion, if applicable), absolute lymphocyte count (K/μL), the patient's primary oncologic diagnosis, and their ability to perform daily tasks (as measured by the Karnofsky Performance Scale) [17, 21, 22].

By estimating the likelihood of short-, intermediate-, and long-term survival, one can determine whether surgery is likely to be beneficial as well as whether a more durable implant may be required [27]. This can help avoid over- or undertreatment of disease, where the chosen surgical intervention does not align with the life expectancy of the patient. Overtreatment results when a patient's expected survival is too short for the patient to recover sufficiently and experience the benefits of surgical intervention. Undertreatment results when a patient's expected survival is longer than the durability of the chosen operative treatment, limiting the ability for the operation to improve the patient's quality of life and functional independence until end of life. Given that surgical intervention of MBD aims to improve pain, quality of life, and physical functioning, it is important for the patient to be able to experience these benefits after postoperative rehabilitation and for these benefits to persist until the end of life [28, 29].

Nonsurgical and Surgical Indications

Decision-making on the appropriate surgical intervention for metastatic bone disease must involve the patient, surgeon, and any additional medical teams contributing to their oncologic care. Particularly, a team-based approach and interdisciplinary management of MBD plays a crucial role in designing an optimal treatment plan with appropriate surgical intervention. The medical oncology team can help determine the timing of surgery with respect to the administration of systemic therapies. The radiation oncology team can offer preoperative stereotactic body radiation therapy (SBRT) or postoperative palliative radiotherapy. Additionally, palliative and rehabilitation teams can help with management of postoperative recovery or nonsurgical interventions. Overall, this team-based approach has shown to be helpful in designing effective treatment plans in oncology care. A systematic review of the impact of multidisciplinary teams (MDTs) in oncology treatment showed that patients who were discussed were more likely to receive complete preoperative staging and neoadjuvant and adjuvant treatments and that 4–50% of patients had changes to their diagnostic reports following MDT meetings [30, 31].

Within these MDTs, the goal of treatment must be decided as palliation or curative intent, after determining the local and systemic progression of disease. In general, the goal of treatment in patients with MBD is to restore function, relieve pain, and improve quality of life. With a high incidence of skeletal-related events in MBD patients, surgical intervention can be necessary to provide stabilization or reconstruction after events such as pathologic fractures.

There is a wide variety of approaches for surgical intervention for MBD based on each patient's estimated life expectancy. However, the goals of surgery always remain as improving the quality of life of patients by maximizing patient function immediately after surgery and providing pain relief. Control of pain is often achieved by local tumor control and restoring the patient's functional independence. The goal of palliative treatment for MBD typically includes controlling pain, preventing or treating fractures, maintaining independence and preventing tumor progression, and generally improving the quality of remaining life [32]. Orthopedic surgeons should be involved early for treatment of any impending fractures to ensure limb functionality and spine stability [33]. Additionally, any surgical fixation must be robust enough for early weight-bearing and minimize potential complications, even with a range of fixation durability according to the patient's estimated survival [32]. Finally, although surgical treatment of metastatic disease is typically palliative in nature, wide excision of some solitary lesions due to renal cell, thyroid, or melanoma can have a survival benefit and should be considered.

Nonsurgical Palliative Care

Nonsurgical palliative care options should be considered for metastatic bone lesions with extremely short survival estimates to avoid unsuccessful surgery at the end of life [16]. Most patients will be receiving or will have received cytotoxic chemotherapy, targeted therapy, immunotherapy, and radionuclide therapy, which have variable effects on metastatic lesions [34, 35]. Most patients will benefit from a comprehensive pain management strategy involving acetaminophen, nonsteroidal anti-inflammatory drugs (NSAIDS) including cyclooxygenase-2 (COX-2) inhibitors, and judicious use of opioid analgesics. Other systemic therapies such as bone-modifying agents (e.g., zoledronic acid, denosumab, and romosozumab) should be strongly considered unless there is a contraindication. Treatment with these agents preserves bone stock and creates a less hospitable environment for bone lesions [36–38].

Radiation therapy is often the most common treatment of symptomatic lesions due to MBD [38]. It is an effective palliative intervention to improve quality of life by relieving pain from bone metastases and preventing pathologic fractures or spinal cord compression. This can help reduce analgesic requirements and maintain or improve skeletal function [39]. External beam radiation therapy (EBRT) is most frequently utilized with indications including pain from bone metastases, which are difficult to control with oral analgesics, malignant spinal cord compression, and prevention of pathologic fractures [40]. Stereotactic body radiation therapy (SBRT) is a newer method that delivers higher doses of radiation to the target area while sparing surrounding healthy tissues. It has proven to be effective in pain management for bone metastases, as shown by a meta-analysis of 18 studies and 1685 MBD patients [41]. In this study, SBRT and EBRT showed a similar response in pain management of bone metastases. However, MBD patients treated with SBRT had a higher rate of complete pain response, defined as a pain score of zero without increased opioid use, compared to patients treated with EBRT [41]. Often, treatment regimens may require use of chemotherapy and radiation therapy for successful management of oncologic disease. Concurrent use of chemotherapeutic agents with radiotherapy must be avoided in some cases due to increased risk of toxicities such as radiation-induced heart disease or peripheral neuropathy [42].

Other less invasive approaches can also be used to treat symptomatic bone metastases that are nonstructural or are not associated with impending or pathologic fractures. These include thermal ablation, embolization, and percutaneous cementation [43]. Particularly, patients with short life expectancies greatly benefit from these less invasive procedures that can provide significant pain palliation, prevent skeletal-related events, and provide durable local tumor control. Thermal ablation involves using extreme heat (radio-frequency ablation) or cold (cryoablation), resulting in coagulative necrosis and cell death in the targeted tissue [44]. Percutaneous cementation involves injecting polymethyl methacrylate (PMMA) cement to consolidate osteolytic lesions [45]. Selective embolization is another procedure often performed for bone metastases due to their commonly hypervascular nature [33, 46].

Surgical Indications

After a skeletal-related event (SRE), surgical intervention is often necessary to improve or preserve functional independence and quality of life. Surgeons have a variety of approaches to treat impending or displaced pathologic fractures, such as intramedullary nailing, endoprosthesis reconstruction, and open reduction–internal fixation [47]. It is critical to choose an implant and surgery that most aligns with the patient's estimated life expectancy. For instance, intramedullary femoral implants used to treat impending or completed pathologic fractures have a median survival of approximately 1 year [47]. As such, less durable implants such as osteosynthesis or intramedullary implants have a greater risk of failure and should be considered for patients with shorter estimated survival [48].

Minimally Invasive, Less Durable Surgery

Patients with relatively short life expectancies of 1–6 months may be candidates for less invasive surgical interventions. These include intramedullary implants used to stabilize impending pathologic fractures or treat completed pathologic fractures (Fig. 11.1a and b) [16]. Additional techniques include balloon-assisted osteoplasty with expandable intramedullary devices that are filled with a monomer and hardened after exposure to blue light, stabilizing the bone segment for impending pathologic fractures, and balloon-assisted osteoplasty with a plate for completed fractures [49, 50]. Furthermore, periacetabular bone metastases may be treated with percutaneous cementoplasty with or without internal fixation using screws or pins for all periacetabular bone [51].

More Invasive, More Durable Surgery

Patients who are likely to survive an excess of 6 months post-surgery are believed to benefit from more durable reconstruction options, such as conventional arthroplasty implants, or modular "tumor" prostheses, particularly for lesions involving the peritrochanteric or subtrochanteric femur (Fig. 11.1c and d) [16].

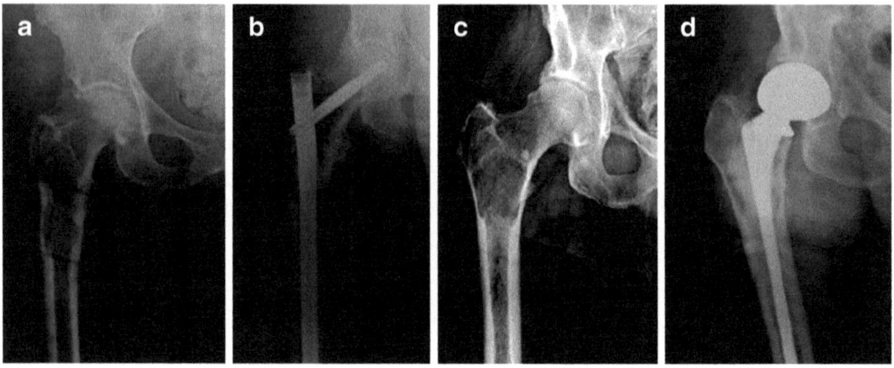

Fig. 11.1 Two implant options for patients with impending pathologic proximal femur fractures are shown. Relatively sick patients with a high likelihood of perioperative complications and survival estimates <6 months may benefit from intramedullary stabilization. (**a**) Preoperative and (**b**) postoperative radiographs are shown. Healthier patients who have a lower likelihood of perioperative complications and more favorable survival estimates of >6 months may require more durable implants, such as the prosthesis shown here. (**c**) Preoperative and (**d**) postoperative radiographs are shown. Patients with very short life expectancies of <1 month may be candidates for palliative radiotherapy, without surgery. (Source: *CORR Journal*, April 2017 Issue 4) [16]

Longer estimates of survival further justify the need for durable and, at times, complex custom reconstructive options. These options are particularly attractive in patients with periacetabular lesions due to renal cell, thyroid, or melanoma [16]. Among patients treated for impending or displaced pathologic proximal femur fractures, data were analyzed for 93 patients treated between 1990 and 2010 in the Karolinska Skeletal Metastasis Registry and 298 patients treated between 1993 and 2008 at the Memorial Sloan Kettering Cancer Center (MSKCC). Both datasets suggest greater durability and lower failure rates with endoprosthetic reconstruction. In the MSKCC dataset, the number of treatment failures was significantly lower in the endoprosthetic reconstruction group (3.1%) compared to intramedullary nailing (6.1%) and open reduction–internal fixation (42.1%) groups. Overall, these data support the use of endoprosthetic reconstruction as the salvage treatment of choice even at end of life. With greater durability and lower implant failure rates, these benefits are valuable for MBD patients with longer life expectancies at 12 months post-surgery [47, 48].

Outcomes

The prognosis of patients with metastatic bone disease (MBD) is strongly tied to their primary cancer type, which plays an important role in selecting the appropriate surgical treatment [32]. A review of survival, prognostic factors, and outcomes after surgical intervention for appendicular skeleton bone metastases indicated the following important factors predicting prognosis: the primary tumor histotype, the presence of visceral metastasis, pathologic fractures, and multiple metastases [52].

In the following section, the outcomes of patients with MBD will be detailed according to their primary cancer diagnosis.

The Scandinavian Sarcoma Group (SSG) Skeletal Metastases Registry is a multicenter prospective registry of patients surgically treated for non-spinal skeletal metastases at one of eight major Scandinavian referral centers. As the world's largest registry of surgically treated skeletal metastases, it includes 1107 patients that were operated on between June 1999 and October 2009. This registry gathered data including demographic characteristics, the primary tumor, the site of metastasis, the presence of other metastases, the type of surgery, complications, and the date and cause of death. These data indicated that metastatic bone disease can develop from a wide range of malignancies and can metastasize to many different bones. Overall, more than 20 different primary cancers led to fractures requiring surgical intervention, with 78% of cases originating from breast, prostate, renal, lung, and myeloma cancers. Skeletal metastases were most commonly found in the femur, humerus, and pelvis, regardless of the primary cancer diagnosis [53].

Over the 10 years of SSG registry data, the overall patient survival post-surgery was 58% at 6 months, 41% at 12 months, and 2% after 5 years. The median survival based on patients' primary cancer diagnosis was the longest for myeloma (26.3 months), then thyroid cancer (22.7 months), followed by breast cancer (12 months), and kidney cancer (10 months). Additionally, the longest survival times were seen in patients that had primary breast and prostate cancers. Some patients with primary renal cancer, myeloma, and lymphoma survived more than 9.5 years post-surgery. Lastly, the primary cancer with the shortest median survival was melanoma at 2.3 months [53].

Across the literature, not all studies agree with the predictive role of each primary cancer diagnosis in MBD patient survival. Hansen et al. found that MBD patients with myeloma, lymphoma, breast, and kidney primary cancers had a higher survival rate at 12 months compared to patients with prostate, lung, or unknown primary cancer. Univariate analysis identified lung, myeloma, prostate, and unknown primary cancers as significant predictors. Further multivariate analysis found primary lung cancer to be a negative predictor of survival and myeloma as positive. Meanwhile, multivariate analysis by Ratasvuori et al. identified breast, renal, thyroid, myeloma, and lymphoma primary cancers as positive predictors of survival. Additionally, Schneiderbauer et al. found MBD patients with renal, lung, and other primary cancers not including breast and prostate to have poorer survival than other primary cancers. Given the contrast among these results, it is important to note differences in patient populations for these studies. The studies by Ratasvuori et al. and Hansen et al. were conducted on a larger scale to investigate both upper and lower extremities, while Schneiderbauer et al. only focused on survival after hip replacement surgery. The main difference is seen in the predictive nature of primary renal cancer. Ratasvuori et al. found renal cell carcinoma to be a positive predictor of survival, which was supported by Hansen et al., while Schneiderbauer et al. identified renal cell carcinoma to be a poorer predictor [52, 54, 55].

More recently, an international database has been developed to study trends in life expectancy after patients were treated for MBD between 2000 and 2022.

This included 3353 patients in the International Bone Metastasis Registry treated for MBD between 2015 and 2022 as well as historical data from the validation of PATHFx models. Patients were adults aged 18 years or older, underwent treatment (surgery and/or radiation therapy) for MBD to the axial and/or appendicular skeleton, and had records of the date of surgery and survival data. Among the 3353 patients, 2156 or 64% had record of a tissue-specific diagnosis for their primary malignancy. The five most common primary cancer diagnoses included breast cancer (n = 377, 17%), prostate cancer (n = 353, 16%), lung cancer (n = 349, 16%), multiple myeloma (n = 239, 11%), and renal cancer (n = 232, 11%). Patients were grouped by their treatment date: period 1 (2000–2009), period 2 (2010–2019), and period 3 (2020–2022). Among the five most common primary cancer diagnoses, the median survival time, defined as the time at which 50% of patients survived, was determined for each period, as seen in Table 11.1 [27].

There was a notable increase in median survival time from period 1 to period 2 in patients with breast, prostate, lung, and renal primary cancers. Additionally, median survival was only reached for MBD patients with primary lung cancer by period 3, likely due to the lower median survival times compared to other primary cancers [27]. Overall, these data suggest that median survival time has increased across most common primary cancer diagnoses, indicating that the life expectancy of patients with radiotherapy and/or surgery for MBD has increased over time. This finding was supported across nearly every cancer type subgroup.

Finally, with advancements in systemic therapies, patients with MBD are living longer. Table 11.1 highlights data from an international MBD database with 3353 patients who underwent surgery for MBD and/or radiation therapy between 2000 and 2022. Among the five most common primary cancer diagnoses with MBD, the median survival time had a notable increase over time for breast, prostate, lung, and renal cancers [27].

Table 11.1 The median survival time in three periods for patients with the five most common primary cancer diagnoses with MBD [27]

	Median survival time (months)		
Primary cancer	Period 1	Period 2	Period 3
Breast	16.5 (95% CI, 7–29)	34 (95% CI, 25–49)	NR[a]
Prostate	6 (95% CI, 4–8)	11 (95% CI, 8–17)	NR
Lung	2 (95% CI, 1–5)	4 (95% CI, 3–4)	6 (95% CI, 4–13)
Multiple myeloma	31 (95% CI, 23–60)	38 (95% CI, 24–62)	NR
Renal	5 (95% CI, 3–12)	35 (95% CI, 18–51)	NR

MBD metastatic bone disease; 95% CI, 95% confidence interval
[a]*NR* not reported; period 1 (2000–2009), period 2 (2010–2019), period 3 (2020–2022)

Conclusions

Caring for patients with metastatic bone disease requires an intimate understanding of projected disease progression, complications requiring medical or surgical intervention, and how treatment plans involve multiple oncologic fields. Given the complexity of this, it is crucial that physicians leverage multidisciplinary teams and technological advancements in prognostic modeling to make the best choices for their patients. Over time, advancements in cancer care have significantly improved our ability to treat metastatic disease and have extended life expectancy in this population. With that comes a greater need to appropriately treat different complications in accordance with patients' estimated survival. This chapter details the necessary steps to vigilantly monitor MBD disease progression, identify complications, and effectively treat them according to patients' estimated survival time.

References

1. Chen W, Hoffmann AD, Liu H, Liu X. Organotropism: new insights into molecular mechanisms of breast cancer metastasis. npj Precision Oncol. Nature Publishing Group. 2018;2
2. Piccioli A, Maccauro G, Spinelli MS, Biagini R, Rossi B. Bone metastases of unknown origin: epidemiology and principles of management. J Orthop Traumatol. Springer-Verlag Italia srl. 2015;16:81–6.
3. Gao Y, Bado I, Wang H, Zhang W, Rosen JM, Zhang XHF. Metastasis organotropism: redefining the congenial soil. Develop Cell. Cell Press. 2019;49:375–91.
4. DiCaprio MR, Murtaza H, Palmer B, Evangelist M. Narrative review of the epidemiology, economic burden, and societal impact of metastatic bone disease. Ann Joint. AME Publishing Company. 2022;7
5. Zhang J, Cai D, Hong S. Prevalence and prognosis of bone metastases in common solid cancers at initial diagnosis: a population-based study. BMJ Open. 2023;13(10)
6. Tsukamoto S, Kido A, Tanaka Y, Facchini G, Peta G, Rossi G, et al. Current overview of treatment for metastatic bone disease. Curr Oncol. MDPI. 2021;28:3347–72.
7. Coleman RE, Brown J, Holen I. CAUSES initiation of bone metastases. 2020.
8. Gomis RR, Gawrzak S. Tumor cell dormancy. Mol Oncol. 2017;11(1):62–78.
9. Endo H, Inoue M. Dormancy in cancer. Cancer Sci. Blackwell Publishing Ltd. 2019;110:474–80.
10. Zhang XHF, Wang Q, Gerald W, Hudis CA, Norton L, Smid M, et al. Latent bone metastasis in breast cancer tied to Src-dependent survival signals. Cancer Cell. 2009;16(1):67–78.
11. Rougraff BT, Kneisl JS, Simon MA. Skeletal metastases of unknown origin. A prospective study of a diagnostic strategy. J Bone Joint Surg Am. 1993;75(9):1276–81. https://doi.org/10.2106/00004623-199309000-00003.
12. Zamani-Siahkali N, Mirshahvalad SA, Farbod A, Divband G, Pirich C, Veit-Haibach P, et al. SPECT/CT, PET/CT, and PET/MRI for response assessment of bone metastases. Semin Nucl Med. W.B. Saunders. 2024;54:356–70.
13. Ouvrard E, Kaseb A, Poterszman N, Porot C, Somme F, Imperiale A. Nuclear medicine imaging for bone metastases assessment: what else besides bone scintigraphy in the era of personalized medicine? Front Med. Frontiers Media SA. 2023;10:1320574.
14. Broder MS, Gutierrez B, Cherepanov D, Linhares Y. Burden of skeletal-related events in prostate cancer: unmet need in pain improvement. Support Care Cancer. Springer Verlag. 2015;23:237–47.
15. Oster G, Lamerato L, Glass AG, Richert-Boe KE, Lopez A, Chung K, et al. Natural history of skeletal-related events in patients with breast, lung, or prostate cancer and metastases to bone: a 15-year study in two large US health systems. Support Care Cancer. 2013;21(12):3279–86.

16. Forsberg JA, Wedin R, Boland PJ, Healey JH. Can we estimate short- and intermediate-term survival in patients undergoing surgery for metastatic bone disease? Clin Orthop Relat Res. 2017;475(4):1252–61.
17. Piccioli A, Spinelli MA, Forsberg JA, Wedin R, Healey JH, Ippolito V, et al. How do we estimate survival? External validation of a tool for survival estimation in patients with metastatic bone disease-decision analysis and comparison of three international patient populations. BMC Cancer. 2015;15(1)
18. Ogura K, Gokita T, Shinoda Y, Kawano H, Takagi T, Ae K, et al. Can a multivariate model for survival estimation in skeletal metastases (PATHFx) be externally validated using Japanese patients? Clin Orthop Relat Res. 2017;475(9):2263–70.
19. Shah AA, Schwab JH. Predictive modeling for spinal metastatic disease. Diagnostics. Multidisciplinary Digital Publishing Institute (MDPI). 2024;14:962.
20. Meares C, Badran A, Dewar D. Prediction of survival after surgical management of femoral metastatic bone disease—a comparison of prognostic models. J Bone Oncol. 2019;15:100225.
21. Willeumier JJ, van der Linden YM, van der Wal CWPG, Jutte PC, van der Velden JM, Smolle MA, et al. An easy-to-use prognostic model for survival estimation for patients with symptomatic long bone metastases. J Bone Joint Surg (Am Vol). 2018;100(3):196–204.
22. Anderson AB, Wedin R, Fabbri N, Boland P, Healey J, Forsberg JA. External validation of PATHFx version 3.0 in patients treated surgically and nonsurgically for symptomatic skeletal metastases. Clin Orthop Relat Res. 2020;478(4):808–18.
23. Forsberg JA, Sjoberg D, Chen QR, Vickers A, Healey JH. Treating metastatic disease: which survival model is best suited for the clinic? Tumor. In: Clinical orthopaedics and related research. Springer New York LLC; 2013. p. 843–50.
24. Gulshan V, Peng L, Coram M, Stumpe MC, Wu D, Narayanaswamy A, et al. Development and validation of a deep learning algorithm for detection of diabetic retinopathy in retinal fundus photographs. JAMA. 2016;316(22):2402–10.
25. Esteva A, Kuprel B, Novoa RA, Ko J, Swetter SM, Blau HM, et al. Dermatologist-level classification of skin cancer with deep neural networks. Nature. 2017;542(7639):115–8.
26. Menden MP, Iorio F, Garnett M, McDermott U, Benes CH, Ballester PJ, et al. Machine learning prediction of cancer cell sensitivity to drugs based on genomic and chemical properties. PLoS One. 2013;8(4):e61318.
27. Rogers DL, Raad M, Rivera JA, Wedin R, Laitinen M, Sørensen MS, et al. Life expectancy after treatment of metastatic bone disease: an international trend analysis. J Am Acad Orthop Surg. 2024;32(6):E293–301.
28. Malviya A, Gerrand C. Evidence for orthopaedic surgery in the treatment of metastatic bone disease of the extremities: a review article. Palliat Med. 2012;26:788–96.
29. Errani C, Mavrogenis AF, Cevolani L, Spinelli S, Piccioli A, Maccauro G, et al. Treatment for long bone metastases based on a systematic literature review. Eur J Orthop Surg Traumatol. Springer-Verlag France. 2017;27:205–11.
30. Pillay B, Wootten AC, Crowe H, Corcoran N, Tran B, Bowden P, et al. The impact of multidisciplinary team meetings on patient assessment, management and outcomes in oncology settings: a systematic review of the literature. Cancer Treat Rev. W.B. Saunders Ltd. 2016;42:56–72.
31. Shao J, Rodrigues M, Corter AL, Baxter NN. Multidisciplinary care of breast cancer patients: a scoping review of multidisciplinary styles, processes, and outcomes. Curr Oncol. 2019;26(3):e385–97.
32. Soeharno H, Povegliano L, Choong PF. Multimodal treatment of bone metastasis - a surgical perspective. Front Endocrinol. Frontiers Media S.A. 2018;9:518.
33. Mavrogenis AF, Angelini A, Vottis C, Pala E, Calabro T, Papagelopoulos PJ, et al. Modern palliative treatments for metastatic bone disease awareness of advantages, disadvantages, and guidance. Clin J Pain. Lippincott Williams and Wilkins. 2016;32:337–50.
34. Abarrategi A, Tornin J, Lucia MC, Hamilton A, Enrique MC, Rodrigo JP, et al. Osteosarcoma: cells-of-origin, cancer stem cells, and targeted therapies. Stem Cells Int. Hindawi Limited. 2016;2016:3631764.

35. Nathenson MJ, Conley AP, Sausville E. Immunotherapy: a new (and old) approach to treatment of soft tissue and bone sarcomas. Oncologist. 2018;23(1):71–83.
36. Gdowski AS, Ranjan A, Vishwanatha JK. Current concepts in bone metastasis, contemporary therapeutic strategies and ongoing clinical trials. J Exp Clin Cancer Res. 2017;36(1):108.
37. Kobayashi T, Hara M, Shimanoe C, Morimoto T, Masaaki M, Ito K, et al. Efficacy and safety of romosozumab: a meta-analysis of placebo-controlled trials. J Bone Miner Metab. Springer. 2024;42(5):492–502.
38. Siegel GW, Biermann JS, Calinescu AA, Spratt DE, Szerlip NJ. Surgical Approach to Bone Metastases. Curr Osteoporos Rep. 2018;16(4):512–18. https://doi.org/10.1007/s11914-018-0463-7.
39. De Felice F, Piccioli A, Musio D, Tombolini V. The role of radiation therapy in bone metastases management [Internet]. Oncotarget. 2017;8. Available from: www.impactjournals.com/oncotarget/.
40. Garbar V, Richter S. Palliative radiation therapy for bone metastases. 2025.
41. Bindels BJJ, Mercier C, Gal R, Verlaan JJ, Verhoeff JJC, Dirix P, et al. Stereotactic body and conventional radiotherapy for painful bone metastases: a systematic review and meta-analysis. JAMA Netw Open. 2024:E2355409.
42. Nagata Y. Stereotactic body radiation therapy: principles and practices. 2nd ed. Springer Nature; 2023. p. 1–325.
43. Khanmohammadi S, Noroozi A, Yekaninejad MS, Rezaei N. Cryoablation for the palliation of painful bone metastasis: a systematic review. In: Cardiovascular and interventional radiology, vol. 46. Springer; 2023. p. 1469–82.
44. Moynagh MR, Kurup AN, Callstrom MR. Thermal ablation of bone metastases. Semin Intervent Radiol. 2018;35(4):299–308.
45. Katsanos K, Sabharwal T, Adam A. Percutaneous cementoplasty. Semin Intervent Radiol. 2010;27(2):137–47.
46. Gatto L, Facchini G, Saponara M, Nannini M, Rossi G, Di Scioscio V, et al. Successful selective arterial embolizations for bone metastases in renal cell carcinoma integrated with systemic therapies: a case report. Radiol Case Rep. 2017;12(4):775–9.
47. Steensma M, Boland PJ, Morris CD, Athanasian E, Healey JH. Endoprosthetic treatment is more durable for pathologic proximal femur fractures. Clin Orthop Relat Res. 2012;470(3):920–6.
48. Forsberg JA, Wedin R, Bauer H. Which implant is best after failed treatment for pathologic femur fractures? Tumor. In: Clinical orthopaedics and related research. Springer New York LLC; 2013. p. 735–40.
49. Terek RM; IlluminOss Study Group; McGough R, Fabbri N, Cheung F, Brigman B, et al. IlluminOss photodynamic bone stabilization system improves pain and function in the treatment of humeral metastatic disease. Bone Joint J. 2024;106-B(12):1485–92. https://doi.org/10.1302/0301-620X.106B12.BJJ-2023-1089.R2.
50. Zoccali C, Attala D, Pugliese M, Scotto di Uccio A, Baldi J. The IlluminOss® photodynamic bone stabilization system for pathological osteolyses and fractures of the humerus: indications, advantages and limits in a series of 12 patients at 24 months of minimum follow-up. BMC Musculoskelet Disord. 2021;22(1):63. https://doi.org/10.1186/s12891-020-03927-6.
51. Imanishi J, Yang R, Kawano H, Lee FY. Recent advances in minimally invasive local cancer control and skeletal stabilization of periacetabular osteolytic metastases under C-arm imaging guidance. J Am Acad Orthop Surg. 2025;33(3):e136–e150. Epub 2024 Sep 27. https://doi.org/10.5435/JAAOS-D-24-00077.
52. Kirkinis MN, Lyne CJ, Wilson MD, Choong PFM. Metastatic bone disease: a review of survival, prognostic factors and outcomes following surgical treatment of the appendicular skeleton. Eur J Surg Oncol. W.B. Saunders Ltd. 2016;42:1787–97.
53. Ratasvuori M, Wedin R, Keller J, Nottrott M, Zaikova O, Bergh P, et al. Insight opinion to surgically treated metastatic bone disease: Scandinavian sarcoma group skeletal metastasis registry report of 1195 operated skeletal metastasis. Surg Oncol. 2013;22:132–8.

54. Hansen BH, Keller J, Laitinen M, Berg P, Skjeldal S, Trovik C, et al. The Scandinavian sarcoma group skeletal metastasis register: survival after surgery for bone metastases in the pelvis and extremities. Acta Orthop Scand Suppl. 2004;75(311):11–5.
55. Schneiderbauer MM, Von Knoch M, Schleck CD, Harmsen WS, Sim FH, Scully SP. Patient survival after hip arthroplasty for metastatic disease of the hip [Internet]. J Bone Joint Surg. 2004. Available from: http://journals.lww.com/jbjsjournal.

“Oops…This Is About to Break!”: Impending Pathologic Fractures of Long Bones

12

Joseph J. Connolly, Neha Jejurikar, Sarah Ballatori, Megan H. Goh, and Santiago A. Lozano Calderón

Introduction

Despite improvements in survival for patients with advanced malignancies, bone metastases portend a poor prognosis for cancer patients and present a significant clinical challenge for multidisciplinary care teams [1]. These lesions frequently cause pain, requiring intervention with multimodal analgesic regimens, radiation therapy (RT), and/or surgery. As bone involvement worsens, pain and instability often result in significant functional impairment. In more severe cases, osseous metastases can compromise the structural integrity of bone and put patients at risk of a pathologic fracture. Pathologic fractures occur when innate disease, such as osteoporosis, metabolic bone disease, inherited bone disorders, and infection, weakens bone and causes it to break under normal, atraumatic mechanical loads or even spontaneosly [2, 3].

In the cancer setting, pathologic bone primarily arises from metastatic bone disease [4, 5]. Bone metastases weaken bone and can precipitate a fracture, which has significant implications for patient quality of life and overall delivery of care. Complete pathologic fractures are significantly disabling due to the pain and functional impairment associated with a fractured bone. Surgical management is complex; operative fixation of a displaced fracture requires reduction and extended operative time. Recovery from a complete pathologic fracture fixation is also highly involved for the patient due to the inflammation associated with the injury. In addition, pathologic fractures often cause unexpected and potentially harmful interruptions to life-sustaining adjuvant therapy, thereby complicating multidisciplinary management of disease.

J. J. Connolly · N. Jejurikar · S. Ballatori · M. H. Goh · S. A. Lozano Calderón (✉)
Division of Musculoskeletal Oncology, Department of Orthopedic Surgery / Massachusetts General Hospital - Harvard Medical School, Boston, MA, USA
e-mail: jconnolly28@mgh.harvard.edu; njejurikar@mgh.harvard.edu; sballatori@mgb.org; mgoh3@mgh.harvard.edu; slozanocalderon@mgh.harvard.edu

J. Pretell-Mazzini, S. A. Lozano Calderón (eds.), *Multidisciplinary Approach of Musculoskeletal Metastasis*, https://doi.org/10.1007/978-3-032-09498-8_12

As such, identifying impending pathologic fractures—bone lesions at a high risk of fracture without intervention—prior to injury is crucial to palliate symptoms and sustain quality of life, enhance multidisciplinary coordination of care, and, at times, improve oncologic outcomes [6–11]. In fact, some studies suggest a survival benefit for patients undergoing prophylactic fixation versus post-fracture management [7, 10, 11]. However, identifying and defining true impending fractures is challenging, and clinical decision-making tools are controversial [12]. Treatment decisions require a multidisciplinary approach, considering the patient's clinical presentation and imaging characteristics, as well as the patient's overall health, prognosis, and the biologic behavior of the primary tumor.

Epidemiology and Pathophysiology

The five most common cancers that metastasize to bones are prostate, breast, lung, kidney, and thyroid cancers [13–15]. Prostate and breast cancers show a particular affinity for bone, with more than two-thirds of patients with advanced disease having bone metastases [15, 16]. However, a significant 30–40% of lung, kidney, and thyroid cancers develop bone metastases [15, 17]. In hematologic malignancies, patients with multiple myeloma characteristically present with bone involvement in >80% of cases, while bone can be a primary or—more frequently—secondary site of disease for lymphoma [18, 19]. Each cancer type brings unique tumor biology and, with that, differing prognoses and variable efficacies of multimodal anticancer therapy in achieving disease control.

Metastatic bone disease is typically characterized by its effect on the surrounding bone architecture. Osteolytic lesions cause excessive bone breakdown via osteoclast upregulation or osteoblast downregulation, while osteoblastic lesions work in opposite and result in abnormal bone formation; both types of lesions disrupt normal osseous homeostasis and weaken the biomechanical stability of bone. Lung, kidney, and thyroid carcinomas most often—and multiple myeloma exclusively—produce lytic lesions [15]. Patients with prostate cancer are typically afflicted with blastic osseous lesions, while breast cancer frequently forms mixed lytic and blastic lesions [15]. Because of the bone loss associated with lytic lesions, they are at the highest risk of fracture, followed by mixed lesions and then blastic lesions. However, it is important to highlight that blastic lesions still pose a risk of fracture; osteoblastic activity increases the brittleness of bone and alters its normal mechanical properties [6].

Similarly, the distribution of skeletal metastases also varies depending on the primary tumor, but, generally, the spine is the most common site, followed by the pelvis, ribs, femur, and humerus [1, 15]. Axial skeleton involvement is pronounced in prostate cancer, which primarily metastasizes to the spine and pelvis [16]. In contrast, breast cancer often spreads to the proximal long bones, ribs, and sternum [20]. Lung cancer metastasizes broadly and is typically the suspected cancer of origin when acral metastases are discovered [21, 22]. The risk of pathologic fracture varies depending on the anatomic location and is largely related to the weight-bearing nature of each bone. The bones of the lower extremity, pelvis, and spine are under more stress on average and are thus more likely to fracture than the bones of the upper extremities and ribs [13, 23].

Diagnosis

Diagnosis of an impending pathologic fracture requires a thorough clinical assessment to properly ascertain fracture risk and guide multidisciplinary treatment decisions. The workup includes a targeted history and physical exam, imaging, laboratory testing, and, in some cases, biopsy. Importantly, the extent of the workup depends on whether the patient has a known diagnosed primary tumor. When the primary tumor is unknown, additional testing to identify the site of origin is essential, and, because tumor histology informs both fracture risk and clinical decision-making, it should be performed before treatment decisions are made. A more detailed description of the workup for an unknown primary can be found in Chap. 38.

History and Physical Exam

Although the clinical presentation of skeletal metastases varies with location and extent of disease, in most cases, the initial and most prominent symptom of skeletal metastases is pain [15]. Pain can come in one of two major forms: neoplastic and mechanical. Neoplastic pain occurs as a direct result of oncologic processes in bone: tumor-induced inflammation, periosteal stretching, cortical microfracture, and neurovascular involvement [24]. Mechanical pain, on the other hand, is functional in nature and worsens with activity and weight-bearing. Mechanical pain suggests a loss in structural integrity of the bone and indicates a higher risk of fracture. Despite this, neoplastic pain often precedes mechanical pain and subsequent fracture and, as such, still demands a thorough workup and a watchful eye.

Neurologic changes signal additional treatment considerations beyond fracture risk, particularly in cases of spinal metastases, in which neurologic findings correlate with the level(s) involved and help establish the disease extent (see Chap. 13). In both the spine and extremities, neurovascular involvement can lead to numbness or weakness in the distal limbs. Palpable soft tissue components can be indicative of cortical breakthrough and provide information on the overall aggressiveness of the lesion. Hypercalcemia often manifests as fatigue, nausea and vomiting, constipation, polyuria and polydipsia, confusion, and/or cardiac arrhythmias.

Overall, history and physical examination should be tailored to the site in question. Particular attention should be paid to assessing whether pain is functional in nature. Ascertaining extremity dominance is also important for clinical decision-making, as patients often bear more weight through a dominant limb compared to the contralateral side. Other important examination findings are tenderness, swelling, range-of-motion limitations, neurologic changes, palpable soft tissue components, and outward symptoms of hypercalcemia.

Imaging

Imaging is critical to the evaluation of an impending pathologic fracture, most often beginning first with plain film radiographs. Radiographs are useful to establish baseline lesion characteristics and rule out the risk of fracture. For example, up to 50% bone loss is needed to be detectable on X-ray imaging, meaning that small metastases that are not visible are rarely at risk of a pathologic fracture [25]. In addition, plain radiographs are an apt imaging modality to distinguish between lytic and blastic lesions (Fig. 12.1).

When lesions are concerning for a pathologic fracture on X-ray, further imaging may not be required; however, when plain radiographs incompletely visualize the area of concern, are inconclusive, or do not correlate with the patients' clinical presentation, advanced imaging is recommended, most commonly in the form of computed tomography (CT). CT imaging provides a higher level of detail to assess the three-dimensional (3D) extent of bony destruction. CT is particularly relevant for evaluating pathologic fracture risk, as it helps define cortical involvement, which is directly related to biomechanical stability and the likelihood of a fracture [26, 27] (Fig. 12.2). As with plain radiographs, CT imaging of the entire affected region of bone is recommended, and, in cases of an unknown tumor primary, staging studies of the chest, abdomen, and pelvis should be conducted.

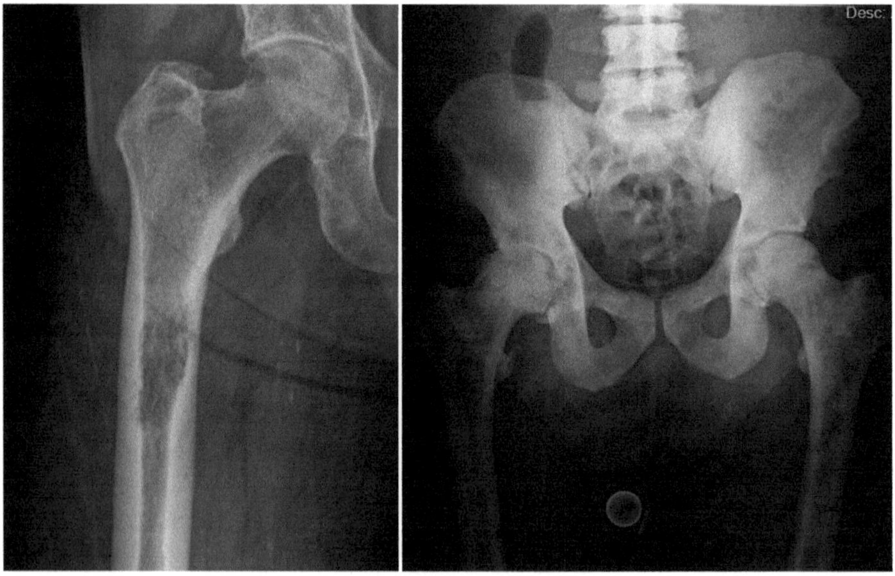

Fig. 12.1 Representative images of lytic and blastic lesions on plain radiographs. The left image demonstrates a lytic lesion of the subtrochanteric region of the right femur in a patient with metastatic lung cancer. The right image shows diffuse blastic lesions of the pelvis and bilateral femurs in a patient with metastatic prostate cancer

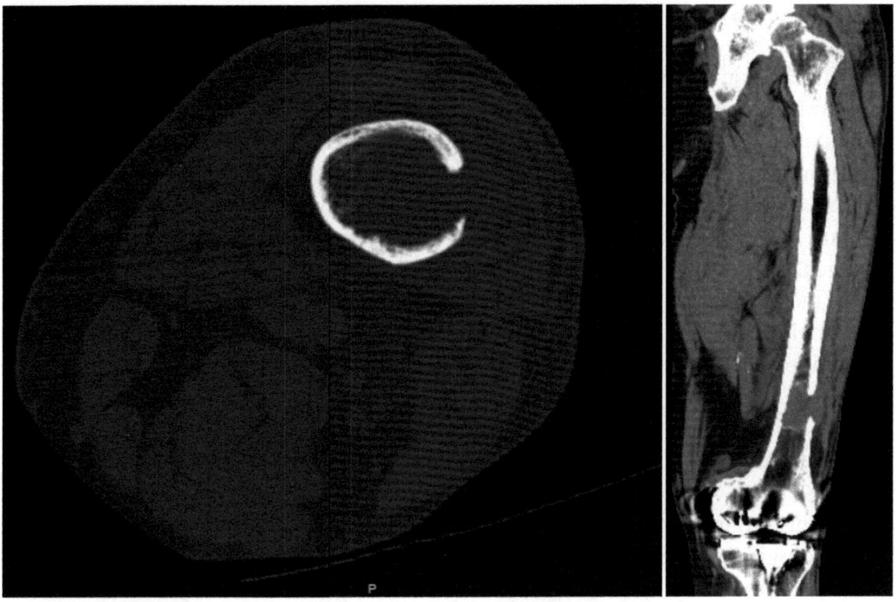

Fig. 12.2 Computed tomography shows cortical breakthrough on axial (left) and coronal (right) views in a patient with metastatic renal cell carcinoma of the left distal femur

Other forms of advanced imaging, such as magnetic resonance imaging (MRI) and positron emission tomography (PET), are useful diagnostic tools. MRI most accurately visualizes soft tissue involvement and spinal cord compression. PET imaging detects metabolically active tumors, which can be used to assess treatment response and risk for continued progression [28, 29]. Bone scans highlight areas of decreased bone density, which are at a higher risk of fracture. Taken together, imaging studies effectively characterize the lesion type and extent of disease, which factor in heavily in fracture risk assessment.

Laboratory Testing and Biopsy

In most cases, history, physical exam, and imaging are sufficient to establish a primary diagnosis and evaluate the risk of pathologic fractures [30, 31]. However, particularly in cases in which there is an unknown primary tumor, additional laboratory testing is important to fully characterize the disease status. Lab testing can be a surrogate marker for disease burden, responsiveness to treatment, and tumor activity. Baseline testing should include a complete blood count and a comprehensive metabolic panel. Specifically, serum calcium quantifies hypercalcemia, a measure of bone resorption, while alkaline phosphatase is a marker of bone turnover. Tumor-specific disease markers include serum and urine protein electrophoresis for multiple myeloma, prostate-specific antigen for prostate cancer, cancer antigen 15-3 for breast cancer, and carcinoembryonic antigen for colorectal cancer.

While biopsy is not necessarily required to evaluate fracture risk, treatment decisions hinge heavily on the tumor histology of origin. First and foremost, biopsy can definitively rule out a primary malignancy of bone, in which case treatment would be curative rather than palliative. Even among the metastatic solid and hematologic malignancies, differences in treatment paradigms for different primary histologies affect the utility of systemic therapy and RT and, sometimes, the surgical intervention of choice [32, 33]. A biopsy can also be useful to evaluate the molecular evolution of metastatic disease compared to the primary tumor, which may inform changes in systemic therapy. The goal of a biopsy is to obtain a representative sample of the pathologic tissue; importantly, intramedullary reaming in the setting of a pathologic fracture with an unknown primary does not replace a well-performed open biopsy to establish a diagnosis and is not recommended.

Risk Assessment

Following a thorough clinical evaluation, care teams must carefully evaluate the risk of pathologic fractures, which has significant implications for treatment decision-making. In summary, lytic lesions, arising from lung, renal, and thyroid carcinomas, as well as multiple myeloma, are at a higher risk of fractures than mixed and blastic lesions, which arise mainly from breast and prostate cancers, respectively. Anatomic location plays an important role; weight-bearing bones including the spine, pelvis, and lower extremity are at a higher risk of fracture than those in the upper extremities and ribs. Tumor histology provides less direct information about fracture risk but heavily impacts the application of adjuvant therapy.

A detailed characterization of pain is highly informative; mechanical, or functional, pain is more indicative of a fracture risk than neoplastic pain. Imaging studies provide, arguably, the most sensitive information about fracture risk. Plain radiographs help determine lesion type and establish a minimum threshold for substantive bone loss. CT imaging reveals 3D lesion size and cortical involvement; large lesions with significant cortical destruction are at a higher risk of fracture because of the loss of biomechanical stability. Risk assessment is multifactorial and challenging, and newer stratification strategies seek to overcome current limitations in positive predictive value [34]. Regardless, clinical decision-making should be multidisciplinary and consider the patient's overall health, prognosis, and goals of care.

Current Risk Stratification Systems

Several risk assessment tools, with varying clinical utility, have been proposed to evaluate impending pathologic fractures. In the extremities, the earliest attempts at quantifying pathologic fracture risk came from Harrington and Carnesale [35, 36]. The Harrington criteria for prophylactic fixation of the femur include lesions $\geq$25 mm, lytic involvement of $\geq$50% of the cortex, and significant, RT-refractory pain upon weight-bearing. Carnesale's definition is similar and indicates for surgery lesions that cause significant pain and have either >50% cortical involvement or size >30 mm. Both sets of criteria incorporate commonly accepted risk factors but were based on retrospective data and have not since been validated prospectively.

Today, the most widely used clinical decision-making tool for impending pathologic fractures of the long bones is the Mirels' scoring system [37]. This system incorporates four clinical and radiographic factors—anatomic location, type of pain, radiographic appearance, and size—with each category assigned a score of 1–3 (Table 12.1). With regard to location, lesions in the weight-bearing bones of the lower extremity, particularly in the peritrochanteric region, are at a higher risk than those of the upper extremity. Functional pain renders the highest score, while moderate or mild neoplastic pain scores lower. Radiographically, as previously discussed, lytic lesions are considered high risk, followed by mixed and then blastic lesions. The variable size incorporates cortical involvement; lesions at the highest risk involved greater than two-thirds of the cortex.

According to Mirels, who quoted 90% accuracy in correctly predicting pathologic fractures, a score ≥8 indicated a patient for prophylactic fixation, while scores ≤7 were deemed low risk. Current practice considers scores ≥9 high risk and scores ≤7 low risk; recommendations for patients with a score of 8 are controversial and require clinical judgment. However, application of the Mirels' score to independent patient series has shown high sensitivity but low specificity [12, 38, 39]. If followed strictly, the Mirels' scoring system would lead to overtreatment with unnecessary prophylactic fixation, which must be taken into account when assessing the risk of fracture.

In the setting of spinal metastases, the Spinal Instability Neoplastic Score (SINS) is the primary scoring system for impending pathologic fractures [40]. The score is in some ways like Mirels', taking into account location, pain, and lesion type; however, spine-specific criteria include radiographic spinal alignment, percentage vertebral body collapse, and bilateral vs. unilateral posterolateral involvement (Table 12.2). Aggregate scores are considered stable (0–6), possibly unstable [7–12], and unstable [13–18]. Unstable lesions are definitively indicated for prophylactic fixation, while possibly unstable lesions may benefit from prophylactic fixation but require clinical judgment. SINS has been validated by several retrospective series and shows high rates of reproducibility and acceptable sensitivity and specificity [41].

Table 12.1 Mirels' scoring system for evaluating impending pathologic fractures

	Score		
Variable	1	2	3
Site	Upper extremity	Lower extremity	Peritrochanteric
Pain	Mild	Moderate	Functional
Lesion type	Blastic	Mixed	Lytic
Size	<1/3 of the cortex	≥1/3 to ≤2/3 of the cortex	>2/3 of the cortex

Table 12.2 Spinal Instability Neoplastic Score scoring system for evaluation of spinal metastases at risk of pathologic fractures

Variable	Score
Location	
Junctional (O–C2, C7–T2, T11–L1, L5–S1)	3
Mobile (C3–6, L2–4)	2
Semirigid (T3–10)	1
Rigid (S2–5)	0
Pain	
Mechanical pain	3
Occasional pain but not mechanical	1
Painless	0
Lesion type	
Lytic	2
Mixed	1
Blastic	0
Radiographic spinal alignment	
Subluxation/translation present	4
De novo deformity	2
Normal alignment	0
Vertebral body collapse	
>50% collapse	3
<50% collapse	2
>50% vertebral involvement, with no collapse	1
None of the above	0
Posterior spinal element involvement	
Bilateral	3
Unilateral	1
Neither	0

Alternative Risk Assessment Strategies

Imaging-Based Strategies

Three-dimensional advanced imaging is an essential element of risk assessment—cortical involvement is arguably the most predictive risk factor [26, 27, 42]—and, as such, several groups have created imaging-based models to improve fracture prediction. For example, studies incorporating scintigraphy into risk stratification tools have shown promise in improving specificity. Riaz et al. created a single-photon emission computed tomography–CT (SPECT–CT)-modified Mirels' score, which used SPECT–CT in lieu of radiographs to categorize lesion type [29]. Another retrospective study of breast cancer patients with proximal femur lesions showed that the glycolysis rate of a lesion could predict subsequent fracture with 85% sensitivity and 80% specificity [28].

CT-based biomechanical methods better represent axial and circumferential cortical involvement and allow for patient- and lesion-specific stiffness and stress modeling. This tailored analysis is important; one cadaveric study found that inferior medial femoral neck lesions had half the strength compared to superior lateral lesions [43]. One strategy, CT-based rigidity analysis (CT-RA), uses Hounsfield units to identify weaker areas of low bone density, which correlates well with load-bearing capacity [44, 45]. In a prospective study comparing the predictive value of CT-RA with Mirels' score, CT-RA outperformed Mirels' in both sensitivity and specificity, even after controlling for baseline confounding variables [46]. Lesions with a 35% reduction in axial, bending, or torsional rigidity were considered high risk for fracture.

Finite element analysis (FEA) similarly uses CT imaging and Hounsfield units to estimate bone quality but considers bone geometry by creating 3D computer models that can be subjected to various physiologic loading conditions [47, 48]. CT–FEA is particularly useful in areas of complex 3D anatomy, such as the femoral neck and trochanteric regions, and in cases in which muscular forces may impact fracture risk [47, 49]. As with CT-RA, CT–FEA has shown to improve fracture risk prediction compared to Mirels' score and other conventional risk metrics [50–52]. However, studies with head-to-head comparisons of CT-RA and CT–FEA show marginal improvement in accuracy using CT–FEA [53]. Despite this, CT–FEA has yet to be prospectively validated, and, along with CT-RA, faces challenges with imaging quality variability and risk-cutoff standardization, which hampers its clinical application [49].

Machine Learning Models

Clinical and imaging-based risk stratification tools are highly useful for evaluating the risk of pathologic fractures, but, as previously discussed, when deciding whether to treat operatively or nonoperatively, clinicians must simultaneously evaluate patient factors including quality of life, prognosis, and tumor responsiveness to chemotherapy and RT. Machine learning models represent a promising adjunct to clinical expertise, and multiple tools have been created to calculate prognosis in patients with impending or complete pathologic fractures. The PATHFx tool (https://www.pathfx.org), which has been validated in several continents, demonstrated excellent survival prediction at several follow-up time points between 3 and 24 months in multiple studies [54, 55]. Another model developed by the Skeletal Oncology Research Group (https://sorg-apps.shinyapps.io/extremitymetssurvival/) assessed 90-day and 1-year survival in patients undergoing operative management for metastatic bone disease [56–59].

Thus far, no machine learning algorithm has been validated to predict fracture risk in patients with impending pathologic fractures. Given machine learning's ability to provide personalized recommendations based on patient-specific inputs, fracture risk algorithms hold immense potential in improving multidisciplinary management of skeletal metastases. Ideally, contemporary models should capitalize on strengths from previous models by combining clinical, radiographic, and biomechanical variables. One proposed model demonstrated improved performance of a CT-based prediction tool when combined with clinical features [60]. Machine learning algorithms can be easily disseminated via online tools but require large training sets and multiple external validation cohorts before they can be applied clinically.

Treatment

The goal of treatment for impending pathologic fractures is to maintain or improve quality of life by preventing fracture and allowing stable, pain-free weight-bearing. Treatment should be minimally disruptive to the oncologic management of disease, but, in many cases, impending fracture treatment and oncologic management can overlap (e.g., RT, radio-frequency ablation, curettage, and metastasectomy). Multidisciplinary treatment decisions primarily hinge on whether management should be operative or nonoperative. Considerations include likelihood of complete pathologic fracture, chemo- and radiosensitivity of the lesion, life expectancy, and surgical morbidity. Subsequent chapters will provide more detailed outlines of surgical treatment options for common sites of skeletal metastases, but this section will provide a general overview of the principles of management for impending pathologic fractures.

When fracture risk is incontrovertible, surgery is almost always the first-choice treatment because it quickly diminishes the risk of fracture and allows for early weight-bearing and return to activities of daily living. However, in cases of highly chemo- and radiosensitive primary cancers—multiple myeloma, lymphoma, and, less often, breast and prostate cancers—palliative RT, systemic treatment changes, and the addition of bone-targeted agents (BTAs) may be sufficient to reconstitute bone quality. Life expectancy factors in heavily; patients with a poor prognosis are more often managed nonoperatively via protected weight-bearing, analgesic regimens, RT, and/or systemic therapy. In contrast, durable reconstructive constructs are paramount for patients with long life expectancies to avoid future revision surgery. Perhaps what is most critical is that patients must be surgical candidates to be considered for operative management; those with multiple comorbidities or diffuse, uncontrolled metastatic disease may benefit better from nonoperative treatment..

Nonoperative Treatment

Radiation Therapy

RT is frequently used in the setting of impending pathologic fractures in combination with bone-targeted agents and/or surgery. It is primarily effective at providing pain relief by shrinking tumor size and reducing local inflammation, particularly in radiosensitive histologies such as multiple myeloma, lymphoma, breast, prostate, and lung cancers [61]. In addition, RT is more often considered the primary treatment for lesions in the upper extremities or non-weight-bearing regions of the pelvis, which are generally considered at a lower risk of fracture. In cases in which patients are undergoing active chemotherapy to control life-threatening visceral metastases, RT can also be a bridging therapy to manage at-risk skeletal metastases until surgery can be considered.

In the setting of skeletal metastases, RT is most often delivered with external beam radiation therapy (EBRT). Current guidelines support the use of single-fraction (8 Gy in 1 fraction) or hypofractionated (20 Gy in 5 fractions or 30 Gy in 10 fractions) schemes, which provide significant benefit while reducing hospital visits [62]. When used in combination with surgery, RT it generally administered postoperatively about 2–3 weeks after surgery; it is generally recommended in these cases that the treatment field include the entire fixation device to mitigate potential tumor seeding and reduce the risk of local recurrence [63, 64]. However, when short-stem fixation devices are used, RT field planning should consider whether reamers were used in surgery, as this may contaminate the entire shaft and warrant RT delivery far below the tip of the stem. RT may also be administered preoperatively in cases in which rapid palliation is required, although studies have variably associated preoperative RT with an increased rate of implant failure [65–67]. Late effects of RT are also not trivial and include radiation-associated fractures, which are debilitating and have high nonunion rates, and radiation-associated cancers, which have a poor prognosis and limited treatment efficacy [68–70].

Systemic Therapy

Systemic anticancer therapies, including cytotoxic chemotherapy and hormone therapy, have historically had poor or temporary response rates in bone. However, advances in immunotherapy and targeted therapies have improved outcomes in patients with bone metastases. In the setting of metastatic disease, most patients are on systemic therapy for the treatment of their overall disease, not just for the treatment or palliation of a metastatic lesion. Therefore, when surgical intervention is indicated, open communication between orthopedic surgery and medical oncology teams is critical to understand systemic therapy schedules and plan surgery accordingly. Systemic anti-neoplastic therapy often needs to be held during the perioperative period, and, as such, poorly planned surgical interventions may cause harmful disruptions in life-prolonging treatment [71].

Bone-targeted agents (BTAs), including bisphosphonates and denosumab, inhibit osteoclastic activity and are commonly employed for skeletal metastases to prevent new fracture and strengthen bone after treatment with surgery or RT [33]. Used in tandem with systemic anticancer therapies, BTAs are effective in delaying and reducing the incidence of skeletal-related events and can be employed at the time of bone metastasis diagnosis to prevent future fractures [72, 73]. Several studies have demonstrated noninferior or even improved outcomes with denosumab versus bisphosphonates, along with a reduced toxicity profile [74–76]. Caution should be exercised when considering stopping BTAs, as subsequent disease progression can be rapid in the absence of the BTA's protective effect. Other risks of BTAs include hypocalcemia, atypical fractures, osteonecrosis of the jaw, and renal insufficiency.

Operative Treatment

Operative treatment must balance the need to avoid additional surgical morbidity with the need for a stable, durable construct that obviates future reoperation. For patients with shorter life expectancy, chemo- or radiosensitive malignancies, and/or diffuse metastatic disease, operative treatment should be as minimally invasive as possible, allowing for shorter recovery and a faster return to systemic anticancer therapy. For patients with longer life expectancy, particularly those with oligometastatic disease, a more aggressive surgical approach—curative-intent metastasectomy followed by reconstruction—may be beneficial [32]. Minimally invasive surgery is relatively straightforward in the extremities, but, until recently, it has been challenging in complex areas such as the pelvis and spine. Establishing a primary diagnosis prior to surgery is recommended, but, when not feasible, an open biopsy should be conducted before fixation to confirm a diagnosis of metastatic disease. For patients with highly vascular skeletal metastases, such as those with thyroid or renal cell carcinoma, multiple myeloma, or vascularized soft tissue components, preoperative embolization is important to limit intraoperative blood loss.

In the extremities, options for surgical management include intramedullary nailing, plate and screw constructs, photodynamic intramedullary nails, and/or cementation. Bone cement (polymethylmethacrylate) and bone substitutes can be used to fill bone defects and provide a rigid substrate for screw fixation. In the femur, dynamic hip screws or cephalomedullary nails are critical for protecting the integrity of the femoral neck [77]. These fixation techniques may be used alone or in combination with one another and may be paired with techniques such as curettage or radio-frequency ablation, which can be useful for local control. When disease involves or abuts articular surfaces, or there is inadequate bone stock for fixation, resection and reconstruction with joint arthroplasty and endoprostheses is typically recommended. Regardless of the surgical strategy, fixation should span the entire length of disease to mitigate the risks of disease progression and allow fixation into uninvolved bone. Newer nonconventional implants, including photodynamic balloons (IlluminOss Medical, Inc.) and carbon fiber implants (CarboFix Orthopaedics, Ltd.), are radiolucent and aid multidisciplinary disease management by improving disease surveillance and radiation therapy planning [78, 79].

Surgical management of the spine and pelvis has historically been challenging; however, the recent adoption of minimally invasive techniques has changed the risk–benefit calculus. Pelvis metastases—particularly those in the periacetabular region—can be managed effectively with percutaneous screw fixation, photodynamic balloon stabilization, or cementoplasty [80–83] Open surgery is generally avoided, if possible, but, in cases of oligometastatic disease or where minimally invasive strategies provide insufficient stability, surgery may consist of en bloc resection and reconstruction with endoprostheses or custom-made implants.

Post-treatment Considerations

After surgery, the goal is to get patients back to weight-bearing, ambulation, and other activities of daily living. Patients should start with physical therapy on postoperative day one to encourage early ambulation. As previously discussed, the fixation strategy should be durable enough to prevent further fractures despite potential disease progression, but patients should still be regularly monitored for implant failure. Standard infection prevention precautions should be taken, as downstream disruptions in oncologic care because of infection can be catastrophic. One study showed an increased risk of thromboembolic events in patients undergoing prophylactic fixation versus after-fracture care [84]; regardless, patients should be put on standard postoperative anticoagulation for prevention of deep vein thrombosis and pulmonary embolism. Despite improved outcomes for prophylactic fixation, complications still occur, which highlights the need for careful preoperative optimization and postoperative monitoring.

Conclusions

Management of impending pathologic fractures due to skeletal metastases is complex and requires a multidisciplinary approach. Several risk factors for fractures have been identified, but the overall assessment of risk is challenging, and, despite being a significant research focus over the past two decades, current risk stratification tools have limitations. The goal of treatment, which includes combinations of surgery, radiation therapy, and systemic treatments, is to prevent a complete fracture, allow weight-bearing, and improve quality of life.

References

1. DeVita VT Jr, Lawrence TS, Rosenberg SA. DeVita, Hellman, and Rosenberg's cancer: principles & practice of oncology. 10th ed. Lippincott Williams & Wilkins; 2015.
2. Azar FM, Canale S, Beaty JH. In: Campbell WC, editor. Campbell's operative orthopaedics. 14th ed. Philadelphia: Elsevier; 2020.
3. Bryson DJ, Wicks L, Ashford RU. The investigation and management of suspected malignant pathological fractures: a review for the general orthopaedic surgeon. Injury. 2015;46(10):1891–9. PMID: 26254572.
4. Rajani R, Quinn RH, Throckmorton TW, Fischer SJ. Metastatic bone disease [Internet]. AAOS. 2021 [cited 2025 Apr 28]. Available from: https://orthoinfo.aaos.org/en/diseases%2D%2Dconditions/metastatic-bone-disease/.
5. Coleman RE. Clinical features of metastatic bone disease and risk of skeletal morbidity. Clin Cancer Res. 2006;12(20 Pt 2):6243s–9s. PMID: 17062708.
6. Damron TA, Mann KA. Fracture risk assessment and clinical decision making for patients with metastatic bone disease. J Orthop Res. 2020;38(6):1175–90. PMID: 32162711.
7. Yanagisawa S, Uchiyama Y, Tanzawa Y, Watanabe T, Wasai S, Suzuki T, Watanabe M. Comparison of the clinical outcomes of intramedullary nailing between impending and completed pathological fractures caused by metastatic femoral tumors from solid cancers. J Orthop Sci. 2024;29(4):1119–24. PMID: 37507316.
8. Nooh A, Goulding K, Isler MH, Mottard S, Arteau A, Dion N, Turcotte R. Early improvement in pain and functional outcome but not quality of life after surgery for metastatic long bone disease. Clin Orthop Relat Res. 2018;476(3):535–45. PMID: 29529637.
9. Blank AT, Lerman DM, Patel NM, Rapp TB. Is prophylactic intervention more cost-effective than the treatment of pathologic fractures in metastatic bone disease? Clin Orthop Relat Res. 2016;474(7):1563–70. PMID: 27020430.
10. Philipp TC, Mikula JD, Doung YC, Gundle KR. Is there an association between prophylactic femur stabilization and survival in patients with metastatic bone disease? Clin Orthop Relat Res. 2020;478(3):540–6. PMID: 32168065.
11. Groot OQ, Lans A, Twining PK, Bongers MER, Kapoor ND, Verlaan JJ, Newman ET, Raskin KA, Lozano-Calderon SA, Janssen SJ, Schwab JH. Clinical outcome differences in the treatment of impending versus completed pathological long-bone fractures. J Bone Joint Surg Am. 2022;104(4):307–15. PMID: 34851323.
12. Damron TA, Morgan H, Prakash D, Grant W, Aronowitz J, Heiner J. Critical evaluation of Mirels' rating system for impending pathologic fractures. Clin Orthop Relat Res 2003;(415 Suppl):S201–7. PMID: 14600611.

13. Christ AB, Piple AS, Gettleman BS, Duong A, Chen M, Wang JC, Heckmann ND, Menendez L. Prevalence of primary malignant tumours, rates of pathological fracture, and mortality in the setting of metastatic bone disease. Bone Jt Open. 2023;4(6):424–31. PMID: 37272304.
14. Huang JF, Shen J, Li X, Rengan R, Silvestris N, Wang M, Derosa L, Zheng X, Belli A, Zhang XL, Li YM, Wu A. Incidence of patients with bone metastases at diagnosis of solid tumors in adults: a large population-based study. Ann Transl Med. 2020;8(7):482. PMID: 32395526.
15. Coleman RE. Metastatic bone disease: clinical features, pathophysiology and treatment strategies. Cancer Treat Rev. 2001;27(3):165–76. PMID: 11417967.
16. Bubendorf L, Schöpfer A, Wagner U, Sauter G, Moch H, Willi N, Gasser TC, Mihatsch MJ. Metastatic patterns of prostate cancer: an autopsy study of 1,589 patients. Hum Pathol. 2000;31(5):578–83. PMID: 10836297.
17. Galasko CS. Skeletal metastases. Clin Orthop Relat Res. 1986;(210):18–30. PMID: 3757360.
18. Kyle RA, Gertz MA, Witzig TE, Lust JA, Lacy MQ, Dispenzieri A, Fonseca R, Rajkumar SV, Offord JR, Larson DR, Plevak ME, Therneau TM, Greipp PR. Review of 1027 patients with newly diagnosed multiple myeloma. Mayo Clin Proc. 2003;78(1):21–33. PMID: 12528874.
19. Demircay E, Hornicek FJ, Mankin HJ, Degroot H. Malignant lymphoma of bone: a review of 119 patients. Clin Orthop Relat Res. 2013;471(8):2684–90. PMID: 23591933.
20. Nie H, Yuan Y, Li J, Xu F, Wang T, Bian L, Jiang Z, Zhang S. Occurrence and distribution of bone metastases in 984 metastatic breast cancer patients. Transl Breast Cancer Res. 2021;2:4–4.
21. Roato I. Bone metastases: when and how lung cancer interacts with bone. World J Clin Oncol. 2014;5(2):149–55. PMID: 24829862.
22. Fones L, Schroeder JT, Beredjiklian PK, Gallant GG. Acral metastasis to the hand as the primary presentation of malignancy. J Hand Surg Glob Online. 2024;6(4):593–6. PMID: 39166202.
23. Miller BJ, Soni EEC, Gibbs CP, Scarborough MT. Intramedullary nails for long bone metastases: why do they fail? Orthopedics. 2011;34(4). PMID: 21469628.
24. Zajączkowska R, Kocot-Kępska M, Leppert W, Wordliczek J. Bone pain in cancer patients: mechanisms and current treatment. Int J Mol Sci. 2019;20(23). PMID: 31801267.
25. Isaac A, Dalili D, Dalili D, Weber MA. State-of-the-art imaging for diagnosis of metastatic bone disease. Radiologe. 2020;60(Suppl 1):1–16. PMID: 32211916.
26. Van der Linden YM, Dijkstra PDS, Kroon HM, Lok JJ, Noordijk EM, Leer JWH, Marijnen CAM. Comparative analysis of risk factors for pathological fracture with femoral metastases. J Bone Joint Surg Br. 2004;86(4):566–73. PMID: 15174555.
27. van der Wal CWPG, Eggermont F, Fiocco M, Kroon HM, Ayu O, Slot A, Snyers A, Rozema T, Verdonschot NJJ, Dijkstra PDS, Tanck E, van der Linden YM. Axial cortical involvement of metastatic lesions to identify impending femoral fractures; a clinical validation study. Radiother Oncol. 2020;144:59–64. PMID: 31733489.
28. Ulaner GA, Zindman AM, Zheng J, Kim TWB, Healey JH. FDG PET/CT assesses the risk of femoral pathological fractures in patients with metastatic breast cancer. Clin Nucl Med. 2017;42(4):264–70. PMID: 28166159.
29. Riaz S, Bashir H, Niazi IK, Butt S, Qamar F. 99mTc MDP SPECT-CT-based modified Mirels classification for evaluation of risk of fracture in skeletal metastasis: a pilot study. Clin Nucl Med. 2018;43(6):e180–3. PMID: 29561523.
30. Rosenthal DI. Radiologic diagnosis of bone metastases. Cancer. 1997;80(8 Suppl):1595–607. PMID: 9362427.
31. Rougraff BT, Kneisl JS, Simon MA. Skeletal metastases of unknown origin. A prospective study of a diagnostic strategy. J Bone Joint Surg Am. 1993;75(9):1276–81. PMID: 8408149.
32. Langerhuizen DWG, Janssen SJ, van der Vliet QMJ, Raskin KA, Ferrone ML, Hornicek FJ, Schwab JH, Lozano-Calderón SA. Metastasectomy, intralesional resection, or stabilization only in the treatment of bone metastases from renal cell carcinoma. J Surg Oncol. 2016;114(2):237–45. PMID: 27156495.
33. Soares do Brito J, Lopes-Brás R, Abrunhosa-Branquinho A, Fernandes I, Gomes I, Casimiro S, Costa L. A tailored approach for appendicular impending and pathologic fractures in solid cancer metastases. Cancers (Basel). 2022;14(4). PMID: 35205641.

34. Benca E, Patsch JM, Mayr W, Pahr DH, Windhager R. The insufficiencies of risk analysis of impending pathological fractures in patients with femoral metastases: a literature review. Bone Rep. 2016;5:51–6. PMID: 28326347.
35. Harrington KD. New trends in the management of lower extremity metastases. Clin Orthop Relat Res. 1982;169:53–61. PMID: 7105590.
36. Carnesale P. Malignant tumours of bone. In: St Canale J, editor. Campbell's operative orthopaedics. 10th ed. Saint Louis: Mosby-Year Book Inc.; 2003. p. 551.
37. Mirels H. Metastatic disease in long bones. A proposed scoring system for diagnosing impending pathologic fractures. Clin Orthop Relat Res. 1989;249:256–64. PMID: 2684463.
38. Jawad MU, Scully SP. In brief: classifications in brief: Mirels' classification: metastatic disease in long bones and impending pathologic fracture. Clin Orthop Relat Res. 2010;468(10):2825–7. PMID: 20352387.
39. Ramsey DC, Lam PW, Hayden J, Doung YC, Gundle KR. Mirels scores in patients undergoing prophylactic stabilization for femoral metastatic bone disease in the veterans administration healthcare system. J Am Acad Orthop Surg Glob Res Rev. 2020;4(9):e20.00141.
40. Fourney DR, Frangou EM, Ryken TC, Dipaola CP, Shaffrey CI, Berven SH, Bilsky MH, Harrop JS, Fehlings MG, Boriani S, Chou D, Schmidt MH, Polly DW, Biagini R, Burch S, Dekutoski MB, Ganju A, Gerszten PC, Gokaslan ZL, Groff MW, Liebsch NJ, Mendel E, Okuno SH, Patel S, Rhines LD, Rose PS, Sciubba DM, Sundaresan N, Tomita K, Varga PP, Vialle LR, Vrionis FD, Yamada Y, Fisher CG. Spinal instability neoplastic score: an analysis of reliability and validity from the spine oncology study group. J Clin Oncol. 2011;29(22):3072–7. PMID: 21709187.
41. Murtaza H, Sullivan CW. Classifications in brief: the spinal instability neoplastic score. Clin Orthop Relat Res. 2019;477(12):2798–803. PMID: 31764353.
42. Tatar Z, Soubrier M, Dillies AF, Verrelle P, Boisgard S, Lapeyre M. Assessment of the risk factors for impending fractures following radiotherapy for long bone metastases using CT scan-based virtual simulation: a retrospective study. Radiat Oncol. 2014;9:227. PMID: 25319635.
43. Cheal EJ, Hipp JA, Hayes WC. Evaluation of finite element analysis for prediction of the strength reduction due to metastatic lesions in the femoral neck. J Biomech. 1993;26(3):251–64. PMID: 8468338.
44. Whealan KM, Kwak SD, Tedrow JR, Inoue K, Snyder BD. Noninvasive imaging predicts failure load of the spine with simulated osteolytic defects. J Bone Joint Surg Am. 2000;82(9):1240–51. PMID: 11005515.
45. Hong J, Cabe GD, Tedrow JR, Hipp JA, Snyder BD. Failure of trabecular bone with simulated lytic defects can be predicted non-invasively by structural analysis. J Orthop Res. 2004;22(3):479–86. PMID: 15099624.
46. Damron TA, Nazarian A, Entezari V, Brown C, Grant W, Calderon N, Zurakowski D, Terek RM, Anderson ME, Cheng EY, Aboulafia AJ, Gebhardt MC, Snyder BD. CT-based structural rigidity analysis is more accurate than Mirels scoring for fracture prediction in metastatic femoral lesions. Clin Orthop Relat Res. 2016;474(3):643–51. PMID: 26169800.
47. Sas A, Tanck E, Sermon A, van Lenthe GH. Finite element models for fracture prevention in patients with metastatic bone disease. A literature review Bone Rep 2020;12:100286. PMID: 32551337.
48. Keyak JH, Kaneko TS, Tehranzadeh J, Skinner HB. Predicting proximal femoral strength using structural engineering models. Clin Orthop Relat Res. 2005;437:219–28. PMID: 16056052.
49. Nguyễn MV, Carlier C, Nich C, Gouin F, Crenn V. Fracture risk of long bone metastases: a review of current and new decision-making tools for prophylactic surgery. Cancers (Basel). 2021;13(15). PMID: 34359563.
50. Sternheim A, Traub F, Trabelsi N, Dadia S, Gortzak Y, Snir N, Gorfine M, Yosibash Z. When and where do patients with bone metastases actually break their femurs? Bone Joint J. 2020;102-B(5):638–45. PMID: 32349590.
51. Eggermont F, van der Wal G, Westhoff P, Laar A, de Jong M, Rozema T, Kroon HM, Ayu O, Derikx L, Dijkstra S, Verdonschot N, van der Linden Y, Tanck E. Patient-specific finite

element computer models improve fracture risk assessments in cancer patients with femoral bone metastases compared to clinical guidelines. Bone. 2020;130:115101. PMID: 31655223.
52. Goodheart JR, Cleary RJ, Damron TA, Mann KA. Simulating activities of daily living with finite element analysis improves fracture prediction for patients with metastatic femoral lesions. J Orthop Res. 2015;33(8):1226–34. PMID: 25761000.
53. Oftadeh R, Karimi Z, Villa-Camacho J, Tanck E, Verdonschot N, Goebel R, Snyder BD, Hashemi HN, Vaziri A, Nazarian A. Curved beam computed tomography based structural rigidity analysis of bones with simulated lytic defect: a comparative study with finite element analysis. Sci Rep. 2016;6:32397. PMID: 27585495.
54. Anderson AB, Wedin R, Fabbri N, Boland P, Healey J, Forsberg JA. External validation of PATHFx version 3.0 in patients treated surgically and nonsurgically for symptomatic skeletal metastases. Clin Orthop Relat Res. 2020;478(4):808–18. PMID: 32195761.
55. Carrwik C, Tsagkozis P, Wedin R, Robinson Y. Predicting survival of patients with spinal metastatic disease using PathFx 3.0 – a validation study of 668 patients in Sweden. Brain & spine. 2022;2:101669. PMID: 36506283.
56. Lee CC, Chen CW, Yen HK, Lin YP, Lai CY, Wang JL, Groot OQ, Janssen SJ, Schwab JH, Hsu FM, Lin WH. Comparison of two modern survival prediction tools, SORG-MLA and METSSS, in patients with symptomatic long-bone metastases who underwent local treatment with surgery followed by radiotherapy and with radiotherapy alone. Clin Orthop Relat Res. 2024;482(12):2193–208. PMID: 39051924.
57. de Groot TM, Ramsey D, Groot OQ, Fourman M, Karhade AV, Twining PK, Berner EA, Fenn BP, Collins AK, Raskin K, Lozano S, Newman E, Ferrone M, Doornberg JN, Schwab JH. Does the SORG machine-learning algorithm for extremity metastases generalize to a contemporary cohort of patients? Temporal validation from 2016 to 2020. Clin Orthop Relat Res. 2023;481(12):2419–30. PMID: 37229565.
58. Thio QCBS, Karhade AV, Bindels BJ, Ogink PT, Bramer JAM, Ferrone ML, Calderón SL, Raskin KA, Schwab JH. Development and internal validation of machine learning algorithms for preoperative survival prediction of extremity metastatic disease. Clin Orthop Relat Res. 2020;478(2):322–33. PMID: 31651589.
59. Tseng TE, Lee CC, Yen HK, Groot OQ, Hou CH, Lin SY, Bongers MER, Hu MH, Karhade AV, Ko JC, Lai YH, Yang JJ, Verlaan JJ, Yang RS, Schwab JH, Lin WH. International validation of the SORG machine-learning algorithm for predicting the survival of patients with extremity metastases undergoing surgical treatment. Clin Orthop Relat Res. 2022;480(2):367–78. PMID: 34491920.
60. Oh E, Seo SW, Yoon YC, Kim DW, Kwon S, Yoon S. Prediction of pathologic femoral fractures in patients with lung cancer using machine learning algorithms: comparison of computed tomography-based radiological features with clinical features versus without clinical features. J Orthop Surg (Hong Kong). 2017;25(2):2309499017716243. PMID: 28659051.
61. Rich SE, Chow R, Raman S, Liang Zeng K, Lutz S, Lam H, Silva MF, Chow E. Update of the systematic review of palliative radiation therapy fractionation for bone metastases. Radiother Oncol. 2018;126(3):547–57. PMID: 29397209.
62. Alcorn S, Cortés ÁA, Bradfield L, Brennan M, Dennis K, Diaz DA, Doung YC, Elmore S, Hertan L, Johnstone C, Jones J, Larrier N, Lo SS, Nguyen QN, Tseng YD, Yerramilli D, Zaky S, Balboni T. External beam radiation therapy for palliation of symptomatic bone metastases: an ASTRO clinical practice guideline. Pract Radiat Oncol. 2024;14(5):377–97. PMID: 38788923.
63. Rosen DB, Haseltine JM, Bartelstein M, Flynn JR, Zhang Z, Kohutek ZA, Yamada Y, Schmitt A, Higginson DS, Vaynrub M, Yang JT, Gillespie EF. Should postoperative radiation for long bone metastases cover part or all of the orthopedic hardware? Results of a large retrospective analysis. Adv Radiat Oncol. 2021;6(6):100756. PMID: 34585024.
64. Summers AR, Philipp T, Mikula JD, Gundle KR. The role of postoperative radiation and coordination of care in patients with metastatic bone disease of the appendicular skeleton. Orthop Rev (Pavia). 2017;9(4):7261. PMID: 29564074.

65. Vijayakumar G, Tang A, Vance D, Yoon RS, Sweeney K, Blank AT. Does perioperative radiation affect implant survivorship of primary Total hip arthroplasty in the setting of metastatic bone disease? Arthroplast Today. 2024;25:101296. PMID: 38292148.
66. Willeumier JJ, Kaynak M, van der Zwaal P, Meylaerts SAG, Mathijssen NMC, Jutte PC, Tsagozis P, Wedin R, van de Sande MAJ, Fiocco M, Dijkstra PDS. What factors are associated with implant breakage and revision after intramedullary nailing for femoral metastases? Clin Orthop Relat Res. 2018;476(9):1823–33. PMID: 30566108.
67. Kraus RD, Weil CR, Wells S, Tward JD, Groundland JS, Jones KB, Cannon DM. Radiation therapy in conjunction with surgical stabilization of impending or pathologic fractures secondary to metastasis: is there a difference between single and multifraction regimens? Adv Radiat Oncol. 2022;7(2):100795. PMID: 35128177.
68. Shimoyama T, Katagiri H, Harada H, Murata H, Wasa J, Hosaka S, Suzuki T, Takahashi M, Asakura H, Nishimura T, Yamada H. Fracture after radiation therapy for femoral metastasis: incidence, timing and clinical features. J Radiat Res. 2017;58(5):661–8.
69. Inchaustegui ML, Kon-Liao K, Ruiz-Arellanos K, Silva GAE, Gonzalez MR, Pretell-Mazzini J. Treatment and outcomes of radiation-induced soft tissue sarcomas of the extremities and trunk-a systematic review of the literature. Cancers (Basel). 2023;15(23). PMID: 38067287.
70. Omurzakov A, Rampam S, Gonzalez MR, Lozano-Calderon SA. What is the incidence and non-union rate of radiation-associated fractures? – a systematic review of the literature. Radiother Oncol. 2025;202:110617. PMID: 39510139.
71. Okay E, Connolly JJ, Gonzalez MR, Lozano-Calderon SA. Wound-healing effects of common antineoplastic agents and perioperative considerations for the orthopaedic surgeon. J Am Acad Orthop Surg. 2024;32(14):e671–82. PMID: 38713761.
72. Coleman RE, Croucher PI, Padhani AR, Clézardin P, Chow E, Fallon M, Guise T, Colangeli S, Capanna R, Costa L. Bone metastases. Nat Rev Dis Primers. 2020;6(1):83. PMID: 33060614.
73. Stopeck AT, Lipton A, Body JJ, Steger GG, Tonkin K, de Boer RH, Lichinitser M, Fujiwara Y, Yardley DA, Viniegra M, Fan M, Jiang Q, Dansey R, Jun S, Braun A. Denosumab compared with zoledronic acid for the treatment of bone metastases in patients with advanced breast cancer: a randomized, double-blind study. J Clin Oncol. 2010;28(35):5132–9. PMID: 21060033.
74. Fizazi K, Carducci M, Smith M, Damião R, Brown J, Karsh L, Milecki P, Shore N, Rader M, Wang H, Jiang Q, Tadros S, Dansey R, Goessl C. Denosumab versus zoledronic acid for treatment of bone metastases in men with castration-resistant prostate cancer: a randomised, double-blind study. Lancet. 2011;377(9768):813–22. PMID: 21353695.
75. Raje N, Terpos E, Willenbacher W, Shimizu K, García-Sanz R, Durie B, Legieć W, Krejčí M, Laribi K, Zhu L, Cheng P, Warner D, Roodman GD. Denosumab versus zoledronic acid in bone disease treatment of newly diagnosed multiple myeloma: an international, double-blind, double-dummy, randomised, controlled, phase 3 study. Lancet Oncol. 2018;19(3):370–81. PMID: 29429912.
76. Henry DH, Costa L, Goldwasser F, Hirsh V, Hungria V, Prausova J, Scagliotti GV, Sleeboom H, Spencer A, Vadhan-Raj S, von Moos R, Willenbacher W, Woll PJ, Wang J, Jiang Q, Jun S, Dansey R, Yeh H. Randomized, double-blind study of denosumab versus zoledronic acid in the treatment of bone metastases in patients with advanced cancer (excluding breast and prostate cancer) or multiple myeloma. J Clin Oncol. 2011;29(9):1125–32. PMID: 21343556.
77. O'Sullivan MB, Saha D, Clement JM, Dowsett RJ, Pacheco RA, Balach T. Team approach: the treatment of metastatic tumors of the femoral diaphysis. JBJS Rev. 2017;5(2). PMID: 28248740.
78. Lam FC, Guru S, AbuReesh D, Hori YS, Chuang C, Liu L, Wang L, Gu X, Szalkowski GA, Wang Z, Wohlers C, Tayag A, Emrich SC, Ustrzynski L, Zygourakis CC, Desai A, Hayden Gephart M, Byun J, Pollom EL, Rahimy E, Soltys S, Park DJ, Chang SD. Use of carbon fiber implants to improve the safety and efficacy of radiation therapy for spine tumor patients. Brain Sci. 2025;15(2). PMID: 40002531.
79. Depauw N, Pursley J, Lozano-Calderon SA, Patel CG. Evaluation of carbon fiber and titanium surgical implants for proton and photon therapy. Pract Radiat Oncol. 2023;13(3):256–62. PMID: 36738918.

80. Lozano-Calderon SA, Gonzalez MR, Werenski JO, Quinn K, Freiberger D, Raskin KA. A prospective study of the illuminoss photodynamic nail system for pelvic stabilization: treatment of impending and actual fractures from metastatic bone disease, multiple myeloma, and primary bone lymphoma. JB JS Open Access. 2024;9(3). PMID: 39281295.
81. Yang R, Goch A, Murphy D, Wang J, Charubhumi V, Fox J, Sen M, Hoang B, Geller D. A novel tripod percutaneous reconstruction technique in periacetabular lesions caused by metastatic cancer. J Bone Joint Surg Am. 2020;102(7):592–9. PMID: 32079881.
82. English DI, Lea WB, King DM, Tutton SM, Neilson JC. Minimally invasive stabilization with or without ablation for metastatic Periacetabular tumors. J Bone Joint Surg Am. 2021;103(13):1184–92. PMID: 34038393.
83. Lee FY, Latich I, Toombs C, Mungur A, Conway D, Alder K, Ibe I, Lindskog D, Friedlaender G. Minimally invasive image-guided ablation, osteoplasty, reinforcement, and internal fixation (AORIF) for osteolytic lesions in the pelvis and Periarticular regions of weight-bearing bones. J Vasc Interv Radiol. 2020;31(4):649–658.e1. PMID: 32139256.
84. Aneja A, Jiang JJ, Cohen-Rosenblum A, Luu HL, Peabody TD, Attar S, Luo TD, Haydon RC. Thromboembolic disease in patients with metastatic femoral lesions: a comparison between prophylactic fixation and fracture fixation. J Bone Joint Surg Am. 2017;99(4):315–23. PMID: 28196033.

Management of Spinal Metastatic Disease

13

Spencer M. Richardson, Zayd Almaya, and Motasem Al Maaieh

Epidemiology

The spine is the most common site of metastatic disease to the skeletal system and is estimated to occur in up to one-third of patients with cancer based on cadaveric studies; however, spinal metastases are diagnosed in the clinical setting at a much lower rate (Fig. 13.1) [2, 3], Cancers most commonly found to metastasize to the spine include breast, prostate, and lung carcinoma, while gastrointestinal cancers have the lowest rate of metastasis to the spine [4]. Other common malignancies metastasizing to the spine include thyroid cancer, renal cancer, and lymphoma. The thoracic spine is responsible for the majority (~70%) of symptomatic lesions [5]. Metastatic epidural spinal cord compression (MESCC) is a sequela of metastatic disease to the spine that occurs in approximately 5% of cancer patients and can precede diagnosis of malignancy in up to 20% of cases [6]. Approximately 10% of patients with known skeletal metastatic disease will develop MESCC or pathologic vertebral compression fracture [3]. Improved systemic treatment options for many cancers have increased survival, and thus there is an increasing need for treatment of symptomatic spinal metastases [4].

S. M. Richardson
Department of Orthopaedic Surgery, Indiana University School of Medicine, Indianapolis, IN, USA
e-mail: spenrich@iu.edu

Z. Almaya
Butler University, Indianapolis, IN, USA
e-mail: zalmaya@butler.edu

M. Al Maaieh (✉)
Associate Professor of Orthopaedic Surgery, Department of Orthopaedic Surgery, Indiana University School of Medicine, Indianapolis, IN, USA
e-mail: malmaaieh@iuhealth.org

J. Pretell-Mazzini, S. A. Lozano Calderón (eds.), *Multidisciplinary Approach of Musculoskeletal Metastasis*, https://doi.org/10.1007/978-3-032-09498-8_13

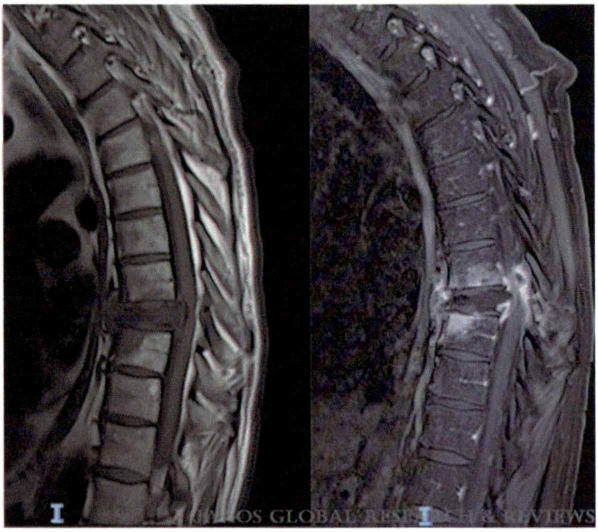

Fig. 13.1 T1 without (left) and with (right) contrast sagittal MRI demonstrating metastatic disease of thoracic spine. (From Ryan et al. [1]. Used under Creative Commons CC BY-ND)

Surgical Indications

Indications for surgical treatment of metastatic disease to the spine can include pathological fracture, tissue biopsy for diagnosis, pain, progressive neurologic symptoms, and instability. Patient factors such as comorbidities and prognosis are important in determining whether surgery is indicated or not. The Neurologic, Oncologic, Mechanical, and Systemic (NOMS) decision framework was developed as a multidisciplinary approach to spinal metastases (Fig. 13.2) [8]. Neurologic assessment involves both the presence of neurologic deficits and the extent of epidural spinal cord compression (ESCC). ESCC is graded on the degree of spinal cord compression and has been shown to have high inter- and intra-observer reliability [9]. Compression rated Grade 2 or 3 is considered high-grade ESCC and requires surgical decompression prior to radiation therapy unless the tumor is highly radiosensitive (Fig. 13.3). Patients with high grade compression are at risk of rapidly progressive paralysis [11]. The oncologic decision point is made based on the responsiveness of a tumor histology to current treatments. Metastatic lesions are radiosensitive or radioresistant based on response to conventional external beam radiotherapy (cEBRT). Radiosensitive histologies include lymphoma, seminoma, multiple myeloma, breast, prostate, ovarian, and neuroendocrine carcinomas. Histologies with unfavorable or intermediate response to radiotherapy include gastrointestinal, non-small cell lung carcinoma, melanoma and sarcoma [12]. Conventional radiotherapy is increasingly being replaced by stereotactic body radiotherapy therapy (SBRT); which allows ablative doses of radiation to be delivered to the site of disease without significant toxicity to surrounding critical structures such as the spinal cord [13]. Patients with high grade ESCC have metastatic disease compressing the spinal cord. Separation surgery is indicated in patients with

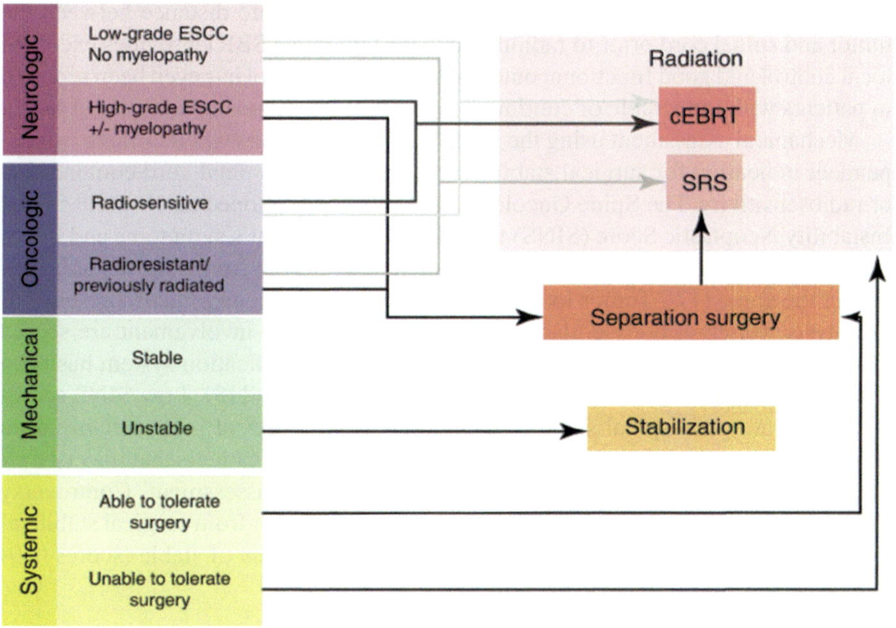

Fig. 13.2 The NOMS Framework. (Zuckerman et al. [7]. Used with permission)

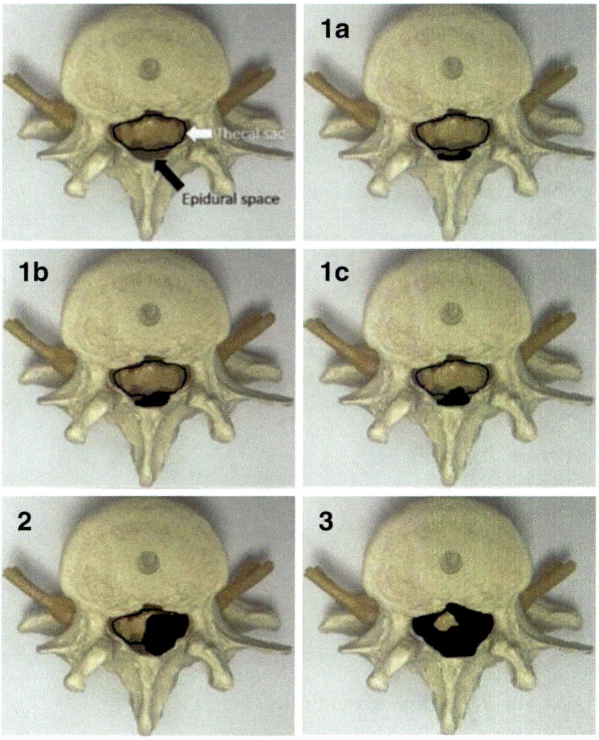

Fig. 13.3 Epidural spinal cord compression (ESCC) is rated on a grade of 1–3 based on the amount of compression and displacement of the spinal cord. The thecal sac and epidural space is labeled in the top left image. Grade 1 compression (1a-c) includes compression in the epidural space without deformity of thecal sac. Grade 2 (2 bottom left) shows compression and deformity of the thecal sac. Grade 3 (3 bottom right) involves compression of the thecal sac and no visible cerebrospinal fluid. (Used with permission. From Goodwin et al. [10])

high-grade ESCC and radioresistant tumors to create a safe distance between the tumor and spinal cord prior to radiotherapy. Postoperative SBRT provides excellent local control and good functional outcomes, and local control has even been reported in patients with inoperable or continued high-grade ESCC postoperatively [14–16].

Mechanical assessment using the NOMS criteria considers whether there is independent indication for surgical stabilization, regardless of spinal cord compression or radiosensitivity. The Spine Oncology Study Group developed an 18-point Spinal Instability Neoplastic Score (SINS) that incorporates patient's symptoms and radiographic findings to determine whether a patient may benefit from surgical stabilization of the spine [17]. Tumor location, patient's pain, spinal alignment, presence of osteolysis, vertebral body collapse, and posterior element involvement are scored for a maximum of 18 points (Table 13.1). The SINS classification system has been found to have excellent inter- and intra-observer reliability [18]. Low SINS scores (0–6) indicate good spinal stability while high SINS scores (13–18) reliably need surgical stabilization. Intermediate scores (7–12) represent patients that may or may not benefit from surgical stabilization and need further assessment. Controversy exists on whether patients with a SINS score of 7–12 benefit from surgical stabilization with some authors advocating for a binary classification of stable (scores 0–9) and unstable (scores 10–18) [19–23].

Table 13.1 The SINS Classification System

Spinal Instability Neoplastic Score (SINS)	
Location	
Junctional (Occiput-C2, C7-T2, T11-L1, L5-S1)	3
Mobile Spine (C3-C6, L2-L4)	2
Semi-rigid (T3-T10)	1
Rigid (S2-S5)	0
Pain relief with recumbency and/or pain with movement/loading of the spine	
Yes	3
No (occasional pain but not mechanical)	1
Pain free lesion	0
Bone lesion	
Lytic	2
Mixed (lytic/blastic)	1
Blastic	0
Radiographic spinal alignment	
Subluxation/translation present	4
de novo deformity (kypohsis/scoliosis)	2
Normal alignment	0
Vertebral body collapse	
>50% Collapse	3
<50% Collapse	2
No collapse with >50% body involved	1
None of the above	0
Posterolateral involvement of the spinal elements (facet, pedicle, or CV joint fracture)	
Bilateral	3
Unilateral	1
None of the above	0

Used with Permission. From Fisher et al. [17]

The systemic assessment considers the patient's comorbidities and overall tumor burden to determine what intervention a patient can tolerate without unfavorable morbidity or mortality. Patient's functional status is commonly measured by the Karnofsky Performance Status (KPS) and the Eastern Cooperative Oncology Group (ECOG) scale [24, 25]. Consideration for nonsurgical or non-instrumented intervention such as laminotomy or non-destabilizing laminectomies should be given to patients with low performance status. Expected patient survival is also considered, as any surgical intervention should allow patient improvement prior to death. A scoping review of emerging evidence related to systemic considerations in surgical treatment of spinal metastatic disease examined 61 articles accounting for 22,335 patients and identified various patient demographics, medical comorbidities, laboratory abnormalities, poor functional status, and systemic disease burden as negatively associated with outcomes [26]. Further work is needed to better define these variables and allow improved decision-making.

The risks of surgery may outweigh the benefits if a patient does not survive long after surgery. Various scoring systems have been developed to predict prognosis of patients with metastatic disease to the spine, though all have accuracy less than 90% [27, 28]. Primary tumor histology and the presence of visceral metastases are the most commonly employed variable in prognostic models [29]. The Skeletal Oncology Research Group (SORG) Nomogram has the highest accuracy at predicting 30-day, 90-day, and 1-year mortality [30–32]. The following factors were used to create a nomogram to determine patient-specific risk of death at 30-days, 90-days, and 1- year (Fig. 13.4).

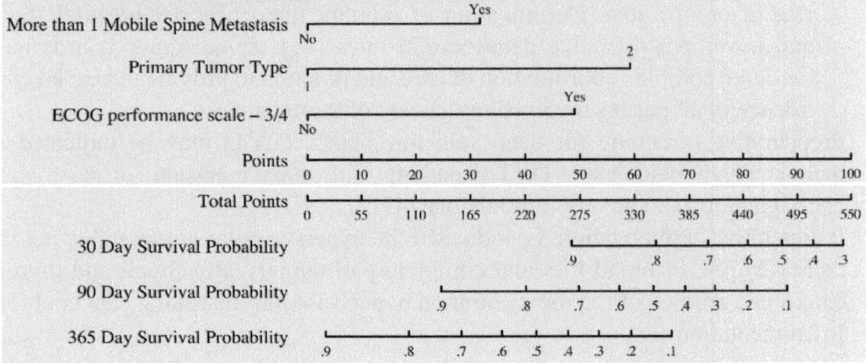

Fig. 13.4 The SORG Nomogram. (Used with permission. From Zuckerman et al. [7])

Further iterations of the SORG algorithms have been developed and externally validated. These models have been calibrated to multiple time points and are available for clinicians as web-based risk calculators that produce patient-specific probability of mortality [29, 33–35]. Traditionally, a patient's lifespan of at least 90 days has been desired before any surgical intervention. However, a recent review of 253 patients who underwent surgery for spinal metastasis found that health-related quality of life (HRQOL) was independent of whether the patient survived 3 months postoperatively. The authors of this study advocate for baseline performance status to be prioritized over predicted survival when making surgical decisions for these patients [36]. The SORG group further developed a machine learning model to predict 6-week mortality which has been externally validated [33]. The New England Spinal Metastasis Score (NESMS) is a relatively simple scoring system that is applicable to patients regardless of treatment strategy and has been prospectively validated for 6-month and 1-year mortality [37]. Additionally, palliative surgery may be increasing appropriately for patients with a short lifespan with the increased focus on minimally invasive percutaneous instrumentation and navigation in spine surgery for some [38, 39].

Preoperative Planning

- Imaging
  - MRI of entire spine with and without contrast.
  - Upright radiographs of area of interest.
  - Obtain CT of chest, abdomen, and pelvis if unknown primary. Consider Bone Scan or PET/CT for identification of other lesions for easier biopsy if primary tumor is unknown.
- Multidisciplinary tumor boards [40, 41].
  - Multidisciplinary spinal tumor boards consisting of radiologists, radiation oncology, spine surgeons, oncologist, and other subspecialist are increasingly recognized as vital to the proper workup and management of patients with metastatic disease to the spine. Institutional reviews of treatment pathways prior to and after implementation of a spine tumor board found an increased rate of preoperative identification of primary tumor, shorter hospital stays, and lower postoperative deterioration rates [41]. Spine tumor boards help facilitate complex coordination of care and function to provide increased confidence to all parties involved in the care of these patients.
- Preoperative screening for deep vein thrombosis (DVT) may be indicated as there is a high incidence of DVT in patients with spinal metastatic disease, especially patients who are non-ambulatory [42].
- Preoperative embolization is indicated in hyper-vascular tumors due to the increased risk of blood loss and complexity of surgery. Renal cell and thyroid carcinoma are two of the most common hyper-vascular metastatic cancers [43].
- Instrumentation
  - Cement augmentation can enhance the durability of instrumentation in patients with metastatic disease to the spine [13, 44].

  - MRI compatible instrumentation may allow easier monitoring for recurrence.
  - Bone graft grafting is likely to benefit patients with a life expectancy greater than 1 year as pseudoarthrosis is a major risk in oncologic spine surgery. More research is needed to quantify the impact of radiation therapy and chemotherapy on fusion rates and how to minimize their effects [45].
- Consultation with plastic surgery should be performed in the setting of revision metastatic spine tumor surgery as these patients are high risk for infection and wound complications [46]. This may also be considered in initial treatment of patients at risk for wound complications, such as those receiving chemotherapy, radiation treatment or obese patients [47].

Surgical Options

Separation Surgery

Separation surgery is indicated in patients with high-grade epidural spinal cord compression (ESCC) from radioresistant malignancies. Separation surgery differs from aggressive gross total or en bloc resections, as only the epidural space is removed of tumor with a target of 2–3 mm of separation. High intraoperative blood loss and resulting anemia may occur as intralesional resection of the metastatic lesion is performed. Durotomies may occur during dissection and should be repaired primarily if feasible. Vertebral body fractures, pseudoarthrosis, and hardware failure may occur as delayed complications of high radiation dosing [48]. Radiation myelopathy is a rare but devastating complication of radiotherapy that usually presents within 3 years of treatment [49]. Separation surgery that accomplishes downgrading of ESCC grade to 0 or 1 shows improved local control and survival in patients receiving stereotactic body radiotherapy (SBRT) (Fig. 13.5) [51]. Patients with high-grade ESCC even after attempted separation surgery still benefit from SBRT [15].

Minimally Invasive Surgery

Growing interest in minimally invasive approaches to spinal metastatic disease has led to exploration of percutaneous instrumentation, mini-open approaches, tubular retractors, and endoscopic surgery [52]. Benefits to MIS found in the treatment of metastatic spine disease are like those found in the treatment of other spine pathologies. Several authors have reported decreased blood loss and shorter hospitalization with MIS for metastatic disease [53, 54]. A review of Level III studies of MIS vs. open approaches for symptomatic spinal metastases found comparable outcomes [55]. Percutaneous instrumentation for spinal stabilization after vertebral compression fracture has been reported to improve pain and function in patients with metastatic thoraco-lumbar disease [56]. Instrumentation may be placed by fluoroscopic guidance, image-guided navigation, or robotic-assisted navigation. Indications for

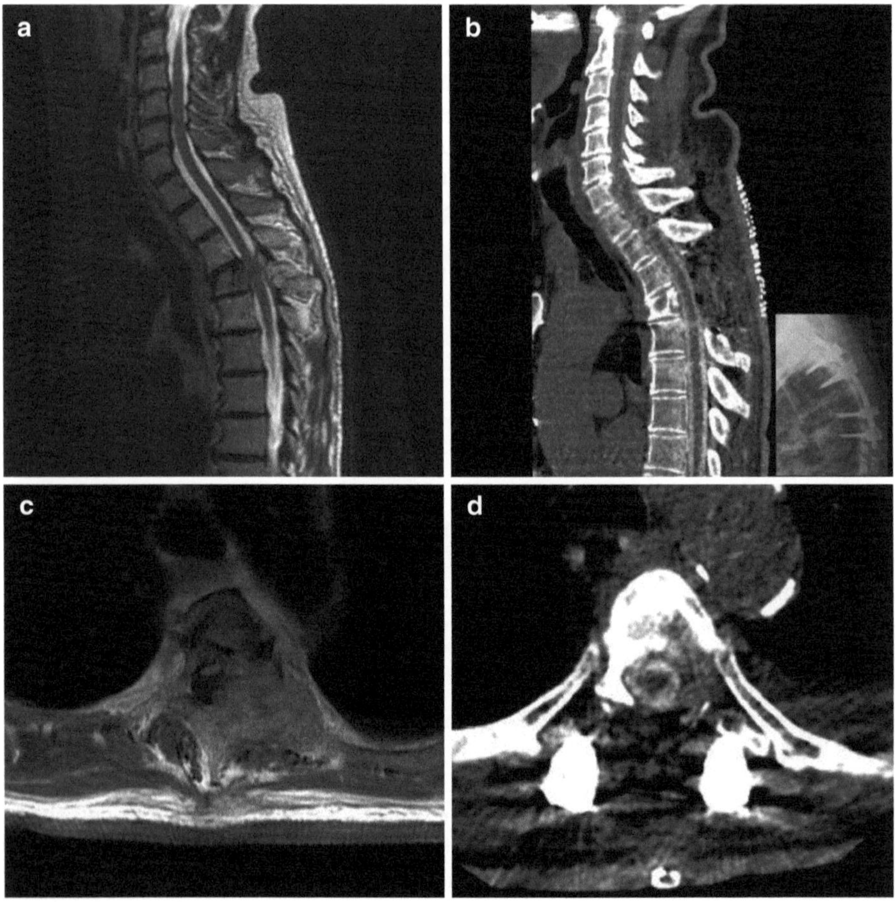

Fig. 13.5 Separation surgery in a 76-year-old male with renal cell carcinoma metastasis to thoracic spine. Preoperative and postoperative MRI T2 sagittal sequences showing a T4 tumor (**a, b**). Preoperative MRI T1 axial sequence showed compression and deformation of thecal sac from tumor (**c**). Postoperative CT myelography shows complete decompression of thecal sac (**d**). (Used with permission. Barzilai et al. [50])

percutaneous instrumentation are currently being investigated. A study of 50 patients treated with fluoroscopic-guided percutaneous stabilization for metastatic disease reported improved pain and all patients ambulating postoperatively. There was one case of implant failure [57]. Cement augmentation with fenestrated pedicle screw fixation has allowed short-segment constructs to be investigated and report good outcomes and similar hardware failure rates to traditional long-constructs, even with percutaneous instrumentation [44, 58]. Intraoperative image-guided percutaneous instrumentation is advantageous due to reduced radiation exposure and faster operative times, while allowing minimally invasive separation surgery [59]. Interest in the use of robotics for instrumentation and even tumor resection is growing in spinal oncology [60, 61]. Indications for MIS include separation surgery that is amendable to MIS approaches as this may decrease recovery time and allow faster delivery of radiotherapy. MIS surgery may be a good option for medially frail patients as well who cannot tolerate larger procedures. Contraindications to MIS include the need for circumferential decompression (difficult to achieve with MIS), highly vascular tumors that may be better decompressed open, and hemostasis controlled through open approaches, and those requiring anterior column support in addition to posterior instrumentation.

Open Decompression

Decompression is indicated in patients with high-grade ESCC and the ideal approach depends on the location of compression and spinal level. The thoracolumbar spine is most often approached from posterior and may also be approached posterolateral in the thoracic spine or variations of lateral in the lumbar spine. Transpedicular decompression allows access anterior to the cord from a posterior approach. The anterior approach to the thoracic or lumbar spine requires coordination with an access surgeon. There is a lack of consensus for the need for anterior column reconstruction after partial corpectomy for decompression of the thoracolumbar spine. Reconstruction of the cervical spine after anterior decompression with a single or multilevel corpectomy is performed with a metal cage. Posterior instrumentation should be performed if greater than two level cervical corpectomy is performed. En bloc resection and vertebrectomy may be considered for patients with a single, isolated metastasis with an unfavorable response to radiation such as renal cell carcinoma [62].

Kyphoplasty/Vertebroplasty

Compression fractures are a common presentation of metastatic disease to the spine. A collapsed vertebral body from pathologic compression fracture may be augmented with PMMA cement through a minimally invasive kyphoplasty. Vertebroplasty involves the creation of a cavity in a vertebral body with injection of PMMA cement. Vertebral compression fractures from metastatic disease to the spine may be treated with kyphoplasty with rapid reduction in pain and improved function [63]. The decision for vertebral augmentation should be based on patient symptoms and imaging findings. Patients with pain interfering with daily life, worsening pain, edema on advanced imaging (STIR sequences), kyphotic deformity, and loss of vertebral height are more likely to benefit from vertebral augmentation. Surgical outcomes are improved in patients with early augmentation (2–8 weeks) compared to longer time to intervention [64]. Complications of cement extravasation include neurological injury and cement embolism. An incompetent posterior vertebral wall or burst fracture is a contraindication to vertebroplasty or kyphoplasty due to risk of cement extravasation into the spinal canal.

Surgical Technique

Separation Surgery [50, 65]

- Intraoperative neurophysiological monitoring is typically used.
- Prone positioning and fluoroscopic imaging to localize levels of interest for skin incision.
- Standard midline incision with sub-periosteal muscle dissection of posterior elements and transverse process.
  - *MIS approach to separation surgery is increasing being reported [52, 66].
- Excessive downward pressure on pathologic posterior elements is avoided due to risk of further compression of spinal cord.
- Instrumentation of a minimum of two levels cranial and caudal with pedicle or lateral mass screws.
- Decompression of the spinal cord must occur circumferentially.
  - Access to dorsal cord is accomplished through thinning of posterior elements with burr to expose ligamentum flavum and dura.
  - Tenotomy scissors and Woodson Dissector are used to dissect normal dural planes and resect tumor and ligamentum flavum from dura.
  - Facet joints removal and drilling of the pedicles to the level of the body is performed unilaterally or bilaterally to allow access to the ventral dural space.
  - A Woodson dissector is used to delineate the posterior longitudinal ligament (PLL) and tenotomy scissors are used to cut across the PLL and release fibrous connects to the dura. The epidural venous plexus will require hemostasis.
  - Resection of a portion of the vertebral body may be required. If greater than 50% of the vertebral body is resected, polymethylmethacrylate (PMMA) augmentation should be used for anterior column support.
- Hemostasis is achieved and a drain is placed prior to standard multilayered closure.

Tips and Tricks

- Ultrasonography may be used to confirm adequate circumferential separation has been achieved.
- Partial vertebrectomy of approximately 20% is typically sufficient to allow ventral decompression.
- Laminectomy alone is typically inadequate to provide circumferential decompression.

Pearls and Pitfalls

- Separation surgery is part of the "hybrid therapy" in combination with radiosurgery to treat metastatic disease causing high-grade ESCC.
- Gross-total resection of metastatic lesion with favorable response to radiotherapy is not required since postoperative radiotherapy will treat remaining tumor and increases operative morbidity. Patients with metastatic cancers with unfavorable response to radiotherapy may also benefit from separation surgery and radiotherapy [67].

- Avoid blunt dissection of the PLL which can place traction on the dura leading to a spinal cord injury.

Percutaneous Instrumentation- Fluoroscopic Guided [68, 69]

- Prone positioning and fluoroscopic imaging to localize levels of interest
- AP fluoroscopic imaging is used to obtain a crisp, single superior endplate with pedicles equidistant from the spinous process.
- Pedicles are identified, and a Kirschner wire (K-wire) is placed longitudinally over the skin, and a skin marking pen is used to mark the cranial-caudal location of the lateral margin of the pedicle.
- Skin incisions made be made as paraspinal stab incisions, parallel longitudinal paraspinal incisions, or a single midline incision after localization of trajectory.
- The midpoint of the lateral pedicle border serves as the entry point for a Jamshidi needle
- This needle is advanced to a depth of 2 cm under AP fluoroscopic view without breaching the medial cortex.
- A blunt guidewire is inserted through the needle and the needle removed.
- Screws can be placed over the guidewire and should be advanced until meeting resistance against the lateral wall of the facet.
- Screw heads must be aligned properly in the coronal and sagittal planes to place percutaneous rods.
- Rods are passed percutaneously through the screw heads and locked in place. Reduction towers may be used to allow locking screws to be placed.

Tips and Tricks

- Proper rod placement through the cannula and screw can be assess by attempted rotation of the cannula and screw. The rod is properly through the cannula and screw if unable to rotate.

Pearls and Pitfalls

- Proper angulation medial-to-lateral is essential as this orientation changes throughout the thoracolumbar spine
- Surgeons should be familiar with the MIS approach to spinal instrumentation and prepared for conversion to an open procedure if inadequate fluoroscopic views are available or anatomy is distorted.

Percutaneous Instrumentation—Image-Guided and Robotic-Assisted [70]

- Prone positioning and fluoroscopic imaging to localize levels of interest. If the docking location is a palpable landmark, placement may be performed without imaging.

- Dissection is performed down to the docking location, and the reference frame is secured.
  - The reference frame may be docked either to the table or to the patient, commonly to the spinous process. A reference frame may be placed in the posterior superior iliac spine (PSIS) for lower lumbar or sacral surgery.
  - Robotic-assisted instrumentation requires mounting a stable base to the patient, which may be the spinous process, PSIS, or other landmark, and is system specific.
- Intraoperative imaging or verification and mapping to preoperative imaging is next performed.
  - Intraoperative CT is performed with the mobile CT scanner (O-arm). Long segments may require more than one reference docking location.
  - If navigation is performed based on pre-operative images without intraop imaging, registration of instruments is performed with three or more exposed points matched with preop imaging.
- CT guided instrumentation is completed as follows:
  - Pedicle trajectories and entry point is selected with a navigated probe.
  - A pilot hole is burred and navigated Lenke probe is used to cannulate the pedicle in the predetermined path and this trajectory is saved.
  - A navigated tap is used and follows the saved trajectory that can be monitored in real-time on axial and sagittal planes.
  - The path for pedicle screw placement is probed with a ball tip guide to look for pedicle breach.
  - A navigated screwdriver is used to place screw with confirmation of depth and trajectory on navigation.
- Robotic assisted instrumentation is performed after accuracy is assessed and confirmed.
  - After selection of a vertebral level, the system software guides the robotic arm into position.
  - Skin incision and deep dissection is performed down to posterior elements.
  - A cannulated guide is advanced through the soft tissues and the pedicle is drilled.
  - Some systems allow real-time verification of instrumentation via navigation.
  - Pedicle screws are placed, and a smooth transition of cannulas should be seen to allow for appropriate rod contouring.

Tips and Tricks

- If using more than one reference station, cover the second reference frame with a sterile drape to minimize interference.
- Proper rod placement through the cannula and screw can be assess by attempted rotation of the cannula and screw. The rod is properly through the cannula and screw if unable to rotate.

Pearls and Pitfalls

- Place the reference station and infrared camera to allow maximum ergonomics and continuous monitoring without the need for readjustment.
- Inaccurate registration may cause inadvertent hardware malposition.

Open Decompression

- The surgical approach will dictate the position of the patient.
- Posterior decompression may be performed at all spinal levels
  - Standard midline incision after confirmation of levels of interest by fluoroscopy
  - The fascia is incised in line with the incision and a subperiosteal dissection performed.
  - Decompression may be performed through a laminotomy, partial laminectomy, or complete laminectomy.
    - Complete laminectomy is performed using a burr the lamina beginning cranially 5 mm medial to the pars and moving caudally. Once the bone is thinned, a Kerrison rongeur is used to complete the osteotomy before the complete lamina is removed *en bloc*. A Woodson elevator is used to detach any adherent tissue as gentle upward traction is placed on the posterior elements.
      - Care must be taken not to remove bone too laterally into the pars articularis due to risk of pars fracture.
      - Laminectomies that are taken laterally into the facet joints risk iatrogenic instability and great care should be taken if no instrumentation is planned.
  - A laminotomy or partial laminectomy may be performed with the burr and Kerrison rongeur if enough access is allowed to remove any dorsal tumor.
- Anterior cervical corpectomy
  - The standard anterior approach to the cervical spine can typically expose the anterior vertebral bodies from C2 to T1.
  - A subperiosteal dissection of the vertebral body is performed and annulotomy of the disc space after confirmation of vertebral body level on fluoroscopy.
    - Wide exposure is often needed beyond the uncinate process to allow adequate visualization and dissection.
  - Distraction is performed with Caspar pins or Gardner-Wells tongs.
  - Discectomy is performed at the level above and below the cervical vertebrae of interest.
  - The corpectomy width is marked with a high-speed burr longitudinally. A large Leksell rongeur can be used to remove large pieces of bone while the high-speed burr can provide fine dissection down the posterior wall
  - Kerrison rongeurs are used to remove remaining thin cortical shell and small size curettes may be needed to separate from dura or tumor.

  - The superior endplate of the caudal vertebral body and inferior endplate of cranial vertebral body should be removed of cartilage.
  - A graft or mesh cage is placed after sizing with implant template or calipers.
  - Distraction is released and an anterior plate with locking screws is placed to span the involved levels.
- Anterior lumbar corpectomy
  - An access surgeon familiar with the anterior approach to the lumbar spine should be used.
  - Confirm approach angle on lateral fluoroscopy to ensure segment of interest can be accessed.
  - A midline incision or oblique paramedial incision may be used. This is followed by dissection down to the anterior rectus sheath. Fascial incision of the rectus is performed 1–2 cm from the midline to allow later closure.
  - The rectus muscle may be mobilized laterally, and blunt dissection is performed superficial to the transversalis fascia until the retroperitoneal fat is seen. There will not be a posterior rectal sheath if dissecting inferior to the arcuate line.
  - Careful mobilization of the common iliac vessels depends on the vertebral segment of interest. Segmental vessels, the iliolumbar vein, and middle sacral vessels may need to be ligated depending on the level as well.
  - The levels are confirmed with fluoroscopy after placement of needle in the disc space.
  - Perform annulotomy and discectomy of the disc space cranial and caudal to the targeted vertebral level.
    - Extreme care is needed when performing annulotomy with sharp knife and the great vessels are protected. A cobb elevator and curettes are used to perform discectomy to the level of the posterior longitudinal ligament.
  - Piecemeal corpectomy with large Leksell rongeurs is then performed. Kerrison rongeurs are used more posteriorly when the bone has been sufficiently thinned. Curved curettes are used to separate the posterior vertebral wall from thecal sac and/or tumor when resecting this bone.
  - Graft sizing can be performed with measuring a cotton tip applicator to measure depth and length.
  - Reconstruction of anterior column is performed with bone graft, cage, or combination.
- Tips and Tricks
  - The anterior lip of the cranial endplate may be flattened to allow graft insertion after anterior cervical corpectomy.
- Pearls and Pitfalls
  - Anterior cervical corpectomies greater than 3-levels should be augmented with posterior instrumentation due to high risk of failure.
  - Reconstruction of the anterior column in the lumbar spine should be performed with a graft that is large enough to contact the ring apophysis of the cranial and caudal vertebrae.

Kyphoplasty/Vertebroplasty

- The patient is positioned prone and correct vertebral level is confirmed with fluoroscopy
- A 1 cm incision is made 1 cm lateral to the pedicle entry point. This is typically 2–3 cm lateral to midline
- A Jamshidi needle is advanced through the soft tissues to the lateral margin of the pedicle and then until the medial wall of the pedicle is reached on fluoroscopic imaging.
  - Ensure no medial wall breaching on fluoroscopic imaging.
- The needle is advanced 2 mm beyond the posterior vertebral wall on the lateral view.
- The stylet is removed from the needle, and a drill is placed through the cannula.
- The drill is advanced until 2 mm of the anterior vertebral wall.
- The balloon system is inserted and inflated with contrast dye. Confirm appropriate positioning of the cannulas and contrast prior to full inflation of the balloon.
- Deflate the balloon after the appropriate volume is achieved within the vertebral body. This is determined by cortical apposition, realignment of the endplate, or maximum volume achieved.
- PMMA cement is injected into the cavity under fluoroscopy to confirm proper placement

Tips and Tricks

- Position of the Jamshidi needle is at the 9-o'clock position for the left pedicle and 3-o'clock position for the right pedicle.

Pearls and Pitfalls

- Do not remove cannulas before the cement has hardened.

Postoperative Care

- Prophylactic anticoagulation on postop day 1 unless high drain output or high risk of uncontrolled bleeding.
- Immediate mobilization with physical therapist on postoperative day 1.
- Bracing for patient comfort may allow easier mobilization of patient.
- Multimodal pain control and may consider consultation with pain service if available.
- Foley catheter should be removed as soon as the patient is ambulatory or on postoperative day 1.
- Patients are at high risk of decubitus ulcers or pressure sores. An air mattress may be indicated in patients who have neurological deficits or are unable to mobilize.
- Radiation and/or chemotherapy should be delayed 2–3 weeks postop or until wounds have healed.

References

1. Ryan SE, Hall JRL, Humble R, Olinger CR. Basal cell carcinoma with metastasis to the thoracic spine: an uncommon case report and review of the literature. JAAOS: Glob Res Rev. 2024;8(2):e23.00214. https://doi.org/10.5435/JAAOSGlobal-D-23-00214.
2. Ortiz Gomez JA. The incidence of vertebral body metastases. Int Orthop. 1995;19(5):309–11. https://doi.org/10.1007/BF00181116.
3. Van den Brande R, Cornips EM, Peeters M, Ost P, Billiet C, Van de Kelft E. Epidemiology of spinal metastases, metastatic epidural spinal cord compression and pathologic vertebral compression fractures in patients with solid tumors: a systematic review. J Bone Oncol. 2022;35:100446. https://doi.org/10.1016/j.jbo.2022.100446.
4. Shakil H, Malhotra AK, Badhiwala JH, Karthikeyan V, Essa A, He Y, et al. Contemporary trends in the incidence and timing of spinal metastases: a population-based study. Neuro-Oncol Adv. 2024;6(1):vdae051. https://doi.org/10.1093/noajnl/vdae051.
5. Ecker RD, Endo T, Wetjen NM, Krauss WE. Diagnosis and treatment of vertebral column metastases. Mayo Clin Proc. 2005;80(9):1177–86. https://doi.org/10.4065/80.9.1177.
6. Cole JS, Patchell RA. Metastatic epidural spinal cord compression. Lancet Neurol. 2008;7(5):459–66. https://doi.org/10.1016/S1474-4422(08)70089-9.
7. Zuckerman SL, Laufer I, Bilsky M. NOMS. In: Marco R, editor. Metastatic spine disease. Cham: Springer; 2018. https://doi.org/10.1007/978-3-319-76252-4_4.
8. Laufer I, Rubin DG, Lis E, Cox BW, Stubblefield MD, Yamada Y, et al. The NOMS framework: approach to the treatment of spinal metastatic tumors. Oncologist. 2013;18(6):744–51. https://doi.org/10.1634/theoncologist.2012-0293.
9. Bilsky MH, Laufer I, Fourney DR, Groff M, Schmidt MH, Varga PP, et al. Reliability analysis of the epidural spinal cord compression scale. J Neurosurg Spine. 2010;13(3):324–8. https://doi.org/10.3171/2010.3.SPINE09459.
10. Goodwin ML, Buchowski JM, Schwab JH, Sciubba DM. Spinal tumors: diagnosis and treatment. J Am Acad Orthopaedic Surg. 2022;30(17):e1106–21. https://doi.org/10.5435/JAAOS-D-21-00710.
11. Uei H, Tokuhashi Y, Maseda M. Analysis of the relationship between the Epidural Spinal Cord Compression (ESCC) scale and paralysis caused by metastatic spine tumors. Spine (Phila Pa 1976). 2018;43(8):E448–e55. https://doi.org/10.1097/brs.0000000000002378.
12. Gerszten PC, Mendel E, Yamada Y. Radiotherapy and radiosurgery for metastatic spine disease: what are the options, indications, and outcomes? Spine (Phila Pa 1976). 2009;34(22 Suppl):S78–92. https://doi.org/10.1097/BRS.0b013e3181b8b6f5.
13. Newman WC, Laufer I, Bilsky MH. Neurologic, oncologic, mechanical, and systemic and other decision frameworks for spinal disease. Neurosurg Clin N Am. 2020;31(2):151–66. https://doi.org/10.1016/j.nec.2019.11.005.
14. Blakaj DM, Palmer JD, Dibs K, Olausson A, Bourekas EC, Boulter D, et al. Postoperative stereotactic body radiotherapy for spinal metastasis and predictors of local control. Neurosurgery. 2021;88(5):1021–7. https://doi.org/10.1093/neuros/nyaa587.
15. Patel PP, Cao Y, Chen X, LeCompte MC, Kleinberg L, Khan M, et al. Oncologic and functional outcomes after stereotactic body radiation therapy for high-grade malignant spinal cord compression. Adv Radiat Oncol. 2024;9(1):101327. https://doi.org/10.1016/j.adro.2023.101327.
16. Palacio Giraldo A, Sohm D, Neugebauer J, Leone G, Bergovec M, Dammerer D. Stereotactic radiosurgery in metastatic spine disease-a systemic review of the literature. Cancers (Basel). 2024;16(16):2787. https://doi.org/10.3390/cancers16162787.
17. Fisher CG, DiPaola CP, Ryken TC, Bilsky MH, Shaffrey CI, Berven SH, et al. A novel classification system for spinal instability in neoplastic disease: an evidence-based approach and expert consensus from the Spine Oncology Study Group. Spine (Phila Pa 1976). 2010;35(22):E1221–9. https://doi.org/10.1097/BRS.0b013e3181e16ae2.

18. Fourney DR, Frangou EM, Ryken TC, Dipaola CP, Shaffrey CI, Berven SH, et al. Spinal instability neoplastic score: an analysis of reliability and validity from the spine oncology study group. J Clin Oncol. 2011;29(22):3072–7. https://doi.org/10.1200/JCO.2010.34.3897.
19. Vargas E, Shabani S, Mummaneni PV, Park C, Rechav Ben-Natan A, Rivera JJ, et al. Does surgery for metastatic spinal tumors improve functional outcomes in patients without spinal cord compression but with potentially unstable spines (SINS 7-12)? J Neurosurg Spine. 2023;39(2):287–94. https://doi.org/10.3171/2023.3.SPINE221120.
20. Pennington Z, Ahmed AK, Westbroek EM, Cottrill E, Lubelski D, Goodwin ML, et al. SINS score and stability: evaluating the need for stabilization within the uncertain category. World Neurosurg. 2019;128:e1034–e47. https://doi.org/10.1016/j.wneu.2019.05.067.
21. Hussain I, Barzilai O, Reiner AS, DiStefano N, McLaughlin L, Ogilvie S, et al. Patient-reported outcomes after surgical stabilization of spinal tumors: symptom-based validation of the Spinal Instability Neoplastic Score (SINS) and surgery. Spine J. 2018;18(2):261–7. https://doi.org/10.1016/j.spinee.2017.07.008.
22. Lenschow M, Lenz M, von Spreckelsen N, Ossmann J, Meyer J, Kessling J, et al. Impact of spinal instrumentation on neurological outcome in patients with intermediate spinal instability neoplastic score (SINS). Cancers (Basel). 2022;14(9) https://doi.org/10.3390/cancers14092193.
23. Versteeg AL, Sahgal A, Rhines LD, Sciubba DM, Schuster JM, Weber MH, et al. Health related quality of life outcomes following surgery and/or radiation for patients with potentially unstable spinal metastases. Spine J. 2021;21(3):492–9. https://doi.org/10.1016/j.spinee.2020.10.017.
24. Bollen L, van der Linden YM, Pondaag W, Fiocco M, Pattynama BP, Marijnen CA, et al. Prognostic factors associated with survival in patients with symptomatic spinal bone metastases: a retrospective cohort study of 1,043 patients. Neuro-Oncol. 2014;16(7):991–8. https://doi.org/10.1093/neuonc/not318.
25. Switlyk MD, Kongsgaard U, Skjeldal S, Hald JK, Hole KH, Knutstad K, et al. Prognostic factors in patients with symptomatic spinal metastases and normal neurological function. Clin Oncol (R Coll Radiol). 2015;27(4):213–21. https://doi.org/10.1016/j.clon.2015.01.002.
26. MacLean MA, Touchette CJ, Georgiopoulos M, Brunette-Clement T, Abduljabbar FH, Ames CP, et al. Systemic considerations for the surgical treatment of spinal metastatic disease: a scoping literature review. Lancet Oncol. 2022;23(7):e321–e33. https://doi.org/10.1016/S1470-2045(22)00126-7.
27. Tokuhashi Y, Uei H, Oshima M, Ajiro Y. Scoring system for prediction of metastatic spine tumor prognosis. World J Orthop. 2014;5(3):262–71. https://doi.org/10.5312/wjo.v5.i3.262.
28. Ahmed AK, Goodwin CR, Heravi A, Kim R, Abu-Bonsrah N, Sankey E, et al. Predicting survival for metastatic spine disease: a comparison of nine scoring systems. Spine J. 2018;18(10):1804–14. https://doi.org/10.1016/j.spinee.2018.03.011.
29. Shah AA, Schwab JH. Predictive modeling for spinal metastatic disease. Diagnostics (Basel). 2024;14(9):962. https://doi.org/10.3390/diagnostics14090962.
30. Paulino Pereira NR, Janssen SJ, van Dijk E, Harris MB, Hornicek FJ, Ferrone ML, et al. Development of a prognostic survival algorithm for patients with metastatic spine disease. J Bone Joint Surg Am. 2016;98(21):1767–76. https://doi.org/10.2106/JBJS.15.00975.
31. Smeijers S, Depreitere B. Prognostic scores for survival as decisional support for surgery in spinal metastases: a performance assessment systematic review. Eur Spine J. 2021;30(10):2800–24. https://doi.org/10.1007/s00586-021-06954-6.
32. Paulino Pereira NR, McLaughlin L, Janssen SJ, van Dijk CN, Bramer JAM, Laufer I, et al. The SORG nomogram accurately predicts 3- and 12-months survival for operable spine metastatic disease: external validation. J Surg Oncol. 2017;115(8):1019–27. https://doi.org/10.1002/jso.24620.
33. Karhade AV, Fenn B, Groot OQ, Shah AA, Yen HK, Bilsky MH, et al. Development and external validation of predictive algorithms for six-week mortality in spinal metastasis using 4,304 patients from five institutions. Spine J. 2022;22(12):2033–41. https://doi.org/10.1016/j.spinee.2022.07.089.

34. Karhade AV, Thio Q, Ogink PT, Bono CM, Ferrone ML, Oh KS, et al. Predicting 90-day and 1-year mortality in spinal metastatic disease: development and internal validation. Neurosurgery. 2019;85(4):E671–E81. https://doi.org/10.1093/neuros/nyz070.
35. Karhade AV, Thio Q, Ogink PT, Shah AA, Bono CM, Oh KS, et al. Development of machine learning algorithms for prediction of 30-day mortality after surgery for spinal metastasis. Neurosurgery. 2019;85(1):E83–91. https://doi.org/10.1093/neuros/nyy469.
36. Dea N, Versteeg AL, Sahgal A, Verlaan JJ, Charest-Morin R, Rhines LD, et al. Metastatic spine disease: should patients with short life expectancy be denied surgical care? An International Retrospective Cohort Study. Neurosurgery. 2020;87(2):303–11. https://doi.org/10.1093/neuros/nyz472.
37. Schoenfeld AJ, Ferrone ML, Schwab JH, Blucher JA, Barton LB, Tobert DG, et al. Prospective validation of a clinical prediction score for survival in patients with spinal metastases: the New England spinal metastasis score. Spine J. 2021;21(1):28–36. https://doi.org/10.1016/j.spinee.2020.02.009.
38. Kim CH, Chung CK, Sohn S, Lee S, Park SB. Less invasive palliative surgery for spinal metastases. J Surg Oncol. 2013;108(7):499–503. https://doi.org/10.1002/jso.23418.
39. Schupper AJ, Patel S, Steinberger JM, Germano IM. The role of minimally invasive surgery within a multidisciplinary approach for patients with metastatic spine disease over a decade: a systematic review. Neuro-Oncology. 2024;26(3):417–28. https://doi.org/10.1093/neuonc/noad206.
40. Mann M, Annable N, Emch TM, Wu J, Chao ST, Benzel E, et al. Management of patients with spine tumors strengthened by a dedicated multidisciplinary spine tumor board: a 15-year single-institutional experience. World Neurosurg. 2023;175:e397–405. https://doi.org/10.1016/j.wneu.2023.03.110.
41. Miyazaki K, Kanda Y, Sakai Y, Yoshikawa R, Yurube T, Takeoka Y, et al. Effect of bone metastasis cancer board on spinal surgery outcomes: a retrospective study. Medicina (Kaunas). 2023;59(12) https://doi.org/10.3390/medicina59122087.
42. Zacharia BE, Kahn S, Bander ED, Cederquist GY, Cope WP, McLaughlin L, et al. Incidence and risk factors for preoperative deep venous thrombosis in 314 consecutive patients undergoing surgery for spinal metastasis. J Neurosurg Spine. 2017;27(2):189–97. https://doi.org/10.3171/2017.2.SPINE16861.
43. Nair S, Gobin YP, Leng LZ, Marcus JD, Bilsky M, Laufer I, et al. Preoperative embolization of hypervascular thoracic, lumbar, and sacral spinal column tumors: technique and outcomes from a single center. Interv Neuroradiol. 2013;19(3):377–85. https://doi.org/10.1177/159101991301900317.
44. Barzilai O, McLaughlin L, Lis E, Reiner AS, Bilsky MH, Laufer I. Utility of cement augmentation via percutaneous fenestrated pedicle screws for stabilization of cancer-related spinal instability. Oper Neurosurg (Hagerstown). 2019;16(5):593–9. https://doi.org/10.1093/ons/opy186.
45. Elder BD, Ishida W, Goodwin CR, Bydon A, Gokaslan ZL, Sciubba DM, et al. Bone graft options for spinal fusion following resection of spinal column tumors: systematic review and meta-analysis. Neurosurg Focus. 2017;42(1):E16. https://doi.org/10.3171/2016.8.FOCUS16112.
46. Mesfin A, Sciubba DM, Dea N, Nater A, Bird JE, Quraishi NA, et al. Changing the adverse event profile in metastatic spine surgery: an evidence-based approach to target wound complications and instrumentation failure. Spine (Phila Pa 1976). 2016;41(Suppl 20):S262–s70. https://doi.org/10.1097/brs.0000000000001817.
47. Jarvers JS, Lange M, Schiemann S, Pfränger J, Heyde CE, Osterhoff G. Risk factors for wound-related complications after surgical stabilization of spinal metastases with a special focus on the effect of postoperative radiation therapy. BMC Surg. 2021;21(1):423. https://doi.org/10.1186/s12893-021-01431-9.
48. Jawad MS, Fahim DK, Gerszten PC, Flickinger JC, Sahgal A, Grills IS, et al. Vertebral compression fractures after stereotactic body radiation therapy: a large, multi-institutional, multinational evaluation. J Neurosurg Spine. 2016;24(6):928–36. https://doi.org/10.3171/2015.10.SPINE141261.

49. Ong WL, Wong S, Soliman H, Myrehaug S, Tseng CL, Detsky J, et al. Radiation myelopathy following stereotactic body radiation therapy for spine metastases. J Neuro-Oncol. 2022;159(1):23–31. https://doi.org/10.1007/s11060-022-04037-0.
50. Barzilai O, Laufer I, Robin A, Xu R, Yamada Y, Bilsky MH. Hybrid therapy for metastatic epidural spinal cord compression: technique for separation surgery and spine radiosurgery. Oper Neurosurg (Hagerstown). 2019;16(3):310–8. https://doi.org/10.1093/ons/opy137.
51. Al-Omair A, Masucci L, Masson-Cote L, Campbell M, Atenafu EG, Parent A, et al. Surgical resection of epidural disease improves local control following postoperative spine stereotactic body radiotherapy. Neuro-Oncology. 2013;15(10):1413–9. https://doi.org/10.1093/neuonc/not101.
52. Zuckerman SL, Laufer I, Sahgal A, Yamada YJ, Schmidt MH, Chou D, et al. When less is more: the indications for MIS techniques and separation surgery in metastatic spine disease. Spine (Phila Pa 1976). 2016;41(Suppl 20):S246–S53. https://doi.org/10.1097/BRS.0000000000001824.
53. Hansen-Algenstaedt N, Kwan MK, Algenstaedt P, Chiu CK, Viezens L, Chan TS, et al. Comparison between minimally invasive surgery and conventional open surgery for patients with spinal metastasis: a prospective propensity score-matched study. Spine (Phila Pa 1976). 2017;42(10):789–97. https://doi.org/10.1097/BRS.0000000000001893.
54. Kumar N, Malhotra R, Maharajan K, Zaw AS, Wu PH, Makandura MC, et al. Metastatic spine tumor surgery: a comparative study of minimally invasive approach using percutaneous pedicle screws fixation versus open approach. Clin Spine Surg. 2017;30(8):E1015–E21. https://doi.org/10.1097/BSD.0000000000000400.
55. Pennington Z, Ahmed AK, Molina CA, Ehresman J, Laufer I, Sciubba DM. Minimally invasive versus conventional spine surgery for vertebral metastases: a systematic review of the evidence. Ann Transl Med. 2018;6(6):103. https://doi.org/10.21037/atm.2018.01.28.
56. Barzilai O, McLaughlin L, Amato MK, Reiner AS, Ogilvie SQ, Lis E, et al. Minimal access surgery for spinal metastases: prospective evaluation of a treatment algorithm using patient-reported outcomes. World Neurosurg. 2018;120:e889–901. https://doi.org/10.1016/j.wneu.2018.08.182.
57. Kwan MK, Lee CK, Chan CY. Minimally invasive spinal stabilization using fluoroscopic-guided percutaneous screws as a form of palliative surgery in patients with spinal metastasis. Asian Spine J. 2016;10(1):99–110. https://doi.org/10.4184/asj.2016.10.1.99.
58. Newman WC, Amin AG, Villavieja J, Laufer I, Bilsky MH, Barzilai O. Short-segment cement-augmented fixation in open separation surgery of metastatic epidural spinal cord compression: initial experience. Neurosurg Focus. 2021;50(5):E11. https://doi.org/10.3171/2021.2.FOCUS217.
59. Nasser R, Nakhla J, Echt M, De la Garza RR, Kinon MD, Sharan A, et al. Minimally invasive separation surgery with intraoperative stereotactic guidance: a feasibility study. World Neurosurg. 2018;109:68–76. https://doi.org/10.1016/j.wneu.2017.09.067.
60. Sayari AJ, Pardo C, Basques BA, Colman MW. Review of robotic-assisted surgery: what the future looks like through a spine oncology lens. Ann Transl Med. 2019;7(10):224. https://doi.org/10.21037/atm.2019.04.69.
61. Barzilai O, Goh AC, Park B, Rusch V, Weiser M, Leitao MM Jr, et al. Robotic resection of spinal and paraspinal tumors. Oper Neurosurg (Hagerstown). 2024; https://doi.org/10.1227/ons.0000000000001333.
62. Howell EP, Williamson T, Karikari I, Abd-El-Barr M, Erickson M, Goodwin ML, et al. Total en bloc resection of primary and metastatic spine tumors. Ann Transl Med. 2019;7(10):226. https://doi.org/10.21037/atm.2019.01.25.
63. Berenson J, Pflugmacher R, Jarzem P, Zonder J, Schechtman K, Tillman JB, et al. Balloon kyphoplasty versus non-surgical fracture management for treatment of painful vertebral body compression fractures in patients with cancer: a multicentre, randomised controlled trial. Lancet Oncol. 2011;12(3):225–35. https://doi.org/10.1016/s1470-2045(11)70008-0.

64. Papanastassiou ID, Filis A, Aghayev K, Kokkalis ZT, Gerochristou MA, Vrionis FD. Adverse prognostic factors and optimal intervention time for kyphoplasty/vertebroplasty in osteoporotic fractures. Biomed Res Int. 2014;2014:925683. https://doi.org/10.1155/2014/925683.
65. Rothrock R, Pennington Z, Ehresman J, Bilsky MH, Barzilai O, Szerlip NJ, et al. Hybrid therapy for spinal metastases. Neurosurg Clin N Am. 2020;31(2):191–200. https://doi.org/10.1016/j.nec.2019.11.001.
66. Martins Coelho Junior VP, Dhaliwal JS, Chakravarthy VB. MIS technique for separation surgery in lumbar spine metastatic disease. Neurosurg Focus Video. 2024;10(2):V11. https://doi.org/10.3171/2024.1.FOCVID23222.
67. Rades D, Huttenlocher S, Bajrovic A, Karstens JH, Adamietz IA, Kazic N, et al. Surgery followed by radiotherapy versus radiotherapy alone for metastatic spinal cord compression from unfavorable tumors. Int J Radiat Oncol Biol Phys. 2011;81(5):e861–8. https://doi.org/10.1016/j.ijrobp.2010.11.056.
68. Rose PS, Clarke MJ, Dekutoski MB. Minimally invasive treatment of spinal metastases: techniques. Int J Surg Oncol. 2011;2011:494381. https://doi.org/10.1155/2011/494381.
69. Moussazadeh N, Rubin DG, McLaughlin L, Lis E, Bilsky MH, Laufer I. Short-segment percutaneous pedicle screw fixation with cement augmentation for tumor-induced spinal instability. Spine J. 2015;15(7):1609–17. https://doi.org/10.1016/j.spinee.2015.03.037.
70. Amin AG, Barzilai O, Bilsky MH. CT-based image-guided navigation and the DaVinci robot in spine oncology: changing surgical paradigms. HSS J. 2021;17(3):294–301. https://doi.org/10.1177/15563316211032009.

Management of Pelvic Metastasis – Minimally Invasive Pelvic Surgery

14

Barlas Goker, Joseph J. Connolly, Jichuan Wang, Bang H. Hoang, David S. Geller, and Rui Yang

Introduction

Second to the spine, the pelvis is one of the most common locations for bone metastases; prevalences range between 9.4 and 38% in cancer patients [1–3]. Overall estimated incidence of metastatic bone disease are anywhere between 300,000 and 600,000 people, meaning that between 28,000 and 228,000 people are living with pelvic metastases [4]. Metastases of the pelvis can be highly morbid, as they cause severe pain with ambulation and put patients at high risk of functional limitations secondary to impending or actual pathologic fracture. A sizeable percentage of these patients are elderly, frail, and suffer from multiple comorbidities; given the significant impact of pelvic metastases on quality of life and the potential to impact the delivery of critical systemic therapies, surgery is often indicated to relieve pain, improve function, and, less commonly, attain local tumor control. Traditionally, open surgery was the gold standard for the management of pelvic metastases. However, more recently, minimally invasive procedures have gained traction as a viable surgical option because of the decreased complication profile and faster time to recovery. Widespread adoption of these

B. Goker · B. H. Hoang · D. S. Geller
Department of Orthopaedic Surgery, Montefiore Medical Center, Albert Einstein College of Medicine, Bronx, New York, NY, USA

J. J. Connolly
Division of Musculoskeletal Oncology, Department of Orthopaedic Surgery, Massachusetts General Hospital/Harvard Medical School, Boston, MA, USA

J. Wang
Musculoskeletal Tumor Center, Beijing Key Laboratory for Musculoskeletal Tumors, Peking University People's Hospital, Beijing, China

R. Yang (✉)
Department of Orthopaedic Surgery, University of Texas, Houston, Houston, USA

J. Pretell-Mazzini, S. A. Lozano Calderón (eds.), *Multidisciplinary Approach of Musculoskeletal Metastasis*, https://doi.org/10.1007/978-3-032-09498-8_14

methods may benefit patients who receive treatment with the appropriate indications, offering shorter recovery periods and improved function. Comparative research is necessary to understand and better delineate the indications and role of these surgical techniques.

Anatomy of the Pelvis

The pelvic ring is a unique and complex structure composed of multiple bones and articulations, as well as various vessels, nerves, and internal organs. The os coxae are formed by the fusion of the ilium, ischium, and pubis and articulate with each other anteriorly through the pubic symphysis and posteriorly with the sacrum via the sacroiliac joints. The pelvis bridges the axial and appendicular skeletons, acting as a crucial axis point for locomotion and weight transfer from the trunk to the lower extremities, with some of the strongest muscle attachments in the human body. It functions as a key conduit for major nerves and vessels serving the lower abdomen and extremities and supports and protects several internal organs, including the intestines, urinary bladder, and genitalia. The anatomy of the pelvis is remarkable; however, its anatomy is precisely what makes the pelvis particularly challenging to treat surgically. The complex 3-dimensional (3D) structure muddies pelvic fixation and reconstruction; the numerous neurovascular structures increase the risk of bleeding and functional deficits; and the proximity to the gastrointestinal and genitourinary systems predisposes patients to infectious complications.

Multiple classification systems have been designed to better understand the implications of tumor involvement on the local anatomy. The first of those classification systems, described by Enneking, categorizes a tumor involving the innominate bone as Type I, the periacetabular region as Type II, the pubis as Type III, and the sacrum as Type IV [5]. Harrington et al. divided periacetabular tumors even further according to the extent of bone destruction: isolated cavitary defects with intact acetabular walls as Class I; medial wall or dome involvement as Class II; superior or lateral cortical destruction as Class III [6]. The latest and most comprehensive system is the metastatic acetabular classification (MAC), which describes the lesions based on their anatomic involvement by improving upon the previously reported Harrington classification [7, 8]. According to the MAC classification, cavitary dome lesions with an intact acetabular roof are considered Type I; medial wall defects are Type II; single column (anterior or posterior) defects are Type III; and defects in both the anterior and posterior columns are considered Type IV (Table 14.1).

Table 14.1 Classifications of pelvic, and peri-acetabular metastatic disease

Classification system	Location of tumor involvement
Enneking classification	
Type I	Innominate bone
Type II	Periacetabular region
Type III	Pubic bone
Type IV	Sacrum
Harrington periacetabular classification	
Class I	Isolated cavitary defect with intact acetabular walls
Class II	Medial wall or dome involvement
Class III	Superior or lateral cortical destruction
Metastatic acetabular classification (MAC)	
Type I	Cavitary dome lesion with intact acetabular roof
Type II	Medial wall defect
Type III	Single column defect in either the anterior or posterior column
Type IV	Defects in both anterior and posterior columns

Surgical Indications

Surgery for metastatic disease of the pelvis is primarily indicated for two main reasons: pain and structural instability. Pain is most often managed conservatively at first; combinations of radiation therapy, systemic anti-cancer therapy, and bisphosphonates attempt to reduce pain by inducing local regression of the tumor and reconstitution of normal bone matrix. When pain is refractory to conservative management or mechanical in nature, surgery can be used to overcome those limitations and provide pain relief to the patient. Structural instability caused by destructive bone lesions can put patients at risk for pathologic fracture; patients presenting with impending or actual pathologic fractures are generally indicated for surgery. Both pain and structural instability can contribute to burdensome functional limitations that impact quality of life and pose a barrier to successful delivery of life-sustaining systemic therapy.

Treatment of a pelvic metastatic tumor is typically guided by the disease location and extent of bone destruction, as well as the patient's functional status, life expectancy, comorbidities, and goals of care. Historically, surgeons have opted for open procedures such as curettage and cementation, pelvic resection, total hip arthroplasty (THA), or Harrington reconstruction. According to the MAC classification, Type I periacetabular lesions have been treated by bipolar hemiarthroplasty or THA with or without cement augmentation; type II lesions required reinforcement of the medial wall with THA using flanged cups, antiprotrusio cages or tantalum cup on cup constructs; type III lesions were treated with THA with cement augmentation to achieve sufficient stability; and type IV lesions were often treated even more aggressively due to the extensive bone loss, and management ranged from THA with cement and cup/cage or cup on cup constructs to pelvic resections and reconstructions. However, open surgery is inherently morbid, with the added difficulty of operating on a diseased pelvis with poor bone quality. These surgeries are associated with complications including bleeding, infection, pulmonary embolization, dislocation, neurovascular injuries, and aseptic loosening [8–11].

To reduce the complication risk in an especially vulnerable population, several minimally invasive techniques have been described as alternative surgical management options for patients with all lesion types. Although minimally invasive surgeries may not eliminate the risks of surgery, they are considerably lower than open surgeries. The surgical duration is often shorter in minimally invasive surgeries, mitigating the risks associated with prolonged anesthesia and minimizing the length of hospital stay. As such, where possible, minimally invasive surgery is generally indicated over open surgery whenever possible; open surgery is now only preferred in cases where minimally invasive surgery is insufficient in managing the patient's extent of disease. Minimally invasive surgeries for pelvic metastases are contraindicated in patients with less than 1 month of expected survival. Advanced periacetabular disease involvement (Harrington III, MAC III/IV) is a relative contraindication to some percutaneous interventions due to increased risk of failure without reconstruction; however, this is controversial, and selected cases may benefit from percutaneous fixation that can traverse the pelvic discontinuity line.

Preoperative Planning

As per the current standard of care, minimally invasive surgery is performed in the operating theater under the administration of either general anesthesia or local anesthesia with sedation. Radiolucent tables and image intensification are the mainstays of minimally invasive pelvic surgery. Either C-arm imaging or O-arm navigation can be utilized depending on surgeon preference and experience as well as availability. Company representatives should be involved if the plan includes the use of implants.

A profound knowledge of different radiographic views of the pelvis is imperative, as obtaining clear images is vital for a successful and safe operation. Patients should undergo pelvic imaging with direct radiographs and either computed tomography (CT) or magnetic resonance imaging (MRI). In addition to anterior-posterior (AP) pelvis views, radiographs should also include pelvic inlet/outlet and Judet oblique views to highlight the anatomical extent of disease, which is fundamentally important in diagnosing a pathologic fracture and understanding its stability. A thorough review of the patient's preoperative imaging is crucial, and it is often useful to draw or template the trajectory of implants and the entry portals. In available centers, studying the individual anatomy with 3D-printed models may be helpful. It is even possible to print guides or frames to direct the surgeon and facilitate the procedure [12, 13].

Patients can be placed in the supine, lateral, or prone position for these procedures. Positioning generally depends on the anatomy of the defect, the patient's comorbidities, surgeon preference, and equipment capabilities. In general, fixation of the anterior column is best approached in the supine position, while the posterior column and transcolumnar trajectories are best approached in the lateral or prone position. In some cases, patient positioning may be switched mid-procedure to optimize the approach for each stage of the surgery. The authors do not prefer placing a bump under the sacrum or the hips. Abduction of the upper extremities into at least 90 degrees is often helpful to have a better angle for anterior column and gluteal pillar paths. After preparation of the skin, sterile draping should be done widely with attention given to each entry point.

Surgical Options

The options for minimally invasive surgery range from radiofrequency ablation (RFA), cryoablation, cementation with polymethyl methacrylate (PMMA), photodynamic nails, screw fixation, and their combinations [14–18]. Generally, these techniques are associated with successful reduction in pain scores with complications and morbidity lower than that of open surgeries; however, the ideal method has not yet been determined due to the scarcity of comparative high-quality evidence and lack of controlled studies in the literature. Regardless, all these techniques should be part of the orthopaedic oncologist armamentarium, as open techniques are. Surgeons can also opt for different combinations of these techniques based on the specific characteristics of individual cases as well as their personal preferences. The advantages and disadvantages of each technique should be kept in mind when combination methods are utilized.

Cementoplasty

Cementoplasty is a well-described technique, originally used for the treatment of spine lesions, that is now widely used for augmentation of lytic bone defects in the pelvis [19]. Cementoplasty most often uses PMMA, which consists of liquid and powder components that form an exothermic reaction when mixed. The liquid contains the monomer, accelerator, and inhibitor, while the powder contains the polymer, radio-opacifier, and initiator. PMMA has a Young's modulus between cortical and cancellous bone and has excellent strength against compressive forces. The viscosity of cement allows it to be injected percutaneously and conform to the bone defect before solidifying. Injection of PMMA can be done either directly into the bone or inside an inflatable balloon [20].

Cementoplasty with PMMA is a relatively simple technique that is often combined with ablation or internal fixation. Alone, it is an effective short-term intervention for patients with painful lesions who have limited life expectancy. It has the advantages of providing immediate stability to allow early weight-bearing, lower cost compared to more sophisticated implants, and an easier learning process for surgeons. Heat released from the cement polymerization can induce necrosis in the surrounding tumoral tissues, although it is not currently accepted as an anticancer treatment. Complications related to cement leakage are the primary drawbacks of cementoplasty. As PMMA fills the defect, bone reconstitution is not possible following surgery. The use of an additional "venting" port facilitates filling of the defect and decreases the risk of fat embolism associated with high intraosseous pressure during injection. It also improves filling as existing air within the lesion is evacuated.

Ablative Techniques

Ablation methods provide local control of tumors by targeting them with probes that induce drastic temperature changes. RFA generates extreme heat by an electric current that flows through electrodes, coagulating the tissues and causing necrosis as a result. Cooled systems add a circulating irrigation system to avoid charring the tissues and limiting the area of necrosis. Cryoablation utilizes extreme cold temperatures with pressurized argon to achieve local tumor control. The cold forms an "ice ball" that causes destruction of cellular cytoplasm and cycles of freezing and thawing causes osmotic damage. RFA and cryoablation are commonly used for pelvic metastases and are frequently combined with cement application [21].

Ablation methods target the lesion and necrose the area of disease involvement and can degrade the matrix prior to cement injection. Cryoablation and RFA are relatively easier to learn and adopt than the internal fixation techniques, as advancing the probe to the center of the lesion is sufficient in many cases without the need to traverse multiple pathways. Their effectiveness can be limited in larger tumors, and multiple points of ablation are sometimes needed. RFA is preferred in lesions measuring less than 1 cm, whereas cryoablation is preferred in lesions less than 5 cm in size. Thermal damage to neighboring neurovascular structures and internal organs is a specific concern of ablation procedures with RFA being the preferred method when close to neurovascular structures.

Percutaneous Fixation

Cannulated Screws

Percutaneous screw fixation was first introduced in metastatic pelvic bone lesions in 2015 [22]. Initially, the procedures were limited to screw placement through lytic masses. Recently, the novel "Tripod" technique revolutionized the concept to include the management of periacetabular metastases by creating a rebar structure around the acetabulum [17]. This allowed patients to start bearing weight early in the postoperative course and act as a bridging treatment until radiation would provide local control of the lesion. The construct is built by fixing screws in three different corridors: the anterior column, the posterior column, and the transcolumnar (sciatic corridor) trajectories (Fig. 14.1). Depending on the size of the lesion additional screws including a second transcolumnar or a gluteal pillar screw can be placed to reinforce the construct.

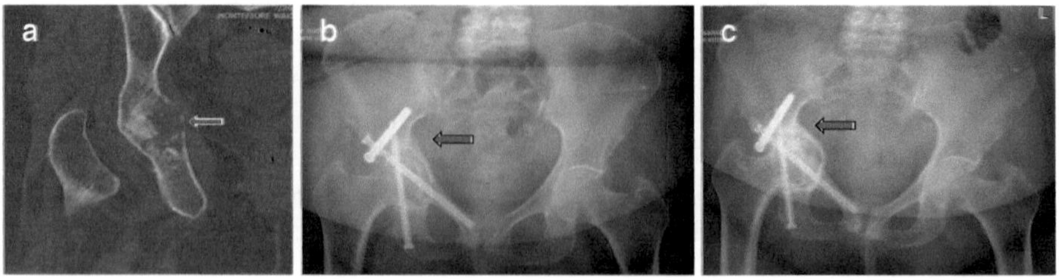

Fig. 14.1 73-year-old female with metastatic thyroid cancer to bone who presented with severe right hip pain. CAT scan (**a**, arrow) showed periacetabular disease with posterior column pathological fracture. After biopsy-proven diagnosis, the patient underwent percutaneous Tripod screw fixation (**b**) with significant pain relief. At a one-year follow-up, radiographs showed a completely healed fracture (**c**). She was able to ambulate independently with occasional use of an assisting device

Percutaneous screw fixation has the advantages of early weight-bearing and avoidance of complications associated with cementoplasty (extravasation) and ablation (thermal injuries). The host bone can be preserved more as lesions can heal with bone formation after radiation [17]. The setup requires only a radiolucent table, fluoroscopy, and the instrumentation tray, obviating the need for an interventional radiology suite. The main disadvantage of percutaneous screw fixation is the steep learning curve for surgeons who are not accustomed to and experienced in the procedure.

Photodynamic Nails

Photodynamic nails are novel implants that were originally developed to provide fixation in long bones without the need for locking screw fixation [23]. They have since been FDA-approved for use in the pelvis, where they are especially useful in navigating the complex anatomic corridors. The system consists of a smooth balloon catheter that can successfully accommodate into curvy and complex bony structures. The catheter is then filled with monomer to pervade the entire canal or defect. The monomer then responds to blue light by polymerizing and converting into a stiff construct that fixes the defect and fracture (Fig. 14.2). Photodynamic nails have shown a significant improvement in pain and physical function scores by the 6-week postoperative visit [24, 25].

Photodynamic nails share many similarities with the percutaneous screw placement technique as another form of internal fixation. The nails provide great fixation in anatomically complicated bone and serve well in challenging cases of pelvic metastases with notably distorted trajectories or uncontained defects. Internal fixation methods can also be incorporated into THA to augment the stability of the construct [23]. Since the monomer is injected into a balloon, there is no risk of extravasation, but the ability to fill a large defect is sometimes limited by the

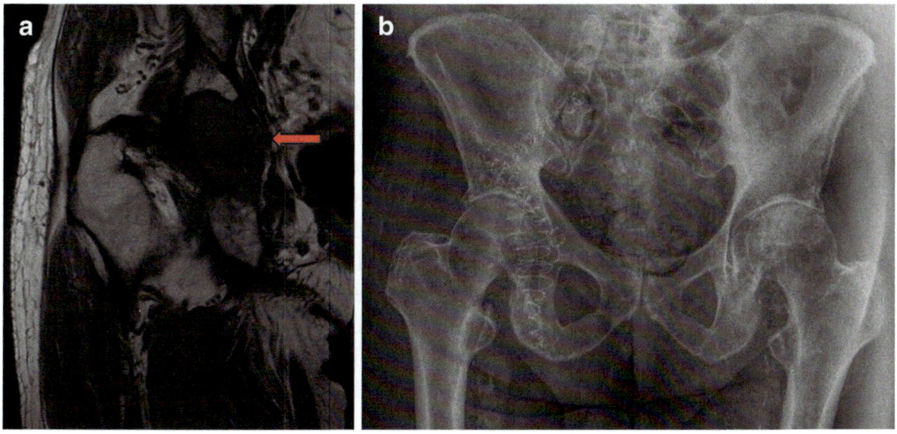

Fig. 14.2 84-year-old male with multiple myeloma presenting with right hip pain exacerbated by axial loading. MRI of the hip (**a**) demonstrates an expansile marrow-replacing lesion of the posterior acetabulum with extraosseous extension. The patient underwent percutaneous fixation with photodynamic nails (**b**), which resulted in significant improvement in pain. At six-months follow-up, the patient is now able to ambulate without assistive devices

available balloon sizes. In addition, given the radiolucency of photodynamic nails, they allow for easier radiation planning and postoperative oncologic surveillance. Potential increases in costs and limited accessibility of these implants are disadvantages of this technique.

Ablation, Osteoplasty, Reinforcement, and Internal Fixation (AORIF)

Ablation, osteoplasty, reinforcement, and internal fixation (AORIF) is a combination strategy that utilizes multiple of the previously discussed surgical techniques to provide local disease control and stable fixation [15, 26]. In this technique, a cannulated screw is partially inserted through intact bone up to the lesional border and is subsequently used as a common entry port for ablation, followed by osteoplasty and PMMA cementation. Osteoplasty ensures that space is created for cement injection and creates an improved tissue boundary to prevent against cement extravasation. A second screw is often used as ventilation for this portion of the procedure. Prior to PMMA curing, the original cannulated screw is advanced through the now-filled defect and fixed within the cement.

AORIF has several advantages. Like photodynamic nails, the use of cementation in combination with percutaneous screw fixation allows the construct to fill the osteolytic defect and provide increased biomechanical support [27]. The use of ablation provides local tumor control, which likely decreases the risk of construct failure due to disease progression. However, AORIF has a learning curve and inherits the risk of each of the individual techniques that comprise the procedure, namely cement extravasation and embolization. Additionally, the screw fixation within the PMMA cement makes revision difficult, should it be necessary.

Surgical Technique

The entry point of the cannulated screw or photodynamic nail is determined by positioning the guidewire over the skin under image intensification. A stab incision is made to insert a 3.2-mm guidewire with a power instrument. A biopsy sample can be taken through the stab incision for histopathological analysis before the placement of guidewires. Drilling the near cortex is often sufficient, and in cases with poor bone quality, this can also be omitted. When using screw fixation, 8.0-mm fully threaded cannulated screws should be used whenever possible; however, 6.5-mm cannulated screws can be fixed for curvy and narrow trajectories. Lengths for photodynamic nails typically range from 100 to 120 mm in length and can vary in diameter from 9 to 22 mm, depending on the size of the defect. For periacetabular metastases, the typical screw or nail paths include the anterior column, posterior column, and transcolumnar corridors (sciatic corridor).

As anterior column screws or nails are generally more difficult to place, it may be best to start with them (Fig. 14.3). The entry point of an anterior column screw or nail is in the supraacetabular region of the ilium. The screw or nail should pass from above the acetabulum into the medial aspect of the ipsilateral pubic ramus. Anterior column screws or nails should be placed using obturator oblique views for

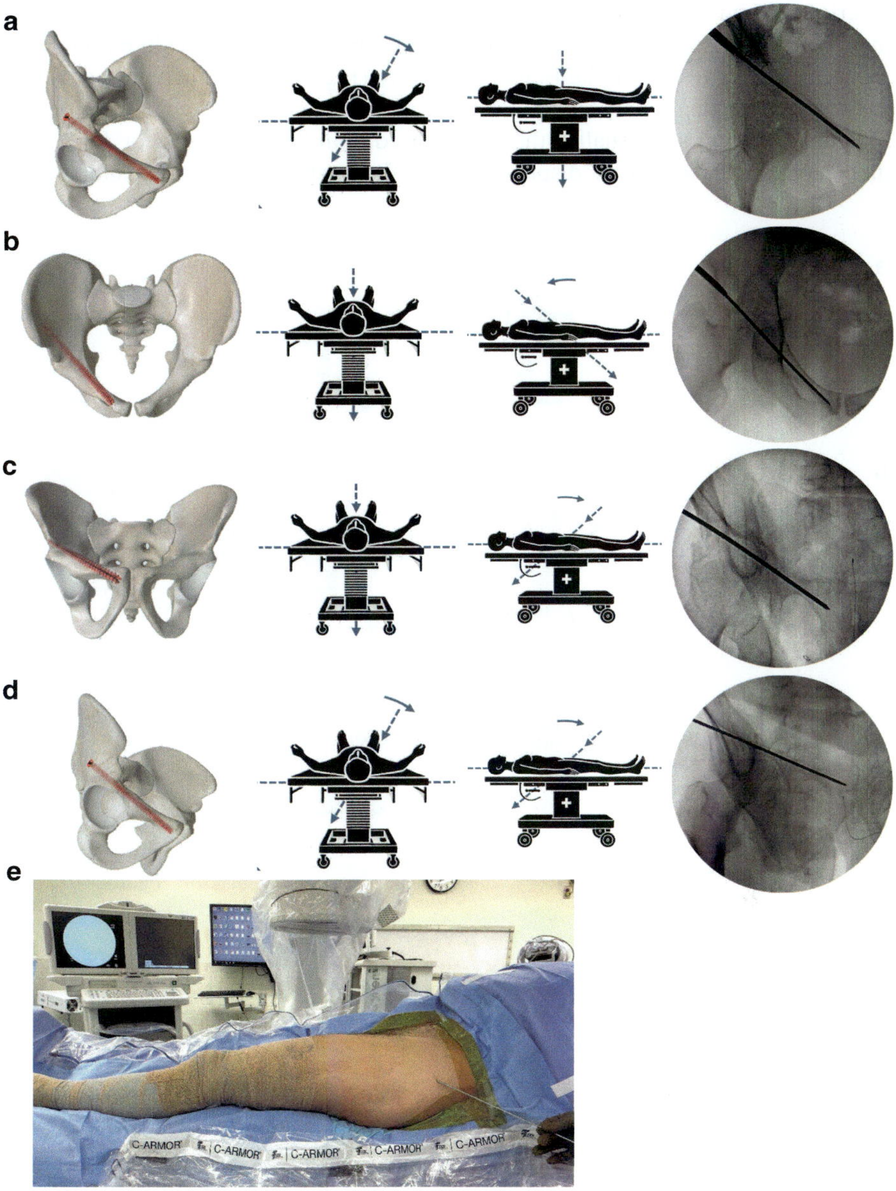

Fig. 14.3 Anterior column screw: (**a**) The obturator oblique view is utilized to obtain the starting point for a 3.2mm guidewire. (**b**) The inlet and (**c**) outlet views allow the surgeon to place the guidewire appropriately. (**d**) The obturator outlet view ensures that the hip joint is not violated. 3D illustrations (left), corresponding projection trajectory of C-arm (middle) and intraoperative fluoroscopic images (right) are also presented. (E) Clinical picture shows the starting point of the anterior column screw

anterior column visualization, pelvic inlet views for anterior-posterior (AP) adjustment, and outlet views for superior-inferior adjustment. Posterior column screws or nails can be inserted from the ischial tuberosity (Fig. 14.4). AP pelvis view is best used for the starting point, and the lower extremity should be flexed at the hip. Iliac oblique views, which visualize the posterior column best, are used for AP adjustment and are excellent in displaying the relationship to the acetabulum joint.

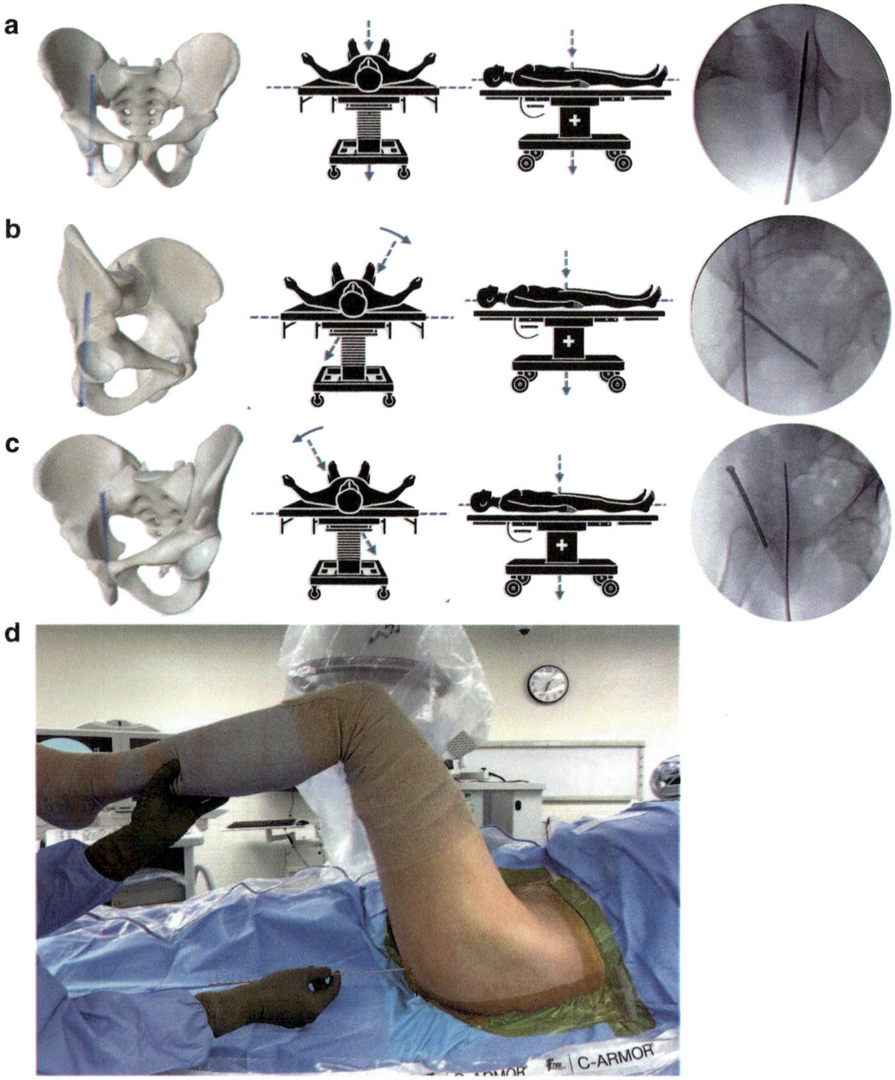

Fig. 14.4 Retrograde posterior column screw: (**a**) The AP pelvis view is utilized to find the starting point and trajectory. (**b**) The obturator oblique and (**c**) iliac oblique view are used to ensure that the guidewire remains in the posterior column, particularly as it is advanced past the sciatic notch. 3D illustrations (left), corresponding projection trajectory of C-arm (middle) and intraoperative fluoroscopic images (right) are also presented. (**d**) Clinical picture shows hip in flexion and abduction to provides access to the ischial tuberosity for the starting point

Obturator oblique views are helpful in mediolateral placement of the screw or nail. When the patient is in the supine position, the transcolumnar screw or nail is inserted through the anterior inferior iliac spine (AIIS)–which can be located below the teardrop on an obturator outlet view–and aimed at the posterior inferior iliac spine (PIIS) (Fig. 14.5). When the patient is in the prone position, the screw or nail is

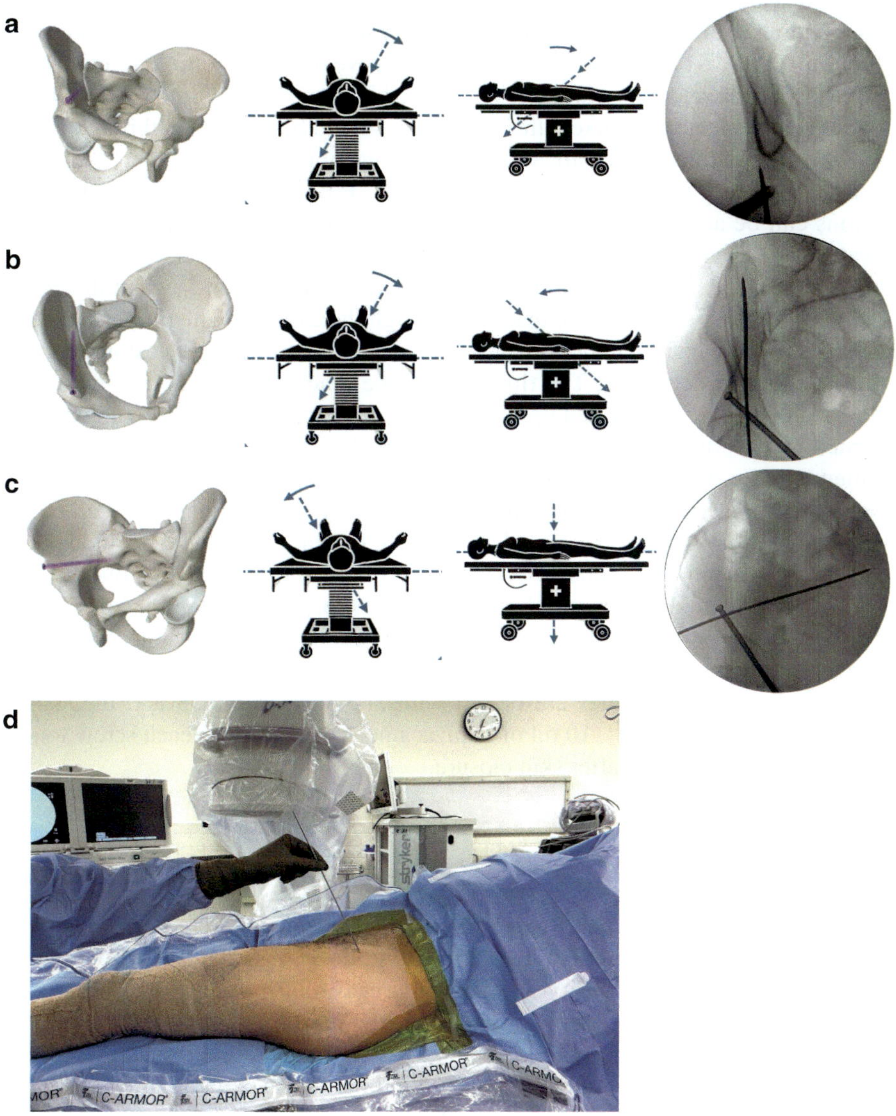

Fig. 14.5 Trans-columnar screw: (**a**) The obturator outlet (teardrop) view demonstrates the starting point for trans-columnar screw placement, typically the AIIS. Medial/lateral and cranial/caudal positioning are determined by the (**b**) obturator inlet and (**c**) iliac oblique views, respectively. 3D illustrations (left), corresponding projection trajectory of C-arm (middle) and intraoperative fluoroscopic images (right) are also presented. (**d**) Clinical picture shows the starting point of the trans-columnar screw

inserted through the PIIS and aimed toward the AIIS. Iliac oblique views are great at displaying the wire's position in the iliac wing and its relationship to the sciatic notch and allow for superior/inferior adjustments. The inlet and/or obturator inlet views are utilized for medial/lateral placement between the inner and outer tables of the ilium.

Gluteal pillar screws or nails through the iliac crest or inferior pubic ramus screws or nails can be placed based on the location for pelvic tumors localized in other regions. For lesions impacting the iliosacral region, screws or nails should be placed along the S1 corridor. In cases with extensive disease, multiple screws parallel in the same trajectory can increase the stability of the construct. After all the screws or nails have been fixed, their positions should be confirmed through different views under image intensification or CT-navigation.

Ablation and cementation techniques are relatively straightforward. The lesions can be accessed through entry points based on their locations. The iliac crest can be used as an entry point for lesions in the ilium. After a stab incision, the cortex of the iliac crest can be perforated with a trocar, and a cannula can be inserted. An RFA probe can be inserted through the cannula. The probe should be centered in the lesion under image guidance; however, large lesions can be ablated in multiple points. The tip of the probe reaches 95 °C during RFA, and the zone of ablation is heated up to 70 °C for 15 mins. Cryoablation can be performed with a similar approach, as tumor necrosis is achieved at −40 °C. PMMA cement can be injected either directly through a cannula, or a kyphoplasty balloon can be inserted for cement application. Injection of cement is often easier when it is still in the viscous liquid state, since the cannula can be occluded with cement in a semi-solid phase. As previously mentioned, when ablation and cementation techniques are combined with screw or photodynamic nail fixation, the same entry point can be used for the RFA probe or cement injection as for screw or nail fixation [15].

The subcutaneous tissues are closed with 2–0 Vicryl or PDS and the skin is closed with 3–0 Monocryl. 10 ml of bupivacaine is injected into each screw trajectory as a local anesthetic after skin closure.

Postoperative Care

Patients can be weight-bearing as tolerated for lesions located in Enneking zones I and III. Periacetabular (Enneking zone II) lesions depend on the procedure chosen and the stability of the construct. It is possible to allow early weight-bearing with Tripod percutaneous screw fixation or photodynamic nail stabilization as the internal fixation construct works as a rebar. Ablative procedures do not add stability unless they are combined with cementoplasty or internal fixation. Surgeons should consider the extent of destruction and the ability of the construct to tolerate loading for weight-bearing restrictions. The use of assistive devices depends on the individual's capacity, which is sometimes needed especially during the first 4–6 weeks. Patients should be mobilized as early as possible to improve functional outcomes and prevent thrombotic events. Rehabilitation with physical therapy exercises can provide a great benefit for recovery in the first 3 months.

Complications

Percutaneous procedures are generally considered safer compared to open surgeries; however, there is still a risk of complications such as superficial/deep infection, thrombosis/embolism, and neurovascular injuries [28]. Minimizing the risks of some of these complications is possible by preemptive measures such as administering antibiotics to prevent infection, mechanical or pharmacological thromboprophylaxis as well as early mobilization for lowering rates of thrombosis. Surgeons are obligated to thoroughly inform patients about the risks. Healthy communication is key to building a relationship of trust should complications occur, and facilitates further management, ensures continuity of care as well as lowers the risk of litigation [29]. Each percutaneous treatment option has been reported to have specific complications in addition to the risks of minimally invasive pelvic surgery.

Percutaneous screw fixation has been associated with intraarticular screw placement and a 20 percent rate of conversion to THA, although these complications are less frequent in non-periacetabular metastases [17, 30]. Photodynamic nails can rupture, and removal can be difficult when revision surgery is needed [19]. Cryoablation and RFA have been associated with risks secondary to thermal damage to adjacent nerves, and there has been a case report of chondrolysis and osteonecrosis of the femoral head [31, 32]. Cementoplasty can cause leakage of the PMMA cement, leading to a range of complications from asymptomatic extravasation to pulmonary embolization, a devastating complication that can even lead to patient mortality [19, 33]. Balloon-assisted osteoplasty can mitigate extravasation by containing the PMMA inside the augmentation balloon, however, does not eliminate the risk of leakage entirely [20].

Pearls and Pitfalls

Guide-wire positioning can be challenging in cases with extensive bone loss or distorted anatomy. When a wire is not perfectly located in the initial attempt, a second wire can be inserted without removing the first one and using it as a reference line. This is especially useful in circumstances where a wire breaches through cortical bone as it can block the second wire's passage.

Wires that point toward cortical bone can be advanced through the medullary canal with gentle mallet taps instead of using a power tool. This is a particularly helpful technique for traversing the pubic rami during anterior column screw placement. Another way of redirecting the wire is to slightly retract and advance a 4.9 cannulated drill until the wire is swallowed. This widens the path to reposition the wire inside a canal for close distances.

Imaging is the key to success in minimally invasive procedures of the pelvis. Surgeons should be familiar with many different views of the pelvis, which is imperative in understanding the 3D structure of the bone from multiple angles, and the position of wires or implants in relation to critical structures and landmarks.

References

1. Kakhki VRD, Anvari K, Sadeghi R, Mahmoudian A-S, Torabian-Kakhki M. Pattern and distribution of bone metastases in common malignant tumors. Nucl Med Rev Cent East Eur. 2013;16:66–9.
2. Nakamoto Y, Osman M, Wahl RL. Prevalence and patterns of bone metastases detected with positron emission tomography using F-18 FDG. Clin Nucl Med. 2003;28:302–7.
3. Tofe AJ, Francis MD, Harvey WJ. Correlation of neoplasms with incidence and localization of skeletal metastases: an analysis of 1,355 diphosphonate bone scans. J Nucl Med. 1975;16:986–9.
4. DiCaprio MR, Murtaza H, Palmer B, Evangelist M. Narrative review of the epidemiology, economic burden, and societal impact of metastatic bone disease. Ann Jt. 2022;7:28.
5. Enneking WF, Dunham WK. Resection and reconstruction for primary neoplasms involving the innominate bone. J Bone Joint Surg Am. 1978;60:731–46.
6. Harrington KD. The management of acetabular insufficiency secondary to metastatic malignant disease. J Bone Joint Surg Am. 1981;63:653–64.
7. Healey JH, Brown Lenard H. 2.15 Pathological pelvic fractures and acetabular reconstruction in metastatic disease. In: Fractures of the pelvis and acetabulum; 2015. https://doi.org/10.1055/b-0035-121667.
8. Marco RA, Sheth DS, Boland PJ, Wunder JS, Siegel JA, Healey JH. Functional and oncological outcome of acetabular reconstruction for the treatment of metastatic disease. J Bone Joint Surg Am. 2000;82:642–51.
9. Harrington KD, Sim FH, Enis JE, Johnston JO, Diok HM, Gristina AG. Methylmethacrylate as an adjunct in internal fixation of pathological fractures. Experience with three hundred and seventy-five cases. J Bone Joint Surg Am. 1976;58:1047–55.
10. Tillman R, Tsuda Y, Puthiya Veettil M, Young PS, Sree D, Fujiwara T, Abudu A. The long-term outcomes of modified Harrington procedure using antegrade pins for periacetabular metastasis and haematological diseases. Bone Joint J. 2019;101-B:1557–62.
11. Brown TS, Salib CG, RosePS SFH, Lewallen DG, Abdel MP. Reconstruction of the hip after resection of periacetabular oncological lesions: a systematic review. Bone Joint J. 2018;100-B:22–30.
12. Park JW, Kang HG, Kim JH, Kim H-S. The application of 3D-printing technology in pelvic bone tumor surgery. J Orthop Sci. 2021;26:276–83.
13. Wang J, Zhao Z, Liang H, Zhang R, Liu X, Zhang J, Singh S, Guo W, Yan T, Hoang BH, et al. Artificial intelligence assisted preoperative planning and 3D-printing guiding frame for percutaneous screw reconstruction in periacetabular metastatic cancer patients. Front Bioeng Biotechnol. 2024;12:1404937.
14. Hartung MP, Tutton SM, Hohenwalter EJ, King DM, Neilson JC. Safety and efficacy of minimally invasive acetabular stabilization for periacetabular metastatic disease with thermal ablation and augmented screw fixation. J Vasc Interv Radiol. 2016;27:682–688.e1.
15. Lee FY, Latich I, Toombs C, Mungur A, Conway D, Alder K, Ibe I, Lindskog D, Friedlaender G. Minimally invasive image-guided ablation, osteoplasty, reinforcement, and internal fixation (AORIF) for osteolytic lesions in the pelvis and periarticular regions of weight-bearing bones. J Vasc Interv Radiol. 2020;31:649–658.e1.
16. Fourman MS, Ramsey DC, Newman ET, Raskin KA, Tobert DG, Lozano-Calderon S. How I do it: percutaneous stabilization of symptomatic sacral and periacetabular metastatic lesions with photodynamic nails. J Surg Oncol. 2021;124:1192–9.

17. Yang R, Goch A, Murphy D, Wang J, Charubhumi V, Fox J, Sen M, Hoang B, Geller D. A novel tripod percutaneous reconstruction technique in periacetabular lesions caused by metastatic cancer. J Bone Joint Surg Am. 2020;102:592–9.
18. Araneta KTS, Rizkallah M, Boucher L-M, Turcotte RE, Aoude A. Joint-sparing reconstruction for extensive periacetabular metastases: literature review and a novel minimally invasive surgical technique. J Bone Oncol. 2022;34:100428.
19. Cotten A, Deprez X, Migaud H, Chabanne B, Duquesnoy B, Chastanet P. Malignant acetabular osteolyses: percutaneous injection of acrylic bone cement. Radiology. 1995;197:307–10.
20. Kurup AN, Morris JM, Schmit GD, Atwell TD, Schmitz JJ, Rose PS, Callstrom MR. Balloon-assisted osteoplasty of periacetabular tumors following percutaneous cryoablation. J Vasc Interv Radiol. 2015;26:588–94.
21. Hoffmann RT, Jakobs TF, Trumm C, Weber C, Helmberger TK, Reiser MF. Radiofrequency ablation in combination with osteoplasty in the treatment of painful metastatic bone disease. J Vasc Interv Radiol. 2008;19:419–25.
22. Deschamps F, de Baere T, Hakime A, Pearson E, Farouil G, Teriitehau C, Tselikas L. Percutaneous osteosynthesis in the pelvis in cancer patients. Eur Radiol. 2016;26:1631–9.
23. Zoccali C, Attala D, Pugliese M, di Uccio AS, Baldi J. The IlluminOss® photodynamic bone stabilization system for pathological osteolyses and fractures of the humerus: indications, advantages and limits in a series of 12 patients at 24 months of minimum follow-up. BMC Musculoskelet Disord. 2021;22:63.
24. Lozano-Calderon SA, Gonzalez MR, Werenski JO, Quinn K, Freiberger D, Raskin KA. A prospective study of the IlluminOss photodynamic nail system for pelvic stabilization: treatment of impending and actual fractures from metastatic bone disease, multiple myeloma, and primary bone lymphoma. JB JS Open Access. 2024;9
25. Lozano-Calderon SA, Clunk MJ, Gonzalez MR, Sodhi A, Krueger RK, Gruender AC, Greenberg DD. Assessing pain and functional outcomes of percutaneous stabilization of metastatic pelvic lesions via photodynamic nails: a bi-institutional investigation of orthopaedic outcomes. JB JS Open Access. 2024;9
26. Dussik CM, Toombs C, Alder KD, Yu KE, Berson ER, Ibe IK, Li F, Lindskog DM, Friedlaender GE, Latich I, et al. Percutaneous ablation, osteoplasty, reinforcement, and internal fixation for pain and ambulatory function in periacetabular osteolytic malignancies. Radiology. 2023;307:e221401.
27. Morris MT, Alder KD, Moushey A, Munger AM, Milligan K, Toombs C, Conway D, Lee I, Chen F, Tommasini SM, et al. Biomechanical restoration of metastatic cancer-induced periacetabular bone defects by ablation-osteoplasty-reinforcement-internal fixation technique (AORIF): to screw or not to screw? Clin Biomech (Bristol, Avon). 2022;92:105565.
28. Ibe I, Dussik CM, Callan AK, Barr J, Lee FY. Emerging minimally invasive percutaneous procedures for periacetabular osteolytic metastases. J Bone Joint Surg Am. 2023;105:479–89.
29. Schleiter KE. Difficult patient-physician relationships and the risk of medical malpractice litigation. Virtual Mentor. 2009;11:242–6.
30. Yang R, Singh S, Falk A, Wang J, Thornhill B, Fox J, Sen M, Hoang B, Geller DS. Percutaneous screw stabilization of non-periacetabular pelvic lesions caused by metastatic cancer and multiple myeloma. J Bone Joint Surg Am. 2022;104:577–85.
31. English DI, Lea WB, King DM, Tutton SM, Neilson JC. Minimally invasive stabilization with or without ablation for metastatic periacetabular tumors. J Bone Joint Surg Am. 2021;103:1184–92.
32. Friedman MV, Hillen TJ, Wessell DE, Hildebolt CF, Jennings JW. Hip chondrolysis and femoral head osteonecrosis: a complication of periacetabular cryoablation. J Vasc Interv Radiol. 2014;25:1580–8.
33. Park JW, Lim H-J, Kang HG, Kim JH, Kim H-S. Percutaneous cementoplasty for the pelvis in bone metastasis: 12-year experience. Ann Surg Oncol. 2022;29:1413–22.

Management of Pelvic Metastasis: Open Surgical Techniques

15

Herrick J. Siegel

Epidemiology

The etiology of pelvic bone metastasis involves the spread of cancer cells from a primary tumor to the bones in the pelvic region. The pelvis is only second to the spine as the most frequent site of bone metastasis [1–4]. The cancers that most commonly metastasize to bone include prostate, breast, lung, kidney, thyroid, and gastrointestinal cancers. Hematogenous spread in which cancer cells travel through the bloodstream to distant sites is the most common mode of metastasis. Bones with a high blood supply, such as the pelvis, are often sites for metastatic growth due to increased access. Local Invasion may occur near the pelvis, from such primary sites as the bladder, rectal, or cervical cancers. These may invade nearby the pelvic bones by direct extension. Chemotactic and growth factors are specific proteins secreted from tumors that attract cancer cells to bone tissue and enable their growth in bone [5–8]. The pelvic bone marrow is a rich vascular structure that creates a supportive environment for cancer cell survival and proliferation [9–11].

Anatomic Considerations

The anatomy of the pelvis is a bony ring consisting of the sacrum, coccyx, and bilateral os coxae (ilium, ischium, and pubis), forming the major load-bearing region between the spine and lower extremities. It provides structural support for standing and walking and serves as an attachment site for numerous muscles and ligaments. It is divided into three main regions based on load transfer and function. The degree of bone loss, tumor type, and extent of involvement of pelvic bone are paramount to

H. J. Siegel (✉)
Rothman Orthopaedics/Jefferson Hospital, Philadelphia, PA, USA

J. Pretell-Mazzini, S. A. Lozano Calderón (eds.), *Multidisciplinary Approach of Musculoskeletal Metastasis*, https://doi.org/10.1007/978-3-032-09498-8_15

the decision of whether to surgically treat, which surgical approach and the type of surgical resection and reconstruction.

1. Anterior Pelvis: Formed by the pubic symphysis and superior/inferior pubic rami, it is critical in supporting lower abdominal viscera and plays a minor role in weight transmission.
2. Posterior Pelvis: Comprising the sacrum, sacroiliac joints, and parts of the ilium, the posterior pelvis transmits most of the body weight from the spine to the lower extremities. The sacrum is particularly vulnerable to pathologic fractures in metastatic disease.
3. Lateral Pelvis: Includes the acetabulum and hip joint, where the body weight is distributed through the femoral head during standing and ambulation. Pathologic lesions here can lead to hip instability and fractures, affecting mobility.

Pelvic metastases can involve any of these regions, necessitating a tailored surgical approach based on the specific anatomical location, size of the lesion, and functional compromise.

Surgical Indications/Non-surgical Indications

Surgery for pelvic metastases is primarily indicated in patients with debilitating symptoms or those at risk for complications such as pathologic fractures and/or neurovascular compromise [1–3]. The indications for surgery include:

1. Intractable Pain: When pain is unresponsive to conservative treatments such as analgesics, radiation therapy, or systemic therapy, surgical intervention may be necessary to remove or debulk the tumor, stabilize the bone, or decompress nerves.
2. Pathologic Fractures: Pelvic bones, particularly the acetabulum and sacrum, are susceptible to pathologic fractures due to metastatic bone destruction. Surgical stabilization is indicated when fractures occur or are imminent.
3. Mechanical Instability: Metastatic lesions in load-bearing regions of the pelvis, such as the acetabulum or sacrum, can compromise skeletal integrity, leading to instability and impaired mobility. Surgery is indicated to restore stability.
4. Neurological Compromise: Tumors in the sacrum or near the sacroiliac joint can cause compression of nearby nerves, leading to neurological deficits such as weakness, sensory loss, or bladder and bowel dysfunction. Decompression surgery may be necessary to relieve these symptoms.
5. Failure of Non-Surgical Treatments: When conservative treatments such as chemotherapy, targeted therapy, immunotherapy, or radiation therapy fail to control tumor growth or alleviate symptoms, surgery may be considered as a salvage option.
6. Prevention of Complications: In patients with high-risk lesions, such as large lytic metastases in weight-bearing regions, prophylactic surgery may be performed to prevent fractures or other complications.

Surgical decision-making requires a thorough assessment of the patient's overall oncological prognosis, functional status, life expectancy, and the risks and benefits of surgery [4]. Non-surgical options may include activity and/or weight-bearing restrictions, radiation or systemic treatment. The etiology, location of anatomic involvement and extent of tumor burden at other sites are essential factors in the decision-making process.

Surgical Classifications

The most used classification systems for pelvic bone metastasis are the Enneking Classification and the Harrington Classification. The Enneking divides the pelvis into three zones: Zone I: Iliac wing, between the acetabulum and sacroiliac joint, zone II: Acetabulum and zone III the ischium. The Enneking system may also be used to describe the type of internal hemipelvectomy. The zones may also be combined for large resections, that is, Zones 1–2 would include the iliac wing and acetabulum [5, 6].

The Harrington Classification focuses primarily on Enneking zone II and describes the extent of bone loss of the acetabulum. In class I lesions, the articular surface is disrupted, but the walls and columns are intact. Patients with class I lesions were reconstructed with cemented acetabular and femoral components. In class II lesions, the medial wall and quadrilateral plate are deficient, requiring acetabular reconstruction with a flanged cup to transfer weight-bearing forces away from the medial acetabulum and to prevent medial collapse of the reconstruction. In class III lesions, the roof and acetabular rim are deficient and require reconstruction with Steinmann pins in cement and total hip arthroplasty (THA) using a flanged cup or cage (i.e., cement-reinforced THA) [5].

The metastatic acetabular classification (MAC) describes lesions in four anatomic zones—the acetabular dome, medial wall, anterior column, and posterior column. Specific reconstructions were proposed based on the anatomic pattern of destruction. For all these reconstructions, adequate bone stock in the ipsilateral pelvis and ischium or floor of the acetabulum is needed for Steinmann pin and cage-screw fixation [5].

Preoperative Planning

Preoperative planning for pelvic metastasis surgery is crucial to ensure optimal outcomes and minimize complications. This process involves a multidisciplinary approach, including orthopedic surgeons, oncologists, radiologists, and anesthesiologists. Key considerations in preoperative planning include:

1. Imaging:
   - X-ray: Radiographs are useful for initial evaluation of bone lesions and fractures, though they may not show early metastatic involvement [8].

- CT scan: Provides detailed cross-sectional images of bone and soft tissue structures, enabling precise assessment of bone destruction and tumor involvement. CT angiography may often be useful to assess the adjacent vascular structures [8].
- MRI: Essential for evaluating soft tissue and neural structures, particularly in cases of sacral or nerve root involvement. MR angiogram may be useful to assess the extent of vascular invasion or degree of vessel displacement by the tumor [9]. Also very helpful in identifying occult metastasis [8, 9, 12].
- Bone Scintigraphy: A bone scan can detect multiple metastatic sites throughout the skeleton and assess the overall burden of metastatic disease [9].
- PET-CT: Combines metabolic and anatomical information, useful for identifying tumor activity, treatment response and bone stock as well as distant systemic disease [10].

2. Tumor Staging: The extent of metastatic disease, both in the pelvis and elsewhere in the body, is critical in determining the appropriateness of surgery. Patients with widespread metastases, rapid disease progression, or poor oncological prognosis may not benefit from extensive surgical intervention [10].
3. Functional Assessment: A preoperative evaluation of the patient's functional status, including mobility, neurological function, and overall physical health, is essential to determine the potential benefits of surgery. In patients with limited life expectancy or poor functional status, less invasive options may be more appropriate [11].
4. Neurovascular Considerations: The proximity of the pelvic bones to major blood vessels (iliac vessels) and nerves (lumbosacral plexus) necessitates careful planning to avoid iatrogenic injury during surgery. Preoperative imaging can help identify the relationship between the tumor and these critical structures [11].
5. Anesthetic Considerations: Given the complexity and potential duration of pelvic surgeries, a thorough anesthetic evaluation is required. Patients may have comorbid conditions related to their cancer or treatments that increase perioperative risk. Blood loss can be significant, necessitating appropriate preparation for transfusions and hemodynamic monitoring [13]. Preoperative embolization procedures may be considered if the blood loss is expected to be excessive. Blood bank consultation may be needed to assure that the appropriate type and number of units of packed red blood cells and/or platelets are available.

Surgical Options

The choice of surgical technique depends on the location and extent of the pelvic metastasis, the goals of surgery, and the patient's overall condition. Surgical options can be broadly classified into procedures for stabilization, debulking, resection, and reconstruction. Each technique has its own indications and challenges [14–20].

1. *Curettage and Bone Filling*

   Curettage involves the removal of tumor tissue from within the bone using curettes and/or high-speed burrs. It is primarily indicated for osteolytic lesions that have not caused complete destruction of the bone or mechanical instability. After curettage, cement augmentation or bone substitutes are typically used to fill the resulting cavity and restore structural integrity [21–23].

   - Advantages: It can be performed minimally invasively, which will translate to less morbidity compared to major resections.
   - Disadvantages: Risk of incomplete tumor removal and local progression. Also, inadequate stability of compromised pelvic bone.

2. *Cement Augmentation*

   Cement augmentation, often used in conjunction with curettage, involves the injection or packing of polymethylmethacrylate (PMMA) cement into osteolytic lesions. This technique provides immediate mechanical stability, reduces pain, and is particularly useful in treating acetabular and sacral metastases [22, 23].

   - Advantages: Quick recovery, immediate pain relief, minimally invasive.
   - Disadvantages: Limited to less extensive disease; does not offer long-term tumor control. It may be less than optimal for areas of higher mechanical stresses/weight-bearing regions.

3. *Prophylactic Stabilization with Internal Fixation*

   For patients at risk of pathologic fractures, prophylactic stabilization with internal fixation can prevent fracture-related complications (Fig. 15.1). *Details were discussed in the previous chapter.* Common fixation techniques include:

   - Screws and Plates: Used for smaller lesions or fractures in non-weight-bearing regions of the pelvis.

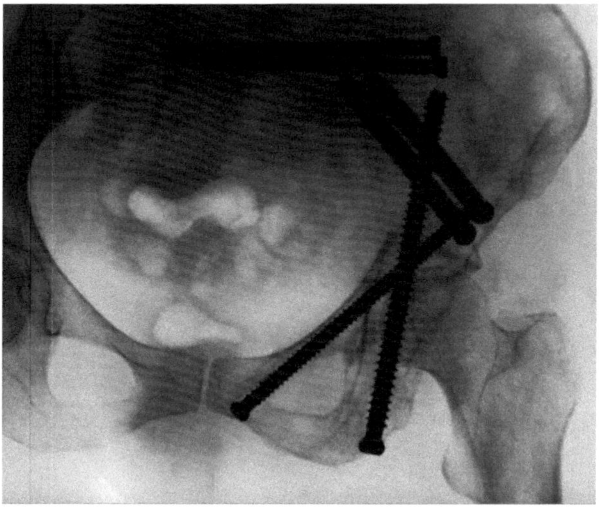

Fig. 15.1 AP pelvis showing multiple percutaneous screws placed to stabilize pelvis with large osteolytic lesion. Cement may be used cautiously to fill a contained defect

- Intramedullary Nails: Provide stabilization for long bone involvement extending to the pelvis, particularly when femoral metastasis is also present.
- Pelvic Reconstruction Plates: Used in cases of acetabular involvement, these plates help stabilize the hip joint and restore weight-bearing capacity.
- Advantages: Prevents fractures and preserves mobility.
- Disadvantages: Potential for hardware failure or infection, limited to structurally compromised bone.

4. *Acetabular Reconstruction*

   Metastatic lesions involving the acetabulum can lead to hip joint instability and mechanical failure. Acetabular reconstruction is indicated for patients with acetabular destruction or impending fracture [24–28] (Fig. 15.2).

   - Total Hip Arthroplasty (THA): In cases where the femoral head or acetabulum is extensively involved, total hip replacement may be necessary. Cemented acetabular components are often preferred due to the poor quality of metastatic bone.
   - Cemented Acetabular Cups: In cases of severe acetabular destruction, the use of cemented acetabular cups can restore joint stability and allow early mobilization. This is often done in conjunction with metallic rebar with Steinman pins or screws (Fig. 15.3a–c).
   - Cage Reconstruction: In cases of extensive bone loss, a cage reconstruction can be performed to support the acetabular component of a total hip replacement. (Fig. 15.4a and b).
   - Advantages: Restores joint function and stability, allows for weight-bearing.
   - Disadvantages: Major surgery with potential for complications, particularly in patients with extensive disease.

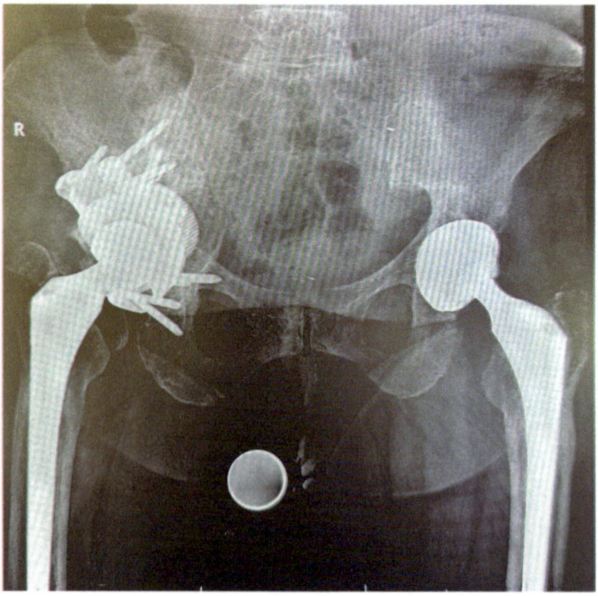

Fig. 15.2 AP pelvis showing a complex reconstruction using a combination of a custom triflange implant with cement augmentation to treat a pathologic fracture of the right acetabulum thyroid carcinoma metastasis

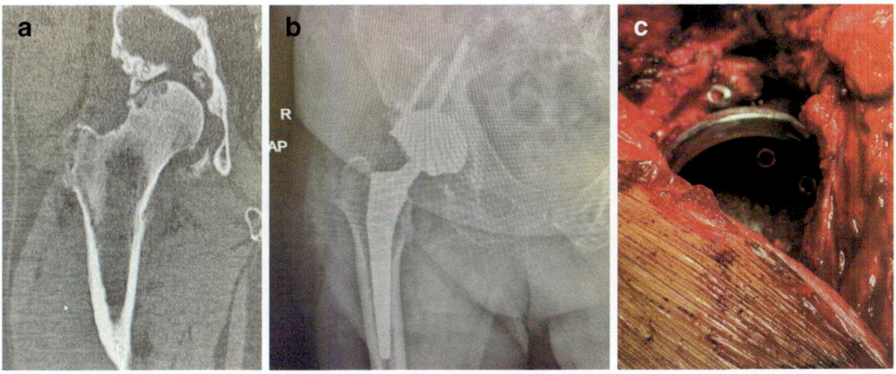

Fig. 15.3 (**a**) Preop AP pelvis, (**b**) Postop hip reconstruction using pelvic screws and cement for rebar. (**c**) Intra operative photo of superior screw acting as buttress and rebar to stabilize cup

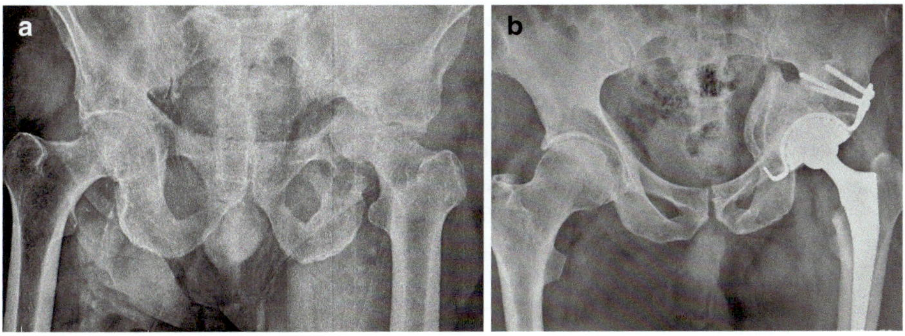

Fig. 15.4 (**a**) Preoperative AP pelvic image. (**b**) 6 months post op image showing anti-protrusio cage and healed periacetabular bone following systemic treatment and bone grafting

Novel Reconstructive Options

Acetabular augments, cup in cup construct using tantalum or titanium revision shells in combination with cemented cups and cups supported internally by cages may also be used. Instability may be a result of changes in the hip center/joint line. While bipolar heads may be used with proximal femoral resections, dual mobility and constrained liners may be required if hip stability cannot be obtained intraoperatively. Containment does add increase stress through the acetabular construct and may lead to early loosening; dual mobility would likely not have a similar negative impact on construct stability (Fig. 15.5a and b).

5. *Sacroiliac Fixation and Sacral Tumor Resection*

   The sacrum is a common site for pelvic metastases, particularly in prostate, breast, and renal cell carcinomas. Tumors in this region can cause significant pain and neurological compromise due to involvement of the sacral nerve roots [29–31]. Surgical options for sacral metastases include:

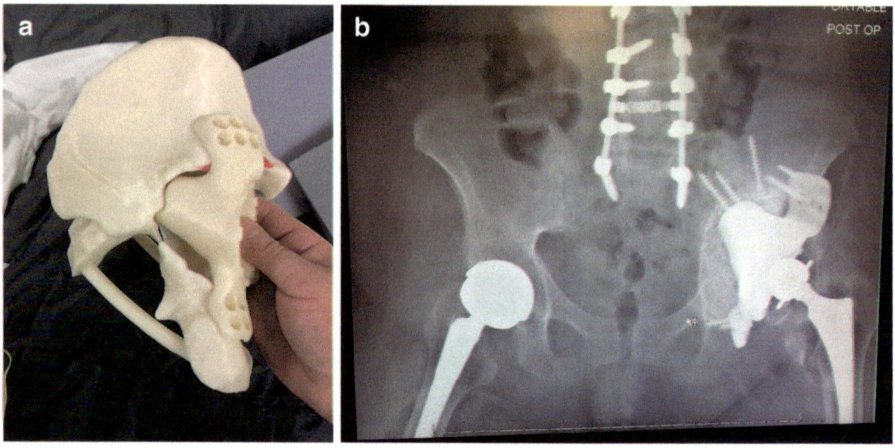

Fig. 15.5 (**a**) Pre op model for a custom designed implant in patient with renal cell metastastsis. (**b**) Post op image showing reconstruction with custom implant and use of constrained Freedom liner to enhance stability due to compromise of gluteal muscles

- Sacral Decompression: In cases of nerve compression, decompression surgery can relieve symptoms and improve neurological function. This may involve removal of part of the sacrum (partial sacrectomy).
- Sacroiliac Fixation: When metastases involve the sacroiliac joint, fixation with screws or rods may be necessary to restore stability and prevent further collapse.
- Advantages: Pain relief and improvement of neurological symptoms, as well as stabilization of the posterior pelvis.
- Disadvantages: Sacral resections can be technically challenging and may result in significant morbidity, including loss of bowel or bladder function in extensive resections.

6. En Bloc Resection

   In selected cases, particularly in patients with isolated or limited metastatic disease en bloc resection of the tumor may be considered. Additionally, disease free time may be considered if the pelvic bone involvement presents after a considerable time of undetectable disease. En Bloc resection involves the complete removal of the tumor along with a margin of healthy bone and soft tissue. Reconstruction with bone grafts, custom implants or off-the-shelf tantalum augments or cages may be required to restore pelvic stability.

   - Advantages: Provides optimal local disease control.
   - Disadvantages: Highly invasive, associated with significant morbidity, and often not feasible in patients with widespread disease.

7. Girdlestone (resection arthroplasty).

   The removal of the femoral head without reconstruction may be an option for palliative care. In patients with significant pain that are poor candidates for

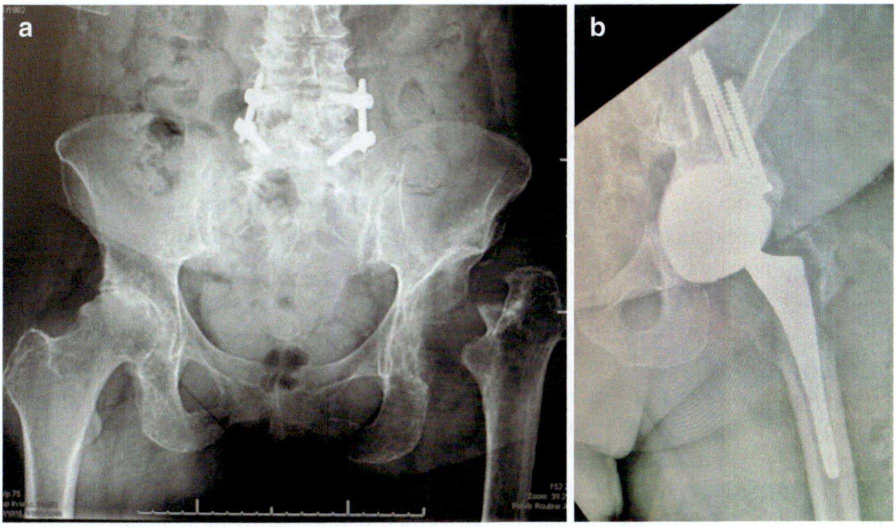

Fig. 15.6 (**a**) AP of pelvis. The patient is s/p a Girdlestone procedure for a pathologic acetabular fracture from metastatic gastrointestinal adenocarcinoma. (**b**) The patient was optimized for hip reconstruction and was treated with a arthroplasty and Osteocentric fasteners, to assist with cup stabilization

reconstruction and/or have failed non-operative management. Soft tissue reconstruction may be performed with suture anchors, synthetic graft and heavy sutures. The procedure avoids the high complications often associated with more complex reconstructions [32–35] (Fig. 15.6a and b).

Surgical Technique

Patient Positioning and Surgical Incision

The patient positioning is highly dependent on the anatomic location of the tumor. The periacetabular region has been regarded as a common location for intervention due to its importance biomechanically for gait and sitting. In general, the patient is positioned in the lateral decubitus with the assistance of a bean bag. This provides easily accessible extensile approach to the hemipelvis, including the greater sciatic notch. While there are multiple tendon/soft tissue releases that may improve acetabular exposure, a trochanteric osteotomy may be performed to mobilize the gluteal muscles. Alternatively, a trochanteric slide may be done, whereby the gluteal tendon insertions and vastus origin or left in continuity. A utilitarian incision may be used in most cases. This may take the shape of a "T" or inverted "L" depending on location of the tumor or more commonly, an iliofemoral approach. A "T shape" incision is not of first choice given the high wound complication rates associated to it. The sciatic notch is important to identify. It helps to identify the gluteal vessels and the sciatic nerve. These structures need to be identified, protected, and carefully mobilized during tumor excision/curetting and the subsequent reconstruction.

Steps for Tumor Resection

Exposures of the pelvis are highly dependent on the location of the tumor and the surgical plan. The amount of periacetabular bone loss dictates the type of surgery. Soft tissue flaps are generally made after incision through the fascia. The posterolateral approach is similar to a standard total hip approach. Once with fascia is opened, the hip may be dislocated posteriorly with the releases of the external rotators and capsule. If more exposure is needed the gluteus maximus insertion on the femur may be tagged and cut. Once the femoral head is removed the acetabulum is exposed with the placement of multiple Homan retractors included a blunt retractor in the sciatic notch. Depending on the need for column exposure, the gluteal muscles may be elevated with the use of a Cobb elevator or trochanteric slide. If the plan is a total hip, the cup may be prepared as needed, and further exposure may be performed if internal fixation and/or augments are needed to supplement fixation. If a press fit cup is used it is recommended that 2 or more screws are placed in the cup. The osteolytic lesion may be curetted and cemented as needed. Alternatively, a resection of the area involved may be performed. With this option it is important to have reconstruction implants available. Neurovascular structures should be identified and carefully protected during the planned resection.

Bone Reconstruction

A commonly used tool to indicate the acetabular destruction is the Harrington Classification organized into four groups [24–26]. In Group I, if the subchondral bone is intact, a simple curettage may be performed with or without an adjuvant such as cryoprobes or an Argon beam laser. Once the lesion is removed, it is filled with bone cement, with or without rebar metal fixation with pins and/or screws. If the articular cartilage is involved, a total hip replacement may be performed with a cemented cup and screws. Group II involves a more extensive involvement with loss of the medial wall, with preservation of the superior dome/roof and the lateral wall. With this pattern of bone loss and an anti-protrusio cup/ring should be considered to prevent protrusion. In Group III, extensive osteolysis not only affects the medial wall but also the roof and lateral rim. This requires a more extensile approach with reconstruction using internal fixation into the iliac wing, a cement augmentation construct followed by placement of the acetabular cup. Harrington described a technique using large threaded Steinman pins placed with in the surrounding hemipelvis to support the cemented acetabular component. The internal fixation is used to transform the weight-bearing stresses from the component to the unaffected bone. This technique often requires elevation of the gluteal muscles off the iliac wing and is technically challenging. Placing a finger in the sciatic notch while drilling helps to orient the pins and protects the nerve [32–34].

Bone reconstruction can be performed as well with tantalum augments with the shape of "orange slices" or augment with iliac fixation to reconstruct the posterior or anterior depending on the size of the defect. If there is protrusion or ring discontinuity, a tantalum mega cup can be secured vertically with custom-made holes based on the remaining bone for purposes of fixation with 6.5 mm divergent screws in the posterior column/ischium, ilium and anterior column. A cemented dual mobility head or constrained liner are added for acetabular reconstruction pending the remaining capsular soft tissues in addition to the abductors and external rotators [32–34].

Soft Tissue Reconstruction

Soft tissue reconstruction should be performed meticulously. If the soft tissue is not involved with the tumor, the normal host tissue around the hip should be repaired. Bone holes around the proximal femur may be used for large non absorbable suture, to repair the capsule and rotators. If there is a soft tissue deficit, an allograft may be used or alternatively a synthetic mesh material may enhance reconstruction. Local and vascularized free flaps may be used to assist with dead space management and wound closure. Preoperative planning and consultation with a plastic surgeon that routinely performs soft tissue management for pelvic defects is highly recommended [36, 37].

Wound Closure

The wound should be closed meticulously in multiple layers. Undermining may be needed for adequate closure without significant tension. Drains may be used either above or below the fascial layer. Copious irrigation is generally used prior to closure. When possible, the external rotators and posterior capsule shoulder be repaired. Additionally, if the gluteal tendon was taken down from the femur, it should be repaired. Deep repair is typically done with a #1 or #0 absorbable suture and more superficial tissue with 2.0 suture. Skin closure may be with nylon, prolene or staples depending on surgeon preference. Incisional VAC system use has grown and popularity and may have several advantages. I prefer to use a silver coated negative pressure dressing followed by application of sponge and sealant dressing.

"Tips and Tricks"

Pelvic bone surgery, often complex due to the anatomy and load-bearing nature of the pelvis, requires precision and technique to ensure successful outcomes. Here are key tips and tricks to consider:

1. Preoperative Planning.

- Comprehensive Imaging: High-resolution CT scans with 3D reconstructions can significantly improve understanding of fracture patterns and guide surgical planning. Preoperative templating helps map implant positions and screw trajectories.
- Assess Patient-Specific Factors: Age, bone quality, and comorbidities play a crucial role in surgical planning. Elderly patients often have osteoporotic bones, impacting fixation choices, while younger patients might need different fixation options for faster recovery.

2. Choosing the Right Approach.
   - Surgical Exposure: The approach—whether anterior (ilioinguinal or Stoppa) or posterior (Kocher-Langenbeck)—depends on fracture type and location. Familiarity with different approaches is essential; combined approaches may be necessary for complex fractures.
   - Minimally Invasive Options: In select cases, minimally invasive techniques, such as percutaneous screw fixation, flexi-nail or photodynamic nails reduce soft tissue damage and postoperative recovery time.
3. Patient Positioning and Stabilization.
   - Positioning: Proper patient positioning is crucial, especially with complex fractures. For example, placing the patient supine or lateral can offer better visualization of the pelvis.
   - Stable Positioning Aids: Use stable pelvic positioning devices or padded tables to ensure the patient remains immobile throughout the procedure, minimizing intraoperative adjustments.
4. Surgical Technique and Implant Selection.
   - Bone Quality Assessment: For osteoporotic bones, use locking plates and longer screws to improve fixation strength. Hybrid fixation techniques combining plates and screws can provide greater stability [37].
   - Use of Fluoroscopy and Navigation: Real-time fluoroscopy or computer-assisted navigation ensures accurate screw placement and implant positioning. This is especially useful for sacroiliac joint stabilization.
   - Contouring Implants: In complex fractures, pre-contour plates fit the unique curvature of each patient's pelvis, reducing operative time and enhancing fit. Recent advances in implant materials and design have impacted on the immediate stability of pathologic fractures. Minimally invasive fracture care with the use of a special "liquid glue" that fills a balloon after insertion within the involved bone. (Illuminoss ™, Providence, RI) (Fig. 15.7). The material hardens within the balloon with the use of a blue light source and provides not only structural support but may assist with the stabilization of plate and screws, since screw fixation may be secured into the hardened material. Another novel technology used to stabilize pelvic bone loss is Osteocentric's UnififMI Fasteners. Although they have a similar appearance to cortical screws, the fasteners have a unique thread geometry that instantly and circumferentially interlocks with bone by entrapping and containing bone between the thread forms. (Osteocentric ™ Logan, UT).

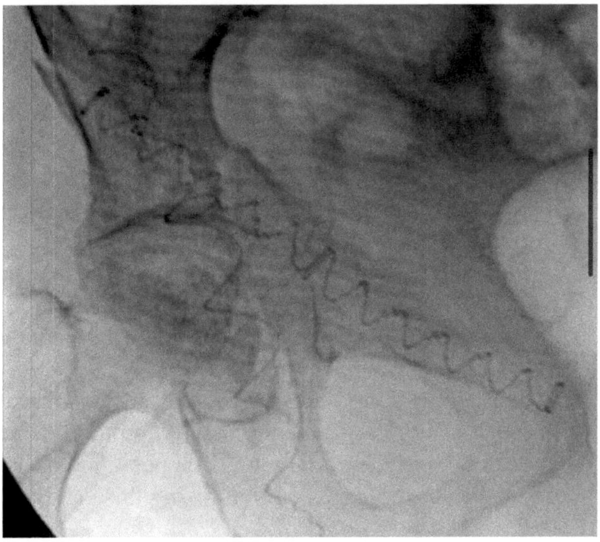

Fig. 15.7 AP of pelvis. Patient treated with Illuminoss implant to stabilize acetabular metastatic lesion without joint reconstruction

5. Avoiding Neurovascular Injury.
   - Mapping Neurovascular Structures: Awareness of critical neurovascular landmarks, such as the sciatic nerve and superior gluteal vessels, is vital. Blunt dissection and careful retraction reduce the risk of damaging these structures.
   - Meticulous Hemostasis: Ensure hemostasis throughout to prevent excessive blood loss, as pelvic surgeries can be prone to significant bleeding.
6. Postoperative Care.
7. Early Mobilization: Encourage early mobilization within limits to prevent complications like deep vein thrombosis (DVT) and promote bone healing.
8. Rehabilitation and Follow-Up: A structured rehabilitation plan and regular follow-ups with imaging monitor healing and ensure functional recovery.

By combining these tips with patient-specific adjustments, surgeons can enhance both the safety and efficacy of pelvic surgeries, improving overall patient outcomes.

"Pearls and Pitfalls"

Pelvic bone surgery is complex, often required in cases of trauma, malignancy, or deformity, and presents unique challenges due to the anatomy and function of the pelvic region. Success in pelvic surgery depends on understanding the "pearls" (best practices) and "pitfalls" (potential complications) that characterize these procedures.

Pearls of Pelvic Bone Surgery

1. Comprehensive Preoperative Planning: Due to the complexity of the pelvic anatomy, detailed preoperative imaging is essential. Techniques like CT and MRI can help identify fracture patterns or tumor extensions, allowing surgeons to plan approaches with precision. Three-dimensional imaging can provide additional clarity, enabling preoperative planning of implant positioning and screw trajectories.
2. Multidisciplinary Approach: Pelvic surgeries often benefit from the involvement of a team including orthopedic surgeons, vascular surgeons, and sometimes plastic surgeons. This collaborative approach can aid in complex cases, especially when resection or reconstruction is involved, reducing risks related to blood loss or soft tissue complications.
3. Patient Positioning and Access: Given the intricate nature of pelvic surgery, proper positioning—whether supine, prone, or lateral—can enhance exposure and access to surgical targets. Each position has specific indications depending on the injury location, and optimizing this can facilitate more precise execution of the surgery.
4. Intraoperative Navigation and Technology: Intraoperative imaging and navigation technologies, such as computer-assisted systems, allow for enhanced accuracy in screw placement and implant positioning. This minimizes the risk of iatrogenic injuries to surrounding neurovascular structures.

Pitfalls of Pelvic Bone Surgery

1. High Risk of Blood Loss: The pelvis is highly vascular, and pelvic fractures or surgeries can lead to significant bleeding. Even with careful planning, blood loss can be substantial, necessitating transfusions and sometimes vascular interventions, such as preoperative embolization.
2. Neurovascular Injury: The proximity of major blood vessels (e.g., iliac vessels) and nerves (e.g., sciatic nerve) make them vulnerable to damage during surgery. Inaccurate drilling or instrumentation can lead to severe complications, including permanent nerve damage or vascular injury.
3. Infection Risks: Infection rates are relatively high in pelvic surgery, particularly with open fractures or cancer resections. Preventive measures, including perioperative antibiotics and meticulous aseptic techniques, are crucial.
4. Poor Healing and Nonunion: Due to limited blood supply to certain areas of the pelvis, such as the acetabulum, pelvic bones are prone to nonunion. This risk is higher in older patients or those with comorbidities affecting bone healing, like diabetes or osteoporosis.
5. Postoperative Mobility Issues: Even with successful surgery, patients may face prolonged recovery periods, and some may develop long-term mobility impairments or chronic pain. Early physical therapy and rehabilitation are essential to improve outcomes but must be carefully balanced with the need for proper healing.

In summary, pelvic bone surgery demands careful planning, expertise, and close attention to detail to avoid significant complications. Employing modern technology, ensuring multidisciplinary involvement, and focusing on patient rehabilitation can enhance outcomes and mitigate the common pitfalls of this challenging surgical field.

Postoperative Care

The postoperative management of patients undergoing surgery for pelvic metastasis is crucial for ensuring optimal outcomes. Key components of postoperative care include:

1. Pain Management: Adequate pain control is essential for early mobilization and recovery. Multimodal analgesia, including opioids, NSAIDs, and nerve blocks, may be used.
2. Rehabilitation: Early mobilization and physical therapy are important for restoring function and preventing complications, such as deep vein thrombosis or pressure ulcers. The type and extent of rehabilitation depend on the surgical procedure performed and the patient's overall condition.
3. Management of Surgical Complications: Complications such as infection, hardware failure, and wound dehiscence may occur, particularly in patients with poor nutritional status or immunosuppression. Close monitoring and prompt intervention are necessary to address these issues.
4. Oncological Follow-Up: Continued oncological treatment is necessary to manage systemic disease. Radiation therapy may be used postoperatively to control local tumor growth or alleviate symptoms. Chemotherapy or targeted therapies may also be resumed as appropriate.

References

1. Aboulafia AJ, Levine AM, Schmidt D, Aboulafia D. Surgical therapy of bone metastases. Seminars in oncology. Elsevier; 2007.
2. Picci P, Manfrini M, Fabbri N, Gambarotti M, Vanel D. Atlas of musculoskeletal tumors and tumorlike lesions: the Rizzoli case archive. Springer; 2014.
3. Hage WD, Aboulafia AJ, Aboulafia DM. Incidence, location, and diagnostic evaluation of metastatic bone disease. Orthop Clin N Am. 2000;31(4):515–28.
4. Li S, Peng Y, Weinhandl ED, Blaes AH, Cetin K, Chia VM, et al. Estimated number of prevalent cases of metastatic bone disease in the US adult population. Clin Epidemiol. 2012;4:87–93.
5. Issack PS, Kotwal SY, Lane JM. Management of metastatic bone disease of the acetabulum. JAAOS J Am Acad Orthop Surg. 2013;21(11):685–95.
6. Enneking W, Dunham W, Gebhardt M, Malawar M, Pritchard D. A system for the classification of skeletal resections. La Chirurgia degli organi di movimento. 1990;75(1 Suppl):217–40.
7. Capanna R, Campanacci DA. The treatment of metastases in the appendicular skeleton. J Bone Joint Surg Br Vol. 2001;83(4):471–81.

8. Dolek BA, Ciliz DS, Ozdemir N, Ozet G, Duran S, Ozdemir N. Comparative analysis of whole-body diffusion-weighted imaging and PET/CT in metastasis detection: a prospective study. Cureus. 2024;16(11)
9. Bosma S, Vriens D, Gelderblom H, van de Sande M, Dijkstra P, Bloem J. 18 F-FDG PET-CT versus MRI for detection of skeletal metastasis in Ewing sarcoma. Skeletal Radiol. 2019;48:1735–46.
10. Yasko AW, Rutledge J, Lewis VO, Lin PP. Disease-and recurrence-free survival after surgical resection of solitary bone metastases of the pelvis. Clin Orthop Relat Res. 2007;459:128–32.
11. Kunisada T, Choong PF. Major reconstruction for periacetabular metastasis: early complications and outcome following surgical treatment in 40 hips. Acta Orthop Scand. 2000;71(6):585–90.
12. Ishiguchi H, Ito S, Kato K, Sakurai Y, Kawai H, Fujita N, et al. Diagnostic performance of 18 F-FDG PET/CT and whole-body diffusion-weighted imaging with background body suppression (DWIBS) in detection of lymph node and bone metastases from pediatric neuroblastoma. Ann Nucl Med. 2018;32:348–62.
13. Kitagawa Y, Ek E, Choong P. Pelvic reconstruction using saddle prosthesis following limb salvage operation for periacetabular tumour. J Orthop Surg. 2006;14(2):155–62.
14. Harrington KD. The management of acetabular insufficiency secondary to metastatic malignant disease. JBJS. 1981;63(4):653–64.
15. Nilsson J, Gustafson P, Fornander P, Ornstein E. The Harrington reconstruction for advanced periacetabular metastatic destruction: good outcome in 32 patients. Acta Orthop Scand. 2000;71(6):591–6.
16. Breau RH, Blute ML. Surgery for renal cell carcinoma metastases. Curr Opin Urol. 2010;20(5):375–81.
17. Nieder E, Elson R, Engelbrecht E, Kasselt M, Keller A, Steinbrink K. The saddle prosthesis for salvage of the destroyed acetabulum. J Bone Joint Surg Br Vol. 1990;72(6):1014–22.
18. Aboulafia AJ, Buck R, Mathews J, Li W, Malawer MM. Reconstruction using the saddle prosthesis following excision of primary and metastatic periacetabular tumors. Clin Orthop Relat Res. 1995;314:203–13.
19. Jansen J, Van de Sande M, Dijkstra P. Poor long-term clinical results of saddle prosthesis after resection of periacetabular tumors. Clin Orthop Relat Res. 2013;471(1):324–31.
20. Tang X, Guo W, Ji T. Reconstruction with modular hemipelvic prosthesis for the resection of solitary periacetabular metastasis. Arch Orthop Trauma Surg. 2011;131:1609–15.
21. Rove KO, Crawford ED. Metastatic cancer in solid tumors and clinical outcome: skeletal-related events. Oncology (Williston Park). 2009;23(14 Suppl 5):21–7.
22. Giurea A, Ritschl P, Windhager R, Kaider A, Helwig U, Kotz R. The benefits of surgery in the treatment of pelvic metastases. Int Orthop. 1997;21:343–8.
23. Nagarajan R, Clohisy D, Neglia J, Yasui Y, Mitby P, Sklar C, et al. Function and quality-of-life of survivors of pelvic and lower extremity osteosarcoma and Ewing's sarcoma: the Childhood Cancer Survivor Study. Br J Cancer. 2004;91(11):1858–65.
24. Harrington KD. Orthopaedic management of extremity and pelvic lesions. Clin Orthop Relat Res (1976–2007). 1995;312:136–47.
25. Oh D, Huh SJ. Insufficiency fracture after radiation therapy. Radiat Oncol J. 2014;32(4):213.
26. Harrington KD. Impending pathologic fractures from metastatic malignancy: evaluation and management. Instr Course Lect. 1986;35:357–81.
27. Bates T. A review of local radiotherapy in the treatment of bone metastases and cord compression. Int J Radiat Oncol Biol Phys. 1992;23(1):217–21.
28. Maher E. The use of palliative radiotherapy in the management of breast cancer. Eur J Cancer Clin Oncol. 1992;28(2–3):706–10.
29. Fini M, Salamanna F, Parrilli A, Martini L, Cadossi M, Maglio M, et al. Electrochemotherapy is effective in the treatment of rat bone metastases. Clin Exp Metastasis. 2013;30:1033–45.
30. Cadossi R, Ronchetti M, Cadossi M. Locally enhanced chemotherapy by electroporation: clinical experiences and perspective of use of electrochemotherapy. Future Oncol. 2014;10(5):877–90.

31. Gangi A, Kastler B, Klinkert A, Dietemann JL. Injection of alcohol into bone metastases under CT guidance. J Comput Assist Tomogr. 1994;18(6):932–5.
32. Melnic CM, Salimy MS, Hosseinzadeh S, Moverman MA, Bedair HS, Lozano-Calderón SA, et al. Trabecular metal augments in severe malignancy-associated acetabular bone loss. Hip Int. 2023;33(4):678–84.
33. Melnic CM, Cohen-Levy WB, Lozano-Calderon SA, Heng M. Treatment of severe acetabular bone loss using a tantalum acetabular shell and a cemented monoblock dual mobility acetabular cup. JAAOS J Am Acad Orthop Surg. 2022;30(3):e301–e6.
34. Salimy MS, Blackburn AZ, Alpaugh K, Lozano-Calderón SA, Bedair HS, Melnic CM. Postoperative outcomes in total hip and total knee arthroplasty for patients who have multiple myeloma. J Arthroplast. 2023;38(11):2269–74.
35. Goetz MP, Callstrom MR, Charboneau JW, Farrell MA, Maus TP, Welch TJ, et al. Percutaneous image-guided radiofrequency ablation of painful metastases involving bone: a multicenter study. J Clin Oncol. 2004;22(2):300–6.
36. Cotten A, Deprez X, Migaud H, Chabanne B, Duquesnoy B, Chastanet P. Malignant acetabular osteolyses: percutaneous injection of acrylic bone cement. Radiology. 1995;197(1):307–10.
37. Li Z, Butala NB, Etheridge BS, Siegel HJ, Lemons JE, Eberhardt AW. A biomechanical study of periacetabular defects and cement filling. J Biomech Eng. 2007;129:129.

Management of Shoulder Gridle Metastatic Disease

16

Alexandra M. Arguello, Leilani Garayua Cruz, Jonathan D. Barlow, and Matthew T. Houdek

Introduction/Background/Prevalence

With the aging population, metastatic bone disease is becoming increasingly common, with 20% of patients with metastatic bone disease having involvement of the upper extremity. In the upper extremity, skeletal-related events often involve the humerus, which is the third most common site overall for metastatic bone disease (behind the spine, including the pelvis, and femur) [1–6]. In addition to the proximal humerus, metastatic disease is the most common malignant tumor to impact the scapula and clavicle [7–9]. Recently the Musculoskeletal Tumor Society published clinical practice guidelines on the management of humeral metastatic disease, however, the evidence and strength of recommendations for all statements were limited due to the lack of high-quality data [10]. In addition, there are currently no guidelines to assist surgeons in the management of scapular or clavicular metastatic disease.

Work-Up and Exam

In order to assist clinicians in the work-up for a patients over the age of 40 years with a new bone lesion, or in those with a remote history of malignancy, the Rougraff criteria is often utilized to identify the origin of the tumor [11]. The Rougraff criteria include obtaining a detailed history about pain and when the pain occurs (i.e., functional pain) and a physical exam, in addition to obtaining labs (complete blood count, comprehensive metabolic panel, erythrocyte sedimentation rate, C-reactive protein, alkaline phosphatase, serum and urine protein electrophoresis) and imaging

A. M. Arguello · L. G. Cruz · J. D. Barlow · M. T. Houdek (✉)
Department of Orthopedic Surgery, Mayo Clinic, Rochester, MN, USA
e-mail: Arguello.alexandra@mayo.edu; Garayuacruz.leilani@mayo.edu; Barlow.jonathan@mayo.edu; Houdek.matthew@mayo.edu

J. Pretell-Mazzini, S. A. Lozano Calderón (eds.), *Multidisciplinary Approach of Musculoskeletal Metastasis*, https://doi.org/10.1007/978-3-032-09498-8_16

(CT of the chest, abdomen, and pelvis with and with-out contrast if possible). Additional tests could be ordered based on the history of the patient (i.e., PSA in males) but are not essential for the initial work-up. The Rougraff criteria has been shown to identify the primary of the tumor in 85% of patients [11]. Even if a potential primary visceral or hematologic malignancy is discovered during this work-up process, in patients who continue to have a solitary bony tumor, it is important to obtain a core-needle biopsy from the bony site of disease prior to operative intervention.

Imaging of the upper extremity often begins with orthogonal radiographs including the joint above and below the area of pain. Malignant tumors frequently cause cortical disruption with a soft-tissue component, periosteal reaction, and have a broad, permeative zone of transition [12]. Following plain radiographs, cross-sectional imaging can then be obtained to better define the anatomic extent of the tumor if necessary. For metastatic disease, a MRI or CT is typically not needed. However, these can be particularly helpful in suspected clavicular metastases due to its location and curvature, as a clavicular lesion can be difficult to detect on plain radiographs [9]. For these patients advanced imaging is often needed to identify a pathologic fracture and determine its anatomic extent. In addition, an MRI of the entire bone with and without gadolinium contrast should be obtained to define the local bony and soft-tissue extent of the tumor for surgical planning when a resection with negative margins is being considered in the setting of oligometastatic disease [13]. This is especially true in patients with aggressive tumors (i.e., renal cell carcinoma) as they often present with large soft tissue components that require resection. It is important to carefully evaluate the critical neurovascular structures (brachial plexus and axillary vessels) of the shoulder girdle, as complete encasement requiring resection of vital structures would be a contraindication to proceeding with limb-sparing surgery, and an amputation may need to be discussed [14, 15]. Involvement of these structures is suspected in the setting of intractable radicular pain and motor deficits. In addition, forequarter amputation should only be utilized as a last resort, palliative option for patients with metastatic disease as there remains a high risk of local relapse, but does have the potential to improve quality of life [16–18].

When the Rougraff criteria fail to identify a primary site of origin for the tumor, or in the setting of an isolated bone lesion even with visceral metastatic disease elsewhere, a well-planned biopsy is a critical step in the management of a patient. An image-guided core needle biopsies is the standard practice to obtain a diagnosis if needed, and the need for an open surgical biopsy is limited. When performing these procedures collaboration between the orthopedic oncologist managing the patient and the radiologist performing the biopsy is necessary to avoid a poorly performed biopsy that could have substantial implications in the care of the patient [19–21]. For tumors of the proximal humerus the biopsy tract should be lateral to the deltopectoral interval, and through the anterior deltoid, in order to avoid the subscapularis and pectoralis muscle bellies and tendons as this could lead to contamination of the underlying major neurovascular structures [14]. For the scapula, anterior biopsies should be avoided, and a posterior biopsy in line with a planned incision should be performed.

Determining Need for Operative Intervention

It is important to understand the functional anatomy of the shoulder girdle prior to discussing surgical options for reconstruction based on the location of the tumor in addition to the goals of surgery with an en-bloc/gross total resection versus an intra-lesional procedure. The shoulder joint is the most mobile joint in the body, with the main function of the shoulder to provide a stable platform for positioning the hand in space [22]. In the upper extremity, most of the metastatic disease that is encountered which requires operative management involves the proximal humerus, and only rarely would a metastatic lesion involving the scapula or clavicle warrant operative management.

Patients with an impending pathologic fracture of the upper extremity will frequently present with pain and limited function. These limitations can have a substantial impact on the patients quality of life, especially when the dominant arm is affected [5, 23]. In order to risk-stratify patients for surgery versus observation, the Mirels criteria is utilized [24]. When compared to the lower extremity, patients with impending pathologic fractures of the upper extremity will warrant surgery with a lower total Mirel's score (≥7 or 8 in the upper extremity versus ≥9 in the lower extremity) or in those with a cortical breach noted on imaging [25, 26]. In addition to the Mirel's scoring system, patient selection for surgery is also important for outcomes a different algorithm have been developed to assist surgeons in determining if a patient's life expectancy and goals of care align for the appropriate treatment [3, 23, 26–28]. Although there are some instances where non-operative management of metastatic disease can be considered, in the setting of a pathologic humerus fracture fixation is often warranted when the patient is able to tolerate surgery as these fractures are unlikely to heal without treatment.

For the scapula and the clavicle, metastatic disease can often be managed without surgical intervention. Relative indications for operative management of metastatic lesions include pathologic clavicle fractures preventing weight bearing, medialization of the glenoid with pain refractory to radiation, and oligometastatic disease warranting resection.

Preoperative Planning

The decision between different options for management of a humeral metastatic lesion involves a careful evaluation of the preoperative imaging and patient factors. Often plain radiographs of the humerus are all that is needed to determine which treatment modality will be used; however, if there is concern for a large soft-tissue component or involvement of the humeral head, cross-sectional imaging can be obtained. One of the most important determining factors is the tumor histology. For all intralesional procedures the use of postoperative radiotherapy is essential to reduce the risk of local recurrence/progression. If the tumor is radioresistant, and the patient has a favorable life expectancy consideration for an en bloc procedure is preferred over an intralesional procedure. Additional features that are important to analyze vary from location to location in the humerus:

- Proximal humerus
  - Determine extent of disease in the humeral head
    - Diffuse involvement of the articular surface or fracture through the humeral head often requires an arthroplasty.
  - Involvement of the humeral head without collapse likely warrants extended curettage and plate fixation.
  - If consideration of intramedullary nail is given there needs to be adequate bone stock to allow for multiple locking screws
  - Use of extended curettage and plate fixation preferred technique
    - Augmentation with cement or photo-dynamic implants
- Scapula:
  - Determine involvement of glenoid, coracoid, and scapular spine
    - Structures essential for shoulder function
      - Consideration for curettage and cement only in the setting of refractory disease with functional pain
      - Resection and reconstruction reserved for oligometastatic disease.
  - Lesions refractory to radiotherapy involving the medial boarder of scapula can be considered for resection with limited morbidity.
- Clavicle
  - Tumor often involve the medial and lateral third of the clavicle.
    - Pathologic fractures of the medial clavicle can be difficult to visualize on plain radiographs.
- Non-operative management
  - Indications
    - Patient requiring commencement of chemotherapy.
    - Tumor histology with biological potential to heal with appropriate systemic treatment.
      - Lymphoma, breast carcinoma, myeloma
    - Ability to maintain weight-bearing restrictions.
    - Limited life expectancy
    - Majority of cases of scapular or clavicular metastatic disease
  - Contraindications
    - Inability to maintain weight-bearing restrictions.
    - Radioresistant histologies
    - Intractable pain
    - Pathologic fracture

Surgical Treatment Options

Once the determination has been made to proceed with operative intervention diverse options exist, without a "gold-standard" treatment. The treating surgeon should choose which ever option they are most comfortable and facile with based on the tumors location as based on the clinical evidence, there does not seem to be a difference in outcomes.

ORIF with Locking Plate and Cementation

Curettage, cementation and locking plate fixation is a primary means of treating the vast majority of metastatic lesions in the humerus and upper extremity (Fig. 16.1). It can be used for disease in all locations of the humerus and provides a rigid construct to allow for weight bearing and use immediately following the procedure [29, 30]. For this surgery to be successful, the intralesional resection is a crucial step. It is important to make a large oval cortical window to have access to entire tumor defect to allow for curettes and a burr to reach the entire cavity to minimize disease burden and decrease the risk of tumor progression [29]. Adjuvants such as hydrogen peroxide and an argon beam coagulator can be utilized to treat areas of potential microscopic tumor contamination [31, 32]. Following the extended curettage, polymethylmethacrylate cement is used to fill the defect to provide further stability [30, 33]. Screws should be placed while the cement is curing to prevent cracking of the cement mantel, however, can also be placed following the cementation. Multiple drill bits maybe needed however if the cement has cured, as the cement can interdigitate in the drill bit. This procedure reliably improves pain and function. However, complications such failure of fixation and pain with abduction are commonly reported [29, 30, 34, 35] In addition, compared to an intramedullary nail, this is a larger surgical procedure in terms of exposure, operative time and blood loss [34].

Intramedullary Nail

Patients with proximal metaphyseal lesions of the proximal humerus and diaphyseal lesions of the humerus can be amenable to nail fixation (Fig. 16.2) [36, 37] Compared to plate fixation, the use of an intramedullary nail is a shorter surgical procedure; however, often these procedures are commonly performed without a concurrent curettage and cementation but this may not lead to higher rates of disease progression [34]. Intramedullary nails can be combined with curettage and cementation [37], and augmentation with cement has been found to improve early pain and functional outcomes compared to patients who did not undergo cement augmentation [38, 39]. Intramedullary nails commonly fail secondary to disease progression, mainly in the setting of renal cell carcinoma or other radioresistant tumors [3]. Although disease progression is a concern, the most common complication following this procedure is shoulder pain and stiffness, which may require manipulation under anesthesia [34].

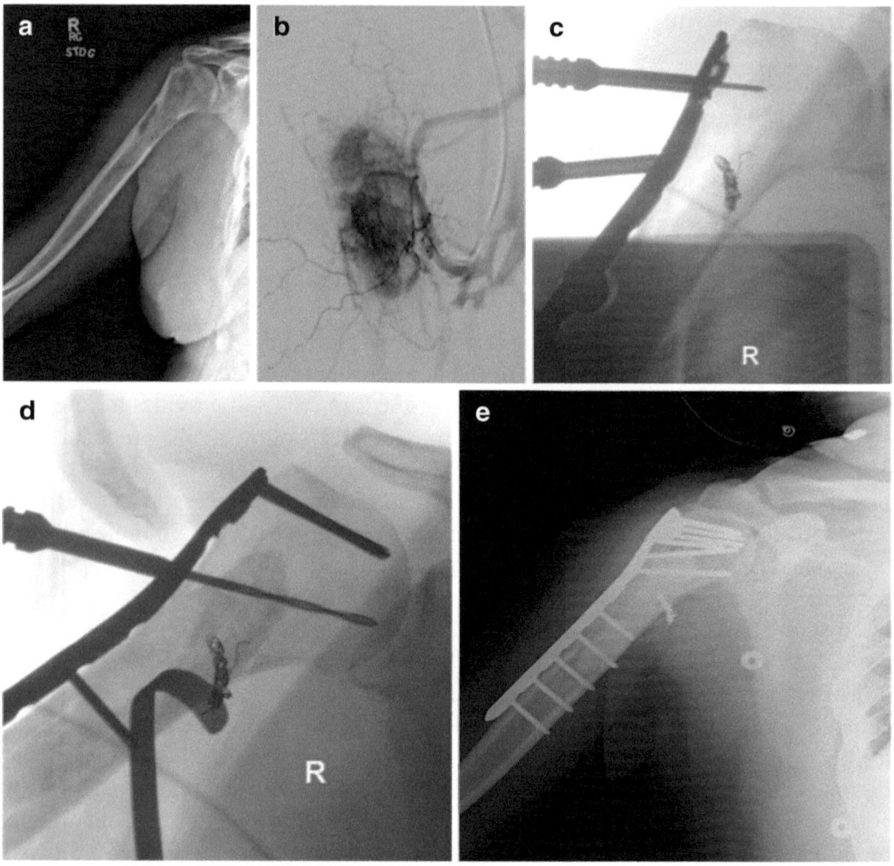

Fig. 16.1 Preoperative AP radiographs showing an impending pathologic fracture of the proximal humerus due to renal cell carcinoma (**a**). The patient was treated with preoperative embolization (**b**) secondary to the planned curettage, cementation, and plate fixation. A locking plate was provisionally fixed through a deltopectoral approach (**c**), and an oval-shaped window was made over the lesion. An extensive curettage was performed, and the defect is filled with cement (**d**) with locking screws placed through the cement while it is curing, with the goal to achieve rigid fixation that allows for unrestricted use following surgery (**e**)

Photodynamic Bone Stabilization

Photodynamic bone stabilization (IlluminOss System®, IlluminOss Medical, Inc. East Providence, RI, USA) had initial promise of providing fixation while allowing for a minimal morbidity [34, 40]. This system is a Dacron catheter balloon that is inflated with a liquid polymer, which hardens when exposed to white light. Since the ballon is inflated with a liquid monomer, it can expand into the defects and unique anatomy of a pathologic or impending pathologic fracture. Once the monomer is cured, screws can be placed through the proximal and distal end, similar to an intramedullary nail. However, when used in this fashion, or without the use of

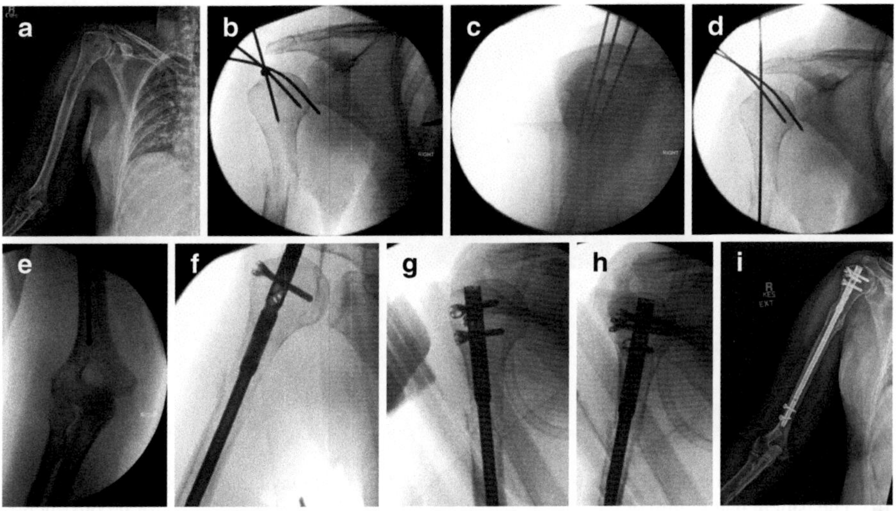

Fig. 16.2 Preoperative AP radiograph showing an impending pathological fracture due to myeloma (**a**). Prior to surgery the patient sustained a pathological fracture (**b**) and planned for an intramedullary nail. An incision was made over the deltoid, and the interval was split. Large K-wires can be placed in the proximal humerus to act as joysticks to help with reduction and as retractors of the soft tissue to expose the proximal humerus and allow for placement of the guide pin on the apex of the proximal humerus in line with the humeral shaft on AP (**c**) and lateral views (**d**). The guide wire is then passed (**e**) and placed as distal as possible (**f**). The humerus is then reamed while holding reduction of the fracture site and the nail is placed which can help with indirect reduction (**g**). Multiple screws are placed in the humeral head, and it is important to evaluate the screw length on multiple rotations and views (**h** and **i**) to be sure screws do not penetrate the joint surface. At 1-year follow-up the fracture is fully healed, and the patient is using the arm as tolerated

additional cross screws, this type of reconstruction is associated with a high rate of failure [34, 40]. As such the use of this implant is now considered to be an adjuvant to plate fixation, rather than a stand-alone treatment. In addition to structural failure, hemodynamic instability can be seen during the inflation of the balloon similar to a fat embolism [41]. A benefit of this system is that it can be utilized in the clavicle as an intramedullary device secondary to its ability to contour with the curvature of the bone (Fig. 16.3). When combined with a locking plate, this provides rigid fixation where previous options for fixation were limited.

Endoprosthesis

Most lesions in the proximal humerus can be treated without the use of an arthroplasty; however, when there is substantial bony destruction, pathologic fracture, or involvement of the articular surface of the proximal humerus, endoprosthetic reconstruction is considered. In addition, in the setting of oligometastatic disease or

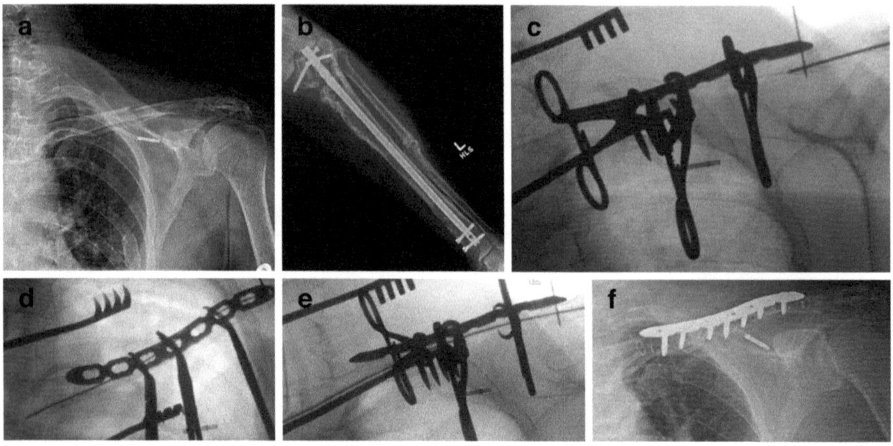

Fig. 16.3 AP radiographs showing a pathologic fracture of the clavicle in the setting of myeloma (**a**). The patient had failure of fixation of a tibial nail (**b**) and was not a candidate for reconstruction or revision and refused amputation. To allow for weight-bearing fixation of the clavicle was performed. An incision was made over the clavicle and the fracture was reduced to a plate. A K-wire was placed through the acromion and acromioclavicular joint (**c**). Reamers were placed over the K-wire and a ball-tipped guide wire was passed into the clavicle for a photodynamic implant, which was confirmed in the appropriate location on multiple views (**d** and **e**). The implant was placed to be sure it did not end in the acromioclavicular joint and was fixed with a locking plate (**f**) to allow for unrestricted weight bearing

histologies which are radioresistant, a negative margin resection and endoprosthetic reconstruction is a viable treatment option.

Reverse shoulder arthroplasty endoprosthesis (Fig. 16.4) has become the preferred means of endoprosthetic reconstruction for the proximal humerus secondary to the improvements in patient function and decrease in complications compared to a traditional hemiarthroplasty endoprosthesis [42–45]. For these procedures it is important to reduce the complications, namely dislocation. In order to do so, it is important to utilize the largest glenosphere for all oncologic reconstructive reverse shoulders [46]. In patients with a favorable life expectancy consideration for a reverse allograft prosthetic composite is an option. This type of reconstruction has been found to have improved functional outcomes compared to patients undergoing a cemented reverse endoprosthesis and a hemiarthroplasty endoprosthesis as this allows for repair of the rotator cuff to the donor rotator cuff on the humeral allograft [42, 47]. It is also felt the APC allows for the deltoid to be adequately tensioned, resulting in improved functional outcomes and stability [48, 49]. One drawback to an APC is that healing of the host/allograft junction could be compromised due to radiation, and as such a reverse endoprosthesis should be considered. Some argue that fixation of the APC with a long-cemented stem and a plate is sufficient for fixation even in the case that the junction does not heal.

Historically a hemiarthroplasty endoprosthesis was utilized for a vast majority of cases. However, they are associated with a high rate of complications [42, 50–52].

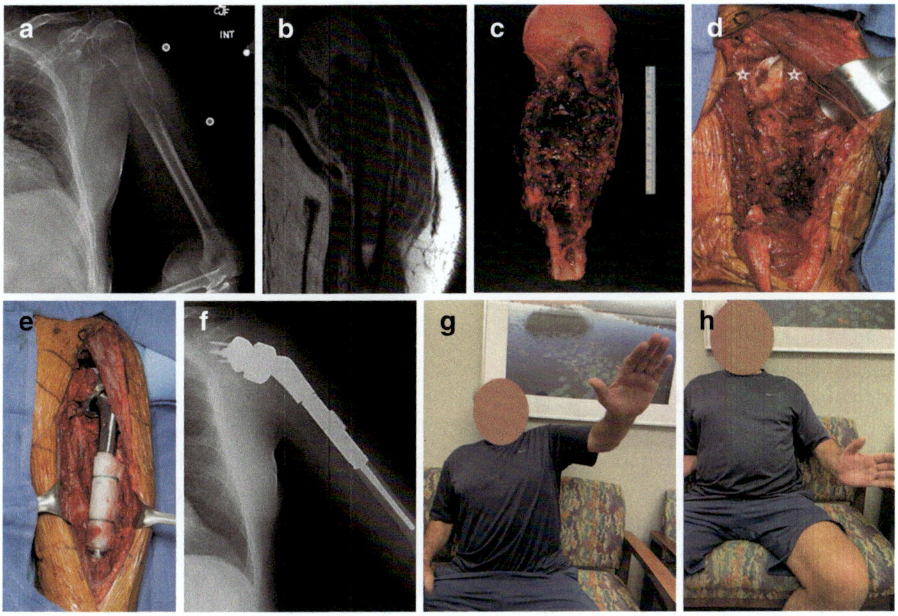

Fig. 16.4 AP radiographs showing a pathologic fracture through metastatic renal cell carcinoma (**a**) with a coronal MRI showing diffuse involvement of the proximal humerus (**b**). The patient was treated with a resection of the proximal humerus (**c**) through an extended deltopectoral interval (**d**). The rotator cuff tendons were able to be preserved (stars); however, the resection was below the deltoid insertion. The rotator cuff was repaired to the implant, and the deltoid insertion was reconstructed with mesh (**e**) around a reverse total shoulder endoprosthesis (**f**). Six months following surgery, the patient had 90° of active forward elevation (G) however, external rotation was limited (**h**)

This is often secondary to the lack of a soft-tissue envelope which leads to subluxation and proximal migration. In order to reduce this complication, synthetic mesh has been utilized as a scaffold for attachment of soft-tissues [53–56]. Additionally, the use of a mesh has been found to potentially improve function and range of motion compared to patients where a mesh was not used for reconstruction [57]. Following an extra-articular resection of the scapula, a proximal humerus or a suspension endoprosthesis can be utilized [58]. Typically, this implant is suspended from the residual clavicle and wrapped in a mesh to allow for soft-tissue healing. It is important to use a cemented stem as long as possible as shorter cemented stems have been associated with stress shielding [58]. Due to the limited function of the shoulder following these procedures, some authors have advocated for a "less is more" approach and recommend the use of a cement spacer, with similar functional outcomes [59].

There are limited studies which specifically examines the outcome of patients with metastatic disease of the proximal humerus undergoing either a reverse endoprosthesis or a hemiarthroplasty endoprosthesis. A recent study by Sullivan et al. [60] showed improvements in function with the use of a reverse endoprosthesis;

however, patients often require a longer period of immobilization of the shoulder. As such, their institution currently utilizes a reverse endoprosthesis when the patient has over a 3 month life expectancy as that is the time it typically takes to complete the rehabilitation [60]. If they have less than a 3 month life expectancy, then a hemi-arthroplasty endoprosthesis suspended with mesh is utilized as this allows the patient to use the shoulder immediately following surgery [60].

Surgical Techniques

Patient Positioning

Proximal Humerus

- Plan for extended deltopectoral interval for majority of cases
  - Important to have the arm prepped in free and include the ipsilateral chest wall.
    - Can be supine or in a semi-seated vs. beach-chair position depending on surgeon preference.
- Consider the use of a radiolucent table and have C-Arm come in from the head to allow for imaging of proximal humerus.
  - Important for humeral nail
- No matter what approach is used be sure you can image the entire extremity prior to prepping and draping patient.

Scapula

- Incision is based on the location of the tumor.
  - Total scapulectomy or tumors involving the glenoid are done through a utilitarian shoulder incision.
    - Extension of an extended deltopectoral interval carried over the shoulder.
    - Curved along the outer border of the scapula and curved medial.
    - Patient is in the rolling lateral position with access to the anterior and posterior chest wall.
      - Should be prepped past midline on anterior and posterior chest.
      - Arm prepped free.
    - Include the ipsilateral neck.
  - Table placed in reverse Trendelenburg.
  - Partial scapulectomy done through a posterior approach (Fig. 16.5).
    - Curved J incision or medial incision based on tumor location.
      - Similar to posterior portion of utilitarian shoulder
    - Patient is in the rolling lateral position with access to the anterior and posterior chest wall.
      - Should be prepped past midline on anterior and posterior chest.
      - Arm prepped free.
      - Include the ipsilateral neck.

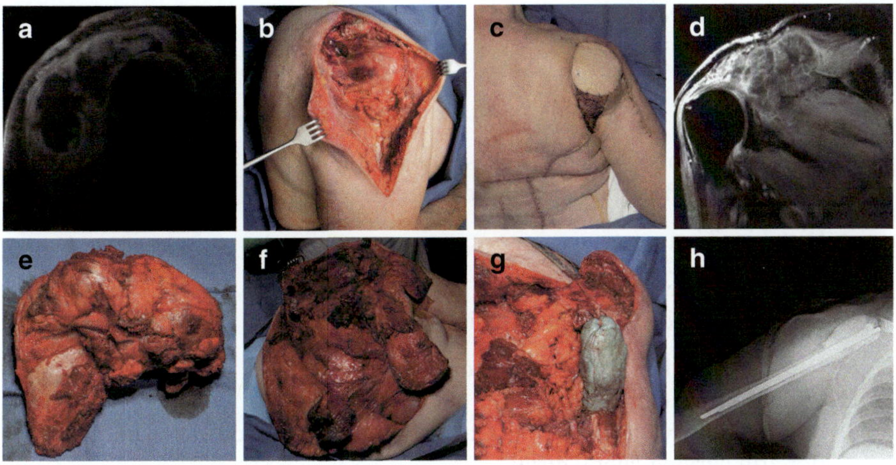

Fig. 16.5 Preoperative sagittal contrast enhanced T1 MRI of the shoulder showing a large soft-tissue mass arising from the acromion involving the posterior deltoid. The mass biopsied as metastatic colorectal carcinoma (**a**). Initially the area was treated with radiotherapy, however, the mass continued to grow and involve the skin, causing severe pain. The patient was treated with partial scapulectomy, with resection of the posterior deltoid and preservation of the rotator cuff and axillary nerve (**b**) at the time of surgery margins were reported as negative. A pedicled latissimus flap was used for coverage of the area (**c**). The patient developed a rapid local recurrence with-in 3-months of the surgery with involvement of the glenohumeral joint (**d**). A total scapulectomy with an extra-articular resection was performed (**e**) leaving a large soft-tissue defect (**f**) The latissimus flap was re-inset, and the humerus was suspended from the residual clavicle utilizing a cement spacer (**g**) fabricated with rush-rods (**h**)

  - Patient rolled forward to allow for visualization of the midline back.
- Imaging of the arm is often not needed.

Anesthesia Considerations

- Postoperative nerve block for pain
- Avoid long-acting paralytics for induction of anesthesia and preoperative nerve blocks.
- Reaming, cementation, inflating photodynamic implants.
  - Ensure adequate volume status.
  - Consider using 100% FiO2.
- Tranexamic acid
  - 1 G IV at incision and 1 G IV at closure
- Blood loss can be considerable while curettage tumor especially with certain histologies.
  - Preoperative embolization
    - Renal, thyroid, solitary fibrous tumors

Incision/Exposure

- Humeral nail
  - Incision along the deltoid to split the fibers of the deltoid.
    - From acromion distal to expose the rotator cuff
    - Rotator cuff is split and tagged for repair.
  - Distal incision
    - Can be anterior or lateral.
    - Care is needed to identify radial nerve lateral.
      - Should directly visualize bone.
- Extended deltopectoral
  - Runs from clavicle, over or just lateral to the coracoid down the midline anterior arm curving lateral distal.
  - Care is taken to preserve the cephalic vein.
  - Pectoralis major is released and tagged exposing the long head of the biceps and axillary space.
    - Tenotomy of long head can be performed if needed for resection of proximal humerus.
  - For plate/cement construct often do not need to release long-head of biceps
    - Can release a portion of the distal anterior deltoid to allow for exposure and plate placement.
      - Should repair to plate at completion of procedure.
  - In setting of a proximal humerus resection should release teres/latissimus from humerus and tag for repair
    - Axillary nerve is running superior.
    - Radial nerve is running inferior.
  - Ligation of the anterior division of the circumflex vessels allows for the vascular bundle to fall medial and exposes the axillary vessels/nerve.
  - Rotator interval is created, and the subscapularis and supraspinatus tendon are released and tagged.
    - This allows for elevation of the humeral head out of the wound and access to the posterior rotator cuff and capsule, which can be divided.
  - Distal to the glenoid can place finger between inferior capsule and axillary nerve.
    - Cut onto finger with cautery to release inferior capsule and protect axillary nerve.
  - Traction can be placed on the humerus allowing sectioning of the triceps to fully release the proximal humerus.
  - Conjoint tendon can often be preserved; however, can be released if further exposure is needed medially.
    - Preservation of the tendon assists with judging your reconstruction length.

  - Brachialis can be split at midline distal to expose the humerus avoid lateral exposure of the radial nerve.
- Utilitarian Shoulder
  - Posterior continuation of the extended deltopectoral
  - Deltoid released off the clavicle, acromion and scapular spine based on tumor extent.
    - Posteriorly the axillary nerve and posterior circumflex vessels can be visualized.
      - Nerve and vessels are kept in continuity with the muscle belly by freeing them from the underlying capsule.
    - Triceps are reflected off the scapula to fully visualize nerve.
  - Trapezius is released off scapular spine.
  - Posterior incision follows the lateral boarder of the latissimus, curving towards the inferior angle of the scapula.
  - Need to create large skin flaps to expose underlying muscle and scapula.
  - Shoulder is internally rotated, and the patients hand placed behind the iliac crest in a "Figure 4" fashion.
    - Elevates the scapula away from the chest wall.
    - Release of medial periscapular muscles from the scapula.
      - Inferior to superior fashion
    - Allows for inspection of the chest wall.
      - Develop interval between serratus anterior muscle and subscapularis.
  - Lateral border of scapula is identified following the latissimus and triceps proximal.
    - Thoracodorsal vessel is preserved in case a latissimus flap is needed.
  - Clavicle can be osteotomized or the acromioclavicular joint is disarticulated, allowing for mobilization removal of the specimen.
    - Allows for retraction of the scapula to ligate perforating vessels into the subscapularis and supraclavicular vessels.
    - Scapula can be lifted off the chest wall posteriorly exposing the axillary vessels from the back.
      - Helpful in setting of a large soft-tissue mass
  - Osteotomy of the proximal humerus can be done at this time or the capsule of the glenohumeral joint/rotator cuff tendons are released based on tumor extent.
    - Difficulty removing the specimen it is often secondary to the preservation of the coracoclavicular ligaments.

Tumor Resection

- Resection of the tumor is based on the goals of surgery.

  - En-bloc/gross total resection in the setting of oligometastatic or radioresistant tumors
    - Goal for at least 1 cm bony margin and cuff of normal soft tissue
  - Harmonic scalpel can be used to make scapular osteotomies.
- Curettage used for majority of metastatic tumors.
  - Thermal, mechanical, and chemical adjuvants to enhance local tumor control.
    - Our preferred technique is to utilize a high-speed burr, hydrogen peroxide and an argon beam electrocoagulator.
  - Defect filled with bone cement with antibiotics (1 g of vancomycin and 1 g of gentamycin per batch), or a photodynamic implant can be placed through defect.

Wound Closure

- Depending on patient skin
  - Nylon sutures (3–0) and an incisional vacuum dressing
    - Poor skin quality
    - Edema of extremity
  - Steri-Strips and topical skin adhesive
    - Majority of incisions

Proximal Humerus Reconstruction Tips and Tricks

- Technique for reconstruction is based on expected patient survival and goals of care.
  - Majority of patients with metastatic disease are treated with a cemented reverse total shoulder endoprosthesis (RTSE)
    - Reverse APC can be considered in the setting of resections below the deltoid insertion, those with a favorable life expectancy (>1 year) and when radiation may not be considered following tumor resection.
  - Cemented hemiarthroplasty endoprosthesis is used the setting of unfavorable life expectancy and need for immediate use of the shoulder for weight bearing.
    - Typically, this is for patients with <6 weeks of life expectancy.
- Joint stability in a RTSE
  - Lateralizing the glenoid as much as possible
  - Using the largest glenosphere
  - Place glenosphere with inferior tilt.
  - Semi-constrained humeral polyethylene insert
- Joint stability in hemiarthroplasty endoprosthesis
  - Proximal humerus component is wrapped in mesh prior to cementation.

- Multiple heavy sutures are passed through the mesh and holes in implant to secure the mesh to the implant.
- Drill is used to create bone tunnels for polyester sutures in acromion, coracoid and glenoid.
  - Sutures are passed through the tunnels and through the mesh/implant.
  - Secured into place with implant centered below acromion.
- Soft tissue can be repaired directly to mesh with locking sutures.
  - Biceps tendon, remaining rotator cuff, pectoralis tendon

- *Postoperative Rehabilitation*
  - All patients with metastatic disease often require some form of venous thromboembolism prophylaxis.
  - *Reverse Endoprosthesis*
    - 6-weeks of immobilization patients with a sling and abduction pillow
      - At 6-weeks patients can start active and passive motion with deltoid
    - If the deltoid insertion is reconstructed with mesh, patients are immobilized for 12-weeks prior to starting therapy.
    - Internal and external rotation/stretching should be avoided.
    - At 3-months (12-weeks) postoperative patients can start unrestricted motion, wall stretches, and light strengthening, progressively increasing the use of the shoulder.
    - 15-pound life-time lifting restriction
  - *Plate and Cement, Humeral Nail, Photodynamic Implant*
    - Weight bearing as tolerated immediately following surgery
    - Simple sling for comfort
  - *Total Scapular Resection or Partial Resections with Joint Involvement*
    - 6-weeks of immobilization patients with a sling and abduction pillow or simple sling
    - Allowed for active assisted and passive ROM of the elbow hand and wrist
    - Internal and external rotation/stretching should be avoided.
    - At 6-weeks patient can stop wearing sling for everyday activities
    - At 3-months (12-weeks) postoperative patients can start unrestricted motion of elbow hand and wrist
      - Motion of the shoulder should be avoided
    - 15-pound life-time lifting restriction
  - *Partial Scapular Resections (Not Involving Glenohumeral Joint)*
    - Simple sling for comfort
    - Avoid internal and external rotation exercises for 4 weeks depending on repair of rotator cuff and serratus muscles
      - Can be activities as tolerated with elbow, hand and wrist.
- *Postoperative Medical Management*

- All patients should be evaluated by medical oncology and radiation oncology for ongoing cancer directed cares
  - Postoperative radiotherapy is important to reduce the risk of local progression
- Wounds should be evaluated by surgical team 2–3 weeks postoperative to be sure they have healed appropriately for either cytotoxic chemotherapy or radiotherapy
  - Some targeted treatments require a longer period prior to commencement of treatment

References

1. Wisanuyotin T, et al. Prognostic and risk factors in patients with metastatic bone disease of an upper extremity. J Bone Oncol. 2018;13:71–5.
2. Voskuil RT, Mayerson JL, Scharschmidt TJ. Management of metastatic disease of the upper extremity. JAAOS J Am Acad Orthop Surg. 2021;29(3):e116–25.
3. Arpornsuksant P, et al. What factors are associated with local metastatic lesion progression after intramedullary nail stabilization? Clin Orthop Relat Res. 2022;480(5):932–45.
4. Frassica FJ, Frassica DA. Evaluation and treatment of metastases to the humerus. Clin Orthop Relat Res. 2003;415 Suppl:S212–8.
5. Bickels J, Dadia S, Lidar Z. Surgical management of metastatic bone disease. J Bone Joint Surg Am. 2009;91(6):1503–16.
6. Hu H, et al. The personalized shoulder reconstruction assisted by 3D printing technology after resection of the proximal humerus tumours. Cancer Manag Res. 2019;11:10665–73.
7. Kaiser CL, et al. Tumors of the scapula: a retrospective analysis identifying predictors of malignancy. Surg Oncol. 2020;32:18–22.
8. Suresh S, Saifuddin A. Unveiling the 'unique bone': a study of the distribution of focal clavicular lesions. Skeletal Radiol. 2008;37(8):749–56.
9. Hussain S, et al. Anatomical distribution, the incidence of malignancy and diagnostic workup in the pathological lesions of the clavicle: a review of 410 cases. Arch Orthop Trauma Surg. 2023;143(6):2981–7.
10. Tedesco NS, et al. The musculoskeletal tumor society clinical practice guideline on the management of metastatic humeral disease. J Am Acad Orthop Surg. 2024;32(10):e482–8.
11. Rougraff BT, Kneisl JS, Simon MA. Skeletal metastases of unknown origin. A prospective study of a diagnostic strategy. J Bone Joint Surg Am. 1993;75(9):1276–81.
12. Costelloe CM, Madewell JE. Radiography in the initial diagnosis of primary bone tumors. AJR Am J Roentgenol. 2013;200(1):3–7.
13. Kager L, et al. Skip metastases in osteosarcoma: experience of the Cooperative Osteosarcoma Study Group. J Clin Oncol. 2006;24(10):1535–41.
14. Malawer MM. Tumors of the shoulder girdle. Technique of resection and description of a surgical classification. Orthop Clin North Am. 1991;22(1):7–35.
15. Wittig JC, et al. Osteosarcoma of the proximal humerus: long-term results with limb-sparing surgery. Clin Orthop Relat Res. 2002;397:156–76.
16. Merimsky O, et al. Is forequarter amputation justified for palliation of intractable cancer symptoms? Oncology. 2001;60(1):55–9.
17. Wittig JC, et al. Palliative forequarter amputation for metastatic carcinoma to the shoulder girdle region: indications, preoperative evaluation, surgical technique, and results. J Surg Oncol. 2001;77(2):105–13; discussion 114.

18. Puhaindran ME, et al. Major upper-limb amputations for malignant tumors. J Hand Surg Am. 2012;37(6):1235–41.
19. Bickels J, et al. Biopsy of musculoskeletal tumors. Current concepts. Clin Orthop Relat Res. 1999;(368):212–9.
20. Mankin HJ, Mankin CJ, Simon MA. The hazards of the biopsy, revisited. Members of the musculoskeletal tumor society. J Bone Joint Surg Am. 1996;78(5):656–63.
21. Welker JA, et al. The percutaneous needle biopsy is safe and recommended in the diagnosis of musculoskeletal masses. Cancer. 2000;89(12):2677–86.
22. Aparisi Gómez MP, et al. Functional and surgical anatomy of the upper limb: what the radiologist needs to know. Radiol Clin North Am. 2019;57(5):857–81.
23. Thai DM, Kitagawa Y, Choong PF. Outcome of surgical management of bony metastases to the humerus and shoulder girdle: a retrospective analysis of 93 patients. Int Semin Surg Oncol. 2006;3:5.
24. Mirels H. Metastatic disease in long bones a proposed scoring system for diagnosing impending pathologic fractures. Clin Orthop Relat Res. 1989;249:256–64.
25. Evans AR, et al. Mirels' rating for humerus lesions is both reproducible and valid. Clin Orthop Relat Res. 2008;466(6):1279–84.
26. Tat J, et al. Predicting pathological fractures at metastatic humeral lesions. JB JS Open Access. 2023;8(2)
27. Miller BJ, et al. Intramedullary nails for long bone metastases: why do they fail? Orthopedics. 2011;34(4)
28. Hoban KA, et al. Mirels' score for upper limb metastatic lesions: do we need a different cutoff for recommending prophylactic fixation? JSES Int. 2022;6(4):675–81.
29. Siegel HJ, et al. Pathological fractures of the proximal humerus treated with a proximal humeral locking plate and bone cement. J Bone Joint Surg Br. 2010;92(5):707–12.
30. Wilson WT, et al. Stabilisation of pathological humerus fractures using cement augmented plating: a case series. J Clin Orthop Trauma. 2021;15:93–8.
31. Chen C, et al. Postoperative complications with cryotherapy in bone tumors. J Bone Oncol. 2017;7:13–7.
32. Yang Y, et al. Advances in limb salvage treatment of osteosarcoma. J Bone Oncol. 2018;10:36–40.
33. Sim FH, Daugherty TW, Ivins JC. The adjunctive use of methylmethacrylate in fixation of pathological fractures. JBJS. 1974;56(1):40–8.
34. Hoellwarth JS, et al. Evaluating the reoperation rate and hardware durability of three stabilizing implants for 105 malignant pathologic humerus fractures. Injury. 2020;51(4):947–54.
35. Weiss KR, et al. Fixation of pathological humeral fractures by the cemented plate technique. J Bone Joint Surg Br. 2011;93(8):1093–7.
36. Alvi HM, Damron TA. Prophylactic stabilization for bone metastases, myeloma, or lymphoma: do we need to protect the entire bone? Clin Orthop Relat Res. 2013;471(3):706–14.
37. Choi ES, et al. Intramedullary nailing for pathological fractures of the proximal humerus. Clin Orthop Surg. 2016;8(4):458–64.
38. Laitinen M, Nieminen J, Pakarinen TK. Treatment of pathological humerus shaft fractures with intramedullary nails with or without cement fixation. Arch Orthop Trauma Surg. 2011;131(4):503–8.
39. Sarahrudi K, et al. Surgical treatment of pathological fractures of the shaft of the humerus. J Trauma. 2009;66(3):789–94.
40. Krumme J, et al. Early experience in pathologic humerus fracture treated with the photodynamic bone stabilization system shows limitations related to patient selection. Orthopedics. 2021;44(3):154–9.
41. Sabharwal S, Boland PJ, Vaynrub M. Severe hemodynamic collapse during humerus stabilization with photodynamic implant: a report of two cases. JBJS Case Connect. 2024;14(2)
42. Houdek MT, et al. Comparison of reconstructive techniques following oncologic intraarticular resection of proximal humerus. J Surg Oncol. 2021;123(1):133–40.

43. Maclean S, et al. Reverse shoulder endoprosthesis for pathologic lesions of the proximal humerus: a minimum 3-year follow-up. J Shoulder Elb Surg. 2017;26(11):1990–4.
44. Chalmers PN, Keener JD. Expanding roles for reverse shoulder arthroplasty. Curr Rev Musculoskelet Med. 2016;9(1):40–8.
45. Ackland DC, Patel M, Knox D. Prosthesis design and placement in reverse total shoulder arthroplasty. J Orthop Surg Res. 2015;10:101.
46. Grosel TW, et al. Oncologic reconstruction of the proximal humerus with a reverse total shoulder arthroplasty megaprosthesis. J Surg Oncol. 2018;118(6):867–72.
47. Sanchez-Sotelo J, Wagner ER, Houdek MT. Allograft-prosthetic composite reconstruction for massive proximal humeral bone loss in reverse shoulder arthroplasty. JBJS Essent Surg Tech. 2018;8(1):e3.
48. Sanchez-Sotelo J, et al. Allograft-prosthetic composite reconstruction for massive proximal humeral bone loss in reverse shoulder arthroplasty. J Bone Joint Surg Am. 2017;99(24):2069–76.
49. King JJ, et al. Allograft-prosthetic composite reverse total shoulder arthroplasty for reconstruction of proximal humerus tumor resections. J Shoulder Elb Surg. 2016;25(1):45–54.
50. Damron TA, et al. Functional laboratory assessment after oncologic shoulder joint resections. Clin Orthop Relat Res. 1998;348:124–34.
51. Fuhrmann RA, Roth A, Venbrocks RA. Salvage of the upper extremity in cases of tumorous destruction of the proximal humerus. J Cancer Res Clin Oncol. 2000;126(6):337–44.
52. Cundy WJ, et al. Constrained or unconstrained shoulder replacement for musculoskeletal tumor resections? J Shoulder Elb Surg. 2020;29(10):2104–10.
53. Takahashi T, et al. Reverse shoulder arthroplasty and synthetic mesh for reconstruction of the shoulder joint after malignant bone tumor resection: a case report. Medicine. 2021;2(2):e0051.
54. Marulanda GA, et al. Proximal and total humerus reconstruction with the use of an aortograft mesh. Clin Orthop Relat Res. 2010;468(11):2896–903.
55. Raiss P, et al. Replacement of the proximal humerus with MUTARS tumor endoprostheses. Eur J Surg Oncol. 2010;36(4):371–7.
56. van de Sande MA, Dijkstra PD, Taminiau AH. Proximal humerus reconstruction after tumour resection: biological versus endoprosthetic reconstruction. Int Orthop. 2011;35(9):1375–80.
57. Tang X, et al. Synthetic mesh improves shoulder function after intraarticular resection and prosthetic replacement of proximal humerus. Clin Orthop Relat Res. 2015;473(4):1464–71.
58. Braig ZV, et al. Humeral stress shielding following cemented endoprosthetic reconstruction: an under-reported complication? J Surg Oncol. 2021;123(2):505–9.
59. Ebeid WA, et al. Outcomes of modular endoprosthesis reconstruction versus cement spacer reconstruction following resection of proximal humeral tumors. BMC Musculoskelet Disord. 2022;23(1):484.
60. Sullivan MH, et al. Comparison of reconstructive techniques for nonprimary malignancies in the proximal humerus. J Surg Oncol. 2024;130(1):64–71.

Management of Diaphyseal Humeral Metastases

17

Sarah Ballatori, Lucy Hederick, Megan Goh, Neha Jejurikar, Joseph Benevenia, and Santiago A. Lozano Calderón

Introduction

Metastatic bone disease (MBD) is on the rise in the United States due to improvements in systemic cancer therapies combined with an aging population. The increased prevalence and longevity of patients with MBD necessitate durable surgical solutions [1]. Humeral lesions, often resulting from cancers such as multiple myeloma or metastatic carcinoma, cause pain, hinder activities of daily living, and preclude mobilization in certain patients. Surgery aims to alleviate pain and restore the function of the upper extremity. This chapter will focus on the surgical management of diaphyseal humeral metastatic lesions.

Epidemiology

The incidence of skeletal-related events secondary to metastatic disease is approaching 800,000 events annually, resulting in morbidity for patients and a substantial financial burden on the health care system [2, 3]. Osseous metastases most commonly affect the axial skeleton, and the upper extremity comprises about 20% of all metastatic bone disease [4]. The humerus is the second-most affected long bone, following the femur [4, 5]. Approximately 16–39% of all impending or completed pathologic fractures occur in the humerus [2].

S. Ballatori · L. Hederick · M. Goh · N. Jejurikar · S. A. Lozano Calderón (✉)
Division of Musculoskeletal Oncology, Department of Orthopedic Surgery / Massachusetts General Hospital - Harvard Medical Schoo, Boston, MA, USA
e-mail: slozanocalderon@mgh.harvard.edu

J. Benevenia
Department of Orthopaedic Surgery, Musculoskeletal Oncology Division, Rutgers Cancer Institute, Rutgers New Jersey Medical School, Newark, NJ, USA

J. Pretell-Mazzini, S. A. Lozano Calderón (eds.), *Multidisciplinary Approach of Musculoskeletal Metastasis*, https://doi.org/10.1007/978-3-032-09498-8_17

Patients with MBD have a wide array of primary cancer diagnoses, from breast and lung carcinomas to multiple myeloma and lymphoma. Primary tumor histology dictates predilection for bony metastasis and subsequent treatment options. For instance, approximately 70% of patients with metastatic breast or prostate cancer develop skeletal metastases, compared to 36% of patients with metastatic lung cancer and only 5% of patients with gastrointestinal cancers [4].

Indications

Nonoperative

Compared to lesions of the lower extremities or pelvis, upper extremity lesions are more frequently treated nonoperatively. This is due to the lower weight-bearing requirement of the upper extremity, particularly in those patients who ambulate without an assistive device. Immobilizing impending or completed diaphyseal humerus fractures with functional bracing such as Sarmiento bracing, coaptation splinting, or a sling is feasible. Nonoperative treatment is best reserved for patients who: (a) have very small lesions not at risk of impending pathologic fracture, with a high probability of favorable response to systemic or radiation therapy, or (b) have a very poor prognosis or a contraindication to surgery [6]. Initial nonoperative management should also be employed in those awaiting a formalized histologic diagnosis [7]. Of note, according to the Musculoskeletal Tumor Society Clinical Practice Guideline (MSTS CPG), there is insufficient data to determine which patients optimally benefit from nonsurgical versus surgical management [2].

Operative

Indications for surgical management of diaphyseal humeral metastatic lesions center around each patient's symptoms, prognosis, and goals of care. While these indications are similar in principle to the management of all skeletal lesions, there are several unique considerations for the humerus. For one, the humerus is the structural foundation of the upper extremity, responsible for positioning and structurally supporting the hand in space. Therefore, surgically stabilizing the humerus can restore a painless arc of motion, which is critical for activities of daily living such as eating, dressing, and hygiene. In addition, unlike MBD of the lower extremities, the humerus is a non-weight-bearing bone. This means treatment is less urgent, as patients can usually maintain their mobility and keep pain controlled with sling immobilization while awaiting surgery [2, 7].

Surgical intervention is indicated for the following: alleviating pain, restoring function, preventing impending fracture, improving prognosis, and for certain primary tumor histologies. The choice of surgical technique will vary based on several factors, such as lesion size, tumor sensitivity to systemic and/or radiation therapy, hand dominance, patient prognosis, and anticipated healing potential [2, 7, 8]. The wide range of surgical techniques, from minimally invasive ablative treatment to en bloc resection, is covered in the following section.

Pain

Most patients with completed or impending fractures of the humeral diaphysis report pain that worsens with activity and improves with immobilization. Although pain in isolation is not an absolute indication to proceed with surgery, it is important to remember that the goal of treatment for most patients is palliative. Surgical treatment is indicated for patients with persistent pain in the setting of failed nonoperative treatment or anticipated unsuccessful nonoperative treatment [2].

Function

In deciding whether to indicate a patient for surgery, it is important to consider the patient's weight-bearing status. If a patient relies on the upper extremity for use of a walker or cane, this may further support the plan for operative management. In all cases of palliative fixation, the postoperative plan should include full weight bearing as tolerated, and this principle is even more important for those patients who mobilize with a walker. Even if the patient does not weight bear through the upper extremity for ambulation, surgical fixation can restore or protect the integrity of the humeral shaft, thereby enabling function of the upper limb for activities of daily living. This is particularly important when the patient's dominant hand is affected [2].

Facture Risk and Prognosis

Untreated metastatic lesions to the diaphyseal humerus can result in pathologic fracture, causing substantial pain and disability [9]. The ability to accurately predict a lesion's fracture risk is essential, as prophylactic fixation decreases morbidity compared to pathologic fracture fixation [10]. Risk factors for progression to pathologic fracture of humeral metastatic lesions include: advanced cancer stage, poor disease control, increased tumor size, rapid tumor growth, lytic composition, weight-bearing bone, pain, female sex, advanced age, poor bone health, and increased fall risk [2].

A Mirels score is commonly calculated to predict the fracture risk of long bones [11]. The Mirels numerical scoring system is covered in other chapters. While the Mirels score is conceptually helpful in assessing fracture risk, more recent studies have altered its use in the humerus. Hoban et al. retrospectively analyzed the use of the Mirels score in upper extremity MBD, including 35 patients with humeral lesions. They report poor sensitivity and specificity in predicting upper extremity fractures and recommend using a lower Mirels threshold of ≥ 7 for considering prophylactic fixation [12]. Novel methods, including computed tomography (CT)-based structural rigidity analysis (CTRA), finite element (FE) analysis, and PET/CT avidity have also been shown to predict fracture risk in metastatic lesions [13].

Prognostic tools can be useful to inform surgical decisions. The PATHFx tool predicts survival in patients with a pathologic fracture in the setting of metastatic disease [14]. The Skeletal Oncology Research Group (SORG)'s machine learning algorithms can predict 90-day and 1-year survival in patients with operatively managed extremity metastatic disease [15].

Generating a more accurate understanding of fracture risk and prognosis using the aforementioned tools can help guide multidisciplinary and patient–provider discussions to decide the extent of therapeutic intervention. Regardless of the scoring system used, these scores must be interpreted within the context of each patient's individual clinical history, lesion characteristics, and goals of care.

Improving Prognosis

The ultimate goal of surgical treatment for diaphyseal humerus lesions is to improve quality of life and palliate pain, without impact on the patient's overall prognosis. However, in select cases of oligometastatic disease, en bloc resection may be considered. Surgical metastasectomy, defined as removal of the cancerous lesion(s) that have spread from the primary tumor, is indicated for two groups of patients: (1) patients with tumors that do not respond well to systemic and radiation therapy and (2) patients with oligometastatic disease for whom wide resection is anticipated to improve prognosis [16, 17].

This concept is most extensively studied in renal cell carcinoma, with multiple primarily retrospective studies demonstrating improved overall survival and improved recurrence-free survival with en bloc resection of oligometastatic disease [17–20]. This phenomenon has also been observed in oligometastatic papillary thyroid carcinoma and metastatic breast carcinoma with late occurrence of osseous metastasis [7, 21, 22]. However, there is insufficient evidence for or against this practice in humeral metastatic lesions according to the MSTS CPG [2].

Tumor Histology

Metastatic bone disease to the diaphyseal humerus comprises a diverse list of histologies, from carcinomas such as breast, lung, and prostate, to blood cancers including multiple myeloma and lymphoma. Regardless of the metastatic lesion's primary histology, pathologic fractures have a diminished ability to heal spontaneously compared to fractures of normal bone [8]. Different tumor histologies demonstrate both a varied healing potential after fracture and a wide range of responses to systemic therapies, such as cytotoxic chemotherapy, targeted or immunotherapy, and anti-resorptive medications. For example, denosumab has been shown to decrease skeletal-related events in breast cancer patients [23]. Similarly, some cancer types are more radiosensitive, while others are notoriously radioresistant. For example, breast cancer, prostate cancer, and multiple myeloma are highly radiosensitive, while renal cell carcinoma is not [7]. Therefore, indications for surgery will depend on the anticipated response to systemic and/or radiation therapy.

In summary, indications for surgical treatment of a humeral shaft metastatic lesion hinge on the following: pain, function, fracture risk, prognosis, and tumor histology, taken together with the patient's baseline functional status and goals of care.

Pre-operative Planning and Surgical Options

There are four major considerations that help guide surgical planning:

1. Is there a confirmed diagnosis?
2. Is surgical metastasectomy indicated?
3. Is fixation or reconstruction indicated?
4. Is there a need for radiolucent implants?

Is There a Confirmed Diagnosis?

It is essential that, prior to moving forward with surgical planning, a comprehensive workup of the humeral lesion is completed. If a confirmed histologic diagnosis is lacking, a biopsy must be performed to avoid inadvertent intralesional fixation through a primary osseous malignancy [7]. General metastatic disease workup and biopsy are covered in prior chapters.

Is Surgical Metastasectomy Indicated?

As discussed above, patients whose primary tumors do not respond well to systemic or radiation therapy or who may have improved prognosis with removal of osseous metastases can be indicated for surgical metastasectomy. For the first category of patients, removal of local tumor burden via metastasectomy can lessen the risk of local recurrence and therefore decrease the likelihood of implant failure [17].

Surgical approaches to metastasectomy include (1) intralesional tumor excision with operative fixation, or (2) wide resection with reconstruction. Curettage, high-speed burring, radiofrequency ablation, cryoablation, and/or bone void fillers (such as polymethyl methacrylate (PMMA)) may be employed as an augment to surgery. Further indications for surgical fixation vs reconstruction are discussed in the following section.

Is Fixation or Reconstruction Indicated?

Whether or not metastasectomy is performed, the general surgical options to stabilize a skeleton affected by MBD include:

1. Fixation, with or without tumor excision, ablation, or other adjuvant treatments. As previously mentioned, the addition of intralesional metastasectomy may be considered to decrease the risk of local recurrence and, therefore, decrease the likelihood of fixation failure [17].
2. Reconstruction, either after wide resection metastasectomy is performed, or secondary to a large, destructive metastatic lesion.

Fixation

It is important to keep in mind the increased load on the fixation construct when fixing pathologic fractures due to their reduced healing potential [8]. Fixation options for lesions of the humeral diaphysis include plate and screw internal fixation, intramedullary nail fixation, intramedullary photodynamic polymer fixation, and a combination of them. As there is limited evidence directly comparing these fixation methods, the decision is based on surgeon preference and patient-specific considerations [2].

Plate-screw internal fixation offers the benefit of greater rigidity than intramedullary nail fixation. Further stability can be achieved with the use of locking screws and bone cementation. Plating is preferred for diaphyseal lesions located near the proximal or distal end of the humeral shaft. The wide surgical exposure required for plate–screw fixation is beneficial for direct visualization of the lesion but heightens soft tissue damage and risk of wound complications [24].

Intramedullary nails (IMNs) are ideal load-sharing devices for proximal third and mid-diaphyseal lesions, particularly those that are sensitive to radiation or systemic therapy. Clinical outcomes after intramedullary nailing of humeral shaft metastatic lesions are quite good, resulting in reduced pain and improved function [25, 26]. Outcomes of humeral intramedullary nail fixation are further improved with the addition of curettage and PMMA packing in less radiosensitive tumors. Therefore, this augment should be considered for patients with primary tumors like renal cell carcinoma [27]. IMN technique can involve reaming of the medullary canal or placing the nail in an unreamed canal; studies tend to favor the unreamed intramedullary nailing technique [28, 29]. In a retrospective comparative study of 55 patients, those who underwent unreamed intramedullary nailing demonstrated reduced need for blood transfusion, with no statistical difference in bone consolidation rate or pain scores, compared to those who underwent reamed intramedullary nailing [28].

Radiolucent intramedullary fixation, including intramedullary photodynamic polymer fixation and carbon fiber implants, will be covered in greater detail in the following subsection on radiolucent implants.

Reconstruction

Intercalary reconstruction is often necessary in patients left with an intercalary bone void after en bloc resection of a diaphyseal humeral lesion or when bony apposition of fracture fragments is not feasible due to the size and destructiveness of a lesion. Reconstruction of a diaphyseal defect can be achieved with an endoprostheses, allograft, autograft, or combination of cement with IMN and/or plate fixation [30].

Endoprosthetic reconstruction requires at least 5 cm of intramedullary canal fixation proximal and distal to the lesion to accommodate the endoprosthetic stems, and other systems of fixation, such as the Zimmer/Biomet Compress require at least 4.5 cm of remaining bone proximal and distal to the lesion (Fig. 17.1) [31]. Endoprosthetic reconstruction immediately restores form and function to the affected extremity. Outcomes in the setting of metastatic disease are favorable, affording pain relief and a low complication rate in the short term [13, 32]. However,

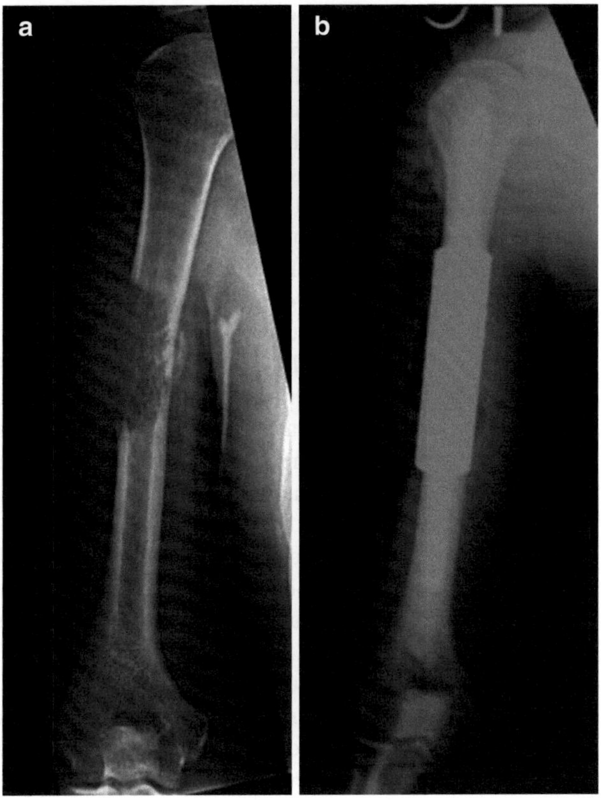

Fig. 17.1 (**a**) Patient with metastatic lung cancer with a pathologic fracture in the right diaphyseal humerus. (**b**) The patient underwent an intercalary prosthetic reconstruction after en bloc resection of the lesion

these implants are susceptible to aseptic loosening, infection, and periprosthetic fracture [33, 34]. A study evaluating the performance of custom-made endoprosthetic replacements for MBD of the humeral diaphysis with an average follow-up of 4.7 years reported relatively high complication rates due to aseptic loosening and periprosthetic fracture [35]. Blocking screws through the stem of the implant or additional plate fixation can be used to account for the torsional forces placed upon humeral reconstructions [36].

Allografts or autografts offer a strong biologic option for reconstructing diaphyseal humeral defects, particularly for younger patients with a favorable prognosis requiring a longer-term surgical solution. The most common complication with allograft reconstruction is nonunion of the graft-host junction [37]. To increase healing potential, a vascularized fibular autograft can be considered; however, these are also associated with graft failure, fracture, host-site morbidity, and high reoperation rate [38].

Is There a Need for Radiolucent Implants?

Unlike standard metal implants, radiolucent implants do not obscure postoperative imaging, enabling more accurate adjuvant radiation therapy planning, better visualization of fracture healing, and improved surveillance of local tumor recurrence [39]. Radiolucent options for diaphyseal humerus lesions include photodynamic nails (PDNs), carbon fiber nails, and carbon fiber plates.

Photodynamic intramedullary nail fixation (IlluminOss System) can be performed in either a standard antegrade fashion or in the less conventional retrograde manner, accessing the medial and lateral columns of the distal humerus (Fig. 17.2). Therefore, this implant offers a fixation solution when a proximal incision should be avoided or when the lesion is located distally in the humeral diaphysis. A PDN can be placed in isolation, often in cases of prophylactic fixation or nondisplaced pathologic fractures, or it can be combined with percutaneous screws or plate-screw constructs (Fig. 17.3). As with standard metal IMNs, the PDN offers pain reduction and improved function for patients with humeral metastatic disease [40]. PDNs are particularly helpful in treating uncontained lesions, as the liquid monomer remains contained within the expandable balloon and therefore does not extravasate into the soft tissues, unlike liquid bone substitutes or bone cement. Biomechanically, the implant conforms in size to match the area and diameter of the targeted metastasis, not only bypassing the defect but also filling it [41].

Similar to PDNs, carbon fiber (CF) nails (Fig. 17.4) and plates (Fig. 17.5) are radiolucent and do not produce a scattering effect on MRI or CT, thus allowing for postoperative surveillance and accurate radiation planning. Carbon fiber implants have a high tensile strength with a modulus of elasticity close to that of native bone [42]. According to a 2023 study, carbon fiber and titanium humeral nails have comparable failure rates and modes. However, carbon fiber nails may be susceptible to failure, particularly in patients requiring large segmental resection and cement spacer reconstruction (Fig. 17.6), if less than 5 cm of residual distal humerus remains for nail fixation. In these cases, the authors

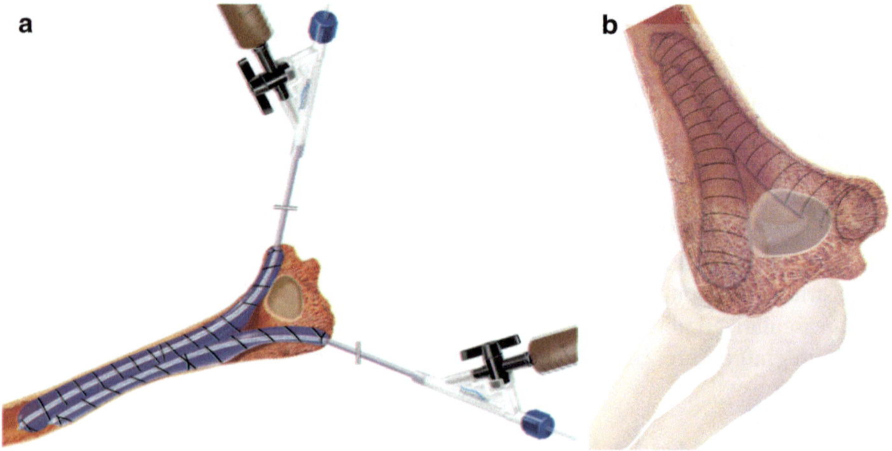

Fig. 17.2 (**a**) Schematic of the retrograde approach to stabilize distal diaphyseal humerus impending and/or pathologic fractures with photodynamic nails. (**b**) Schematic of the anatomic positioning of the balloons

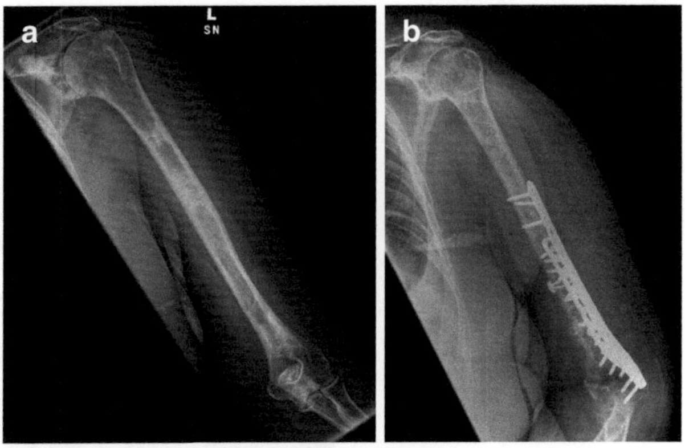

Fig. 17.3 (**a**) Patient with metastatic non-small cell lung cancer with an impending pathologic fracture of the left diaphyseal humerus. (**b**) The patient was treated with an intramedullary photodynamic nail augmented with a titanium plate and screws for prophylactic stabilization

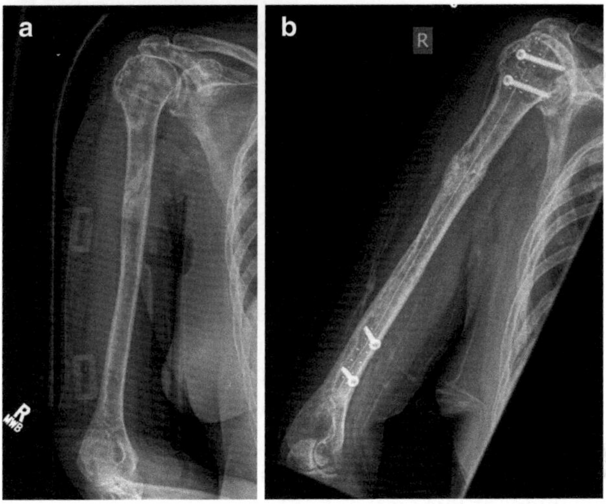

Fig. 17.4 (**a**) Patient with metastatic breast cancer with a pathologic fracture of the right diaphyseal humerus. (**b**) The patient underwent stabilization with a carbon fiber intramedullary nail

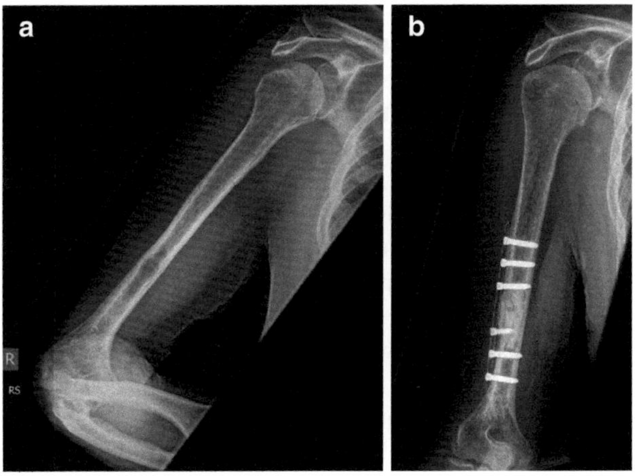

Fig. 17.5 (**a**) Patient with metastatic renal cell carcinoma with an impending pathologic fracture of the right diaphyseal humerus. (**b**) The patient was treated with excision and curettage followed by ablation of the lesion and cement packing with prophylactic stabilization using a carbon fiber plate

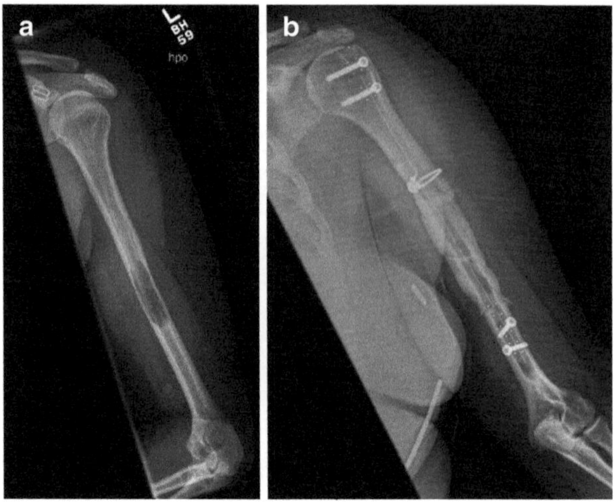

Fig. 17.6 (**a**) Patient with metastatic synovial sarcoma with impending pathologic fracture of the left diaphyseal humerus. (**b**) The patient underwent an en bloc resection and reconstruction with a carbon fiber intramedullary nail and insertion of a cement spacer

suggest fixation augmentation with either an intercalary allograft, addition of a plate distally, or use of a thicker titanium nail (>8.5 mm) [43]. A 2023 systematic review of literature on carbon fiber plates demonstrated no significant differences in clinical or radiologic outcomes between patients treated with CF and metal plates [44].

Overall, many surgical solutions exist for patients with metastatic disease to the diaphyseal humerus. In addition to considering these surgical options, it is important to engage with a multidisciplinary team to determine the best treatment plan for each patient. In addition to systemic cancer therapies, several noninvasive and minimally invasive techniques can be used in combination with surgical intervention. These modalities include radiation therapy, radiofrequency ablation, cryoablation, and embolization, and will be covered in other chapters of this textbook [45].

Surgical Technique

Patient Positioning and Surgical Approach

Surgical positioning and approach are dictated by the diaphyseal humerus lesion location, size, pathologic fracture morphology, presence of associated soft tissue mass, and, in the case of impending fracture, cortical destruction location. Indications for closed versus open surgical treatment were discussed in the "Indications" section.

For most cases, the patient is positioned supine on a radiolucent table, with a folded blanket bump under the scapula to elevate the humerus and facilitate fluoroscopic views. The patient should be shifted towards the non-affected extremity, so the affected arm is fully supported by the radiolucent table. Additional support with blankets under the distal humerus helps keep the arm in a neutral, non-hyperextended position, facilitating appropriate insertion of the initial guide wire, reaming, and ball-tipped guide wire placement if nailing has been planned. Fluoroscopy is located on the contralateral side of the table. Opening the arm board helps create space for C-arm positioning. Depending on body habitus, a radiolucent hand table or plexiglass may be needed.

In the rarer cases requiring an open posterior approach–for example, a patient with a distal humeral shaft lesion requiring posterolateral plating–the patient is positioned in the lateral decubitus. In this case, the table is turned 90 degrees, and fluoroscopy enters from the head of the table.

For treatment of an impending fracture or a closed reduction of a pathologic fracture, a standard IMN surgical approach is used. Skin incision is located at the anterolateral corner of the acromion, extending distally 1–2 cm in line with the deltoid fibers. The deltoid fascia is incised at the raphe between the anterior and middle third of the deltoid muscle, and the muscle is retracted to expose the rotator cuff and supraspinatus tendon. The rotator cuff is sharply split in line with its fibers, and the tendon edges are retracted with sutures. Next, a guidewire is inserted and the entry point in the proximal humerus is confirmed fluoroscopically–just posterolateral to the biceps tendon, medial to the greater tuberosity sulcus on the anteroposterior (AP) view, and collinear to the diaphysis on the lateral view. The remaining surgical steps are in accordance with the implant's surgical technique guide.

Retrograde PDN placement employs the same general principles as the technique used in pediatric patients with supracondylar fractures fixed with Kirshner wires. The lateral condyle and medial epicondyle are palpated, and a longitudinal 1 cm incision is made directly over each bony prominence. Careful blunt dissection is performed to access the condyles, particularly on the medial side to avoid injury to the ulnar nerve, which is immediately adjacent to the medial epicondyle. PDN placement is sequentially performed into the medial and lateral columns of the distal humerus (Fig. 17.2).

If an open approach to the humeral shaft is selected for a procedure involving intralesional curettage or for a wide resection, either an extensile anterolateral approach or a posterior triceps-sparing approach is used based on surgeon preference and lesion characteristics. The anterolateral approach to the humeral shaft is an excellent approach for lesions of the proximal third and mid-humeral shaft, and the posterior approach is best for mid- to distal third shaft lesions.

The anterolateral approach to the humeral shaft begins with the deltopectoral interval proximally, extending distally and laterally along the arm to the interval between the biceps and the brachialis. The fascia is incised between the biceps and the brachialis, retracting the biceps medially. Proximally, the cephalic vein is identified in the deltopectoral interval, and if the dissection is carried distally, the radial nerve is identified between the brachialis and mobile wad distally and traced proximally through the lateral intermuscular septum to the spiral groove. The brachialis is then split in its internervous plane, between the radial and musculocutaneous nerves, affording direct access to the humeral shaft.

In the posterior approach to the humeral shaft, a longitudinal midline incision is made along the dorsal arm, keeping the lesion location in mind. If distal exposure is necessary, the incision can be curved radially along the olecranon process. Next, the triceps fascia is incised in line with the skin incision. The ulnar nerve is identified and protected medially as it travels along the medial intermuscular septum. The radial nerve is identified laterally as it penetrates the lateral intermuscular septum approximately 10 cm proximal to the lateral epicondyle. Distal exposure to the humeral shaft is accomplished by retracting the triceps medially or laterally, and more proximal access is achieved by splitting the long and lateral heads of the triceps. The inferior lateral cutaneous branch of the radial nerve can be used to track the radial nerve proper after releasing this sensory branch from the lateral septum. This also allows mobilization of the triceps medially.

Fixation

Fracture fixation and impending fracture stabilization using intramedullary nails, plate-screw constructs, or photodynamic implants employ the above surgical approaches and basic principles of fracture reduction. On occasion, a combination of intramedullary nail-and-plate or photodynamic nail-and-plate constructs may be used, particularly in the setting of large, destructive lesions. For each of these implants, please refer to the specific company technique guides for current, accurate instrumentation steps.

Metastatectomy and Reconstruction

Surgical removal of tumor can be achieved by either intralesional excision or wide resection. Indications for each of these approaches were previously discussed. Briefly, if the lesion is amenable to wide resection, which may improve local recurrence or overall survival, or too large and destructive to consider simple curettage and packing, wide resection is the optimal choice. Proximal and distal osteotomies are planned based on pre-operative imaging, and a modular intercalary endoprosthesis is cemented into place, recreating the native humeral length and soft tissue tension. If the remaining bone is too short (<5 cm) to accommodate a stemmed prosthesis, either a Compress system or a custom-made implant can be considered. Intercalary allografts with concomitant plate, nail, or combined nail–plate constructs are also an alternative for reconstruction in diaphyseal lesions.

More commonly, intralesional curettage is performed. This technique involves thorough curettage of the lesion, with or without adjuvant therapies such as radiofrequency ablation, cryoablation, phenol or peroxide irrigation, and/or argon beam coagulation. If the humerus is intact, a longitudinal, oval cortical window is made, centered on the lesion. The corticotomy can be established with incrementally larger straight curettes if the lesion has weakened the cortex, or with a high-speed burr. The cortical window must be large enough to sufficiently access the entire lesion.

Gross tumor is meticulously excised with curettes, followed by burring of the tumor cavity with care to maintain the cortical walls. The bone void is then irrigated thoroughly and filled with a bone void filler per the surgeon's preference. Bone void filler options commonly include PMMA, allograft, or synthetic bone substitutes comprised of Calcium Sulfate, Calcium Phosphate, Magnesium Sulfate, bi-phasic hydroxyapatite, or a combination thereof.

Soft Tissue Closure

Minimal to no soft tissue disruption is expected in the surgical approach to the humeral shaft, except for splitting the rotator cuff tendon for the insertion of an intramedullary nail. After repair of the rotator cuff tendon with #1 nonabsorbable suture such as Ethibond or Fiberwire, closure involves fascia closure with 0 absorbable suture such as Vicryl or PDS, subcutaneous and dermis closure with 2–0 absorbable suture, and skin closure with staples, 3–0 Monocryl or Nylon, all per surgeon preference. Occasionally, a surgical drain is left in place if, for example, there is any difficulty achieving hemostasis.

Postoperative Care

Postoperatively, patients are often admitted to the hospital for observation, pain control, and perioperative antibiotics. A sling is provided for comfort, but range of motion and weight bearing are allowed right away. Upon confirmation of wound healing, about 2 weeks after surgery, the patients may begin systemic and radiation therapy.

Tips and Tricks

- Supine, flat Jackson table, move the patient over so the arm is on the edge of the table. Use support in distal humerus to prevent hyperextension.
- For displaced pathologic fractures, it is important to reduce the fracture prior to reaming for placement of an intramedullary device, particularly when considering using a PDN.
- Percutaneous 2.0 guide wires can be used to assist in fracture reduction in cases with coronal or sagittal deformities.
- In lesions extending into the proximal or distal metaphysis, plate fixation is preferred if sufficient fixation for intramedullary devices is not achievable.
- When using photodynamic nails, the use of additional screws for rotational control is based on the surgeon's preference. Our preference is to use them in patients with complete fractures.
- In cases where aggressive curettage of a lesion is performed in the humeral diaphysis, our preference for fixation is plate and screws with or without

supplementation of photodynamic nails, depending on the size of the defect and/or the presence of other lesions in the humerus.
- Embolization prior to intralesional curettage of renal cell carcinoma can aid in visualization intraoperatively and reduce blood loss.
- For intramedullary implants, placement of a cannulated awl over the starting guidewire facilitates exchange to the ball-tipped guide wire.
- For cement packing of bone voids, wait for the cement to partially cure prior to packing to avoid cement extravasation into the soft tissues.

References

1. Siegel RL, Miller KD, Jemal A. Cancer statistics, 2020. CA Cancer J Clin. 2020;70(1):7–30. https://doi.org/10.3322/caac.21590.
2. Tedesco NS, Mesko N, Wodajo F, Management of Metastatic Humeral Disease Work Group, Staff of the American Academy of Orthopaedic Surgeons and the Musculoskeletal Tumor Society, Management of Metastatic Humeral Disease Work Group and Staff of the American Academy of Orthopaedic Surgeons and the Musculoskeletal Tumor Society. The musculoskeletal tumor society clinical practice guideline on the management of metastatic humeral disease. J Am Acad Orthop Surg. 2024;32(10):e482–8. https://doi.org/10.5435/JAAOS-D-24-00053.
3. Schulman KL, Kohles J. Economic burden of metastatic bone disease in the U.S. Cancer. 2007;109(11):2334–42. https://doi.org/10.1002/cncr.22678.
4. Coleman RE. Clinical features of metastatic bone disease and risk of skeletal morbidity. Clin Cancer Res. 2006;12(20 Pt 2):6243s–9s. https://doi.org/10.1158/1078-0432.CCR-06-0931.
5. Rovere G, Meschini C, Piazza P, et al. Proximal humerus fractures treatment in adult patients with bone metastasis. Eur Rev Med Pharmacol Sci. 2022;26(1 Suppl):100–5. https://doi.org/10.26355/eurrev_202211_30288.
6. Flemming JE, Beals RK. Pathologic fracture of the humerus. Clin Orthop Relat Res. 1986;203:258–60.
7. Quinn RH, Randall RL, Benevenia J, Berven SH, Raskin KA. Contemporary management of metastatic bone disease: tips and tools of the trade for general practitioners. J Bone Joint Surg Am. 2013;95(20):1887–95. https://doi.org/10.2106/00004623-201310160-00011.
8. Gainor BJ, Buchert P. Fracture healing in metastatic bone disease. Clin Orthop Relat Res. 1983;178:297–302.
9. Frassica FJ, Frassica DA. Evaluation and treatment of metastases to the humerus. Clin Orthop Relat Res. 2003;(415 Suppl):S212–8. https://doi.org/10.1097/01.blo.0000093052.96273.a7.
10. Piccioli A, Spinelli MS, Maccauro G. Impending fracture: a difficult diagnosis. Injury. 2014;45 Suppl 6:S138–41. https://doi.org/10.1016/j.injury.2014.10.038.
11. Mirels H. Metastatic disease in long bones. A proposed scoring system for diagnosing impending pathologic fractures. Clin Orthop Relat Res. 1989;(249):256–64.
12. Hoban KA, Downie S, Adamson DJA, MacLean JG, Cool P, Jariwala AC. Mirels' score for upper limb metastatic lesions: do we need a different cutoff for recommending prophylactic fixation? JSES Int. 2022;6(4):675–81. https://doi.org/10.1016/j.jseint.2022.03.006.
13. Damron TA, Mann KA. Fracture risk assessment and clinical decision making for patients with metastatic bone disease. J Orthop Res. 2020;38(6):1175–90. https://doi.org/10.1002/jor.24660.
14. Anderson AB, Wedin R, Fabbri N, Boland P, Healey J, Forsberg JA. External validation of PATHFx version 3.0 in patients treated surgically and nonsurgically for symptomatic skeletal metastases. Clin Orthop Relat Res. 2020;478(4):808–18. https://doi.org/10.1097/CORR.0000000000001081.

15. Thio QCBS, Karhade AV, Bindels BJ, et al. Development and internal validation of machine learning algorithms for preoperative survival prediction of extremity metastatic disease. Clin Orthop Relat Res. 2020;478(2):322–33. https://doi.org/10.1097/CORR.0000000000000997.
16. Lin PP, Mirza AN, Lewis VO, et al. Patient survival after surgery for osseous metastases from renal cell carcinoma. J Bone Joint Surg Am. 2007;89(8):1794. https://doi.org/10.2106/JBJS.F.00603.
17. Langerhuizen DWG, Janssen SJ, van der Vliet QMJ, et al. Metastasectomy, intralesional resection, or stabilization only in the treatment of bone metastases from renal cell carcinoma. J Surg Oncol. 2016;114(2):237–45. https://doi.org/10.1002/jso.24284.
18. Fottner A, Szalantzy M, Wirthmann L, et al. Bone metastases from renal cell carcinoma: patient survival after surgical treatment. BMC Musculoskelet Disord. 2010;11:145. https://doi.org/10.1186/1471-2474-11-145.
19. Ratasvuori M, Wedin R, Hansen BH, et al. Prognostic role of en-bloc resection and late onset of bone metastasis in patients with bone-seeking carcinomas of the kidney, breast, lung, and prostate: SSG study on 672 operated skeletal metastases. J Surg Oncol. 2014;110(4):360–5. https://doi.org/10.1002/jso.23654.
20. Zaid HB, Parker WP, Safdar NS, et al. Outcomes following complete surgical metastasectomy for patients with metastatic renal cell carcinoma: a systematic review and meta-analysis. J Urol. 2017;197(1):44–9. https://doi.org/10.1016/j.juro.2016.07.079.
21. Wegener B, Schlemmer M, Stemmler J, Jansson V, Dürr HR, Pietschmann MF. Analysis of orthopedic surgery of bone metastases in breast cancer patients. BMC Musculoskelet Disord. 2012;13:232. https://doi.org/10.1186/1471-2474-13-232.
22. Satcher RL, Lin P, Harun N, Feng L, Moon BS, Lewis VO. Surgical management of appendicular skeletal metastases in thyroid carcinoma. Int J Surg Oncol. 2012;2012:417086. https://doi.org/10.1155/2012/417086.
23. Stopeck AT, Lipton A, Body JJ, et al. Denosumab compared with zoledronic acid for the treatment of bone metastases in patients with advanced breast cancer: a randomized, double-blind study. J Clin Oncol. 2010;28(35):5132–9. https://doi.org/10.1200/JCO.2010.29.7101.
24. Janssen SJ, Teunis T, Hornicek FJ, Bramer JAM, Schwab JH. Outcome of operative treatment of metastatic fractures of the humerus: a systematic review of twenty three clinical studies. Int Orthop. 2015;39(4):735–46. https://doi.org/10.1007/s00264-014-2584-7.
25. Redmond BJ, Biermann JS, Blasier RB. Interlocking intramedullary nailing of pathological fractures of the shaft of the humerus. J Bone Joint Surg Am. 1996;78(6):891–6. https://doi.org/10.2106/00004623-199606000-00012.
26. Pretell J, Rodriguez J, Blanco D, Zafra A, Resines C. Treatment of pathological humeral shaft fractures with intramedullary nailing. A retrospective study. Int Orthop. 2010;34(4):559–63. https://doi.org/10.1007/s00264-009-0833-y.
27. Laitinen M, Nieminen J, Pakarinen TK. Treatment of pathological humerus shaft fractures with intramedullary nails with or without cement fixation. Arch Orthop Trauma Surg. 2011;131(4):503–8. https://doi.org/10.1007/s00402-010-1172-6.
28. Younis M, Barnhill SW, Maguire J, Pretell-Mazzini J. Management of humeral impending or pathological fractures with intramedullary nailing: reaming versus non reaming technique-a retrospective comparative study. Musculoskelet Surg. 2022;106(1):35–41. https://doi.org/10.1007/s12306-020-00668-6.
29. Colello MJ, Hunter MD, Tanner SL, Porter SE. Intramedullary nail fixation for the treatment of pathologic humeral shaft fractures. Orthopedics. 2020;43(5):e389–98. https://doi.org/10.3928/01477447-20200619-08.
30. Fan Y, Zhang B, Guo L, Yao W. Long bone shaft metastasis: a comparative study between cement filling and intercalary prosthesis. World J Surg Oncol. 2023;21:374. https://doi.org/10.1186/s12957-023-03242-z.
31. Abudu A, Carter SR, Grimer RJ. The outcome and functional results of diaphyseal endoprostheses after tumour excision. J Bone Joint Surg Br. 1996;78(4):652–7.

32. Benevenia J, Kirchner R, Patterson F, et al. Outcomes of a modular intercalary endoprosthesis as treatment for segmental defects of the femur, tibia, and humerus. Clin Orthop Relat Res. 2016;474(2):539–48. https://doi.org/10.1007/s11999-015-4588-z.
33. Hennessy DW, Raskin KA, Schwab JH, Lozano-Calderón SA. Endoprosthetic reconstruction of the upper extremity in oncologic surgery. J Am Acad Orthop Surg. 2020;28(8):e319–27. https://doi.org/10.5435/JAAOS-D-19-00219.
34. Ruggieri P, Mavrogenis AF, Bianchi G, Sakellariou VI, Mercuri M, Papagelopoulos PJ. Outcome of the intramedullary diaphyseal segmental defect fixation system for bone tumors. J Surg Oncol. 2011;104(1):83–90. https://doi.org/10.1002/jso.21893.
35. McGrath A, Sewell MD, Hanna SA, et al. Custom endoprosthetic reconstruction for malignant bone disease in the humeral diaphysis. Acta Orthop Belg. 2011;77(2):171–9.
36. Zhao Z, Ye Z, Yan T, Tang X, Guo W, Yang R. Intercalary prosthetic replacement is a reliable solution for metastatic humeral shaft fractures: retrospective, observational study of a single center series. World J Surg Oncol. 2021;19:140. https://doi.org/10.1186/s12957-021-02250-1.
37. Ogink PT, Teunissen FR, Massier JR, Raskin KA, Schwab JH, Lozano-Calderon SA. Allograft reconstruction of the humerus: complications and revision surgery. J Surg Oncol. 2019;119(3):329–35. https://doi.org/10.1002/jso.25309.
38. Rose PS, Shin AY, Bishop AT, Moran SL, Sim FH. Vascularized free fibula transfer for oncologic reconstruction of the humerus. Clin Orthop Relat Res. 2005;438:80–4. https://doi.org/10.1097/01.blo.0000179586.34727.5b.
39. Zoccali C, Attala D, Pugliese M, di Uccio AS, Baldi J. The IlluminOss® photodynamic bone stabilization system for pathological osteolyses and fractures of the humerus: indications, advantages and limits in a series of 12 patients at 24 months of minimum follow-up. BMC Musculoskelet Disord. 2021;22(1):63. https://doi.org/10.1186/s12891-020-03927-6.
40. Terek RM, IlluminOss Study Group, McGough R, et al. IlluminOss photodynamic bone stabilization system improves pain and function in the treatment of humeral metastatic disease. Bone Joint J. 2024;106-B(12):1485–92. https://doi.org/10.1302/0301-620X.106B12.BJJ-2023-1089.R2.
41. Goh MH, Ballatori SE, Gonzalez MR, Lozano-Calderon SA. Management of Impending and/or pathologic fractures of the proximal radius in the setting of metastatic bone disease with an intramedullary photodynamic nail. Tech Hand Up Extrem Surg. 2025; https://doi.org/10.1097/BTH.0000000000000513.
42. Yeung CM, Bhashyam AR, Patel SS, Ortiz-Cruz E, Lozano-Calderón SA. Carbon fiber implants in orthopaedic oncology. J Clin Med. 2022;11(17):4959. https://doi.org/10.3390/jcm11174959.
43. Bhashyam AR, Yeung C, Sodhi A, et al. Titanium vs. carbon fiber-reinforced intramedullary nailing for humeral bone tumors. J Shoulder Elb Surg. 2023;32(11):2286–95. https://doi.org/10.1016/j.jse.2023.04.023.
44. Rijs Z, Weekhout A, Daniel S, et al. Carbon-fibre plates for traumatic and (impending) pathological fracture fixation: where do we stand? A systematic review. J Orthop Traumatol. 2023;24(1):42. https://doi.org/10.1186/s10195-023-00724-4.
45. Hammad A, Ahmed O, Connell PP, Olson D, Balach T. Team approach: management of pathologic fractures. JBJS Rev. 2023;11(1) https://doi.org/10.2106/JBJS.RVW.22.00166.

Management of Metastatic Disease Around the Elbow

18

Niveditta Ramkumar and Dipak B. Ramkumar

Introduction and Epidemiology

The burden of osseous metastatic disease continues to rise, with the increase in the overall incidence of metastatic malignancy [1]. While surgical treatment for metastatic disease is rarely curative, modern advances in systemic and targeted therapies have progressively increased life expectancy in patients with osseous metastatic disease [1]. As a result, surgical treatment, when indicated, needs to be durable and outlive the increasing life expectancies of patients with metastatic bone disease. While the upper extremity, in general, can account for almost 20% of cases of osseous metastatic disease, the distal upper extremities, including the elbow, forearm, wrist, and hand, are an extremely rare site of disease involvement and typically include <1% of the total number of cases [2–4]. As with other sites, the most common histological subtypes present in the upper extremity include lung, breast, prostate, kidney, and thyroid carcinoma, with both lung and renal accounting for the majority of acral metastasis (i.e. osseous involvement distal to the elbow) [5, 6]. Metastatic bone disease in the upper extremities can be an important cause of morbidity and can impart a significant decrease in quality of life. These lesions can cause pain and functional deficits, which in turn can affect activities of daily living, including feeding, hygiene, and overall independence.

N. Ramkumar
Geisel School of Medicine at Darmouth, Lebanon, NH, USA

D. B. Ramkumar (✉)
Dignity Health St. Joseph's Medical Center, Creighton University School of Medicine, Phoenix, AZ, USA

J. Pretell-Mazzini, S. A. Lozano Calderón (eds.), *Multidisciplinary Approach of Musculoskeletal Metastasis*, https://doi.org/10.1007/978-3-032-09498-8_18

Evaluation

Tumors in and around the elbow pose a diagnostic challenge. As with most other conditions, the evaluation of the patient with an elbow lesion begins with a thorough clinical history and physical examination. Most patients presenting with focal elbow lesions usually present with persistent or progressive, often unexplained pain in the affected elbow. This can also be accompanied by focal swelling, non-mechanical pain, and possibly a soft tissue mass. Patients with metastatic disease may also present with additional systemic symptoms, including unexplained weight loss, fevers, chills, or generalized fatigue/malaise. These symptoms are often not associated with any specific antecedent of trauma or inciting event. In these situations, one must maintain a high index of suspicion for an underlying malignancy.

Work-up usually also includes initial diagnostic imaging. This can start with plain film radiographs evaluating the elbow, and the entirety of the humerus and forearm. Patients presenting with lytic lesions with significant cortical erosion, periosteal reactions or other aggressive features, should also have additional advanced, cross-sectional imaging for further work-up [7]. This can include magnetic resonance imaging (MRI), which can help provide additional information regarding the location and size of the lesion, characterizing any associated soft tissue masses, and ascertaining the adjacency and involvement of any associated critical neurovascular structures. In patients with an established diagnosis of osseous metastatic disease and with known, multifocal osseous involvement, further preoperative histopathological tissue sampling may not be necessary. The risk of second malignancy or collision tumor is an exceedingly rare entity, and while this diagnosis should be entertained, the incidence is exceedingly low. In solitary lesions, the risk of primary bone malignancy increases, however, and if these lesions present with aggressive features, appropriate preoperative tissue sampling should be completed. Once a tissue diagnosis has been established, appropriate staging should be completed prior to initiation of any local treatment for the bone lesion. Imaging modalities for staging ultimately depend on the underlying histology and can include computed tomography (CT) of the chest, abdomen pelvis, Tc99m whole body bone scan, low dose whole body CT, or positron emission tomography (PET) CT scans.

Surgical Indications/Nonsurgical Indications

Like in other skeletal locations, the treatment of metastatic bone disease is ultimately targeted at the underlying histology. Typically, available treatment modalities include systemic therapy (i.e., chemotherapy, immunotherapy, and targeted therapy), radiation, and surgical and interventional treatment options. While systemic therapy remains a ubiquitous cornerstone in the management of metastatic carcinomas, the role of surgical treatment is more nuanced. Surgical treatment is often employed in cases of completed or impending pathological fractures. While multiple classification systems exist for predicting the risk of pathological fractures, none are perfect, as most systems are subjective in nature and often have wide inter-rater reliabilities [8]. When an impending fracture is suspected, the general recommendation is to proceed with prophylactic stabilization, as this results in a lower 30-day risk of any major medical complication, increased likelihood of discharge home, shorter hospital stay, and a cost savings of approximately $25,000 per case, compared with fixation after a completed pathological fracture [9].

Mirel's criteria are most commonly used to predict risk of an impending pathological fracture, with mechanical pain (i.e., pain with weight bearing) being the most predictive clinical factor of fracture completion [1]. These criteria are somewhat more challenging to apply in the upper extremity, however, mainly because the upper extremity is non-weight bearing. Further, there are recognized load differences in the upper extremity with force vectors that may convey a divergent fracture risk when compared with the femur [10]. While Evans et al. have demonstrated that Mirels' rating for humerus lesions was both reproducible and valid when comparing the data with patient outcomes, this has not been studied in other upper extremity locations, including for disease in and around the elbow [10]. Additionally, it is also important to consider the forces absorbed by the upper extremity in each individual patients. If patients have concomitant lower extremity lesions that may necessitate the use of a walker or crutches, then the upper extremity effectively becomes partially weight bearing. As such, in these cases, surgical intervention may be preferred, especially if it improves function and allows the patient to appropriately participate and complete activities of daily living.

For the reasons outlined above, the majority of lesions in the upper extremity that have not fractured can usually be managed conservatively. Once again, it is important to account for the location of the lesion, the extent of disease burden, the risk of progression and fracture, completed fracture healing potential, response to systemic and radiation therapy, and overall prognosis when electing to proceed with nonoperative management. Non-operative management can include immobilization of the affected joint or bone as well as external beam radiation. Multiple studies have reported that between 50% and 90% of patients can achieve a period of partial or complete pain relief after palliative radiation [1].

Preoperative Planning

It is important to recall that the distal humerus is the least commonly involved region of the humerus with respect to osseous metastatic disease. It accounts for approximately 7% of all surgical procedures related to metastatic disease, but also has one of the highest revision rates, approaching 33% [11]. In the distal humerus and elbow, the osseous anatomy, adjacency of neurovascular structures, short articular block, and soft tissue envelope all contribute to the challenges of surgical treatment. As such, meticulous preoperative planning is often required in the management of these lesions, including completion of appropriate pre-operative imaging and anticipation of issues related to soft tissue management. Appropriate pre-operative imaging should include plain film radiographs of the elbow in orthogonal planes in addition to cross-sectional imaging, such as magnetic

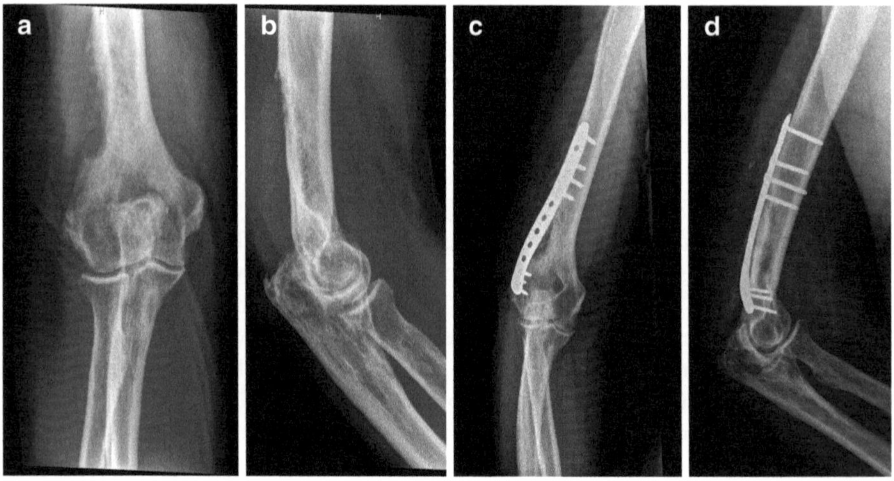

Fig. 18.1 Lytic lesion involving the right distal humerus, with compromise of the posterolateral cortex. (**a**) and (**b**) represent preoperative AP and lateral radiographs, respectively; (**c**) and (**d**) demonstrate post-operative AP and lateral images after curettage, cementoplasty, and prophylactic plate fixation have been performed

resonance imaging (MRI) or CT scans of the elbow, to assess for the extent of bone destruction, the presence of an extra-osseous soft tissue component, and to better ascertain the relationship of critical neurovascular structures to any extra-osseous soft tissue masses (Fig. 18.1). In addition, as delineated above, a thorough neurovascular examination with assessment and documentation of distal neurovascular function is imperative. Additionally, one must also account for the histology of the underlying lesion and its relative sensitivity to adjuvant systemic therapy and radiation. Lesions that are less sensitive to both radiation and systemic therapy may benefit from additional local control treatment, which could include intra-lesional curettage or even surgical resection and reconstruction of the involved area of bone. This can directly impact surgical reconstruction options and must be accounted for ahead of time. Further, if considering intralesional treatment and the histology dictates (i.e., renal cell carcinoma, thyroid carcinoma), consideration of pre-operative embolization would be prudent to help minimize blood loss and decrease the risk of perioperative complications.

Surgical Treatment Options

The surgical treatment options available for the treatment of metastatic disease in the elbow include the following: plate and screw fixation, modular megaprosthesis/distal humeral arthroplasty, and reconstruction with bulk or osteochondral allografts. The goal of surgical treatment is to attain rigid and durable internal fixation and to allow for immediate postoperative use of the upper extremity [12].

Internal Fixation with or without Intralesional Curettage

For lesions that preserve the articular surface of the distal humerus and have sufficient bone stock for plate/screw fixation, internal fixation may be sufficient for both impending and complete fractures [8]. Generally, lesions within 20–40 mm of the distal articular surface of the humerus are best managed with plate fixation. In this region, neither antegrade nor retrograde nails can provide rigid enough fixation, unless also combined with plate fixation [12]. As noted above, some histological subtypes are less sensitive to systemic therapy and radiation therapy options, and as a result, these lesions can also be treated with intra-lesional curettage, cementation, and plate fixation. Further, radiofrequency ablation, an emerging technique, can also be used to achieve additional palliative pain control by thermally necrosing the lesion and helping achieve some degree of local control of the disease [13–16]. Generally, an expansile approach to the distal humerus is recommended, and this could entail a posterior triceps splitting approach or a paratricipital approach (Alonso-Llames) can be undertaken. Occasionally, an olecranon osteotomy may also be necessary to access more distal lesions, adjacent to the articular block, if internal fixation is selected as the surgical treatment option (Fig. 18.1).

Modular Endoprostheses

Modular endoprostheses are generally favored for reconstruction of large bone defects about the elbow joint [17]. Generally, large, uncontained lesions involving the articular block of the distal humerus, or that exhibit intra-articular extension, can be treated more effectively with distal humeral arthroplasty after resection of the tumor [18]. Multiple studies have reviewed functional results after use of these prostheses, and in general, results have been favorable [19–21]. In one retrospective review, Capanna et al. reported an average MSTS score of 22.9 out of 30 in the population of patients undergoing distal humeral megaprosthetic reconstruction, with a complication rate of 16.7% [21]. The main complications were nerve injury, mechanical failure of the articular prosthesis component, and one deep infection. One common theme among most studies is the somewhat high complication rate—ranging from 15% to 25% with common complications including nerve injury, aseptic loosening, vascular injury, and deep infection.

Surgical Technique

Internal Fixation

The approach to internal fixation at the elbow ultimately depends on the location of the lesion being treated and the planned surgical construct. For most plating constructs, we recommend using a posterior approach, which allows for excellent visualization of the entire posterior surface of the distal humerus and elbow joint. Further, the posterior approach allows for direct identification of the radial nerve, when necessary, and is also extensile, allowing for access to the entirety of the humerus, if required. Intra-articular lesions can also be approached in this manner. In rare cases, where visualization of the articular surface is necessary, an olecranon osteotomy can also be performed.

Patient Positioning

When pursuing a posterior approach to the elbow, we typically position patients in the lateral decubitus position. The patient is usually secured in this position with a bean bag positioner or using posts and a pegboard. The operative extremity is then draped over a radiolucent bolster, allowing for elbow flexion over the bolster. The non-operative extremity is then positioned on an arm board. Fluoroscopy is usually brought in parallel to the bed, allowing unimpeded access to the elbow in the antero-posterior plane. Generally, we rotate the upper extremity to visualize the lateral plane. See Fig. 18.2.

Surgical Approach

As delineated above, we typically undertake a posterior approach to the distal humerus and elbow. This approach has multiple advantages, including excellent visualization of the distal humerus and elbow joint, and for identifying and protecting the radial nerve. The approach starts with a posterior midline longitudinal incision that crosses the extension surface of the elbow in an S-shaped pattern. The distal extent of the incision usually ends along the proximal ulna. The superficial dissection should protect the posterior antebrachial cutaneous nerve and other cutaneous branches. Once the fascia is incised in line with the incision, the deeper musculature, including the triceps muscles and aponeurosis, is exposed. Several different deeper dissections can be undertaken, including the triceps-splitting approach, the triceps-reflecting (or Bryan-Morrey) approach, or the paratricipital/triceps-sparing approach. As the name implies, the triceps splitting approach splits the midline of the triceps distally and provides

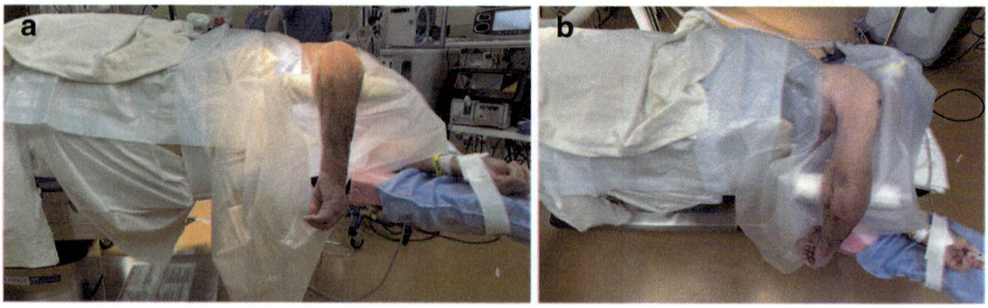

Fig. 18.2 Overview of patient positioning for a posterior approach to the distal humerus and elbow. As demonstrated, the patient is positioned in the lateral decubitus position and secured in this position with a bean bag positioner (**a**). The operative extremity is then draped over a radiolucent roll to allow for adequate flexion of the elbow (**b**). The non-operative upper extremity is positioned on an arm board to allow for unimpeded access to the operative site, with fluoroscopy

direct access to the distal humerus. This can, however, lead to potential triceps functional weakness post-operatively. The triceps-reflecting approach detaches the triceps tendon from the olecranon, allowing for its reflection proximally. This approach again offers excellent visualization of the distal articular surface of the humerus and the elbow, but necessitates repair of the triceps insertion upon closure, and may ultimately necessitate a period of immobilization or limited mobilization through the elbow to protect the repair. Lastly, the triceps sparing approach allows for the creation of medial and lateral windows around the triceps, leaving the triceps completely intact and contiguous. This approach is beneficial when triceps integrity is critical, i.e., after endoprosthetic elbow reconstruction.

Intra-lesional Curettage

Intra-lesional curettage is typically an essential component in the treatment of impending pathological fracture or completed pathological fracture in osseous metastatic disease, and perhaps more critical in histologies that are generally resistant to either radiation therapy or systemic therapy. Curettage allows for improved local control by intralesionally decreasing tumor burden. Access to the lesion is straightforward in cases of a completed pathological fracture or an impending pathological fracture with significant cortical bone loss. We typically use curettes of both straight and angled varieties to access the lesion and perform the curettage. This should be done systematically to ensure that the entirety of the lesion has been accessed and removed. Direct and fluoroscopic visualization help provide complete curettage of the lesion. At our institution, we prefer to separate the curettage specimen into two aliquots—one containing osseous fragments and the other primarily soft tissue. The specimen containing osseous fragments is usually submitted for decalcification by our pathology department. The decalcification process can cause significant denaturing of nucleic acids, and thus, this sample is generally not appropriate for molecular diagnostics, including next-generation sequencing. This is why we submit a separate "soft tissue" specimen, bypassing the decalcification process and allowing for fixation without significant compromise to genetic material.

In cases where the cortical structure and continuity of the bone remain intact, but operative intervention is still indicated for impending pathological fracture risk, the lesional tissue can be accessed through a small cortical window. Depending on the overall integrity of the bone, this can be completed by simply using a straight curette to enter the lesion, or, in our preferred approach, using a small pencil tip or router tip burr to help create a more controlled cortical window. The window should be large enough to allow for the passage of multiple curettes and ensure successful and complete intralesional curettage. In situations where a cementoplasty is also planned, ensuring that the cortical window is large enough to pack the cement material manually is prudent.

Implant Selection

In the setting of a completed or impending pathological fracture involving the elbow, implant selection is guided by several key factors, including underlying histology, extent of osseous destruction, anticipated biological response to adjuvant therapies (i.e., radiation therapy or systemic therapy), and the patient's overall functional status and life expectancy. A myriad of intramedullary nail and plate/screw constructs are commercially available, and most of these constructs offer both standard and locking screw fixation. Peri-articular, pre-contoured plates are also plentiful and often offer variable-angle locking options, which can be helpful for periarticular completed or impending fractures. Ultimately, hardware selection comes down to the surgeon's choice and comfort level, if it still helps address the underlying tumor and compromised bone.

Modular Endoprosthesis

Patient Positioning

For modular endoprosthetic reconstruction procedures, we typically prefer to utilize an anterior approach. This allows for excellent exposure of critical neurovascular structures and their protection during deep dissection and tumor resection. As such, we place the patient in a supine position with the operative extremity abducted and placed on a padded radiolucent table.

Surgical Approach

Generally, we employ an anterior approach to the elbow and distal humerus. An S-shaped incision beginning in the middle of the arm, along the medial border of the biceps muscle, extending distally to the level of the antecubital fossa, is utilized. At the distal extent, the incision curves laterally along the elbow's volar aspect, towards the brachioradialis muscle's volar margin. Ideally, full-thickness flaps are raised, protecting both the medial and lateral antebrachial cutaneous nerves. Once exposed, we prefer to move to the more proximal aspect of the incision and approach and develop the interval between the biceps and triceps musculature. Generally, in this area, more native tissue planes are preserved, allowing for easier identification of the neurovascular structures. With these structures identified, the overlying investing fascia is opened in a proximal-to-distal direction through the entire course of the exposure. The brachial artery, median and ulnar nerves, and the medial antebrachial cutaneous nerves are dissected individually and isolated with a vessel loop proximally. The neurovascular structures are then meticulously dissected off the underlying tumor and humerus. Once complete, attention is turned to the isolation of the biceps muscle. Usually, the biceps is not involved with the tumor, and as such can be preserved. The biceps or the brachialis muscle must be maintained to allow elbow flexion. Next, the pronator teres and common flexor muscles are released from their medial origins on the distal humerus. Similarly, the common extensor muscles are also released from the lateral origins on the distal humerus. Care must be taken to preserve the neural innervation, including branches of the median nerve on the flexor side and the posterior interosseous nerve and its branches on the lateral side, should be preserved where feasible.

Attention is then turned to muscular dissection—anteriorly, the brachialis muscle is elevated off the distal humerus, if there is no soft tissue involvement.

Posteriorly, the triceps is elevated off the distal humerus. As noted, on occasions with soft tissue infiltration, the brachialis and the short head of the triceps may need to be resected, depending on the extent of the tumor. Care must be taken to ensure that the triceps insertion on the olecranon is preserved, and the olecranon itself is not osteotomized. Next, the elbow joint is opened anteriorly, and the capsule is released from the distal humerus, radial head, and ulna. This allows for disarticulation of the elbow joint. The humerus is then osteotomized proximally at the appropriate level based on the extent of the tumor. At this point, preparation on the humeral and ulnar side can continue, as outlined by the specific prosthesis being utilized.

Implant Selection

Generally, modular endoprostheses are recommended. Depending on the extent of osseous resection and soft tissue tensioning, these can often be customized to the desired length. Occasionally, the length may need to be shortened by one to two centimeters to help facilitate soft tissue closure. The stems of the humerus and ulnar components are usually cemented, and preoperative planning to ascertain the maximal stem diameter and length is helpful. Generally, we cement both stems to maximize fixation at the time of implantation. Reaming on both humeral and ulnar sides is also completed with hand reamers. Care must be taken to ensure that the ulnar implant is placed as deep as possible within the olecranon fossa without damaging the posterior cortex of the bone. We also recommend the utilization of linked prostheses to maximize stability.

Postoperative Care

The post-operative care of patients undergoing surgical treatment of metastatic disease about the elbow is critical. After operative fixation, in our practice, the elbow and forearm are usually splinted in a long posterior slab splint for soft tissue rest and protection, to allow for adequate healing of the skin incisions. Typically, we keep patients in a splint for at least 2 weeks postoperatively, at which point we transition patients into a hinged elbow brace with free range of motion from full extension to 120 degrees of flexion, as tolerated. Patients are encouraged to mobilize the elbow as tolerated and are often referred to physical and occupational therapy to aid with range of motion, strengthening, and edema control. Once patients can consistently demonstrate adequate control in flexion/extension arcs, they are weaned out of the hinged elbow brace and are allowed to use the elbow and upper extremity as tolerated.

For patients undergoing megaprosthetic reconstruction, generally, the post-operative care is more conservative. Early in the post-operative period, edema control and skin/soft tissue rest are critical, especially if additional soft tissue reconstructions are employed to ensure adequate coverage of the prosthesis. Patients are usually splinted with a bulky dressing and a posterior slab splint that maintains the elbow in approximately 60 degrees of flexion. The extremity is elevated for at least the first 3–4 days after surgery to aid with edema control. Any drains and nerve catheters that are employed for perioperative seroma and pain control, respectively, are usually discontinued during this time frame. In our practice, we have patients undergo dressing and splint change on post-operative day #4, to allow for assessment of the surgical wound after superficial re-epithelialization should have already occurred. The elbow is then re-splinted at 60 degrees of flexion. Generally, the elbow is immobilized in a splint for at least 6 weeks post-operatively to allow for adequate soft tissue healing and scarring. During this time, elbow range of motion is strictly prohibited. At the 6-week mark, the patient is transitioned to a hinged elbow brace and permitted to advance elbow motion from 30 to 130 degrees of flexion through active, active-assisted, and passive range of motion, under the supervision of a physical or occupational therapist. At 12 weeks postoperatively, the brace is then adjusted to allow for a full range of motion of the elbow and strengthening of the muscle groups in and around the elbow are initiated at this time. Weight-bearing restriction of a maximum of two pounds is also introduced at the 12-week mark. We continue use of the hinged elbow brace for at least 6 more weeks thereafter. At approximately week 18 after surgery, the patient is transitioned out of the elbow brace and can use a sling as needed for comfort. At this point, resistance strengthening is begun, and the patient's weight restriction is advanced to 5 pounds. Ultimately, we continue a 5-pound weight restriction for the rest of the patient's life.

References

1. Johnson CN, Gurich RW Jr, Pavey GJ, Thompson MJ. Contemporary management of appendicular skeletal metastasis by primary tumor type. JAAOS J Am Acad Orthop Surg. 2019;27(10):345–55.
2. Savvidou OD, Koutsouradis P, Chloros GD, et al. Bone tumours around the elbow: a rare entity. EFORT Open Rev. 2019;4(4):133–42.
3. Kendal JK, Abbott A, Kooner S, Johal H, Puloski SK, Monument MJ. A scoping review on the surgical management of metastatic bone disease of the extremities. BMC Musculoskelet Disord. 2018;19:1–10.
4. Muramatsu K, Ihara K, Iwanagaa R, Taguchi T. Treatment of metastatic bone lesions in the upper extremity: indications for surgery. Orthopedics. 2010;33(11):1–7.
5. Mavrogenis AF, Mimidis G, Kokkalis ZT, et al. Acrometastases. Eur J Orthop Surg Traumatol. 2014;24:279–83.
6. Stomeo D, Tulli A, Ziranu A, Perisano C, De Santis V, Maccauro G. Acrometastasis: a literature review. Eur Rev Med Pharmacol Sci. 2015;19(15):2906–15.
7. Miller TT. Bone tumors and tumorlike conditions: analysis with conventional radiography. Radiology. 2008;246(3):662–74.
8. Voskuil RT, Mayerson JL, Scharschmidt TJ. Management of metastatic disease of the upper extremity. JAAOS J Am Acad Orthop Surg. 2021;29(3):e116–25.
9. El Abiad JM, Raad M, Puvanesarajah V, Rao SS, Morris CD, Levin AS. Prophylactic versus postfracture stabilization for metastatic lesions of the long bones: a comparison of 30-day postoperative outcomes. JAAOS J Am Acad Orthop Surg. 2019;27(15):e709–16.
10. Evans AR, Bottros J, Grant W, Chen BY, Damron TA. Mirels' rating for humerus lesions is both reproducible and valid. Clin Orthop Relat Res. 2008;466(6):1279–84.
11. Wedin R, Hansen BH, Laitinen M, et al. Complications and survival after surgical treatment of 214 metastatic lesions of the humerus. J Shoulder Elb Surg. 2012;21(8):1049–55.

12. Frassica FJ, Frassica DA. Metastatic bone disease of the humerus. JAAOS J Am Acad Orthop Surg. 2003;11(4):282–8.
13. Levy J, Hopkins T, Morris J, et al. Radiofrequency ablation for the palliative treatment of bone metastases: outcomes from the multicenter OsteoCool Tumor Ablation Post-Market Study (OPuS One Study) in 100 patients. J Vasc Interv Radiol. 2020;31(11):1745–52.
14. Thanos L, Mylona S, Galani P, et al. Radiofrequency ablation of osseous metastases for the palliation of pain. Skeletal Radiol. 2008;37:189–94.
15. Deschamps F, Farouil G, Ternes N, et al. Thermal ablation techniques: a curative treatment of bone metastases in selected patients? Eur Radiol. 2014;24:1971–80.
16. Gennaro N, Sconfienza LM, Ambrogi F, Boveri S, Lanza E. Thermal ablation to relieve pain from metastatic bone disease: a systematic review. Skeletal Radiol. 2019;48:1161–9.
17. Hennessy DW, Raskin KA, Schwab JH, Lozano-Calderón SA. Endoprosthetic reconstruction of the upper extremity in oncologic surgery. JAAOS J Am Acad Orthop Surg. 2020;28(8):e319–27.
18. Funovics PT, Schuh R, Adams SB Jr, Sabeti-Aschraf M, Dominkus M, Kotz RI. Modular prosthetic reconstruction of major bone defects of the distal end of the humerus. JBJS. 2011;93(11):1064–74.
19. Hanna S, David L, Aston W, et al. Endoprosthetic replacement of the distal humerus following resection of bone tumours. J Bone Joint Surg Br Vol. 2007;89(11):1498–503.
20. Tang X, Guo W, Yang R, Tang S, Yang Y. Custom-made prosthesis replacement for reconstruction of elbow after tumor resection. J Shoulder Elb Surg. 2009;18(5):796–803.
21. Capanna R, Muratori F, Campo FR, et al. Modular megaprosthesis reconstruction for oncological and non-oncological resection of the elbow joint. Injury. 2016;47:S78–83.

Management of Metastatic Disease Around the Forearm, Wrist, and Hand

19

Melinda J. Choi and Seth D. Dodds

Epidemiology

Bone metastases to the hand and wrist are rare, accounting for 0.1–0.42% of all bony metastases, with a reported 6.1% of malignant hand tumors attributed to metastatic disease [1, 2]. The relative rarity of hand metastases is attributed to the small amount of bone marrow in the small bones of the hand [2]. However, underdiagnosis is likely as staging may not examine the distal extremities or may be missed, especially if patients are asymptomatic [3–5]. Some believe the development of acrometastasis to be initiated by trauma, which can trigger an inflammatory pathway, attracting tumor cells to the area with subsequent adherence and growth [6–8]. Common primary cancers resulting in bony metastases to the upper extremity include skin, renal, breast, prostate, and lung cancer, with lung cancer being the most common primary tumor [9–11]. Typically, the appearance of bony metastases in the hand, wrist, or forearm is a poor prognostic sign and is associated with more advanced disease, with a median patient survival of approximately 6 months after the appearance of the metastases [1, 3, 6, 9, 12]. While acrometastases may develop at any age, they commonly present in the middle to late decades, directly correlating with the age of cancer diagnoses.

Metastatic disease in the hand has been observed everywhere between the distal phalanx and carpus, *though the distal phalanx is the most commonly involved bone*, followed by the metacarpal and the proximal phalanx [2, 3, 6]. The predilection of tumor cells to the distal phalanx is thought to be related to the dense capillary network present at the nail matrix [10]. Interestingly, adjacent articular surfaces tend to

M. J. Choi
Divison of Plastic Surgery, University of Miami, Miami, FL, USA
e-mail: mjc316@miami.edu

S. D. Dodds (✉)
Department of Orthopaedics, University of Miami, Miami, FL, USA
e-mail: sdodds@med.miami.edu

J. Pretell-Mazzini, S. A. Lozano Calderón (eds.), *Multidisciplinary Approach of Musculoskeletal Metastasis*, https://doi.org/10.1007/978-3-032-09498-8_19

be spared from the metastatic lesion [2]. Metastases to the hand or forearm may initially present in the soft tissue with subsequent spread or invasion to underlying bone [11]. Men are more likely to develop acrometastasis, thought to be due to the higher prevalence of smoking-related lung cancer in men [6]. Patients may present with pain, swelling, and occasionally occult fracture [1]. Metastases to the hand have been mistaken for felon, infections, and gout [10]. Additionally, the symptoms associated with upper extremity metastases may lead to the discovery of the primary cancer or even be the first indicator of metastatic disease [3, 7, 12]. The work-up of a hand lesion has led to the diagnosis of malignancy in a reported 16–36% of patients [1, 2, 10].

Surgical Indications/Nonsurgical Indications

Patient selection for surgery is made on an individual basis. The primary goal of surgery in these cases is typically palliative, aiming to improve debilitating pain and preserve upper extremity function. Additional indications may include large, fungating tumors involving the soft tissue and/or chronic infection [13]. Given the typical short survival time upon discovery of bony metastases in these locations, care should be taken to reduce any post-operative morbidity and minimize risks of post-operative complications. In cases of renal cell carcinoma, consideration may be taken towards wide local excision of a solitary metastatic lesion, as it has been demonstrated to improve long-term survival [14, 15].

Patients with a significant tumor burden bilaterally may not be eligible for surgical intervention given the potential loss of bilateral upper extremity function with surgical resection. Patients who are unable to tolerate excision, even without reconstruction, due to significant deconditioning or disease should consider nonsurgical management if tolerated. Use of bisphosphonates, specifically for breast cancer or multiple myeloma, or anti-RANKL antibody (denosumab) may mitigate the progression of bone lysis [16, 17]. Additional nonsurgical interventions for upper extremity metastases include radiotherapy, chemotherapy, immunotherapy, or medical pain management [10]. These therapies are more frequently employed for more proximal lesions [10]. Radiotherapy has been demonstrated to improve pain associated with lesions, resulting in an improvement in function [6]. Failure of radiotherapy or medical pain management, however, may necessitate surgical intervention [18].

Preoperative Planning

When preparing for surgical intervention on a patient with hand, wrist, or forearm bony metastasis, an appropriate work-up, including imaging and labs, should be performed. Imaging, including X-ray, CT, MRI, and bone scans, may identify the lesion(s) of interest. Hand and upper extremity metastasis will typically present as a lytic lesion on radiographs, though blastic or mixed lesions may also be seen (Figs. 19.1 and 19.2) [7, 8, 10]. However, radiographs may only reveal lesions that

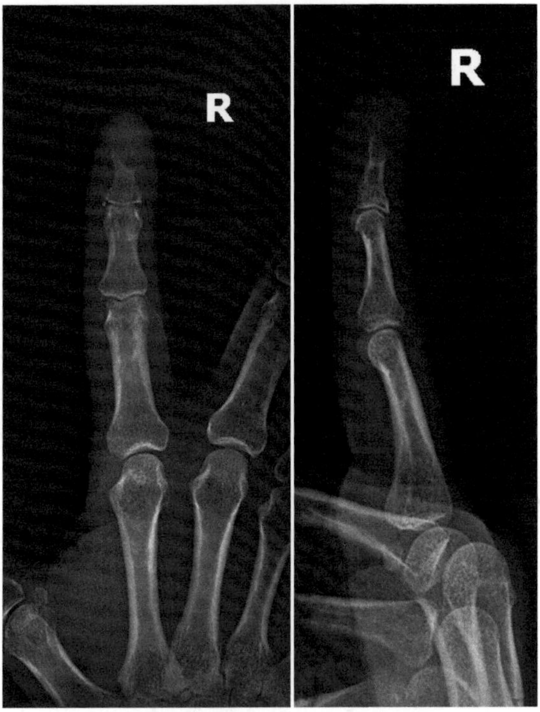

Fig. 19.1 A 64-year-old left-hand dominant female presented with a 2-month history of right index finger pain at the distal phalanx. Radiographs demonstrate the characteristic lytic lesion seen in distal phalanx metastasis

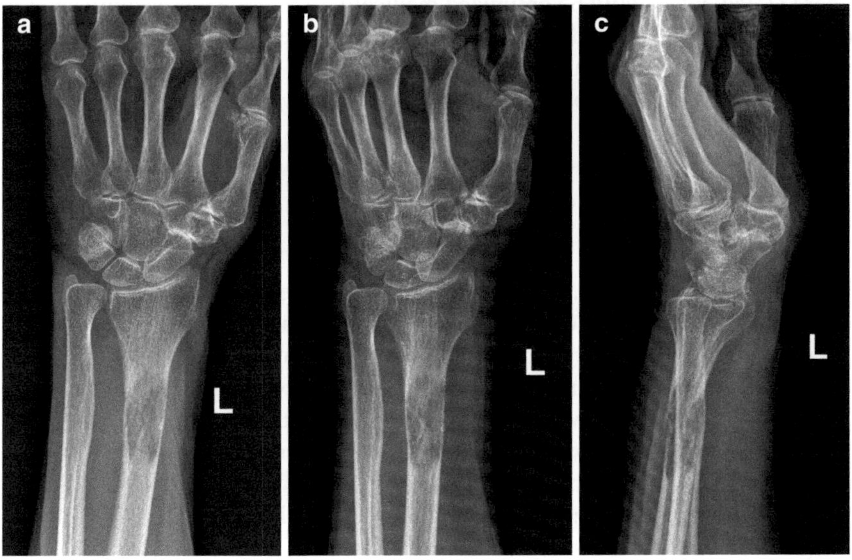

Fig. 19.2 A 76-year-old right-hand dominant female presented with a 5-week history of severe left distal forearm pain. She had a known history of renal cell carcinoma diagnosed 25 years prior to presentation. Radiographs demonstrate a characteristic lytic lesion at the distal third of the radius on the (**a**) posteroanterior and (**b**) oblique views. A pathologic fracture of the dorsal cortex is seen on the (**c**) lateral view

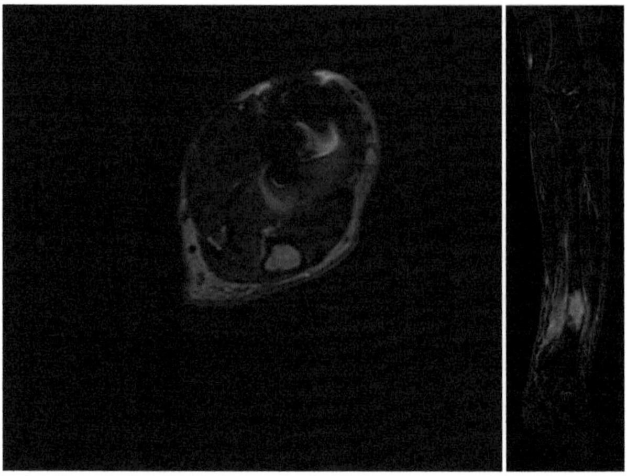

Fig. 19.3 Right upper extremity MRI of the 76-year-old patient demonstrates an intramedullary lesion with cortical breakthrough and extension to the soft tissue. The pronator quadratus was noted to be involved with the lesion intra-operatively

have replaced over half of the cancellous bone, making an MRI more effective for early detection [4]. MRI has a high sensitivity and specificity for the identification of metastases [1, 5, 10]. On MRI, lesions can be seen with associated bony destruction and any additional soft tissue involvement (Fig. 19.3) [8]. PET scans have also demonstrated great sensitivity and may reveal distant, asymptomatic metastases earlier [12]. The resolution of CT scans can be limited.

Definitive diagnosis is performed with a biopsy (CT guided or open) to confirm the diagnosis of metastatic disease and verify that the lesion is not a new primary tumor [8]. Patients should be optimized medically to tolerate a procedure and minimize postoperative morbidity. While most procedures may be performed under tourniquet control to minimize blood loss in an already anemic population, surgeons should aim for a platelet count >50,000 μL and hemoglobin >7 g/dL prior to surgical intervention.

Surgical Options

The surgical plan for hand, wrist, or forearm bony metastasis initially begins with excision of the metastatic lesion. Treatment options include intralesional curettage with or without adjuvant treatment, wide local excision, or amputation [3, 4, 19]. Amputation is less commonly performed in more proximal lesions [10]. However, a study of three patients with metastases to the radius who underwent palliative transhumeral amputation demonstrated reduced pain in all patients, no local recurrence, and the ability to obtain a prosthesis in two of the patients [13]. Surgeons should consider performing an amputation under regional anesthesia, if possible, for significantly ill patients unable to tolerate general anesthesia [13]. Specifically for renal cancer, *en bloc* excision of a solitary or oligometastatic lesion may be associated with increased survival [1].

Amputation is the frequently preferred treatment method for acrometastasis, especially if involving the distal phalanx [3]. Amputation through the next proximal joint level will rapidly relieve pain and preserve as much hand function as possible

[10]. Additionally, amputation offers a relatively short recovery time that may be optimal for ill or debilitated patients. For proximal or cutaneous, isolated lesions, local excision may be more appropriate to minimize morbidity and again, optimize as short of a healing time as possible [3]. Local excision may be combined with radiotherapy, though caution should be taken with regards to the development of wounds associated with radiation.

Reconstruction may or may not be indicated depending on the level of surgery and the patient's prognosis. For patients undergoing palliative excisions, reconstruction will subject patients to greater surgical risks, longer surgeries, or longer recovery periods, which may not outweigh the severity of illness or a patient's limited life expectancy [13]. However, for patients with less aggressive primary tumors and better prognosis, reconstruction may be necessary to maintain upper extremity function and stability. Targeted muscle reinnervation (TMR) or regenerative peripheral nerve interface (RPNI) could be considered for patients undergoing more proximal upper extremity amputations to augment pain management. These procedures may add between 30 min to a few hours of additional operative time, depending upon the familiarity of the surgeon. Depending on the patient, this additional operative time may be tolerated to minimize the risk of formation of painful neuromas. Simple forms of reconstruction such as skin grafting over excision of involved soft tissue at the web space is feasible and does not add significant operative time or morbidity [2]. Additionally, local flaps such as a V-Y advancement may be performed for wide local excision of larger metastases [19].

Specifically in the forearm, the resected segmental defect may be reconstructed with autograft, allograft, or poly methyl methacrylate (PMMA). Risks associated with the use of allograft include nonunion, fracture, and infection [20]. The use of vascularized autograft to reconstruct large segmental defects, in particular a free vascularized fibula requiring microvascular anastomosis, is typically not advised for these patients given the morbidity associated with extensive surgery and the required time for union potentially exceeding the relatively short lifespan of these immunocompromised patients [20]. Some advocate for use of cemented intercalary prostheses to reconstruct segmental defects, as it allows for early range of motion and weight bearing, thus improving quality of life [20].

In cases of bony metastases to the carpal bones, surgical options are again individualized. In some cases, excision of the singular carpal bone may be sufficient [21, 22]. Adjunctive procedures such as four-corner fusion might not be attempted due to the additional surgical morbidity and desire for quick recovery [22]. However, additional procedures to treat pain, such as wrist denervation, fall under the palliative goals of surgical intervention and do not add any additional recovery time.

Contraindications

Patients unwilling to undergo amputation, unable to tolerate surgery, or those insistent upon limb-sparing procedures may not be eligible for surgery depending on the extent of metastatic disease. Limb or digit-sparing procedures may not be appropriate for tumors that have invaded the soft tissue. In these cases, patients should be counseled that amputation may be the best option to improve pain, retain as much function as possible, and prevent the development of any future, time-consuming infections or wound care.

Complications

Immediate complications of amputation include infection, wound dehiscence, or incomplete resection. Persistent pain may still arise, especially if the metastatic lesion recurs. Neuromas and phantom limb pain may present after amputation if peripheral nerves are not managed at the time of amputation. Reconstructive efforts using allograft or autograft are at risk for infection, non-union, and fracture. The risks of these complications should be weighed for each patient, as not every patient may be able to tolerate revision or salvage procedures. Additional complications include incomplete excision or local recurrence/progression of metastatic disease.

Surgical Technique

Patient Positioning and Surgical Incision

An arm table is secured to the operating table. A tourniquet is placed on the upper arm. The arm may be exsanguinated by gravity, but is traditionally not exsanguinated with an Esmarch or elastic wrap due to the theoretical risk of forcing malignant cells into adjacent, clear tissue [23]. The surgical incision varies based on the location of the lesion. For amputations of the digits, a traditional fish-mouth incision may be designed to provide adequate soft tissue coverage. Amputations through the level of the metacarpophalangeal joint are traditionally designed with a curvilinear incision on the volar palm and a Y-incision dorsally, which creates a dorsal skin flap for web-space coverage. At the forearm, a fish mouth incision can be used, with skin flaps designed 2 cm distal to the planned bone cut to ensure adequate soft tissue coverage. For the carpal bones, a dorsal longitudinal incision ulnar to Lister's tubercle is typically made, which will allow for wide exposure. To access the distal radius or radial shaft, a longitudinal incision is designed with a volar FCR approach, or a dorsal approach centralized over Lister's tubercle, depending on the location of the tumor. At the more proximal radius and ulna, incisions are planned longitudinally or curvilinear for adequate exposure. A volar or dorsal approach may be used depending on the exact location of the metastases or to avoid critical neurovascular structures.

Steps for Tumor Resection

For digital amputations, the skin and soft tissue are incised. The flexor and extensor tendons are transected and allowed to retract into the forearm. Any remaining soft tissue and periosteum are elevated off the bone. When amputating through a joint, the distal cartilage at the remaining, proximal phalanx should be debrided with a rongeur to prevent continued production of synovial fluid [23]. When amputating through the bone, a small oscillating saw is used, along with rongeurs, to ensure that the remaining stump does not have sharp edges.

For border digits including the index finger and small finger, a modified ray resection may be preferable to disarticulation to eliminate the metacarpal head prominence while preserving the width of the hand [23]. Soft tissue is elevated circumferentially off the metacarpal, and the interossei are elevated off the metacarpal. Transmetacarpal ligaments are divided and then later repaired for the third and fourth metacarpals. Using an oscillating saw, the metacarpal osteotomy should be angled 45 degrees volarly to maintain the natural contour of the hand. Digital nerves should be identified and treated to prevent neuroma formation.

In cases of amputation at the level of the metacarpals for the index and middle fingers, the metacarpal base should be preserved to maintain ECRL and ECRB attachments respectively. Central digits are typically approached with a complete ray amputation. For middle and ring finger amputations, it is recommended to transpose the adjacent ray border to close down the dead space [23]. Overall, at the level of the metacarpal, the associated neurovascular bundle should be preserved, as the common vessels will supply adjacent digits.

For lesions involving the carpus, wide local excision is suitable. The carpal bone of interest is excised. Particularly invasive lesions may require larger *en bloc* resection, involving resection of surrounding tendons, wrist capsule, or adjacent carpal bones that may be involved [5]. Margins will ideally be disease-free [5]. Forearm amputations are treated similarly, incising skin and soft tissue, with structures divided systematically by layers. In cases of segmental resection involving the radius or ulna shaft, dissection proceeds either to the level of the bone where the lesion is or in a wide approach if soft tissue is involved. The radius or ulna is then exposed circumferentially, with dissection extending 2–3 cm both proximal and distal to the metastatic lesion. A sagittal saw is used to perform *en bloc* resection, with care taken to protect underlying deep structures. Frozen sections may be sent intra-operatively to confirm negative margins [24]. Alternatively, for a wide local excision, a corticotomy may be performed and the bone lesion can be curetted. In cases where there is already a pathological fracture, a volar-based radial shaft plate is placed and secured with screws proximally and distally to maintain the radius in a reduced position prior to bone excision. During dissection, hemostasis should be a priority to allow easier dissection. Care should be taken during dissection around peripheral nerves. In cases of amputation, nerves are sharply divided and are either buried into adjacent muscle or reconstructed with TMR or RPNI to prevent neuroma formation.

Reconstruction

For amputations, careful design of the skin incision will avoid the need for extensive soft tissue reconstruction and ensure adequate coverage. In amputations requiring nerve transection, care should be taken to attempt to mitigate the formation of neuroma and additional stump pain. Methods such as nerve implantation into muscle, RPNI, or TMR in more proximal amputations may be employed. For implantation into muscle, the distal end of the nerve is dissected free and buried into the adjacent muscle. The muscle can be secured loosely around the nerve with a 4-0 Prolene stitch. In RPNI, a 1 cm × 1 cm muscle graft is harvested and wrapped loosely around the end of the nerve with a 6-0 Prolene stitch.

When *en bloc* resection at the carpus is performed, the remaining space is typically left open without placement of graft. Even with the bony defect, patients may not progress to carpal instability after resection [5]. Additionally, invasive lesions may require excision of the extensor retinaculum. Care should be taken to repair the extensor retinaculum to prevent tendon subluxation [5]. Intercarpal fixation or arthrodesis may not be warranted depending on the patient's activity level, and should be evaluated on a case-by-case basis, given additional healing time required and any additional revision procedures that may arise [5].

For resections involving the distal radius, techniques for reconstruction include wrist arthrodesis, allograft, autograft, or prosthetic replacement. Total wrist arthrodesis will resolve pain and aims to preserve the patient's grip strength [15]. The disadvantage of this approach is the potential need for allograft or autograft to bridge the bony gap. Autograft and allograft may be used alone for reconstruction, as well [25]. Use of allograft is quick, requires no donor site, and can completely fill any bony defect. Disadvantages of allograft include fracture, non-union, and resorption [15, 26, 27]. Specifically in the distal radius, however, the risk of allograft-related infection is rare given its small area [26]. A locking compression plate to span the allograft may mitigate the risk of fracture (Fig. 19.4). The plate is placed on the volar radial shaft and contoured to fit the distal radius. A combination of non-locking and locking screws is used, with locking screws placed distally. The distal radial ulnar joint (DRUJ) should be evaluated for stability after fixation, pinning as needed.

In the case of autograft, this frequently requires a donor site for harvest, which may not be feasible for deconditioned patients [26]. Cadaveric autograft may be used instead as a strut graft for large defects of the radius or ulna. Additional risks include non-union and fracture of the graft [15]. Autograft may be taken from the fibula, iliac crest, or adjacent distal ulna. The fibula head can be used to reconstruct the articular surface of the distal radius defect. This may be harvested as a vascularized graft for large defects or as a non-vascularized graft, though this has associated risks of non-union, resorption, and collapse [15, 26]. Disadvantages of vascularized bone grafts include long, invasive surgery and recovery times that patients may not be able to tolerate. As an alternative to autograft, ulnar translocation of the distal ulna to the defect site may circumvent the need for a distal donor site, however, the load-bearing across the ulna may be insufficient [26]. Finally, prosthetic reconstruction avoids the need for union, donor-site, or resorption and may be suitable for individuals with a short life expectancy [24].

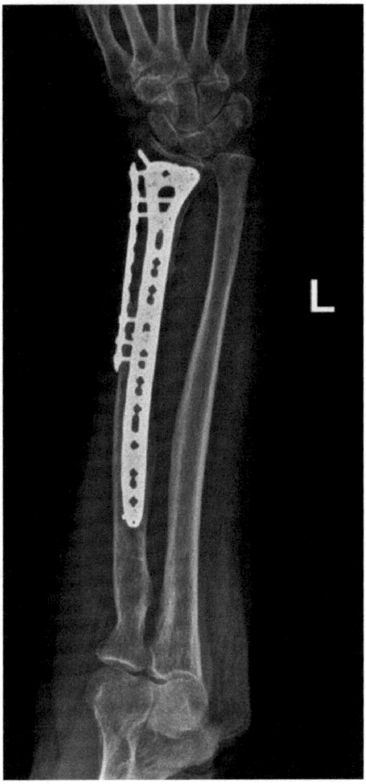

Fig. 19.4 The 76-year-old patient was treated with curettage of the metastatic lesion. Before curettage, internal fixation of the pathologic fracture was performed along the radial aspect of the radial shaft to maintain alignment during excision of the lesion. The bone defect was then reconstructed with cancellous allograft, followed by a variable-angle locking distal radius plate along the volar aspect. Radiograph demonstrates postoperative fixation

Soft Tissue Reconstruction

In cases of the hand or digit where tumors may have invaded soft tissue, resection of all areas of soft tissue should be undertaken, which may involve a more proximal bony resection. Margins will ideally be cleared. In cases where the proximal bone can be spared, soft tissue reconstruction should be as simple as possible, especially in deconditioned patients. If paratenon is still present, wounds may be treated with a simple skin graft. Thigh or groin skin may be used as a donor for skin grafts. If amputations are proximal enough and the soft tissue has not been violated by the lesion, donor tissue for the skin graft may be used from the resected skin. In cases of exposed bone or tendon, local flaps may be required. For hand or forearm reconstruction, these might include the reverse radial forearm flap, lateral arm flap, or the posterior interosseous artery flap. If the wound defect is large enough to require a free tissue transfer, clinicians may consider instead performing a more proximal level amputation to obtain sufficient soft tissue flaps for coverage, given the palliative nature of these surgeries. While free tissue transfer may certainly be used to reconstruct defects, this decision should be made on an individual basis and after extensive discussion with the patient.

Tendon or nerve transfers may be considered in cases where the radial, median, or ulnar nerve must be sacrificed due to involvement with metastases. In healthier patients with good prognoses, this may provide additional functional benefits. When large segmental defects in the nerve are present after wide local excision, nerve allograft may be used for reconstruction. These procedures will add additional operative time that should be considered, though recovery time is not significantly affected.

Wound Closure

Long incisions should be closed in layers to optimize healing and minimize the risk of dehiscence, especially in the setting of future radiation. For amputations, permanent sutures may be used to close the skin. When tendons are involved, care should be taken to ensure there is still adequate excursion with closure. Flexor tendons should not be tacked down as it may lead to quadrigia effect due to the shared muscle belly. Extensor tendons may also have intact juncturae, so an adhered residual extensor tendon might prevent remaining digits from fully extending. Dead spaces should be minimized and may require the use of buried, tacking sutures or drain placement to prevent seroma formation.

Tips and Tricks

In the forearm, segmental resection of the metastatic lesion is preferred in order to spare adjacent joints, if feasible. Surgeons might consider placing a plate across the radial or ulnar shaft prior to resection to minimize the risk of fracture of an area of thinned cortex. If reconstruction is to be performed, a graft or PMMA may then be placed with subsequent plating.

An important goal of reconstruction is to minimize any additional complications. The use of autograft or allograft, which will require time for union and graft take, should be evaluated on a case-by-case basis (Table 19.1).

Table 19.1 Common approaches for tumor resection, bone reconstruction, and soft tissue reconstructive options for each region of the upper extremity

Region	Approaches for tumor resection	Bone reconstructive options	Soft tissue reconstructive options
Phalange	Curettage, amputation	Allograft, PMMA	Local flaps, skin graft
Metacarpal	Curettage, ray resection, amputation	Allograft, PMMA	Local flaps, skin graft
Carpus	Wide local excision, amputation	Allograft, arthrodesis	Pedicled and local flaps, skin graft, wrist denervation
Radius/Ulna	Curettage, wide local excision, amputation	Allograft, autograft, PMMA, arthrodesis, prosthetic	Pedicled and local flaps, skin graft, free tissue, TMR/RPNI/nerve implantation into muscle

Pearls and Pitfalls

When placing grafts, consideration should be taken to place hardware that will not require removal. Returning to the operating room for removal of hardware may not be practical for many of these patients. Additionally, allograft should be used with caution prior to confirmation of clear margins, since the presence of positive margins may necessitate adjuvant radiation, which could potentially jeopardize healing.

In cases of recurrence of the metastasis, revision procedures may need to be performed, especially if there are additional fractures, hardware loosening, or bone gaps. Reconstruction after excision of a recurrent lesion may require use of a more robust structural graft (Fig. 19.5). These grafts may take a longer amount of time to heal, but can be quite reliable and stable in the postoperative period (Figs. 19.6 and 19.7).

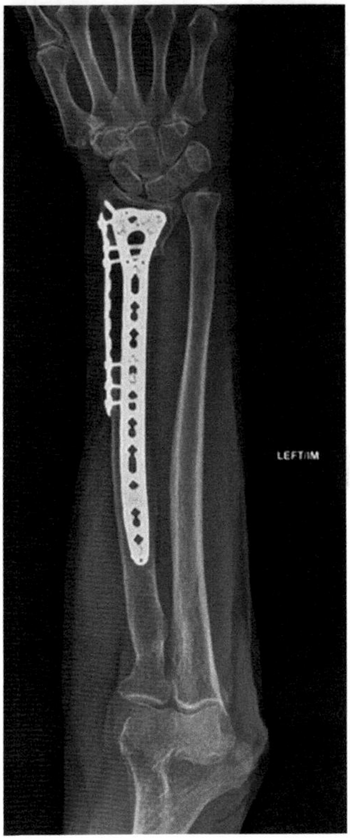

Fig. 19.5 Approximately 1 year after the index procedure, the patient developed a recurrence of the metastatic lesion with bony resorption and a subsequent 5 cm bone loss

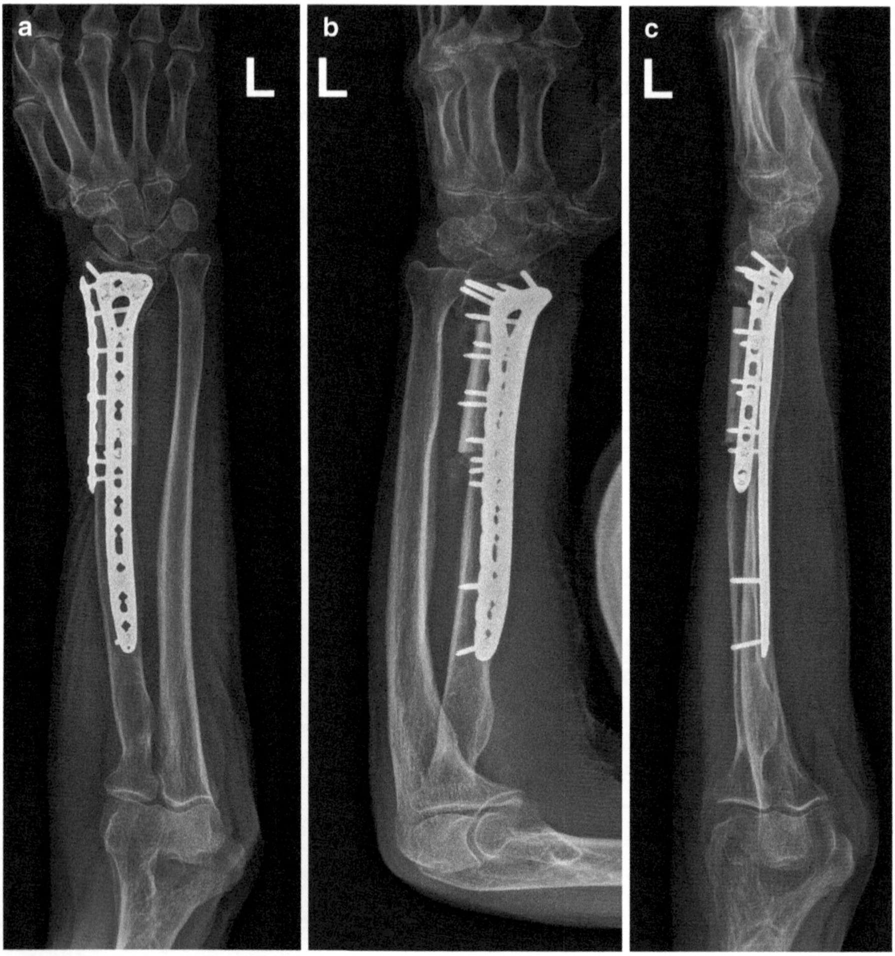

Fig. 19.6 The previous hardware was removed. A structural freeze-dried allograft was placed in the bony defect as a structural strut graft. Radial and volar plates were revised. Radiographs demonstrate postoperative fixation

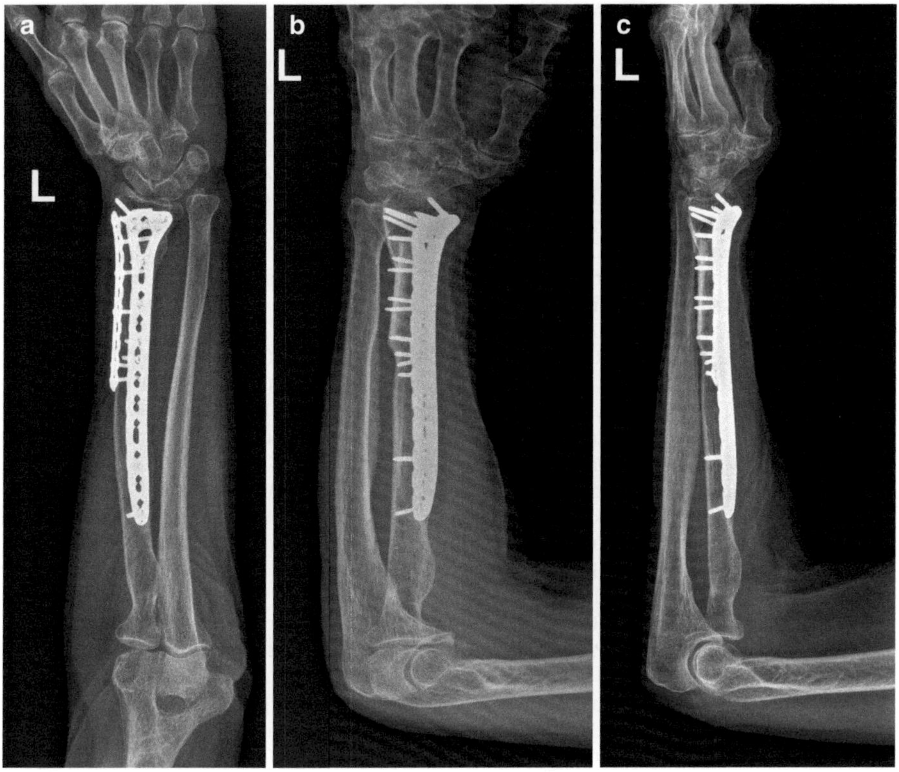

Fig. 19.7 The patient demonstrated interval healing of her radial strut graft and well-seated hardware at 30-months postoperatively

Postoperative Care

Postoperatively, patients are placed in a soft dressing or splint for 5–7 days. Amputation stumps can be dressed with a moist, occlusive dressing followed by dry gauze, cast padding, and an ace wrap. The soft dressing may be taken down in 3 days with the patient beginning to perform washes and keeping the incision clean and dry. Splinting and/or casting may be necessary for cases where there is concern for instability of the wrist, fragile remaining bone, or when graft or hardware is placed in the forearm. Depending on the stability of the reconstructed construct, splints may be removed as soon as 1 week post-operatively with early active range of motion protocols, or patients may be transitioned to casting to ensure adequate fixation and union. Casts are typically in place for 6–8 weeks with radiograph surveillance to assess for bone healing. Follow-up clinic visits are performed at regular intervals once any splints or casts are removed.

References

1. Morris G, Evans S, Stevenson J, et al. Bone metastases of the hand. Ann R Coll Surg Engl. 2017;99(7):563–7. https://doi.org/10.1308/rcsann.2017.0096.
2. Kerin R. The hand in metastatic disease. J Hand Surg Am. 1987;12(1):77–83. https://doi.org/10.1016/s0363-5023(87)80164-8.

3. Amadio PC, Lombardi RM. Metastatic tumors of the hand. J Hand Surg Am. 1987;12(2):311–6. https://doi.org/10.1016/s0363-5023(87)80299-x.
4. Sur YJ, Kang YK, Bahk WJ, Chang DK, Rhee SK. Metastatic malignant tumour in the hand. J Plast Surg Hand Surg. 2011;45(2):90–5. https://doi.org/10.3109/2000656X.2011.556224.
5. Tolo ET, Cooney WP, Wenger DE. Renal cell carcinoma with metastases to the triquetrum: case report. J Hand Surg Am. 2002;27(5):876–81. https://doi.org/10.1053/jhsu.2002.34368.
6. Flynn CJ, Danjoux C, Wong J, Christakis M, Rubenstein J, Yee A, Yip D, Chow E. Two cases of acrometastasis to the hands and review of the literature. Curr Oncol. 2008;15(5):51–8.
7. Tan KT, Simpson C, Chandrasekar CR. Hand metastasis: an unusual presentation of renal cell carcinoma. Hand (NY). 2012;7(2):204–6. https://doi.org/10.1007/s11552-012-9393-5.
8. Fones L, Schroeder JT, Beredjiklian PK, Gallant GG. Acral metastasis to the hand as the primary presentation of malignancy. J Hand Surg Glob Online. 2024;6(4):593–6. https://doi.org/10.1016/j.jhsg.2024.04.006.
9. Hernandez RK, Wade SW, Reich A, Pirolli M, Liede A, Lyman GH. Incidence of bone metastases in patients with solid tumors: analysis of oncology electronic medical records in the United States. BMC Cancer. 2018;18(1):44. https://doi.org/10.1186/s12885-017-3922-0.
10. Dow T, Davis C, ElAbd R, Lalonde D, Williams J. Cancer metastases to the hand: a systematic review and meta-analysis. Hand. 2024;19(6):865–74.
11. Kerin R. Metastatic tumors of the hand. J Bone Joint Surg. 1983;65-A(9):1331–5.
12. Afshar A, Farhadnia P, Khalkhali H. Metastases to the hand and wrist: an analysis of 221 cases. J Hand Surg Am. 2014;39(5):923–32 e17. https://doi.org/10.1016/j.jhsa.2014.01.016.
13. LiBrizzi CL, Levin AS, Strike SA, Morris CD. Indications and outcomes of palliative major amputation in patients with metastatic cancer. Surg Oncol. 2022;40:101700. https://doi.org/10.1016/j.suronc.2021.101700.
14. Kavolius JP, Mastorakos DP, Pavlovich C, Russo P, Burt ME, Brady MS. Resection of metastatic renal cell carcinoma. J Clin Oncol. 1998;16(6):2261–6.
15. Zoccali C, Formica VM, Sperduti I, et al. Wide resection for giant-cell tumor of the distal radius: which reconstruction? A systematic review of the literature and pooled analysis of 176 cases. Hand Surg Rehabil. 2022;41(5):552–60. https://doi.org/10.1016/j.hansur.2022.07.002.
16. Coleman RE. Metastatic bone disease: clinical features, pathophysiology and treatment strategies. Cancer Treat Rev. 2001;27:165–76. https://doi.org/10.1053/ctrv.2001.0210.
17. Puhaindran ME, Athanasian EA. Malignant and metastatic tumors of the hand. J Hand Surg Am. 2010;35(11):1895–900; quiz 1900. https://doi.org/10.1016/j.jhsa.2010.08.014.
18. Cegla P, Konstanty E, Fundowicz M, Pietrasz K, Matuszewski K, Heydrych A. First case of acrometastases to the wrist reported from penile cancer. J Cancer Res Ther. 2021;17(2):599.
19. Bubau M, Georgiu C, Fodor L. Bilateral hand metastases from a pleomorphic/dedifferentiated leiomyosarcoma: a case report. Am J Dermatopathol. 2022;44(2):118–20.
20. Gibson PD, Ippolito JA, Benevenia J. Radial shaft reconstruction with an intercalary endoprosthesis following resection of metastatic tumor. Orthopedics. 2017;40(2):e242–7. https://doi.org/10.3928/01477447-20160901-03.
21. Song Y, Yao J. Trapezial metastasis as the first indication of primary non-small cell carcinoma of the lung. J Hand Surg Am. 2012;37(6):1242–4. https://doi.org/10.1016/j.jhsa.2012.03.006.
22. Stahl S, Santos Stahl A, Lotter O, Pfau M, Perner S, Schaller HE. Palliative surgery for skeletal metastases from melanoma in the scaphoid–a critical case report appraisal. J Plast Reconstr Aesthet Surg. 2012;65(8):1111–5. https://doi.org/10.1016/j.bjps.2012.01.019.
23. Plate AM, Steiner G, Posner MA. Malignant tumors of the hand and wrist. J Am Acad Orthop Surg. 2006;14(12):680–92.
24. Aljuhani WS, Alanazi AM, Edrees AO. Single bone forearm reconstruction of proximal ulna metastatic lesion: a case report. Int J Surg Case Rep. 2023;106:108259. https://doi.org/10.1016/j.ijscr.2023.108259.
25. Althausen P, Althausen A, Jennings LC, Mankin HJ. Prognostic factors and surgical treatment of osseous metastases secondary to renal cell carcinoma. Cancer. 1997;80(6):1103–9. https://doi.org/10.1002/(sici)1097-0142(19970915)80:6<1103::Aid-cncr13>3.0.Co;2-c.
26. Liu W, Wang B, Zhang S, Li Y, Hu B, Shao Z. Wrist reconstruction after En bloc resection of bone tumors of the distal radius. Orthop Surg. 2021;13(2):376–83. https://doi.org/10.1111/os.12737.
27. Duncan SF, Athanasian EA, Healey JH. Radius neck-to-humerus trochlea transposition for elbow reconstruction after resection of the proximal ulna: report of 2 cases. J Hand Surg Am. 2008;33(8):1384–7. https://doi.org/10.1016/j.jhsa.2008.04.004.

Management of Metastatic Disease to the Proximal Femur

20

Costantino Errani

Epidemiology

Bone is the third most common site of metastatic disease, after the lungs and the liver [1–3]. Long bone metastases are a common presentation in patients with advanced cancer, occurring in up to 70% of patients [1–3]. The prevalence of metastatic bone disease in the U.S.A. is estimated between 300,000 and 600,000 patients per year [1–3]. The femur is the third most frequent site for bone metastases after the spine and pelvis [4, 5]. The most common anatomic locations requiring surgery are the spine and the femur [6].

Surgical Indications/Nonsurgical Indications

Surgical indications for proximal femoral metastases include pathological fractures and impending fractures, unless the general condition of the patient precludes any surgical intervention, such as bone marrow carcinosis or respiratory failure [5]. The surgical indication of proximal femur metastasis depends as well on the patient's general condition, tumor type, and the extent of the lesion. Our goal is to palliate the pain and symptoms of bone metastasis, and to stabilize the bone for immediate weight bearing so that the patient's quality of life is not impaired [7].

The indications for surgical treatment of an impending fracture seem to differ between countries: in the U.S.A. up to 71% of patients with an impending fracture are treated surgically compared to only 18% in the Nordic countries [8, 9]. The life expectancy of patients with bone metastases seems to be the most important factor in determining the surgical treatment, avoiding over- or under-treatment [2, 10].

C. Errani (✉)
Orthopaedic Service, Musculoskeletal Oncology Department, IRCCS Istituto Ortopedico Rizzoli, Bologna, Italy
e-mail: costantino.errani@ior.it

J. Pretell-Mazzini, S. A. Lozano Calderón (eds.), *Multidisciplinary Approach of Musculoskeletal Metastasis*, https://doi.org/10.1007/978-3-032-09498-8_20

In metastatic patients with a limited expected survival, the indications for surgical treatment are limited [11]. An expected survival of less than 6 weeks is a relative contraindication for surgical treatment of any impending or pathological fractures. For patients with a life expectancy of 3–12 months, it is recommended to a nail fixation. For patients with life expectancy longer than 12 months, en bloc resection of a metastasis with a durable reconstruction, such as prosthetic reconstruction, could be the best indication [2, 10]. In addition, for patients with solitary bone metastasis without concomitant organ metastases at the initial diagnosis of renal cell carcinoma, resection of bone metastasis should be considered because local tumor control can be achieved and overall survival may be extended by resection [12].

Preoperative Planning

Treatment of the proximal femur requires the careful analysis of the preoperative imaging studies. An accurate preoperative planning of the bone metastasis by antero-posterior and latero-lateral 1:1 radiographs is mandatory, including the view of bone joints (hip and knee) to prepare appropriate sizes of the prosthesis. Magnetic resonance imaging is useful to determine the surgical approach and the type of reconstruction by evaluating the exact extent of the lesion (extraosseous extension) and its relationship with neurovascular bundles, which is critical for planning surgical margins. It is mandatory to plan the length of resection and check the implant availability. The prosthesis must be able to reconstruct the skeletal defect, and the stem needs to match the size of the medullary canal of the residual femur.

Surgical Options

Intramedullary fixation or prosthetic reconstruction should be chosen on the basis of the location of the lesion, the extent of bone destruction, and the ability of the construct to last the life expectancy of the patients [2].

In patients with bone metastases limited to the head and neck area of the proximal femur, hemiarthroplasty or total arthroplasty may be indicated without difference in the risk of complications, reoperations, or revisions between patients treated with hemiarthroplasty and total hip arthroplasty [13].

In patients with bone metastases in the inter- or subtrochanteric area, intramedullary nailing, hemiarthroplasty, and megaprosthesis could be considered, without difference in the risk of complications between patients treated with intramedullary nailing, hemiarthroplasty, total hip arthroplasty and megaprosthesis [13].

Advantages and Disadvantages of These Previous Options

Nail fixation is a less invasive surgical treatment that do not need prolonged rehabilitation, by contrast, prosthetic reconstruction is associated with longer rehabilitation times [2]. Most studies reported that intramedullary nail fixation and prosthetic reconstruction seem to have a similar risk of complications [14–17]. Both intramedullary nail fixation and prosthetic reconstructions could be associated with complications: intramedullary nail fixation could break or cut out secondary to disease progression, while prosthetic reconstruction could be associated with dislocation or deep infection [14]. Most complications after intramedullary nail fixation occur one year after surgery; in contrast, prosthetic reconstruction seems to be a more durable surgical procedure but can cause complications earlier, within a year after surgery [2, 10, 12].

Contraindications (If Any) to Use Either One of These Options

- Megaprosthesis is usually contraindicated in a very fragile patients due to the risk of infection or dislocation [13].
- Malnutrition, skin disorders with active lesions or ulcers, and significant medical comorbidities may preclude prosthetic reconstructions.
- Femoral vessel compromise by tumor is a relative contraindication because it is always possible an arterial reconstruction; however a palliative therapy or an amputation (disarticulation) should be considered to minimize the risk of interruption to systemic therapy.
- Systemic sepsis is a contraindication to prosthetic reconstruction.

Complications

The risk of hardware breakage, screw cutout, nonunion is higher in patients treated with intramedullary nailing, compared to the risk of periprosthetic fractures and loosening of patients treated with prosthetic reconstructions [13]. The risk of revision due to disease progression is higher in patients treated with intramedullary nailing compared to prosthetic reconstructions [13].

The risk of reoperation due to infection is higher in patients treated with megaprosthesis compared to patients treated with intramedullary nailing, hemiarthroplasty, and total hip arthroplasty [13]. The risk of dislocation is higher in patients treated with megaprosthesis with acetabular component compared to those treated without an acetabular component [13].

Surgical Technique

Patient Positioning and Surgical Incision

A standard operating room table can be utilized as well as a radiolucent table if intra operative imaging is desired. The patient is placed in lateral decubitus position using table pegs. The bony prominences of the legs and pelvis must be protected with pads under the lateral malleolus and knee of the bottom leg and a pillow between the knees, leaving the feet free so they can be palpated for later assessment of limb length. The entire limb is sterilely prepared and included in the sterile field. The limb should be draped leaving free room for movement during the operation. The drapes should leave uncovered the lateral aspect of the femur as the incision may be extended distally.

The incision is a lateral hip approach ("Hardinge" or "Trans-gluteal") with distal extension as needed based on the length of planned resection. Some surgeons may consider using a postero-lateral approach; however, the lateral approach helps to visualize the hip both anteriorly and posteriorly. Longitudinal incision through the skin and subcutaneous tissue, with its proximal end directed slightly posteriorly. The incision should start 3 cm proximal to the anterior-middle third of great trochanter and extended distally in line with the femur with an extension based on the planned length of the femoral resection (Fig. 20.1a, b, c). A gluteal extension of the incision passes between the tensor of fascia lata and the gluteus maximus muscle.

Steps for Tumor Resection

- After dissection of the subcutis, the fascia is incised longitudinally in the same line as the skin, 3 cm proximally to the great trochanter and extending distally (Fig. 20.2a). The fascial sleeve is retracted with a Charnley retractor, exposing gluteus maximus, gluteus Medius and vastus lateralis.

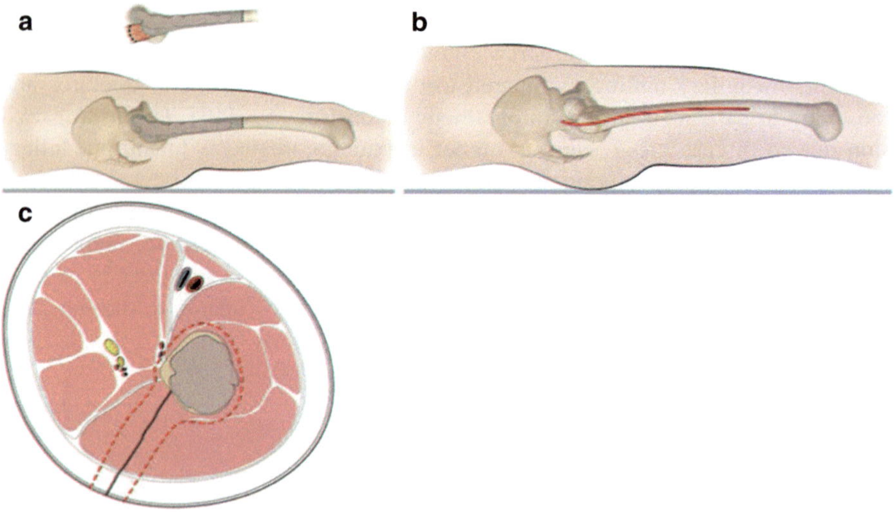

Fig. 20.1 (**a**) Surgical planning of a bone metastasis of the proximal femur. (**b**) A lateral surgical approach, including a biopsy incision (**c**), should start 3 cm proximal to the great trochanter and extend distally in line with the femur with an extension based on the planned length of the femoral resection

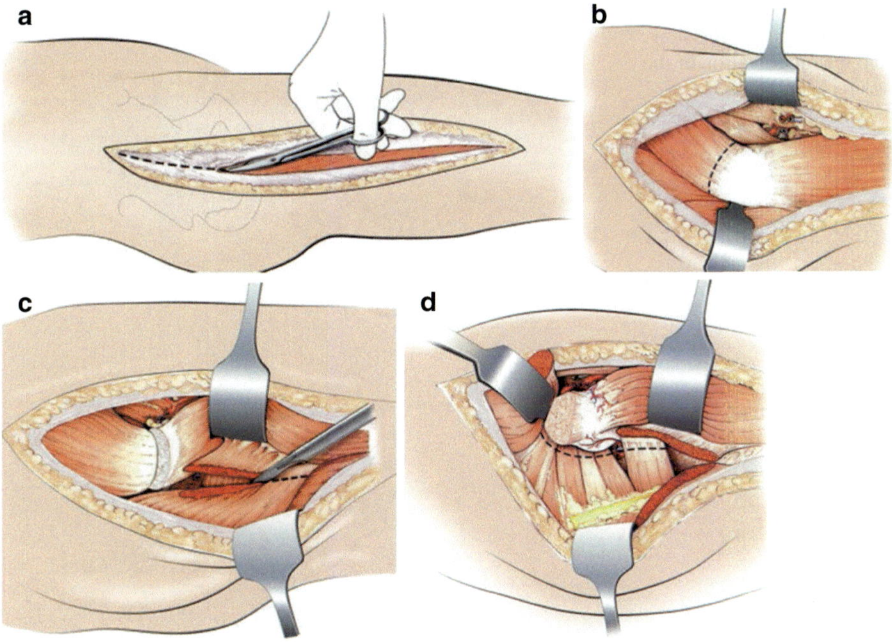

Fig. 20.2 Following the fascia incision (**a**), the gluteus medius and minimus muscles (**b**), the gluteus maximus (**c**) and the short external rotators of the hip are transected from their insertion from the proximal femur (**d**)

- The gluteus Medius and minimus muscles are divided at their insertion to the greater trochanter (Fig. 20.2b). Some surgeons will consider detaching the abductor muscles and the vastus lateralis from the greater trochanter as a soft tissue sleeve.
- The plane underneath the gluteus maximus is entered and completely opened, dividing the tendon of the gluteus maximus from the femur and transecting the distal fibers of the muscle from its insertion in the lateral surface of the femoral bone (Fig. 20.2c).
- Internally rotating the hip, the short external rotators of the hip (the piriformis, gemellus, obturator internus and externus, and quadratus femoris) are transected from their insertion from the proximal femur (Fig. 20.2d).
- Mobilize the vastus lateralis and intermedius muscles distally and anteriorly to expose the anterior, lateral and posterior aspects of the femur (Fig. 20.3). The vastus lateralis and intermedius may be lifted using the coagulation diathermy from its proximal attachment and gently detached from the lateral intermuscular septum as the dissection goes distally.

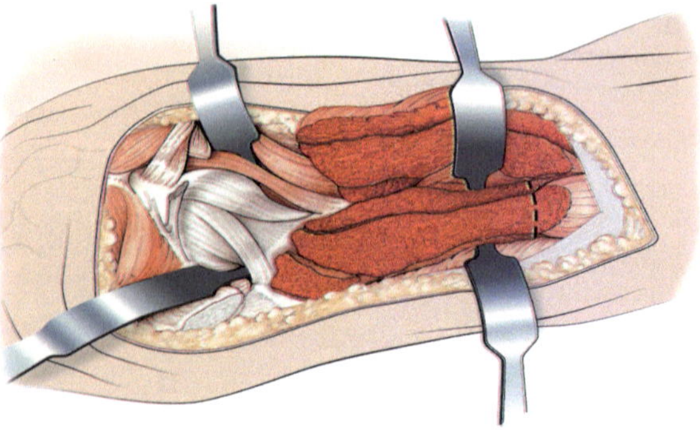

Fig. 20.3 Following the previous muscles incision and the mobilization of the vastus lateralis, the proximal femur is exposed

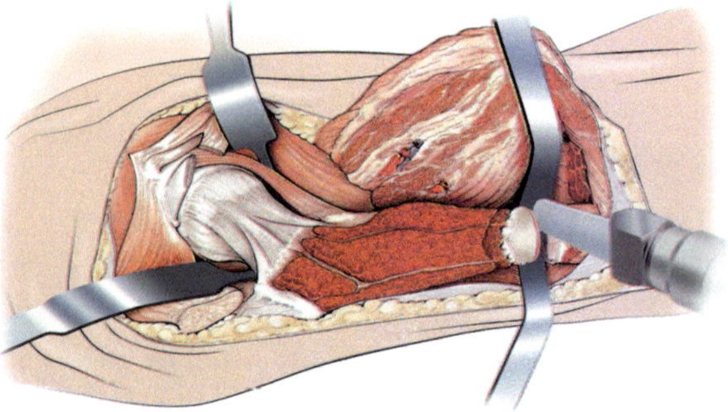

Fig. 20.4 Femur osteotomy 2 cm distal to the tumor

- The osteotomy of the femur can be performed at least 2 cm distal to the tumor using the preoperative measurements determined by Magnetic Resonance Imaging (Fig. 20.4). Before performing the transverse osteotomy of the distal femur, it is important to determine the length and rotation of the host femur, as well as the length of the resected proximal femur.
- The vastus intermedius and medialis muscles can be dissected around the medial side of the femur to reach the medial intermuscular septum, and care should be taken to avoid injury to the femoral vessels within the adductor canal (Fig. 20.5). Once the intermuscular septum is incised, the posterior compartment structures can be mobilized anteriorly in line with detachment of the tendon of the gluteus maximus.
- The muscles anterior to the hip joint (the sartorius and the rectus femoralis) are mobilized. With the hip in full abduction, flexion, and external rotation (gynecologic position), the iliopsoas is divided at its insertion to the lesser trochanter and proximally reflected (care should be taken to ligate and divide the vessels passing on the anterior surface of the iliopsoas).
- Release the capsule antero-inferiorly and antero-superiorly to expose the femoral head and neck and permit free external rotation of the femur. Make a T-shaped capsulotomy to expose the joint but preserve the acetabular labrum unless a total hip arthroplasty is planned (Fig. 20.6).

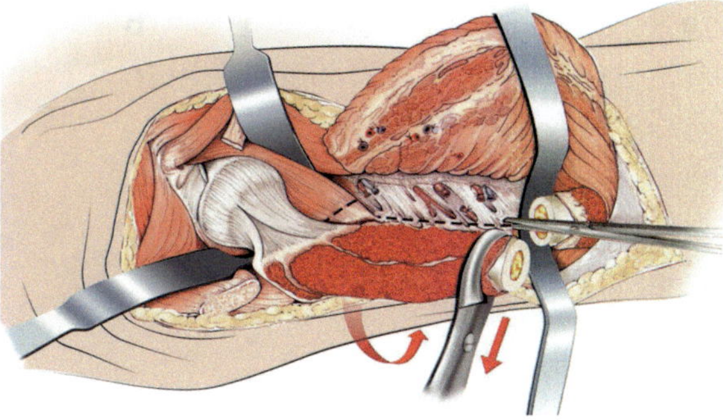

Fig. 20.5 Incision of the medial intermuscular septum

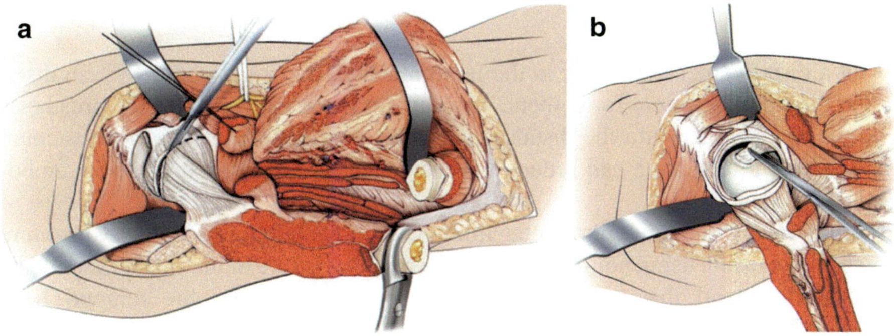

Fig. 20.6 After the capsular incision, (**a**), the ligamentum teres is sectioned to permit the hip dislocation (**b**)

- Following posterior capsular incision, a dislocation of the hip completes the resection of the proximal femur (Fig. 20.7a).

Proximal Femur Reconstruction with Megaprosthesis

- The distal residual femur is prepared by reaming the canal for better cement osteo-integration. In these patients, cemented stems are preferable to be used over press-fit ones due to bone quality, life expectancy and possible need for radiation therapy.
- Determination of the size that best fits and insertion of the trial components to check the stability of the hip and the length of the limb.
- The prosthesis can be assembled based on the size and length of the trial components, that can be compared also directly with the resected specimen. Introduction

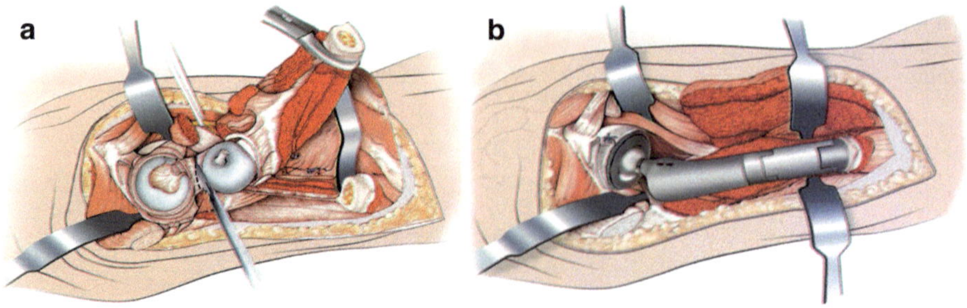

Fig. 20.7 (**a**) Capsular incision completes the proximal femur resection. (**b**) Reconstruction

of a distal cement restrictor in the femoral canal, leaving it about 1–2 cm distal to the tip of the stem for cement. The cement is pressurized, and the prosthesis can be cemented into the distal part of the femur. Intraoperative bone-vacuum cementing technique can reduce the incidence of fat and bone marrow embolism [18].

- If the reconstruction of the acetabulum is not indicated, it is mandatory to suture the hip capsule using non-adsorbable sutures. The suture of the capsule should be passed around the periphery of the capsule to capture the femoral head (Fig. 20.7b).
- If the reconstruction of the acetabulum is needed, three retractors are then utilized to gain exposure of the acetabulum: anterior, superior, and inferior. Positioning the hip in flexion to relax the neurovascular structures during this maneuver is also helpful. Releasing the inferior hip capsule between the anterior and posterior retractors helps improve acetabular visualization and assists with insertion/removal of acetabular reamers and components. The labrum, pulvinar, and any loose soft tissues are excised, followed by routine acetabular reaming. The acetabular component is inserted in a press-fit manner with screw augmentation if necessary.

Soft Tissue Reconstruction as Needed

- Heavy sutures, typically placed through holes in the prosthesis, are used to reattach the gluteus medium and minimus to the intertrochanteric region of the prosthesis with non-absorbable suture. This reattachment can be further reinforced by suturing the gluteus medius to the vastus lateralis muscle.
- Soft tissue reattachment of the iliopsoas to the prosthesis can be done with non-absorbable suture, improving stability of the prosthesis and the flexion of the hip.
- Close also the gluteus maximus and fascia proximally, and the vastus lateralis fascia distally.

Wound Closure

- Hemostasis is secured and must be continued until the blood pressure is at least 100 mm Hg.
- Close the fascia lata with either interrupted or running sutures followed by routine closure of the subcutaneous tissues and skin.
- One suction drain may be placed underneath the gluteal fascia before closure of the wound.

"Tips and Tricks"

- To prevent rotational malposition, it may be useful to mark the rotation with electrocautery one the trial component prosthesis is positioned, with particular attention to avoid an internal rotation (an externally rotated limb is better than an internally rotated limb). The linea aspera may be also used as a guide for femoral rotation after tumor resection.
- Proper leg length can be obtained using the trial component prosthesis, avoiding limb-length discrepancy with particular attention to avoid elongation (a shorter limb is better than a longer limb).
- If the acetabular reconstruction is needed, the transverse acetabular ligament, reamer position relative to the floor, and cup-positioning guides are used to establish proper version and inclination during both reaming and cup insertion, with particular attention to avoid a vertical cup (a horizontal cup is better than a vertical cup).
- If the acetabular reconstruction is not needed, the hip capsule should be detached from the femur as close to the femoral bone, because this facilitates the reconstruction of the capsule, reducing the risk of dislocation of the megaprosthesis.
- If it is not possible to reattach the gluteus medium and minimus to the intertrochanteric region of the prosthesis due to the sacrifice of soft tissue, the fascia may be sutured through holes in the prosthesis to stabilize the prosthesis, reducing the risk of dislocation.
- Graphic of proximal femur showing the reconstruction with megaprosthesis (Fig. 20.8)

"Pearls and Pitfalls"

- Communicate with the patient and his or her family the possible risk of complications such as infection and dislocation that are not uncommon. The overall risk of complications is 11%. The risk of infection is 3%, while the risk of dislocation is 2% in patients treated without an acetabular component and 7% in patients treated with acetabular component [13].
- Detailed preoperative templating is mandatory.

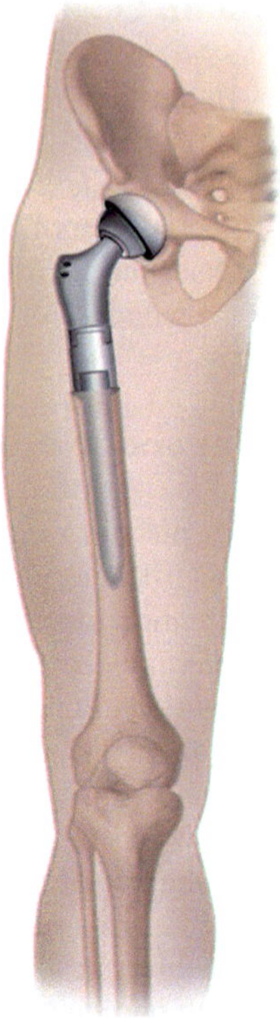

Fig. 20.8 A proximal femur megaprosthesis

- The length of the prosthesis should not exceed the length of the resected bone. Adequate restoration of the limb length is mandatory to avoid excessive tension on the sciatic nerve
- A bipolar acetabular reconstruction may be needed in patients with poor soft tissue tension and instability, reducing the risk of dislocation of the megaprosthesis.
- The leg is brought to abduction during the sutures through the holes in the prosthesis or around the proximal portion of the prosthesis with non-adsorbable sutures.

Postoperative Care

- Intravenous prophylactic antibiotic are given during the surgery, and there is no evidence suggesting that extented postoperative antibiotic therapy may reduce the risk of infection [19].
- Thromboembolic prophylaxis for 4 weeks is usually administered.
- The patient is encouraged to get out of bed during postoperative day 1 and no later than postoperative day 3.

- The drain if placed, is removed when its output is below 30 mL for 24 hours, usually within 3 days.
- Daily physical therapy for ambulation with a removable abduction hip orthosis and exercise of the knee is recommended.
- The abduction hip orthosis should be applied for 3 months with a flexion range from 0 to 30 degrees in the first month, then 0–60 the second month, and 0–90 the third month.
- If staples or non-absorbable sutures are used for skin closure these should be left for 3 weeks to avoid wound disruption. With intensive physical rehabilitation, patients may become ambulatory with a cane at approximately 4 weeks.

Acknowledgments Thanks to the graphic designer Silvia Bassini who designed the illustrations of the chapter.

References

1. DiCaprio MR, Murtaza H, Palmer B, Evangelist M. Narrative review of the epidemiology, economic burden, and societal impact of metastatic bone disease. Ann Joint. 2022;7:28.
2. Errani C, et al. Treatment for long bone metastases based on a systematic literature review. Eur J Orthop Surg Traumatol Orthop Traumatol. 2016; https://doi.org/10.1007/s00590-016-1857-9.
3. Coleman RE. Clinical features of metastatic bone disease and risk of skeletal morbidity. Clin Cancer Res. 2006;12:6243s–9s.
4. Damron TA, Mann KA. Fracture risk assessment and clinical decision making for patients with metastatic bone disease. J Orthop Res. 2020;38:1175–90.
5. Tanaka A, et al. Surgery for femoral metastases. Bone Joint J. 2020;102-B:285–92.
6. Behnke NK, et al. Risk factors for same-admission mortality after pathologic fracture secondary to metastatic cancer. Support Care Cancer. 2017;25:513–21.
7. Lam Y-L. CORR insights®: what factors are associated with implant breakage and revision after intramedullary nailing for femoral metastases? Clin Orthop. 2018;476:1834–6.
8. Forsberg JA, Eberhardt J, Boland PJ, Wedin R, Healey JH. Estimating survival in patients with operable skeletal metastases: an application of a bayesian belief network. PLoS One. 2011;6:e19956.
9. Ratasvuori M, et al. Prognostic role of en-bloc resection and late onset of bone metastasis in patients with bone-seeking carcinomas of the kidney, breast, lung, and prostate: SSG study on 672 operated skeletal metastases. J Surg Oncol. 2014;110:360–5.
10. Errani C, et al. C-reactive protein and tumour diagnosis predict survival in patients treated surgically for long bone metastases. Int Orthop. 2021;45:1337–46.
11. Errani C, et al. What's new in management of bone metastases? Eur J Orthop Surg Traumatol. 2019;29:1367–75.
12. Errani C, Mavrogenis AF, Tsukamoto S. What's new in musculoskeletal oncology. BMC Musculoskelet Disord. 2021;22:704.
13. Solou K, et al. Risk of complications and implant survival of surgical treatment of proximal femoral metastasis: a meta-analysis. Eur J Orthop Surg Traumatol. 2024;34:1779–94.
14. Harvey N, Ahlmann ER, Allison DC, Wang L, Menendez LR. Endoprostheses last longer than intramedullary devices in proximal femur metastases. Clin Orthop. 2012;470:684–91.
15. Wedin R, Bauer HCF. Surgical treatment of skeletal metastatic lesions of the proximal femur: endoprosthesis or reconstruction nail? J Bone Joint Surg Br. 2005;87:1653–7.

16. Hunt KJ, Gollogly S, Randall RL. Surgical fixation of pathologic fractures: an evaluation of evolving treatment methods. Bull Hosp Joint Dis. 2006;63:77–82.
17. van Doorn R, Stapert JW. Treatment of impending and actual pathological femoral fractures with the long Gamma nail in The Netherlands. Eur J Surg Acta Chir. 2000;166:247–54.
18. Pitto RP, Koessler M, Kuehle JW. Comparison of fixation of the femoral component without cement and fixation with use of a bone-vacuum cementing technique for the prevention of fat embolism during total hip arthroplasty. A prospective, randomized clinical trial. J Bone Joint Surg Am. 1999;81:831–43.
19. Zeitlinger L, Wilson M, Randall RL, Thorpe S, on behalf of the PARITY Investigators. Surgical duration is independently associated with an increased risk of surgical site infection and may not be mitigated by prolonged antibiotics: a secondary analysis of the PARITY trial of infection after lower-extremity endoprosthetic reconstruction for bone tumors. J Bone Joint Surg Am. 2023;105:79–86.

Management of Metastatic Disease to the Femoral Diaphysis

21

Wesley M. Durand and Adam S. Levin

Introduction

The evaluation and management of metastatic disease occupies a large proportion of musculoskeletal oncology practice. As medical management of metastatic disease has advanced, the survival of patients with metastases has increased [1–3]. This has led not only to an increase in the number of patients who may potentially benefit from surgery for metastatic disease, but also has influenced the manner of surgical treatment that they receive. This chapter will focus on the surgical management of femoral diaphyseal metastatic lesions. Lesions extending from the diaphysis into either the proximal or distal femur are addressed separately, in respective sections.

Epidemiology

Treatment of metastatic lesions requires an understanding of tumor type and biology. Ward et al. conducted a 2003 retrospective analysis of a single surgeon's 12-year experience treating metastatic disease (including both fractured and non-fractured lesions) at an academic medical center in the Southeastern United States [4]. Of 323 total cases, 56% were in the femur. Of these, 23% were lung, 21% breast, and 15% were myeloma, with other primaries such as renal, prostate, lymphoma, gastrointestinal, and endometrial all accounting for <10% of the total. In a large retrospective review of 142 consecutive patients with metastatic pathologic fractures of the femur, Sarahrudi et al. observed the most common primary tumors to be breast (47%), lung (10%), prostate (8%), and renal (5%) [5]. This distribution may be changing over time, as the duration of survival from each of these diseases evolves over time.

W. M. Durand · A. S. Levin (✉)
Department of Orthopaedic Surgery, Johns Hopkins Hospital, Baltimore, MD, USA
e-mail: wdurand@jhmi.edu; alevin25@jhmi.edu

J. Pretell-Mazzini, S. A. Lozano Calderón (eds.), *Multidisciplinary Approach of Musculoskeletal Metastasis*, https://doi.org/10.1007/978-3-032-09498-8_21

Metastases to the femoral diaphysis are observed less frequently than in other regions of the femur. Sarahrudi et al. found the trochanteric region to be the most common lesion location, followed by the diaphysis [5]. In a 2011 retrospective single-center review of patients undergoing intramedullary rod fixation for skeletal metastasis, Miller et al. found 81 patients with lesions of the femur, 21 of whom (26%) had lesions of the femoral shaft (proximal femur lesions accounted for 58% of patients) [6].

A large proportion of the heterogeneity of this data is undoubtedly attributable, in part, to differences in retrospective study technique, viz., whether the study included only patients with completed pathologic fracture, or also those with impending fracture. It is likely that the population reflected in studies of patients with completed fractures may have a higher proportion of tumors with a more aggressive biology, lytic metastases, and metastases located in bony regions experiencing greater biomechanical stresses. This is also likely, however, to reflect the population most likely to be treated by musculoskeletal oncologists.

Indications

Predicted survival is among the most critical considerations when deciding between operative vs. nonoperative management of metastatic lesions. As a general guideline, patients with greater than 6 weeks predicted survival should be considered for operative management [7], though patient- and disease-specific considerations should be taken into account in surgical decision-making. There is an abundance of manuscripts and predictive algorithms for survival, ranging from metastatic disease broadly to specific tumor and anatomic location instruments. A study by Kotian et al. demonstrated that, among patients with pathologic femur fractures, patients with lower hemoglobin, lower albumin, and primary tumor subtype including sarcomas, had worse survival [8]. The PATHFx tool is an externally validated algorithm capable of predicting survival following a pathologic fracture [9]. Other studies have utilized a variety of mathematical and machine learning methods with varying levels of success, including logistic regression, Bayesian Belief Network, and random forest models, to identify just a selection [10–12]. The Skeletal Oncology Research Group (SORG) produced machine-learning algorithms with very good predictive capacity for survival at both 90 days and 1 year [12]. This model was subsequently externally validated on a Taiwanese cohort, with maintenance of good discriminative capability [13]. Models have further been developed to predict surgery after radiotherapy, such as the METSSS model: *M*etastasis location, *E*lderly (age), primary *T*umor, *S*ex, *S*ickness (i.e., comorbidity burden), and radiotherapy *S*ite [14].

The Mirels' score is one of the most commonly utilized systems for operative vs. nonoperative decision-making in skeletal metastases [15]. This is calculated based on the lesion radiographic pattern (1 = blastic, 2 = mixed, 3 = lytic), size of the lesion as a proportion of the diameter of the bone (1 = <1/3, 2 = 1/3–2/3, 3 = >2/3), lesion site (1 = upper extremity, 2 = lower extremity, 3 = peritrochanteric), and the

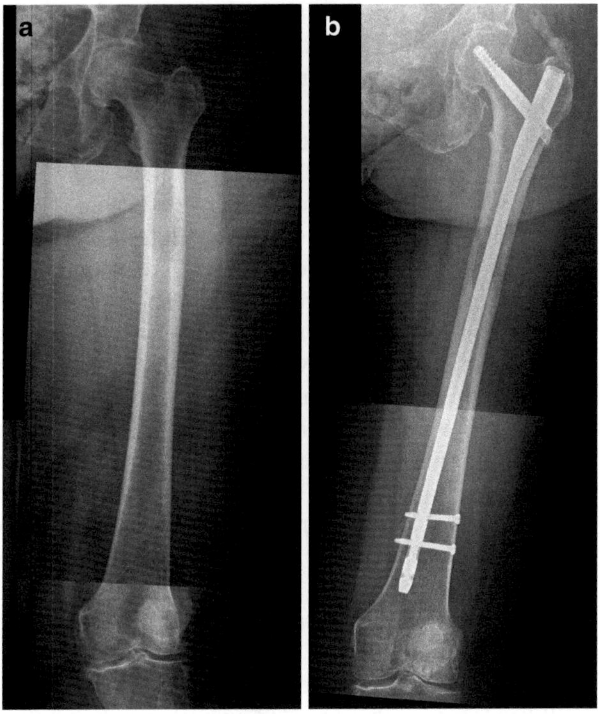

Fig. 21.1 (**a**) A 69-year-old female with metastatic breast carcinoma presented with right lower extremity pain and was found to have a lytic proximal diaphyseal femur lesion (Mirels' score 10). (**b**) She underwent prophylactic intramedullary rod fixation

presence of pain (1 = mild, 2 = moderate, 3 = mechanical). Scores ≤7 were recommended for non-operative management, 8 for individualized decision-making, and ≥9 for operative management. Despite its widespread use, it is important to interpret the Mirels' score in the context of a patients' predicted survival, functional capacity, and available surgical treatment options. (Fig. 21.1).

Preoperative Planning and Surgical Options

Surgical Metastatectomy

The first major decision in developing a surgical strategy for the operative management of femoral diaphyseal lesions is whether the patient would benefit from resection of the lesion. Much controversy exists around metastatectomy, with a significant body of literature addressing renal cell carcinoma, specifically. Ratasvuori et al. published a retrospective analysis of the Scandinavian Sarcoma Group registry with 672 patients with metastatic disease, and found that en bloc vs. intralesional resection of kidney cancer was associated with an approximately four-fold increase in survival [16]. Fottner et al. reported on 116 surgeries in 101 patients with renal cell carcinoma, and found that patients exhibited longer survival with solitary metastases, the absence of pathologic fracture, and wide/radical metastatectomy [17]. In a large study of 183 patients with renal cell carcinoma, Langerguizen et al. found that those with negative resection margins had improved overall survival and recurrence rates [18]. Lin et al. similarly observed superior survival rates in patients with solitary renal cell metastases—35% at 5-years for those with solitary lesions, compared to 11% for those with multiple bony lesions, and 8% for those with both bone and visceral involvement [19]. Higuchi et al. observed that those undergoing en bloc metastatectomy vs. intralesional curettage had no statistically significant difference in 10-year overall survival (51.3% vs. 30.3%, $p = 0.29$), but had a markedly significantly higher rate of recurrence-free 10-year survival (97.6% vs. 18.0%, $p < 0.001$) [20]. Evenski et al. similarly observed no significant difference in overall survival, but lower rates of recurrence with en bloc vs. intralesional resection [21]. It is important to note that adjuvant therapy protocols differed greatly across these studies. Given this abundant and, at times, conflicting data, Zaid et al. conducted a systematic review and meta-analysis, finding that complete surgical metastatectomy leads to improved overall survival vs. incomplete resection, even after adjusting for performance status [22]. This has traditionally been attributed, at least in part, to relative chemo- and radio-resistance of renal cell carcinoma, though recent studies have demonstrated similar response of osseous vs. visceral metastases to receptor tyrosine kinase inhibition [23]. Growing evidence suggests that select patients with skeletal lesions from other radioresistant primary tumors may also benefit from surgical metastatectomy (e.g., urothelial carcinoma, melanoma, etc.), though further study is needed in these indications [24, 25].

Taken together, these results support resection with a goal of negative margins in select patients with renal cell carcinoma, particularly those with solitary or oligometastatic disease with a longer predicted survival. Patients with other malignancies that are relatively radio- or chemo-resistant and with longer predicted survival may also potentially benefit, though this has been less well-studied.

Reconstructive Options

Intramedullary Rod Fixation

The choice of reconstructive options differs primarily based on whether metastatectomy is pursued. For patients with femoral diaphyseal metastases not undergoing resection, the mainstay of treatment is typically intramedullary rod fixation. This is, by far, the most selected treatment option for these lesions across the literature. In a retrospective study of 53 patients with femoral diaphyseal lesions, Kask et al. found that 94.4% of patients were treated with intramedullary rods [26]. This is expected, given the relatively restrictive criteria for metastasis resection and reconstruction. Intramedullary rod fixation is a broadly used technique in orthopedic surgery, and its outcomes and complication profile for femoral diaphyseal lesions are overall favorable. Miller et al. studied 112 consecutive patients undergoing intramedullary rod fixation of long-bone metastases, including 21 femoral diaphyseal lesions [6]. There were only two failures in this group, one from a broken nail that was revised to a megaprosthesis in a patient with breast cancer, and another from tumor progression in a patient with renal cell carcinoma that was revised to an intercalary allograft. Sarahrudi et al. observed a similar failure rate of 9.5% of closed reduction and intramedullary rod fixation of femoral lesions [5]. Ward et al. observed failure in approximately 2% of femoral lesions treated with intramedullary rod fixation [4]. Vermesan et al.'s study included 6 patients with femoral diaphyseal pathologic fractures, all treated with intramedullary rod fixation, two of whom required additional surgery [27]. Kask et al.'s series showed complications in 9 of 53 patients with femoral diaphyseal lesions, including five with osteosynthesis complications and three with tumor progression [26]. Arpornsuksant et al. found that local progression occurred in 4.9% of patients treated with an intramedullary rod for metastatic lesions, but that renal cell carcinoma and younger age were associated with increased likelihood of progression [28]. (Fig. 21.2) Carbon fiber intramedullary implants (CFI) have grown in popularity in recent years, in part due to favorable radiologic characteristics. Lozano-Calderon et al. performed a multi-center review of 239 patients undergoing intramedullary long-bone fixation, including 132 (55%) with femoral surgeries [29]. In this series, only 5 patients with femoral implants experienced mechanical failure (3.8% of patients with femoral surgeries). Herzog et al. retrospectively compared patients with impending or completed pathologic

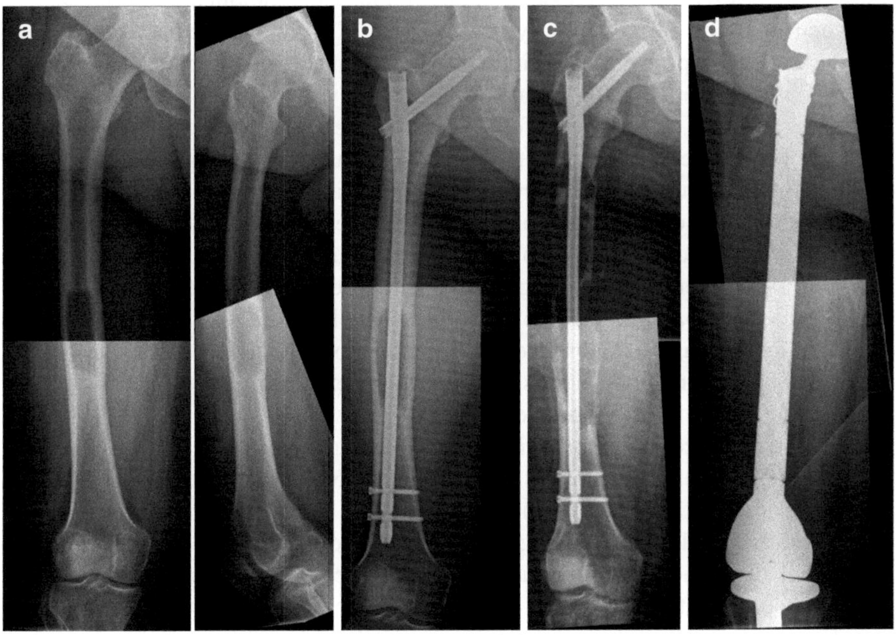

Fig. 21.2 (**a**) A 49-year-old male with a history of obesity and widely metastatic renal cell carcinoma including brain, spine, and extremity metastases, with a large right femur metastatic lesion (Mirels' score 11). (**b**) He underwent prophylactic stabilization with a locked, antegrade cephalomedullary rod. (**c**) By 18 months postoperatively, he experienced marked tumor progression despite chemotherapy, with significant pain in his right femur. (**d**) He was ultimately revised to a total-femur-replacing hip/knee arthroplasty

fractures, treated with CFI (n = 53 fractures, 40 of which were femoral) vs. titanium (n = 47 fractures, 29 of which were femoral) intramedullary implants [30]. The complication profiles were overall not statistically different between the carbon fiber (19% total complication rate, 15% implant-related complication rate) and titanium groups (19% total complication rate, 10% implant-related complication rate). Yeung et al. performed a retrospective matched case-control study comparing 72 patients undergoing treatment of impending or completed pathologic long-bone fractures with 36 patients in both CFI and titanium groups [31]. Implant-related complications occurred in 7 patients in the titanium group (19%) and 1 patient in the CFI group (3%), though this did not reach statistical significance ($p = 0.055$). Gonzalez et al. retrospectively compared patient-reported outcomes for patients undergoing treatment of impending or completed pathologic fractures, using propensity-score matching to include 156 patients in the titanium group and 51 patients in the CFI group. No significant differences were observed between the two groups at 3-months, 6-months, and 1-year postoperatively for PROMIS and pain VAS scores [32]. While select studies have reported anecdotal differences in the cost of CFI vs. traditional implants, these values will be highly variable based on location and often subject to institution-specific purchasing patterns and contracting [30]. Cost-effectiveness studies comparing CFI vs. traditional implants are highly warranted, as are further, large-scale studies comparing complication profiles.

When planning intramedullary rod fixation, it is imperative that the surgeon ensure that the appropriate size implant is available for their patient's anatomy. Patients with blastic lesions may pose difficulty in passing instruments and implants through the lesion safely. Many surgeons elect to place additional implants into the femoral neck out of concern for future pathologic fracture in this location—this is accomplished readily in antegrade implants with head screw options, though it can also be done for retrograde implants with additional cannulated screws placed adjacent to the proximal aspect of the rod. In Moon et al.'s study of 145 patients undergoing femoral intramedullary rod fixation for metastatic disease, no patients were found to have developed femoral neck lesions at a mean of 13 months postoperative radiographic follow-up [33]. Further study is necessary before any conclusions can be drawn regarding the long-term safety of leaving the femoral neck unprotected.

It is also not uncommon for patients with metastatic disease to require fixation of multiple long bones. Data on this topic is sparse. Miller et al. reported on 11,427 patients from a database of trauma care, with 118 undergoing simultaneous tibia/femoral nailing and 28 undergoing staged tibia/femoral nailing [34]. They found no significant difference in cardiopulmonary complications for patients undergoing simultaneous vs. staged fixation. Raasck et al. completed a systemic review on simultaneous vs. staged bilateral femur intramedullary rod fixation for metastatic disease, finding no significant difference in survival or other postoperative complications between the two approaches [35]. There is likely, however, some element of confounding in these studies based on patient health status. Pending more definitive data, surgeons should individualize this decision-making to the clinical status of each patient.

Resection and Intercalary Reconstruction

The technique of resection in patients deemed eligible for wide surgical metastatectomy is highly individualized to the tumor, patient, and surgeon's unique characteristics and is beyond the scope of this chapter. When performed, however, reconstruction of the bony defect is required. A large variety of techniques have been described for the reconstruction of diaphyseal bony defects, and the comparison of these techniques is an area of active investigation and controversy. The choice of technique will depend not only on patient characteristics, but also surgeon's comfort level and experience, as well as the availability of the required components.

The primary methods of bony reconstruction for femoral diaphyseal lesions are intercalary prosthesis and allograft reconstruction. While a variety of other techniques have been developed for reconstruction after diaphyseal resections, including vascularized bone graft, combined vascularized bone graft and allograft, and segmental bone transport, these techniques generally require prolonged recovery and rehabilitation, making them often less ideal for patients with metastatic disease [36]. In many such patients, early return to function and weight-bearing is preferable in light of their estimated shorter survival. Select studies have reported on allograft use for intercalary femoral reconstruction, though scarce evidence exists to directly compare the two approaches [7, 37, 38]. Lun et al. retrospectively compared 16 patients undergoing intercalary prosthetic reconstruction vs. 18 patients with segmental allograft reconstruction for intercalary resection of femoral diaphyseal metastases with pathologic fracture [38]. Primary tumors included lung, breast, gastric, liver, renal, and prostate. All intercalary prosthetics were cemented, and segmental allografts were stabilized with either intramedullary rod or plate/screw constructs. At final follow-up (mean 9.6 months overall), patients with intercalary prosthesis had significantly lower total complications (18.8% vs. 66.7%) and implant-related complications (12.5% vs. 55.6%). This was driven by trends in lower allograft fracture (n/a for the prosthetic group vs. 16.7% for the allograft group), infection (6.3% vs. 27.8%), and non-union (0% vs. 38.9%). These results, while not definitive, suggest that intercalary endoprosthesis may be a more suitable option than segmental allograft for many patients with metastatic disease, and are in line with the authors' clinical practice. (Fig. 21.3).

Intercalary endoprosthetic reconstruction of diaphyseal defects has received increasing attention over recent years, and select reports are available to guide expectations regarding outcomes and implant survival [39–41]. Johnson et al. studied 19 patients with cemented intercalary endoprosthetic reconstruction of diaphyseal defects from resection of metastasis for a median 24-month follow-up, including 10 femur lesions and 9 humerus lesions [40]. Of the femur lesions, 5 were from renal cell primaries, two metastatic sarcomas (clear cell sarcoma and leiomyosarcoma), one melanoma, one colon, and one unknown. At 3-month follow-up, 9 patients with femur prostheses were ambulatory, and at 1-year follow-up, 5 patients remained alive and were ambulatory without assistance. The most common complication was aseptic loosening, occurring with primarily humeral prostheses—only one of 10 cemented femoral intercalary endoprostheses experienced aseptic loosening. Two patients with femoral prostheses experienced recurrence (one with renal cell primary, one with leiomyosarcoma primary), and one demonstrated

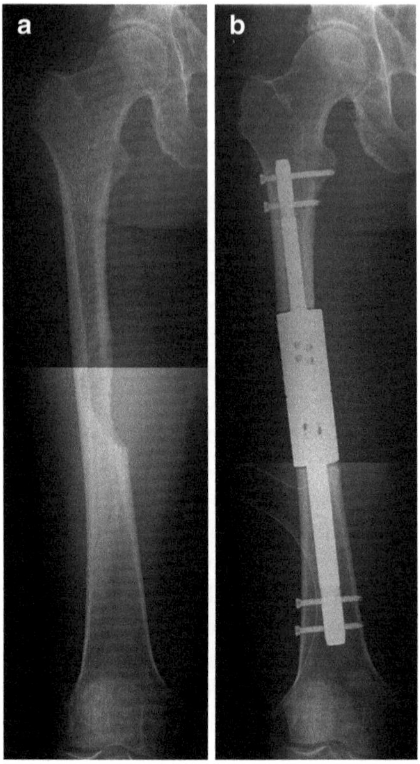

Fig. 21.3 (**a**) A 59-year-old male presented with oligometastatic clear cell renal cell carcinoma status-post nephrectomy, and a right femur lesion (Mirels' score = 11). He was indicated for prophylactic stabilization given the risk of pathologic fracture, and given his metastatic status and primary, was indicated for metastatectomy with reconstruction. (**b**) He underwent wide resection of his right femur lesion with negative margins, with reconstruction with a non-cemented intercalary prosthesis, followed by radiation therapy

wound dehiscence. Benevenia et al. retrospectively reviewed patients undergoing modular intercalary endoprosthetic reconstruction, including 21 patients with femoral reconstruction, 9 of whom received cemented implants, with the remaining 11 patients receiving non-cemented implants [39]. Primary tumors among patients with femoral lesions included renal cell, breast, prostate, lymphoma, leukemia, melanoma, myxoid malignant fibrous histiocytoma, osteosarcoma, and leiomyosarcoma, in addition to several patients indicated for failed previous prostheses. Of 12 non-oncologic complications observed, all occurred in patients with femoral prostheses. There was no significant difference in complications observed overall between cemented vs. non-cemented prostheses (21% cemented vs. 33% non-cemented, $p = 0.39$). When present, however, failure of non-cemented implants occurred via aseptic loosening, whereas cemented implants failed at the modular clamp/rod interface.

Cementation of the bony defect resulting from either wide resection or intralesional curettage is also feasible, typically paired with plate-and-screw fixation to the remnant proximal and distal femur. This technique may be highly variable in implementation, potentially involving intramedullary/intracement wires as well as variable plate selection based on the location and morphology of the lesion. Fan et al. studied 15 patients undergoing intralesional curettage, cementation, and plate-screw fixation of long-bone metastatic lesions, and compared to 13 patients undergoing resection and reconstruction with a non-cemented intercalary prosthesis [42]. They observed significantly lower rates of complications (including failure of fixation and tumor recurrence) in the group undergoing resection with intercalary prosthetic (15%) vs. curettage, cementation, and plate–screw fixation (33%).

Short intramedullary segment compressive osteointegration is another, newer technique for bone-prosthetic anchoring that has been increasingly studied in recent years, though reports of use in diaphyseal resections are sparse [43–46]. Tedesco et al. studied 6 patients with lower extremity skeletal malignancies or previous failed reconstruction undergoing diaphyseal resection and double-ended short segment compressive osteointegration with a modular intercalary endoprosthetic, and observed three mechanical implant failures requiring revision surgery. The authors concluded that, while this technique may facilitate intercalary resection in patients who might otherwise have required arthroplasty, patients should be advised regarding the high risk of implant failure. The need for prolonged rehabilitation and high risk of implant failure should make this technique particularly less attractive for patients with metastatic disease unless long survival expectancy is assured. This is, again, an area of active research, and more insights are required prior to definitive conclusions regarding the optimal methods of bone-prosthetic anchoring.

Surgical Technique

The techniques of surgical wide resection and intramedullary rod fixation of the femur have been thoroughly treated in previous works and are beyond the scope of this chapter. Here, we will describe the technique for cemented intercalary prosthetic reconstruction of a segmental bony defect following resection of a metastatic lesion with transverse osteotomies, illustrated via a case example.

Surgical Metastatectomy with Cemented Intercalary Prosthetic Reconstruction

A 24-year-old patient with a history of metastatic adamantinoma-like Ewing sarcoma of the thyroid presented with a metastasis to the right femoral shaft. Past medical history was notable for obesity (BMI >50) and myositis. The patient had initially presented to an outside hospital with worsening back pain and weight loss and was ultimately found to have multiple bony lesions and a thyroid mass, with biopsy demonstrating a small-round blue cell tumor with morphology and immunohistochemistry supporting the diagnosis. Work-up ultimately revealed multiple bony metastases, including lesions in the spine, ribs, and right femur. Following thyroidectomy, chemotherapy, and radiotherapy to sites of metastatic disease, follow-up imaging demonstrated persistence of FDG-avidity in the right femoral diaphyseal lesion on PET/CT. A subtle fracture was also appreciable on preoperative imaging. The patient was indicated for surgical metastatectomy with intercalary prosthetic reconstruction. (Fig. 21.4).

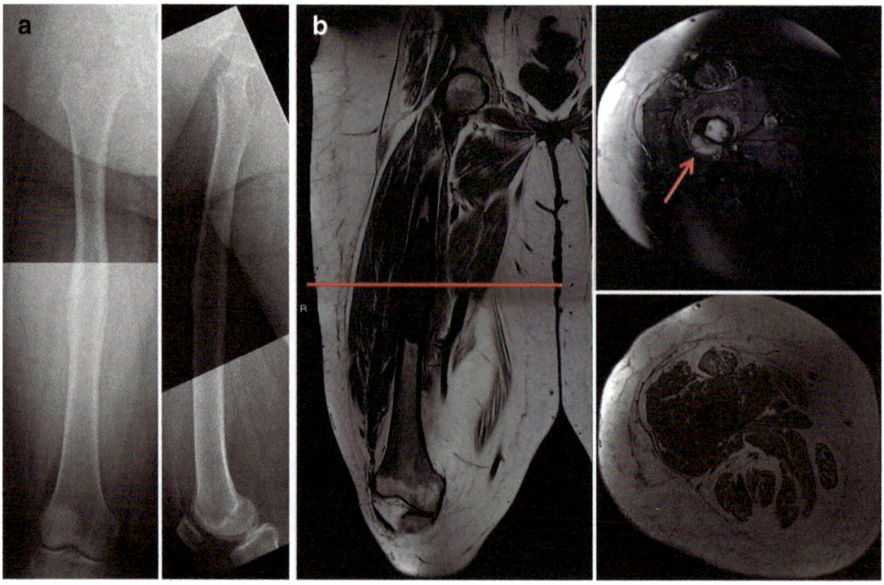

Fig. 21.4 (**a**) Preoperative radiographs, demonstrating a lytic, expansile lesion in the mid-diaphyseal right femur, with subtle cortical lucency on the lateral view indicating pathologic fracture. (**b**) MRI demonstrating a mid-diaphyseal, T1-hypointense, T2-heterogeneously hyperintense lesion with cortical thinning and breakthrough, with an associated soft-tissue component (red arrow)

Positioning, Surgical Approach, and Resection

The patient was positioned in the right-side-up lateral decubitus position using a beanbag. Specific attention was taken to ensure that the contralateral leg would be palpable through the drapes for limb length comparison. The surgical incision was made laterally in line with the femur and was extended through the skin and subcutaneous tissues to the level of the iliotibial band. The iliotibial band was split, and a lateral subvastus approach was performed, keeping the vastus intermedius intact as a surgical margin. The lesion was then localized fluoroscopically and, via both fluoroscopic guidance as well as palpation of the underlying mass, a circumferential wide margin of muscle was maintained around the tumor, carefully developing the plane between the tumor and the profunda femoris artery and vein. The intramedullary longitudinal extent of the tumor was measured preoperatively to be 13 cm, and additional margins proximally and distally were planned. The rotation of the femur was marked, first with electrocautery followed by a surgical marker, proximally and distally, to ensure proper rotational orientation following reconstruction. Transverse cuts were made under direct visualization. A frozen section biopsy was sent from the intramedullary canal of both proximal and distal remnant femur, and bone wax was placed in both ends of the canal. The intercalary resection was passed off as a specimen, and frozen section analyses were negative at the proximal and distal margins. (Fig. 21.5).

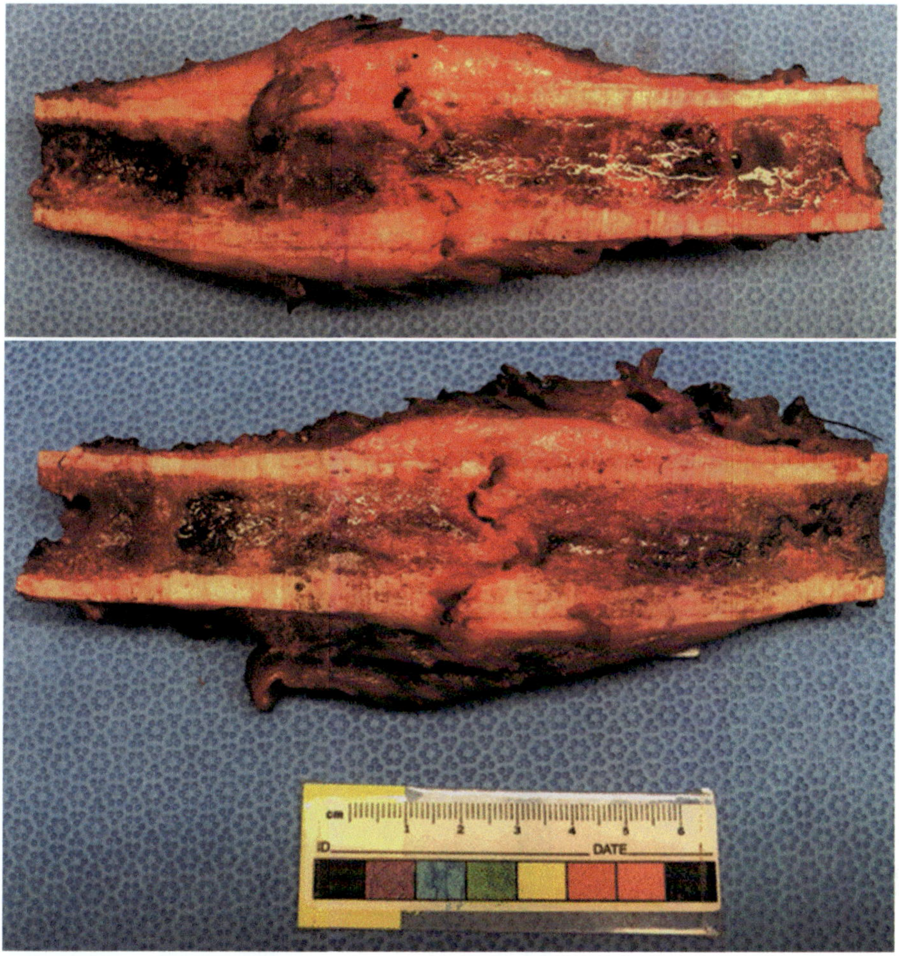

Fig. 21.5 Bivalved resection specimen demonstrating intramedullary tumor, pathologic fracture, and soft tissue extension. Final margins and intramedullary frozen sections were negative

Intercalary Reconstruction

Both proximal and distal femoral canals were prepared by insertion of a ball-tipped guidewire, and subsequent serial reaming up to size 15 mm. The proximal and distal bone stock were both sufficient to accommodate at least 90 mm length stems. After preparation of the proximal and distal bone, trial stems were inserted, and intercalary trials were assembled to recreate the length of bone resected. Leg length, rotation, and alignment were verified. The decision was made for cementation of the implant stems, and both the proximal and distal canals were prepared for cementation. Cement was injected into the canal using a standard cement gun. After the stems were each cemented into position, and following curing of the PMMA, the intercalary implant was assembled and positioned within the defect, taking care to appropriately align rotation with our previous markings. All connecting screws were then locked. Limb length and rotation were checked again and found to be satisfactory. (Fig. 21.6).

Tips and Tricks

- Marking of femoral rotation before osteotomy—the markings for rotation on the remaining femur on both the proximal and distal cuts must be aligned with each other prior to osteotomy. It is the authors' practice to make the marking first with electrocautery, followed by the surgical marker. Failure to mark or failure to align the proximal and distal marks along the same axis, may lead to malrotation.
- Remaining bone stock—ensure that sufficient remaining diaphyseal bone stock will remain following resection to accept the intramedullary component of the device.
- Surgical approach—while the illustrated case utilized a lateral approach to the femur, the selection of the surgical approach should be made based on the anatomy of the resection, i.e., if the tumor abuts the superficial femoral artery, an approach facilitating direct visualization of these structures may be warranted.

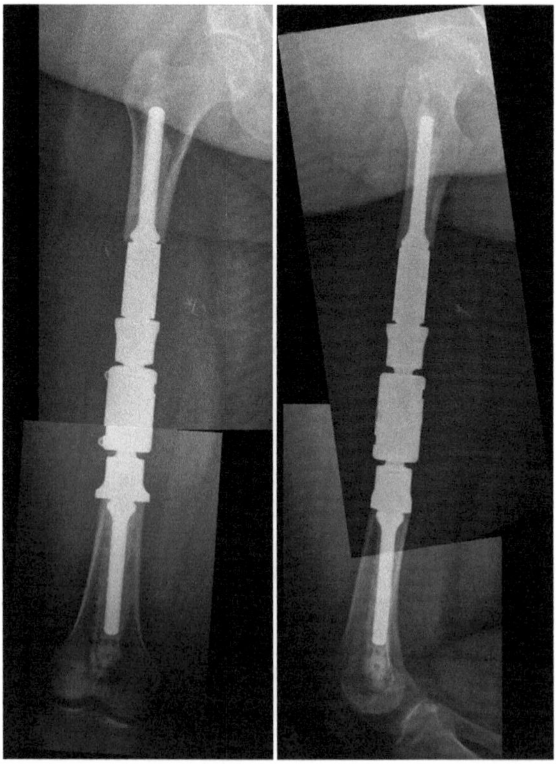

Fig. 21.6 Postoperative radiographs, demonstrating a cemented intercalary prosthetic reconstruction

- Extent of the proximal implant—it is the authors' practice to avoid terminating the proximal stem of the implant distal to the lesser trochanter, to avoid a stress riser in the intertrochanteric region.
- Knowledge of implant system—existing modular intercalary implant systems have minimum resection lengths; knowledge of these devices and appropriate preoperative planning is paramount to ensure appropriate matching of the implant system to the bony defect. If cementing, it is imperative to ensure that the available stem diameters will allow for a cement mantle of appropriate thickness. Finally, surgeons must ensure that stems of appropriate length are available to allow for optimal proximal and distal extent.

Conclusion

Metastatic disease of the femoral diaphysis is commonly encountered in orthopedic oncologic practice. In the vast majority of cases, surgical management involves intramedullary rod fixation. A minority of patients with renal cell carcinoma, papillary thyroid carcinoma, and some breast cancers with later osseous metastasis—particularly those with solitary or oligometastatic disease with a longer predicted survival—may be appropriate for surgical metastatectomy and intercalary reconstruction. The latter has been accomplished using a variety of techniques, including endoprosthetic and allograft reconstruction.

References

1. Sundquist M, Brudin L, Tejler G. Improved survival in metastatic breast cancer 1985–2016. Breast. 2017;31:46–50.
2. Gallagher DJ, Kemeny N. Metastatic colorectal cancer: from improved survival to potential cure. Oncology. 2010;78(3–4):237–48.
3. Luyendijk M, Visser O, Blommestein HM, de Hingh IHJT, Hoebers FJP, Jager A, et al. Changes in survival in de novo metastatic cancer in an era of new medicines. JNCI J Natl Cancer Inst. 2023;115(6):628–35.
4. Ward WG, Holsenbeck S, Dorey FJ, Spang J, Howe D. Metastatic disease of the femur: surgical treatment. Clin Orthop Relat Res. 2003;(415 Suppl):S230–44.
5. Sarahrudi K, Greitbauer M, Platzer P, Hausmann J-T, Heinz T, Vécsei V. Surgical treatment of metastatic fractures of the femur: a retrospective analysis of 142 patients. J Trauma Acute Care Surg. 2009;66(4):1158–63.
6. Miller BJ, Soni EEC, Gibbs CP, Scarborough MT. Intramedullary nails for long bone metastases: why do they fail? Orthopedics. 2011;34(4)
7. Errani C, Mavrogenis AF, Cevolani L, Spinelli S, Piccioli A, Maccauro G, et al. Treatment for long bone metastases based on a systematic literature review. Eur J Orthop Surg Traumatol. 2017;27:205–11.
8. Kotian RN, Puvanesarajah V, Rao S, El Abiad JM, Morris CD, Levin AS. Predictors of survival after intramedullary nail fixation of completed or impending pathologic femur fractures from metastatic disease. Surg Oncol. 2018;27(3):462–7.
9. Anderson AB, Wedin R, Fabbri N, Boland P, Healey J, Forsberg JA. External validation of PATHFx version 3.0 in patients treated surgically and nonsurgically for symptomatic skeletal metastases. Clin Orthop Relat Res. 2020;478(4):808–18.
10. Sørensen MS, Gerds TA, Hindsø K, Petersen MM. Prediction of survival after surgery due to skeletal metastases in the extremities. Bone Jt J. 2016;98-B(2):271–7.

11. Forsberg JA, Eberhardt J, Boland PJ, Wedin R, Healey JH. Estimating survival in patients with operable skeletal metastases: an application of a bayesian belief network. PLoS One. 2011;6(5):e19956.
12. Thio QCBS, Karhade AV, Ogink PT, Bramer JAM, Ferrone ML, Calderón SL, et al. Development and internal validation of machine learning algorithms for preoperative survival prediction of extremity metastatic disease. Clin Orthop Relat Res. 2020;478(2):322.
13. Tseng T-E, Lee C-C, Yen H-K, Groot OQ, Hou C-H, Lin S-Y, et al. International validation of the SORG machine-learning algorithm for predicting the survival of patients with extremity metastases undergoing surgical treatment. Clin Orthop Relat Res. 2022;480(2):367.
14. Zaorsky NG, Liang M, Patel R, Lin C, Tchelebi LT, Newport KB, et al. Survival after palliative radiation therapy for cancer: the METSSS model. Radiother Oncol. 2021;158:104–11.
15. Mirels H. Metastatic disease in long bones a proposed scoring system for diagnosing impending pathologic fractures. Clin Orthop Relat Res (1976–2007). 1989;249:256–64.
16. Ratasvuori M, Wedin R, Hansen BH, Keller J, Trovik C, Zaikova O, et al. Prognostic role of en-bloc resection and late onset of bone metastasis in patients with bone-seeking carcinomas of the kidney, breast, lung, and prostate: SSG study on 672 operated skeletal metastases. J Surg Oncol. 2014;110(4):360–5.
17. Fottner A, Szalantzy M, Wirthmann L, Stähler M, Baur-Melnyk A, Jansson V, et al. Bone metastases from renal cell carcinoma: patient survival after surgical treatment. BMC Musculoskelet Disord. 2010;11:1–6.
18. Langerhuizen DW, Janssen SJ, van der Vliet QM, Raskin KA, Ferrone ML, Hornicek FJ, et al. Metastasectomy, intralesional resection, or stabilization only in the treatment of bone metastases from renal cell carcinoma. J Surg Oncol. 2016;114(2):237–45.
19. Lin PP, Mirza AN, Lewis VO, Cannon CP, Tu S-M, Tannir NM, et al. Patient survival after surgery for osseous metastases from renal cell carcinoma. J Bone Joint Surg Am. 2007;89(8):1794–801.
20. Higuchi T, Yamamoto N, Hayashi K, Takeuchi A, Abe K, Taniguchi Y, et al. Long-term patient survival after the surgical treatment of bone and soft-tissue metastases from renal cell carcinoma. Bone Joint J. 2018;100(9):1241–8.
21. Evenski A, Ramasunder S, Fox W, Mounasamy V, Temple HT. Treatment and survival of osseous renal cell carcinoma metastases. J Surg Oncol. 2012;106(7):850–5.
22. Zaid HB, Parker WP, Safdar NS, Gershman B, Erwin PJ, Murad MH, et al. Outcomes following complete surgical metastasectomy for patients with metastatic renal cell carcinoma: a systematic review and meta-analysis. J Urol. 2017;197(1):44–9.
23. Tai KY, El Abiad JM, Morris CD, Markowski MC, Levin AS. Comparing the responses of osseous versus soft-tissue metastases of renal cell carcinoma to receptor tyrosine kinase inhibitors and immunotherapy. Kidney Cancer. 2020;4(3):151–8.
24. Lemke E, Sahasrabudhe D, Guancial E, Bylow K, Johnson S, Messing E, et al. The role of metastasectomy in urothelial carcinoma: where are we in 2020? Clin Genitourin Cancer. 2020;18(4):e478–e83.
25. Colman MW, Kirkwood JM, Schott T, Goodman MA, McGough RL III. Does metastasectomy improve survival in skeletal melanoma? Melanoma Res. 2014;24(4):354–9.
26. Kask G, Nieminen J, Parry MC, van Iterson V, Pakarinen T-K, Ratasvuori M, et al. Revision rate of reconstructions in surgically treated diaphyseal metastases of bone. Eur J Surg Oncol. 2019;45(12):2424–30.
27. Vermesan D, Prejbeanu R, Haragus H, Dema A, Oprea MD, Andrei D, et al. Case series of patients with pathological dyaphiseal fractures from metastatic bone disease. Int Orthop. 2017;41:2199–203.
28. Arpornsuksant P, Morris CD, Forsberg JA, Levin AS. What factors are associated with local metastatic lesion progression after intramedullary nail stabilization? Clin Orthop Relat Res. 2022;480(5):932–45.
29. Lozano-Calderon SA, Rijs Z, Groot OQ, Su MW, Werenski JO, Merchan N, et al. Outcomes of long bones treated with carbon-fiber nails for oncologic indications: international multi-institutional study. J Am Acad Orthop Surg. 2024;32(3):e134–e45.

30. Herzog LN, Traven SA, Walton ZJ, Leddy LR. The use of carbon fiber implants for impending or existing pathologic fractures. J Orthop Trauma. 2022;36(7):e260–e4.
31. Yeung CM, Bhashyam AR, Groot OQ, Merchan N, Newman ET, Raskin KA, et al. Comparison of carbon fibre and titanium intramedullary nails in orthopaedic oncology. Bone Joint Open. 2022;3(8):648–55.
32. Gonzalez MR, Xu RF, Sodhi A, Fang V, Kim C, de Groot TM, et al. Patient-reported outcomes after intramedullary nailing of oncologic impending or pathologic fractures with carbon fiber or titanium implant. J Am Acad Orthop Surg Global Res Rev. 2024;8(6)
33. Moon B, Lin P, Satcher R, Bird J, Lewis V. Intramedullary nailing of femoral diaphyseal metastases: is it necessary to protect the femoral neck? Clin Orthop Relat Res. 2015;473:1499–502.
34. Miller B, Phillips M, Krech L, Biberstein B, Parker J, Pounders S, et al. Outcomes of simultaneous versus staged intramedullary nailing fixation of multiple long bone lower extremity fractures. Injury. 2023;54(7):110831.
35. Raasck K, Al Farii H, Alnusif N, Wang X, Wang C. The effects of simultaneous versus staged bilateral femoral intramedullary nailing for metastatic disease on survival time and perioperative complications: a systematic review. J Orthop Oncol. 2018;4(2)
36. Zekry KM, Yamamoto N, Hayashi K, Takeuchi A, Alkhooly AZA, Abd-Elfattah AS, et al. Reconstruction of intercalary bone defect after resection of malignant bone tumor. J Orthop Surg. 2019;27(1):2309499019832970.
37. Liu Q, He H, Duan Z, Zeng H, Yuan Y, Wang Z, et al. Intercalary allograft to reconstruct large-segment diaphysis defects after resection of lower extremity malignant bone tumor. Cancer Manag Res. 2020;12:4299–308.
38. Lun D-X, Hu Y-C, Yang X-G, Wang F, Xu Z-W. Short-term outcomes of reconstruction subsequent to intercalary resection of femoral diaphyseal metastatic tumor with pathological fracture: comparison between segmental allograft and intercalary prosthesis. Oncol Lett. 2018;15(3):3508–17.
39. Benevenia J, Kirchner R, Patterson F, Beebe K, Wirtz DC, Rivero S, et al. Outcomes of a modular intercalary endoprosthesis as treatment for segmental defects of the femur, tibia, and Humerus. Clin Orthop Relat Res. 2016;474(2):539–48.
40. Johnson JD, Satcher RL, Feng L, Lewis VO, Moon BS, Bird JE, et al. What is the prosthetic survival after resection and intercalary endoprosthetic reconstruction for diaphyseal bone metastases of the humerus and femur? Clin Orthop Relat Res. 2022; https://doi.org/10.1097/CORR.0000000000002669.
41. Hamada K, Naka N, Tamiya H, Ozaki R, Outani H, Fujimoto T, et al. Intercalary endoprosthetic reconstruction for impending pathological fractures in patients with femoral diaphyseal bone metastases. Eur J Orthop Surg Traumatol. 2009;19:547–51.
42. Fan Y, Zhang B, Guo L, Yao W. Long bone shaft metastasis: a comparative study between cement filling and intercalary prosthesis. World J Surg Oncol. 2023;21(1):374.
43. Pedtke AC, Wustrack RL, Fang AS, Grimer RJ, O'Donnell RJ. Aseptic failure: how does the Compress® implant compare to cemented stems? Clin Orthop Relat Res. 2012;470(3):735–42.
44. Farfalli GL, Boland PJ, Morris CD, Athanasian EA, Healey JH. Early equivalence of uncemented press-fit and Compress® femoral fixation. Clin Orthop Relat Res. 2009;467(11):2792–9.
45. Zimel MN, Farfalli GL, Zindman AM, Riedel ER, Morris CD, Boland PJ, et al. Revision distal femoral arthroplasty with the Compress® prosthesis has a low rate of mechanical failure at 10 years. Clin Orthop Relat Res. 2016;474(2):528–36.
46. Tedesco NS, Van Horn AL, Henshaw RM. Long-term results of intercalary endoprosthetic short segment fixation following extended diaphysectomy. Orthopedics. 2017;40(6):e964–e70.

Management of Metastatic Disease Around the Knee

22

Marcos Galli Serra, Bruno Terrarrossa Retegui, and Walter Parizzia

Skeletal metastases can develop in two-thirds of cancer patients. Although almost any carcinoma can metastasize to bone, those that do so most frequently are prostate, breast, renal, lung, or thyroid. Bone metastases are often the most symptomatic and disabling manifestation of secondary cancer, and it is essential to define appropriate service delivery levels to provide high-quality clinical care. The improvement in the average survival after recognizing bone metastases has been most marked in breast and prostate cancer, with breast cancer accounting for most cases requiring orthopedic treatment. Despite advances in both orthopedic oncology and spinal surgery, there is relatively low awareness in hospital and primary care settings of what can be achieved. Studies and clinical experience suggest significant variation in the standards of care for patients with bone metastatic disease and the surgical techniques and implants used. Even though guidelines were introduced over 20 years ago, errors in managing patients with secondary bone lesions remain common [1]. There are clinical surgical situations that can lead to poor outcomes, especially when these lesions are around the knee:

A. Not intervening prophylactically when it is appropriate.
B. Using inappropriate surgical implants that depend on long-term fracture consolidation. Failure of union commonly seen in pathologic fractures leads to high failure rates in such devices.
C. Lack of appreciation of the available treatment options, particularly in more advanced cases of the disease, where concurrent clinical service opinions are often not considered.

M. Galli Serra (✉) · B. Terrarrossa Retegui · W. Parizzia
Hospital Universitaro Austral, Facultad de Ciencias Biomedicas, Universidad Austral, Buenos Aires, Argentina
e-mail: Mgalli@cas.austral.edu.ar; BTERRARO@cas.austral.edu.ar; WPARIZZI@cas.austral.edu.ar

J. Pretell-Mazzini, S. A. Lozano Calderón (eds.), *Multidisciplinary Approach of Musculoskeletal Metastasis*, https://doi.org/10.1007/978-3-032-09498-8_22

Secondary bone lesions typically present with one of these three clinical manifestations:

A. Acute admission with pathological fracture.
B. Referral from the oncologist/surgical oncology team (surgeon, oncologist, and radiologist).
C. Referral to an orthopedic clinic with unexplained musculoskeletal pain with or without a history of oncologic disease.

Pain is the most common clinical symptom and varies from a dull ache to deep pain, exacerbated by weight-bearing and sometimes worsens at night. A transverse or short oblique fracture that occurs after a period of prior pain and a relatively low-energy trauma should raise suspicion of a pathological fracture.

Although femur is a common site for metastasis, distal femoral metastases are less common than those from the proximal end. Metastases usually originate from breast, lung, thyroid, kidney, and prostate primary carcinomas. Condylar metastases are easily treated by curettage or resection and reconstruction with cement packing with additional plate and screw fixation or arthroplasty reconstruction ranging from primary joint replacements with or without augments, stems, and /or cones and ultimately, cemented mega prostheses when necessary. The decision between joint sparing and joint reconstructive procedures depends on the location of the metastatic lesion in relation to the joint, the amount of healthy bone between the knee joint and the lesion, and the presence of coexistent osteoarthritic disease. Extraosseous soft tissue extension is uncommon except for cortical metastases secondary to lung cancer and renal cell carcinoma. The former are radiosensitive and therefore, subjective of intralesional treatment followed by radiation while the latter is not, and therefore, more commonly treated with resections that require sacrifice of soft tissue attachments to attain local control. In this setting hinged arthroplasty reconstruction or megaprostheses is required. Additional procedures for local control in renal cell carcinoma of less invasive nature are gaining traction. These include radiofrequency ablation, cryoablation and/or microwave ablation. However, data in the oncologic outcomes of this type of interventions remains limited.

Isolated lesions are uncommon but may occur in relation to thyroid and renal metastases. Occasionally, late presentation of breast carcinoma metastases many years after the treatment of the primary may present as a solitary bone lesion as well. In all these cases, en bloc resection may be attempted with local curative intent and aiming for overall survival increase. More often, however, disease arises in multiple sites within a single bone and progression of disease is to be expected. In planning surgical management, protection of the whole bone should be sought in anticipation of further development of metastatic disease. Long-stemmed cemented intramedullary fixation is preferred as a method for protecting the femoral diaphysis, whereas the distal metaphysis is more commonly treated with retrograde nailing, plating or arthroplasty reconstruction.

Currently, there is a lack of auditable standards for the treatment of this pathology. This is a topic where, until recently, there has been a relative scarcity of scientific data. Prospective controlled trials of different treatments for bone metastasis are scarce. Our goal is to reflect the clinical consensus on best practices in the surgical area. There is very little evidence from randomized controlled trials for surgical intervention; most studies are retrospective case series or individual case reports [2]. Many studies are from the Scandinavian Sarcoma Group, which has the world's largest registry of surgically treated skeletal metastases [3]. There is additional evidence that replacement with oncologic reconstruction prostheses of the femur is beneficial for patients with proximal and distal femoral metastases [4, 5].

Surgical Indications/Non-surgical Indications of Metastasis Disease Around the Knee

The prognosis for many patients with secondary bone lesions, particularly those without visceral lesions, has significantly improved in recent years, mainly due to advances in medical therapy, including hormonal blocking treatment, bisphosphonates, chemotherapy, immunotherapy and targeted agents.

Orthopedic surgeons responsible for treating secondary bone lesions around the knee must ensure that oncologists and other members of the multidisciplinary team treating cancer patients are aware of the benefits of orthopedic interventions for these patients. Considering that any bone lesion in the lower limb can cause high rates of transient or permanent disability and keep the patient, who already has a shortened life expectancy due to their underlying disease, away from their usual life for a long time.

The role of the orthopedic oncology surgeon in managing this pathology is divided into four main categories:

A. Establish the diagnosis. Biopsy or any procedure leading to diagnosis.
B. Medical treatment and orthopedic advice for bone lesions with Mirels score <7 [6].
C. Surgical treatment of painful bone lesions and prevention of fractures (mostly Mirels score >8).
D. Stabilization or reconstruction after a pathological fracture.

The orthopedic surgeon will also assess the suitability of surgery and can help coordinate care and involve other specialties as appropriate, such as palliative care, primary care and oncology.

The objectives of surgery are to relieve pain and maintain or restore function. There is evidence that appropriate and timely treatment of metastatic bone disease can improve survival [7]. Treating lower limb lesions can be crucial for maintaining mobility and performance status, and thus, access to other treatments such as chemotherapy or radiotherapy. The general orthopedic principles underlying the management of the pathological fracture are as follows:

A. The procedure should provide immediate mechanical stability, allowing rapid return to pre-event function as soon as possible.
B. The fixation should be durable (therefore, implant selection and an estimation of life expectancy are essential).
C. All lesions in the affected bone should be stabilized if reasonable. Treatments should always be appropriate for the stage of the disease and the patient's general condition and should reflect the patient's treatment preferences.
D. All patients requiring surgery should be under the care of a specialist in the field.
   Patients should benefit from a multidisciplinary discussion of their care situation. In most cases, the discussion will be communicated to the patient and their family, documenting the benefits and potential complications derived from it.

Pre-operative Planning

Patients with oncological disease and bone metastases are complex patients with associated morbidity and mortality. All patients should be evaluated and assessed in an MDT, where the decisions of treatment are made: Firstly, if the patient is a candidate for surgery. Secondly, the decision of what type of reconstruction is going to be used. For this evaluation, the patient must have all their staging studies complete. The evaluation of these lesions must be thorough and multi-modal, utilizing a combination of imaging studies to determine the extent of disease, evaluate bone integrity, and guide the choice of appropriate surgical intervention. The surgical evaluation of metastatic lesions in the distal femur and proximal tibia requires plain radiographs, CT, and/or MRI. These imaging modalities provide critical information about the extent of disease, the involvement of adjacent structures, and the overall surgical feasibility. A clear understanding of how these imaging studies interrelate and inform surgical decisions is essential for achieving optimal outcomes, balancing oncologic control with functional preservation.

Metastatic bone disease in the distal femur and proximal tibia presents a unique challenge for orthopedic surgeons due to the complex anatomy, functional importance of these regions, and the need for precise surgical planning.

Everything regarding assessment and staging studies has already been explained in other chapters, so in this section, we will only focus on the results of local studies around the knee. They should include at least the following.

Plain Radiographs (X-ray)

Plain radiographs are still the first-line imaging modality for suspected bone metastasis. They provide a first assessment of the skeletal integrity, the presence of lytic or blastic lesions, and the overall distribution of the disease. Complete x-rays of the entire bone segment affected should be included. When evaluating the distal femur and proximal tibia, radiographs should include the entire femur and tibia, from the hip to the ankle, to assess for any possible involvement of adjacent joints, especially the knee. This is crucial for figuring out whether joint preservation is possible or if more extensive surgical resection is needed. In the case of distal femur, the distal diaphysis and metaphysis of the bone should be evaluated for coexistent lesions in addition to the integrity of the knee joint itself. The cortical integrity, periosteal reaction, and soft tissue involvement should be carefully assessed, as these can influence surgical decisions, particularly with respect to fixation vs. reconstruction options. For this reason, it is helpful to have additional imaging studies when radiographs are not sufficient to evaluate appropriately the bones, or they look suspicious. Computerized axial tomography and MRI can provide additional helpful information are about to resect or reconstruct in cases of oligometastatic disease.

Computed Tomography (CT) Scan

Detailed Bone Assessment: A CT scan is particularly useful for evaluating the bony architecture and extent of metastatic lesions, especially in regions of complex anatomy such as the distal femur and proximal tibia. CT provides detailed cross-sectional imaging, allowing for the assessment of cortical involvement, the extent of bone destruction, and any fractures that may require stabilization. CT is indicated when there is uncertainty about the extent of the lesion on plain radiographs, particularly when planning for bone resection or reconstructive procedures. It is also beneficial for evaluating the involvement of the bone marrow and surrounding structures that may not be well visualized on X-rays. CT scans help in planning for the margins of resection in oligometastatic disease, the selection of appropriate surgical approaches, and the design of prosthetic reconstruction, if needed. It aids in the assessment of the distal femur and proximal tibia's mechanical axis, which is crucial when considering limb salvage.

Magnetic Resonance Imaging (MRI)

Soft Tissue and Bone Marrow Involvement: MRI is invaluable in evaluating the soft tissue components, including the involvement of the muscles, ligaments, and tendons around the knee, as well as the bone marrow. It provides high-contrast resolution for detecting bone marrow edema and soft tissue infiltration, both of which may influence the surgical approach. MRI is particularly helpful in assessing the extent of the tumor within the metaphyseal region and evaluating the knee joint for any intra-articular extension, which may influence the decision to preserve the joint or proceed with joint replacement. MRI is often required when the clinical presentation suggests soft tissue involvement or when there is concern for cortical breach and soft tissue invasion. Additionally, MRI is essential for assessing the spinal cord and neural structures when there is suspicion of metastatic involvement in the spinal axis.

Role of Imaging in Surgery Planning

Preoperative Planning: Comprehensive imaging using plain radiographs, CT, and MRI is essential to define the extent of the metastasis, plan the surgical approach, and assess the feasibility of various reconstruction options, including the use of prostheses or allografts.

Surgical Decision-Making: The goal of surgery in metastatic bone disease is generally to relieve pain, prevent fractures, and restore function, where possible. The decision to proceed with limb salvage or amputation is guided by the extent of the lesion, the mechanical integrity of the bone, and the involvement of surrounding tissues, which are well-documented in imaging studies. If the lesion is localized with adequate bone stock, resection with reconstruction (e.g., prosthetic replacement or bone grafting) may be feasible. In cases of extensive soft tissue or joint involvement, amputation may be considered.

Intraoperative Guidance: Intraoperative imaging, including fluoroscopy, may also be used for real-time guidance, particularly when performing osteotomies or placing fixation devices.

Preoperative Evaluation

It is crucial that some patients who may not wish to consider surgery in a palliative context engage in open discussions with their families and healthcare providers. In patients with low physical performance and limited life expectancy, management strategies should be discussed with the oncologist prior to proceeding with surgery [8–11]. The orthopedic surgeon must evaluate the suitability of any surgical intervention. The most critical factors affecting post-operative survival include the primary tumor, the presence of metastases in organs, overall health status, and the number of bone metastases. Patients with bone metastases arising from primary breast or kidney cancers tend to have a better prognosis than those with metastatic lung cancer. Estimating the burden of disease and survival is crucial in determining the procedure and implant selected. For patients with favorable prognosis, procedures that allow for durable reconstruction should be considered, and en-bloc resection should be evaluated in solitary metastases with favorable prognosis carcinomas, particularly in renal cases [12, 13].

It is essential to address the overall status of the patient prior to surgery, which necessitates a comprehensive medical history and thorough physical examination. Comorbidities should be optimized, and the evaluation should pay particular attention to nutritional status, respiratory complications related to malignancy (such as infections and pleural effusions), and pulmonary and myocardial toxicity secondary to chemotherapeutic agents (especially anthracyclines, trastuzumab, bleomycin, busulfan, and mitomycin). Both the disease process and treatment may have affected bone marrow function and coagulation. A complete blood count and coagulation assessment should be performed, and hematological support and appropriate bone marrow growth factors should be considered to minimize the risks of infection and bleeding. Additionally, electrolyte imbalances, including hypercalcemia, should be assessed and, if possible, corrected prior to surgery, along with monitoring fluid balance.

Imaging studies of the entire affected bone are a minimum requirement. Staging should involve appropriate studies and investigations based on the clinical situation. If an extensive reconstructive procedure is being considered, systemic restaging is recommended. Early engagement with the anesthesia service is crucial for planning analgesia in opioid-tolerant patients undergoing major surgery, as well as for perioperative management, which includes potential significant blood loss, vascular access, and postoperative care (HDU, epidurals, etc.). Patients with active thoracic compromise may have particular risks during surgical interventions involving cemented implants.

Tumors associated with bleeding risk (such as renal, thyroid tumors and in general tumors with associated large soft tissue components) should be considered for preoperative embolization, as this has been shown to significantly reduce blood loss, packed red blood cell transfusion volume, and operative time. Ideally, embolization should be performed less than 48 hours prior to surgery [14].

Surgical Options

Minimally Invasive Techniques for Patella, Distal Femur, and Proximal Tibia (Radiofrequency Ablation, Cementoplasty, Cryotherapy)

These options have demonstrated pain relief and decreased opioid consumption in the short term. Patients with painful bone metastasis could benefit more from combining cryotherapy or RFA with other therapies, such as radiotherapy and bisphosphonates, than from single therapies.

In metastatic bone disease around the knee, these minimally invasive treatments are reserved for small lesions, outside of weight-bearing areas, and patellar lesions.

In our experience, we consider radiofrequency as the best option for this type of metastases due to the low complication rate and its availability.

Patella

The incidence of metastatic lesions to the patella is lower than 0.12% [15, 16]. A Medline search from studies published between 1960 and 2016 yielded 44 reported cases of patellar metastases. Primary tumors that have been reported to metastasize to the patella include carcinomas of the lung, kidney, breast, esophagus, uterine cervix, skin (malignant melanoma), bowel, larynx, head, and neck. The most common symptom for presentation is anterior knee pain as the first manifestation. Surgery or minimally invasive techniques are optimal methods for most symptomatic patients who possess a patellar metastasis.

Depending on the characteristics of the lesion and the clinical status of the patient, different techniques may be performed. Alternatives include Lesion excision for small marginal lesions, curettage with cement packing if the peripheral shell of bone is appropriate. When metastatic disease affects more than 50% of the patella or involves its articular surface, we suggest performing a patellectomy with reconstruction of the extensor mechanism. Alternatively, resurfacing of the patella with a cemented patellar button may be sufficient in some cases. To preserve motional function, the intact extensor apparatus should be maintained or the extensor structure reconstructed. After performing a patellectomy, we suggest performing an adequate reconstruction of the extensor apparatus, using, if necessary, a Prolene mesh. Radiotherapy is recommended in cases of partial resection in order to avoid progression of the disease [17, 18].

The contraindication for these procedures is when the patient is not fit to undergo surgery. In such cases, isolated radiotherapy remains a valid option.

Complications that may occur include rupture of the extensor apparatus and surgical site infection, potentially progressing to septic arthritis.

Distal Femur and Proximal Tibia

In metastatic bone disease of the distal femur and proximal tibia, the treatment will depend on the type of primary tumor, the size of the lesion, and the patient's age and functional demand.

In cases of renal-origin tumors, prior to surgical resection, we always perform vascular embolization to reduce the risk of bleeding if an intralesional procedure is planned.

Internal fixation with a plate combined with cement augmentation is indicated for lesions involving less than 30% of the total metaphysis or that do not invade the articular surface. The advantages of this procedure are shorter surgical time, low surgical site infection rates, and the possibility of converting to a prosthetic replacement if needed.

In borderline-sized lesions, away from the articular surface, where primary arthroplasty would require complete metaphyseal resection, especially in younger patients, we opt for tumor resections using cutting guides based on 3D CT planning and reconstruction with 3D printed metal/polyethylene implants.

However, the gold standard for metastatic bone disease of the distal femur and proximal tibia is non-conventional total knee arthroplasty, using modular and hinged systems depending on the length of the lesion. The use of arthroplasty revision cones has facilitated the management of metaphyseal lesions, especially when used in combination with long stems. Ultimately, if bone loss is significant and not sufficient for any revision construct, cemented distal femoral or proximal tibia replacements are strong tools for reconstruction. These provide immediate weight bearing despite the extent of the surgery. Recent studies and series are demonstrating that for individuals with good cortical bone, uncemented stems may be used safely [19, 20].

In cases where metastatic lesions involve the anterior tibial tubercle, reconstruction of the extensor apparatus must be planned to avoid loss of function and preserve the individual's quality of life. For these situations, reconstructions with attachment of the patellar tendon to the prosthesis are our preference. In some cases, a gastrocnemius flap can be added over this reconstruction. Due to eventual laxity of the soft tissues, some advocate for prolene mesh augmentations, while others prefer to use extensor mechanism allografts or reinforcement with Achilles tendon allografts. Our preference is the latter [21–25].

Beyond the improvement in pain and early ambulation with a joint replacement, the disadvantage of these therapeutic options is that they are major surgeries performed on patients who often do not have an optimal clinical status.

Complications include prosthetic dislocation, infection, periprosthetic fracture, soft tissue coverage problems and component loosening due to disease progression or infection.

Surgical Technique: [26–28]

Patella

(a) Patient positioning and Surgical Incision: The patient could be in the supine or lateral decubitus position. Our preference is supine positioning.
(b) Steps for tumor resection: Depending on the location of the lesion, the approach may be anterior, medial, or lateral. In all cases, access is made through the area of cortical weakness to the cavitary lesion, and treatment is performed. In the case of partial resection or complete patellectomy, an attempt is made to perform through a medial arthrotomy.
(c) Reconstruction: Contained cavitary lesions: Fill with cement and possibly augment with cannulated screws. Uncontained cavitary lesions: Partial or complete resection of the lesion followed by reconstruction of the extensor apparatus. Reconstruction can be performed by suturing both ends of the opening, or it may be associated with the augmentation of the same using synthetic mesh or tendon allografts for reinforcement of the extensor apparatus.
(d) soft tissue reconstruction as needed: In cases of partial resection of the extensor apparatus in its soft tissues, the defect can be closed using mesh with end-to-end suturing, or tendon grafts from the patient or cadaveric sources can be used for reconstruction and reinforcement of the extensor apparatus using known techniques.
(e) Wound closure: The wound is closed in layers without major complications. It may occur that in reconstructions that increase the size of the knee (e.g., extensor apparatus) or in patients with multiple previous surgeries, it becomes necessary to use a rotational flap from the inner gastrocnemius and a free skin graft.
(f) Tips and Tricks: The key is to identify areas of weakness and potential failure during the surgery to reinforce and prevent ruptures of the extensor apparatus.

(g) Pearls and Pitfalls: Lesions in the patella are very exposed in the anterior region of the knee and very close to the skin and bony structures (especially the femur). It is a region not commonly affected by metastatic disease, requiring a high degree of suspicion as patients with metastatic disease may experience knee pain due to many other causes. The patient must remain with a knee extension splint for at least 6 weeks. After this period, they will begin progressive rehabilitation.

Distal Femur

Depending on the type of reconstruction, the position of the patient can change.

Retrograde IM Nailing

(a) Patient positioning and Surgical Incision: The patient should be in a supine position with a support under the thigh to maintain the necessary semi-flexed position of 30°. The incision should be made according to the surgical technique for the selected retrograde intramedullary nail insertion. In general, this involves a parapatellar approach that allows access to the intercondylar area where, under radioscopic control, the precise entry point for the nail is determined [6].
(b) Steps for tumor resection: In most cases, when using this reconstruction technique, extensive tumor resections are not performed, as this technique largely relies on the external structure of the distal femur for support and fixation of the reconstruction.
(c) Reconstruction: Depending largely on the anatomopathological diagnosis of the lesion, the technique used will vary. In the case of using intramedullary nails, this is generally done when one decides to preserve the bony structure and "bypass" the lesion, aiming to obtain fixation proximally and distally.
(d) In cases of extensive lytic lesions or cortical defects, careful consideration should be given to the nail size to avoid stress risers and to distribute load appropriately. Cement augmentation may also be considered in conjunction with reaming to fill voids and improve stability.
(e) Soft tissue reconstruction as needed: In this type of reconstruction, soft tissue reconstruction is generally not needed, as it is not usually associated with significant resections.
(f) Wound closure: The wound is closed in the usual manner.
(g) Tips and Tricks:
   - Correct patient positioning.
   - Proper imaging to facilitate the placement of nails and locking screws without difficulty. Check the fluoroscopy C-arm before incision. It is possible that dual radiation beam might be used.

- Use several surgical drapes to cover de c-arm to have full movement without risk of contamination.

(h) Pearls and Pitfalls.
- Correctly localize the entry point of the nail. Is mandatory the correct view of Blumensaat's line.
- Use a triangular radiolucent bolster under the knee to maintain its position during surgery.
- When placing the distal locking screws, use proper imaging in both the anteroposterior and lateral views.
- Ensure thorough preoperative planning, including the assessment of the femoral neck and knee joint stability, to avoid complications during the procedure.

Controlled reaming can increase bone-biomaterial contact and enhance the stability of the nail, especially in areas of cortical thinning or lytic lesions caused by metastasis

- Avoid excessive reaming, which can weaken the bone and increase the risk of fracture propagation.
- The diameter of the intramedullary nail should typically be 1–2 mm smaller than the final reamed diameter of the canal. This ensures a snug fit without excessive pressure, which could compromise the bone or impede insertion. (If the canal is reamed to 12 mm, the nail diameter should generally be 10–11 mm.)
- Oversized nails relative to the reamed diameter can cause cortical damage or increase insertional forces, while undersized nails may result in suboptimal stabilization.

Intratumoral Resection and Cement Reconstruction with Plate

(a) Patient Positioning and Surgical Incision: The patient should be in a supine, lateral position, or medial with leg in the frog leg position if accessing the medial portion of the knee.

(b) Steps for Tumor Resection: The site to be operated on and the approach, whether medial or lateral, is defined preoperatively. If the approach is lateral, a subvastus approach is attempted, ensuring not to damage the lateral collateral ligament at its insertion on the epicondyle. In the case of a medial approach, if a slightly more posterior resection is decided, the medial collateral ligament should be detached and can be re-anchored with simple sutures.

Once access to the tumor site is achieved, a window of sufficient size is created to progressively resect the lesion using curettes until healthy, bleeding bone is encountered. Using electrocautery at 100 and with spray or fulgurate mode is a good option to control bleeding but also as an adjuvant in addition or alternative to high-speed burr and subsequent electrofulguration with an argon electrocautery can be used. The use of advanced hemostatic tools, such as the

Aquamantys® System, can significantly reduce intraoperative bleeding, improve surgical visibility, and enhance patient outcomes. The Aquamantys is a bipolar electrosurgical device that combines radiofrequency energy with saline irrigation to achieve hemostasis through a process called transcollation. This technology is particularly effective for managing bleeding from cancellous bone and vascularized lesions, where traditional methods like packing or coagulation may be insufficient. The Aquamantys system may be more expensive than conventional hemostatic methods, so its use should be prioritized for cases where bleeding control is critical. Surgeons unfamiliar with the device may require training to optimize its application and efficiency

(c) Reconstruction: Once cleaned, the site is filled with cement if the lesion is small. If the lesion is very large, stability can be increased with prophylactic osteosynthesis, and after reconstruction, the cavity is filled with cement, ensuring the screws are trapped within the cement. The osteosynthesis should be locked, as the primary fixation will be provided by its attachment to the cement and the osteosynthesis material, given the low bone quality.

(d) Soft Tissue Reconstruction as Needed: In both approaches, the reconstruction should be covered by advancing the medial or lateral vastus, as appropriate, to cover the osteosynthesis material. In the case of the lateral approach, the fascia lata should be closed in flexion, ensuring that the screws and plate do not contact it to avoid IT friction syndrome.

(e) Wound Closure: In our experience, we attempt to close the wounds in knee flexion as much as possible to allow for quicker mobility.

(f) Tips and Tricks:

Ensure thorough preoperative planning, including imaging studies to define the extent of the tumor and the best surgical approach.

Use intraoperative imaging to confirm the correct placement of instruments and implants.

(g) Pearls and Pitfalls:

Correctly localize the entry point of the nail.

Use a triangular radiolucent bolster under the knee to maintain its position during surgery.

When placing the distal locking screws, use proper imaging in both the anteroposterior and lateral views.

Avoid excessive reaming, which can weaken the bone and increase the risk of fracture propagation. Ensure the osteosynthesis material does not contact the fascia lata to prevent IT friction syndrome.

Intratumoral Resection and Reconstruction Using Arthroplasty Revision Techniques

Intratumoral resection, a palliative approach often used in cases of metastatic bone disease, involves removing the bulk of the tumor while sparing critical structures. Reconstruction after such resections in the metaphyseal and diaphyseal regions of long bones can be challenging, particularly when addressing extensive bone loss or structural instability. Arthroplasty revision techniques, including the use of metaphyseal cones and long-stem prostheses (cemented or non-cemented), provide durable solutions for stabilization and functional restoration.

Metaphyseal Cones: Provide structural support in the setting of metaphyseal bone loss by transferring load from the implant to the remaining intact bone. Enhance implant fixation in compromised bone. They are typically made of porous metal (e.g., tantalum or titanium) to promote osseointegration and biological fixation.

The surgical technique includes, after intratumoral resection, the bone cavity is debrided to remove necrotic or infiltrated tissue. The cone is press-fit or lightly impacted into the remaining metaphyseal bone. The cone serves as a foundation for the femoral or tibial component of the prosthesis. It provides excellent primary stability, promotes long-term osseointegration, and can be used with both cemented and non-cemented stems.

In some cases, Long-Stem Prostheses are needed also Used for diaphyseal reconstruction when bone loss extends beyond the metaphysis, requiring additional support. The Cemented Stems: Provides immediate stability by anchoring the prosthesis to the bone using polymethylmethacrylate (PMMA) cement. It is ideal for patients with poor bone quality or extensive lytic lesions. It Allows for quick weight-bearing postoperatively. Cemented stems are inserted into a prepared canal filled with cement, ensuring uniform distribution for optimal fixation.

In some specific patients, a non-Cemented Stems can be used (Typically preferred in younger patients or those with limited metastatic burden). It relies on press-fit fixation and subsequent osseointegration, is suitable for patients with adequate cortical bone stock and expected long-term survival. The stem is selected based on the length and diameter required to achieve diaphyseal stability as they are inserted into the canal with a tight press-fit, often aided by preoperative templating and intraoperative trialing. The advantages of Long-stem prostheses is that they distribute load evenly, minimizing stress risers and also provide robust stabilization for extensive resections.

In some cases, these techniques can be combined: As "Cone and Cemented Stem". As it combines the structural support of a metaphyseal cone with the immediate fixation of a cemented stem. It is used in cases with extensive bone loss and poor-quality diaphyseal bone.

Preoperative planning is essential in this type of reconstruction: Imaging studies (X-rays, CT, MRI) are critical for determining the extent of bone loss, the tumor's location, and the quality of remaining bone and templating for cone size and stem length ensures proper fit and alignment.

Arthroplasty revision techniques using metaphyseal cones and long-stem prostheses (cemented or non-cemented) are effective strategies for reconstructing bone defects after intratumoral resection. These approaches provide durable stability, optimize functional outcomes, and allow for early mobilization in patients with metastatic bone disease. With careful planning and technique, these methods can address the challenges of extensive bone loss while improving the quality of life for affected patients.

Resection and Prothesis Reconstruction

(a) Patient Positioning and Surgical Incision: The patient should be in a supine or lateral position based on surgeon's preference. If the resection is distal, some may prefer to use a tourniquet. In our experience, we prefer not to use it for larger resections. The incision might include the resection of the biopsy tract. Our preference is a straight anterior incision with medial parapatellar arthrotomy if possible. Extended distally according to preoperative planning. If the proximal resection is wide or planned, we choose to cross the approach laterally and perform a lateral parapatellar arthrotomy. This results in a lateral thigh approach with the possibility of continuing proximally with a posterolateral hip approach.

The segment of the femur to be resected should be measured.

(b) Steps for Tumor Resection: We perform the medial parapatellar incision with extension over the medial region of the quadriceps tendon and muscle extension. The approach is deepened to the distal femur. If the lesion to be resected extends higher and the approach needs to be extended proximally, we must consider releasing the adductor ring and identifying the femoral neurovascular bundle in that region.

A key step when performing the anterior approach is to remember that once the gastrocnemius muscles are detached from the posterior condyles, whether medially or laterally, the popliteal vessels appear. This is a critical moment in surgery to pay attention and to avoid collateral vascular and nerve damage.

In cases where the lesion has a more posteromedial soft tissue component, a more medial and thus posterior approach can be used, anatomically utilizing the space between the sartorius muscle and the vastus medialis. We generally use this type of approach more for primary tumors.

Once the bone is free from all surrounding soft tissues, we perform the osteotomy using an oscillating saw with irrigation. The cut is transverse (Figs. 22.1 and 22.2). Although it is controversial in some cases, we take a sample from the remaining medullary canal, which is sent for frozen section analysis to determine if a larger bone cut is needed and to obtain histological certification to identify any residual disease in the canal. This helps in evaluating the need for subsequent radiotherapy.

(c) Reconstruction: We perform the reconstruction with a modular oncologic reconstruction prosthesis of the knee with a rotational hinge. We try to have curved and straight stems available, depending on the magnitude and length of the resection. The gold standard for this type of patient and reconstruction is a cemented stem. Before cementing the stem, a plastic cement restrictor is

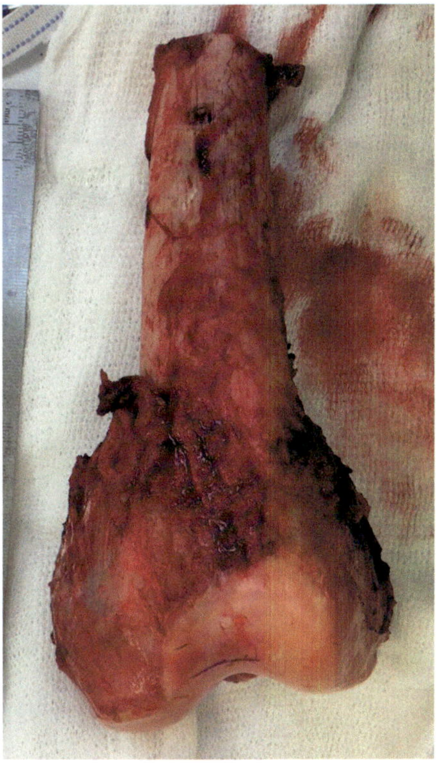

Fig. 22.1 Distal femur resection anterior. Note that there is a small cuff of soft tissue resection in the lateral condyle, due to tumor cortical disruption

always used in the medullary canal, and third or fourth-generation cementing is performed to achieve proper fixation of the stem, leaving a 2 mm of cement mantle between the bone and the stem. In some special situations, an uncemented stem can be used (good cortical width, younger patients with oligometastatic disease), in these situations we try to place the widest stem possible (Fig. 22.3).

It is crucial when placing the femoral stem to have a reference, either from the implant or by marking the bone, for correct femoral orientation and rotation in relation to the line aspera (Figs. 22.3 and 22.4), ensuring proper patellofemoral tracking (Fig. 22.5). This can be verified during the placement of the trial components of the implant by testing knee flexion-extension tracking and making the necessary markings before the final implantation.

(d) Soft Tissue Reconstruction as Needed: We must ensure and perform the necessary muscle releases so that the implant is fully covered along its entire length.
(e) Wound Closure: We close the wound with the knee in almost 90° flexion in all planes. If the approach has been extended more medially or laterally, a lateral or medial release can be performed as needed to ensure proper patellofemoral tracking. In all cases, we use a deep drain, which we remove once its output is no more than 30 cc/24 h.

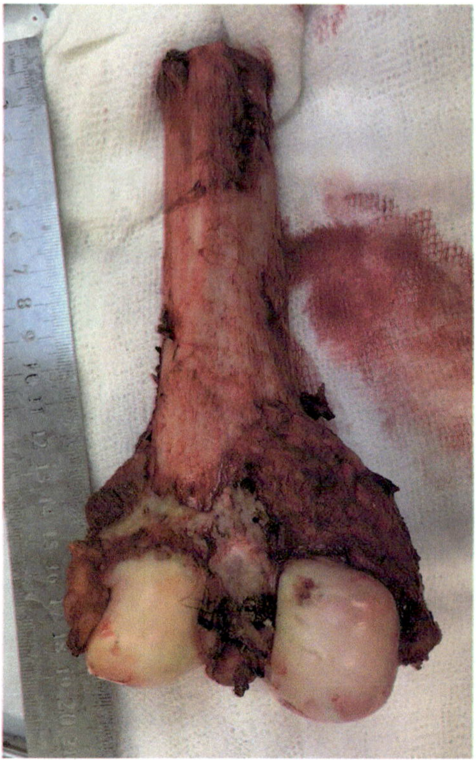

Fig. 22.2 Distal femur resection posterior. Note that there is a small cuff of soft tissue resection in the lateral condyle, due to tumor cortical disruption

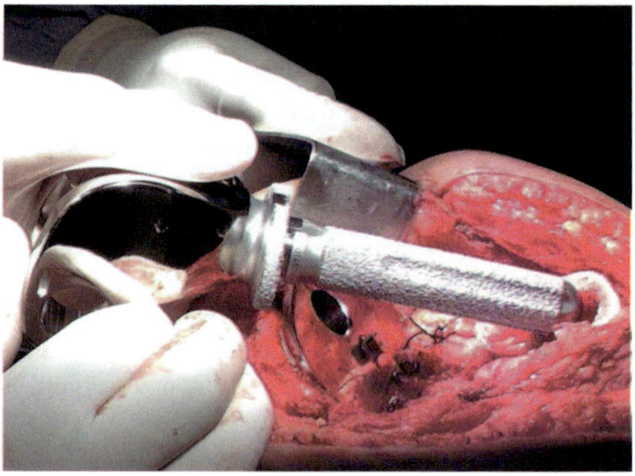

Fig. 22.3 Distal femur straight non-cemented stem

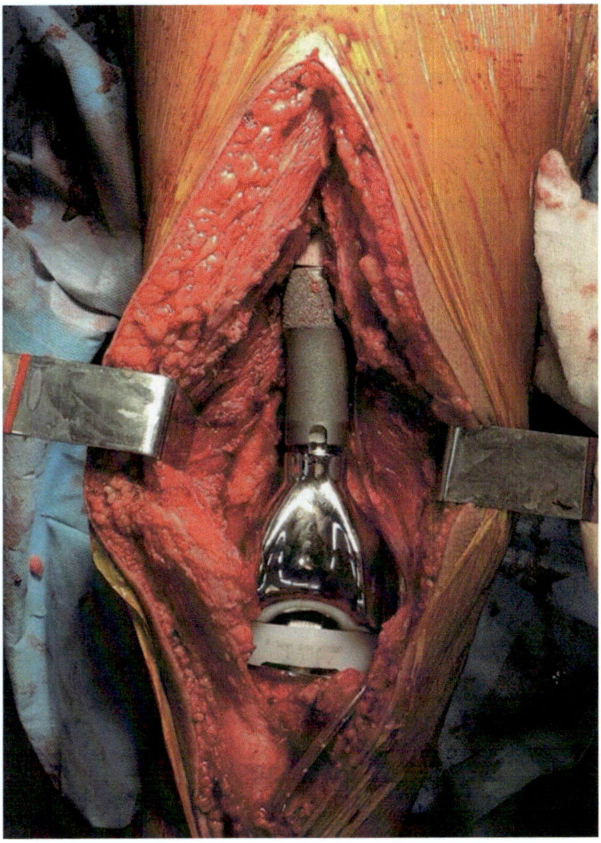

Fig. 22.4 It is crucial to have the correct orientation and rotation of the implant

Fig. 22.5 Always test, patellofemoral tracking in full flexion

(f) Tips and Tricks:
- Ensure thorough preoperative planning, including imaging studies to define the extent of the tumor and the best surgical approach.
- If needed, use intraoperative imaging to confirm the correct placement of instruments and implants.
- During the placement of trial components, test knee flexion-extension to ensure proper tracking before final implantation.
- If you are planning to go medial and posterior identification of the mayor vessels is important to avoid injuries of them.
- Be aware of the rotation of the femoral component when you are impacting or cementing it. After the implantation and before the cement hardens test the flexion—extension of the knee to confirm adequate patellar tracking. You can even keep the knee in about 45 degrees flexion while the cement curing process happens making sure the patella stays in adequate position.

(g) Pearls and Pitfalls:
- For resections extending proximally, releasing the vascular elements should be a consideration.
- Proper orientation of the femoral component is crucial to avoid incorrect implant rotation and subsequent patellofemoral maltracking.
- Choose the appropriate stem for each patient (cemented vs. uncemented).
- We should aim to provide muscular soft tissue coverage for the implant along its entire length. Whenever possible, use implants with rotational hinges.

Proximal Tibia

(a) Patient Positioning and Surgical Incision**:** The patient should be in a supine position. A bolster is typically used to maintain the knee flexed at 90 degrees during surgery.

(b) Steps for Tumor Resection: In patients with metastatic disease, the priority should be joint function, and the proximal tibia with the anterior tibial tuberosity (ATT) and the patellar tendon attachment are fundamental. If the pathology allows, the anterior knee incision is continued with an anterior medial arthrotomy to the anterior tibial tuberosity and patellar tendon. In this case, the patellar tendon is stripped from the tuberosity and left attached to the entire lateral flap of the tibia during resection. If the resection is only of a proximal segment above the ATT, an osteotomy can be performed, which is sufficient without extensive tissue release, and often a tibiofibular arthrotomy is not necessary. However, if the resection is larger, the recommendation is to start medially, entering the space beneath the gastrocnemius muscles, and to locate and release the neurovascular elements from the popliteal region to the tibiofibular trunk. These elements are released and protected along their course to the tibiofibular region. Once protected, the dissection of soft tissues in the posterior tibia continues. At this point, the decision is made whether to leave the popliteus muscle as a posterior margin. The tibiofibular arthrotomy is performed from proximal and posterior to distal and anterior, and the neurovascular elements of the leg are fully released, often including the common peroneal nerve. Once these important structures are released, the proximal tibial osteotomy is performed according to preoperative planning, and the segment is resected en bloc. It is crucial before resection to establish a clear reference for the height of the joint line. The parameniscal structures can be used (Fig. 22.6), and if resection is not necessary, the fibular head can serve as a reference, understanding that the joint line should be 1.5 cm above it.

(c) Reconstruction: As with the femur, there are various reconstruction techniques ranging from less invasive to more invasive, which will be described. The difference regarding the distal femur is that they will not be treated separately, as the surgical position, soft tissue reconstruction, wound closure, tips and tricks, and pearls and pitfalls are very similar.

Proximal Tibial Intramedullary Nailing: (It has been included in Chap. 23.) This can be performed for proximal metaphyseal-diaphyseal lesions using nails with multi-locked fixation systems to provide stability to the reconstruction. In some cases, it can be augmented with cement.

Proximal Tibial Osteosynthesis Plate: (it has been included in Chap. 23) In juxta-articular or metaphyseal-epiphyseal cavitary lesions, mixed reconstructions can be performed with cement and osteosynthesis or cannulated screws. It is important to note that for this type of reconstruction to be durable over time, a proper analysis of the remaining bone structure must be performed. A cortical involvement greater than 30% of the bone's circumference is not a good indication for using this technique.

Oncologic Reconstruction Prosthesis: This is the recommended type of reconstruction for extensive lesions in the articular or metaphyseal region of the proximal tibia. The surgical technique was described above in terms of the tibial cut and the type of implant to be used. The recommendation is to use modular prostheses that allow replacement of the affected bone segment with a rotational hinge joint surface and the option of cemented and

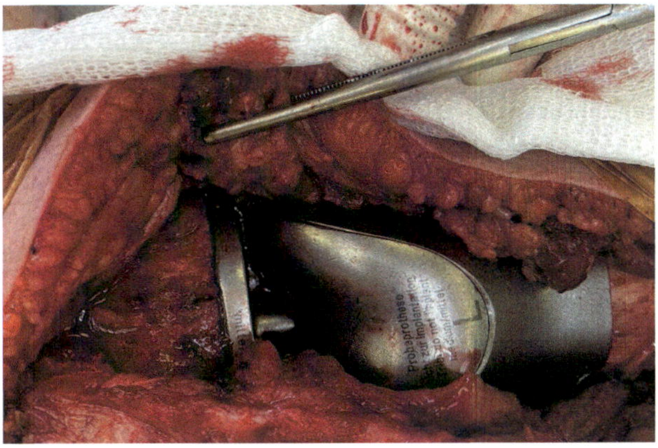

Fig. 22.6 Parameniscal structure as reference for joint line height

uncemented stems. A significant challenge in these cases is reconstructing the extensor mechanism to maintain knee extension and functional outcomes. Techniques that preserve the anterior cortex and extensor mechanism while sacrificing the remaining tibia offer the advantage of improved soft tissue preservation, reduced risk of complications, and better functional recovery.

Techniques for Anterior Cortex and Extensor Mechanism Preservation: Fournan et al. [29]. Technique (JBJS): Preserve the anterior tibial cortex along with the patellar tendon insertion while performing wide resection of the proximal tibia.

Surgical Steps: The tibial resection is planned to spare the anterior tibial cortex as a flap, ensuring that the patellar tendon remains intact and in continuity with the tibial tuberosity. After tumor resection, a proximal tibial endoprosthesis is implanted. The preserved anterior cortex is sutured securely to the soft tissue or the prosthetic components to maintain alignment and support. Advantages: Preservation of the extensor mechanism maintains active knee extension and the anterior cortex provides additional soft tissue coverage over the prosthesis, reducing implant exposure risk.

Pedicle Flap Technique: The patellar tendon and tibial tuberosity are preserved as a pedicle flap with part of the anterior tibial cortex during the proximal tibial resection. The flap is rotated or sutured directly to the proximal end of the endoprosthesis or a soft tissue reconstruction component (e.g., mesh or synthetic ligament) to maintain extensor mechanism continuity.

(d) Soft Tissue Reconstruction: In the proximal tibia, there are several options for soft tissue reconstruction if necessary. Several approaches can be taken, routinely performing a medial gastrocnemius rotational flap with patellar tendon reconstruction during the primary surgery or leaving the flap option for a secondary operation in case of infectious complications or coverage defects (Fig. 22.7).

    Other flap options include rotational locoregional coverage flaps, such as soleus or peroneus flaps, or even the possibility of distant flaps (lateral thigh or latissimus dorsi) in more complex cases. It is understood that we are dealing with patients with metastatic disease, who have a higher risk of complications when performing these complex surgeries.

    In some cases, mesh of different materials (depending on the availability at the center) can also be used to reconstruct the extensor apparatus and provide some fixation to the surrounding soft tissues near the implant (Fig. 22.8).

(e) Wound Closure: The wound should be closed in a conventional manner at 30 degrees of flexion to maintain the integrity of the extensor mechanism, covering the implant or reconstruction. In some cases, leg swelling may prevent full skin closure; in these situations, it is essential to be prepared to use a rotational gastrocnemius flap and free skin graft for coverage. It is important to remember that the tibia is rapidly exposed to complications in the anterior leg region due to limited soft tissue coverage.

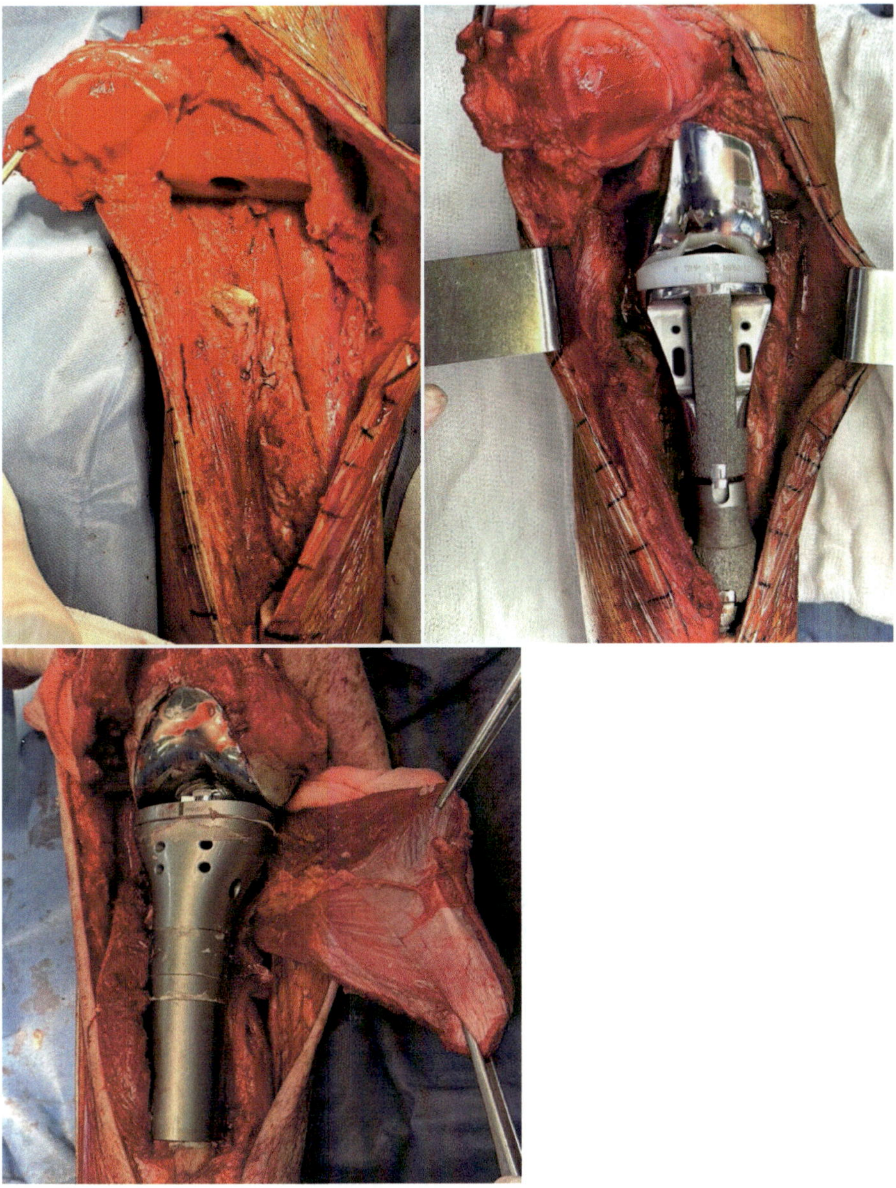

Fig. 22.7 resection and reconstruction of the proximal tibia. The patellar tendon can be sutured to prothesis or to the gastrocnemius flap

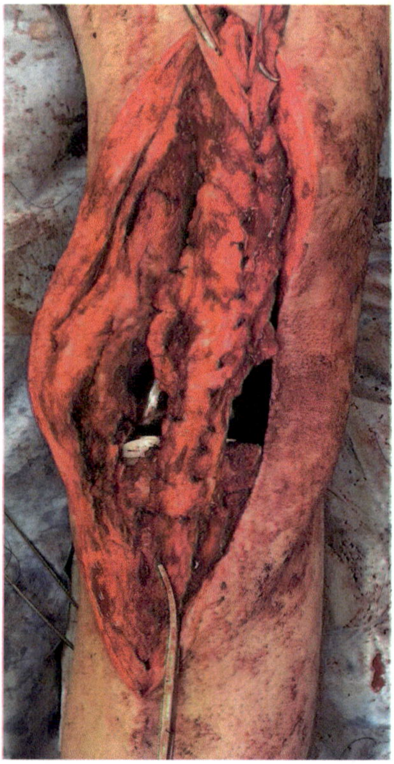

Fig. 22.8 Extensor mechanism reconstruction

(f) Tips and Tricks:
   - Proper preoperative planning of the resection to define the implant to be used.
   - Keep in mind the need for coverage flaps to cover the prosthetic component.
   - Identify the vascular and nerve structures in cases of resections requiring tibiofibular joint arthrotomy and release the popliteal neurovascular bundle as well as the common peroneal nerve.

(g) Pearls and Pitfalls.
   - Proper preoperative planning to ensure the availability of suitable tibial stems.
   - Keep in mind the reconstruction of the extensor apparatus and have the necessary tools ready for proper anchoring.
   - Be familiar with the surgical technique or have a plastic surgeon available in the operating room to perform the gastrocnemius flap if needed.
   - Provide full coverage of the implant.
   - If the skin is under tension, it is preferable to perform a flap immediately rather than waiting for complications to arise.

Postoperative Care

Immobilization

In cases where we perform a reconstruction of the extensor apparatus in any of its forms, we immobilize the patient with a splint in knee extension with full extension for 6 weeks. After this period, we remove the splint and begin a comprehensive and supervised rehabilitation plan, incorporating progressive passive and active flexion and extension protocols.

In cases where we do not work on the extensor mechanism, we do not routinely use a knee brace. Each case should be evaluated individually based on the specific surgical approach and patient factors.

Weight Bearing

Depending on the patient's condition and physical capabilities, we recommend the use of assistive devices, such as a walker, crutches, or Canadian cane, during the first's day after surgery, for swelling and edema.

- In general, prosthetic reconstructions of the distal femur and proximal tibia can have full weight bear immediately after surgery, as tolerated by the patient's pain level.
- For reconstructions involving nails, plates, and cement, weight-bearing and new pain should be followed up with regular X-rays to assess the reconstruction and prevent mechanical complications as implant or hardware failure.

Thromboembolic Prophylaxis

Decisions regarding the appropriate thromboembolic prophylaxis regimen should be made after considering the patient's comorbidities and the results of coagulation studies and complete blood counts. In general, patients with metastatic cancer undergoing surgery have an increased risk of thromboembolic complications. Blood chemistry and the mechanics of injury should be considered to establish prophylaxis. Platelet function may be abnormal in patients with widespread metastatic disease, necessitating potential variations in the regimen during the perioperative period to minimize bleeding risk. The appropriate use of analgesics is an essential part of patient management. This not only enhances the patient's quality of life but also aids in improving mobility and the potential for significant recovery following surgery. Patients may require high doses of opioids.

Antibiotic Prophylactic

Antibiotic prophylactic should be 24 h of cefazolin 30 min prior to incision and 3 more doses, to complete 24 h. In the Parity randomized clinical trial [30], the 5-day regimen did not reduce the rate of surgical site infection compared with the 1-day regimen. Although the results of this randomized clinical trial do not definitively exclude an important benefit of post-operative prophylactic antibiotic therapy longer than 24 h, the significantly higher risk of antibiotic-related complications in the 5-day regimen represents a critical finding. Clinicians should consider the uncertainty of the benefits and the relative confidence in findings of harm to make an informed decision on antibiotic duration after surgical resection and endoprosthesis reconstruction for lower extremity bone tumors.

References

1. Harvie P, Whitwell D. Metastatic bone disease. Bone Joint Res. 2013;2(6):96–101. https://doi.org/10.1302/2046-3758.26.2000154.
2. Willeumier JJ, van der Wal CWPG, Schoones JW, van der Wal RJ, Dijkstra PDS. Pathologic fractures of the distal femur: current concepts and treatment options. Wiley; 2018. https://doi.org/10.1002/jso.25218.
3. Ratasvuori M, et al. Insight opinion to surgically treated metastatic bone disease: Scandinavian Sarcoma Group Skeletal Metastasis Registry report of 1195 operated skeletal metastasis. Surg Oncol. 2013;22(2):132–8. https://doi.org/10.1016/J.SURONC.2013.02.008.
4. Ashford RU, et al. Proximal femoral replacements for metastatic bone disease: financial implications for sarcoma units. Int Orthop. 2010;34(5):709–13. https://doi.org/10.1007/s00264-009-0838-6.
5. Chandrasekar CR, Grimer RJ, Carter SR, Tillman RM, Abudu AT. Modular endoprosthetic replacement for metastatic tumours of the proximal femur. J Orthop Surg Res. 2008;3(1) https://doi.org/10.1186/1749-799X-3-50.
6. H. Mirels, "Metastatic disease in long bones: a proposed scoring system for diagnosing impending pathologic fractures. 1989.," Clin Orthop Relat Res, 415, Suppl, 2003, https://doi.org/10.1097/01.blo.0000093045.56370.dd.
7. Groot OQ, et al. Clinical outcome differences in the treatment of impending versus completed pathological long-bone fractures. J Bone Joint Surg Am. 2022;104(4):307–15. https://doi.org/10.2106/JBJS.21.00711.
8. Forsberg JA, Wedin R, Boland PJ, Healey JH. Can we estimate short- and intermediate-term survival in patients undergoing surgery for metastatic bone disease? Clin Orthop Relat Res. 2017;475(4):1252–61. https://doi.org/10.1007/s11999-016-5187-3.
9. Forsberg JA, Eberhardt J, Boland PJ, Wedin R, Healey JH. Estimating survival in patients with operable skeletal metastases: an application of a bayesian belief network. PLoS One. 2011;6(5):e19956. https://doi.org/10.1371/journal.pone.0019956.
10. Forsberg JA, Sjoberg D, Chen Q-R, Vickers A, Healey JH. Treating metastatic disease: which survival model is best suited for the clinic? Clin Orthop Relat Res. 2013;471(3):843–50. https://doi.org/10.1007/s11999-012-2577-z.
11. Piccioli A, et al. How do we estimate survival? External validation of a tool for survival estimation in patients with metastatic bone disease—decision analysis and comparison of three international patient populations. BMC Cancer. 2015:1–8. https://doi.org/10.1186/s12885-015-1396-5.

12. Ratasvuori M, et al. Prognostic role of en-bloc resection and late onset of bone metastasis in patients with bone-seeking carcinomas of the kidney, breast, lung, and prostate: SSG study on 672 operated skeletal metastases. J Surg Oncol. 2014;110(4):360–5. https://doi.org/10.1002/jso.23654.
13. Laitinen M, et al. Survival and complications of skeletal reconstructions after surgical treatment of bony metastatic renal cell carcinoma. Eur J Surg Oncol. 2015;41(7):886–92. https://doi.org/10.1016/j.ejso.2015.04.008.
14. Pazionis TJC, Papanastassiou ID, Maybody M, Healey JH. Embolization of hypervascular bone metastases reduces intraoperative blood loss: a case-control study. Clin Orthop Relat Res. 2014;472(10):3179–87. https://doi.org/10.1007/s11999-014-3734-3.
15. Khanmohammadi S, Noroozi A, Yekaninejad MS, Rezaei N. Cryoablation for the palliation of painful bone metastasis: a systematic review. Cardiovasc Intervent Radiol. 2023;46(11):1469–82. https://doi.org/10.1007/S00270-022-03356-Z.
16. Ehne J, Tsagozis P. Current concepts in the surgical treatment of skeletal metastases. World J Orthop. 2020;11(7):319–27. https://doi.org/10.5312/wjo.v11.i7.319.
17. Zhao J, Zhao Y, Ma X, Feng H, Gao X. Case report comprehensive imaging and pathology of solitary patellar metastasis from breast cancer with literature review on its diagnosis and prognosis. 2022. [Online]. Available: www.ajtr.org.
18. Mercuri M, Casadei R. Patellar tumors. Clin Orthop Relat Res. 2001;389:35–46. https://doi.org/10.1097/00003086-200108000-00007.
19. Yu A, et al. Outcomes of distal femoral replacement versus internal fixation for metastatic bone diseases of the distal femur. J Orthop. 2024;58:117–22. https://doi.org/10.1016/J.JOR.2024.06.031.
20. Bonnevialle P, et al. Surgical treatment of tibial metastases: retrospective, multicenter, observational study of 25 patients. Orthop Traumatol Surg Res. 2020;106(6):1039–45. https://doi.org/10.1016/J.OTSR.2019.07.017.
21. Abdel MP, Salib CG, Mara KC, Pagnano MW, Perry KI, Hanssen AD. Extensor mechanism reconstruction with use of marlex mesh: a series study of 77 total knee arthroplasties. J Bone Joint Surg Am. 2018;100(15):1309–18. https://doi.org/10.2106/JBJS.17.01165.
22. Stephen SR, Burnett J, Berger RA, Della Valle CJ, Sporer SM. Extensor mechanism allograft reconstruction after total knee arthroplasty. J Arthroplast. 2005;23:35.
23. Burnett RSJ, et al. Extensor mechanism allograft reconstruction after total knee arthroplasty. J Bone Joint Surg Am. 2005;87. Suppl 1, no. Pt 2:175–94. https://doi.org/10.2106/JBJS.E.00442.
24. Vyas P, Cui Q. Management options for extensor mechanism discontinuity in patients with total knee arthroplasty. Cureus. 12:e9225. https://doi.org/10.7759/cureus.9225.
25. Jaureguito JW, Dubois CM, Smith SR, Gottlieb LJ, Finn HA. Medial gastrocnemius transposition flap for the treatment of disruption of the extensor mechanism after total knee arthroplasty. J Bone Joint Surg Am. 1997;79(6):866–73, [Online]. Available: http://www.ncbi.nlm.nih.gov/pubmed/9199384.
26. Sim FH, Choong PFM, Weber KL, W. Kluwer Lippincott Health Williams. Master techniques in orthopedic surgery. 1st ed. Philadelphia: Lippincott Williams & Wilkins, A Wolters Kluwer; 2011.
27. Martin M, Paul S. Musculoskeletal cancer surgery treatment of sarcomas and allied diseases, vol. ebook. 1st ed. Kluwer Academics Publishers; 2004.
28. Martin M, James W, Jacob B. Operative techniques in orthopedic surgical oncology. Philadelphia: Lippincott Williams & Wilkins; 2012.
29. Fourman MS, Ramsey DC, Newman ET, Heng M, Raskin KA, Lozano-Calderon SA. Tubercle-sparing proximal tibial reconstruction in patients with primary and metastatic bone disease: a case report. JBJS Case Connect. 2022;12(1) https://doi.org/10.2106/JBJS.CC.21.00483.
30. Ghert M, et al. Comparison of prophylactic intravenous antibiotic regimens after endoprosthetic reconstruction for lower extremity bone tumors: a randomized clinical trial. JAMA Oncol. 2022;8(3):345–53. https://doi.org/10.1001/JAMAONCOL.2021.6628.

Management of Metastatic Disease to the Tibial Shaft

23

Jestin Williams, Irfan A. Khan, Christopher Bloise, and Matthew G. Cable

Epidemiology

Tibial metastatic lesions comprise a small proportion of metastatic disease to bone [1, 2]; with the axial skeleton, proximal femur, and proximal humerus more commonly affected [3–5]. When lesions are found in the tibia, they likely spread hematogenously through arterial tree metastasis, frequently involving the most proximal segment, as with other peripheral long bones. This can signify advanced metastatic disease, with the majority of patients diagnosed with tibial metastases also having visceral metastases and polyostotic metastatic disease [6, 7]. Tibial metastatic lesions can cause significant pain, functional limitations, and risk for pathologic fracture. Patients treated for tibial metastatic lesions often have an impending pathologic fracture, with a smaller proportion of patients having displaced pathologic tibial fractures at the time of presentation.

The average age of patients diagnosed with tibial metastatic lesions is 65 years old, with a similar proportion of men and women being affected [8]. Metastatic lesions in the tibia most commonly affect the proximal tibia (44%), followed by the tibial shaft (35%), with the distal tibia being least affected (11%) (Table 23.1, Fig. 23.1) [8]. The primary tumor for suspected tibial metastases is known in approximately 85% of cases, which mirrors the primary tumor identification rate when utilizing the diagnostic strategy proposed by Rougraff et al. [9]. The most common lytic metastases to the tibia are renal cell carcinoma (24%), lung carcinoma (13%), and hematologic malignancies (10%) (Table 23.1) [8]. While less

J. Williams
Department of Orthopaedic Oncology, MD Anderson Cancer Center, Houston, TX, USA
e-mail: Jmwilliams6@mdanderson.org

I. A. Khan · C. Bloise · M. G. Cable (✉)
Department of Orthopedics, Louisiana State University Health Science Center, New Orleans, LA, USA
e-mail: ikhan6@lsuhsc.edu; cblois@lsuhsc.edu; mcabl1@lsuhsc.edu

J. Pretell-Mazzini, S. A. Lozano Calderón (eds.), *Multidisciplinary Approach of Musculoskeletal Metastasis*, https://doi.org/10.1007/978-3-032-09498-8_23

Table 23.1 Tibial location and primary tumor of tibial metastatic lesions

Item	Percentage (%)
Tibial location of metastases	
Proximal	44
Shaft	34
Distal	11
Not specified	10
Primary tumor	
Renal cell	24.1
Lung	13.0
Hematologic	10.2
Lymphoma	6.5
Myeloma	3.7
Colorectal	8.3
Breast	7.4
Endometrial	5.6
Skin	3.7
Prostate	2.8
Other	7.4
Unknown/ not specified	17.5

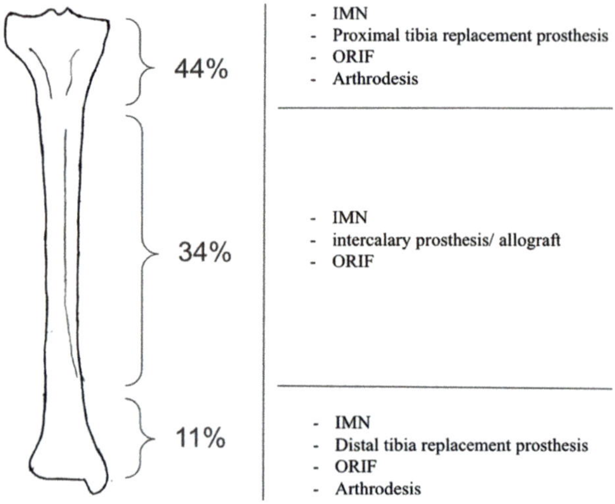

Fig. 23.1 Frequency of tibial metastases and associated fixation options based on lesion location

common, sclerotic or osteoblastic metastasis of the tibia is also possible from breast, ovarian, prostate, carcinoid, and neurogenic tumors. Of note, giant cell tumors and infections are nonmalignant entities that can present with lytic lesions in the tibia, and should be considered with isolated lesions.

Amongst patients with tibial shaft metastases, treatment options comprise curettage, intramedullary nail (IMN) placement, intercalary allograft (IA) or intercalary endoprosthesis (IE), and open reduction internal fixation (ORIF) with plates and screws (Fig. 23.2). A systematic review by Greco et al. showed that more than 80% of patients with tibial shaft metastases are treated with IMN placement, with 13% of patients treated with intercalary prostheses, and 5% of patients treated with ORIF (Fig. 23.1) [8]. Metadiaphyseal lesions were not included in this specific data analysis, and are commonly stabilized with ORIF with proximal tibial plating. A small

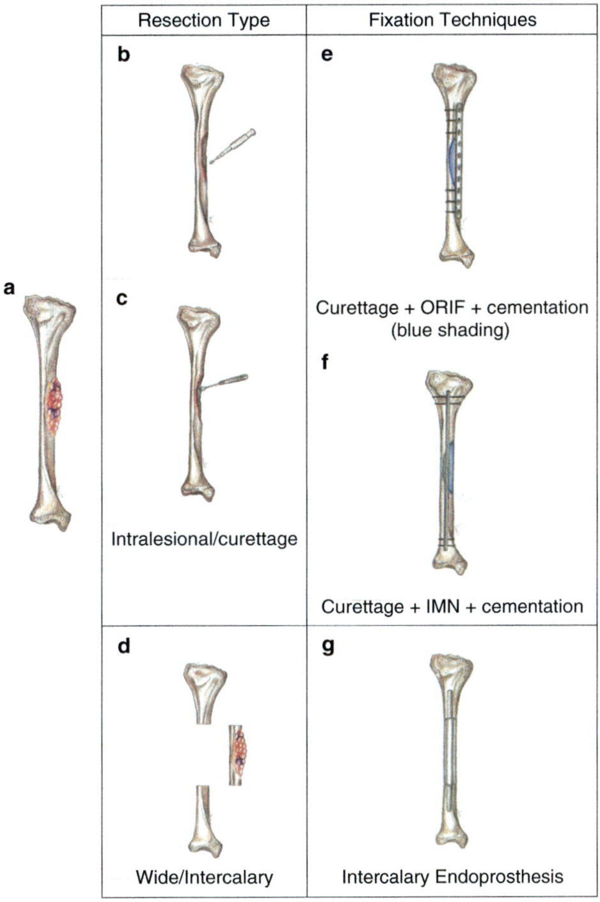

Fig. 23.2 (**a**–**g**) Model of diaphyseal tibial lesion, various resections and fixation techniques

number of patients may be treated with amputation. In poor prognostic settings or nonambulatory patients, tibial metastases can be treated nonoperatively with radiation and immobilization.

Nonsurgical & Surgical Indications

Treatment aims at improving pain, function, and quality of life. A multidisciplinary approach involving, considering concurrent systemic therapies and radiation. Treatment considerations are multifactorial, including location of the tumor within the bone, additional skeletal metastasis, impending or displaced pathological fracture, tumor histology, pain, patient comorbidities, and patient prognosis. General indications for nonsurgical management include life expectancy less than 6 weeks and extensive comorbidities that increase the risk of perioperative morbidity and mortality. The expected recovery should not exceed estimated life expectancy.

Predicting individual prognosis is imperative in determining surgical management. Willeumier et al. developed a life expectancy prognostic model for patients with long bone metastasis treated with radiotherapy or surgery [10]. This model takes into account the primary tumor, Karnofsky performance score, and the presence of visceral and/or brain metastases [10]. PathFX is another model to predict life expectancy, developed using Bayesian machine learning techniques [11]. PathFX incorporates objective quantifiable variables, including primary tumor type, sex, age, Eastern Cooperative Oncology Group performance status score, presence of visceral metastases, presence of polyostotic skeletal metastases, pathological fracture, hemoglobin, and lymphocyte count [11]. A relative contraindication for surgical intervention is widespread metastatic disease (extensive axial skeleton involvement, spinal cord compression, heavy visceral burden, and intracranial disease), as this relates to the patient's ability to recover from surgery and future quality of life.

Metastatic bone pain can be structural, in the form of impending and/or pathological fractures, or cancer-related pain due to intraosseous swelling and inflammation. Surgical indications for tibial metastasis are impending and actual pathological fractures. Snell and Beal were of the first to predict fracture risk by analyzing the rates of femur fractures in those with breast cancer [12, 13]. They showed an increased fracture risk in patients with lytic lesions of the cortex larger or equal to 2.5 cm [12, 13]. Harrington et al. evaluated fracture risk in skeletal metastasis and considered patients at high risk when they had destructive lytic or blastic lesions involving more than 50% of the circumferential cortical bone or when weight-bearing pain persists, recurs, or worsens after local radiation [14]. They highlighted the proximal femur, with a higher fracture risk with lesions greater than 2.5 cm in any dimension or if there was an associated avulsion fracture of the lesser trochanter [14]. The Mirel's criteria, one of the most highly used systems, predicts pathologic fracture based on lesion size on plain radiographs, location, radiographic appearance (lytic, blastic, or mixed), and the presence/severity of pain [15]. CT-based rigidity analysis (CTRA), a more novel technique, defines impending pathologic fractures based on more than 35% reduction in axial, bending, or torsional rigidity [16]. Damron et al. demonstrated more favorable predictions of fracture risk with the CTRA versus the Mirel's criteria [16].

Preoperative Planning

Forgoing general preoperative considerations that have been addressed elsewhere in this volume, specific considerations for tibial metastasis include orthogonal plain films of the tibia, knee, and ankle to rule out articular involvement. Plain films provide a good description of the lesion, extent of involvement, pathological fracture, and bony destruction. Magnetic resonance imaging (MRI) with and without contrast is usually the modality of choice to further describe the lesion. Pathological tissue is generally seen as hypo- (black) or isointense with muscle on T1-weighted images and hyperintense (white) on T2-weighted images. T1 weighted images best highlight differences in the fat, marrow, and give details of the mass' anatomical position and relationship to surrounding tendinous and nervous structures. T2-weighted images are best when evaluating tumor edema and vascular structures. Fat-saturated images aid in delineating the mass as fat versus other tissue and highlight areas of necrosis. CT scans provide high-resolution information about the cortical bone and associated bony destruction, which is particularly useful in lytic lesions and multiple myeloma. Bone scans can detect additional peripheral lesions as it visualizes the entire skeleton, which can also be used in multiple myeloma, being aware of up to 30% false negative rate., Also, be aware of false positives in patients with osteopenia or degenerative changes and correlate findings with plain films. Positron emission tomography (PET)/CT scan is an efficient way to identify both osseous and soft tissue/lymphatic involvement. For osseous lesions in patients with a known cancer diagnosis but no previous osseous involvement, a tissue biopsy must still be obtained.

Surgical Treatments

Surgical treatment for tibial shaft metastasis ranges from prophylactic curettage procedures with cement augmentation with or without hardware, to management of completed fractures with open or closed reduction and internal fixation, to more involved lesions that require resection with IA or IE reconstruction or amputation. The cutaneous soft tissue envelope covering the tibia diaphysis is usually a deciding factor, allowing both easy approach for lesions warranting resection, or supporting less invasive methods like intramedullary fixation if there is concern for wound healing or soft tissue compromise in the setting of systemic chemotherapy and radiation.

Literature is sparse regarding surgical management of tibial shaft metastasis, mostly limited to case reports, case series, and retrospective cohorts [2, 6–8]. This likely stems from patients being grouped within a heterogenous population in studies including primary bone tumors of the tibia and anatomical locations (epiphyseal and metaphyseal), because of the rarity of tibial shaft metastasis.

Lesions of the tibial shaft, including the metadiaphyseal area, can be treated by curettage with the addition of adjuvants and/or stabilization. A thorough curettage performed manually or with the use of a high-speed burr should aim to remove as much pathological bone as possible. After macroscopic lesional material is removed, use of adjuvant therapies such as cryotherapy, radiofrequency ablation (RFA), hydrogen peroxide, phenol, ethanol, argon beam, cautery, and/or bone cement is highly encouraged to ensure microscopic disease is reduced and extend the physiologic treatment margin. After curettage, the tibia can be stabilized with plate and screw fixation or IMN (Figs. 23.2, 23.3, 23.4, and 23.5). There is no consensus supported by literature regarding which adjuvant is most effective. The most widely used and supported adjuvant is bone cement (PMMA). Cement allows immediate fixation, does not rely on bony healing, can be used for plating or IMN, and can be used to elute antibiotics or cytotoxic drugs. Proper cementing techniques include

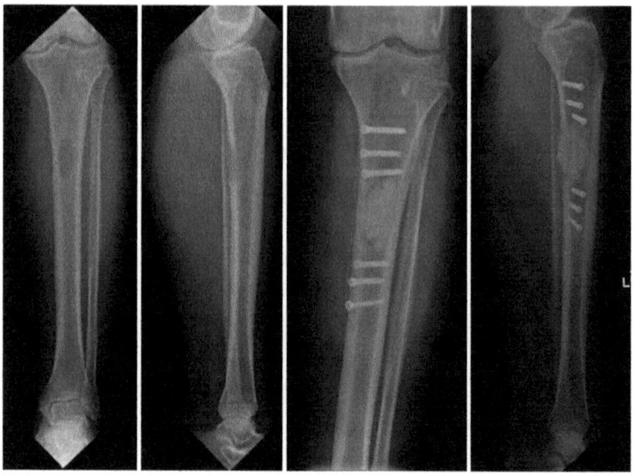

Fig. 23.3 Pre- and postoperative plain films of a 60-year-old male with recurrent metastatic thyroid cancer with left tibia proximal diaphyseal metastatic lesion treated with open curettage, ablation with electrocautery, cementoplasty with gentamycin-containing cement, and carbon fiber broad plate with a combination of locking and cortical screws

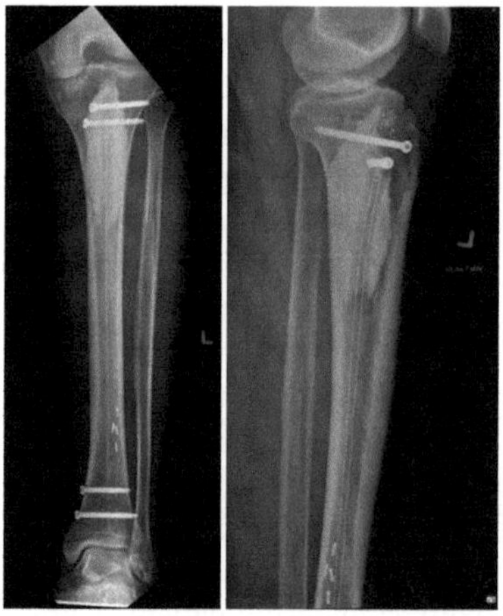

Fig. 23.4 Postoperative plain films of a 72-year-old male with metastatic renal cell carcinoma with left proximal tibial metadiaphyseal lesion treated with open curettage, ablation with electrocautery, followed by ethanol, carbon fiber intramedullary nail, and cementoplasty

digital pressurization, which allows for interdigitation, and to avoid drilling through hardened cement, which can affect compression strength. If plating prior to cementation, such as in the setting of realized pathologic fractures, there must be careful attention to plate position, ensuring an adequate working window for cement placement or injection. Plate length is often much longer than with standard ORIF techniques to avoid stress risers due to poor bone quality. If a curetted defect must be traversed by a screw, hardware should be placed first and the cement injected into the defect. The preferred cementation method with IMN is to first place the nail for shaft lesions, then cement around the nail. If there is metadiaphyseal involvement that may interfere with IMN interlock screws, one can predrill the interlock screws and inject cement through the drill holes, then place the interlock screws prior to cement curing.

Indications for curettage, adjuvant, and fixation are metastatic lesions that allow stable fixation after curettage with or without cementation, namely, lesions without

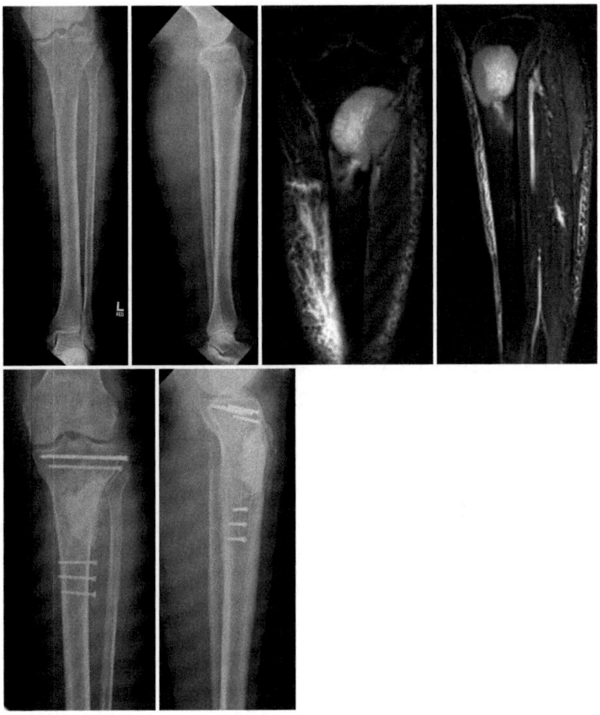

Fig. 23.5 Preoperative plain films and T2-weighted MRI cuts with postoperative plain films of a 65-year-old male with multiple myeloma with left proximal tibial metadiaphyseal lesion treated with open curettage, ablation with electrocautery, followed by irrigation with hydrogen peroxide, allograft placement, and carbon proximal tibial plate with combination of locking and cortical screws. Allograft can be considered in lieu of bone cement when radiation is not needed or anticipated, as with some hematologic malignancies and with patients exhibiting good response to systemic therapies

circumferential bone loss Locked plating is favored over non locked plating due to its increased pull-out strength especially in the setting poor bone quality [17]. An advantage of plating is that it does not require a separate incision. Disadvantages for plating tibial shaft lesions are the limited soft tissue envelope over the lower leg, potential risks of skin compromise from surgery and adjuvant radiation, and weight bearing limitations postoperatively.

IMN for fixation is the favored treatment for most tibial shaft metastatic lesions [18]. Advantages of IMN include smaller incisions and possibility for full weight bearing immediately. Disadvantages are the need for a separate incision for curettage, if performed, and the theoretical potential of seeding the entire tibial canal and increasing planned radiation treatment zones. In a systematic review of surgical treatment for tibia lesions with 38 patients having diaphyseal lesions, 29 were treated with IMN with or without cementation, and 2 were treated with ORIF with or without cementation [8]. A retrospective review by Kelly et al. showed 6 of 25 patients with tibia metastases having diaphyseal lesions who were treated with curettage and IMN; 4 of 6 underwent cementation and 2 without [19]. Miller et al. describe IMN failures for the treatment of long bone metastatic lesions in a retrospective review including lesions of the humerus, femur, and tibia; there were 112 constructs and only 2 tibial shaft lesions [20]. The failure mechanisms included surgeon error, tumor progression, nonunion, and hardware failure. There were ten failures, and none were of the tibia shaft. The review showed low likelihood of failure in those with short survival times and metastatic lung cancer lesions. The likelihood of failure was higher in those with solitary lesions, lymphoma, renal cell carcinoma, and radiation-associated fractures [20].

Another consideration for fixation is the use of carbon fiber (CF) versus stainless steel (Ss)/titanium (Tt) implants. The most well-known advantage of CF-implants is its radiolucency (Figs. 23.2, 23.3, 23.4, and 23.5). This is particularly advantageous in orthopaedic oncology with the use of CT and MRI scans for disease surveillance, while Ss/Tt implants create significant artifacts, distorting the image quality and reliability [21]. Radiolucency also applies with postoperative radiation, with metal implants creating interference in the planning imaging, and scattering the radiation particle beam, leading to errors in mapping and dosing, while carbon fiber negates this risk [22–24]. Compared to metallic implants, CF has a closer elastic modulus to bone, yet has similar strength in four-point bending, torsional, and bending fatigue, and lower particle generation in wear testing [25].

Intercalary resection and reconstruction involves circumferential removal of an involved diaphyseal segment of bone. Transverse cuts are made above and below the lesion where there is no longer circumferential bone involvement, and the bone is deemed structurally sound to support the reconstruction. Soft tissue margins are not as critical in these resections, and the surrounding soft tissue envelope can be preserved if there is not significant extraperiosteal tumor extension. Radiation-resistant tumors (e.g., thyroid and renal cell carcinomas) deserve particular attention, as intercalary resections usually involve lesions that have failed other fixation techniques and have not responded to adjuvant radiation. Regarding reconstruction, there are diverse options including: vascularized fibula autograft (VFAG), IA, combination of VFAG and IA, IE, extracorporeal irradiated autograft, bone transport either with external fixation or IMN, innovative techniques using intercalary scaffolds augmented with growth factors (tissue engineering techniques, masquelet technique), or simply using cement circumferentially around a large diameter intramedullary nail. Of these, IE or cement augmented nails are almost exclusively used for metastatic tibial lesions warranting intercalary resection. The reasons are the low healing capacity and inability to immediately weight bear after the reconstruction.

Potential advantages of VFAG include the ability to reconstruct soft tissue defects with an osteocutaneous flap of the fibula [26]. VFAG also has the ability to act as a hemicortical graft. The main disadvantages of the VFAG are its weak structural support and need for extended periods of immobilization for healing, the morbidity of harvesting, and longer surgical time. Bony union for the lower extremity usually takes about 5–7 months, and most surgeons advocate protected weight bearing until signs of radiographic union [27]. Thus, the risks and benefits of this treatment strategy and patient life expectancy must be considered. VFAG alone is not advocated. In the tibia, a combination of VFAG and IA can be considered, which allows structural support by the IA and active biology by the VFAG [28, 29]. Complications at the recipient site include hardware failure, graft fracture, anastomotic thrombus and flap necrosis, nonunion/delayed union, and infection [28–30]. A frequent complication of VFAG is fatigue fractures of the graft, occurring in approximately 32–35%

of patients [31, 32]. Donor site complications include valgus ankle deformity, ankle instability, peroneal nerve palsy, and sensory deficits of the foot, skin graft failure and resultant exposed underlying structures, and flexor hallucis longus contracture [30, 32].

IA reconstruction is another form of biologic material used for reconstruction of segmental defects. It may also be used as a structural hemicortical graft in near-circumferential defects that don't warrant a full intercallary resection. Contraindication for IA reconstruction is inability to obtain stable fixation secondary to insufficient native bone stock. Factors associated with adverse outcomes are age >18 and length of resection [33]. There is an increased fracture risk in reconstruction of bone defects >17 cm with IA alone [33, 34]. As stated earlier, VFAG is commonly used with IA, but both are uncommon for metastasis. The main advantage of IA graft is its almost infinite sizing; it can be used to reconstruct near full bone lengths, which is not always possible with intercalary endoprosthesis stems, and the ability to provide initial mechanical strength [27, 35, 36]. The allograft to be used is based on preoperative planning and imaging to obtain the best anatomical match, and can be readily available as frozen femur shaft segments or custom-ordered through a bone bank. After the pathologic bone has been resected, it can be used as a template to create the IA, or patient-specific cutting guides can be used. The IA is fixated either with plates and screws, an IMN, or both [27]. There is no consensus on which fixation is superior in native bone reconstruction; one study demonstrated no difference in union while another favored plating with a 8% non-union rate compared to 22% by IMN, [35, 37] however, intramedullary fixation tends to be favored for metastatic reconstruction due to immediate weight bearing.

The main complications of IA are infection, fracture, and nonunion, with almost 70% of patients needing additional surgery (Fig. 23.6a) [36]. These characteristics are largely due to the inactive nature of the graft, but studies demonstrate 10-year graft survival approaching 80% (50–80%) [35, 38]. Various studies demonstrate that most complications occur in the first 3–4 years after initial surgery, with the reconstruction becoming significantly stable and functional after this time period [35, 36, 39–41]. The inactive bone may predispose it to infections, reported rates are 12–14% [36, 42]. Allograft fracture prevalence is approximately 9–19% [38, 43]. It has been shown that fracture rates are higher with plate fixation, as the screws inserted through the allograft act as a stress riser [37, 43]. To decrease stress risers and reduce fracture risk, some advocate spanning the entire allograft to provide extra-cortical support or fixating with IMN [40]. Gerrand et al. described filling the intramedullary canal with cement and using a plate and screw fixation to span the entire bone and supplementing the host-allograft junction with cancellous autograft [44]. In his series of 27 IA, they had no fractures. The incidence of allograft-host junction nonunion has been reported between 17% and 50%, with a predilection of higher risk with IMN [36, 42, 45]. This is likely due to the IMN not allowing compression across the host-allograft site; however, it is a load-sharing device and permits immediate weight bearing. There may also be a negative impact by chemotherapy on fracture healing [46]. It should be noted that the majority of published IA cases do not involve metastatic disease.

Extracorporeal irradiated autograft provides another option for intercalary reconstruction. This technique is used more in countries where it is against certain beliefs, where allograft bone banks are not as prominent, such as Asia. The advantages are restoring large defects, no sizing mismatch, and ease of procurement without using a bone bank. A disadvantage of this technique is if there is already substantial cortical bone loss, the irradiated autograft provides little structural support. There are no guidelines for irradiation dosing. Uyttendaele et al. and Chen et al. reported low rates of recurrence using a dosage of 30,000 rads [46, 47]. There have been other methods used for devitalization, including autoclaving, pasteurization, and liquid nitrogen [48–57]. Complications range up to 52% including infection, nonunion, fracture, bone resorption, and graft necrosis [48, 50–52].

IE is a frequently utilized intercalary reconstruction method for diaphyseal tibial metastasis (Fig. 23.6). Indications for use are the same as VFAG and IA, with a predisposition in patients with metastasis, myeloma, or lymphoma [40]. The advantage of IE is the immediate stability conferred, ideal for metastatic bone disease, elderly, those on systemic therapies with poor healing capacity, and those with limited life expectancy, where immediate weight bearing and full function outweigh construct durability. Ruggieri et al., in a retrospective review of 24 patients treated with primary or metastatic bone tumors in the diaphysis of the femur, tibia, or humerus treated with segmental resection and endoprosthetic reconstruction,

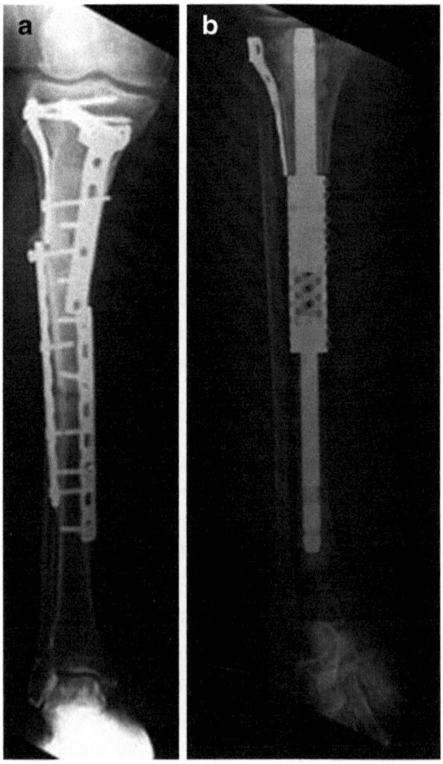

Fig. 23.6 Preoperative (**a**) and postoperative (**b**) plain films of a patient with failed intercalary allograft of the right tibia diaphysis, treated with an intercalary cemented endoprosthesis device as a limb salvage device. Courtesy of Merete Technologies, Inc

describe metastatic surgical indications as extensive bone loss and destruction that precluded standard reconstruction with IMN, with defect size greater than 5 cm in the tibia [53].

The approach and resection for IE are standard as described prior. Accurately measuring residual native bone stock is pivotal for preoperative planning and intra-operative assessment. There must be approximately 5 cm of intramedullary canal remaining to accept most stems and provide stable fixation [40]. Shorter resection lengths can be "built up" with porous diaphyseal cones to accept intramedullary stem lengths, but this adds to the complexity of the procedure. After resection, the canal is prepared with rigid reamers to ensure proper fit and under fluoroscopy to ensure no intra-articular breach. Planing reamers are used to create a flush docking site for the implant. Next, a trial prosthesis is assembled and inserted to ensure adequate reduction. IE are often fixated with cement as an adjuvant to prevent canal spread, particularly with radioresistant tumors. Some stem designs also accommodate transfixion interlocking pins. As with all cemented stems, canal preparation is key for stability. The canal is irrigated with pulsatile lavage, packed for hemostasis, then optional restrictors are placed, and finally, cement is impacted.

Complications of IE are infection, prosthetic or periprosthetic fracture, aseptic loosening, or mechanical wear. Complications associated with implant failure range up to 33% [41, 53]. In contrast to IA gaining stability after the first couple of years, IE failure increases over time. The 10-year survival for IE ranges from 63% to 80% (Tables 23.2, 23.3, 23.4, and 23.5) [55, 56].

Table 23.2 Technique—open curettage and cementation of tibial diaphyseal metastases

Radiolucent OR table, operative limb elevated on foam to facilitate lateral fluoroscopy views. Tourniquet used with gravity exsanguination rather than Esmark.
Resection is intralesional, so incision first made ONLY over planned bony cortical window to limit tumor seeding.
Incision lengthened to accommodate plate/screw placement (if needed) AFTER curettage.
Direct medial approach used to avoid anterior and lateral compartment contamination.
Minimize cautery use on soft tissues. Sharply develop full-thickness flaps to bone and incise periosteum longitudinally and elevate.
Mark out oval bone window with Bovie using fluoroscopy, incorporating thinned cortical bone when possible. Bone quality is typically too poor to preserve the bone window.
A 5 mm fluted or fast-cutting burr is optimal for making the cortical window.
A diamond-coated round burr is optimal for decorticating the lesion periphery, removing a few millimeters of bone without accidental cortical breach. This burr also cauterizes bone as it is removed.
All lesional contents are grossly removed using curettes, pituitary rongeurs, burr, and suction.
Contrast (i.e., Isovue) injected into the cavity to verify complete tumor removal. Filling defects on fluoroscopy should be addressed with further curettage. This also confirms if the lesion is contained, allowing safe use of adjuvants like ethanol or phenol without extraosseous leakage.
Appropriate plate selected that allows at least 3 holes above and below the lesion. The entire diaphysis should be spanned if possible. The most proximal and distal screws can be placed percutaneously.
Middle third diaphyseal lesions typically can be filled with cement first and then a straight locking plate applied.
For proximal or distal third diaphyseal lesions, an anatomic tibial plate may be required.

(continued)

Table 23.2 (continued)

If overlap of planned screw trajectories through the curetted lesion, if possible, the plate and screws are applied first, and cement injected second.
A cortical screw first reduces the plate to bone. Additional locking screws are placed, and the first cortical screw can be exchanged for a locking screw.
Bone cement is manually packed and pressurized into the lesion. Medium or high-viscosity cement facilitates manual handling, packing, and pressurization.
If the bone window is partly obstructed by the plate, medium or low viscosity cement can be injected with a 50 cc Foley-tipped syringe through screw holes or partly exposed window.
The cement should fill both the lesion and the cortical window.
Tourniquet deflated before wound closure to ensure hemostasis.
For patients on cytotoxic therapy or tyrosine kinase inhibitors (frequently used in renal cell cancers), nylon suture is used.
Patients on hormone-based therapy or checkpoint inhibitors are at lower risk for delayed healing, and monocryl or resorbable sutures can be used.
Tips and tricks
For gaunt or malnourished patients, intramedullary nail is favored to avoid plate/hardware prominence and wound healing complications. Anterolateral approach can help with hardware coverage and is considered if the cortical defect is not medial or posterior.
If Hohmann retractors are used, ensure surrounding cortical bone is thick enough to support them without accidentally penetrating the bone. A Weitlaner retractor is typically sufficient.
The cortical window does not need to be wider than 1 cm to accommodate most curettes.
Window length does not need to span the entire lesion, since curved curettes can easily access 2 cm above and below the window.
At the surgeon's discretion, a bisphosphonate (4 mg zolendronate in 1 cc) can be added to the cement to prevent further tumor-induced osteolysis.

Table 23.3 Technique—intramedullary stabilization of tibial diaphyseal metastases

Open lesion curettage performed as outlined in Table 23.2, then nail insertion, and lastly cementation.
Nail insertion should be extra-articular (parapatellar or tendon splitting). Suprapatellar insertion is avoided to prevent tumor seeding into the knee.
For prophylactic nailing, starting point and insertion technique are critical. There is less "wiggle room" as compared with completed fractures, and it is easy to iatrogenically fracture the tibia during nail insertion.
Tendon splitting technique with the knee fully flexed facilitates precise and direct starting point visualization, and allows easy irrigation and removal of canal contents. This position also elevates the patella to prevent nail impingement that could result in accidental posterior cortical breach or iatrogenic fracture during insertion.
Canal reaming is minimized to accommodate a 9 mm or 10 mm nail diameter. A reamer irrigation aspiration (RIA) device can permit canal preparation in a single pass.
At least two interlock screws, both proximally and distally are preferred.
Tips and tricks
For metadiaphyseal metastases (Fig. 23.3), cement can be injected through an interlock drill hole using a Jamshidi bone biopsy canula with attached 20 cc syringe. The drill tract should be unicortical to prevent extraosseous cement extravasation. The tract can then be drilled bicortically before the cement hardens, and the interlock screw fully inserted.
Minimally invasive methods are possible. First percutaneous radiofrequency ablation of appropriately sized lesions, reamer irrigation/aspiration of the canal, nail placement, followed by percutaneous balloon osteoplasty and cement injection using widely available kyphoplasty instrumentation. This avoids making a second incision for lesion removal.

Table 23.4 Technique—intercalary endoprosthesis

Wide soft tissue margins are not required in the setting of metastasis, but wide bony margins are encouraged.
The shortest stems for most intercalary implants are ~6 cm in length. For very proximal tibia lesions, intercalary allograft or long-stemmed tibia component of a total knee arthroplasty should be considered.
Near circumferential loss of bone is present. The resection length should continue until (1) circumferential bone is present and (2) cortical thickness is near normal.
Intercalary prostheses come in segmental lengths that can be interconnected. Bone resection length should be planned accordingly. For example, an implant design may employ 4 cm intercalary segments, so a 6 cm bony defect would require an 8 cm resection length to accommodate two stacked 4 cm intercalary segments.
Transverse bone cuts made, taking care not to accidentally splinter or crack the diseased bone.
The often-used technique of starting with a saw and completing the osteotomy with an osteotome does NOT work well here. Full circumferential protection of the diaphysis with two Hohmann or malleable retractors is not particularly difficult after gross removal of an intercalary lesion. This facilitates using a small saw blade to completely make necessary osteotomies. A bone planar should be used if available.
Straight reamers are employed proximally and distally to the physeal scars if possible.
Cement restrictors are usually not required. Canals are thoroughly cleaned prior to cementation.
While press-fit stems are available, they are NOT recommended in the oncologic setting.
Proximal and distal stems should be cemented separately, with separate cement mixes.
Many stems offer interlock screws, inserted using an outrigger jig that attaches to the stem. Using an interlock screw after cementing a stem is "off-label" for some vendors, but is considered standard of care by many orthopaedic oncologists.
The intercalary body is connected last to both stems. Many are clam-shell in design, with sequential tightening of many screws.
Incision closure with nylon suture.
Tips and tricks
Radiation does not work well here. The large size of the metal endoprosthesis causes significant external radiation beam scattering, further compromising the limited soft tissue envelope over the tibia. With this in mind, to decrease risk of local metastasis progression, Aquamantis™ or similar is used to aggressively ablate the peripheral soft tissues around the removed bone segment without causing tissue carbonization/charring.
Unlike with intercalary resections of the humerus and femur, care should be taken NOT to fracture the fibula during canal preparation.
Cement mantle thickness is usually 1–2 mm, but a 0 mm French paradox technique can be used for very narrow canals.
Consider medial gastrocnemius or soleus flap to help with implant coverage in severely malnourished patients. Any wound complication with exposed hardware can result in amputation in these patients.

Table 23.5 Comparison of surgical management techniques for tibial metastasis

Reconstruction	Advantages	Disadvantages
Curettage, cementation, plate, and screw fixation	Best for large solitary lesions Single incision Immediate fixation with cementation Lesion can be contained or uncontained	Marginal resection Larger incision needed for plating, may delay restarting chemotherapy Thin soft tissue envelope over tibia Possible weight-bearing limitations Not well suited for multifocal metastasis
Curettage, cementation, and intramedullary nail	Best for multiple lesions throughout the tibia Immediate fixation with cementation Less disruption of soft tissue envelope Immediate weight bearing	Marginal resection Theoretic seeding of intramedullary canal Adjuvant radiation often needed over entire bone.
Intercalary allograft	Wide resection Reconstruction of large defects (>6 cm and circumferential) Accommodates large proximal and distal lesions while avoiding arthroplasty	Larger incision Prolonged weight bearing limitations Higher nonunion risk, with poor bone healing capacity Risk of graft fracture
Intercalary endoprosthesis	Wide resection Addresses large uncontained defects (>6 cm and circumferential) Immediate weight bearing Single incision	Higher risk of infection, prosthetic, periprosthetic fracture, and aseptic loosening Nonbiologic fixation

Postoperative Care

After undergoing surgical intervention for tibial metastatic lesions, patients are at an increased risk of medical and surgical complications. They are at an increased risk for deep vein thrombosis (DVT) or pulmonary embolism (PE) due to having all three risk factors for Virchow's Triad [57], an increased risk for wound complications and infections due to having weakened immunity from impaired function of antigen-presenting cells due to having cancer [58], and undergoing preoperative/postoperative cancer treatments. Additionally, many patients with metastatic cancer are malnourished and sarcopenic, which can impair wound and fracture healing. [59–61] Patients may have significant postoperative pain from increased tolerance to traditionally utilized opioid pain medications and being contra-indicated from utilizing some non-opioid medications in traditional multimodal pain regimens [62]. Therefore, multidisciplinary management of patients incorporating effective anticoagulation, infection prevention principles, nutritional optimization, and optimal postoperative analgesia can prevent complications and improve patient outcomes.

The primary goals for undergoing surgical intervention for tibial shaft metastasis are to improve pain and preserve/restore function. The majority of patients undergoing IMN or ORIF can begin weight-bearing immediately postoperatively, facilitating improved functional ability and functional outcomes. Greco et al. reported in their meta-analysis that amongst 38 patients treated for tibial shaft metastases, 18 patients had preoperative and postoperative pain levels compared, with all patients having significant improvement in their pain postoperatively [8]. Additionally, amongst 9 patients with functional outcomes that were reported, all patients had improved functional outcomes [8]. Amongst 18 patients whose ambulation status was reported, 17/18 (94%) were able to ambulate postoperatively, with 3/4 (75%) patients who received an intercalary prosthesis being able to ambulate [8]. Bonnevaille et al. also reported pain and functional outcomes for 21 patients who were alive 6 months postoperatively, but for all patients with tibial metastasis, not stratified by tibial shaft metastases [6]. Bonnevaille et al. reported that 17/21 (80%) of patients had pain reduction, 4/21 (19%) of patients had no pain, and 1/21 (1%) of patients had no change in pain [6]. When subjectively assessing ambulation and knee/ankle range of motion postoperatively, 3/21 (14%) of patients had normal function, 11/21 (52%) had acceptable function, and 7/21 (33%) had insufficient function [6]. Therefore, surgical intervention for tibial metastases appears to be associated with excellent pain relief and good functional outcomes for the majority of patients.

Patients undergoing surgical fixation of metastatic tibial shaft lesions should undergo postoperative radiation therapy (PORT) to facilitate improved pain control, decrease the risk of local recurrence and subsequent skeletal-related events (SREs), and improve functional outcomes. Many studies have found that PORT is associated with decreased risk for SREs. [63–69] After allowing for adequate time for postoperative wound healing, PORT should be started within 2–5 weeks postoperatively [70]. PORT consists of external-beam radiation therapy that is typically given as multiple fractions with lower doses (30 Gray [Gy] in 10 fractions or 20 Gy in 5 fractions), or one fraction with a higher dose (8 Gy in 1 fraction) [71]. All hardware should be covered when delivering SBRT to minimize the risk of recurrence [67]. Therefore, in patients who undergo tibia IMN fixation, the entire tibia should be covered to minimize the risk of tibia local recurrence [67, 68].

Multi-disciplinary discussion regarding utilizing bisphosphonates should be conducted. Bisphosphonates are effective at inhibiting osteoclast-mediated bone resorption, which has been found to decrease the risk for developing SREs and improve pain, which can improve patient outcomes [72–74]. Additionally, evidence is being published regarding the antitumor properties of bisphosphonates, which may strengthen the immune system reaction towards tumor cells, causing inhibition of tumor growth and metastasis, and induce tumor cell apoptosis [75]. Studies have been published demonstrating increased survival benefit when utilizing bisphosphonates in patients with breast and prostate cancer. However, further study on this topic is warranted, and the benefits/risks of bisphosphonates, including osteonecrosis of the jaw must be considered [76].

Survivorship in patients with tibial metastatic disease has been reported by a small number of studies. As reported in a meta-analysis by Greco et al., amongst 64 patients, 18 (28%) were alive at a mean follow-up of two years postoperatively, while 46 (72%) patients passed away at a median of 15 months postoperatively [8]. A retrospective study by Bonnevaille et al. found that amongst 25 patients, the median survival was 14 months postoperatively (range 0–49 months) [6]. They found that the type of primary cancer plays a large role in survival, with patients diagnosed with a lung primary tumor having a median survival of 4 months (none lived past one year), while patients with renal primary tumor had a median survival of 32 months (3/10 patients were alive after 3 years) [6]. Additionally, whether surgical intervention was being performed prophylactically versus for a pathologic fracture influenced survival postoperatively. Patients treated prophylactically had a longer median survival of 15 months, while patients treated for a pathologic fracture had a shorter median survival of 5 months [6]. The short median survival times reported likely represent that many patients diagnosed with tibial metastases have advanced metastatic disease.

References

1. De Geeter K, Reynders P, Samson I, Broos PL. Metastatic fractures of the tibia. Acta Orthop Belg. 2001;67(1):54–9.
2. Kelly M, Lee M, Clarkson P, O'Brien P. Metastatic disease of the long bones: a review of the health care burden in a major trauma centre. Can J Surg. 2012;55(2):95–8. https://doi.org/10.1503/cjs.049009.
3. Jayarangaiah A, Kemp AK, Theetha Kariyanna P. Bone metastasis. In: StatPearls. StatPearls Publishing; 2024. http://www.ncbi.nlm.nih.gov/books/NBK507911/. Accessed 25 Sept 2024.
4. Capanna R, Campanacci DA. The treatment of metastases in the appendicular skeleton. J Bone Joint Surg Br. 2001;83-B(4):471–81. https://doi.org/10.1302/0301-620X.83B4.0830471.
5. Leeson MC, Makley JT, Carter JR. Metastatic skeletal disease distal to the elbow and knee. Clin Orthop. 1986;206:94–9.
6. Bonnevialle P, Descamps J, Niglis L, et al. Surgical treatment of tibial metastases: retrospective, multicenter, observational study of 25 patients. Orthop Traumatol Surg Res. 2020;106(6):1039–45. https://doi.org/10.1016/j.otsr.2019.07.017.
7. Greenbaum SL, Thornhill BA, Geller DS. Characterization and surgical management of metastatic disease of the Tibia. Am J Orthop Belle Mead NJ. 2017;46(6):E423–8.
8. Greco T, Fulchignoni C, Cianni L, Maccauro G, Perisano C. Surgical management of tibial metastases: a systematic review. Acta Bio-Medica Atenei Parm. 2022;92(S3):e2021552. https://doi.org/10.23750/abm.v92iS3.12540.
9. Rougraff BT, Kneisl JS, Simon MA. Skeletal metastases of unknown origin. A prospective study of a diagnostic strategy. J Bone Jt Surg. 1993;75(9):1276–81. https://doi.org/10.2106/00004623-199309000-00003.
10. Willeumier JJ, Van Der Linden YM, Van Der Wal CWPG, et al. An easy-to-use prognostic model for survival estimation for patients with symptomatic long bone metastases. J Bone Jt Surg. 2018;100(3):196–204. https://doi.org/10.2106/JBJS.16.01514.
11. Piccioli A, Spinelli MS, Forsberg JA, et al. How do we estimate survival? External validation of a tool for survival estimation in patients with metastatic bone disease—decision analysis and comparison of three international patient populations. BMC Cancer. 2015;15(1):424. https://doi.org/10.1186/s12885-015-1396-5.

12. Snell W, Beals RK. Femoral metastases and fractures from breast cancer. Surg Gynecol Obstet. 1964;119:22–4.
13. Beals RK, Lawton GD, Snell WE. Prophylactic internal fixation of the femur in metastatic breast cancer. Cancer. 1971;28(5):1350–4. https://doi.org/10.1002/1097-0142(1971)28:5<1350::aid-cncr2820280539>3.0.co;2-6.
14. Harrington KD. Impending pathologic fractures from metastatic malignancy: evaluation and management. Instr Course Lect. 1986;35:357–81.
15. Mirels H. Metastatic disease in long bones. A proposed scoring system for diagnosing impending pathologic fractures. Clin Orthop. 1989;(249):256–64.
16. Damron TA, Nazarian A, Entezari V, et al. CT-based structural rigidity analysis is more accurate than mirels scoring for fracture prediction in metastatic femoral lesions. Clin Orthop. 2016;474(3):643–51. https://doi.org/10.1007/s11999-015-4453-0.
17. Tejwani NC, Guerado E. Improving fixation of the osteoporotic fracture: the role of locked plating. J Orthop Trauma. 2011;25(Supplement 2):S56–60. https://doi.org/10.1097/BOT.0b013e31821b8a52.
18. Scolaro JA, Lackman RD. Surgical management of metastatic long bone fractures: principles and techniques. J Am Acad Orthop Surg. 2014;22(2):90–100. https://doi.org/10.5435/JAAOS-22-02-90.
19. Kelly CM, Wilkins RM, Eckardt JJ, Ward WG. Treatment of metastatic disease of the tibia. Clin Orthop. 2003;(415 Suppl):S219–29. https://doi.org/10.1097/01.blo.0000093843.72468.3a.
20. Miller BJ, Soni EEC, Gibbs CP, Scarborough MT. Intramedullary nails for long bone metastases: why do they fail? Orthopedics. 2011;34(4):01477447-20110228-12. https://doi.org/10.3928/01477447-20110228-12.
21. Zimel MN, Hwang S, Riedel ER, Healey JH. Carbon fiber intramedullary nails reduce artifact in postoperative advanced imaging. Skeletal Radiol. 2015;44(9):1317–25. https://doi.org/10.1007/s00256-015-2158-9.
22. Huang JY, Followill DS, Howell RM, et al. Approaches to reducing photon dose calculation errors near metal implants. Med Phys. 2016;43(9):5117–30. https://doi.org/10.1118/1.4960632.
23. Tedesco G, Gasbarrini A, Bandiera S, Ghermandi R, Boriani S. Composite PEEK/carbon fiber implants can increase the effectiveness of radiotherapy in the management of spine tumors. J Spine Surg. 2017;3(3):323–9. https://doi.org/10.21037/jss.2017.06.20.
24. Nevelsky A, Borzov E, Daniel S, Bar-Deroma R. Perturbation effects of the carbon fiber-PEEK screws on radiotherapy dose distribution. J Appl Clin Med Phys. 2017;18(2):62–8. https://doi.org/10.1002/acm2.12046.
25. Golish SR, Mihalko WM. Principles of biomechanics and biomaterials in orthopaedic surgery. J Bone Jt Surg-Am. 2011;93(2):207–12. https://doi.org/10.2106/00004623-201101190-00013.
26. Kamei Y, Nakayama B, Toriyama K, et al. Combined fibular osteocutaneous and omental flaps. Plast Reconstr Surg. 2007;119(5):1499–504. https://doi.org/10.1097/01.prs.0000256065.62382.89.
27. Malawer MM, Wittig JC, Bickels J, Wiesel SW. Operative techniques in orthopaedic surgical oncology. Wolters Kluwer Health; 2015. https://books.google.com/books?id=1u80CwAAQBAJ
28. Capanna R, Campanacci DA, Belot N, et al. A new reconstructive technique for intercalary defects of long bones: the association of massive allograft with vascularized fibular autograft. Long-term results and comparison with alternative techniques. Orthop Clin North Am. 2007;38(1):51–60. https://doi.org/10.1016/j.ocl.2006.10.008.
29. Panagopoulos GN, Mavrogenis AF, Mauffrey C, et al. Intercalary reconstructions after bone tumor resections: a review of treatments. Eur J Orthop Surg Traumatol. 2017;27(6):737–46. https://doi.org/10.1007/s00590-017-1985-x.
30. Malizos KN, Zalavras CG, Soucacos PN, Beris AE, Urbaniak JR. Free vascularized fibular grafts for reconstruction of skeletal defects. J Am Acad Orthop Surg. 2004;12(5):360–9. https://doi.org/10.5435/00124635-200409000-00010.
31. Minami A, Kasashima T, Iwasaki N, Kato H, Kaneda K. Vascularised fibular grafts. An experience of 102 patients. J Bone Joint Surg Br. 2000;82(7):1022–5. https://doi.org/10.1302/0301-620x.82b7.10332.

32. de Boer HH, Wood MB, Hermans J. Reconstruction of large skeletal defects by vascularized fibula transfer. Factors that influenced the outcome of union in 62 cases. Int Orthop. 1990;14(2):121–8. https://doi.org/10.1007/BF00180115.
33. Frisoni T, Cevolani L, Giorgini A, Dozza B, Donati DM. Factors affecting outcome of massive intercalary bone allografts in the treatment of tumours of the femur. J Bone Joint Surg Br. 2012;94-B(6):836–41. https://doi.org/10.1302/0301-620X.94B6.28680.
34. Bus MPA, Dijkstra PDS, Van De Sande MAJ, et al. Intercalary allograft reconstructions following resection of primary bone tumors: a nationwide multicenter study. J Bone Jt Surg. 2014;96(4):e26. https://doi.org/10.2106/JBJS.M.00655.
35. Muscolo DL, Ayerza MA, Aponte-Tinao L, Ranalletta M, Abalo E. Intercalary femur and tibia segmental allografts provide an acceptable alternative in reconstructing tumor resections. Clin Orthop. 2004;426:97–102. https://doi.org/10.1097/01.blo.0000141652.93178.10.
36. Ortiz-Cruz E, Gebhardt MC, Jennings LC, Springfield DS, Mankin HJ. The results of transplantation of intercalary allografts after resection of tumors. A long-term follow-up study. J Bone Jt Surg Am. 1997;79(1):97–106. https://doi.org/10.2106/00004623-199701000-00010.
37. Vander Griend RA. The effect of internal fixation on the healing of large allografts. J Bone Jt Surg. 1994;76(5):657–63. https://doi.org/10.2106/00004623-199405000-00005.
38. Mankin HJ, Springfield DS, Gebhardt MC, Tomford WW. Current status of allografting for bone tumors. Orthopedics. 1992;15(10):1147–54. https://doi.org/10.3928/0147-7447-19921001-05.
39. Mankin HJ, Gebhardt MC, Jennings LC, Springfield DS, Tomford WW. Long-term results of allograft replacement in the management of bone tumors. Clin Orthop. 1996;324:86–97. https://doi.org/10.1097/00003086-199603000-00011.
40. Fuchs B, Ossendorf C, Leerapun T, Sim FH. Intercalary segmental reconstruction after bone tumor resection. Eur J Surg Oncol J Eur Soc Surg Oncol Br Assoc Surg Oncol. 2008;34(12):1271–6. https://doi.org/10.1016/j.ejso.2007.11.010.
41. San-Julian M, Cañadell J. Fractures of allografts used in limb preserving operations. Int Orthop. 1998;22(1):32–6. https://doi.org/10.1007/s002640050203.
42. Donati D, Di Liddo M, Zavatta M, et al. Massive bone allograft reconstruction in high-grade osteosarcoma. Clin Orthop. 2000;377:186–94. https://doi.org/10.1097/00003086-200008000-00025.
43. Thompson RC, Garg A, Clohisy DR, Cheng EY. Fractures in large segment allografts. Clin Orthop. 2000;370:227–35. https://doi.org/10.1097/00003086-200001000-00023.
44. Gerrand CH, Griffin AM, Davis AM, Gross AE, Bell RS, Wunder JS. Large segment allograft survival is improved with intramedullary cement. J Surg Oncol. 2003;84(4):198–208. https://doi.org/10.1002/jso.10316.
45. Hornicek FJ, Gebhardt MC, Tomford WW, et al. Factors affecting nonunion of the allograft-host junction. Clin Orthop. 2001;382:87–98. https://doi.org/10.1097/00003086-200101000-00014.
46. Uyttendaele D, De Schryver A, Claessens H, Roels H, Berkvens P, Mondelaers W. Limb conservation in primary bone tumours by resection, extracorporeal irradiation and re-implantation. J Bone Joint Surg Br. 1988;70-B(3):348–53. https://doi.org/10.1302/0301-620X.70B3.3163694.
47. Chen TH, Chen WM, Huang CK. Reconstruction after intercalary resection of malignant bone tumours: comparison between segmental allograft and extracorporeally-irradiated autograft. J Bone Joint Surg Br. 2005;87-B(5):704–9. https://doi.org/10.1302/0301-620X.87B5.15491.
48. Böhm P, Springfeld R, Springer H. Re-implantation of autoclaved bone segments in musculoskeletal tumor surgery. Arch Orthop Trauma Surg. 1998;118(1-2):57–65. https://doi.org/10.1007/s004020050312.
49. Sabo D, Brocai DR, Eble M, Wannenmacher M, Ewerbeck V. Influence of extracorporeal irradiation on the reintegration of autologous grafts of bone and joint. Study in a canine model. J Bone Joint Surg Br. 2000;82(2):276–82.
50. Khattak MJ, Umer M, Haroon-ur-Rasheed UM. Autoclaved tumor bone for reconstruction: an alternative in developing countries. Clin Orthop. 2006;447:138–44. https://doi.org/10.1097/01.blo.0000205876.05093.80.

51. Manabe J, Ahmed AR, Kawaguchi N, Matsumoto S, Kuroda H. Pasteurized autologous bone graft in surgery for bone and soft tissue sarcoma. Clin Orthop. 2004;419:258–66. https://doi.org/10.1097/00003086-200402000-00042.
52. Tsuchiya H, Wan SL, Sakayama K, Yamamoto N, Nishida H, Tomita K. Reconstruction using an autograft containing tumour treated by liquid nitrogen. J Bone Joint Surg Br. 2005;87-B(2):218–25. https://doi.org/10.1302/0301-620X.87B2.15325.
53. Ruggieri P, Mavrogenis AF, Bianchi G, Sakellariou VI, Mercuri M, Papagelopoulos PJ. Outcome of the intramedullary diaphyseal segmental defect fixation system for bone tumors. J Surg Oncol. 2011;104(1):83–90. https://doi.org/10.1002/jso.21893.
54. Benevenia J, Kirchner R, Patterson F, et al. Outcomes of a modular intercalary endoprosthesis as treatment for segmental defects of the femur, tibia, and humerus. Clin Orthop. 2016;474(2):539–48. https://doi.org/10.1007/s11999-015-4588-z.
55. Hanna SA, Sewell MD, Aston WJS, et al. Femoral diaphyseal endoprosthetic reconstruction after segmental resection of primary bone tumours. J Bone Joint Surg Br. 2010;92-B(6):867–74. https://doi.org/10.1302/0301-620X.92B6.23449.
56. Aldlyami E, Abudu A, Grimer RJ, Carter SR, Tillman RM. Endoprosthetic replacement of diaphyseal bone defects. Long-term results. Int Orthop. 2005;29(1):25–9. https://doi.org/10.1007/s00264-004-0614-6.
57. Kushner A, West WP, Khan Suheb MZ, Pillarisetty LS. Virchow Triad. In: StatPearls. StatPearls Publishing; 2024. http://www.ncbi.nlm.nih.gov/books/NBK539697/. Accessed 25 Sept 2024.
58. Allen BM, Hiam KJ, Burnett CE, et al. Systemic dysfunction and plasticity of the immune macroenvironment in cancer models. Nat Med. 2020;26(7):1125–34. https://doi.org/10.1038/s41591-020-0892-6.
59. Arends J. Malnutrition in cancer patients: causes, consequences and treatment options. Eur J Surg Oncol. 2024;50(5):107074. https://doi.org/10.1016/j.ejso.2023.107074.
60. Bossi P, De Luca R, Ciani O, D'Angelo E, Caccialanza R. Malnutrition management in oncology: an expert view on controversial issues and future perspectives. Front Oncol. 2022;12:910770. https://doi.org/10.3389/fonc.2022.910770.
61. Karpouzos A, Diamantis E, Farmaki P, Savvanis S, Troupis T. Nutritional aspects of bone health and fracture healing. J Osteoporos. 2017;2017:1–10. https://doi.org/10.1155/2017/4218472.
62. Ramirez M, Strang A, Roland G, Lasala J, Owusu-Agyemang P. Perioperative pain management and cancer outcomes: a narrative review. J Pain Res. 2023;16:4181–9. https://doi.org/10.2147/JPR.S432444.
63. Lanier CM, Johnson AG, Razavian NB, Farris JC, Hughes RT. Skeletal-related events after surgery with or without radiotherapy for bone metastases to weight-bearing bones. Cureus. Published online December 21, 2022. https://doi.org/10.7759/cureus.32778.
64. Van Geffen E, Wobbes T, Veth RP, Gelderman WA. Operative management of impending pathological fractures: a critical analysis of therapy. J Surg Oncol. 1997;64(3):190–4. https://doi.org/10.1002/(sici)1096-9098(199703)64:3<190::aid-jso3>3.0.co;2-5.
65. Drost L, Ganesh V, Wan BA, et al. Efficacy of postoperative radiation treatment for bone metastases in the extremities. Radiother Oncol. 2017;124(1):45–8. https://doi.org/10.1016/j.radonc.2017.05.010.
66. Townsend PW, Smalley SR, Cozad SC, Rosenthal HG, Hassanein RES. Role of postoperative radiation therapy after stabilization of fractures caused by metastatic disease. Int J Radiat Oncol. 1995;31(1):43–9. https://doi.org/10.1016/0360-3016(94)E0310-G.
67. Epstein-Peterson ZD, Sullivan A, Krishnan M, et al. Postoperative radiation therapy for osseous metastasis: Outcomes and predictors of local failure. Pract Radiat Oncol. 2015;5(5):e531–6. https://doi.org/10.1016/j.prro.2015.02.006.
68. Rosen DB, Haseltine JM, Bartelstein M, et al. Should postoperative radiation for long bone metastases cover part or all of the orthopedic hardware? Results of a large retrospective analysis. Adv Radiat Oncol. 2021;6(6):100756. https://doi.org/10.1016/j.adro.2021.100756.
69. Lutz S, Balboni T, Jones J, et al. Palliative radiation therapy for bone metastases: update of an ASTRO evidence-based guideline. Pract Radiat Oncol. 2017;7(1):4–12. https://doi.org/10.1016/j.prro.2016.08.001.

70. Ramadan S, Arifin AJ, Nguyen TK. The role of post-operative radiotherapy for non-spine bone metastases (NSBMs). Cancers. 2023;15(13):3315. https://doi.org/10.3390/cancers15133315.
71. Van Der Velden J, Willmann J, Spałek M, et al. ESTRO ACROP guidelines for external beam radiotherapy of patients with uncomplicated bone metastases. Radiother Oncol. 2022;173:197–206. https://doi.org/10.1016/j.radonc.2022.05.024.
72. Fukuokaya W, Mori K, Urabe F, et al. Bone-modifying agents in patients with high-risk metastatic castration-sensitive prostate cancer treated with abiraterone acetate. JAMA Netw Open. 2024;7(3):e242467. https://doi.org/10.1001/jamanetworkopen.2024.2467.
73. O'Carrigan B, Wong MH, Willson ML, Stockler MR, Pavlakis N, Goodwin A. Bisphosphonates and other bone agents for breast cancer. Cochrane Breast Cancer Group, ed. Cochrane Database Syst Rev. 2017;2018(11) https://doi.org/10.1002/14651858.CD003474.pub4.
74. Van Broekhoven DL, Dootjes LW, Van Der Veldt A, Zillikens C, Van Oldenrijk J. Effect of bisphosphonates on skeletal related events in long bone metastases of renal cell carcinoma: a systematic review. Clin Genitourin Cancer. 2023;21(3):e190–7. https://doi.org/10.1016/j.clgc.2022.12.010.
75. Sun M, Iqbal J, Singh S, Sun L, Zaidi M. The crossover of bisphosphonates to cancer therapy. Ann NY Acad Sci. 2010;1211(1):107–12. https://doi.org/10.1111/j.1749-6632.2010.05812.x.
76. Drake MT, Clarke BL, Khosla S. Bisphosphonates: mechanism of action and role in clinical practice. Mayo Clin Proc. 2008;83(9):1032–45. https://doi.org/10.4065/83.9.1032.

24 Management of Metastatic Disease to the Foot and Ankle

Taha Aksoy and H. Thomas Temple

Introduction

Acral metastases are rare, and even the terms (acral and acro) and the definition of a distal metastasis are ill-defined. As a result, there is considerable variability in how acral metastatic disease is reported in the literature. Some articles include acral metastases in both the upper and lower extremities [1–4]. Acral metastases in the foot are clearly different from those in the hand in how patients present, the types of disability that result from the disease, and the treatment necessary to mitigate pain and restore function. Some authors include all lesions below the knee [3–6]; others include lesions in the foot and ankle exclusively [7–12]. Metastases to the foot are very uncommon and are reported to be one-half to one-third the incidence of metastatic lesions observed in the hand [2, 13, 14]. Others, however, have reported acral metastases to be more common in the lower extremity [3, 5].

Most articles on acral metastases are case reports, while others are limited to retrospective case series over several years from either large institutional or public databases, both fraught with inherent limitations due to data loss and selection bias. For this chapter, the term *acral metastasis* will be used, and the incidence, diagnosis, and treatment of this condition will be confined to lesions in the *ankle and foot only*.

In both sexes combined, lung cancer is the most diagnosed cancer (11.6% of the total cases) and the leading cause of cancer death (18.4% of the total cancer deaths), closely followed by female breast cancer (11.6%), prostate cancer (7.1%), and

T. Aksoy
Musculoskeletal Oncology, University of Miami Miller School of Medicine, Miami, FL, USA

H. T. Temple (✉)
Orthopaedic Surgery and Pathology, University of Miami Miller School of Medicine, Miami, FL, USA
e-mail: Htemple@med.miami.edu

J. Pretell-Mazzini, S. A. Lozano Calderón (eds.), *Multidisciplinary Approach of Musculoskeletal Metastasis*, https://doi.org/10.1007/978-3-032-09498-8_24

colorectal cancer (6.1%) [15]. Bone is one of the most common target sites of metastases in patients with advanced cancer [16], but metastases to the foot and hand (acral metastasis) are very rare (0.07–0.3%) [11]. With advances in diagnosis and treatment, patient survival with different cancers has significantly improved [17]. Because of this, the number of patients with bone metastases has been increasing in recent decades [15].

In general, metastatic bone disease may occur in all age groups but is more commonly encountered in males and older individuals [18]. Similarly, acral metastases may also occur in a wide range of ages and are also more frequent in males than females and in general, older individuals. In patients with acral metastases, the lung is the most common primary malignancy, followed by colorectal, breast, and genitourinary (GU) [4–6] (Table 24.1). The rarity of acral metastases is believed to be due to the paucity of red marrow in the bones of the foot and ankle and the further distance from the primary tumor and Batson's plexus that allows retrograde flow through the valveless network of veins [6, 9, 13, 19]. Because acral metastases are so rare, it is difficult to estimate the true incidence and to define the epidemiologic characteristics of this disease. For this reason, the location of acral metastases in the foot varies, but in general, the hindfoot and collective bones of the midfoot are the most common sites, followed by the bones in the forefoot [7, 12].

In most cases, acral metastases are seen in patients with widespread metastatic disease [7, 12]. However, in a few reviews, 31.5–44.5% of patients presented with a metastatic deposit in the foot, representing the first manifestation of metastatic disease [5, 9, 11, 19]. This is contrary to a study by Thompson et al., where only one patient in 38 included in the series had isolated metastatic disease of the foot at the time of diagnosis [12]. Delays in diagnosis in this subgroup of patients are common [4, 7]. The mean duration of symptoms to time of diagnosis of the acral metastasis was 3.2–22.3 months [5, 7, 11, 12, 20]

Delays occur because symptoms of metastatic disease in the foot and ankle are thought to be due to other common conditions, such as tendinopathy, gout, arthritis, overuse, and trauma [7]. A higher index of suspicion should be entertained in patients with known primary cancers; however, as noted in some series of patients with acral metastases, 37–80% of patients had no known primary cancer at the time of diagnosis [7, 9]. Aside from being uncommon and unsuspected, in those patients with known primary disease, acral metastases that are clinically asymptomatic may be missed because routine positron emission tomography (PET) imaging typically extends only to the knees. Therefore, unless the PET scan includes the whole body or a technetium bone scan or other focused imaging study is performed, the acral metastasis will go undetected. This raises the possibility that acral metastases may be under-reported.

Table 24.1 Literature Review for Acral Metastases

Study	Publication year	*N*	Age (years)	Symptom duration to diagnosis (months)	Survival (months)	Most common primary tumor	First sign of malignancy in diagnosis (%)	Type
El Abiad et al.[a] [1]	2019	16	61.5	N/A	9.7	Lung	N/A	Single institution
Ciftdemir et al.[a] [5]	2021	48	61.5	19.1	24.9	Lung	43.5	Single institution
Evans et al. [7]	2014	15	64.5	4	20.1	Infra-diaphragm[b]	20	Single institution
Greco et al. [9]	2020	46	63.2	N/A	N/A	Lung	37	Single institution
Thompson et al. [12]	2022	44	73	21.3	12	Prostate	2	Literature review
Meheshwari et al. [11]	2008	14	66.6	3.2	14.8	Renal	31.5	Single institution
Bahrabadi et al. [20]	2020	38	59.4	3.2	5.8	Lung	44.5	Literature review

[a]Includes both upper and lower extremity cases
[b]Includes colon and genitourinary malignancies
N/A not reported

Diagnosis

All patients with pain, especially intractable foot pain, should undergo a careful clinical history and physical examination. A prior history of a known primary tumor is very important. A history of a persistent cough, weight loss, change in bowel habits, blood in the stool and urine, urinary frequency and hesitancy, tobacco and alcohol consumption should be noted.

A focused physical examination of the foot and ankle should include careful inspection for deformity and swelling, erythema, limited range of foot and ankle motion, gait alteration, tenderness to palpation, loss of strength, and inability to perform a single heel raise [7].

Orthogonal views of the affected foot and ankle should be obtained. Acral metastases may be solitary, but in patients with advanced disease, they are more often multifocal [12]. Depending on the primary tumor, the appearance of metastasis in bone may be lytic, mixed, or blastic. Since most acral tumors arise from the lung and GU primary sites, the usual appearance is lytic [20] (Fig. 24.1).

For patients who are seen with suspected acral metastasis as the initial presentation, a biopsy is necessary, especially for those individuals with a solitary lesion. Typically, a needle biopsy under image guidance is undertaken. Once a diagnosis is made, and for those individuals with multifocal disease in the foot and the presence of other potential sites of skeletal disease, further imaging is indicated to ascertain the primary site of disease. Staging with appropriate imaging should be performed expeditiously. A diagnostic strategy described by Rougraff et al. is valuable in determining the primary site of disease in most cases [21]. PET imaging is especially helpful in determining the primary site and extent of disease. Technetium^{99} pyrophosphate (Tc^{99} pyrophosphate) may be employed to assess other sites of skeletal disease where bone is being rapidly turned over, as is the case in metastatic bone disease. For patients with myeloma, however, Tc^{99} pyrophosphate imaging is not as effective as radiographs in detecting myeloma deposits in bone [22]. 18-Fluorodeoxyglucose positron emission tomography with computed tomography

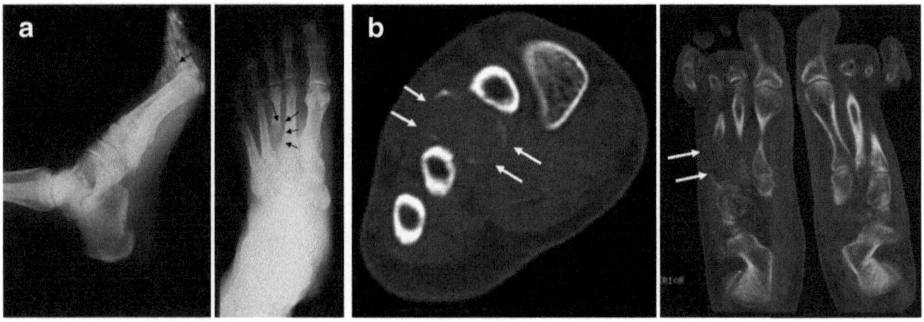

Fig. 24.1 (**a**) 80-year-old patient with a known history of renal cell carcinoma; antero-posterior and lateral radiographs of a radiolucent lesion in the base of the left third metatarsal (black arrows), (**b**) Corresponding axial and coronal computed tomography demonstrating cortical destruction extending to subchondral bone (white arrows)

(FDG PET/CT) has been shown to be superior to whole body x-ray surveillance in evaluating disease extent in patients with myeloma. FDG PET/CT and magnetic resonance imaging (MRI) have higher sensitivity and specificity than whole-body x-ray survey in evaluating disease extent in patients with multiple myeloma [23].

Advanced imaging, CT or MRI, can be valuable in assessing local disease extent, cortical destruction, marrow replacement, and soft tissue extension. All these factors may be important in developing a treatment strategy.

Radiographically, solitary acral metastases may have an indolent appearance and can be confused with benign tumors, dysplasia, or even Paget's disease [24, 25]. Locally destructive metastases can also mimic osteomyelitis [26].

Patients with bone metastases to the foot and ankle are older, often have medical co-morbidities related or unrelated to their primary cancer, and are sometimes significantly disabled by pain and or fracture(s). They require urgent evaluation and expeditious treatment, but unfortunately, care, like for most patients with metastatic disease, is often fragmented and delayed. These patients are best referred to a center where multidisciplinary musculoskeletal oncologic care is available. Ideally, these patients should be presented in a multidisciplinary conference where medical oncologists, orthopedic musculoskeletal surgeons, radiologists, pathologists, nurse navigators, and physical and occupational therapists can discuss and devise diagnostic and therapeutic strategies in an expeditious and coordinated fashion [27]. For many of these patients, end-of-life decisions are in order, and referral to hospice care may be in the patient's and their family's best interest in select cases.

Treatment

Surgical Indication

Treatment is predicated on the type of primary cancer, patient age, co-morbid conditions, available systemic treatment, degree of pain and disability after radiotherapy, and pre-morbid functional status.

Historically, patients with bone metastases were considered incurable, and although the prognosis for most is generally poor, especially those with acral metastases and diffuse skeletal disease, each patient should be approached individually. Targeted therapies for some forms of cancer have improved dramatically, and survival, in general, has improved for patients with metastatic disease [11, 22].

The orthopedic goals of treating metastatic disease are to relieve pain, restore function, and facilitate nursing care. To that end, and in consideration of the treatment for patients with acral metastases, the therapeutic options are specific to the individual's overall health, disease extent, and overall prognosis. In general, these patients have a poor outcome [12], and for many patients, acral metastases represent advanced or terminal disease. In others, though, with solitary tumors or tumors for which effective systemic therapies exist, survival can be prolonged. Treatment may be different for a disease in the foot versus the ankle and even for different sites in the foot (forefoot versus mid or hindfoot).

Treatment Options

Non-operative Treatment

Patients with extensive local disease with diffuse metastases are not reasonable surgical candidates. Treatment for most patients with acral metastases is non-operative. Analgesics, shoe modifications, orthotics, immobilization, and modified weight bearing (cane, crutches, and wheelchair) are employed. External beam radiotherapy may be used in select cases, although there are some tumors, notably renal cell carcinoma, that are relatively radioresistant [28]. Twenty percent of patients who present with renal cell carcinoma (RCC) will have metastases at diagnosis, and 40–50% of those with localized advanced disease will ultimately progress to metastatic disease. Although new targeted therapy paradigms have changed the treatment of patients with advanced RCC and offer prolonged survival, a cure is uncommon in the absence of surgical resection. For this reason, patients with solitary or oligometastases with RCC may be good candidates for surgical resection.

Bisphosphonates or other agents that reduce bone resorption by inhibiting osteoclast formation via the receptor activator nuclear factor-κ B ligand (RANKL) pathway should be utilized in all patients with metastatic bone disease, especially those with lytic metastases.

Surgical Procedures

In general, surgical management focuses on symptomatic control rather than curative intent, except in select cases with limited disease (solitary or oligo metastases). Procedures should aim to relieve symptoms and improve function and quality of life in patients who may already have less than 6 months of mean survival [6]. The ultimate goal should be to allow immediate weight-bearing whenever feasible.

Acral metastases are usually managed with marginal or intralesional resections. When the metastatic deposit is confined to a small area and removal will provide functional or prognostic benefit, wide resection may be considered.

Surgical Options

Depending on the extent, biology, and behavior of the lesion, local control can be achieved with limb salvage or amputation. *Limb salvage* should be considered for cases where the residual tissue after planned excision and reconstruction will be viable and lead to painless function. *Amputation* is commonly performed when extensive tumor involvement is present, tumor recurrences and when persistent infection supervenes. The neurovascular status should be evaluated carefully with non-invasive vascular studies in patients with borderline inflow. Ideally, venous studies should be included in the standard preoperative workup in patients with advanced metastatic cancer to identify venous thrombotic disease.

Resection and Bone Defect Filling

Metastases affecting the soft tissue in the foot and ankle can present with swelling, deformity, discomfort with shoe wear and skin ulceration depending on the site and extent of the disease. Resection can be undertaken with the goals of pain palliation, deformity correction, and restoration of function by restoring foot anatomy and improving biomechanics. Vacuum-assisted closure devices, local flaps, skin grafts, or skin substitutes can be employed, especially for the management of large soft tissue defects.

Resection followed by filling with polymethylmethacrylate (PMMA) or bone graft (particulate and/or structural) should be the principal treatment for lesions with impending or existing pathological fractures. PMMA provides immediate stability by providing mechanical support. In addition to structural properties, bone grafts may also promote bone formation. Plate and screw fixation can be added for additional stability in appropriate locations. Structural integrity is very important and requires careful preoperative planning.

Arthrodesis

Arthrodesis is another effective modality for providing stability in individuals with destructive periarticular disease. However, the status of adjacent joints should be considered as they may require inclusion in the arthrodesis. In the hindfoot, retrograde nails may be useful to stabilize the subtalar and tibiotalar joints, augmented with either structural and particulate grafts, as well as polymethylmethacrylate. Similarly, plate and screw constructs with appropriate grafts may be used to reconstruct and stabilize painful defects in the tarsal joints and first ray.

Endoprosthetic Reconstruction

Endoprosthetic reconstruction can be considered in select patients with oligometastatic and periarticular disease. This is especially true in the distal tibia and talus, where emerging technologies in custom reconstruction utilizing 3-D printed implants may play a significant role.

Minimally Invasive Procedures

Minimally invasive approaches can be beneficial for certain patients, especially with smaller, localized symptomatic lesions who are poor surgical candidates because of the widespread disease or associated co-morbidities. Percutaneous fixation, cementoplasty, cryoablation, and radiofrequency ablation are some useful minimally invasive techniques. Percutaneous fixation or cementoplasty may be indicated for small defects or fractures that do not require open resection. Minimally

Table 24.2 General Outline of Surgical Strategies for Acral Metastases

Location	Deformity or ulceration ± infection	Intractable pain without fracture	Impending fracture	Impending fracture with at-risk joint	Pathological fracture
Forefoot	Amputation Intralesional excision	Amputation Curetting + PMMA MIS	Amputation PMMA/ BG ± IF	Amputation IF + PMMA/BG Arthrodesis	Amputation IF + PMMA/ BG
Midfoot	Intralesional excision Amputation	MIS Curetting + PMMA	PMMA/ BG ± IF Amputation MIS	IF + PMMA/BG Amputation Arthrodesis	Amputation IF + PMMA/ BG
Hindfoot	Intralesional excision Amputation	MIS Curetting + PMMA	PMMA/ BG ± IF MIS Amputation	Arthrodesis IF + PMMA/BG Amputation	IF + PMMA/ BG Arthrodesis Amputation
Distal tibia	Intralesional excision Amputation	MIS Curetting + PMMA	PMMA/ BG ± IF MIS	IF + PMMA/BG Arthrodesis Amputation	IF + PMMA/ BG Arthrodesis Amputation
Distal fibula	Intralesional excision Amputation	MIS Curetting + PMMA	PMMA/ BG ± IF MIS	Distal fibulectomy IF + PMMA/BG Arthrodesis	Distal fibulectomy Arthrodesis

MIS Minimally invasive surgery (Percutaneous fixation or cementoplasty, cryoablation, radiofrequency ablation), *PMMA* Polymethylmethacrylate, *BG* Bone graft, *IF* Internal fixation

invasive procedures are advantageous insofar as they can be done in an ambulatory setting, are cost-effective, and return patients to function sooner.

Procedures can be modified or used in combination to provide optimum care for this sensitive patient population. A general outline is given in Table 24.2.

Surgical Management for Metastases to the Foot

Patients with limited disease and pain after radiotherapy may benefit from surgical intervention. Depending on the degree of bone destruction, and the location in the foot, options include: amputation, wide excision, intralesional excision with chemical or thermal adjuncts, and cementation or grafting, along with plate and screw or nail stabilization.

Forefoot

For isolated disease in the forefoot, amputation of a digit or ray in whole or part may be a prudent oncologic and functional decision; one that removes the tumor in a complete way and restores function quickly. Limited involvement of metatarsals can be managed with local interventions such as curetting and reinforcement with

PMMA. For more extensive involvement of the forefoot, curetting and graft placement with or without stabilization of one or more metatarsals may be useful. For non-reconstructable defects involving multiple digits, a tarsometatarsal amputation (Lisfranc) may be needed.

Ray Amputations

Ray amputations, (involving the central toes) result in near-normal function. On the other hand, border ray amputations may result in significant foot imbalance, especially the first ray. For this reason, if a first ray amputation is considered, saving 8–10 mm of the proximal phalanx can result in substantially improved function [29]. Otherwise, a prosthetic device inside the shoe may be necessary to restore normal ambulation.

Transmetatarsal Amputations

Patients with transmetatarsal amputations can be fit with shoe fillers with reasonable anticipated function.

Tarsometatarsal Amputations (Lisfranc)

Tarsometatarsal amputations, however, result in a significantly shortened residual foot and cause significant foot imbalance, resulting in drop foot and equinus contracture due to the removal of the insertions of the anterior tibialis and peroneus tendons. This, in turn, causes more stress on the lateral foot (cuboid). For these reasons, this amputation requires tendon balancing to create a plantigrade and balanced residual foot.

Surgical technique

- Patient in supine position with tourniquet available.
- Make incisions in the midfoot creating dorsal and plantar flaps.
- Dissect and tag tendons (especially tibialis anterior) for later transfer.
- Transfer the tibialis anterior tendon from the first metatarsal to the medial cuneiform to avoid future drop-foot deformity, preserving mobility of the ankle. This can be fixed by using bone tunnels or suture anchors.
- Disarticulate the metatarsals along the Lisfranc joint.
- Amputate the metatarsals.
- Trim the skin flaps in a fish-mouth pattern. If possible, create a longer plantar flap with a dorsally based suture line.
- Close the skin and apply a well-padded posterior splint—drains should be used if clinically indicated.

Tips: use an atraumatic surgical technique. Use skin hooks to manipulate the flaps. Make direct incisions to the bone to create the thickest possible flaps. Maintain a longer and thicker posterior flap.

Complications: Possible flap necrosis, chronic pressure ulcers, and osteomyelitis if deep infection develops.

Midfoot

Midfoot disease that is limited may also be amenable to either complete or intralesional resection and grafting with plate and screw stabilization (Fig. 24.2). Extensive involvement of the midfoot resulting in significant pain after radiotherapy and chemotherapy may be mitigated by a Chopart amputation through the tarsal navicular and calcaneocuboid joints, leaving the heel and talus intact.

Tarsal Navicular and Calcaneocuboid Joints Amputation (Chopart)

This amputation also results in a very short residual foot and requires rerouting the tibialis anterior through the neck of the talus or the posterior tibialis or a strip of the Achilles tendon through the interosseous membrane, then through the neck of the talus to prevent significant equinus contracture.

Surgical technique

- Patient in supine position with tourniquet available.
- Make incisions in the midfoot, creating dorsal and plantar flaps.
- Transfer of the tibialis anterior to the talar neck is necessary to control the deformity of the hindfoot—the tendon of the tibialis anterior is detached from its insertion and is passed through a hole drilled in the neck of the talus. This is sutured upon itself, and the extensor tendons are carefully sutured to the fascia and soft tissues of the sole of the foot. Rather than the tibialis transfer or in addition to this transfer, the posterior tibial tendon, after release, can be passed through the interosseous membrane and anchored to the talar neck with the tibiotalar joint in dorsiflexion. This can be secured the sutures passed through drill tunnels or suture anchors.
- Rupture of this tendon is common after many years of use, and for that reason, some authors recommend fusion of the ankle joint.
- Amputation is done through the talo-navicular and calcaneocuboid joints.

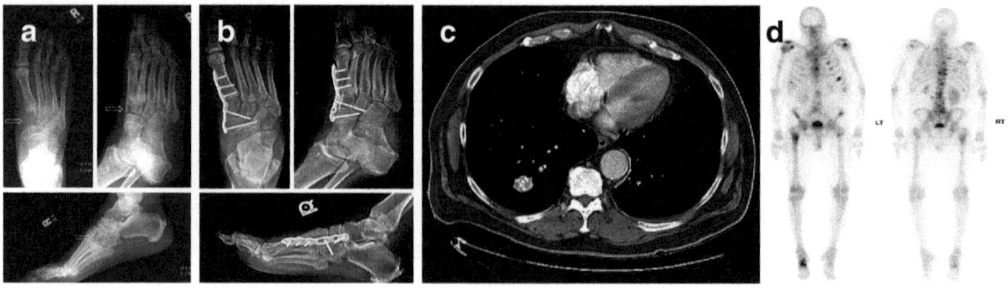

Fig. 24.2 (**a**) 68-year-old patient with bladder carcinoma; antero-posterior, oblique, and lateral radiographs demonstrating a radiolucent abnormality medial cuneiform with extension into the metatarsal-cuneiform joint best seen on the lateral radiograph (hollow arrows), (**b**) Antero-posterior, oblique, and lateral radiographs following marginal excision, structural allograft placement, and fusion of adjacent proximal and distal joints, (**c**) Axial computed tomography showing pulmonary metastasis, (**d**) Technetium^{99} pyrophosphate scan demonstrating multiple bone metastases

- Close the skin and apply a posterior splint—drains should be used as clinically indicated.

Tips: Rebalancing is required to prevent equinus and varus deformities, and can be accomplished by Achilles tenotomy, anterior tibialis, or posterior tibial tendon transfer to the talus, and postoperative casting

Postoperative care: Postoperative casting with the foot in a neutral to slightly dorsiflexed position for 8 weeks is necessary. Postoperatively, prosthetic fit is also necessary to restore normal gait function.

Complications: Progressive equinovarus deformity; gait alterations, difficult prosthetic fitting.

Hindfoot

Tumors of the hindfoot without significant bone loss can be treated with extended curetting and grafting or cementation with or without plate or screw fixation. This is usually performed before or after radiotherapy in an attempt to control local tumor spread and to prevent fractures. Patients with radio-resistant tumors and massive bone loss with intractable pain and inability to ambulate are best served with a Syme amputation and immediate postoperative cast immobilization.

Syme Amputation

A Syme amputation results in a durable end-bearing residual limb on which the patient can negotiate short distances without a prosthesis. With appropriate prosthetic fit, these patients can achieve a near-normal gait.

Surgical technique

- Patient is placed in supine position—Use of tourniquet based on surgeon's preference.
- A transverse incision is made anteriorly from the medial and lateral malleolus, then a posterior flap is fashioned around the heel pad.
- The key landmarks for creating this flap are the inferior aspects of the malleoli. A point 1 cm inferior and 1 cm anterior to the tip of the lateral malleolus is marked. Next, a point 1.5 cm inferior and 1 cm anterior to the tip of the medial malleolus is marked. These points are then connected with a straight line drawn across the anterior of the ankle. It is important that the incision is not proximal to the distal aspect of the tibia. The plantar incision is oriented 90 degrees from the dorsal incision and drawn out across the plantar portion of the foot, extending from the two points below the malleoli. The plantar incision should be carried out to the level of the calcaneocuboid joint to ensure adequate length of the plantar flap.
- First, the anterior incision is used, and dissection by planes is carried down. The skin is incised, followed by dissection through the subcutaneous tissue. Once the extensor tendons are identified (tibialis anterior, extensor hallucis longus,

extensor digitorum longus, and, if present, the peroneus tertius), they are sharply transected with distal traction to allow the tendons to retract proximally. The tibialis anterior artery is identified and ligated, and the deep peroneal nerve is sharply transected under tension. The anterior capsule of the ankle is identified and divided.

- The vertical incisions are then made with slight anterior obliquity from the tips of the lateral and medial malleoli, then another transverse incision just distal to the plantar heel crease. The dissection is deepened through the subcutaneous tissues of the heel. On the lateral aspect, the peroneal tendons are identified, placed under traction, cut sharply, and allowed to retract proximally. The lateral dorsal cutaneous nerve and lesser saphenous vein are also transected and allowed to retract proximally following ligation of the vein. Dissection of the plantar flap is complete at this point once the plantar fascia is visualized.
- It is of utmost importance to be careful not to damage the posterior tibialis neurovascular bundle since this is the only feeder to the heel pad (lateral plantar and calcaneal arteries).
- The foot is then plantarflexed, and the posterior ankle joint capsule and periarticular structures are transected.
- The insertion of the Achilles tendon is identified and released from the calcaneus. A Crego elevator may be useful in freeing the insertion of the Achilles tendon. Following transection of the tendon, further plantarflexion of the talus with the Steinmann pin allows subperiosteal dissection of the calcaneus from its underlying soft tissue attachments.
- At this point, the plantar fascia and underlying muscles are still attached to the calcaneus. Careful dissection is performed, and the posterior tibial artery is identified and traced as distally as possible. Ideally, the artery will be dissected out into its medial and lateral branches. The two branches of the posterior tibial artery are transected at the distal aspect of the flap and ligated. The posterior tibial nerve is placed under traction, transected sharply, and allowed to migrate proximally.
- Then resection of the tibia and fibula is done. These bone cuts are made parallel to the ground-supporting surface. The cartilage of the distal tibia can be removed, which could allow better adherence of the plantar fat pad. Further remodeling of the distal end of the tibia and fibula will be required to square off the osseous component of the stump. Remodeling the lateral aspect of the fibular malleolus and the medial aspect of the medial malleolus will provide a narrower, more cosmetic stump that facilitates optimal prosthetic fit.
- Prior to closure, drill holes are made in the distal anterior aspect of the tibia, and the Achilles tendon and other remaining soft tissues are secured with non-absorbable suture.
- Finally, the plantar flap is advanced, and debulking of the residual intrinsic musculature is performed as necessary. The subcutaneous tissues are reapproximated. The skin is closed with a simple interrupted suture of 3–0 synthetic.
- The wound is dressed in a well-molded modified Jones compression dressing to minimize edema and hematoma, or seroma formation.

Tips: Preserve the longest possible length of the posterior tibialis artery since this is the anatomical structure vascularizes the fat pad used for weight-bearing.

Postoperative care: The patient is non-weight-bearing for a period of 3–6 weeks until the wound has completely healed. Sutures are generally removed around 3 weeks postoperatively. The patient can be placed in a short leg cast, and weight bearing as tolerated is permitted. The cast is changed at two-week intervals until resolution of all soft tissue edema and shrinkage of the stump has occurred. After this, the patient can be referred for prosthetic fitting.

Complications: Heel pad migration, infection, ulceration, flap necrosis, wound dehiscence, residual limb pain, and possible re-amputation to a higher level.

Calcanectomy

An alternative to amputation is calcanectomy that has been described for an isolated destructive acral metastasis of the calcaneus. This procedure has been reported to result in acceptable postoperative function [30]. Since most patients with acral metastases have poor overall survival, resection of the calcaneus and reconstruction with a custom metal implant should only be considered for patients with solitary or oligometastatic disease with good prospects for systemic targeted therapy. If a custom 3-D metal reconstruction of the calcaneus is considered, the implant should be smaller than the resected calcaneus to facilitate wound closure. The surface should be rough (porous coated) to allow for soft tissue adhesion and to prevent fat pad migration, and there should be porous coating at the calcaneotalar and calcaneocuboid joint surfaces to allow integration at the bone metal interfaces. Hydroxyapatite has been used to facilitate integration at these interfaces. The implant is anchored by a conical post that is porous-coated inside and out, which is inserted into the talus, as well as screws that anchor the implant to the talus and cuboid. The Achilles tendon is reattached through holes in the back of the implant. Sometimes, the peroneus brevis is detached during the course of the dissection and can also be reattached to the implant. The holes in the implant should be highly polished to prevent suture breakage. This procedure can be performed in one or two stages. If soft tissue closure is difficult, a fasciocutaneous flap based on the saphenous vessels can be performed. The flap can be raised at the time of the resection and used immediately or allowed to mature and used at a second stage. A bulky dressing is applied along with a posterior plaster splint with the foot in neutral to slight dorsiflexion.

Surgical Management of the Distal Tibia

Surgical management of metastatic tumors of the distal tibia requires careful consideration of the tumor type, degree of bone destruction, response to adjuvant radiation and chemotherapy, age, and degree of dysfunction relative to pre-morbid function. For patients with extensive bone loss of the distal tibia, options include a transtibial amputation or a wide excision. Following wide excision, some authors have described retrograde nailing through the hindfoot into the tibia and replacing the resected bone segment with a PMMA spacer [31]. Hsu and Lindsay [32]

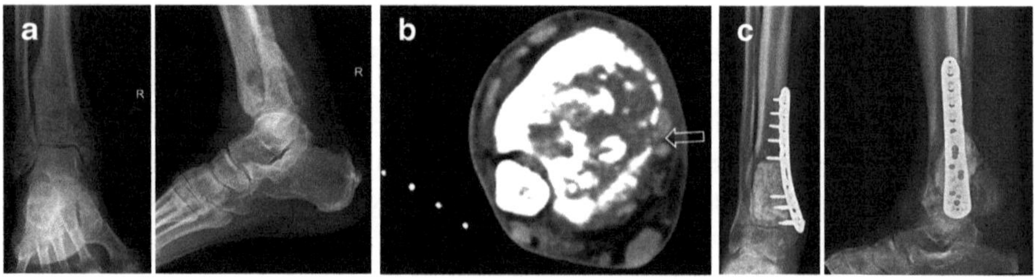

Fig. 24.3 (**a**) 72-year-old patient with prostate carcinoma; AP and lateral views of a mixed radiolucent and radiodense abnormality in the distal tibia meta-epiphysis without obvious joint involvement. (**b**) Medial cortical destruction with local soft tissue extension (hollow arrow), (**c**) Antero-posterior and lateral radiograph of the right ankle following extended curetting with cementation with plate and screw fixation. Note the incomplete pathological fracture in the posterior tibial cortex

reported a one-year follow-up of patients undergoing tibio-talar arthrodesis using cement and a retrograde nail. Patients were able to bear full weight painlessly without assistive devices. For patients with mild to moderate degrees of bone destruction plates and screws supplemented with cement can be effective in relieving pain and restoring function [33, 34] (Fig. 24.3). Other options include: intercalary allograft arthrodesis, the use of a cage construct with cement or graft supplemented with a structural allograft, autograft, vascularized fibulae, endoprosthetic replacement [35], and custom implants. The use of structural allografts, autografts, and vascularized fibulae should be utilized only in selected patients in whom prolonged survival is anticipated.

Surgical Management of the Distal Fibula

Isolated acral metastases to the fibula are very rare and the subject of scattered case reports [36–38]. Individuals with acral metastases in the distal fibula with limited bone destruction are best managed by radiotherapy and possibly plate and screw fixation to limit pain and prevent pathologic fracture. Based on the limited cases available for review, bone destruction is usually significant and requires excision of the distal fibula. Following fibulectomy, there are several options:

1. Reconstruction of the soft tissues around the tibiotalar joint and retaining the peroneal tendons, followed by casting for 8–12 weeks, can result in a stable tibiotalar joint [39].
2. Another group advocated using a mesh to reconstruct the soft tissue over the lateral tibiotalar joint, reporting acceptable function [40].
3. Other authors recommend fusion of the tibiotalar joint to prevent deformity and instability [41]. This can be accomplished using cannulated

screws, screws and plates, and retrograde nails through the hindfoot. This option is best in patients in whom the peroneal tendons are absent.

Again, it is important to understand the patient's disease, type and extent, premorbid function, and anticipated survival before embarking on more complicated reconstructions such as fibula allografts or autografts and tibiotalar fusion.

Summary of Surgical Management

It is important to recognize that bone metastases often occur in conjunction with diffuse metastatic disease and, therefore, a short life expectancy is anticipated. Therefore, treatment should be tailored to pain control and nonoperative optimization of function. When deciding on a surgical intervention, benefits should always outweigh the risks, and to this end, the risk of complications should be minimized. Thus, amputation often provides pain relief and restoration of function with minimal complications. The decision-making process should take all aspects of treatment into account, including but not limited to preoperative optimization, nutrition, wound care, and rehabilitation.

For individuals with limited local bone destruction, tumors that are not responsive to systemic or radiotherapy, and oligometastatic disease, resection and reconstruction may be considered. As previously discussed, survival is improving for patients with metastatic bone disease, and those individuals with solitary acral metastases for which targeted systemic strategies exist are very good candidates for surgical excision and reconstruction.

Awareness, better diagnostic techniques, tailored surgical management, and goal-oriented rehabilitation are crucial for those patients with disease that are refractory to radiation and systemic therapy.

Conclusions

Acral metastases in the foot and ankle are rare, occur in older individuals, and are often unsuspected. This may lead to delays in diagnosis that, in turn, may result in decreased function, increased pain, diminished quality of life, or even shortened survival. Many of these patients, however, present with advanced disease at the time of diagnosis and therefore have a poor prognosis. Treatment is based on the type of tumor and disease extent, both locally and distantly. For patients with limited disease and certain primary tumors, survival can be prolonged, thus highlighting and underscoring the need for prompt and correct diagnosis and expeditious treatment.

References

1. El Abiad JM, Aziz K, Levin AS, McCarthy EM, Morris CD. Osseous metastatic disease to the hands and feet. Orthopedics. 2019;42(2):e197–201. https://doi.org/10.3928/01477447-20181227-04. Epub 2019 Jan 3.
2. Healey JH, Turnbull AD, Miedema B, Lane JM. Acrometastases. A study of twenty-nine patients with osseous involvement of the hands and feet. J Bone Joint Surg Am. 1986;68:743–6.
3. Machado V, San-Julian M. Prognosis and treatment of acrometastases: observational study of 35 cases treated in a single institution. Rev Esp Cir Ortop Traumatol (Engl Ed). 2019;63(1):49–55. https://doi.org/10.1016/j.recot.2018.05.001. English, Spanish. Epub 2018 Jul 7. PMID: 30612612.
4. Stomeo D, Tulli A, Ziranu A, Perisano C, De Santis V, Maccauro G. Acrometastasis: a literature review. Eur Rev Med Pharmacol Sci. 2015;19(15):2906–15. PMID: 26241547.
5. Çiftdemir M, Ustabaşıoğlu FE, Çölbe SA, Üstün F, Usta U, Çiçin İ. Clinicopathological and prognostic characteristics of acral metastases in patients with malignant disease: a retrospective study. Acta Orthop Traumatol Turc. 2021;55(1):67–72. https://doi.org/10.5152/j.aott.2020.20052. PMID: 33650515; PMCID: PMC7932733.
6. Mavrogenis AF, Mimidis G, Kokkalis ZT, Karampi ES, Karampela I, Papagelopoulos PJ, Armaganidis A. Acrometastases. Eur J Orthop Surg Traumatol. 2014;24(3):279–83. https://doi.org/10.1007/s00590-013-1311-1. Epub 2013 Sep 8. PMID: 24013815.
7. Evans S, Ramasamy A, Jeys L, Grimer R. Delayed diagnosis in metastatic lesions of the foot. Ann R Coll Surg Engl. 2014;96(7):536–8. https://doi.org/10.1308/003588414X14055925058111. PMID: 25245734; PMCID: PMC4473441.
8. Gall RJ, Sim FH, Pritchard DJ. Metastatic tumors to the bones of the foot. Cancer. 1976;37(3):1492–5. https://doi.org/10.1002/1097-0142(197603)37:3<1492::aid-cncr2820370335>3.0.co;2-q. PMID: 1260668
9. Greco T, Cianni L, De Mauro D, Dughiero G, Bocchi MB, Cazzato G, Ragonesi G, Liuzza F, Maccauro G, Perisano C. Foot metastasis: current knowledge. OrthopRev (Pavia). 2020;12(Suppl 1):8671. https://doi.org/10.4081/or.2020.8671. PMID: 32913603; PMCID: PMC7459381.
10. Hattrup SJ, Amadio PC, Sim FH, Lombardi RM. Metastatic tumors of the foot and ankle. Foot Ankle. 1988;8(5):243–7. https://doi.org/10.1177/107110078800800503. PMID: 3366429.
11. Maheshwari AV, Chiappetta G, Kugler CD, Pitcher JD Jr, Temple HT. Metastatic skeletal disease of the foot: case reports and literature review. Foot Ankle Int. 2008;29(7):699–710. https://doi.org/10.3113/FAI.2008.0699. PMID: 18785420.
12. Thompson RP, Raina P, Downie S, Jariwala AC. Acrometastasis of the lower limb carries a poorer prognosis: a retrospective cohort study. Foot Ankle Spec. 2022:19386400221136373. https://doi.org/10.1177/19386400221136373. Epub ahead of print. PMID: 36416410.
13. Libson E, Bloom RA, Husband JE, Stoker DJ. Metastatic tumours of bones of the hand and foot. A comparative review and report of 43 additional cases. Skeletal Radiol. 1987;16(5):387–92. https://doi.org/10.1007/BF00350965. PMID: 3306939.
14. Sebghati J, Khalili P, Tsagkozis P. Surgical treatment of metastatic bone disease of the distal extremities. World J Orthop. 2021;12(10):743–50. https://doi.org/10.5312/wjo.v12.i10.743. PMID: 34754830; PMCID: PMC8554352.
15. Bray F, Ferlay J, Soerjomataram I, Siegel RL, Torre LA, Jemal A. Global cancer statistics 2018: GLOBOCAN estimates of incidence and mortality worldwide for 36 cancers in 185 countries. CA Cancer J Clin. 2018;68(6):394–424. https://doi.org/10.3322/caac.21492. Epub 2018 Sep 12. Erratum in: CA Cancer J Clin. 2020 Jul;70(4):313. doi: 10.3322/caac.21609. PMID: 30207593.
16. Hernandez RK, Adhia A, Wade SW, O'Connor E, Arellano J, Francis K, Alvrtsyan H, Million RP, Liede A. Prevalence of bone metastases and bone-targeting agent use among solid tumor patients in the United States. Clin Epidemiol. 2015;7:335–45. https://doi.org/10.2147/CLEP.S85496. PMID: 26229504; PMCID: PMC4514316.

17. Soeharno H, Povegliano L, Choong PF. Multimodal treatment of bone metastasis-a surgical perspective. Front Endocrinol (Lausanne). 2018;9:518. https://doi.org/10.3389/fendo.2018.00518. PMID: 30245668; PMCID: PMC6137681.
18. Xu DL, Zhang XT, Wang GH, Li FB, Hu JY. Clinical features of pathologically confirmed metastatic bone tumors--a report of 390 cases. Ai Zheng. 2005;24(11):1404–7. Chinese. PMID: 16552972.
19. Fornetti J, Welm AL, Stewart SA. Understanding the bone in cancer metastasis. J Bone Miner Res. 2018;33(12):2099–113. https://doi.org/10.1002/jbmr.3618. Epub 2018 Nov 26. PMID: 30476357.
20. Bahrabadi M, Otoukesh B, Moghtadaei M, Hoseinzadeh S, Amiri S. Foot metastasis: review of 38 cases. Arch Bone Jt Surg. 2021;9(1):122–5. https://doi.org/10.22038/abjs.2020.49792.2474. PMID: 33778125; PMCID: PMC7957098.
21. Rougraff BT, Kneisl JS, Simon MA. Skeletal metastases of unknown origin. A prospective study of a diagnostic strategy. J Bone Joint Surg Am. 1993;75(9):1276–81. https://doi.org/10.2106/00004623-199309000-00003. PMID: 8408149.
22. Lindström E, Lindström FD. Skeletal scintigraphy with technetium diphosphonate in multiple myeloma--a comparison with skeletal x-ray. Acta Med Scand. 1980;208(4):289–91. https://doi.org/10.1111/j.0954-6820.1980.tb01196.x. PMID: 6449830.
23. Mesguich C, Fardanesh R, Tanenbaum L, Chari A, Jagannath S, Kostakoglu L. State of the art imaging of multiple myeloma: comparative review of FDG PET/CT imaging in various clinical settings. Eur J Radiol. 2014;83(12):2203–23. https://doi.org/10.1016/j.ejrad.2014.09.012. Epub 2014 Sep 28. PMID: 25308249.
24. Pino AE, Temple HT. Three cases of melorheostosis with foot and ankle involvement. Am J Orthop (Belle Mead NJ). 2012;41(8):E115–9. PMID: 22900256.
25. Smith SE, Murphey MD, Motamedi K, Mulligan ME, Resnik CS, Gannon FH. From the archives of the AFIP. Radiologic spectrum of Paget disease of bone and its complications with pathologic correlation. Radiographics. 2002;22(5):1191–216. https://doi.org/10.1148/radiographics.22.5.g02se281191. PMID: 12235348.
26. Youn L, Kuta A, Shendrik KS. Metastatic disease mimicking osteomyelitis in the foot. Cureus. 2024;16(6):e61608. https://doi.org/10.7759/cureus.61608. PMID: 38962613; PMCID: PMC11221879.
27. Kimura T. Multidisciplinary approach for bone metastasis: a review. Cancers (Basel). 2018;10(6):156. https://doi.org/10.3390/cancers10060156. PMID: 29795015; PMCID: PMC6025143.
28. Ljungberg B. The role of metastasectomy in renal cell carcinoma in the era of targeted therapy. Curr Urol Rep. 2013;14(1):19–25. https://doi.org/10.1007/s11934-012-0293-6. PMID: 23212738.
29. Hakim-Zargar M, Aronow MS, Gibson L, Obopilwe E. Implications for the anatomy of the flexor hallucis brevis insertion. Foot Ankle Int. 2010;31(1):65–8. https://doi.org/10.3113/FAI.2010.0065. PMID: 20067725.
30. Tonogai I, Nishisho T, Miyagi R, Sairyo K. Total calcanectomy for metastasis of renal cell carcinoma in the calcaneus: a case report. Foot Ankle Surg. 2018;24(1):e7–e12. https://doi.org/10.1016/j.fas.2017.06.007. Epub 2017 Jul 6. PMID: 29413783.
31. Futani H, Kamae S, Atsui K, Yoh K, Tateishi H, Maruo S. Successful limb salvage of pathological fracture of the distal tibia caused by cancer metastasis. J Orthop Sci. 2002;7(2):262–6. https://doi.org/10.1007/s007760200044. PMID: 11956990.
32. Hsu AR, Lindsay AD. Management of a tibial plafond lung cancer metastasis using an intramedullary hindfoot fusion nail. Foot Ankle Spec. 2018;11(1):82–7. https://doi.org/10.1177/1938640017733098. Epub 2017 Sep 27. PMID: 28952794.
33. Bonnevialle P, Descamps J, Niglis L, Lebaron M, Falguieres J, Mericq O, Fabre T, Reina N, Sailhan F, members of the SoFCOT. Surgical treatment of tibial metastases: retrospective, multicenter, observational study of 25 patients. Orthop Traumatol Surg Res. 2020;106(6):1039–45. https://doi.org/10.1016/j.otsr.2019.07.017. Epub 2019 Oct 14. PMID: 31624032.

34. De Geeter K, Reynders P, Samson I, Broos PL. Metastatic fractures of the tibia. Acta Orthop Belg. 2001;67(1):54–9. PMID: 11284272.
35. Shekkeris AS, Hanna SA, Sewell MD, Spiegelberg BG, Aston WJ, Blunn GW, Cannon SR, Briggs TW. Endoprosthetic reconstruction of the distal tibia and ankle joint after resection of primary bone tumours. J Bone Joint Surg Br. 2009;91(10):1378–82. https://doi.org/10.1302/0301-620X.91B10.22643. PMID: 19794176.
36. Hekmat S, Ghaedian T, Barati H, Movahed M. Solitary metastasis of gastric cancer to fibula: a case report. Iran J Radiol. 2012;9(3):161–4. https://doi.org/10.5812/iranjradiol.3564. Epub 2012 Sep 17. PMID: 23329984; PMCID: PMC3522377.
37. Karanjia H, Abraham JA, O'Hara B, Shallop B, Daniel J, Taweel N, Schick FA. Distal fibula metastasis of cholangiocarcinoma. J Foot Ankle Surg. 2013;52(5):659–62. https://doi.org/10.1053/j.jfas.2013.02.017. Epub 2013 Apr 9. PMID: 23578566.
38. Luther L, McGlone PJ, Hardacker KD, Alsoof D, Hayda RA, Terek RM. Distal fibular metastasis of colorectal carcinoma: a case report. Orthop Rev (Pavia). 2024;16:91505. https://doi.org/10.52965/001c.91505. PMID: 38469575; PMCID: PMC10927312.
39. Khal AA, Zucchini R, Giannini C, Sambri A, Donati DM, De Paolis M. Distal fibula reconstruction in primary malignant tumours. Curr Oncol. 2021;28(5):3463–73. https://doi.org/10.3390/curroncol28050299. PMID: 34590617; PMCID: PMC8482106.
40. Prajapati A, Gulia A, Hegde P, Puri A. Is minimal reconstruction (meshplasty) adequate to restore ankle function after excision of distal fibula tumors? J Clin Orthop Trauma. 2020;11(3):467–70. https://doi.org/10.1016/j.jcot.2020.03.023. Epub 2020 Apr 9. PMID: 32405211; PMCID: PMC7211805.
41. Dieckmann R, Ahrens H, Streitbürger A, Budny TB, Henrichs MP, Vieth V, Gebert C, Hardes J. Reconstruction after wide resection of the entire distal fibula in malignant bone tumours. Int Orthop. 2011;35(1):87–92. https://doi.org/10.1007/s00264-009-0931-x. Epub 2009 Dec 29. PMID: 20039038; PMCID: PMC3014490.

Part IV

Principles of Medical Treatment

Medical Management of Musculoskeletal Metastasis from Breast Cancer

25

Naomi Dempsey, Yolcar Chamorro, Carlos Hernandez, Priya Bhatt, and Reshma Mahtani

Introduction

Despite advances in the treatment of early-stage breast cancer (BC), 20–30% of patients relapse with distant metastasis [1–3]. Although not considered curable, with current therapeutic options, metastatic breast cancer (MBC) patients can live for several years. The goal of treatment is palliative, and quality of life (QOL) considerations are paramount. At presentation with metastatic disease, it is standard practice to perform a biopsy to confirm the diagnosis and characterize hormone receptor (HR) and human epidermal growth factor receptor 2 (HER2) status, which impacts treatment options. Treatment recommendations are based on receptor subtype, as well as patient and tumor factors (age, comorbidities, toxicities from prior treatment for BC, and sites of disease). Bone is a frequent site of involvement, with bone metastasis (BM) present in approximately 70% of patients across all subtypes of MBC [4, 5]. Although typically associated with an improved prognosis as compared to patients who present with visceral metastasis (i.e., liver, lungs, or brain), the presence of BM can significantly impact QOL. Approximately 25% of MBC patients present with bone-only metastases (BOM), portending a more indolent disease course, and associated with a prolonged median overall survival (OS) [2, 6, 7].

In a retrospective study of patients with BC who developed recurrence in bone, 72.2% developed BM within the first 5 years after initial BC diagnosis [6]. Most had predominantly osteolytic lesions in axial bones, including ribs (13.4%), thoracic vertebrae (12.4%), pelvis (12.2%), lumbar vertebrae (12.1%), and sternum (10.5%) [6]. BC BM are most commonly osteolytic and are associated with bone destruction, but 15–20% are predominantly osteoblastic lesions, where the tumor causes

N. Dempsey · Y. Chamorro · P. Bhatt · R. Mahtani (✉)
Miami Cancer Institute, Baptist Health South Florida, Miami, FL, USA
e-mail: reshma.mahtani@baptisthealth.net

C. Hernandez
Florida International University, Miami, FL, USA

J. Pretell-Mazzini, S. A. Lozano Calderón (eds.), *Multidisciplinary Approach of Musculoskeletal Metastasis*, https://doi.org/10.1007/978-3-032-09498-8_25

increased bone growth [5, 8]. Risk factors for the development of BM from BC include older age, larger tumor size, lymph node involvement, HR positivity, and lower tumor grade [2, 4]. While the most common site of musculoskeletal metastasis from BC is bone, less frequently (<1%), BC can also metastasize to muscle and subcutaneous tissue [9].

BC BM significantly increases the risk of skeletal-related events (SREs), including bone pain requiring local therapy, pathologic fractures, hypercalcemia, and spinal cord compression [5, 10]. SREs affect over 50% of patients with MBC and negatively impact QOL [10]. Patients with BOM at initial MBC diagnosis have an 80% risk of SREs over their lifetime, while those with mixed bone and visceral involvement have a 60% risk. Patients without BM at initial MBC diagnosis have a high likelihood of developing BM subsequently, and according to one series, a 21% risk of developing subsequent SREs [11].

Bone-modifying agents (BMAs), including bisphosphonates and denosumab, significantly reduce the risk of SREs and are recommended by the National Comprehensive Cancer Network (NCCN) guidelines in all patients with BM from MBC [12]. As with all patients with MBC, the management of those with BM aims to control symptoms. Treatments are based on the subtype of disease and options include hormonal therapy, molecularly targeted treatments, HER2-directed therapy, and chemotherapy (+/− immunotherapy). Additional factors taken into consideration for those with BM include the use of local therapies involving surgery and radiation where appropriate, with the ultimate goal of stabilizing bone, preserving and restoring function, and minimizing risk of SREs [5].

Diagnosis and Staging

Patients with a new BC diagnosis with signs or symptoms suggestive of MBC, such as bone pain with an elevated alkaline phosphatase, right upper quadrant abdominal pain, dry cough, or neurologic symptoms, should undergo symptom-directed imaging [12]. Newly diagnosed early-stage (stage I-III) BC patients with lymph node involvement (as documented by tissue biopsy and/or by clinical exam) should have staging imaging regardless of symptoms [13]. If suspicious findings are noted on imaging, a tissue biopsy (preferably from a metastatic site) should be performed to confirm the diagnosis of MBC and assess receptor subtype.

A challenge associated with performing core needle biopsies (CNB) on bone is the potential risk of inaccurate characterization of the breast cancer subtype, due to frequent issues with decalcification of bone, which can impact immunohistochemical staining for HR and HER2 status, resulting in false negatives [14]. Therefore, the internal control for receptors is important for accurate interpretation. One study comparing CNB and fine needle aspiration of BC BM showed that loss of ER expression was more common in CNB when crush artifact was present, signifying that biopsy technique may also influence reliability of results [15]. When an alternative site of metastasis is safe and feasible for biopsy, this is preferred. In cases of BOM, bone biopsy should be performed and interpreted with caution.

At presentation with metastatic disease, standard evaluation includes imaging to assess the extent and sites of metastatic involvement. NCCN guidelines endorse standard imaging with computed tomography (CT) of the chest, abdomen, and pelvis with and without IV contrast in addition to a nuclear bone scan. PET/CT may be useful in certain situations when standard imaging is equivocal [12]. The characterization of BC BM and interpretation of response to therapy by imaging can be challenging due to the low sensitivity of radiographs and CT, difficulty in interpretation of results of functional imaging like PET/CT and bone scan, and the expense and limited availability of MRI. For assessment of response to therapy, the NCCN guidelines recommend repeating the same imaging modality that was used in the baseline assessment [12].

Significant bone mineral density must be lost before a lesion is visible on a plain radiograph, making this is a relatively insensitive modality for detection of BM. Plain radiographs can be used to evaluate focal bone pain in a patient with a history of BC, but are not used for screening in asymptomatic patients at high risk. Although CT can identify BC BM, according to one series, sensitivity is only 74% and specificity is 56% [16]. Despite this, it is standard practice to perform CT scans to evaluate for potential sites of visceral and soft tissue involvement and to add a bone scan.

Bone scintigraphy, or "bone scan" uses nuclear imaging to evaluate areas of high bone turnover, which may be associated with a metastatic lesion. However, there are challenges with the routine use of bone scans in MBC when patients have primarily lytic lesions, for which bone scans have a low sensitivity [17]. Additionally, a prior fracture can cause a false positive, and a scintigraphy flare or "bone flare" may occur when treated BM heal, appearing brighter on follow-up studies. Early on, this phenomenon can mimic progressive disease during effective treatment and can cause clinicians to prematurely discontinue a treatment that is potentially beneficial [18].

18-F-Fluorodeoxyglucose (FDG) PET/CT is a nuclear scan that detects areas of high glucose uptake and can be useful when conventional staging yields non-diagnostic or equivocal results. It is superior to bone scintigraphy in the detection of osteolytic BM and evaluates for metastases in visceral organs with the same exam [19]. The sensitivity and specificity of 18-F-FDG-PET/CT for the detection of bone metastasis are 98% and 56%, respectively [16]. 18-F-FDG PET/CT alone can be used for systemic staging of BC.18-F-Fluoroestradiol (FES) PET is a nuclear scan which uses a radiolabeled form of estradiol to evaluate for uptake in sites of ER expression. Although FES-PET is effective in detecting ER+ BM, this study would miss ER-lesions. Due to the extensive physiologic uptake of this radioisotope in the liver, it is of limited value in the detection of liver metastasis [20]. NCCN guidelines indicate that this modality of imaging may be considered for ER+ MBC [12].

Whole body MRI has a high sensitivity (90%) and specificity (92%) for BM, but is not currently widely available or cost-effective [21]. However, MRI is the imaging exam of choice to evaluate spinal metastases and diagnose spinal cord compression.

In the metastatic setting for monitoring of response to therapy, NCCN guidelines recommend repeating the same imaging modality at regular intervals. For patients with bone-dominant metastasis and primarily lytic lesions that are difficult to assess on imaging, NCCN supports the assessment of tumor markers such as carcinoembryonic antigen (CEA), CA 15-3, and CA 27-29 as an alternative method to assess response [12]. In cases of a mixed response to therapy (one site of metastasis is responding radiographically while other site(s) are documenting progression, a repeat biopsy of the progressing lesion may be indicated to clarify ER/PR/HER2 status at this site, given reports of possible heterogeneity of subtypes in various sites of metastases [22].

Liquid biopsy is a blood-based assay to obtain circulating tumor DNA (ctDNA), which can be used to perform next-generation sequencing (NGS) for the identification of targetable mutations. Patients with BOM and/or a low burden of disease may not shed significant ctDNA into the blood, which may result in false-negative NGS analysis. If no somatic mutations are detected on liquid biopsy, tissue analysis for targetable mutations is recommended.

Management

MBC remains as incurable disease, and the goal of treatment for MBC is palliative in nature. Systemic therapy, such as chemotherapy, endocrine therapy, targeted therapy, and immunotherapy are used to extend life in patients with MBC while also maintaining QOL. Systemic therapy for MBC is selected in a shared decision model based upon receptor subtype, previous treatments received, anticipated adverse events, performance status, and organ function. An assessment of QOL is an important part of the response assessment to ensure that the side effects of treatment do not outweigh the survival benefits. All patients with MBC should undergo germline genetic testing, given the availability of two PARP inhibitors as a treatment option. All patients with BM should be treated with a BMA such as zoledronic acid or denosumab to reduce the risk of SREs.

HR+/HER2- and HR+/HER2 Low

For patients with HR+/HER2- MBC, standard first-line therapy is a cyclin-dependent kinase 4/6 inhibitor (CDK4/6i) (ribociclib, abemaciclib, or palbociclib) in combination with endocrine therapy (ET) such as an aromatase inhibitor (AI) or fulvestrant. In those with a PIK3CA mutation and distant recurrence while on or within 12 months of completing adjuvant endocrine therapy, the INAVO120 trial showed a progression-free and overall survival benefit for the combination of inavolisib, palbociclib, and fulvestrant versus placebo, palbociclib, and fulvestrant [23]. Within pivotal trials evaluating first-line CDK4/6i + ET for HR+/HER2- MBC, the interaction tests between treatment effect and metastatic sites were not statistically significant, including for patients with BOM [24, 25]. Additionally, another meta-analysis demonstrated a statistically significantly improved progression-free survival (PFS) for CDK4/6-inhibitor–based therapies as first line in bone-only metastatic disease,

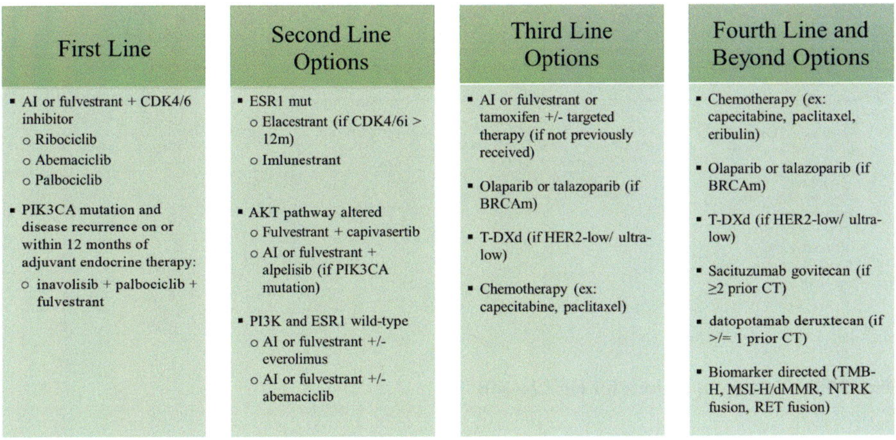

Fig. 25.1 Treatment options for HR+/HER2- MBC

further confirmed by a patient-level pooled analysis from the US Food and Drug Administration [26, 27].

Guidelines recommend continued treatment with ET +/− targeted therapy for up to 3 lines of therapy, or until patients are considered endocrine refractory. Although patients may respond to hormonal therapy for many years, invariably endocrine resistance develops, necessitating the use of chemotherapy. Single-agent sequential chemotherapy represents standard treatment, with the recent incorporation of antibody-drug conjugates (ADCs), which have significantly improved both PFS and OS compared to cytotoxic chemotherapy. Large, randomized trials of patients with HR+/HER2- and HER2-low (HER2 IHC 1+/ 2+ and in situ hybridization negative) MBC show the superiority of Sacituzumab Govitecan (SG) in the TROPiCS-02 trial, Trastuzumab Deruxtecan (T-DXd) in the DESTINY-Breast04 trial, and Datopotomab Deruxtecan (Dato-DXd) in the TROPION-Breast01 trial [28] respectively, over cytotoxic chemotherapy [29, 30] (Fig. 25.1).

An additional consideration in the management of patients with HR+ MBC includes bone density. Many patients with MBC are post-menopausal, and many received adjuvant AI for early-stage breast cancer, both risk factors for decreased bone density. In the metastatic setting, HR+ patients may receive AI or fulvestrant in combination with other agents for years. Steroids used as premedication for chemotherapy also have negative impacts on bone density. Bone mineral density (BMD) should be monitored at baseline and every 2 years while on therapy. Patients should take calcium, vitamin D and be encouraged to engage in weight-bearing exercise.

HER2+

Patients with HER2+ MBC have an increased risk of developing brain metastasis, and when present, this factor tends to drive prognosis and management. Patients with HR+/HER2+ MBC are more likely to have bone metastasis, and patients with

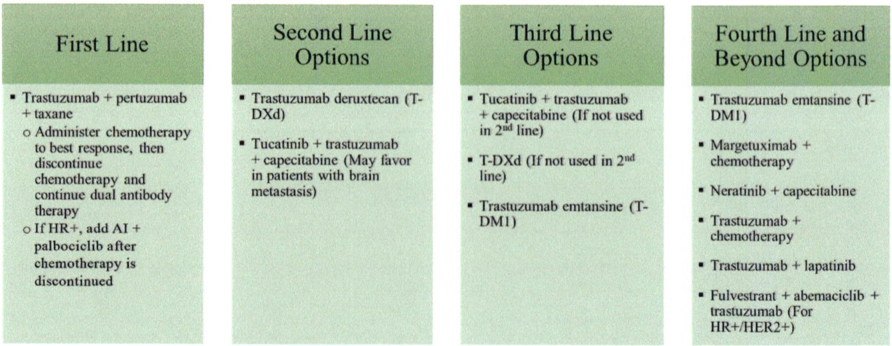

Fig. 25.2 Treatment options for HER2+ MBC

HR-/HER2+ MBC are more likely to have visceral involvement. Standard first-line treatment for HER2+ MBC is evolving with data from DESTINY-Breast09 and PATINA trials, but has historically been taxane chemotherapy in combination with dual anti-HER2 monoclonal antibodies (trastuzumab and pertuzumab), which is still reflected by the NCCN guidelines at the publication of this manuscript. In the second line setting, the ADC T-DXd is preferred, followed by other anti-HER2 therapies such as small molecule oral tyrosine kinase inhibitors in combination with chemotherapy and HER2 monoclonal antibodies (Fig. 25.2). Approximately 50% of patients with HER2+ MBC have HR+/HER2+ (triple positive) breast cancer, and the use of ET in combination with HER2-directed therapy may provide a chemotherapy-free option in certain cases.

TNBC

Patients with metastatic TNBC (mTNBC) represent the greatest unmet need in MBC. At initial diagnosis, tumors are tested for PD-L1, which, if positive (CPS > 10%), chemotherapy in combination with immunotherapy is considered standard first-line treatment. For PD-L1-negative tumors, chemotherapy alone is recommended. ADC therapy is preferred in the second line, given the documented overall survival benefit as compared to chemotherapy. The ASCENT trial demonstrated the superiority of SG over conventional chemotherapy in patients with mTNBC, and the DESTINY-Breast04 trial demonstrated superiority of T-DXd over conventional chemotherapy in patients with HER2-low MBC, including and exploratory subgroup analysis which included HR-patients. In the third line and beyond, additional cytotoxic chemotherapy is considered (Fig. 25.3).

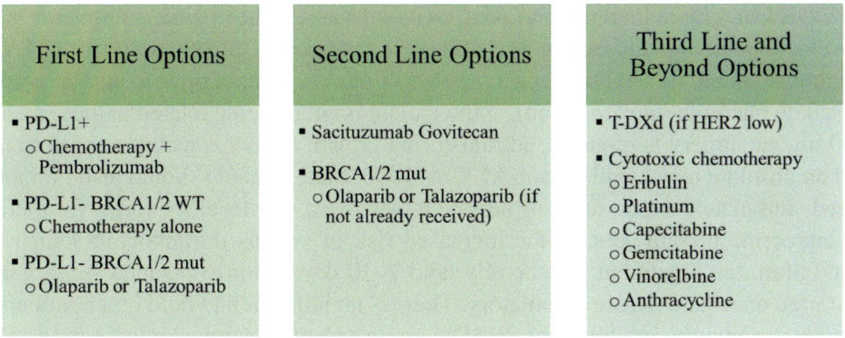

Fig. 25.3 Treatment options for mTNBC

Management of Skeletal-Related Events (SREs) from BC BM

Approximately half to two thirds of patients with MBC with BM will develop an SRE [11, 31]. SREs include bone pain requiring local therapy, pathological fractures, spinal cord compression, and hypercalcemia of malignancy. Because most patients with MBC have BM, which are primarily osteolytic, causing bone destruction, BC is one of the most common cancers causing SREs [32].

SREs often result in hospitalization and may cause chronic pain, neurologic deficits, and negative impacts on QOL. Risk factors for SREs related to MBC include osteoporosis, occurrence of BM as the first metastasis, multiple BM, disease-free interval from early BC of less than 3 years, presence of BM for longer than 2 years, and elevations in CA 15-3 and alkaline phosphatase [33]. Use of BMAs such as denosumab and bisphosphonates in patients with BM due to MBC significantly reduces the risk of SREs, but does not prolong survival [34]. The decision regarding the timing of institution of this supportive therapy is influenced by the extent of bone involvement, the presence of lesions in weight-bearing bones, and patient preference. Effective systemic therapy for control of MBC is key to preventing recurrent SREs and maximizing overall survival.

Pathological Fractures

Up to 50% of patients with BM from BC experience pathological fractures, which are most common in weight-bearing bones such as the proximal femur, pelvis, and spine [35]. Symptoms of bone pain should be investigated by plain radiograph, where the integrity of the cortex must be evaluated. Destruction of cortex or inability to bear weight are warning signs requiring referral to orthopedic surgery [36]. If an orthopedic oncologist is accessible, they should be included in the multidisciplinary team along with radiation oncology, medical oncology, and pain management or palliative care as needed. BMAs, which block bone resorption, are an integral part of the management of BM for fracture prevention.

CDK4/6i, chemotherapy, and ADCs can all cause neutropenia. In patients who develop a pathological fracture while on these treatments, surgery is the priority, and treatment should be held (the exact number of days to be held prior to surgery varies based on chemotherapy regimen). If the patient is neutropenic related to these agents and urgent surgery is required, administering growth factors is recommended to target an absolute neutrophil count (ANC) > 1000. Platelet count should also be monitored, and platelets transfused as needed to achieve a safe level for surgery. In terms of endocrine therapy, due to the increased risk of venous thromboembolism with tamoxifen, this treatment is generally held 7–10 days prior to planned surgery and restarted once patients are ambulatory. There is no indication to hold other endocrine therapy (aromatase inhibitors or SERDs) around planned surgical intervention.

Chemotherapy impairs wound healing, and as such, surgical wounds should be fully healed without signs of infection prior to chemotherapy initiation. Although similar wound healing issues are not of concern with the use of immunotherapy, this treatment can rarely cause adrenal insufficiency, and therefore, there are potential surgical implications. Specifically, an AM cortisol should be checked prior to surgery in patients receiving these agents. It is essential to work in a multidisciplinary collaborative approach to optimize patient outcomes.

Spinal Cord Compression

Spinal cord compression (SCC), a severe SRE which constitutes an oncologic emergency, occurs when BM in the vertebrae causes structural instability and/or compression of the spinal cord due to soft tissue expansion into the canal [37]. Signs of SCC include lower extremity paralysis, bowel and bladder incontinence, and saddle paresthesia. In one review of patients with BM from MBC, 8% developed spinal cord compression (SCC), and SCC was more common in patients with BOM (15%) vs when there were bone and visceral metastasis (4%) [11]. The diagnosis and management of SCC with regard to local therapies are discussed in Chaps. 13 and 31. A multidisciplinary team, including neurosurgery or orthopedic spine surgery, radiation oncology, and medical oncology, is required. BMAs are indicated, and systemic therapy for MBC is started after surgery and radiation have been completed. The perioperative considerations discussed above apply.

A series of 70 cases of SCC due to MBC revealed a median OS of 4 months after the diagnosis of SCC, and the ability to ambulate after treatment for SCC was the best predictor of survival after SCC [38]. Systemic therapy has greatly improved since the time of this series, and perhaps with current treatment options, MBC survival after SCC is improving. However, patients with SCC represent a group of patients with a particularly aggressive type of MBC with poor outcomes.

Patients with vertebral BM that are at risk for SCC are also at risk for leptomeningeal carcinomatosis (LC), which can occur due to direct extension of vertebral BM into the subarachnoid space, causing neurologic deficits that do not clearly associate with a single neurologic location. Breast cancer is the tumor type most often associated with LC, with mTNBC being the most common subtype [39].

Hypercalcemia of Malignancy

Approximately 10–15% of patients with MBC experience hypercalcemia of malignancy, which is more common with bone-dominant MBC, most frequently in those with HR+/HER2+ disease [40]. Parathyroid hormone-related peptide (PTHrP), which is released by BC cells, increases osteoclast activity and causes excessive calcium release from bones, resulting in hypercalcemia. Rarely, MBC can cause hypercalcemia without the presence of bone metastasis in <1% [41]. Patients may present with nausea, fatigue, severe dehydration, renal failure, and coma [42]. Aggressive IV hydration and IV bisphosphonates are urgently indicated. Acute renal dysfunction secondary to hypercalcemia is not a contraindication to the use of bisphosphonates (in contrast to those with chronic renal dysfunction). In cases of hypercalcemia resistant to bisphosphonates, denosumab may be used. To prevent *hypo*calcemia on BMAs, serum calcium levels must be monitored, and calcium should be supplemented [43].

Bone Pain from Breast Cancer Bone Metastasis

BC BM can cause pain related to either structural weakness or inflammatory pain from chemical stimuli [44]. Despite the mechanism, pain from BC BM can be debilitating, resulting in loss of independence and negative impacts on QOL. HER2+ BC and luminal "B" BC (a more aggressive type of high-grade HR+/HER2- BC), can cause particularly severe bone pain, especially in weight-bearing areas such as the pelvis, femur, and vertebrae [45]. Radiation therapy can be extremely effective for the treatment of bone pain from BC BM. Chemotherapy should be held when radiation is administered due to its radio-sensitizing effects. BMAs alone are not sufficient pain control for BC BM, resulting in pain. Radiation and analgesics should be offered, and BMAs should be incorporated. Referral to pain management may be indicated. In some cases, pain may persist even after control of underlying cancer, highlighting the need for multidisciplinary management.

Use of BMAs in BC with BM

BMAs such as bisphosphonates and denosumab are commonly used in patients with BC BM to treat and prevent SREs by inhibiting bone resorption. Pamidronate was the first BMA studied in BC BM, significantly delaying the time to first SRE [34]. Zoledronic acid (ZA) 4 mg IV every 3–4 weeks was then compared to pamidronate in patients with BC BM and decreased the risk of SREs by 20% ($p = 0.037$) in the entire cohort and 30% ($p = 0.01$) in those with osteolytic lesions [46]. Results of three clinical trials, the OPTIMIZE-2 trial, ZOOM trial, and CALGB trial, support extended every 12 weeks dosing of ZA, compared to every 4 weeks administration [47, 48, 56]. The incidence of osteonecrosis of the jaw (ONJ), a rare but serious side effect of long-term bisphosphonate therapy, was low in both groups in the ZOOM trial [47]. The RANKL inhibitor denosumab was studied in a phase 3 trial of patients with BC BM who were randomized to denosumab or ZA. Denosumab delayed the first SRE compared to ZA (HR 0.82; $p = 0.01$). OS was not significantly different, and rates of ONJ were 2.0% for denosumab and 1.4% for ZA [49].

Atypical femur fractures (AFF) are fractures that occur spontaneously or after minimal trauma in the subtrochanteric region or the proximal diaphysis of the femur. AFF are rare but can result from suppression of bone remodeling with BMAs when these agents are used for a long period of time. Retrospective reviews report an AFF incidence of 0.4–12.5% for patients with BM receiving BMAs [50, 51]. It is important to distinguish AFF from active BM on imaging. AFF should be treated by holding BMA and referral to orthopedic surgery for fixation, whereas active BM would be treated with BMA and radiation therapy if symptomatic, with or without surgical stabilization.

The 2017 ASCO guidelines for the treatment of BC BM recommend that all patients be treated with an osteoclast inhibitor. These guidelines do not recommend one agent over another, but offer options to include denosumab, ZA at either a 4-week or 12-week interval, or pamidronate [52]. NCCN endorses ZA every 12 weeks or pamidronate. Patients with severe renal dysfunction should preferentially receive denosumab [12]. Dental evaluation should be done prior to initiation of BMAs so that any necessary dental work that involves the mandibular bone, such as an extraction, dental implant, or root canal, can be completed prior to administration of BMA. Supplemental calcium and vitamin D are recommended.

With extended dose intervals of ZA showing similar benefits for reduction of SREs, "treatment holiday" from BMAs is considered when BMs have healed and are inactive. Discontinuation of denosumab increases the risk of spontaneous fractures, so NCCN guidelines recommend administration of one dose of ZA six months after the final dose of denosumab [12].

Overall Prognosis

MBC patients with BOM are a subgroup with particularly good outcomes. In a Danish registry trial, MBC patients with BOM had the lowest 1-year mortality rate (32%) of any pattern of metastasis. Combined sites had a 1-year mortality of 34%, liver-only 43%, and brain-only 62% [53]. In a German cohort study, at the beginning of first-line therapy for MBC, 36% of patients with HR+ MBC had BOM, and 20% of patients with HR- MBC had BOM. Median OS was 54 months for patients with BOM, 38 months for those with non-visceral metastasis, and 28 months with visceral metastasis [54]. In a SEER analysis of patients with dnMBC with BM at diagnosis, similar negative impacts of added visceral involvement were seen compared to BOM for OS (HR 1.83 with brain metastasis, 1.63 with liver metastasis) [55]. Furthermore, those with a single BM had improved median OS compared to those with multiple BM (7.5 years vs 4.8 years, $p < 0.0001$) [56]. A bone-only pattern of metastasis, particularly when disease burden is low, portends particularly long survival among MBC patients.

When a patient with BOM develops visceral metastasis, this often represents a turning point in the clinical course where the disease becomes more aggressive. Often, this is the time when chemotherapy and ADCs are needed, if not previously used. The goal of systemic therapy for MBC is to extend life while QOL is preserved; therefore, QOL related to the underlying cancer and side effects of treatment is assessed regularly. Once treatment options are exhausted or the patient no longer has preserved QOL on systemic therapy, hospice is discussed.

Conclusion

Bone is the most common site of metastasis in BC and can cause significant morbidity when SREs develop. Multidisciplinary management of patients with BC BM is key for prevention, early identification, and treatment of SREs related to MBC. When warning signs of SREs are present, early evaluation of symptoms and management within a multidisciplinary team helps to reduce the morbidity related to BC SREs. Although MBC remains an incurable disease with treatment that is palliative in nature, *mortality rates from MBC are declining secondary to improvements in novel therapies. Patients with BOM represent a subtype of MBC who may have a particularly indolent disease course with improved overall survival.*

References

1. DeSantis C, Siegel R, Bandi P, Jemal A. Breast cancer statistics, 2011. CA Cancer J Clin. 2011;61:409–18.
2. Luo Y, Ma A, Huang S, Yu Y. Invasive lobular carcinoma has worse outcome compared with invasive ductal carcinoma in stage IV breast cancer with bone-only metastasis. Breast Care (Basel). 2022;17:296–305.
3. Othman A, Winogradzki M, Lee L, et al. Bone metastatic breast cancer: advances in cell signaling and autophagy related mechanisms. Cancers (Basel). 2021;13:13.
4. Pang L, Gan C, Xu J, et al. Bone metastasis of breast cancer: molecular mechanisms and therapeutic strategies. Cancers (Basel). 2022;14:14.
5. Venetis K, Piciotti R, Sajjadi E, et al. Breast cancer with bone metastasis: molecular insights and clinical management. Cells. 2021;10:10.
6. Nie H, Yuan Y, Li J, et al. Occurrence and distribution of bone metastases in 984 metastatic breast cancer patients. Translational Breast Cancer Research. 2021;2:4.
7. Parkes A, Clifton K, Al-Awadhi A, et al. Characterization of bone only metastasis patients with respect to tumor subtypes. NPJ Breast Cancer. 2018;4:2.
8. Woolf DK, Padhani AR, Makris A. Assessing response to treatment of bone metastases from breast cancer: what should be the standard of care? Ann Oncol. 2015;26:1048–57.
9. Di Micco R, Santurro L, Gasparri ML, et al. Rare sites of breast cancer metastasis: a review. Transl Cancer Res. 2019;8:S518–52.
10. Cook GJR. Imaging of bone metastases in breast cancer. Semin Nucl Med. 2022;52:531–41.
11. Domchek SM, Younger J, Finkelstein DM, Seiden MV. Predictors of skeletal complications in patients with metastatic breast carcinoma. Cancer. 2000;89:363–8.
12. National Comprehensive Cancer Network. NCCN Clinical Practice Guidelines in Oncology: Breast Cancer Version 4.2024. In. 2024.

13. Arnaout A, Varela NP, Allarakhia M, et al. Baseline staging imaging for distant metastasis in women with stages I, II, and III breast cancer. Curr Oncol. 2020;27:e123–45.
14. Gertych A, Mohan S, Maclary S, et al. Effects of tissue decalcification on the quantification of breast cancer biomarkers by digital image analysis. Diagn Pathol. 2014;9:213.
15. Zeng J, Piscuoglio S, Aggarwal G, et al. Hormone receptor and HER2 assessment in breast carcinoma metastatic to bone: a comparison between FNA cell blocks and decalcified core needle biopsies. Cancer Cytopathol. 2020;128:133–45.
16. O'Sullivan GJ, Carty FL, Cronin CG. Imaging of bone metastasis: an update. World J Radiol. 2015;7:202–11.
17. Montilla-Soler JL, Makanji R. Skeletal scintigraphy. Cancer Control. 2017;24:137–46.
18. Vogel CL, Schoenfelder J, Shemano I, et al. Worsening bone scan in the evaluation of antitumor response during hormonal therapy of breast cancer. J Clin Oncol. 1995;13:1123–8.
19. Cook GJ, Houston S, Rubens R, et al. Detection of bone metastases in breast cancer by 18FDG PET: differing metabolic activity in osteoblastic and osteolytic lesions. J Clin Oncol. 1998;16:3375–9.
20. Piccardo A, Fiz F, Treglia G, et al. Head-to-head comparison between (18)F-FES PET/CT and (18)F-FDG PET/CT in oestrogen receptor-positive breast cancer: a systematic review and meta-analysis. J Clin Med. 2022;11
21. Wu LM, Gu HY, Zheng J, et al. Diagnostic value of whole-body magnetic resonance imaging for bone metastases: a systematic review and meta-analysis. J Magn Reson Imaging. 2011;34:128–35.
22. Li A, Keck JM, Parmar S, et al. Characterizing advanced breast cancer heterogeneity and treatment resistance through serial biopsies and comprehensive analytics. NPJ Precis Oncol. 2021;5:28.
23. Tuner N, Im S-A, Saura C, et al. Inavolisib-based therapy in pik3ca-mutated advanced breast cancer. N Engl J Med. 2024;391:1584–96.
24. Burris HA, Chan A, Campone M, et al. Abstract P4-22-16: first-line ribociclib + letrozole in patients with HR+, HER2– advanced breast cancer (ABC) presenting with visceral metastases or bone-only disease: a subgroup analysis of the MONALEESA-2 trial. Cancer Res. 2017;77:P4-22-16.
25. Goetz MP, Toi M, Campone M, et al. MONARCH 3: Abemaciclib as initial therapy for advanced breast cancer. J Clin Oncol. 2017;35:3638–46.
26. Gao JJ, Cheng J, Bloomquist E, et al. CDK4/6 inhibitor treatment for patients with hormone receptor-positive, HER2-negative, advanced or metastatic breast cancer: a US Food and Drug Administration pooled analysis. Lancet Oncol. 2020;21:250–60.
27. Toss A, Venturelli M, Sperduti I, et al. First-line treatment for endocrine-sensitive bone-only metastatic breast cancer: systematic review and meta-analysis. Clin Breast Cancer. 2019;19:e701–16.
28. Bardia A, Jhaveri K, Im S-A, et al. Datopotamab deruxtecan versus chemotherapy in previously treated inoperable/metastatic hormone receptor-positive human epidermal growth factor receptor 2-negative breast cancer: primary results From TROPION-Breast01. J Clin Oncol. 2025;43:285–96.
29. Modi S, Jacot W, Yamashita T, et al. Trastuzumab deruxtecan in previously treated HER2-low advanced breast cancer. N Engl J Med. 2022;387:9–20.
30. Rugo HS, Bardia A, Marme F, et al. Sacituzumab Govitecan in hormone receptor-positive/human epidermal growth factor receptor 2-negative metastatic breast cancer. J Clin Oncol. 2022;40:3365–76.
31. Tanaka R, Yonemori K, Hirakawa A, et al. Risk factors for developing skeletal-related events in breast cancer patients with bone metastases undergoing treatment with bone-modifying agents. Oncologist. 2016;21:508–13.
32. Jensen AO, Jacobsen JB, Norgaard M, et al. Incidence of bone metastases and skeletal-related events in breast cancer patients: a population-based cohort study in Denmark. BMC Cancer. 2011;11:29.

33. Zhang H, Zhu W, Biskup E, et al. Incidence, risk factors and prognostic characteristics of bone metastases and skeletal-related events (SREs) in breast cancer patients: a systematic review of the real world data. J Bone Oncol. 2018;11:38–50.
34. Hortobagyi GN, Theriault RL, Lipton A, et al. Long-term prevention of skeletal complications of metastatic breast cancer with pamidronate. Protocol 19 Aredia Breast Cancer Study Group. J Clin Oncol. 1998;16:2038–44.
35. Coleman RE, McCloskey EV. Bisphosphonates in oncology. Bone. 2011;49:71–6.
36. Rosen LS, Gordon D, Tchekmedyian NS, et al. Long-term efficacy and safety of zoledronic acid in the treatment of skeletal metastases in patients with nonsmall cell lung carcinoma and other solid tumors: a randomized, Phase III, double-blind, placebo-controlled trial. Cancer. 2004;100:2613–21.
37. Fizazi K, Carducci M, Smith M, et al. Denosumab versus zoledronic acid for treatment of bone metastases in men with castration-resistant prostate cancer: a randomised, double-blind study. Lancet. 2011;377:813–22.
38. Hill ME, Richards MA, Gregory WM, et al. Spinal cord compression in breast cancer: a review of 70 cases. Br J Cancer. 1993;68:969–73.
39. Coleman RE. Clinical features of metastatic bone disease and risk of skeletal morbidity. Clin Cancer Res. 2006;12:6243s–9s.
40. Coleman RE, Smith P, Rubens RD. Clinical course and prognostic factors following bone recurrence from breast cancer. Br J Cancer. 1998;77:336–40.
41. Hashizume S, Shimoda M, Yoshinami T, et al. Hypercalcemia affected in metastatic breast cancer patients without bone metastasis: report of three cases. Surg Case Rep. 2022;8:210.
42. Coleman R, Gnant M, Morgan G, Clezardin P. Effects of bone-targeted agents on cancer progression and mortality. J Natl Cancer Inst. 2012;104:1059–67.
43. Kristensen B, Ejlertsen B, Mouridsen HT, Loft H. Survival in breast cancer patients after the first episode of hypercalcaemia. J Intern Med. 1998;244:189–98.
44. Falk S, Dickenson AH. Pain and nociception: mechanisms of cancer-induced bone pain. J Clin Oncol. 2014;32:1647–54.
45. Akay CL, Ueno NT, Chisholm GB, et al. Primary tumor resection as a component of multimodality treatment may improve local control and survival in patients with stage IV inflammatory breast cancer. Cancer. 2014;120:1319–28.
46. Rosen LS, Gordon DH, Dugan W Jr, et al. Zoledronic acid is superior to pamidronate for the treatment of bone metastases in breast carcinoma patients with at least one osteolytic lesion. Cancer. 2004;100:36–43.
47. Amadori D, Aglietta M, Alessi B, et al. Efficacy and safety of 12-weekly versus 4-weekly zoledronic acid for prolonged treatment of patients with bone metastases from breast cancer (ZOOM): a phase 3, open-label, randomised, non-inferiority trial. Lancet Oncol. 2013;14:663–70.
48. Himelstein AL, Qin R, Novotny PJ, et al. CALGB 70604 (Alliance): a randomized phase III study of standard dosing vs. longer interval dosing of zoledronic acid in metastatic cancer. J Clin Oncol. 2015;33:9501–9.
49. Stopeck AT, Lipton A, Body JJ, et al. Denosumab compared with zoledronic acid for the treatment of bone metastases in patients with advanced breast cancer: a randomized, double-blind study. J Clin Oncol. 2010;28:5132–9.
50. Ota S, Inoue R, Shiozaki T, et al. Atypical femoral fracture after receiving antiresorptive drugs in breast cancer patients with bone metastasis. Breast Cancer. 2017;24:601–7.
51. Yang SP, Kim TW, Boland PJ, Farooki A. Retrospective review of atypical femoral fracture in metastatic bone disease patients receiving denosumab therapy. Oncologist. 2017;22:438–44.
52. Van Poznak C, Somerfield MR, Barlow WE, et al. Role of bone-modifying agents in metastatic breast cancer: an American Society of Clinical Oncology-Cancer Care Ontario focused guideline update. J Clin Oncol. 2017;35:3978–86.

53. Ording AG, Heide-Jorgensen U, Christiansen CF, et al. Site of metastasis and breast cancer mortality: a Danish nationwide registry-based cohort study. Clin Exp Metastasis. 2017;34:93–101.
54. Schroder J, Fietz T, Kohler A, et al. Treatment and pattern of bone metastases in 1094 patients with advanced breast cancer—results from the prospective German tumour registry breast cancer cohort study. Eur J Cancer. 2017;79:139–48.
55. Liu D, Wu J, Lin C, et al. Breast subtypes and prognosis of breast cancer patients with initial bone metastasis: a population-based study. Front Oncol. 2020;10:580112.
56. Parkes A, Warneke CL, Clifton K, et al. Prognostic factors in patients with metastatic breast cancer with bone-only metastases. Oncologist. 2018;23:1282–8.

Medical Management of Musculoskeletal Metastasis from Prostate Cancer

26

Chinmay T. Jani and Janaki Neela Sharma

Introduction

Prostate cancer (PC) is the most diagnosed malignancy among men in the United States and is the second leading cause of cancer-related mortality [1]. Various epidemiological studies have identified a complex interplay of individual biology and lifestyle factors contributing to the development of this disease [2]. Key risk factors include advancing age, family history, toxic exposures, and lifestyle influences such as diet, smoking, and obesity [2]. Additionally, genome-wide association studies (GWAS) have identified over 180 genetic risk loci associated with a predisposition to PC [3, 4].

PC is classified into localized, locally advanced, or metastatic stages, guiding treatment decisions at each stage [5]. Early metastatic disease, termed metastatic castration-sensitive prostate cancer (mCSPC), initially responds to androgen deprivation therapy (ADT) as PC cells are exquisitely dependent on the androgen receptor for growth signals. Eventually, the disease progresses to metastatic castration-resistant prostate cancer (mCRPC) that requires treatment approaches targeting alternative pathways [5].

PC exhibits a strong preference for hematogenous spread to the bone. An autopsy study indicated that about 90% of men who succumbed to metastatic PC had bone involvement [6]. Bone metastases can present as osteoblastic, osteolytic, or mixed bone lesions, with osteoblastic being the most common [7–9]. The presence of musculoskeletal metastasis profoundly impacts prognosis and quality of life (QoL),

C. T. Jani
Division of Medical Oncology, Sylvester Comprehensive Cancer Center, University of Miami, Coral Gables, FL, USA
e-mail: ctj36@miami.edu

J. N. Sharma (✉)
Sylvester Comprehensive Cancer Center, University of Miami, Coral Gables, FL, USA
e-mail: jsharma@miami.edu

J. Pretell-Mazzini, S. A. Lozano Calderón (eds.), *Multidisciplinary Approach of Musculoskeletal Metastasis*, https://doi.org/10.1007/978-3-032-09498-8_26

often leading to skeletal-related events (SREs) such as pathological fractures and spinal cord compression [10, 11]. Additional complications include hypocalcemia, myelosuppression, and loss of bone mineral density (BMD) due to ADT [10].

The pathophysiology of PC bone metastasis is best explained by the "seed and soil" hypothesis, initially described by Paget in 1889, which posits that metastatic cells ("seeds") thrive in the bone microenvironment ("soil"), facilitating their growth [12]. The bone microenvironment supports PC cell proliferation by disrupting the balance between osteolytic and osteoblastic activity, with paracrine signaling further promoting tumor expansion [13]. This process begins as cancer cells enter the bone through wide-channel sinusoids, facilitated by the SDF-1–CXCR-4 pathway for homing and invasion [14]. The adhesion molecule integrin $\alpha v\beta 3$, along with the receptor activator of NF-κB ligand (RANK–RANKL) signaling pathway, further aids in colonization [11]. Within bones, PC cells acquire characteristics of bone cells, interact with bone marrow stem cells, and create a cyclic loop involving growth factors and osteogenic/osteolytic factors [11]. In osteoblastic metastases, commonly seen in prostate cancer, tumor cells secrete osteotropic factors—including bone morphogenetic proteins (BMPs), vascular endothelial growth factor (VEGF), and endothelin-1 (ET-1)—that promote osteoblast differentiation and activity. This enhances aberrant bone formation and further supports tumor growth within the osseous microenvironment [15, 16].

Understanding the biological and molecular mechanisms of PC bone metastasis has driven the development of novel therapeutic agents [11]. Managing musculoskeletal metastasis requires a multidisciplinary approach, integrating systemic therapies with localized treatments to minimize bone-related complications.

Diagnosis and Staging

Accurate diagnosis and staging of PC, particularly when assessing musculoskeletal metastasis, requires a combination of clinical assessment, imaging techniques, and molecular analysis [17]. The TNM classification system correlates with prognoses and is used for treatment decisions. In earlier editions, prognostic grouping of metastatic disease (M1) was based on the site of spread. According to the AJCC 8th edition, this includes three categories: M1a (non-regional lymph nodes), M1b (bone), and M1c (other sites, such as soft tissue) [18, 19].

Clinically, pain is the most common symptom of bone metastasis in PC, affecting approximately 75% of patients [20]. Its early identification is essential for formulating an appropriate therapeutic approach to prevent or delay SREs [17]. Prostate-specific antigen (PSA) remains the cornerstone for screening and monitoring PC, particularly during the castration-sensitive period. While elevated PSA levels can indicate the presence of PC, they do not specifically confirm metastatic disease. In advanced disease or suspected metastatic cases, PSA doubling time can help predict progression and the likelihood of bone involvement [17, 21]. A 50% reduction in PSA levels from baseline indicates a good metabolic response [22]. In addition to PSA, biochemical markers of bone turnover can provide valuable insights into the response of bone metastases to treatment. Changes in alkaline phosphatase (ALP) are up to 72% sensitive and 88% specific for the progression of skeletal metastases [23]. Furthermore, a rapid increase in bone ALP velocity (>6.3 U/L/year) has been identified as an independent predictor of overall survival (OS) in mCRPC, with a fivefold mortality increase [24]. However, interpreting bone turnover markers in cancer patients requires caution, as age, vitamin D deficiency, and adjuvant hormonal therapy can affect levels [25].

Imaging modalities further support the evaluation of bone involvement in metastatic PC. Conventional imaging techniques include plain radiography, 99mTc-diphosphonate bone scans, CT, and MRI. While plain radiography and CT effectively detect bone lesions at advanced stages, MRI can identify early changes in the bone marrow that precede osteoblastic and osteoclastic responses [17]. MRI sequences, such as T1-weighted, STIR, and diffusion-weighted imaging, are particularly useful in detecting tumor infiltration [17]. Because of the superior sensitivity and specificity of PSMA-PET for detecting micro-metastatic disease, NCCN currently recommends it as a frontline imaging tool, eliminating the need for conventional imaging [26]. Additionally, whole-body MRI (WBMRI), with improved resolution and faster acquisition, has shown higher sensitivity than bone scan/X-ray and equivalent accuracy to CT for nodal assessment, supporting its potential role as a future one-stop staging modality in high-risk prostate cancer [27].

In addition to conventional imaging, Prostate-specific membrane antigen PET/CT (PSMA) and 18F-fluoromethylcholine PET/CT (18F-FCH) have emerged as highly sensitive modalities for detecting metastatic PC, including recurrence beyond the prostate fossa [28]. PSMA PET consistently outperforms other imaging techniques in visualizing distant disease [28–30]. However, planar bone scans using 99mTc-diphosphonates remain a standard for detecting skeletal metastasis due to their availability, cost-effectiveness, and high sensitivity [17]. A systematic review of 37 articles found that while PSMA PET outperformed bone scans in primary staging of hormone-sensitive disease, its superiority in biochemical recurrence and mCRPC is yet to be confirmed [31]. According to the Appropriate Use Criteria (AUC) for PSMA PET, it is indicated for staging in a new PC diagnosis of unfavorable intermediate, high, or very-high risk, particularly if CT and bone scan show none to 5 distant lesions. If an extended metastatic burden has already been confirmed, there is no indication for PSMA PET as it will not change treatment (apart from identifying eligibility for PSMA-targeted therapy in CRPC. Furthermore, no clear evidence exists for PSMA PET in staging PC with a Gleason Score < 7 [32, 33]. Also, false-positive findings on PSMA PET-CT are an important diagnostic consideration, as uptake is not entirely specific to prostate cancer. Benign conditions such as vertebral hemangiomas, splenic hemangiomas, meningiomas, and sarcoidosis, as well as non-prostatic malignancies like melanoma, myeloma, and certain sarcomas, can exhibit PSMA uptake, potentially mimicking metastatic prostate cancer. Neurogenic tumors like neurofibromas have also demonstrated PSMA avidity [34–38]. Therefore, despite limitations like low specificity and potential flare phenomena during systemic treatment, bone scans remain the standard for evaluating skeletal metastasis, given their availability, cost-effectiveness, and high sensitivity [17].

An integrated approach that combines clinical evaluation, PSA monitoring, biochemical markers, and advanced imaging techniques is essential for accurate staging and assessing bone metastasis. For bone-dominant disease, bone-targeted techniques like bone scans, SPECT, or 18F-FCH PET are recommended. PET/CT, combined with MRI, aids in better lesion characterization, especially for indeterminate findings. In high-grade cases, 18F-FDG PET/CT can offer predictive insights into treatment response and prognosis. Still, it should be used alongside other specific agents like radiolabeled choline or PSMA due to PC's metabolic heterogeneity [17].

Biopsy is crucial in confirming the diagnosis and guiding subsequent treatment. However, bone biopsies in PC often yield suboptimal results due to the challenging nature of obtaining sufficient tumor tissue from sclerotic lesions [39, 40]. Studies have shown that bone biopsies suffer from low cellular yield, resulting in inadequate material for histopathologic or molecular analysis [39–41]. Given these limitations, it is generally recommended to obtain biopsy samples from the primary tumor (prostate) or more accessible sites of visceral metastasis [41]. Alternatively, in isolated bone metastasis, dedicated PC bone biopsy protocols can be utilized that can ensure a valuable source of high-quality DNA and RNA for tumor sequencing [39].

The Gleason score, a histopathological grading system based on glandular architecture, has long served as a cornerstone for prostate cancer prognosis [26]. It was later refined into the ISUP Grade Group system to improve clarity and predictive accuracy [42]. This system stratifies tumors into five Grade Groups, with higher groups correlating with worse outcomes [42]. Studies have consistently shown that these Grade Groups predict the risk of biochemical recurrence after radical prostatectomy, as well as all-cause and prostate cancer-specific mortality [43, 44]. For example, patients in Grade Group 1 have an excellent prognosis, while those in Grade Group 5 face significantly higher risks of recurrence and death, highlighting the strong prognostic value of Gleason-based grading [45].

Management

Systemic treatment is key in managing metastatic PC with bone metastasis, focusing on controlling progression, alleviating symptoms, and improving survival, with distinct strategies for mCSPC and mCRPC [46].

Metastatic Castration-Sensitive Prostate Cancer

In mCSPC, the backbone of treatment is ADT to achieve testosterone levels <50 ng/dL. ADT includes bilateral orchiectomy or the use of luteinizing hormone-releasing hormone (LHRH) agonists or antagonists [46, 47]. Significant biochemical response is obtained in almost all patients with a median duration of action of 18 months with single-agent therapy [46, 47].

In addition to ADT, all except the frailest of patients should receive additional systemic therapy with oral anti-androgen agents with or without chemotherapy. Systemic therapy selection depends on metastasis volume and timing (synchronous or metachronous), with high-volume disease defined by CHAARTED criteria as having visceral metastasis or ≥4 bone lesions, including at least one beyond the vertebral bodies and pelvis [26]. Low-volume disease includes all patients who do not meet the criteria for high-volume.

For patients with low-volume metastases, the preferred regimens include abiraterone, apalutamide, enzalutamide, or darolutamide [48–54]. The STAMPEDE and LATITUDE trials showed that adding abiraterone with low-dose prednisone to ADT significantly improved OS compared to ADT alone in newly diagnosed metastatic or high-risk, node-positive disease [48–50]. The TITAN trial reported that combining apalutamide with ADT improved rPFS and OS versus ADT alone [51, 52]. Similarly, the ENZAMET and ARCHES trials demonstrated that enzalutamide plus ADT significantly improved OS and rPFS compared to ADT alone [53, 54]. Finally, the ARANOTE trial demonstrated a 46% reduction in the risk of rPFS or death in patients receiving darolutamide combined with ADT versus ADT alone [55].

Additionally, external beam radiotherapy (EBRT) to the primary tumor should be considered for low-volume metastatic disease, either alone or alongside abiraterone or darolutamide. EBRT to the primary tumor has been associated with an OS benefit in patients with low-volume metastatic disease [56]. EBRT to sites of bone metastases can also be considered if metastases are located in weight-bearing bones or if the patient is symptomatic [26].

In high-volume mCSPC, triplet therapy involving ADT, an ARPI (abiraterone or darolutamide), and docetaxel has shown improved outcomes over ADT plus docetaxel alone. The PEACE-1 trial demonstrated that adding abiraterone to ADT and docetaxel significantly improved rPFS and OS compared to ADT and docetaxel alone [57]. Similarly, the ARASENS trial reported that the addition of darolutamide to ADT and docetaxel led to a 32% reduction in the risk of death, along with improvements in secondary endpoints, such as time to CRPC, skeletal event-free survival, and time to initiation of subsequent therapy [58]. Triplet therapy is now the preferred option for patients with high-volume de novo mCSPC who are fit for chemotherapy [57, 58]. See Table 26.1.

Metastatic Castrate-Resistant Prostate Cancer (mCRPC)

mCRPC is defined as disease progression despite maintaining castrate levels of testosterone (<50 ng/dL) [26, 46, 47]. Management in this setting is tailored based on prior therapies and the results of molecular testing [26, 63].

For patients without prior exposure to docetaxel or novel hormonal therapies, treatment options include abiraterone, enzalutamide, or docetaxel. In the chemotherapy-naïve population, the COU-AA-302 trial demonstrated that abiraterone improved rPFS and OS [64]. Enzalutamide was evaluated in chemotherapy-naïve patients through the PREVAIL trial, which also demonstrated improvements in PFS and OS [65, 66]. For docetaxel, two randomized

Table 26.1 Treatment Modalities for Metastatic Castration-Sensitive Prostate Cancer (mCSPC)

Drug/Combination	Indication	Dosing	Common side effects	Work-up
ADT [59–62] LHRHAgonists: Leuprolide (Inj) Goserelin (Inj) Triptolrelin (Inj) LHRH Antagonists: Degarelix (Inj) Relugolix (oral)	mCSPC, as monotherapy (rarely) or combined with ARPI +/− docetaxel	typical dosing varies by agent, e.g., daily, monthly or quarterly	Hot flashes, fatigue, loss of libido, osteoporosis, metabolic syndrome, cardiovascular disease	Baseline PSA, testosterone, liver/kidney function; DEXA scan for bone health
ADT + ARPI				
Abiraterone + ADT [48–50]	Low- and high-volume mCSPC	1000 mg/day abiraterone + 5 mg prednisone/day	Hypertension, hypokalemia, edema, liver toxicity, fatigue	Monitor BP, potassium, liver function; PSA monitoring; DEXA scan
Apalutamide + ADT [51, 52]	Low- and high-volume mCSPC	240 mg/day	Rash, hypothyroidism, fatigue, ischemic heart disease	Baseline thyroid function, liver/kidney function, PSA levels
Enzalutamide + ADT [53, 54]	Low- and high-volume mCSPC	160 mg/day	Fatigue, hypertension, seizures, cognitive impairment	Baseline BP, liver/kidney function, PSA monitoring, falls & mental status assessment
Darolutamide + ADT [55]	Low- and high-volume mCSPC	600 mg twice daily	Fatigue, rash, hypertension, falls	Baseline BP, liver/kidney function, PSA monitoring, falls risk assessment
Triplet Therapy				
ADT + Docetaxel + Abiraterone) [57]	High-volume mCSPC	Docetaxel: 75 mg/m^2 IV every 3 weeks, 6 cycles; Abiraterone: 1000 mg/day + 5 mg prednisone/day	Febrile neutropenia, hypertension, fatigue, liver toxicity	CBC before each cycle, liver/kidney function, PSA, BP monitoring
ADT + Docetaxel + Darolutamide) [58]	High-volume mCSPC	Docetaxel: 75 mg/m^2 IV every 3 weeks, 6 cycles; Darolutamide: 600 mg twice daily	Febrile neutropenia, fatigue, hypertension, anemia, rash	CBC before each cycle, liver/kidney function, PSA, BP monitoring

ADT Androgen Deprivation Therapy, *ARPI* Androgen Receptor Pathway Inhibitor, *BP* Blood Pressure, *CBC* Complete Blood Count, *CRPC* Castrate-Resistant Prostate Cancer, *DEXA* Dual-Energy X-ray Absorptiometry, *IV* Intravenous, *LHRH* Luteinizing Hormone-Releasing Hormone, *mCSPC* Metastatic Castration-Sensitive Prostate Cancer, *PSA* Prostate-Specific Antigen phase 3 studies, TAX 327 and SWOG 9916, showed that docetaxel-based regimens provided significant survival benefits in symptomatic or rapidly progressing mCRPC [67–69].

For patients progressing after prior docetaxel, treatment expands to include abiraterone, enzalutamide, or cabazitaxel. The COU-AA301 trial established abiraterone's efficacy in the post-docetaxel setting, demonstrating a median OS benefit of 15.8 months compared to 11.2 months with placebo [70, 71]. Enzalutamide was initially approved for mCRPC after docetaxel, based on the AFFIRM trial, which showed a median OS of 18.4 months versus 13.6 months with placebo [72, 73]. Cabazitaxel was approved based on the TROPIC trial and showed a 2.4-month OS improvement compared to mitoxantrone, though it was associated with a higher incidence of febrile neutropenia and diarrhea [74–76].

For patients with BRCA mutations, combinations like olaparib/abiraterone or niraparib/abiraterone are effective when no prior ARPi has been used. For patients with all HRR (homologous recombination repair) mutations, talazoparib/enzalutamide can be used after previous exposure to abiraterone, docetaxel or both. Finally, single-agent olaparib is also recommended for HRR-mutated disease following ARPi exposure [77–83].

Immunotherapy options are also available in mCRPC patients with pembrolizumab recommended in cases of MSI-high (MSI-H) or mismatch repair-deficient (dMMR) tumors [84–86]. Additionally, Sipuleucel-T was the first FDA-approved therapeutic cancer vaccine for solid tumor malignancies based on the results of the IMPACT trial, which showed a 4.1-month median OS benefit in patients with no or minimal symptoms from mCRPC [84].

Radioligand therapy remains an actively evolving treatment option for mCRPC. Radium-223 (Ra-223) continues to be a crucial option for patients with bone-only metastases, demonstrating improved OS and time to first SRE in the ALSYMPCA trial [87]. More recently, Lu-177-PSMA-617 has emerged as a key treatment for PSMA-positive mCRPC at any site, showing significant survival benefits in the VISION trial for patients previously treated with ARPi and taxane-based chemotherapy [88]. See Table 26.2.

Management of Skeletal-Related Events

SREs are common and serious complications in patients with advanced PC with the most common SREs in this population include pathological fractures, spinal cord compression, hypercalcemia of malignancy, and the need for radiotherapy or surgery for pain control or stabilization [10, 89, 90]. Among these, spinal cord compression (SCC) is considered the most urgent, as it can lead to irreversible neurological damage and loss of function if not treated promptly [91, 92]. Effective management of SREs is crucial to maintain quality of life, improve functional outcomes, and reduce morbidity in PC patients [10, 89, 90].

Table 26.2 Treatment Modalities for Metastatic Castration-Resistant Prostate Cancer (mCRPC)

Drug/Combination	Indication	Dosing	Side effects	Work-up
Abiraterone + prednisone [64, 70, 71]	All mCRPC	1000 mg/day abiraterone + 5 mg prednisone twice daily	Hypertension, hypokalemia, edema, liver toxicity, fatigue	Monitor BP, liver function, potassium; PSA monitoring; symptom-directed cardiac assessment
Enzalutamide [65, 66, 72, 73]	All mCRPC	160 mg/day	Fatigue, hypertension, seizures, cognitive impairment	Baseline BP, liver/kidney function, PSA monitoring, mental status assessment
Docetaxel [67–69]	All mCRPC	75 mg/m² IV every 3 weeks, 6 cycles	Neutropenia, febrile neutropenia, fatigue, peripheral neuropathy	CBC before each cycle, liver/kidney function, PSA monitoring
Cabazitaxel [74–76]	Post-docetaxel progression	20 mg/m² IV every 3 weeks, with concurrent steroid	Febrile neutropenia, diarrea, fatigue, anemia	CBC before each cycle, liver/kidney function, PSA monitoring
Olaparib [77, 79, 80]	BRCA/HRR mutations	300 mg twice daily	Nausea, anemia, fatigue, pneumonitis	Genetic testing for BRCA/HRR mutations; CBC, liver/kidney function, PSA monitoring
Olaparib + Abiraterone [78]	BRCA mutations	Olaparib: 300 mg twice daily + Abiraterone: 1000 mg/day	Hypertension, hypokalemia, anemia, fatigue	Genetic testing for BRCA mutations; CBC, liver function, PSA monitoring

(continued)

Table 26.2 (continued)

Drug/Combination	Indication	Dosing	Side effects	Work-up
Niraparib + abiraterone [82, 83]	BRCA mutations	Niraparib: 200 mg daily + abiraterone: 1000 mg/day	Hypertension, thrombocytopenia, anemia, fatigue	Genetic testing for BRCA/HRR mutations; CBC, liver function, PSA monitoring
Talazoparib + Enzalutamide [81]	HRR mutations	Talazoparib: 1 mg daily + Enzalutamide: 160 mg/day	Fatigue, anemia, thrombocytopenia, hypertension	Genetic testing for HRR mutations; CBC, liver/kidney function, PSA monitoring
Radium-223 [87]	Symptomatic bone metastases, post-docetaxel	50 kBq/kg IV every 4 weeks, 6 cycles	Bone marrow suppression, GI upset, peripheral edema	CBC monitoring before each dose; symptom assessment
Sipuleucel-T [84]	Minimally symptomatic mCRPC	3 doses, 2 weeks apart	Chills, pyrexia, headache, flu-like symptoms	CBC, liver/kidney function; monitor for infusion-related reactions
Lu-177-PSMA-617 [88]	PSMA-positive, post-ARPi & taxane therapy	7.4 GBq IV every 6 weeks, 4–6 cycles	Anemia, thrombocytopenia, lymphopenia, fatigue	PSMA PET/CT imaging, CBC, liver/kidney function, PSA monitoring
Pembrolizumab [85, 86]	MSI-H/dMMR tumors	200 mg IV every 3 weeks or 400 mg IV every 6 weeks	Immune-related adverse events (e.g., colitis, hepatitis, pneumonitis)	MSI/MMR testing, liver function, CBC, symptom-directed assessment

ADT Androgen Deprivation Therapy, *ARPi* Androgen Receptor Pathway Inhibitor, *BP* Blood Pressure, *CBC* Complete Blood Count, *dMMR* Mismatch Repair-Deficient, *GI* Gastrointestinal, *HRR* Homologous Recombination Repair, *IV* Intravenous, *mCRPC* Metastatic Castration-Resistant Prostate Cancer, *MSI-H* Microsatellite Instability-High, *OS* Overall Survival, *PET/CT* Positron Emission Tomography/Computed Tomography, *PFS* Progression-Free Survival, *PSA* Prostate-Specific Antigen, *PSMA* Prostate-Specific Membrane Antigen

Osteoclast Inhibitors

Bisphosphonates

Bisphosphonates function by inhibiting bone matrix resorption, a process mediated by osteoclasts. They are categorized into three generations based on potency, with the third-generation agents, including zoledronic acid, ibandronate, neridronate, and risedronate, being the most potent [93]. Beyond their osteoclast inhibition, bisphosphonates exhibit additional effects that may contribute to reduced cancer progression, such as immunomodulation, anti-angiogenesis, induction of cancer cell apoptosis, and inhibition of cancer cell adhesion to bone stroma [94, 95].

Earlier trials investigated the use of zoledronic acid in non-metastatic PC to prevent the development of bone metastases, but results indicated that adjuvant zoledronic acid did not effectively prevent bone metastasis in patients with high-risk localized PC [10, 96, 97]. However, zoledronic acid has demonstrated efficacy in mCRPC, demonstrating a reduction in SREs at 15 and 24 months and significantly prolonging the median time to the first SRE. Among patients who experienced on-study SRE, treatment with zoledronic acid lowered the risk of a subsequent SRE by 36%. Additionally, patients reported pain relief, as indicated by changes in the Brief Pain Inventory (BPI) scores, and reductions were observed in urinary markers of bone resorption and serum bone ALP levels [98].

Denosumab

Denosumab, a fully humanized monoclonal antibody, specifically targets and neutralizes the RANKL inhibiting osteoclast development and function [99]. The HALT 138 trial evaluated denosumab against a placebo in non-metastatic PC patients at high risk for fragility fractures, defined by age ≥70 or a history of fragility fractures. Denosumab demonstrated a significant increase in BMD at the lumbar spine, with a 5.6% increase compared to 1.0% in the placebo. While there was no significant difference in overall fracture incidence or time to first fracture, denosumab reduced new vertebral compression fractures (1.5% vs. 3.9%) and fractures at multiple sites (1.2% vs. 4.7%) [100]. This led to regulatory approval for denosumab to increase BMD in men with non-metastatic PC undergoing ADT who are at high risk for fragility fractures [10, 95, 101].

In the bone microenvironment, RANKL-mediated osteoclast activation increases bone turnover, releasing growth factors from the bone matrix that can promote PC establishment in the skeleton [102]. Preclinical studies demonstrated that osteoclast inhibition prevents bone metastasis [103, 104]. Subsequently, a randomized Phase 3 trial found that denosumab improved bone metastasis-free survival compared to placebo and reduced symptomatic bone metastasis [101]. However, no PFS or OS benefits were observed, and the denosumab group had a 5% incidence of ONJ. Due to the lack of survival benefit and increased ONJ risk, the FDA did not approve denosumab for delaying bone metastasis in non-metastatic CRPC [10].

Clinical trials have compared zoledronic acid and denosumab in reducing SREs in mCRPC. While no significant differences in PFS or OS were found, the time to first on-study SRE was longer with denosumab. Denosumab also showed greater suppression of bone turnover markers (uNTx and BALP), with similar ONJ rates

[10, 105]. In two additional Phase 3 studies involving other solid tumors with bone metastasis, denosumab was found to be more effective than zoledronic acid in preventing SREs [106–108]. This led to FDA approval of denosumab for reducing SREs in PC with bone metastasis [10, 105].

The optimal timing for initiating bisphosphonates or denosumab is early in the course of metastatic castration-resistant disease, as soon as bone metastases are identified, even if asymptomatic, to prevent SREs [109]. Treatment should continue as long as the patient is responding to systemic therapy and deriving clinical benefit, with discontinuation only in cases of unacceptable toxicity or lack of further benefit. Hypocalcemia is a common side effect, so patients should receive calcium and vitamin D supplements to prevent it [109, 110]. ONJ is a severe risk; therefore, patients require preventive dental care before starting therapy, along with regular dental monitoring during treatment [106, 111]. If ONJ develops, therapy may need to be paused. In cases of renal impairment, zoledronic acid may require dose adjustments or might need to be avoided altogether. In such situations, denosumab is preferred since it is not renally excreted [106, 111]. See Table 26.3.

Table 26.3 Overview of Therapeutic Agents for prevention and management of Skeletal-Related Events

Drug	Indication	Dosing	Side effects	Work-up	Duration
Zoledronic Acid [98]	To prevent pathological fractures and SREs in mCRPC with bone metastases	4 mg IV every 3–4 weeks	Hypocalcemia, ONJ, renal impairment	Bone-dedicated imaging for the extent of bone metastasis Oral examination and preventive dentistry	Duration of metastatic disease; consider a holiday after 1–2 years to reduce ONJ risk
Denosumab [101, 105]	To prevent pathological fractures and SREs in mCRPC with bone metastases	120 mg subcutaneously every 4 weeks	Hypocalcemia, hypophosphatemia, ONJ	Bone-dedicated imaging for the extent of bone metastasis Oral examination and preventive dentistry Calcium and phosphate repletion before initiating denosumab	No drug holiday is required
	To decrease fracture risk in men on ADT in non-metastatic PC	60 mg subcutaneously every 6 months.			

ADT Androgen Deprivation Therapy, *CRPC* Castrate Resistant Prostate Cancer, *ONJ* Osteonecrosis of Jaw, *RANK* receptor activator of NF-kB ligand, *SRE* Skeletal Related Events

Radio-Pharmaceuticals

Several radiopharmaceuticals are approved for the treatment of mCRPC. Amongst these, Ra-223 is the only radiopharmaceutical that prolongs OS and decreases symptomatic skeletal events (SSEs) in selected men with mCRPC, while Beta-emitting radio-isotopes (Strontium-89 and Samarium-153) may provide palliation of pain but do not significantly prolong OS [10, 87, 95, 112, 113].

Radium-223

Ra-223 is an alpha-emitting radiopharmaceutical that selectively targets bone metastases in patients with mCRPC. It mimics calcium and binds to areas of increased bone turnover, delivering targeted radiation that induces double-strand DNA breaks in adjacent tumor cells [10, 87, 113]. The ALSYMPCA trial demonstrated significant survival benefits with Ra-223, which improved OS by 3.6 months compared to placebo in mCRPC patients with symptomatic bone metastases and no known visceral metastases. Additionally, Ra-223 delayed the time to the first SRE, with lower incidences of pathologic fractures and SCC [87]. A phase-II trial also showed that Ra-223 palliated pain, reduced fatigue, and improved pain interference [114]. Ra-223 has an acceptable side effect profile, including mild myelosuppression, particularly thrombocytopenia, gastrointestinal symptoms (nausea, diarrhea), and injection site reactions [87]. Ra-223 should be considered for mCRPC patients with symptomatic bone-only metastases. Due to its potential myelosuppressive effects, baseline hematologic evaluation and regular monitoring are recommended during treatment [113].

Strontium-89 and Samarium-153

Strontium-89 and Samarium-153 are beta-emitting radiopharmaceuticals primarily used for the palliative treatment of painful bone metastases. They deliver targeted radiation to sites of increased bone turnover, reducing pain and improving QoL [115, 116]. In clinical trials, strontium-89 has shown significant pain relief and a reduction in analgesic use, with effects lasting several months [115]. However, due to its broader beta-emission, it is associated with greater myelosuppression, limiting its repeated use with a primary role in patients with widespread painful metastases who are not candidates for radium-223 or other systemic therapies [115]. Samarium-153 has demonstrated similar efficacy to strontium-89 in pain relief. It also emits gamma rays, allowing for imaging-verified treatment localization. Samarium-153 provides faster pain relief, often within one week, but like strontium-89, it carries a significant risk of myelosuppression [117–119]. Hematologic monitoring is crucial for both, with dose adjustments or treatment delays in cases of severe cytopenia. While strontium-89 and samarium-153 are generally less preferred than radium-223 due to higher toxicity and lack of survival benefit, they remain important options for pain palliation in patients with mCRPC who experience significant bone pain refractory to other interventions [115–119]. See Table 26.4.

Analgesics

Bone pain is one of the most common and debilitating complications of bone metastases in PC [120, 121]. The pain can be constant or intermittent, exacerbated by movement, and is typically localized to areas of bone involvement, such as the

Table 26.4 Overview of radiopharmaceuticals for pain management and treatment of bone metastases in mCRPC

Drug	Indication	Dosing	Side Effects	Work-up
Radium-223 [87]	CRPC with symptomatic bone metastases and no known visceral metastasis	55 kBq/kg IV every 4 weeks for 6 cycles	Mild myelosuppression (thrombocytopenia, neutropenia), nausea, diarrhea, injection site reactions	Baseline CBC, renal, and liver function tests Regular hematologic monitoring during treatment
Strontium-89 [115]	Painful bone metastases in mCRPC	148 MBq IV (single dose)	Myelosuppression (neutropenia, thrombocytopenia), transient pain flare	Baseline CBC Regular monitoring of hematologic parameters
Samarium-153 [117–119]	Painful bone metastases in mCRPC	1 mCi/kg IV (single dose)	Myelosuppression (neutropenia, thrombocytopenia), nausea, transient pain flare	Baseline CBC Renal function evaluation Regular hematologic monitoring

CBC Complete blood count, *mCRPC* metastatic castrate-resistant prostate cancer

spine, pelvis, and ribs. If untreated, it can decrease mobility, disturb sleep, and deteriorate QoL [121]. Involving palliative care specialists for analgesics from earlier in the disease course is known to be beneficial [122]. When treating pain, hormonal manipulation, radiotherapy, chemotherapy, and surgery should always be considered alongside analgesic therapy [122, 123].

Opioids are the primary choice for moderate to severe pain and can be rapidly titrated to achieve pain relief. In cases of bone-only metastasis, they are generally combined with non-opioid analgesics to enhance efficacy and minimize opioid doses [124]. Short-acting opioids, like morphine, oxycodone, or hydromorphone, are often used for acute episodes, while long-acting formulations, such as extended-release morphine or fentanyl patches, help maintain consistent chronic pain control [120, 125]. Care should be taken to monitor potential opioid-related side effects like constipation, nausea, and sedation. Combinations apart from short-acting with extended-release should be avoided [122].

NSAIDs, such as ibuprofen, naproxen, and diclofenac, are also frequently used, especially in mild to moderate pain or in combination with opioids [126, 127]. However, due to potential side effects like gastrointestinal irritation, renal impairment, and cardiovascular risks, careful monitoring and dose adjustments are necessary [126, 127]. Other analgesics, such as acetaminophen, gabapentinoids (e.g., gabapentin and pregabalin), and steroids (e.g., dexamethasone), may be used as adjuncts to enhance pain relief [128]. Acetaminophen can be combined with opioids for moderate pain, while gabapentinoids are beneficial for neuropathic pain components often associated with advanced cancer [128, 129]. Corticosteroids can reduce inflammation, swelling, and pain around metastatic lesions, offering rapid symptomatic relief [130]. Along with these, randomized trials have also shown that medical cannabinoid drugs can be considered for multi-purpose analgesics in patients with advanced cancer pain not fully relieved by strong opioids [131, 132].

Supportive Measures

Supportive care is essential to improve QoL, maintain functional independence, and address physical and psychological needs. Physical therapy is critical in maintaining mobility, strength, and overall function. It helps improve balance, reduces fall risk, enhances pain control, and improves lean mass through tailored exercise programs [133–135]. Mobility aids, such as walkers, canes, or wheelchairs, may be necessary to ensure safety and prevent falls, especially in patients with compromised bone integrity or pathological fractures [134, 135]. Along with physical care, psychosocial support is highly crucial to addressing emotional and mental health challenges. This support can include counseling, support groups, or referrals to mental health professionals. Studies have shown that psychological interventions can reduce anxiety, depression, and distress while improving coping skills and adherence to treatment regimens [136, 137]. Patients with PC often experience reduced appetite, weight loss, or nutritional deficiencies due to the disease or its treatments. Nutritional interventions should focus on high-protein diets, adequate caloric intake, and calcium and vitamin D supplementation to promote bone health. Registered dietitians can provide personalized nutritional counseling to address specific dietary needs, manage side effects of treatments, and prevent malnutrition [138, 139]. See Table 26.5.

Table 26.5 Overview of non-bisphosphonate and non-radiopharmaceutical strategies for managing cancer-induced bone pain

Treatment modality	Primary indications	Outcomes	Additional considerations
Analgesics			
Opioids [120, 124, 125]	Moderate to severe pain	Effective for short-term relief; potential side effects like constipation, nausea, and sedation require management	Titrate for optimal relief; combine with non-opioids to reduce dose and side effects
NSAIDs [126, 127]	Mild to moderate bone pain	Reduces inflammation and pain; can be combined with opioids	Monitor for GI, renal, or cardiovascular side effects
Acetaminophen [128, 129]	Mild to moderate pain	Effective for mild pain relief; can be combined with opioids	Safe for most patients; avoid in those with liver dysfunction
Gabapentinoids [128]	Neuropathic pain components in bone metastasis	Effective for neuropathic pain; beneficial in combination with opioids	Start at low doses and titrate gradually; monitor for sedation and dizziness
Corticosteroids [130]	Inflammatory pain and swelling around metastatic lesions	Rapid symptomatic relief reduces inflammation and edema	Use short-term due to potential side effects; adjust doses for response

GI Gastrointestinal, *NSAID* Non-Steroidal anti-inflammatory drugs

Prognosis

Survivorship of Patients with Musculoskeletal Metastasis

The prognosis for patients with PC and musculoskeletal metastasis is shaped by several factors, including the extent of metastasis—whether involving visceral organs—as well as the patient's overall health and therapy response [18, 140, 141]. Patients with bone-only metastases generally have a more favorable prognosis than those with visceral involvement. A meta-analysis of nine trials encompassing nearly 9000 patients found that the median OS for men with bone metastases, whether isolated or combined with lymph node involvement, was 21.3 months [141]. In contrast, median OS decreased to 19.4 and 13.5 months for patients with lung and liver metastases, respectively [141]. This variation in survival underscores the importance of considering the metastasis location when making treatment decisions and enrolling patients in clinical trials, as recommended by the Prostate Cancer Clinical Trials Working Group [142–144].

SREs, including pathological fractures, hypercalcemia, and bone pain, affect nearly 50% of patients with osteoblastic metastasis [145]. Predicting and managing SREs is critical as they impact prognosis and complicate treatment options [146]. In a study involving 915 PC patients with bone metastasis, long-term survival was linked to the absence of cancer-related pain, a good ECOG status, and minimal analgesic use. Additional factors associated with better outcomes included localized tumors (T0–T2), PSA < 231 μg/L, and a Soloway score of 1, indicating limited disease spread to fewer than three vertebral bodies [18]. These findings suggest that preventing SREs and symptom management are crucial for improving survival outcomes in this patient population.

Quality of Life Assessment and Supportive Interventions

The presence of musculoskeletal metastasis in PC significantly impacts QoL, with major concerns including bone pain, decreased mobility, and psychological burden. Patients with spinal metastasis are particularly at risk for SCC, which can lead to severe neurological deficits, greatly affecting functional independence and overall well-being [18, 147, 148]. Regular QoL assessment is critical for tailoring treatment and providing adequate supportive care. Several tools are commonly used to evaluate QoL in metastatic PC, focusing on physical, emotional, and functional aspects:

- EORTC QLQ-C30: Assesses physical, emotional, and social functioning, along with symptoms like pain and fatigue [149];
- FACT-P (Functional Assessment of Cancer Therapy—Prostate): A PC-specific QoL tool measuring symptoms related to both the disease and its treatments [150];
- BPI (Brief Pain Inventory): Evaluates pain severity and its impact on daily activities [151].

While SREs are often measured in clinical trials, SSEs provide more clinically relevant insights into patient outcomes and QoL [18]. In a study of 832 patients, the 207 who developed at least one SSE had increased emergency room and inpatient visits, lower mean FACT-P scores, and worse pain scores [18]. This underscores the substantial burden of SSEs on QoL and health resource utilization [148]. Additionally, SRE often signifies a permanent downturn in the disease, contributing to patients' perception of worsening health and possibly depression and anxiety [152]. Therefore, in advanced PC, particularly in cases of combined metastases with limited treatment options and a declining patient condition, the focus may shift from curative to palliative care [125]. Hospice care should be considered when:

- Treatment options are no longer effective or desired by the patient.
- There has been a significant deterioration in physical well-being and performance status (e.g., ECOG 3 or 4).
- SSEs severely affect QoL and are challenging to manage with standard treatments.
- A shift in goals of care toward comfort, focusing on symptom management and psychological support rather than aggressive treatments, is desired by the patient

To summarize, musculoskeletal metastases in PC pose significant challenges to patients' QoL, including physical, functional, and emotional burdens. Effective management of SREs and SSEs is vital to improving patient outcomes. Despite treatment advances, persistent pain, reduced mobility, and psychological distress highlight the need for symptom relief, patient comfort, and psychological support, with hospice care as an option in refractory cases. Ultimately, a patient-centered approach remains key to enhancing the QoL and survival in patients with prostate cancer and bone metastasis.

References

1. Jani C, Mouchati C, Abdallah N, Mariano M, Jani R, Salciccioli JD, et al. Trends in prostate cancer mortality in The United States of America, by state and race, from 1999 to 2019: estimates from the centers for disease control WONDER database. Prostate Cancer Prostatic Dis. 2023;26(3):552–62.
2. Pernar CH, Ebot EM, Wilson KM, Mucci LA. The epidemiology of prostate cancer. Cold Spring Harb Perspect Med. 2018;8(12).
3. Eeles RA, Olama AA, Benlloch S, Saunders EJ, Leongamornlert DA, Tymrakiewicz M, et al. Identification of 23 new prostate cancer susceptibility loci using the iCOGS custom genotyping array. Nat Genet. 2013;45(4):385–91, 91e1–2.
4. Al Olama AA, Kote-Jarai Z, Berndt SI, Conti DV, Schumacher F, Han Y, et al. A meta-analysis of 87,040 individuals identifies 23 new susceptibility loci for prostate cancer. Nat Genet. 2014;46(10):1103–9.
5. Litwin MS, Tan HJ. The diagnosis and treatment of prostate cancer: a review. JAMA. 2017;317(24):2532–42.
6. Bubendorf L, Schopfer A, Wagner U, Sauter G, Moch H, Willi N, et al. Metastatic patterns of prostate cancer: an autopsy study of 1,589 patients. Hum Pathol. 2000;31(5):578–83.
7. Coleman RE. Skeletal complications of malignancy. Cancer. 1997;80(8 Suppl):1588–94.

8. Westendorf JJ, Kahler RA, Schroeder TM. Wnt signaling in osteoblasts and bone diseases. Gene. 2004;341:19–39.
9. Ibrahim T, Flamini E, Mercatali L, Sacanna E, Serra P, Amadori D. Pathogenesis of osteoblastic bone metastases from prostate cancer. Cancer. 2010;116(6):1406–18.
10. Gartrell BA, Saad F. Managing bone metastases and reducing skeletal related events in prostate cancer. Nat Rev Clin Oncol. 2014;11(6):335–45.
11. Body JJ, Casimiro S, Costa L. Targeting bone metastases in prostate cancer: improving clinical outcome. Nat Rev Urol. 2015;12(6):340–56.
12. Paget S. The distribution of secondary growths in cancer of the breast. Lancet. 1889;133:571–3.
13. Casimiro S, Guise TA, Chirgwin J. The critical role of the bone microenvironment in cancer metastases. Mol Cell Endocrinol. 2009;310(1–2):71–81.
14. Wang J, Loberg R, Taichman RS. The pivotal role of CXCL12 (SDF-1)/CXCR4 axis in bone metastasis. Cancer Metastasis Rev. 2006;25(4):573–87.
15. Feeley BT, Gamradt SC, Hsu WK, Liu N, Krenek L, Robbins P, et al. Influence of BMPs on the formation of osteoblastic lesions in metastatic prostate cancer. J Bone Miner Res. 2005;20(12):2189–99.
16. Kitagawa Y, Dai J, Zhang J, Keller JM, Nor J, Yao Z, et al. Vascular endothelial growth factor contributes to prostate cancer-mediated osteoblastic activity. Cancer Res. 2005;65(23):10921–9.
17. Evangelista L, Bertoldo F, Boccardo F, Conti G, Menchi I, Mungai F, et al. Diagnostic imaging to detect and evaluate response to therapy in bone metastases from prostate cancer: current modalities and new horizons. Eur J Nucl Med Mol Imaging. 2016;43(8):1546–62.
18. Klaff R, Berglund A, Varenhorst E, Hedlund PO, Jonler M, Sandblom G, et al. Clinical characteristics and quality-of-life in patients surviving a decade of prostate cancer with bone metastases. BJU Int. 2016;117(6):904–13.
19. Roberts MJ, Teloken P, Chambers SK, et al. Prostate Cancer Detection. [Updated 2018 Jun 11]. In: Feingold KR, Ahmed SF, Anawalt B, et al., editors. Endotext [Internet]. South Dartmouth (MA): MDText.com, Inc.; 2000. Table 8, TNM Staging Classifications [per American Joint Committee on Cancer (AJCC) 8th Edition 2016)(198). Available from: https://www.ncbi.nlm.nih.gov/books/NBK279042/table/prostate-cancer-det.primarytum/.
20. van den Beuken-van Everdingen MH, de Rijke JM, Kessels AG, Schouten HC, van Kleef M, Patijn J. Prevalence of pain in patients with cancer: a systematic review of the past 40 years. Ann Oncol. 2007;18(9):1437–49.
21. Oudard S, Banu E, Scotte F, Banu A, Medioni J, Beuzeboc P, et al. Prostate-specific antigen doubling time before onset of chemotherapy as a predictor of survival for hormone-refractory prostate cancer patients. Ann Oncol. 2007;18(11):1828–33.
22. Smith DC, Dunn RL, Strawderman MS, Pienta KJ. Change in serum prostate-specific antigen as a marker of response to cytotoxic therapy for hormone-refractory prostate cancer. J Clin Oncol. 1998;16(5):1835–43.
23. Zafeirakis AG, Papatheodorou GA, Limouris GS. Clinical and imaging correlations of bone turnover markers in prostate cancer patients with bone only metastases. Nucl Med Commun. 2010;31(3):249–53.
24. Metwalli AR, Rosner IL, Cullen J, Chen Y, Brand T, Brassell SA, et al. Elevated alkaline phosphatase velocity strongly predicts overall survival and the risk of bone metastases in castrate-resistant prostate cancer. Urol Oncol. 2014;32(6):761–8.
25. Michaelson MD, Marujo RM, Smith MR. Contribution of androgen deprivation therapy to elevated osteoclast activity in men with metastatic prostate cancer. Clin Cancer Res. 2004;10(8):2705–8.
26. NCCN clinical practice guidelines in oncology: prostate cancer. Version 4.2024.
27. Lecouvet FE, El Mouedden J, Collette L, Coche E, Danse E, Jamar F, et al. Can whole-body magnetic resonance imaging with diffusion-weighted imaging replace Tc 99m bone scanning and computed tomography for single-step detection of metastases in patients with high-risk prostate cancer? Eur Urol. 2012;62(1):68–75.

28. Emmett L, Metser U, Bauman G, Hicks RJ, Weickhardt A, Davis ID, et al. Prospective, multisite, international comparison of (18)F-fluoromethylcholine PET/CT, multiparametric MRI, and (68)Ga-HBED-CC PSMA-11 PET/CT in men with high-risk features and biochemical failure after radical prostatectomy: clinical performance and patient outcomes. J Nucl Med. 2019;60(6):794–800.
29. Afshar-Oromieh A, Zechmann CM, Malcher A, Eder M, Eisenhut M, Linhart HG, et al. Comparison of PET imaging with a (68)Ga-labelled PSMA ligand and (18)F-choline-based PET/CT for the diagnosis of recurrent prostate cancer. Eur J Nucl Med Mol Imaging. 2014;41(1):11–20.
30. Morigi JJ, Stricker PD, van Leeuwen PJ, Tang R, Ho B, Nguyen Q, et al. Prospective comparison of 18F-fluoromethylcholine versus 68Ga-PSMA PET/CT in prostate cancer patients who have rising PSA after curative treatment and are being considered for targeted therapy. J Nucl Med. 2015;56(8):1185–90.
31. Zhou J, Wu R, Wang W, Zhao Y, Liu X. (68)Ga-PSMA PET/CT for the evaluation of metastasis in patients with prostate cancer: a systematic review and meta-analysis. Hell J Nucl Med. 2022;25(3):297–311.
32. Jadvar H, Calais J, Fanti S, Feng F, Greene KL, Gulley JL, et al. Appropriate use criteria for prostate-specific membrane antigen PET imaging. J Nucl Med. 2022;63(1):59–68.
33. Tsechelidis I, Vrachimis A. PSMA PET in imaging prostate cancer. Front Oncol. 2022;12:831429.
34. Snow HA, Hofman MS, Mitchell CA, Gyorki DE, Smith MJF. Incidental metastatic melanoma identified on 68Ga-prostate-specific membrane antigen PET/CT for metastatic prostate cancer. Clin Nucl Med. 2018;43(7):509–11.
35. Sager MS, Bilgic S, Uslu L, Asa S, Sager G, Sonmezoglu K. Active giant cell vasculitis diagnosis with (68)Ga PSMA PET/CT imaging. Mol Imaging Radionucl Ther. 2021;30(1):57–9.
36. Aggarwal P, Ambalavanan N, Sood A, Gupta K, Dahiya D, Mittal BR. Prostate-specific membrane antigen-avid neurofibroma mimicking cutaneous metastasis in metastatic castration-resistant prostate cancer on 18 F-PSMA-1007 PET/CT. Clin Nucl Med. 2024;49(6):e286–e7.
37. Filizoglu N, Oksuzoglu K, Ozguven S. Distinguishing meningioma from metastasis of prostate cancer on 68Ga-PSMA PET/CT. Clin Nucl Med. 2021;46(11):e553–e5.
38. Merrild EH, Baerentzen S, Bouchelouche K, Buus S. Vertebral myeloma mimicking prostatic carcinoma metastasis in 68Ga-PSMA PET/CT. Clin Nucl Med. 2017;42(10):790–2.
39. Sailer V, Schiffman MH, Kossai M, Cyrta J, Beg S, Sullivan B, et al. Bone biopsy protocol for advanced prostate cancer in the era of precision medicine. Cancer. 2018;124(5):1008–15.
40. Ross RW, Halabi S, Ou SS, Rajeshkumar BR, Woda BA, Vogelzang NJ, et al. Predictors of prostate cancer tissue acquisition by an undirected core bone marrow biopsy in metastatic castration-resistant prostate cancer--a Cancer and Leukemia Group B study. Clin Cancer Res. 2005;11(22):8109–13.
41. Zheng G, Lin MT, Lokhandwala PM, Beierl K, Netto GJ, Gocke CD, et al. Clinical mutational profiling of bone metastases of lung and colon carcinoma and malignant melanoma using next-generation sequencing. Cancer Cytopathol. 2016;124(10):744–53.
42. Epstein JI, Egevad L, Amin MB, Delahunt B, Srigley JR, Humphrey PA, et al. The 2014 International Society of Urological Pathology (ISUP) consensus conference on Gleason grading of prostatic carcinoma: definition of grading patterns and proposal for a new grading system. Am J Surg Pathol. 2016;40(2):244–52.
43. Epstein JI, Zelefsky MJ, Sjoberg DD, Nelson JB, Egevad L, Magi-Galluzzi C, et al. A contemporary prostate cancer grading system: a validated alternative to the Gleason score. Eur Urol. 2016;69(3):428–35.
44. Loeb S, Folkvaljon Y, Robinson D, Lissbrant IF, Egevad L, Stattin P. Evaluation of the 2015 Gleason grade groups in a nationwide population-based cohort. Eur Urol. 2016;69(6):1135–41.
45. Ham WS, Chalfin HJ, Feng Z, Trock BJ, Epstein JI, Cheung C, et al. New prostate cancer grading system predicts long-term survival following surgery for Gleason score 8-10 prostate cancer. Eur Urol. 2017;71(6):907–12.
46. Gravis G. Systemic treatment for metastatic prostate cancer. Asian J Urol. 2019;6(2):162–8.

47. Seidenfeld J, Samson DJ, Hasselblad V, Aronson N, Albertsen PC, Bennett CL, et al. Single-therapy androgen suppression in men with advanced prostate cancer: a systematic review and meta-analysis. Ann Intern Med. 2000;132(7):566–77.
48. Fizazi K, Tran N, Fein L, Matsubara N, Rodriguez-Antolin A, Alekseev BY, et al. Abiraterone plus prednisone in metastatic, castration-sensitive prostate cancer. N Engl J Med. 2017;377(4):352–60.
49. Fizazi K, Tran N, Fein L, Matsubara N, Rodriguez-Antolin A, Alekseev BY, et al. Abiraterone acetate plus prednisone in patients with newly diagnosed high-risk metastatic castration-sensitive prostate cancer (LATITUDE): final overall survival analysis of a randomised, double-blind, phase 3 trial. Lancet Oncol. 2019;20(5):686–700.
50. James ND, de Bono JS, Spears MR, Clarke NW, Mason MD, Dearnaley DP, et al. Abiraterone for prostate cancer not previously treated with hormone therapy. N Engl J Med. 2017;377(4):338–51.
51. Chi KN, Agarwal N, Bjartell A, Chung BH, de Santana P, Gomes AJ, Given R, et al. Apalutamide for metastatic, castration-sensitive prostate cancer. N Engl J Med. 2019;381(1):13–24.
52. Chi KN, Chowdhury S, Bjartell A, Chung BH, de Santana P, Gomes AJ, Given R, et al. Apalutamide in patients with metastatic castration-sensitive prostate cancer: final survival analysis of the randomized, double-blind, phase III TITAN study. J Clin Oncol. 2021;39(20):2294–303.
53. Davis ID, Martin AJ, Stockler MR, Begbie S, Chi KN, Chowdhury S, et al. Enzalutamide with standard first-line therapy in metastatic prostate cancer. N Engl J Med. 2019;381(2):121–31.
54. Armstrong AJ, Szmulewitz RZ, Petrylak DP, Holzbeierlein J, Villers A, Azad A, et al. ARCHES: a randomized, phase III study of androgen deprivation therapy with enzalutamide or placebo in men with metastatic hormone-sensitive prostate cancer. J Clin Oncol. 2019;37(32):2974–86.
55. Saad F, Vjaters E, Shore N, Olmos D, Xing N, de Santana P, Gomes AJ, et al. Darolutamide in combination with androgen-deprivation therapy in patients with metastatic hormone-sensitive prostate cancer from the phase III ARANOTE trial. J Clin Oncol. 2024:JCO2401798.
56. Ali A, Hoyle A, Haran AM, Brawley CD, Cook A, Amos C, et al. Association of Bone Metastatic Burden with survival benefit from prostate radiotherapy in patients with newly diagnosed metastatic prostate cancer: a secondary analysis of a randomized clinical trial. JAMA Oncol. 2021;7(4):555–63.
57. Fizazi K, Foulon S, Carles J, Roubaud G, McDermott R, Flechon A, et al. Abiraterone plus prednisone added to androgen deprivation therapy and docetaxel in de novo metastatic castration-sensitive prostate cancer (PEACE-1): a multicentre, open-label, randomised, phase 3 study with a 2 x 2 factorial design. Lancet. 2022;399(10336):1695–707.
58. Smith MR, Hussain M, Saad F, Fizazi K, Sternberg CN, Crawford ED, et al. Darolutamide and survival in metastatic, hormone-sensitive prostate cancer. N Engl J Med. 2022;386(12):1132–42.
59. Hussain M, Tangen CM, Higano C, Schelhammer PF, Faulkner J, Crawford ED, et al. Absolute prostate-specific antigen value after androgen deprivation is a strong independent predictor of survival in new metastatic prostate cancer: data from Southwest Oncology Group Trial 9346 (INT-0162). J Clin Oncol. 2006;24(24):3984–90.
60. Labrie F, Dupont A, Belanger A, Lachance R. Flutamide eliminates the risk of disease flare in prostatic cancer patients treated with a luteinizing hormone-releasing hormone agonist. J Urol. 1987;138(4):804–6.
61. Duchesne GM, Woo HH, Bassett JK, Bowe SJ, D'Este C, Frydenberg M, et al. Timing of androgen-deprivation therapy in patients with prostate cancer with a rising PSA (TROG 03.06 and VCOG PR 01-03 [TOAD]): a randomised, multicentre, non-blinded, phase 3 trial. Lancet Oncol. 2016;17(6):727–37.
62. Sun M, Choueiri TK, Hamnvik OP, Preston MA, De Velasco G, Jiang W, et al. Comparison of gonadotropin-releasing hormone agonists and orchiectomy: effects of androgen-deprivation therapy. JAMA Oncol. 2016;2(4):500–7.

63. Akhoundova D, Feng FY, Pritchard CC, Rubin MA. Molecular genetics of prostate cancer and role of genomic testing. Surg Pathol Clin. 2022;15(4):617–28.
64. Hussaini A, Olszanski AJ, Stein CA, Bosch B, Nemeth P. Impact of an alternative steroid on the relative bioavailability and bioequivalence of a novel versus the originator formulation of abiraterone acetate. Cancer Chemother Pharmacol. 2017;80(3):479–86.
65. Beer TM, Armstrong AJ, Rathkopf DE, Loriot Y, Sternberg CN, Higano CS, et al. Enzalutamide in metastatic prostate cancer before chemotherapy. N Engl J Med. 2014;371(5):424–33.
66. Beer TM, Armstrong AJ, Rathkopf D, Loriot Y, Sternberg CN, Higano CS, et al. Enzalutamide in men with chemotherapy-naive metastatic castration-resistant prostate cancer: extended analysis of the phase 3 PREVAIL study. Eur Urol. 2017;71(2):151–4.
67. Tannock IF, de Wit R, Berry WR, Horti J, Pluzanska A, Chi KN, et al. Docetaxel plus prednisone or mitoxantrone plus prednisone for advanced prostate cancer. N Engl J Med. 2004;351(15):1502–12.
68. Berthold DR, Pond GR, Soban F, de Wit R, Eisenberger M, Tannock IF. Docetaxel plus prednisone or mitoxantrone plus prednisone for advanced prostate cancer: updated survival in the TAX 327 study. J Clin Oncol. 2008;26(2):242–5.
69. Petrylak DP, Tangen CM, Hussain MH, Lara PN Jr, Jones JA, Taplin ME, et al. Docetaxel and estramustine compared with mitoxantrone and prednisone for advanced refractory prostate cancer. N Engl J Med. 2004;351(15):1513–20.
70. Fizazi K, Scher HI, Molina A, Logothetis CJ, Chi KN, Jones RJ, et al. Abiraterone acetate for treatment of metastatic castration-resistant prostate cancer: final overall survival analysis of the COU-AA-301 randomised, double-blind, placebo-controlled phase 3 study. Lancet Oncol. 2012;13(10):983–92.
71. Logothetis CJ, Basch E, Molina A, Fizazi K, North SA, Chi KN, et al. Effect of abiraterone acetate and prednisone compared with placebo and prednisone on pain control and skeletal-related events in patients with metastatic castration-resistant prostate cancer: exploratory analysis of data from the COU-AA-301 randomised trial. Lancet Oncol. 2012;13(12):1210–7.
72. Fizazi K, Scher HI, Miller K, Basch E, Sternberg CN, Cella D, et al. Effect of enzalutamide on time to first skeletal-related event, pain, and quality of life in men with castration-resistant prostate cancer: results from the randomised, phase 3 AFFIRM trial. Lancet Oncol. 2014;15(10):1147–56.
73. Scher HI, Fizazi K, Saad F, Taplin ME, Sternberg CN, Miller K, et al. Increased survival with enzalutamide in prostate cancer after chemotherapy. N Engl J Med. 2012;367(13):1187–97.
74. de Bono JS, Oudard S, Ozguroglu M, Hansen S, Machiels JP, Kocak I, et al. Prednisone plus cabazitaxel or mitoxantrone for metastatic castration-resistant prostate cancer progressing after docetaxel treatment: a randomised open-label trial. Lancet. 2010;376(9747):1147–54.
75. de Wit R, de Bono J, Sternberg CN, Fizazi K, Tombal B, Wulfing C, et al. Cabazitaxel versus abiraterone or enzalutamide in metastatic prostate cancer. N Engl J Med. 2019;381(26):2506–18.
76. Fizazi K, Kramer G, Eymard JC, Sternberg CN, de Bono J, Castellano D, et al. Quality of life in patients with metastatic prostate cancer following treatment with cabazitaxel versus abiraterone or enzalutamide (CARD): an analysis of a randomised, multicentre, open-label, phase 4 study. Lancet Oncol. 2020;21(11):1513–25.
77. Mateo J, Carreira S, Sandhu S, Miranda S, Mossop H, Perez-Lopez R, et al. DNA-repair defects and olaparib in metastatic prostate cancer. N Engl J Med. 2015;373(18):1697–708.
78. Clarke N, Wiechno P, Alekseev B, Sala N, Jones R, Kocak I, et al. Olaparib combined with abiraterone in patients with metastatic castration-resistant prostate cancer: a randomised, double-blind, placebo-controlled, phase 2 trial. Lancet Oncol. 2018;19(7):975–86.
79. Mateo J, Porta N, Bianchini D, McGovern U, Elliott T, Jones R, et al. Olaparib in patients with metastatic castration-resistant prostate cancer with DNA repair gene aberrations (TOPARP-B): a multicentre, open-label, randomised, phase 2 trial. Lancet Oncol. 2020;21(1):162–74.
80. de Bono J, Mateo J, Fizazi K, Saad F, Shore N, Sandhu S, et al. Olaparib for metastatic castration-resistant prostate cancer. N Engl J Med. 2020;382(22):2091–102.

81. Agarwal N, Azad AA, Carles J, Fay AP, Matsubara N, Heinrich D, et al. Talazoparib plus enzalutamide in men with first-line metastatic castration-resistant prostate cancer (TALAPRO-2): a randomised, placebo-controlled, phase 3 trial. Lancet. 2023;402(10398):291–303.
82. Chi KN, Rathkopf D, Smith MR, Efstathiou E, Attard G, Olmos D, et al. Niraparib and abiraterone acetate for metastatic castration-resistant prostate cancer. J Clin Oncol. 2023;41(18):3339–51.
83. Chi KN, Sandhu S, Smith MR, Attard G, Saad M, Olmos D, et al. Niraparib plus abiraterone acetate with prednisone in patients with metastatic castration-resistant prostate cancer and homologous recombination repair gene alterations: second interim analysis of the randomized phase III MAGNITUDE trial. Ann Oncol. 2023;34(9):772–82.
84. Kantoff PW, Higano CS, Shore ND, Berger ER, Small EJ, Penson DF, et al. Sipuleucel-T immunotherapy for castration-resistant prostate cancer. N Engl J Med. 2010;363(5):411–22.
85. Hansen AR, Massard C, Ott PA, Haas NB, Lopez JS, Ejadi S, et al. Pembrolizumab for advanced prostate adenocarcinoma: findings of the KEYNOTE-028 study. Ann Oncol. 2018;29(8):1807–13.
86. Tucker MD, Zhu J, Marin D, Gupta RT, Gupta S, Berry WR, et al. Pembrolizumab in men with heavily treated metastatic castrate-resistant prostate cancer. Cancer Med. 2019;8(10):4644–55.
87. Parker C, Nilsson S, Heinrich D, Helle SI, O'Sullivan JM, Fossa SD, et al. Alpha emitter radium-223 and survival in metastatic prostate cancer. N Engl J Med. 2013;369(3):213–23.
88. Sartor O, de Bono J, Chi KN, Fizazi K, Herrmann K, Rahbar K, et al. Lutetium-177-PSMA-617 for metastatic castration-resistant prostate cancer. N Engl J Med. 2021;385(12):1091–103.
89. Saad F. Bone-directed treatments for prostate cancer. Hematol Oncol Clin North Am. 2006;20(4):947–63.
90. Coleman RE. Clinical features of metastatic bone disease and risk of skeletal morbidity. Clin Cancer Res. 2006;12(20 Pt 2):6243s–9s.
91. Tazi H, Manunta A, Rodriguez A, Patard JJ, Lobel B, Guille F. Spinal cord compression in metastatic prostate cancer. Eur Urol. 2003;44(5):527–32.
92. Maranzano E, Latini P, Checcaglini F, Ricci S, Panizza BM, Aristei C, et al. Radiation therapy in metastatic spinal cord compression. A prospective analysis of 105 consecutive patients. Cancer. 1991;67(5):1311–7.
93. Lin JH. Bisphosphonates: a review of their pharmacokinetic properties. Bone. 1996;18(2):75–85.
94. Neville-Webbe HL, Coleman RE. Bisphosphonates and RANK ligand inhibitors for the treatment and prevention of metastatic bone disease. Eur J Cancer. 2010;46(7):1211–22.
95. Saylor PJ, Rumble RB, Tagawa S, Eastham JA, Finelli A, Reddy PS, et al. Bone health and bone-targeted therapies for prostate cancer: ASCO Endorsement of a Cancer Care Ontario Guideline. J Clin Oncol. 2020;38(15):1736–43.
96. Smith MR, Kabbinavar F, Saad F, Hussain A, Gittelman MC, Bilhartz DL, et al. Natural history of rising serum prostate-specific antigen in men with castrate nonmetastatic prostate cancer. J Clin Oncol. 2005;23(13):2918–25.
97. Wirth M, et al. Prevention of bone metastases in patients with high-risk nonmetastatic prostate cancer treated with zoledronic acid: efficacy and safety results of the Zometa European Study (ZEUS). Eur Urol. 2014; https://doi.org/10.1016/j.eururo.2014.02.014.
98. Saad F, Gleason DM, Murray R, Tchekmedyian S, Venner P, Lacombe L, et al. A randomized, placebo-controlled trial of zoledronic acid in patients with hormone-refractory metastatic prostate carcinoma. J Natl Cancer Inst. 2002;94(19):1458–68.
99. Gul G, Sendur MA, Aksoy S, Sever AR, Altundag K. A comprehensive review of denosumab for bone metastasis in patients with solid tumors. Curr Med Res Opin. 2016;32(1):133–45.
100. Smith MR, Saad F, Egerdie B, Szwedowski M, Tammela TL, Ke C, et al. Effects of denosumab on bone mineral density in men receiving androgen deprivation therapy for prostate cancer. J Urol. 2009;182(6):2670–5.

101. Smith MR, Saad F, Coleman R, Shore N, Fizazi K, Tombal B, et al. Denosumab and bone-metastasis-free survival in men with castration-resistant prostate cancer: results of a phase 3, randomised, placebo-controlled trial. Lancet. 2012;379(9810):39–46.
102. Dougall WC, Chaisson M. The RANK/RANKL/OPG triad in cancer-induced bone diseases. Cancer Metastasis Rev. 2006;25(4):541–9.
103. Zhang J, Dai J, Qi Y, Lin DL, Smith P, Strayhorn C, et al. Osteoprotegerin inhibits prostate cancer-induced osteoclastogenesis and prevents prostate tumor growth in the bone. J Clin Invest. 2001;107(10):1235–44.
104. Guise TA, Mohammad KS, Clines G, Stebbins EG, Wong DH, Higgins LS, et al. Basic mechanisms responsible for osteolytic and osteoblastic bone metastases. Clin Cancer Res. 2006;12(20 Pt 2):6213s–6s.
105. Fizazi K, Carducci M, Smith M, Damiao R, Brown J, Karsh L, et al. Denosumab versus zoledronic acid for treatment of bone metastases in men with castration-resistant prostate cancer: a randomised, double-blind study. Lancet. 2011;377(9768):813–22.
106. Paller CJ, Carducci MA, Philips GK. Management of bone metastases in refractory prostate cancer--role of denosumab. Clin Interv Aging. 2012;7:363–72.
107. Henry DH, Costa L, Goldwasser F, Hirsh V, Hungria V, Prausova J, et al. Randomized, double-blind study of denosumab versus zoledronic acid in the treatment of bone metastases in patients with advanced cancer (excluding breast and prostate cancer) or multiple myeloma. J Clin Oncol. 2011;29(9):1125–32.
108. Stopeck AT, Lipton A, Body JJ, Steger GG, Tonkin K, de Boer RH, et al. Denosumab compared with zoledronic acid for the treatment of bone metastases in patients with advanced breast cancer: a randomized, double-blind study. J Clin Oncol. 2010;28(35):5132–9.
109. Finianos A, Aragon-Ching JB. Zoledronic acid for the treatment of prostate cancer. Expert Opin Pharmacother. 2019;20(6):657–66.
110. Body JJ, Bone HG, de Boer RH, Stopeck A, Van Poznak C, Damiao R, et al. Hypocalcaemia in patients with metastatic bone disease treated with denosumab. Eur J Cancer. 2015;51(13):1812–21.
111. Yarom N, Shapiro CL, Peterson DE, Van Poznak CH, Bohlke K, Ruggiero SL, et al. Medication-related osteonecrosis of the jaw: MASCC/ISOO/ASCO clinical practice guideline. J Clin Oncol. 2019;37(25):2270–90.
112. Goyal J, Antonarakis ES. Bone-targeting radiopharmaceuticals for the treatment of prostate cancer with bone metastases. Cancer Lett. 2012;323(2):135–46.
113. Terrisse S, Karamouza E, Parker CC, Sartor AO, James ND, Pirrie S, et al. Overall survival in men with bone metastases from castration-resistant prostate cancer treated with bone-targeting radioisotopes: a meta-analysis of individual patient data from randomized clinical trials. JAMA Oncol. 2020;6(2):206–16.
114. McHugh D, Tagawa S, Moryl N, Milowsky M, Heller G, Osborne J, et al. A phase II, non-randomized open trial assessing pain efficacy with Radium-223 in symptomatic metastatic castration-resistant prostate cancer. Clin Genitourin Cancer. 2021;19(5):447–56.
115. Porter AT, McEwan AJ, Powe JE, Reid R, McGowan DG, Lukka H, et al. Results of a randomized phase-III trial to evaluate the efficacy of strontium-89 adjuvant to local field external beam irradiation in the management of endocrine resistant metastatic prostate cancer. Int J Radiat Oncol Biol Phys. 1993;25(5):805–13.
116. Liepe K, Kotzerke J. A comparative study of 188Re-HEDP, 186Re-HEDP, 153Sm-EDTMP and 89Sr in the treatment of painful skeletal metastases. Nucl Med Commun. 2007;28(8):623–30.
117. Sartor O. Overview of samarium sm 153 lexidronam in the treatment of painful metastatic bone disease. Rev Urol. 2004;6 Suppl 10(Suppl 10):S3–S12.
118. Serafini AN, Houston SJ, Resche I, Quick DP, Grund FM, Ell PJ, et al. Palliation of pain associated with metastatic bone cancer using samarium-153 lexidronam: a double-blind placebo-controlled clinical trial. J Clin Oncol. 1998;16(4):1574–81.
119. Etchebehere EC, Pereira Neto CA, Lima MC, Santos Ade O, Ramos CD, Silva CM, et al. Treatment of bone pain secondary to metastases using samarium-153-EDTMP. Sao Paulo Med J. 2004;122(5):208–12.

120. Broder MS, Gutierrez B, Cherepanov D, Linhares Y. Burden of skeletal-related events in prostate cancer: unmet need in pain improvement. Support Care Cancer. 2015;23(1):237–47.
121. Kane CM, Hoskin P, Bennett MI. Cancer induced bone pain. BMJ. 2015;350:h315.
122. Kaya E, Feuer D. Prostate cancer: palliative care and pain relief. Prostate Cancer Prostatic Dis. 2004;7(4):311–5.
123. Thompson JC, Wood J, Feuer D. Prostate cancer: palliative care and pain relief. Br Med Bull. 2007;83:341–54.
124. Caraceni A, Hanks G, Kaasa S, Bennett MI, Brunelli C, Cherny N, et al. Use of opioid analgesics in the treatment of cancer pain: evidence-based recommendations from the EAPC. Lancet Oncol. 2012;13(2):e58–68.
125. Rabow MW, Lee MX. Palliative care in castrate-resistant prostate cancer. Urol Clin North Am. 2012;39(4):491–503.
126. Eisenberg E, Berkey CS, Carr DB, Mosteller F, Chalmers TC. Efficacy and safety of nonsteroidal antiinflammatory drugs for cancer pain: a meta-analysis. J Clin Oncol. 1994;12(12):2756–65.
127. Nabal M, Librada S, Redondo MJ, Pigni A, Brunelli C, Caraceni A. The role of paracetamol and nonsteroidal anti-inflammatory drugs in addition to WHO Step III opioids in the control of pain in advanced cancer. A systematic review of the literature. Palliat Med. 2012;26(4):305–12.
128. Huang R, Jiang L, Cao Y, Liu H, Ping M, Li W, et al. Comparative efficacy of therapeutics for chronic cancer pain: a Bayesian network meta-analysis. J Clin Oncol. 2019;37(20):1742–52.
129. Caraceni A, Zecca E, Bonezzi C, Arcuri E, Yaya Tur R, Maltoni M, et al. Gabapentin for neuropathic cancer pain: a randomized controlled trial from the Gabapentin Cancer Pain Study Group. J Clin Oncol. 2004;22(14):2909–17.
130. Haywood A, Good P, Khan S, Leupp A, Jenkins-Marsh S, Rickett K, et al. Corticosteroids for the management of cancer-related pain in adults. Cochrane Database Syst Rev. 2015;2015(4):CD010756.
131. Karst M, Salim K, Burstein S, Conrad I, Hoy L, Schneider U. Analgesic effect of the synthetic cannabinoid CT-3 on chronic neuropathic pain: a randomized controlled trial. JAMA. 2003;290(13):1757–62.
132. Johnson JR, Burnell-Nugent M, Lossignol D, Ganae-Motan ED, Potts R, Fallon MT. Multicenter, double-blind, randomized, placebo-controlled, parallel-group study of the efficacy, safety, and tolerability of THC:CBD extract and THC extract in patients with intractable cancer-related pain. J Pain Symptom Manag. 2010;39(2):167–79.
133. Cormie P, Newton RU, Spry N, Joseph D, Taaffe DR, Galvao DA. Safety and efficacy of resistance exercise in prostate cancer patients with bone metastases. Prostate Cancer Prostatic Dis. 2015;18(2):196.
134. Galvao DA, Taaffe DR, Cormie P, Spry N, Chambers SK, Peddle-McIntyre C, et al. Efficacy and safety of a modular multi-modal exercise program in prostate cancer patients with bone metastases: a randomized controlled trial. BMC Cancer. 2011;11:517.
135. Galvao DA, Taaffe DR, Spry N, Cormie P, Joseph D, Chambers SK, et al. Exercise preserves physical function in prostate cancer patients with bone metastases. Med Sci Sports Exerc. 2018;50(3):393–9.
136. De Vincentis G, Frantellizzi V, Follacchio GA, Farcomeni A, Pani A, Samaritani R, et al. No evidence of association between psychological distress and pain relief in patients with bone metastases from castration-resistant prostate cancer treated with 223Radium. Eur J Cancer Care (Engl). 2019;28(5):e13112.
137. Colloca G, Colloca P. The effects of social support on health-related quality of life of patients with metastatic prostate cancer. J Cancer Educ. 2016;31(2):244–52.
138. Lin PH, Aronson W, Freedland SJ. Nutrition, dietary interventions and prostate cancer: the latest evidence. BMC Med. 2015;13:3.
139. Hackshaw-McGeagh LE, Perry RE, Leach VA, Qandil S, Jeffreys M, Martin RM, et al. A systematic review of dietary, nutritional, and physical activity interventions for the prevention of prostate cancer progression and mortality. Cancer Causes Control. 2015;26(11):1521–50.

140. Gandaglia G, Karakiewicz PI, Briganti A, Passoni NM, Schiffmann J, Trudeau V, et al. Impact of the site of metastases on survival in patients with metastatic prostate cancer. Eur Urol. 2015;68(2):325–34.
141. Halabi S, Kelly WK, Ma H, Zhou H, Solomon NC, Fizazi K, et al. Meta-analysis evaluating the impact of site of metastasis on overall survival in men with castration-resistant prostate cancer. J Clin Oncol. 2016;34(14):1652–9.
142. Scher HI, Halabi S, Tannock I, Morris M, Sternberg CN, Carducci MA, et al. Design and end points of clinical trials for patients with progressive prostate cancer and castrate levels of testosterone: recommendations of the prostate cancer clinical trials working group. J Clin Oncol. 2008;26(7):1148–59.
143. Ryan CJ, Smith MR, de Bono JS, Molina A, Logothetis CJ, de Souza P, et al. Abiraterone in metastatic prostate cancer without previous chemotherapy. N Engl J Med. 2013;368(2):138–48.
144. Fizazi K, Jones R, Oudard S, Efstathiou E, Saad F, de Wit R, et al. Phase III, randomized, double-blind, multicenter trial comparing orteronel (TAK-700) plus prednisone with placebo plus prednisone in patients with metastatic castration-resistant prostate cancer that has progressed during or after docetaxel-based therapy: ELM-PC 5. J Clin Oncol. 2015;33(7):723–31.
145. Fang J, Xu Q. Differences of osteoblastic bone metastases and osteolytic bone metastases in clinical features and molecular characteristics. Clin Transl Oncol. 2015;17(3):173–9.
146. Baldessari C, Pipitone S, Molinaro E, Cerma K, Fanelli M, Nasso C, et al. Bone metastases and health in prostate cancer: from pathophysiology to clinical implications. Cancers (Basel). 2023;15(5):1518.
147. McKay R, Haider B, Duh MS, Valderrama A, Nakabayashi M, Fiorillo M, et al. Impact of symptomatic skeletal events on health-care resource utilization and quality of life among patients with castration-resistant prostate cancer and bone metastases. Prostate Cancer Prostatic Dis. 2017;20(3):276–82.
148. Hagiwara M, Delea TE, Saville MW, Chung K. Healthcare utilization and costs associated with skeletal-related events in prostate cancer patients with bone metastases. Prostate Cancer Prostatic Dis. 2013;16(1):23–7.
149. EORTC Quality of Life Questionnaire. https://qol.eortc.org/questionnaire/eortc-qlq-c30/. Accessed on 10/21/24.
150. Functional Assessment of Cancer Therapy - Prostate. https://www.facit.org/measures/fact-p. Accessed on 10/21/24.
151. Brief Pain Inventory (BPI) – Symptom Research. Available at: https://www.mdanderson.org/research/departments-labs-institutes/departments-divisions/symptom-research/symptom-assessment-tools/brief-pain-inventory.html. Accessed October 26, 2024.
152. Weinfurt KP, Li Y, Castel LD, Saad F, Timbie JW, Glendenning GA, et al. The significance of skeletal-related events for the health-related quality of life of patients with metastatic prostate cancer. Ann Oncol. 2005;16(4):579–84.

Medical Management of Musculoskeletal Metastasis from Lung Cancer

27

Gavin Wayne Stoel, Erik Geiger, and Estelamari Rodriguez

Introduction

Lung cancer is the leading cause of global cancer incidence and mortality, accounting for an estimated 2.4 million diagnoses and 1.8 million deaths in 2022 [1]. Neoplasms of the lungs are the main cause of cancer death and the second most common cancer diagnosis in men and women (after prostate and breast cancer, respectively). Since 1990, the overall mortality rate of lung cancer has been trending down due to tobacco cessation efforts and significant advances in new immunotherapy and targeted therapies for lung cancer. Overall, 5-year survival from lung cancer has doubled, with a 5-year survival of 25% in the United States [2].

Several risk factors are associated with the development of lung cancer, with smoking being the most significant, accounting for approximately 85% of diagnoses. Other risk factors for lung cancer include radon gas, environmental exposure (asbestos), and long-term exposure to high levels of air pollution, particularly fine particulate matter (PM2.5). Family history also increases risk by 1.7-fold, with a greater risk among first-degree relatives. Despite tobacco exposure being a predominant risk factor, there is a rising trend worldwide of lung cancer diagnoses among never-smokers and young people. In the last 42 years, lung cancer rates in women have risen 84%, while diagnoses in men have decreased 36% [3]. Increased

G. W. Stoel
University of Miami Miller School of Medicine, Miami, FL, USA

E. Geiger
Department of Orthopaedic Surgery, Sylvester Comprehensive Cancer Center and the University of Miami Miller School of Medicine/Jackson Memorial Hospital, Miami, FL, USA

E. Rodriguez (✉)
Department of Medical Oncology, Sylvester Comprehensive Cancer Center, University of Miami Miller School of Medicine, Miami, FL, USA
e-mail: Estelarodriguez@miami.edu

J. Pretell-Mazzini, S. A. Lozano Calderón (eds.), *Multidisciplinary Approach of Musculoskeletal Metastasis*, https://doi.org/10.1007/978-3-032-09498-8_27

susceptibility to lung cancer in women has been attributed to higher proportions of epidermal growth factor receptor mutations and the effects of estrogen [4].

Diagnosis and Staging

Histology and Staging

The two main subtypes of lung cancer are non-small cell lung cancer (80%) and small cell lung cancer (20%). Adenocarcinoma is the most common histologic finding, (54.7%) followed by squamous cell (29.4%) in NSCLC. The proportion of stage I at diagnosis is higher among older patients, whereas the proportion of stage IV is higher among younger patients. Understanding both the histologic and molecular characteristics of NSCLC has important treatment implications. Approximately 50% of NSCLC harbor a molecular mutation that could be defined via tumor next-generation sequencing and, in some instances, circulating tumor DNA (ctDNA).

The current standard in advanced NSCLC is to perform biomarker testing at diagnosis to find specific treatment-related biomarkers like PDL1 expression and actionable mutations that could guide treatment options. Currently, there are ten FDA approved targeted therapies for NSCLC (Fig. 27.1). At the time of diagnosis, proper tissue collection from a core biopsy is favored for adequate molecular testing.

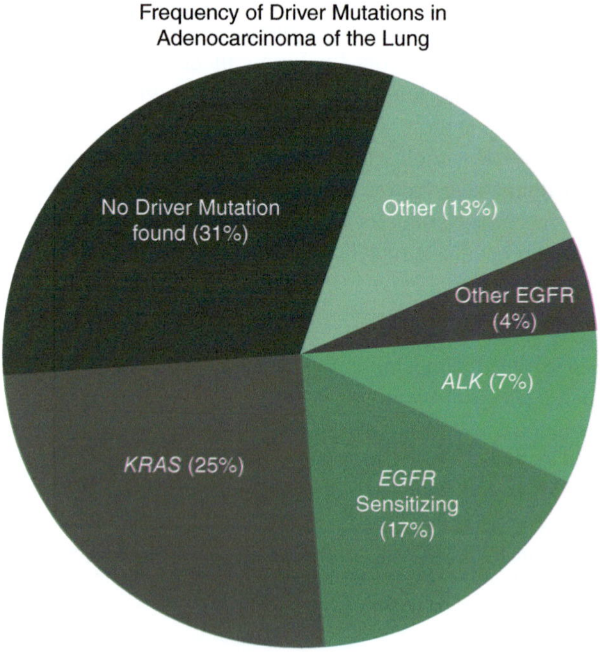

Fig. 27.1 Frequency of driver mutations in non-squamous NSCLC

Bone Metastases in Lung Cancer

Bone metastases are a common complication in advanced lung cancer, particularly non-small cell lung cancer (NSCLC). Approximately 30–40% of lung cancer patients experience bone metastases over the course of their disease. Skeletal metastases often affect the spine, pelvis, and long bones. The presence of bone metastases represents advanced-stage disease classified as Stage IV (M1b) disease in the NSCLC staging criteria. Bone metastases also occur frequently in Extensive Stage SCLC with a generally higher rate of progression. Bone metastases are associated with a significant increase in symptoms due to pain, fractures, spinal cord compression, and other skeletal-related events (SREs).

The incidence of bone metastases in lung cancer can also vary depending on the genetic mutations present in the tumor. Some mutations are associated with a higher likelihood of developing bone metastases, particularly in non-small cell lung cancer (NSCLC). Lung cancer patients with EGFR, ALK, ROS1, and HER2 mutations tend to have a higher likelihood of developing bone metastases, compared to those with KRAS or BRAF mutations.

Pathophysiology of Bone Metastases

Bone metastases in lung cancer occur when cancer cells spread through the bloodstream to the bone, disrupting the normal balance between bone formation and resorption. Tumor cells stimulate osteoclasts, the cells responsible for breaking down bone, leading to increased bone resorption and the formation of osteolytic lesions, which weaken the bone structure (Fig. 27.2). This imbalance can also result in the release of growth factors from the bone matrix that further promote tumor growth. The metastatic process often leads to complications, such as fractures, spinal cord compression, and hypercalcemia, causing significant pain and functional impairment. The destructive cycle of bone remodeling in metastases is driven by interactions between cancer cells, osteoclasts, and osteoblasts, contributing to the progression of skeletal-related events.

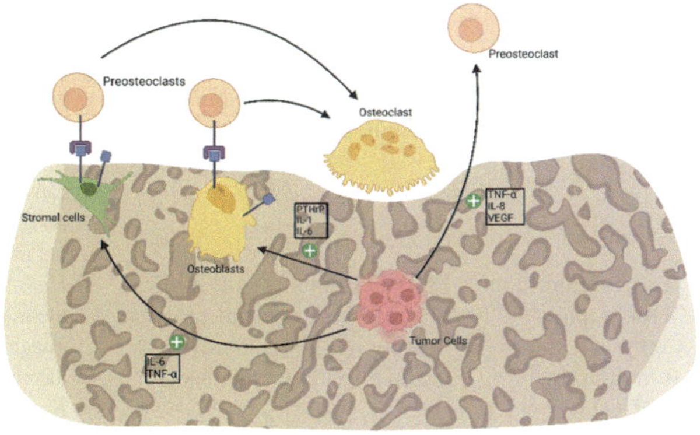

Fig. 27.2 Pathophysiology of bone metastases

Diagnostic Workup for Bone Metastases

The diagnostic workup for bone metastases typically involves a combination of imaging techniques and, in some cases, a biopsy. Each modality has its strengths, providing unique insights into the presence, extent, and characteristics of bone metastases as described in Table 27.1.

Table 27.1 Characteristics of different imaging modalities for skeletal bone metastases in lung cancer

Imaging Modality	Characteristics	Examples
X-rays	Useful for evaluating bone structure Low-cost modality for quick assessment of bone metastases	*Fractures:* X-rays can reveal pathological fractures caused by bone weakening from metastatic involvement. *Osteolytic lesions:* commonly seen in metastases from lung cancer, may appear as areas of bone destruction or thinning on X-rays. *Bone Integrity*: X-rays can assess the overall structural integrity of the bone and help in determining if surgical intervention is necessary.
Magnetic Resonance Imaging (MRI)	MRI is highly sensitive and provides detailed images of bone marrow, soft tissues, and spinal involvement.	*Spinal Metastases:* MRI is the gold standard for assessing metastases in the spine, detecting early-stage bone marrow infiltration, and evaluating the extent of spinal cord compression. *Soft Tissue Involvement:* MRI can also evaluate soft tissue components of metastatic lesions that may not be visible on X-rays or CT scans.
Computed Tomography (CT)	CT scans provide a more detailed view of bone structure and are especially helpful when evaluating complex bone anatomy or when there is a suspicion of bone destruction	*Cortical Bone Involvement:* CT is excellent for visualizing cortical bone integrity and can help plan surgical interventions. *Lesion Characterization:* CT can further characterize the lesion, distinguishing between osteolytic (bone-destroying) and osteoblastic (bone-forming) lesions
Bone Scan (Technetium-99m Scintigraphy	Bone scintigraphy is a nuclear medicine imaging technique used to detect metastatic spread throughout the skeleton	*Whole-Body Evaluation:* Bone scans provide an overview of skeletal involvement, identifying both symptomatic and asymptomatic sites of metastases. *Assessing Treatment Response*: Bone scans can help determine the extent of metastatic bone disease and treatment response. May not image lytic lesions well.
Positron Emission Tomography-Computed Tomography (PET-CT)	PET-CT combines metabolic imaging with anatomical detail, making it highly effective in evaluating metastatic disease.	*High Sensitivity:* PET-CT can detect metastases earlier than conventional imaging techniques by identifying areas of increased metabolic activity in the bone. *Extent of Metastases:* PET-CT provides a full-body assessment for staging and treatment response

Indications for Bone Biopsy

While imaging can usually confirm the presence of bone metastases, a biopsy may be necessary when there is diagnostic uncertainty, or when the pathology could influence treatment decisions (e.g., distinguishing between metastatic lesions and primary bone tumors). Bone biopsies can be done via a percutaneous needle biopsy or a surgical biopsy, guided by imaging (CT or MRI) to ensure accurate sampling of the lesion. The biopsy sample is analyzed to confirm the nature of the metastatic lesion, the histologic classification. Bone biopsies can also help determine if the primary cancer has a specific molecular target, such as EGFR mutations or ALK rearrangements in lung cancer. Molecular testing from bone specimens can be achieved as long as the tissue is not decalcified to preserve tissue DNA. Proper communication with the pathologic teams is needed to maintain tissue integrity for biomarker testing.

In conclusion, the diagnostic workup for bone metastases in lung cancer involves a combination of imaging modalities that provide structural, functional, and metabolic insights, along with biopsy when needed for pathological confirmation. This approach enables accurate diagnosis and guides appropriate treatment strategies.

Management

Systemic Therapies for Lung Cancer with Bone Metastases

Most patients with lung cancer present with advanced disease at the time of diagnosis. Thus, the goals of therapy are of palliative intent for most patients, and managing symptoms from metastatic disease is the primary goal of therapy. There are several important considerations for first-line treatment selection. First is to assess if the disease is broadly metastatic or oligometastatic, and second is to define the histology and biomarkers that can guide treatment selection.

Traditionally, age, burden of disease, ECOG performance status, and weight loss at diagnosis were exclusively predictive of survival. However, since the advancement of targeted therapies and immunotherapy for the treatment of lung cancer, these clinical characteristics are not entirely predictive of prognosis. Significant improvements in survival despite widespread metastatic disease have been seen in selected patients with biomarker-directed therapy and for those whose tumors are more susceptible to immune modulation. Figure 27.3 describes the current treatment algorithm based on histology, PD-L1, and biomarker expression. Actionable mutations are more commonly found in lung adenocarcinomas and patients with a never-smoking or limited-smoking history, but they could be seen in any patients regardless of histology and patient characteristics. This highlights the importance of proper tissue acquisition for molecular testing as part of the diagnostic workup.

Whether a patient has squamous or nonsquamous histology, the amount of PD-L1 expression plays an important role in guiding treatment. With advanced NSCLC expressing PD-L1 on >50% of cells, monotherapy with the PD-L1 inhibitors (i.e., pembrolizumab, cemiplimab, and atezolizumab) has demonstrated better

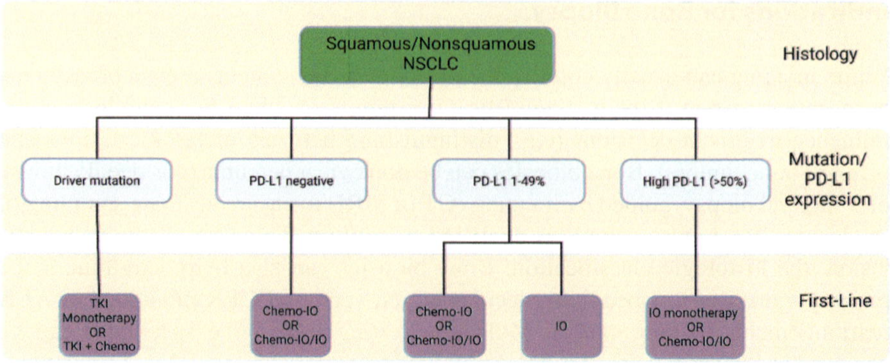

Fig. 27.3 Treatment algorithm for first-line treatment selection in NSCLC

survival outcomes and a more favorable side-effect profile compared to doublet platinum-based chemotherapy. In patients with PD-L1 1–49%, pembrolizumab monotherapy has been approved based on the Keynote 042 trial, although most patients can benefit from the addition of chemotherapy in this setting or treatment with double checkpoint inhibitors to overcome potential intrinsic tumor resistance to immunotherapy. Evidence suggests that combining immunotherapy with platinum-based chemotherapy for managing squamous cell carcinoma (SCC) of the lung with PD-L1 expression <50% may improve OS and PFS versus chemotherapy alone [5]. Table 27.2 describes various cancer therapies used in the management of NSCLC with associated toxicities.

Treatment of Biomarker-Driven NSCLC

A better understanding of the underlying genetic mutations that promote malignant transformation in NSCLC has been beneficial in producing agents that can target the altered protein products of these mutated genes. Some of the more commonly targeted driver mutations will be discussed here.

EGFR-positive non-small cell lung cancer (EGFR+ NSCLC) is a subtype of NSCLC characterized by mutations in the epidermal growth factor receptor (EGFR) gene. These mutations lead to abnormal activation of the EGFR protein, which is involved in cell survival, migration, and proliferation. EGFR mutations are most commonly found in adenocarcinoma, a subtype of NSCLC, and are more frequent in nonsmokers and women, although they can also be found in other demographic groups. The most common EGFR mutations are exon 19 deletions and exon 21 L858R point mutations, which together account for about 85–90% of all EGFR mutations. When EGFR is mutated in cancer, it is constitutively active, driving malignant transformation of tumor cells. EGFR tyrosine kinase inhibitors (EGFR TKIs) take advantage of this mutation by inhibiting EGFR tyrosine kinase and thus slowing growth. EGFR TKI monotherapy has been the preferred first-line option, but recently studies have shown that escalating treatment at diagnosis with the addition of chemotherapy or addition of amivantamab, a bispecific monoclonal antibody targeting the EGFR/MET

Table 27.2 Classification of lung cancer therapies with examples of specific agents

Drug category	Examples	Potential toxicities
Cytotoxic chemotherapy	Carboplatin, paclitaxel, pembrolizumab, gemcitabine	Anemia, thrombocytopenia Nausea, vomiting Renal failure
Immunomodulators	Immune Checkpoint Inhibitors (ICIs) PD1: pembrolizumab, atezolizumab, nivolumab CTLA4: ipilimumab, tremelimumab	Autoimmune conditions Fatigue Colitis Pneumonitis Myositis
Targeted antibodies	VEGF Inhibitors (bevacizumab, ramucirumab) Antibody-Drug Conjugates (trastuzumab deruxtecan)	Poor wound healing Hypertension Proteinuria Infusion reactions
Bispecific antibodies, including Bispecific T-cell engagers (BiTEs)	Amivantamab (Bispecific EGFR/MET) Tarlatamab (BiTE)	Infusion-related reactions Autoimmune syndromes Cytokine release syndrome ICANS
Targeted Tyrosine Kinase Inhibitors	Osimertinib, afatinib Lorlatinib, brigatinib	Drug-related rash, diarrhea QTc prolomgation Thrombotic events

pathways, can lead to improved disease-free survival in these patients [6, 7]. Treatment with EGFR targeted therapy is generally continued until progression of disease, after which a platinum-based chemotherapy or a therapy directed to the acquired molecular resistance pathway, if feasible. Patients with EGFR NSCLC who have oligo-progression (including isolated bone metastases) can be continued on treatment with the EGFR TKI in addition to local radiation [8].

ALK-positive non-small cell lung cancer (ALK+ NSCLC) refers to a subtype of non-small cell lung cancer (NSCLC) that is characterized by a genetic mutation involving the anaplastic lymphoma kinase (ALK) gene. This mutation results from a rearrangement or fusion of the ALK gene with another gene, most commonly EML4, leading to the formation of an abnormal ALK fusion protein. This protein drives cancer cell growth and proliferation. ALK+ NSCLC accounts for about 3–5% of all NSCLC cases and is more commonly seen in younger patients, non-smokers, or light smokers. The first-line treatment for an ALK-positive NSCLC is either a second (alectinib) or third-generation (lorlatinib) ALK tyrosine kinase inhibitor (TKI), which have been shown to improve survival and long-term disease control [9]. In the first-line trial of lorlatinib for ALK+ NSCLC, 60% of patients remained alive without disease progression after 5 years.

KRAS-positive non-small cell lung cancer (KRAS+ NSCLC) refers to a subtype of NSCLC in which mutations occur in the KRAS gene. The KRAS gene plays a key role in regulating cell growth and division by encoding a protein involved in signaling pathways. When mutated, KRAS leads to uncontrolled cell growth and tumor development. KRAS mutations are among the most common genetic alterations in NSCLC, accounting for approximately 25–30% of adenocarcinomas and are most often found in smokers with a high incidence of bone metastases. The management of KRAS+ NSCLC has evolved significantly with the advent of *KRAS G12C inhibitors* like sotorasib and adagrasib, providing targeted treatment options after chemo-immunotherapy for a previously difficult-to-treat subtype of NSCLC [10, 11]. For patients with other KRAS mutations or in cases of resistance, chemotherapy, immunotherapy and clinical trials remain essential components of treatment.

Management of Oligometastatic NSCLC

In non-small cell lung cancer (NSCLC), oligometastatic disease is often considered when there are *1 to 5 metastases* in distant organs (commonly brain, bone, liver, or adrenal glands). These metastases are amenable to *local therapy*, such as surgery or stereotactic body radiotherapy (SBRT). Oligometastatic lung cancer is increasingly seen as a distinct biological state where aggressive local therapies aimed at eradicating the limited metastases may lead to better outcomes, potentially increasing survival when combined with systemic treatments like chemotherapy, immunotherapy, or targeted therapies. In the management of NSCLC oligometastatic to the bone, it is important to tailor individual treatment based on the number of metastases and other factors, such as overall patient health, the primary tumor's response to therapy, and the location of the metastases. Multidisciplinary evaluation is crucial in deciding the extent of aggressive management versus a palliative approach. In summary, patients with oligometastatic NSCLC may undergo radical local therapies like SBRT or surgery for the primary tumor and metastases, in combination with systemic therapies like targeted therapy, chemotherapy, or immunotherapy.

Assessing Bone Metastases Treatment Response

The *RECIST (Response Evaluation Criteria in Solid Tumors)* guidelines are commonly used to assess tumor response to treatment, but they have limitations when evaluating bone metastases. Specifically, bone metastases without a measurable soft tissue component (purely osteoblastic lesions or sclerotic lesions) are generally considered non-measurable under RECIST. These lesions do not typically shrink or change in size in response to treatment, which makes them difficult to quantify using RECIST criteria. Bone lesions with a measurable soft tissue component can be evaluated for response according to RECIST criteria, as long as the soft tissue component is at least 10 mm (1 cm) in the longest diameter on CT or MRI component meets.

To evaluate bone metastasis response to treatment, other criteria such as the *MD Anderson Criteria* or the use of imaging techniques like *PET-CT* or *whole-body MRI* may be more appropriate, as they assess metabolic activity and bone marrow involvement rather than just changes in lesion size. The MDA criteria have been shown to predict survival outcome or failure of treatment earlier than the RECIST criteria [12].

Management of Skeletal Related Events (SREs)

In lung cancer, *skeletal-related events (SREs)* occur commonly as cancer spreads to the bones, leading to significant complications. The most frequent SREs include pathologic fractures, spinal cord compression, hypercalcemia of malignancy, and bone pain. In a prospective cohort of patients with advanced lung cancer, the incidence of bone metastases at initial diagnosis was 48% in stage IV NSCLC and 40% in extensive stage (ED)-SCLC. Forty-five percent of patients who developed bone metastases had SREs [13].

Surgical Management of Pathologic Fractures

Among patients with lung cancer and bone metastases, approximately 8–10% develop pathologic fractures [13]. This risk increases with longer survival, higher tumor burden, and specific sites of metastasis. Almost all patients who suffer from pathologic fractures, or patients deemed high risk for fracture, will receive surgery for stabilization. Fractures commonly occur in bones weakened by metastases, particularly in the spine, pelvis, and long bones (e.g., femur). Surgery for bone metastases in lung cancer is primarily palliative and aims to stabilize fractures, decompress the spinal cord, and alleviate pain, improving the patient's quality of life.

For most patients, *surgery is combined with radiotherapy and systemic therapies* (e.g., bisphosphonates, chemotherapy, targeted therapies, or immunotherapy) to address the overall disease and reduce the risk of skeletal-related events (SREs).

Which Medications Should Be Stopped Before Surgery?

Most systemic therapies can be continued during surgery or radiation to the bone except for vascular endothelial-growth factor (VEGF) monoclonal antibodies like bevacizumab and ramucirumab which should be held at least a month before surgery to avoid poor wound healing.

In general, ICI are believed to be safe in the perioperative period and for many different surgery types, with a low likelihood of grade III or IV Clavien-Dindo complications observed. They likely do not need to be stopped before surgery. In terms of perioperative considerations for the anesthesiologist and surgeon, preoperative assessment must include a review of cardiopulmonary status and inquiry into possible treatment-related toxicities, especially after ICI initiation or titration.

Which Patients Are Good Candidates for Surgery?

Surgical approaches to bone metastasis are largely prophylactic against pathologic fractures and can also provide their palliation of pain, although its effect on the patient's quality of life is variable. Interestingly, prophylactic stabilization may incur a survival benefit as opposed to stabilization after a fracture [14].

Surgical Prediction Models

To determine the need for surgical stabilization, criteria have been proposed to evaluate the risk of pathologic fracture. Mirels' proposed criteria for fracture risk in long bones included the site of the lesion, the level of pain experienced due to the lesion, the type of lesion (lytic, blastic, or mixed), and the size of the lesion [15]. The Mirels' criteria has not been found to be highly sensitive, with a tendency to overestimate the risk of fracture.

One of the first prognostic studies to identify variables associated with different survival outcomes following prophylactic stabilization was in 1995, which demonstrated that the presence of pathologic fracture, the extent of metastasis, and the site of the primary tumor were all prognostic in survival after intramedullary nail fixation. A primary tumor in the lung was statistically determined to be a negative prognostic variable when deciding to prophylactically stabilize, as were pathologic fractures and metastases to the brain or viscera [16]. However, with the development of immunotherapy and molecularly targeted therapies for lung cancer, not every patient may have a lung cancer that is considered a poor prognostic factor for survival, so more recent models have stratified lung cancer based on its treatability and prognosis [17].

Spine-specific scoring systems to predict survival also exist, namely the Tomita and Revised Tokuhashi scores, with the latter being generally accepted as superior for predicting survival [18, 19]. The original Tokuhashi score was created in 1995 and was revised in 2005 [19]. Features of this scoring system include the general condition of the patient based on the Karnofsky performance status (KPS), the number of extraspinal bony metastases, the quantity of bony metastases in the vertebral bodies, the presence of visceral metastases and how resectable these are, spinal cord palsy, and considerations for the primary site of the cancer (Table 27.3). Lower Tokuhashi scores are correlated with a poorer prognosis. Scores of 0–8 correlate to survival of <6 months, scores 9–11 correlate with survival of 6–12 months, and scores 12–15 with survival of >12 months.

Spinal Cord Compression (SCC)

Spinal Cord Compression (SCC) can occur when lung cancer tumor growth or fractures compress the spinal cord, leading to neurological deficits, pain, and even paralysis. Spinal cord compression occurs in approximately 5–14% of patients with bone metastases from lung cancer.

The management of spinal cord compression in lung cancer is a critical medical emergency that requires prompt intervention to prevent neurological damage and improve patient outcomes. The management approach typically involves a combination of medical (corticosteroids, systemic therapies), surgical (spinal decompression or stabilization), and radiation therapies based on the severity of the compression, the patient's overall condition, and the extent of metastatic disease.

For patients with severe neurological deficits or rapidly progressive symptoms, surgical decompression (e.g., laminectomy or tumor removal) may be necessary to relieve pressure on the spinal cord. If there is significant instability or fracture, spinal stabilization may be performed, using instrumentation to maintain vertebral alignment and stability. It is important to consider specific criteria developed for spinal instability when considering surgery for these patients. The Spine Oncology Study Group developed the Spinal Instability Neoplastic Score (SINS) that included factors that overlapped with Mirels' criteria in addition to considerations for spinal alignment and vertebral body collapse [20].

Table 27.3 Revised Tokuhashi Scoring System for predicting survival with spinal bony metastases

Prognostic Factors		Points
Condition (KPS)	Poor (KPS 10–40%)	0
	Moderate (KPS 50–70%)	1
	Good (KPS 80–100%)	2
Quantity of extraspinal bone metastases	≥3	0
	1–2	1
	0	2
Quantity of metastases in vertebral bodies	≥3	0
	2	1
	1	2
Visceral metastases	Unresectable	0
	Resectable	1
	None present	2
Site of primary tumor	Lung or stomach	0
	Kidney, uterus, liver, others, unidentified	1
	Thyroid, prostate, breast, and rectum	2
	Kidney and uterus	3
	Rectum	4
	Thyroid, mammary, and prostate glands	5
Spinal cord palsy	Complete (Frankel A, B)	0
	Partial (Frankel C, D)	1
	None (Frankel E)	2

Management of Skeletal Complications

Cancer-Induced Bone Pain

Bone metastases are a major source of chronic pain in lung cancer patients, often requiring opioids or other pain management strategies. Approximately 70% of patients with bone metastases experience significant pain. Managing pain from bone metastases in lung cancer requires a comprehensive approach that combines pharmacological and non-pharmacological strategies to optimize pain relief and improve the patient's quality of life. Collaborating with a multidisciplinary team, including oncologists, pain specialists, palliative care providers, and rehabilitation therapists, is crucial for comprehensive pain management.

For palliation of painful bone metastases, there are pharmaceutical and radiological options. For mild pain, non-steroidal anti-inflammatory drugs (NSAIDs) or acetaminophen should be used first. The addition of an opioid can help for refractory cases or in cases of more severe pain [21]. In cases of cancer-induced bone pain refractory to oral regimens, external-beam radiation therapy can be used [21, 22]. The American Society for Radiation Oncology recommends the use of a single fraction of radiation, typically at 8 Gy, as opposed to a fractionated treatment regimen [23]. This is simply more economically friendly and convenient for the patient, and single fraction treatment provides the same level of pain relief as does fractionated treatment. However, patients who received single-fraction palliative radiation had higher rates of retreatment, so there may not be as much economic benefit for the patient as is theorized [22, 23]. Whether to use a single fraction or multifractionated approach is case-dependent, and the opinion of the provider is still important in the approach to radiotherapy for bony metastases. For example, patients with a more favorable predicted survival may benefit from fractionated radiation to hopefully prevent the need for retreatment. Table 27.4 describes key strategies for managing bone pain.

Table 27.4 Strategies for management of cancer-related bone pain

Modality		
Pharmacologic Management	Analgesics	*Non-Opioid Analgesics:* Acetaminophen and NSAIDs (e.g., ibuprofen and naproxen) can be effective for mild to moderate pain and help reduce inflammation. *Opioids:* *Weak opioids* (e.g., codeine and tramadol) can be used for moderate pain. *Strong opioids* (e.g., morphine, oxycodone, and fentanyl) are typically needed for severe pain.
	Corticosteroids	*Dexamethasone* can reduce inflammation, swelling, and associated pain, particularly in cases of spinal cord compression or inflammatory bone lesions.
	Bisphosphonates	*Zoledronic Acid, Denosumab* Bisphosphonates can reduce skeletal-related events and bone pain by stabilizing bone metabolism and preventing further complications.
	Antidepressants and Anticonvulsants	*Antidepressants* (e.g., amitriptyline) *Anticonvulsants* (e.g., gabapentin and pregabalin) can be useful for neuropathic pain and may help with overall pain control.
Radiation Therapy	Palliative Radiation	Localized radiation therapy can effectively relieve pain from bone metastases.
	Stereotactic Body Radiation Therapy (SBRT):	SBRT provides targeted radiation to specific areas, offering pain relief while minimizing damage to surrounding healthy tissue.
Surgery	Surgical Stabilization	Surgical stabilization can potentially alleviate pain when bone metastases lead to fractures or spinal instability.
	Decompression Surgery	Surgery may also be indicated for patients with spinal cord compression due to bone metastases.

(continued)

Table 27.4 (continued)

Modality		
Non-Pharmacological Approaches	Physical Therapy	Tailored exercise programs can improve mobility and reduce pain, while also helping maintain strength and function.
	Occupational Therapy	Occupational therapists can help patients adapt their daily activities to minimize pain and improve quality of life.
	Complementary Therapies	Techniques such as acupuncture, massage therapy, or mindfulness meditation may help reduce pain perception and improve emotional well-being.
	Activity Modification	Patients may need assistive devices like crutches, braces, or wheelchairs to protect affected bones and maintain mobility.

Hypercalcemia of Malignancy

Hypercalcemia of Malignancy can be found in 10–20% of patients with bone metastases from advanced cancers, including lung cancer. Bone metastases can cause calcium to be released into the bloodstream, leading to hypercalcemia, which causes nausea, vomiting, confusion, and fatigue. The incidence of hypercalcemia is greater in squamous histology and in advanced stages of disease, except in SCLC, where there is no evidence of hypercalcemia incidence across stages.

Patients with calcium levels >14 mg/dl (3.5 mmol/L) require aggressive treatment with intravenous fluids. If calcium remains high, consider giving intravenous calcitonin with bisphosphonates. Bisphosphonates are effective in treating hypercalcemia but can take 1 to 2 days to decrease calcium levels [24]. In a pooled analysis of two randomized trials, zoledronic acid was superior to pamidronate for the treatment of hypercalcemia. Denosumab is a monoclonal antibody against RANK-L that can also be used to treat hypercalcemia by binding to RANK and reducing osteoclasts, leading to decreased bone resorption. In a randomized trial, denosumab was not inferior to zoledronic acid in delaying time to first-on-study SRE (primary endpoint).

Prognosis and Survivorship

Lung cancer with bone metastases poses significant challenges for prognosis, largely due to the aggressive nature of lung cancer itself and the complications associated with skeletal metastases. Individual patient factors, including overall health, life expectancy, treatment response, and the extent of metastasis, play critical roles in determining outcomes.

Studies indicate that patients with NSCLC and bone metastases can have 1-year survival rates of approximately 30–50%, while those with SCLC often have shorter survival due to the aggressive nature of the disease. Several tools have been developed for preoperative estimation of survival for deciding on the operative management of metastatic bone disease of the extremities. For 90-day survival, the three most important factors associated with poorer survivorship were lower albumin level, higher neutrophil-to-lymphocyte ratio, and rapid growth primary tumor. For 1-year survival, the three most important factors associated with poorer survivorship were lower albumin level, rapid growth primary tumor, and lower hemoglobin level [25]. However, some patients may live longer, particularly with more effective treatments (e.g., targeted therapies) and supportive care. Multidisciplinary management and supportive care are essential for optimizing the quality of life and addressing complications associated with bone metastases.

References

1. Thandra KC, Barsouk A, Saginala K, Aluru JS, Barsouk A. Epidemiology of lung cancer. Contemp Oncol (Pozn). 2021;25(1):45–52. https://doi.org/10.5114/wo.2021.103829.
2. SEER*Explorer: an interactive website for SEER cancer statistics [Internet]. Surveillance Research Program, National Cancer Institute. Accessed at https://seer.cancer.gov/explorer/ on Jan 23, 2024.
3. Jemal A, Schafer EJ, Sung H, Bandi P, Kratzer T, Islami F, Siegel RL. The burden of lung cancer in qomen compared with men in the US. JAMA Oncol. 2023;9(12):1727–8. https://doi.org/10.1001/jamaoncol.2023.4415.
4. Siegfried JM. Women and lung cancer: does oestrogen play a role? Lancet Oncol. 2001;2(8):506–13.
5. Mok TSK, Wu YL, Kudaba I, Kowalski DM, Cho BC, Turna HZ, Castro G Jr, Srimuninnimit V, Laktionov KK, Bondarenko I, Kubota K, Lubiniecki GM, Zhang J, Kush D, Lopes G, KEYNOTE-042 Investigators. Pembrolizumab versus chemotherapy for previously untreated, PD-L1-expressing, locally advanced or metastatic non-small-cell lung cancer (KEYNOTE-042): a randomised, open-label, controlled, phase 3 trial. Lancet. 2019;393(10183):1819–30. https://doi.org/10.1016/S0140-6736(18)32409-7.
6. Planchard D, Jänne PA, Cheng Y, Yang JC, Yanagitani N, Kim SW, Sugawara S, Yu Y, Fan Y, Geater SL, Laktionov K, Lee CK, Valdiviezo N, Ahmed S, Maurel JM, Andrasina I, Goldman J, Ghiorghiu D, Rukazenkov Y, Todd A, Kobayashi K. FLAURA2 Investigators. Osimertinib with or without chemotherapy in *EGFR*-mutated advanced NSCLC. N Engl J Med. 2023;389(21):1935–48. https://doi.org/10.1056/NEJMoa2306434. Epub 2023 Nov 8. PMID: 37937763.
7. Cho BC, Lu S, Felip E, Spira AI, Girard N, Lee JS, Lee SH, Ostapenko Y, Danchaivijitr P, Liu B, Alip A, Korbenfeld E, Mourão Dias J, Besse B, Lee KH, Xiong H, How SH, Cheng Y, Chang GC, Yoshioka H, Yang JC, Thomas M, Nguyen D, Ou SI, Mukhedkar S, Prabhash K, D'Arcangelo M, Alatorre-Alexander J, Vázquez Limón JC, Alves S, Stroyakovskiy D, Peregudova M, Şendur MAN, Yazici O, Califano R, Gutiérrez Calderón V, de Marinis F, Passaro A, Kim SW, Gadgeel SM, Xie J, Sun T, Martinez M, Ennis M, Fennema E, Daksh M, Millington D, Leconte I, Iwasawa R, Lorenzini P, Baig M, Shah S, Bauml JM, Shreeve SM, Sethi S, Knoblauch RE, Hayashi H, MARIPOSA Investigators. Amivantamab plus lazertinib in previously untreated *EGFR*-Mutated advanced NSCLC. N Engl J Med. 2024;391(16). https://doi.org/10.1056/NEJMoa2403614. Epub ahead of print. PMID: 38924756.
8. Harada D, Takigawa N. Oligoprogression in non-small cell lung cancer. Cancer. 2021;13(22):5823. https://doi.org/10.3390/cancers13225823.
9. Shaw AT, Bauer TM, de Marinis F, Felip E, Goto Y, Liu G, Mazieres J, Kim DW, Mok T, Polli A, Thurm H, Calella AM, Peltz G, Solomon BJ, CROWN Trial Investigators. First-line lorlatinib or crizotinib in advanced *ALK*-positive lung cancer. N Engl J Med. 2020;383(21):2018–29. https://doi.org/10.1056/NEJMoa2027187. PMID: 33207094.

10. Jänne PA, Riely GJ, Gadgeel SM, Heist RS, Ou SI, Pacheco JM, Johnson ML, Sabari JK, Leventakos K, Yau E, Bazhenova L, Negrao MV, Pennell NA, Zhang J, Anderes K, Der-Torossian H, Kheoh T, Velastegui K, Yan X, Christensen JG, Chao RC, Spira AI. Adagrasib in non-small-cell lung cancer harboring a *KRASG12C* mutation. N Engl J Med. 2022;387(2):120–31. https://doi.org/10.1056/NEJMoa2204619.
11. Hong DS, Fakih MG, Strickler JH, Desai J, Durm GA, Shapiro GI, Falchook GS, Price TJ, Sacher A, Denlinger CS, Bang YJ, Dy GK, Krauss JC, Kuboki Y, Kuo JC, Coveler AL, Park K, Kim TW, Barlesi F, Munster PN, Ramalingam SS, Burns TF, Meric-Bernstam F, Henary H, Ngang J, Ngarmchamnanrith G, Kim J, Houk BE, Canon J, Lipford JR, Friberg G, Lito P, Govindan R, Li BT. $KRAS^{G12C}$ inhibition with sotorasib in advanced solid tumors. N Engl J Med. 2020;383(13):1207–17. https://doi.org/10.1056/NEJMoa1917239.
12. De Giglio A, Deiana C, Di Federico A. Bone-specific response according to MDA criteria predicts immunotherapy efficacy among advanced non-small cell lung cancer (NSCLC) patients. J Cancer Res Clin Oncol. 2023;149(5):1835–47. https://doi.org/10.1007/s00432-022-04120-z. Epub 2022 Jun 24. PMID: 35750899; PMCID: PMC10097761.
13. Katakami N, Kunikane H, Takeda K, Takayama K, Sawa T, Saito H, Harada M, Yokota S, Ando K, Saito Y, Yokota I, Ohashi Y, Eguchi K. Prospective study on the incidence of bone metastasis (BM) and skeletal-related events (SREs) in patients (pts) with stage IIIB and IV lung cancer-CSP-HOR 13. J Thorac Oncol. 2014;9(2):231–8.
14. Behnke NK, Baker DK, Xu S, Niemeier TE, Watson SL, Ponce BA. Risk factors for same-admission mortality after pathologic fracture secondary to metastatic cancer. Support Care Cancer. 2017;25(2):513–21. https://doi.org/10.1007/s00520-016-3431-8.
15. Mirels H. Metastatic disease in long bones. A proposed scoring system for diagnosing impending pathologic fractures. Clin Orthop Relat Res. 1989;249:256–64.
16. Bauer HC, Wedin R. Survival after surgery for spinal and extremity metastases. Prognostication in 241 patients. Acta Orthop Scand. 1995;66(2):143–6. https://doi.org/10.3109/17453679508995508.
17. Katagiri H, Okada R, Takagi T, Takahashi M, Murata H, Harada H, Nishimura T, Asakura H, Ogawa H. New prognostic factors and scoring system for patients with skeletal metastasis. Cancer Med. 2014;3(5):1359–67. https://doi.org/10.1002/cam4.292.
18. Zhao Y, Liu F, Wang W. Treatment progress of spinal metastatic cancer: a powerful tool for improving the quality of life of the patients. J Orthop Surg Res. 2023;18(1):563. https://doi.org/10.1186/s13018-023-03975-3.
19. Tokuhashi Y, Matsuzaki H, Oda H, Oshima M, Ryu J. A revised scoring system for preoperative evaluation of metastatic spine tumor prognosis. Spine. 2005;30(19):2186–91. https://doi.org/10.1097/01.brs.0000180401.06919.a5.
20. Fisher CG, DiPaola CP, Ryken TC, Bilsky MH, Shaffrey CI, Berven SH, Harrop JS, Fehlings MG, Boriani S, Chou D, Schmidt MH, Polly DW, Biagini R, Burch S, Dekutoski MB, Ganju A, Gerszten PC, Gokaslan ZL, Groff MW, Liebsch NJ, Mendel E, Okuno SH, Patel S, Rhines LD, Rose PS, Sciubba DM, Sundaresan N, Tomita K, Varga PP, Vialle LR, Vrionis FD, Yamada Y, Fourney DR. A novel classification system for spinal instability in neoplastic disease: an evidence-based approach and expert consensus from the Spine Oncology Study Group. Spine. 2010;35(22):E1221–9. https://doi.org/10.1097/BRS.0b013e3181e16ae2.
21. Kvale PA, Simoff M, Prakash UB, American college of chest physicians. Lung cancer Palliative care. Chest. 2003;123:284S–311S. https://doi.org/10.1378/chest.123.1_suppl.284s. PMID: 12527586.
22. Sze WM, Shelley MD, Held I, Wilt TJ, Mason MD. Palliation of metastatic bone pain: single fraction versus multifraction radiotherapy--a systematic review of randomised trials. Clin Oncol (R Coll Radiol). 2003;15(6):345–52. https://doi.org/10.1016/s0936-6555(03)00113-4.
23. Chow E, Zeng L, Salvo N, Dennis K, Tsao M, Lutz S. Update on the systematic review of palliative radiotherapy trials for bone metastases. Clin Oncol (R Coll Radiol). 2012;24(2):112–24. https://doi.org/10.1016/j.clon.2011.11.004.

24. Lutz S, Balboni T, Jones J, Lo S, Petit J, Rich SE, Wong R, Hahn C. Palliative radiation therapy for bone metastases: update of an ASTRO evidence-based guideline. Pract Radiat Oncol. 2017;7(1):4–12. https://doi.org/10.1016/j.prro.2016.08.001.
25. Major P, Lortholary A, Hon J, Abdi E, Mills G, Menssen HD, Yunus F, Bell R, Body J, Quebe-Fehling E, Seaman J. Zoledronic acid is superior to pamidronate in the treatment of hypercalcemia of malignancy: a pooled analysis of two randomized, controlled clinical trials. J Clin Oncol. 2001;19(2):558–67.

Medical Management of Musculoskeletal Metastasis from Renal Cell Carcinoma

28

Palash Asawa and Jaime Rafael Merchan

Introduction

Renal Cancer has a rising worldwide incidence, with an estimated 400,000 new cases annually, and a mortality rate approaching 175,000 deaths per year. Renal cell carcinoma (RCC) accounts for approximately 90–95% of all kidney neoplasms, and 25–30% of all patients have metastatic disease at diagnosis [1, 2]. Established risk factors for RCC include cigarette smoking, obesity, hypertension, and familial cancer syndromes [3]. Diabetes Mellitus, end-stage renal disease, lack of physical activity, alcohol consumption, and industrial exposures are less established risk factors.

Renal cell carcinoma includes a spectrum of different histopathological entities, with clear cell renal cell carcinoma (ccRCC) accounting for the majority (75–80%) of the cases [3]. The remaining are commonly referred to as non-clear cell renal cell carcinoma (nccRCC). These consist of several subtypes of variant histologies with different molecular and cytogenetic profiles. Papillary (10%) [4] and chromophobe renal cell carcinoma (5%) are the most common nccRCC subtypes [5].

RCC is a biologically complex disease that is associated with somatic or germline genomic alterations leading to RCC tumorigenesis. *VHL* silencing, via gene mutation or methylation, is the most common genetic alteration in ccRCC [6]. VHL is a critical tumor suppressor gene, whose loss of function results in the stabilization of its downstream transcription factor, hypoxia-inducible factor (HIF), especially HIF2α. This leads to a "pseudohypoxic" state, leading to HIF-mediated expression

P. Asawa
Department of Hematology and Oncology, Jackson Memorial Hospital and University of Miami Miller School of Medicine, Miami, FL, USA

J. R. Merchan (✉)
Department of Medicine, Division of Medical Oncology, University of Miami Miller School of Medicine and Sylvester Comprehensive Cancer Center, Miami, FL, USA
e-mail: Jmerchan2@med.miami.edu

J. Pretell-Mazzini, S. A. Lozano Calderón (eds.), *Multidisciplinary Approach of Musculoskeletal Metastasis*, https://doi.org/10.1007/978-3-032-09498-8_28

of downstream genes regulating proliferation, survival, metabolism, and angiogenesis, among others [7]. Activating mutations of *MET*, inactivating mutations of *PTEN, TSC1, TSC2*, and *FLCN*, and overactivation of the TFE3, TFEB, and MITF transcription factors are other molecular alterations reported in ccRCC and nccRCC patients [8–10]. Current therapeutic approaches to advanced clear cell RCC target the activated growth factors downstream of the VHL/HIF pathway [11].

The tumor microenvironment plays a critical role in RCC progression. Tumor angiogenesis, driven by genomic alterations, especially the VHL-HIF axis, is a predominant feature in ccRCC [7]. In addition, ccRCC is characterized by infiltration of immune cells, including T cells, B cells, Natural Killer cells, macrophages, and myeloid-derived suppressor cells, among others, which regulate antitumor immune responses as well as immunosuppressive mechanisms [12]. RCC cells express multiple cytokines and immune checkpoints, leading to immune evasion and tumor progression [13]. Knowledge of the complex tumor-stromal interactions has led to the discovery and development of targeted and immune therapy strategies for RCC treatment. As further mechanisms by which these cells cause inhibition of tumor eradication are being elucidated, novel therapies are currently being tested as adjuncts to immune checkpoint inhibitors (CPIs) [14].

Diagnosis and Staging

The extent of local, regional, and distant involvement is determined primarily by radiological imaging and is recommended in all patients with a diagnosis of advanced renal cell carcinoma, or in patients with localized disease, to rule out distant metastases in patients undergoing curative surgery. A computed tomography (CT) of the chest, abdomen, and pelvis with contrast is the most commonly used and preferred method for staging purposes. Alternatively, a chest CT without contrast and an MRI of the abdomen and pelvis with and without gadolinium can be performed for patients with contraindications for iodine contrast. Abdominal imaging helps determine the loco-regional extent of the primary renal tumor, while chest, brain, and bone imaging indicate involvement beyond the abdomen. Evaluation of brain metastases by brain MRI with and without gadolinium (preferred), or alternatively, with a CT scan of the head with contrast, is performed if there is clinical suspicion of brain involvement [15]. Brain imaging to rule out brain metastasis may be pursued in patients with solitary metastasis who are being considered for surgical treatment.

Common imaging modalities used to detect bone metastasis include Technetium-99 m (^{99m}Tc)-diphosphonate bone scintigraphy (bone scan) and conventional radiography. Bone scan is indicated only in patients with bone pain and/or an elevated serum alkaline phosphatase, based on several studies [16–18]. Since bone metastases in RCC and dominantly osteolytic, bone scans have a low sensitivity (around 50% or lower) in detecting bone metastasis in RCC, as bone scans only show osteoblastic reaction to metastatic tissue [19].

The utility of FDG PET–CT for the staging of RCC remains undetermined. It is of limited or no value in the detection of primary renal cell tumors, because FDG is physiologically excreted by the kidneys, obscuring lesions in the kidney [20]. The sensitivity and specificity of FDG–PET–CT for the detection of disease recurrence or metastasis are 89.5% and 83.3%, respectively [21]. On the other hand, PET/CT scan is superior to bone scan for the detection of lytic bone metastases, commonly observed in advanced renal cell carcinoma [22]. This imaging modality can be used as an adjunct when conventional imaging is not conclusive, as early diagnosis of metastatic disease has a significant effect on therapeutic management.

NCCN guidelines suggest that radiological imaging in patients with advanced RCC be performed at baseline, before initiating systemic therapy, and repeated every 6–16 weeks, depending on the patient's clinical status, treatment schedule, and/or MD discretion, to assess response to treatment [15].

RCC Screening

There are no established guidelines for RCC screening in the general population. Patients with an inherited predisposition to RCC require a multidisciplinary approach and routine imaging for lifelong surveillance of renal tumors. Routine screening with CT or MR imaging is recommended for most patients with a genetic predisposition. For patients with hereditary leiomyomatosis and RCC (HLRCC) syndrome, screening begins at an early age, because tumors can be detected as early as 11 years of age and are often aggressive with distant metastases even when small. For individuals with VHL syndrome, biannual surveillance imaging is recommended starting at the age of 20 years and managed with active surveillance until tumors are 3 cm in size [23].

Musculoskeletal Metastases in RCC

The most common metastatic sites in RCC are the lungs (45%), followed by bones, lymph nodes (21% each), liver (20%), adrenal gland (9%), and brain (8%) [24]. An estimated 16–34% of patients with advanced RCC have metastasis to the bone [24, 25, 26]. In these patients, bone metastases are predominantly osteolytic and driven by the predominance of bone resorption over bone formation. The "vicious cycle" hypothesis describes how tumor cells and the bone microenvironment interact to promote both tumor growth and bone destruction. Tumor cells release factors like PTHrP and TGF-β, which stimulate osteoblasts to overproduce RANKL. This leads to increased osteoclast activity, causing bone resorption. In turn, bone degradation releases growth factors that trigger further tumor growth and more PTHrP production, reinforcing the cycle of tumor expansion and bone destruction [27, 28].

Bone metastasis in RCC has been associated with worse overall survival (OS) and contributes to disease burden and significant morbidity through increased risk of skeletal-related events, such as pathological fractures, hypercalcemia of malignancy (HCM), spinal cord, and nerve root compression [29]. It is uncommon to see RCC patients with bone metastases as the only site of metastatic disease. In a retrospective study of 1878 patients with metastatic RCC, 54 patients had bone-only metastasis at diagnosis (2.8%) [30]. The most involved sites include the pelvis, sacrum, spine, and proximal extremities.

A retrospective study with 2027 patients with metastatic RCC from Canada, Denmark, South Korea, and the USA, analyzing data from the IMDC (International Metastatic Renal Cell Carcinoma Database Consortium) showed that patients with bone metastases had significantly shorter mOS and time to treatment failure (TTF) than patients with non-bone metastases (OS: 14.9 vs 25.1 months, respectively, [$p < 0.0001$]; median TTF: 5.7 vs 7.6 months, respectively, [$p < 0.0001$]) [31].

Skeletal muscles comprise nearly 50% of the total human body mass and are well vascularized; however, metastasis to musculature is rare. A painful mass is the most common presenting symptom of skeletal muscle metastasis. An autopsy series of patients with renal cell carcinoma showed that the prevalence of skeletal muscle metastasis ranges from 0.4% to 3.4% of cases [32, 33].

Clinical Prognostic Factors in Advanced RCC

There are certain clinical factors associated with worse overall survival that have been incorporated into prognostic risk models. The most widely used prognostic model, developed by investigators at Memorial Sloan Kettering Cancer Center, considers five factors that have been associated with a decreased rate of survival among patients with advanced renal-cell carcinoma: (1) Poor performance status, (2) A high serum lactate dehydrogenase (LDH) level, (3) A high serum calcium level, (4) A low hemoglobin concentration, and (5) Less than a 1-year interval from diagnosis to treatment. The median survival among patients with none of these risk factors (the favorable-risk group) was significantly better than among patients with 1 or 2 factors (the intermediate-risk group) and among those with 3 or more risk factors (the poor-risk group)—30 months vs. 14 months vs. 5 months ($P < 0.001$) [34]. More recently, the International Metastatic Renal Cell Carcinoma Database Consortium (IMDC) developed risk scoring that incorporated 4 of the 5 adverse prognostic factors from the Memorial Sloan Kettering model, with LDH levels replaced by 2 other factors: high ANC and high platelet count. In a study including 1028 patients, this score was externally validated, where median overall survival (mOS) was 43 months in the favorable-risk group (no prognostic factors), 22 months in the intermediate-risk group (one or two prognostic factors), and 8 months in the poor-risk group (three to six prognostic factors) [35]. IMDC is the currently accepted prognostic score used in the evaluation and management of advanced RCC.

Management of Advanced Renal Cell Carcinoma

Progress in the understanding of the mechanisms of RCC pathogenesis and progression has led to the development of novel treatment options that have changed the RCC treatment landscape and improved patient outcomes. As RCC is a chemotherapy-resistant malignancy, cytotoxic chemotherapy is not used in this disease. Currently, FDA-approved agents for the treatment of advanced RCC include oral receptor tyrosine kinase inhibitors (RTKIs) that target tumor angiogenesis, or checkpoint inhibitors (CPI) immunotherapies, which target PD-1/PDL-1 and CTLA4 pathways (CPIs) [34]. NCCN guidelines recommend the use of combination therapies in the first-line setting. These include CPI combinations of CTLA4 (ipilimumab) and PD1 (nivolumab) inhibitors, or anti-PD-1/PDL-1 + TKI combinations.

As shown in Table 28.1, Ipilimumab-nivolumab, axitinib-pembrolizumab, cabozantinib–nivolumab, and lenvatinib–pembrolizumab are all associated with improved overall survival-measured in years, compared to sunitinib, and are the current standard first-line treatments for advanced/metastatic RCC [36–39]. In patients with contraindications for ICI-based combination therapy, single-agent VEGF-TKIs are acceptable alternative options according to national guidelines [40].

Subsequent therapies for advanced RCC treatment depend on exposure to prior immunotherapy. Options for patients progressing on checkpoint inhibitor immunotherapy can switch to VEGF-TKIs like cabozantinib, axitinib, Lenvatinib, and everolimus, and in certain circumstances, pazopanib or sunitinib [41–44]. Rechallenging with ICIs adds toxicity without benefit, as shown in the CONTACT-03 and the TiNivo2 trials, which found no advantage of combining checkpoint inhibitors to cabozantinib or tivozanib, respectively [45]. The phase 3 TiNivo 2 study compared tivozanib–nivolumab versus tivozanib monotherapy in metastatic renal cell carcinoma patients after one or two prior therapies. Median progression-free survival (PFS) was 5.7 months for the combination and 7.4 months for monotherapy (hazard ratio 1.10, $p = 0.49$). Serious adverse events occurred in 32% of the combination group and 37% of the monotherapy group [46]. This data further showed that ICI rechallenge should be discouraged in patients with advanced renal cell carcinoma. Belzutifan, a hypoxia-inducible factor 2α inhibitor, showed improved progression-free survival over everolimus in the LITESPARK-005 trial for advanced clear-cell renal-cell carcinoma patients previously treated with immune checkpoint and antiangiogenic therapies [47]. In patients who received single-agent VEGF-TKI in the frontline setting, nivolumab is associated with improved OS, PFS, and ORR (compared with Everolimus, CheckMate-025, a phase 3 trial) [48]. Other options include cabozantinib, anti-PD1-RTKI combinations, ipilimumab and nivolumab, Lenvatinib and everolimus, or tivozanib [15].

Treatment selection depends on multiple factors, including patient characteristics, IMDC criteria, anticipated treatment tolerance, presence or absence of medical conditions that contraindicate one agent or another, physician experience, and the patient's wishes. Clinical trials, if available, are preferred and are strongly recommended. While next-generation sequencing is commonly performed in advanced

Table 28.1 Front-line systemic therapies for metastatic renal cell carcinoma

Study	Patient population	Therapy	PFS	OS	Adverse events (Grade 3 or 4)
CheckMate 214 [49]	Locally advanced or metastatic clear cell renal cell carcinoma, $n = 1096$	Ipilimumab (1 mg/kg) plus nivolumab (3 mg/kg) IV once every 3 weeks, followed by nivolumab (3 mg/kg) IV every 2 weeks vs sunitinib 50 mg once daily (oral)	Median PFS: 12 months vs 12 months; HR 0·86 (95% CI 0·73–1·01)	Median OS 53 vs 37 months; HR 0·72 (95% CI 0·62–0·83)	46% vs 63%
CABOSUN [50]	Locally advanced or metastatic renal cell carcinoma with a clear cell component; intermediate or poor risk as per IMDC criteria	Cabozantinib 60 mg once daily oral vs Sunitinib 50 mg once daily oral 4 weeks on, 2 weeks off	Median PFS 8.6 months (95% C 6.8–14.0) vs 5.3 months (95% CI 3.0–8.2)	Median OS was 26.6 months vs 21.2 months; HR 0.80 (95% CI 0.53–1.21)	68% vs 65%.
KEYNOTE-426 [36]	Locally advanced or metastatic clear cell renal cell carcinoma, $n = 862$	Pembrolizumab 200 mg IV once every 3 weeks plus axitinib 5 mg twice daily oral vs sunitinib 50 mg once daily (oral)	Median PFS rates 15·7 months vs 11.1 months; HR 0.69 (95% CI 0.59–0.81)	Median OS 47.2 months vs 40.8; HR 0.84 (95% CI 0.71–0.99)	75.8% vs 70.6%
JAVELIN Renal 101 [51]	Locally advanced or metastatic renal cell carcinoma with a clear cell component ($n = 886$)	Avelumab (10 mg/kg of bodyweight) IV every 2 weeks plus axitinib (5 mg) orally twice daily or sunitinib (50 mg) orally once daily for 4 weeks (6-week cycle)	Median PFS 13·9 months vs 8·5 months HR 0·67 (95% CI 0·568–0·785, $p < 0{\cdot}0001$)	Median OS not reached vs 37·8 months; HR 0·79 (95% CI 0·643–0·969, $p = 0{\cdot}0116$)	71.2% vs 71.5%
CLEAR [37]	Locally advanced or metastatic clear cell renal cell carcinoma, $n = 1069$	Pembrolizumab 200 mg IV once every 3 weeks plus lenvatinib 20 mg once daily orally *vs* sunitinib 50 mg once daily orally	Median PFS 23.9 vs. 9.2 months; HR 0.39 (95% CI 0.32 to 0.49; $P < 0.001$)	53·7 vs 54·3 months; HR 0·79 (95% CI 0·63–0·99)	74.1% vs 60.3%

(continued)

Table 28.1 (continued)

Study	Patient population	Therapy	PFS	OS	Adverse events (Grade 3 or 4)
CheckMate 9ER [39]	Locally advanced or metastatic renal cell carcinoma $N = 651$ (included those with sarcomatoid component)	Nivolumab 240 mg IV every 2 weeks plus cabozantinib 40 mg once daily oral vs sunitinib 50 mg once daily 4 weeks on, 2 weeks off	Median PFS 16·4 months vs 8·4 months; HR 0·58 (95% CI 0·49–0·70)	Median OS 46·5 months vs 36·0 months; HR 0·77 (95% CI 0·63–0·95)	67.5% vs 55.3%

RCC, there are very few actionable mutations in this disease, which is characterized by alterations in tumor suppressor genes, for which there are no available targeted agents. Current efforts are focused on determining the best treatment sequence, as well as predictive biomarkers for response and clinical benefit.

Treatment Options for Musculoskeletal Metastases in Advanced RCC

Systemic Therapy

Patients with advanced RCC and bone-only metastases have been systematically excluded from many clinical trials due, in part, to the absence of measurable disease for response assessment. A meta-analysis reviewing literature in advanced RCC patients and treatment outcomes for populations with and without metastatic bone involvement in the era of immunotherapy and tyrosine kinase inhibitors (TKIs) showed poorer outcomes among patients with bone metastasis. In the various studies examined under this review, the results of the METEOR study, a randomized phase 3 study in patients who have failed prior tyrosine kinase inhibitors, suggested that there may be an overall survival benefit of treatment with cabozantinib compared to everolimus in patients with, compared to patients without bone metastases (HR: 0.54 [95% (CI): 0.34–0.84] vs 0.71 [95% CI: 0.55–0.91]) [52]. However, the small sample size from METEOR and the large and overlapping CIs limit any strict conclusions. In the CABOSUN study, a subgroup analysis in patients with bone metastases versus those without bone metastases observed a PFS benefit with cabozantinib as well. In the Checkmate 9ER trial, nivolumab plus cabozantinib improved survival outcomes compared with sunitinib in this patient population with bone metastases HR 0.63 (95% CI 0.42–0.97) [37]. A similar benefit was not observed in the phase 3 Cosmic-313 trial evaluating cabozantinib in addition to nivolumab and ipilimumab as compared with placebo in addition to nivolumab and ipilimumab in patients with previously untreated advanced renal-cell carcinoma. However, these analyses were limited by the small number of patients in each subgroup [49]. The absence of trials designed to assess the impact of bone metastases on relative treatment benefits limits our ability to architect treatments based on the presence of bone metastasis [29].

Radiotherapy

Radiotherapy remains a major palliative treatment modality for the treatment of bone metastasis in general. Although RCC is considered radioresistant, advances in radiotherapy technologies such as intensity modulation RT (IMRT), stereotactic body RT (SBRT), and stereotactic radiosurgery (SRS) have made it possible to deliver high doses of radiation [53]. A multi-institutional study evaluating the utility of SBRT in 54 patients with bone-only metastatic RCC showed that SBRT resulted in excellent local control rates at 1 (94.9%) and 2 (92%) years [54]. A retrospective study by Rades et al. in 87 patients also demonstrated the efficacy of short-course radiotherapy for the treatment of spinal cord compression [55]. A study conducted by Reichel et al. showed that radiotherapy to osseous sites appeared to provide short-term pain control, prevent fractures, and avoid the need for surgery in patients with RCC and multiple bone metastases [56].

Surgery

Surgery has an important role in the management of RCC bone metastases. In patients with solitary or oligometastatic disease, surgery of the bone lesion can lead to durable, long-term responses. A retrospective study in which 65 patients with RCC underwent surgery for bone metastases showed that patients with solitary bone metastases had a favorable prognosis post-surgery, compared to patients with multiple bone metastases or the presence of concomitant visceral metastases. The presence of single metastases, absence of pathological fractures, and tumor-free resection margins correlated with patient survival post-surgery [57]. However, tumor (safety and feasibility of resection) and patient (performance status, co-morbidities, presence of other sites of metastases) factors limit this option for selected cases.

In patients with metastatic disease in multiple organs, the major contribution of surgical management of bone metastases is in palliation of symptoms, preservation of function, and prevention of skeletal-related events, such as impending pathological fracture or spinal cord compression. These interventions can preserve and restore skeletal integrity and prevent/improve neurological compromise [58].

It is important to recognize the potential risk for postoperative complications in patients receiving tyrosine kinase inhibitors (TKI) in RCC. There is a growing body of literature on the preoperative use of these agents. Studies have reported postoperative complications, including impaired wound healing and hemorrhage perioperatively related to these agents. A retrospective review of all patients undergoing cytoreductive nephrectomy at MD Anderson showed that the use of presurgical systemic targeted therapy with a broad range of VEGF inhibitors, including TKIs, resulted in higher rates of complications >90 days postoperatively, including wound complications. It is recommended that antiangiogenic TKIs be held prior to surgical procedures. Given their different half-lives, the duration of perioperative interruption of RTKIs is different. It is recommended that the clinician consult the drug package's insert for guidance on when to hold and restart TKI for RCC patients requiring surgery.

Radionuclide Therapy

Currently, there is no role of radionuclide therapy in the management of RCC bone metastases. Radium 223 is an alpha-emitting radioisotope with a natural bone-seeking proclivity, which has been shown to prolong survival in men with castration-resistant prostate cancer [59]. Its role in the management of bone-predominant metastatic RCC is being investigated in a randomized phase 2 study investigating cabozantinib, with or without Radium 223, in patients with RCC with bone metastasis is ongoing (NCT04071223) [60].

Supportive Therapy

Bone metastases lead to significant skeletal morbidity, including bone pain, pathologic fractures, spinal cord compression, and HCM. The mechanism of pain secondary to bone metastasis is complex and involves interactions between cancer cells, bone cells, activated inflammatory cells, and bone-innervating neurons. Tumor cells damage bone tissue primarily by activating the RANKL/RANK (receptor activator for nuclear factor kappa B) system. Tumor cells produce RANKL, which binds with RANK on the surface of osteoclasts, initiates their proliferation, and, in consequence, triggers their damaging effect on bones. Osteoclasts resorb bone, which induces acidosis and triggers cancer-induced bone pain by activation of bone-innervating neurons. Also, the structural strength of the skeleton is affected by subsequent mechanical weakness of the bone producing pain. Hypercalcemia of malignancy also occurs because of greatly increased bone resorption and consequent calcium release [61]. The elucidation of these mechanisms has resulted in the use of bone-targeted therapies in the management of bone metastasis and resultant complications in patients with RCC. Treatment goals are the prevention of disease-related skeletal complications, pain palliation, and the maintenance of quality of life.

Bisphosphonates have been extensively studied in the treatment of patients with HCM and in the prevention of skeletal complications associated with bone metastases in patients with breast cancer, prostate cancer, and multiple myeloma. However, studies evaluating their effectiveness in patients with other solid tumors are limited. Zoledronic acid is a third-generation amino-bisphosphonate. A Phase III, placebo-controlled trial demonstrated the efficacy of zoledronic acid in patients with a wide variety of solid tumors, including RCC. A retrospective analysis of the 74 patients with RCC in this study demonstrated that 4 mg zoledronic acid provided a significant treatment benefit. It significantly reduced the percentage of patients who had

SREs (37% vs 75% with placebo, $p = 0.015$), with consistent benefits across all types of SREs. The benefit of zoledronic acid in patients with RCC was also demonstrated clearly in the median time to first SRE (not reached vs 72 days with placebo, $p = 0.006$) and the time to first pathologic fracture (median not reached vs 168 days for the placebo group, $p = 0.003$) [62, 63].

The benefits of bisphosphonate therapy outweigh safety concerns. Their toxicity profile warrants consideration. Acute-phase (flu-like) symptoms are common but can be managed symptomatically. Zoledronic acid is also associated with nephrotoxicity, and dose modification is required in patients with creatinine clearance below normal [64]. Osteonecrosis of the jaw (ONJ) is a rare but potentially serious event reported in patients with cancer who receive bisphosphonates and may be exacerbated by concomitant therapies. There is some signal that concomitant use of bisphosphonates and antiangiogenic therapies (including TKIs such as sunitinib) may cause ONJ more frequently than bisphosphonates alone, but this is not concretely established and needs further elucidation, and hence, close monitoring in this setting is essential. According to current evidence, renal cancer patients should continue to receive bone-targeted therapies for the treatment of bone metastases [65, 66].

Inhibition of RANK-L has been used to reduce the effects of excess bone resorption caused by bone metastases. Denosumab, a human monoclonal antibody specific to RANK-L, has been investigated in multiple trials for the prevention and treatment of bone metastases. A phase III trial of denosumab in patients with skeletal metastasis from a broad range of malignancies, including >100 subjects with RCC, showed that it was non-inferior to zoledronic acid, with overall survival, disease progression, and adverse events being similar in each arm [67]. However, denosumab has the potential advantage of being administered subcutaneously, and patients do not experience the acute phase reaction or the renal toxicity associated with zoledronic acid. Based on the results of this trial, denosumab is now licensed for the treatment of bone metastases in renal cancer. Denosumab, like zoledronic acid, poses a risk for the development of ONJ. The incidence of ONJ between denosumab and zoledronic acid appears to be similar, based on the three major studies directly comparing these drugs [68]. Less common complications from bisphosphonates and denosumab include atypical femoral fractures (AFF) [69]. In a study published by Kaku et al., in a cohort of 529 patients with bone metastases receiving these agents, the overall incidence of AFFs was 0.9% [70].

Conclusions

Bone metastases are a common presentation in advanced renal cell carcinoma. Once present, they are associated with high morbidity and a poor prognosis, compared to other cancers that spread to bone. While recent advances in treatment have improved outcomes for patients with advanced RCC, patients with predominant bone metastatic RCC represent a therapeutic challenge, as response to currently available systemic therapies is limited. The role of bone-directed radionuclide therapy is evolving. Supportive therapies, like bisphosphonates or RANK ligand antibodies, play an important role in treating or preventing bone metastasis complications. Current management of RCC bone metastases is multidisciplinary, involving the medical oncologist, orthopedic oncology surgeon, radiation oncologist, and palliative care. Treatment decisions should be tailored to each patient and clinical situation. The overall poor prognosis of RCC bone metastases and limited systemic therapy options underscore the need for further research to identify novel and effective multidisciplinary strategies to improve the quality of life and outcomes in patients affected by this disease.

References

1. Cirillo L, Innocenti S, Becherucci F. Global epidemiology of kidney cancer. Nephrol Dial Transplant. 2024;39:920–8. https://doi.org/10.1093/ndt/gfae036.
2. Wong MCS, et al. Incidence and mortality of kidney cancer: temporal patterns and global trends in 39 countries. Sci Rep. 2017;7:15698. https://doi.org/10.1038/s41598-017-15922-4.
3. Chow W-H, Dong LM, Devesa SS. Epidemiology and risk factors for kidney cancer. Nat Rev Urol. 2010;7:245–57. https://doi.org/10.1038/nrurol.2010.46.
4. Lopez-Beltran A, Scarpelli M, Montironi R, Kirkali Z. 2004 WHO classification of the renal tumors of the adults. Eur Urol. 2006;49:798–805. https://doi.org/10.1016/j.eururo.2005.11.035.
5. Lopez-Beltran A, et al. 2009 update on the classification of renal epithelial tumors in adults. Int J Urol. 2009;16:432–43. https://doi.org/10.1111/j.1442-2042.2009.02302.x.
6. Comprehensive molecular characterization of clear cell renal cell carcinoma. Nature. 2013;499:43–9. https://doi.org/10.1038/nature12222.
7. Schödel J, et al. Hypoxia, hypoxia-inducible transcription factors, and renal cancer. Eur Urol. 2016;69:646–57. https://doi.org/10.1016/j.eururo.2015.08.007.
8. Gustafson S, Zbuk KM, Scacheri C, Eng C. Cowden syndrome. Semin Oncol. 2007;34:428–34. https://doi.org/10.1053/j.seminoncol.2007.07.009.
9. Nickerson ML, et al. Mutations in a novel gene lead to kidney tumors, lung wall defects, and benign tumors of the hair follicle in patients with the Birt-Hogg-Dubé syndrome. Cancer Cell. 2002;2:157–64. https://doi.org/10.1016/s1535-6108(02)00104-6.
10. Cajaiba MM, et al. The classification of pediatric and young adult renal cell carcinomas registered on the children's oncology group (COG) protocol AREN03B2 after focused genetic testing. Cancer. 2018;124:3381–9. https://doi.org/10.1002/cncr.31578.
11. Linehan WM, et al. The metabolic basis of kidney cancer. Cancer Discov. 2019;9:1006–21. https://doi.org/10.1158/2159-8290.Cd-18-1354.
12. Giraldo NA, et al. Orchestration and prognostic significance of immune checkpoints in the microenvironment of primary and metastatic renal cell cancer. Clin Cancer Res. 2015;21:3031–40. https://doi.org/10.1158/1078-0432.Ccr-14-2926.
13. Monjaras-Avila CU, et al. The tumor immune microenvironment in clear cell renal cell carcinoma. Int J Mol Sci. 2023;24 https://doi.org/10.3390/ijms24097946.
14. Mier JW. The tumor microenvironment in renal cell cancer. Curr Opin Oncol. 2019;31:194–9. https://doi.org/10.1097/cco.0000000000000512.
15. NCCN Clinical Practice Guidelines in Oncology (NCCN Guidelines), https://www.nccn.org/professionals/physician_gls/pdf/kidney.pdf (2024).
16. Seaman E, Goluboff ET, Ross S, Sawczuk IS. Association of radionuclide bone scan and serum alkaline phosphatase in patients with metastatic renal cell carcinoma. Urology. 1996;48:692–5. https://doi.org/10.1016/s0090-4295(96)00236-1.
17. Koga S, et al. The diagnostic value of bone scan in patients with renal cell carcinoma. J Urol. 2001;166:2126–8.

18. Johnson CD, Dunnick NR, Cohan RH, Illescas FF. Renal adenocarcinoma: CT staging of 100 tumors. AJR Am J Roentgenol. 1987;148:59–63. https://doi.org/10.2214/ajr.148.1.59.
19. Sohaib SA, et al. Comparison of whole-body MRI and bone scintigraphy in the detection of bone metastases in renal cancer. Br J Radiol. 2009;82:632–9. https://doi.org/10.1259/bjr/52773262.
20. Liu Y. The place of FDG PET/CT in renal cell carcinoma: value and limitations. Front Oncol. 2016;6:201. https://doi.org/10.3389/fonc.2016.00201.
21. Park JW, Jo MK, Lee HM. Significance of 18F-fluorodeoxyglucose positron-emission tomography/computed tomography for the postoperative surveillance of advanced renal cell carcinoma. BJU Int. 2009;103:615–9. https://doi.org/10.1111/j.1464-410X.2008.08150.x.
22. Wu HC, et al. Comparing whole body 18F-2-deoxyglucose positron emission tomography and technetium-99m methylene diphosphate bone scan to detect bone metastases in patients with renal cell carcinomas—a preliminary report. J Cancer Res Clin Oncol. 2002;128:503–6. https://doi.org/10.1007/s00432-002-0370-1.
23. Diaz de Leon A, Pedrosa I. Imaging and screening of kidney cancer. Radiol Clin North Am. 2017;55:1235–50. https://doi.org/10.1016/j.rcl.2017.06.007.
24. Bianchi M, et al. Distribution of metastatic sites in renal cell carcinoma: a population-based analysis. Ann Oncol. 2012;23:973–80. https://doi.org/10.1093/annonc/mdr362.
25. Huang JF, et al. Incidence of patients with bone metastases at diagnosis of solid tumors in adults: a large population-based study. Ann Transl Med. 2020;8:482. https://doi.org/10.21037/atm.2020.03.55.
26. Santini D, et al. Natural history of malignant bone disease in renal cancer: final results of an Italian bone metastasis survey. PLoS One. 2013;8:e83026. https://doi.org/10.1371/journal.pone.0083026.
27. Mikami S, et al. Invasion and metastasis of renal cell carcinoma. Med Mol Morphol. 2014;47:63–7. https://doi.org/10.1007/s00795-013-0064-6.
28. Brozovich A, et al. All bone metastases are not created equal: revisiting treatment resistance in renal cell carcinoma. J Bone Oncol. 2021;31:100399. https://doi.org/10.1016/j.jbo.2021.100399.
29. Brown J, et al. Implications of bone metastasis on response to systemic therapy in patients with advanced renal cell carcinoma: a systematic literature review. Cancer Treat Rev. 2024;129 https://doi.org/10.1016/j.ctrv.2024.102792.
30. Tang B, et al. Natural history of bone-only metastasis in renal cell carcinoma. Urol Oncol Semin Orig Investig. 2024;42:119.e117–22. https://doi.org/10.1016/j.urolonc.2024.01.030.
31. McKay RR, et al. Impact of bone and liver metastases on patients with renal cell carcinoma treated with targeted therapy. Eur Urol. 2014;65:577–84. https://doi.org/10.1016/j.eururo.2013.08.012.
32. Surov A, et al. Skeletal muscle metastases: primary tumours, prevalence, and radiological features. Eur Radiol. 2010;20:649–58. https://doi.org/10.1007/s00330-009-1577-1.
33. Salman R, et al. Rare skeletal muscle metastasis from renal cell carcinoma: case report and review of the literature. CEN Case Rep. 2018;7:316–9. https://doi.org/10.1007/s13730-018-0350-1.
34. Choueiri TK, Motzer RJ. Systemic therapy for metastatic renal-cell carcinoma. N Engl J Med. 2017;376:354–66. https://doi.org/10.1056/NEJMra1601333.
35. Heng DY, et al. External validation and comparison with other models of the international metastatic renal-cell carcinoma database consortium prognostic model: a population-based study. Lancet Oncol. 2013;14:141–8. https://doi.org/10.1016/s1470-2045(12)70559-4.
36. Powles T, et al. Pembrolizumab plus axitinib versus sunitinib monotherapy as first-line treatment of advanced renal cell carcinoma (KEYNOTE-426): extended follow-up from a randomised, open-label, phase 3 trial. Lancet Oncol. 2020;21:1563–73.
37. Motzer R, et al. Lenvatinib plus Pembrolizumab or Everolimus for advanced renal cell carcinoma. N Engl J Med. 2021;384:1289–300. https://doi.org/10.1056/NEJMoa2035716.
38. Motzer RJ, et al. Nivolumab plus Ipilimumab versus Sunitinib in advanced renal-cell carcinoma. N Engl J Med. 2018;378:1277–90. https://doi.org/10.1056/NEJMoa1712126.

39. Choueiri TK, et al. Nivolumab plus Cabozantinib versus Sunitinib for advanced renal-cell carcinoma. N Engl J Med. 2021;384:829–41. https://doi.org/10.1056/NEJMoa2026982.
40. (eds Parexel & Squibb Bristol-Myers) (2021).
41. Albiges L, et al. CaboPoint: a phase II study of Cabozantinib as second-line treatment in patients with metastatic renal cell carcinoma. Future Oncol. 2022;18:915–26. https://doi.org/10.2217/fon-2021-1006.
42. Powles TB, et al. 718P a phase II study of patients with advanced or metastatic renal cell carcinoma (mRCC) receiving pazopanib after previous checkpoint inhibitor treatment. Ann Oncol. 2020;31:S564. https://doi.org/10.1016/j.annonc.2020.08.790.
43. Ornstein MC, et al. Individualised axitinib regimen for patients with metastatic renal cell carcinoma after treatment with checkpoint inhibitors: a multicentre, single-arm, phase 2 study. Lancet Oncol. 2019;20:1386–94. https://doi.org/10.1016/S1470-2045(19)30513-3.
44. Grande E, et al. INMUNOSUN-SOGUG trial: a prospective phase II study to assess the efficacy and safety of sunitinib as second-line (2L) treatment in patients (pts) with metastatic renal cell cancer (RCC) who received immunotherapy-based combination upfront. J Clin Oncol. 2020;38:5060–70. https://doi.org/10.1200/JCO.2020.38.15_suppl.5060.
45. Pal SK, et al. Atezolizumab plus cabozantinib versus cabozantinib monotherapy for patients with renal cell carcinoma after progression with previous immune checkpoint inhibitor treatment (CONTACT-03): a multicentre, randomised, open-label, phase 3 trial. Lancet. 2023;402:185–95. https://doi.org/10.1016/S0140-6736(23)00922-4.
46. Choueiri TK, et al. Tivozanib plus nivolumab versus tivozanib monotherapy in patients with renal cell carcinoma following an immune checkpoint inhibitor: results of the phase 3 TiNivo-2 study. Lancet. https://doi.org/10.1016/S0140-6736(24)01758-6.
47. Choueiri Toni K, et al. Belzutifan versus Everolimus for advanced renal-cell carcinoma. N Engl J Med. 2024;391:710–21. https://doi.org/10.1056/NEJMoa2313906.
48. Motzer RJ, et al. Nivolumab versus Everolimus in advanced renal-cell carcinoma. N Engl J Med. 2015;373:1803–13. https://doi.org/10.1056/NEJMoa1510665.
49. Tannir NM, et al. Nivolumab plus ipilimumab (NIVO+IPI) vs sunitinib (SUN) for first-line treatment of advanced renal cell carcinoma (aRCC): long-term follow-up data from the phase 3 CheckMate 214 trial. J Clin Oncol. 2024;42:363–73. https://doi.org/10.1200/JCO.2024.42.4_suppl.363.
50. Choueiri TK, et al. Cabozantinib versus sunitinib as initial therapy for metastatic renal cell carcinoma of intermediate or poor risk (Alliance A031203 CABOSUN randomised trial): progression-free survival by independent review and overall survival update. Eur J Cancer. 2018;94:115–25. https://doi.org/10.1016/j.ejca.2018.02.012.
51. Motzer RJ, et al. Avelumab plus Axitinib versus Sunitinib for advanced renal-cell carcinoma. N Engl J Med. 2019;380:1103–15. https://doi.org/10.1056/NEJMoa1816047.
52. Escudier B, et al. Cabozantinib, a new standard of care for patients with advanced renal cell carcinoma and bone metastases? subgroup analysis of the METEOR trial. J Clin Oncol. 2018;36:765–72. https://doi.org/10.1200/jco.2017.74.7352.
53. Umer M, Mohib Y, Atif M, Nazim M. Skeletal metastasis in renal cell carcinoma: a review. Ann Med Surg (Lond). 2018;27:9–16. https://doi.org/10.1016/j.amsu.2018.01.002.
54. Onal C, et al. Bone-only oligometastatic renal cell carcinoma patients treated with stereotactic body radiotherapy: a multi-institutional study. Strahlenther Onkol. 2022;198:940–8. https://doi.org/10.1007/s00066-022-01962-x.
55. Rades D, et al. Short-course radiotherapy (RT) for metastatic spinal cord compression (MSCC) due to renal cell carcinoma: results of a retrospective multi-center study. Eur Urol. 2006;49:846–52. https://doi.org/10.1016/j.eururo.2005.11.004.
56. Reichel LM, Pohar S, Heiner J, Buzaianu EM, Damron TA. Radiotherapy to bone has utility in multifocal metastatic renal carcinoma. Clin Orthop Relat Res. 2007;459:133–8. https://doi.org/10.1097/BLO.0b013e3180616594.
57. Fottner A, et al. Bone metastases from renal cell carcinoma: patient survival after surgical treatment. BMC Musculoskelet Disord. 2010;11:145. https://doi.org/10.1186/1471-2474-11-145.

58. Wood SL, Brown JE. Skeletal metastasis in renal cell carcinoma: current and future management options. Cancer Treat Rev. 2012;38:284–91. https://doi.org/10.1016/j.ctrv.2011.06.011
59. Parker C, et al. Alpha emitter Radium-223 and survival in metastatic prostate cancer. N Engl J Med. 2013;369:213–23. https://doi.org/10.1056/NEJMoa1213755
60. McKay RR, et al. A randomized trial of radium-223 (Ra-223) dichloride and cabozantinib in patients (pts) with advanced renal cell carcinoma (RCC) with bone metastases (RADICAL/Alliance A031801). J Clin Oncol. 2021;39:TPS4593-TPS4593. https://doi.org/10.1200/JCO.2021.39.15_suppl.TPS4593
61. Zajączkowska R, Kocot-Kępska M, Leppert W, Wordliczek J. Bone pain in cancer patients: mechanisms and current treatment. Int J Mol Sci. 2019;20 https://doi.org/10.3390/ijms20236047.
62. van Broekhoven DLM, Dootjes LW, van der Veldt A, Zillikens C, van Oldenrijk J. Effect of bisphosphonates on skeletal related events in long bone metastases of renal cell carcinoma: a systematic review. Clin Genitourin Cancer. 2023;21:e190–7. https://doi.org/10.1016/j.clgc.2022.12.010
63. Rosen LS, et al. Long-term efficacy and safety of zoledronic acid in the treatment of skeletal metastases in patients with nonsmall cell lung carcinoma and other solid tumors. Cancer. 2004;100:2613–21. https://doi.org/10.1002/cncr.20308
64. Shah S, Jean GW, Keisner SV, Gressett Ussery SM, Dowell J. An incidence of renal failure in patients receiving renally adjusted zoledronic acid. J Clin Oncol. 2010;28:e19581–91. https://doi.org/10.1200/jco.2010.28.15_suppl.e19581
65. Brunello A, Saia G, Bedogni A, Scaglione D, Basso U. Worsening of osteonecrosis of the jaw during treatment with sunitinib in a patient with metastatic renal cell carcinoma. Bone. 2009;44:173–5. https://doi.org/10.1016/j.bone.2008.08.132
66. Takahara T, et al. Osteonecrosis of the jaw associated with receiving sunitinib monotherapy: a rare case. J Oral Maxillofac Surg Med Pathol. 2021;33:66–70. https://doi.org/10.1016/j.ajoms.2020.08.005
67. Henry DH, et al. Randomized, double-blind study of Denosumab versus Zoledronic acid in the treatment of bone metastases in patients with advanced cancer (excluding breast and prostate cancer) or multiple myeloma. J Clin Oncol. 2011;29:1125–32. https://doi.org/10.1200/jco.2010.31.3304.
68. Saad F, et al. Incidence, risk factors, and outcomes of osteonecrosis of the jaw: integrated analysis from three blinded active-controlled phase III trials in cancer patients with bone metastases. Ann Oncol. 2012;23:1341–7. https://doi.org/10.1093/annonc/mdr435
69. Takahashi M, et al. Atypical femoral fracture in patients with bone metastasis receiving denosumab therapy: a retrospective study and systematic review. BMC Cancer. 2019;19:980. https://doi.org/10.1186/s12885-019-6236-6
70. Kaku T, et al. Incidence of atypical femoral fractures in the treatment of bone metastasis: an alert report. J Bone Oncol. 2020;23:100301. https://doi.org/10.1016/j.jbo.2020.100301
71. Choueiri TK. et al. TiNivo-2: A phase 3, randomized, controlled, multicenter, open-label study to compare tivozanib in combination with nivolumab to tivozanib monotherapy in subjects with renal cell carcinoma who have progressed following one or two lines of therapy where one line has an immune checkpoint inhibitor. Journal of Clinical Oncology 40, TPS405-TPS405. 2022. https://doi.org/10.1200/JCO.2022.40.6_suppl.TPS405.

Medical Management of Musculoskeletal Metastasis from Thyroid Carcinoma

29

Cesar Ochoa

Introduction

Thyroid cancer is a malignancy of the thyroid parenchymal cells and represents the most common tumor of the endocrine system, representing 2.2% of all new cancer cases. 44,020 new cases of thyroid cancer and 2170 deaths from the disease are estimated in 2024 [1].

Thyroid cancer has a female preponderance, with its incidence being approximately threefold higher in women (22.8 per 100,000 per year in 2014–2018) than in men (8.0 per 100,000 per year) [2]. Incidence varies substantially by geographic location, and in the United States, 1 in 55 U.S. women and one in 149 U.S. men are expected to be diagnosed with thyroid cancer during their lifetime [3].

The incidence of thyroid cancer in the United States increased exponentially in both men and women between 1974 and 2013 [4], reaching a plateau between 2014 and 2016 [5], and has recently started to decrease [6]. The increase in the number of thyroid cancers diagnosed was driven by small and often subclinical thyroid cancers of the papillary subtype, which account for 90% of all cases. The less common subtypes, in descending order of incidence, are follicular, medullary, and anaplastic [7].

The reasons for this increase in incidence remain unknown, as there is no evidence of new risk factors or increased exposure to known contributors to thyroid cancer [8], but has been largely attributed to an overdiagnosis effect, as the growing use of diagnostic techniques, such as ultrasonography, computed tomography and magnetic resonance imaging, combined with an increased access to health care services, results in more fine-needle aspiration biopsies that lead to increased detection

C. Ochoa (✉)
Department of Medical Oncology, Miami Cancer Center, Baptist Health South Florida, Miami, FL, USA
e-mail: Cesar.Ochoa@baptisthealth.net

J. Pretell-Mazzini, S. A. Lozano Calderón (eds.), *Multidisciplinary Approach of Musculoskeletal Metastasis*, https://doi.org/10.1007/978-3-032-09498-8_29

[9]. Recent evidence also points to environmental and lifestyle-related factors and comorbidities such as obesity and excess body weight [10].

Risk Factors

The thyroid gland is highly sensitive to the carcinogenic effects of ionizing radiation, and exposure to ionizing radiation, particularly when the exposure occurs at an early age, is the most established risk factor for thyroid cancer. Ionizing radiation causes DNA strand breaks and somatic mutations, contributing to carcinogenesis [11]. The risk begins to increase after exposure to a mean dose of more than 0.05–0.1 Gy [12], and the latency period between radiation exposure and the appearance of thyroid cancer is at least 5–10 years but can occur decades after exposure [13].

This association has been recognized in children and young adults who received radiation therapy for benign conditions such as acne or tinea capitis [14, 15] or for other cancers such as Hodgkin lymphoma [16], neuroblastoma, or Wilms tumor [17]. In addition, atomic bomb survivors [18] and individuals exposed to radiation fallout from nuclear reactor accidents have shown an increased incidence of thyroid cancer [19].

A family history of thyroid cancer is also a risk factor for the development of this tumor. Approximately 20–25% of medullary thyroid carcinomas are associated with a hereditary syndrome such as multiple endocrine neoplasia 2A (MEN2A), multiple endocrine neoplasia 2B (MEN2B), and familial medullary thyroid carcinoma (FMTC) and are caused by pathogenic germline variants in the RET proto-oncogene [20]. Syndromic familial non-medullary thyroid carcinoma (FNMTC) has been associated with familial adenomatous polyposis, Gardner syndrome, Cowden syndrome, Carney complex, Werner syndrome, and DICER 1 syndrome [21].

Obesity is associated with a chronic and systemic inflammatory state, characterized by an increase in inflammatory factors and production of various cytokines and adipokines, and is an established risk factor for many chronic diseases, including cancer [22]. Observational studies have provided evidence for a potential association between obesity and thyroid cancer [23]. The underlying mechanisms behind this association, however, are not yet clear.

Classification

The thyroid parenchyma consists of two major cell types: the follicular cells, which derive from the endoderm and are responsible for the production of the thyroid hormones tetraiodothyronine (T4) and triiodothyronine (T3), and the parafollicular or C-cells, which derive from the neural crest and secrete calcitonin [24]. The follicular cells give rise to differentiated thyroid cancers, including papillary thyroid cancer, follicular thyroid cancer, and Hurthle cell carcinoma. The parafollicular cells give rise to medullary thyroid carcinoma. Anaplastic thyroid carcinoma, also known as undifferentiated thyroid carcinoma, is a rare and aggressive tumor of the thyroid follicular epithelium that is associated with local invasion and metastatic spread to regional lymph nodes and distant sites.

Papillary thyroid cancers represent approximately 90% of all thyroid malignancies and are the most common variant in the United States [25]. The term encompasses several tumor types that have mutations of genes involved in the mitogen-activated protein kinase (MAPK) signaling pathway. BRAF V600E accounts for most of these mutations, followed by RAS, RET, NTRK, and ALK [26]. There are several histologic variants with different patterns of behavior. Most cases will be clinically indolent and are associated with a benign disease course and a favorable prognosis [27]. Distant metastasis is uncommon for papillary thyroid cancer; only 2–10% of patients will present with metastatic disease, with approximately one fourth of these involving the musculoskeletal system [28]. The tall cell variant is relatively rare and is associated with a more aggressive biologic behavior and has an increased rate of lymph node and distant metastases at presentation [27].

Follicular thyroid cancer is the second most common type of thyroid cancer, accounting for 4% of all thyroid malignancies [25]. It tends to present at an older age than papillary thyroid cancer, peaking at 40–60 years of age [29]. Mutations in the RAS proto-oncogene are common [24] and appear to be associated with a more aggressive disease course. PAX8–PPAR gamma rearrangements have also been described, particularly in younger patients [30]. The invasion of vascular structures is common, resulting in hematogenous dissemination and distant metastases, typically to bones, lungs, and liver. These occur in 7–23% of patients and adversely affect outcomes [31].

Oncocytic carcinoma of the thyroid, previously known as Hurthle cell carcinoma, accounts for 2% of thyroid cancers [25] and was once considered a variant of follicular thyroid cancer. It is now considered genetically distinct follicular cell-derived thyroid cancer. Its clinical presentation and pattern of distant metastases are similar to follicular thyroid cancer [24].

Medullary thyroid cancer is a type of neuroendocrine tumor that constitutes 2% of all thyroid cancers [25] and is associated with a RET proto-oncogene mutation [32]. It can occur in a sporadic or a familial form as part of MEN2A, MEN2B, or in a familial form without a multiple endocrine neoplasia association [24]. Clinically, patients might present with flushing or diarrhea from the tumor secretion of serotonin, histamine, vasoactive intestinal peptide, and prostaglandins. Distant metastases are present in 7–23% of patients at the time of diagnosis and can affect the liver, lung, bones, and, less often, brain and skin [33].

Anaplastic thyroid cancer, which accounts for approximately 1% of thyroid cancers, is a rare and undifferentiated thyroid cancer with an extremely aggressive clinical course and a very poor prognosis [24]. It tends to affect older patients over the age of 60 and is female predominant. It appears to develop from a multistep process with biological transformation from a differentiated thyroid cancer [34]. BRAF and RAS are the predominant drivers and have frequent mutations in TP53, the TERT promoter, and genes involved in epigenetic regulation [24]. Up to 50% of patients will have distant metastases at the time of presentation, mostly involving the lungs. Bone metastases are seen in 5–15% of patients [34]. These tumors are refractory to radioiodine, and traditional chemotherapy and radiotherapy are of limited benefit [35].

Diagnosis and Staging

The most common presentation of thyroid cancer is an asymptomatic nodule or thyroid mass. Rapid growth, fixation to surrounding tissues, cervical adenopathy, or symptoms such as cough, dyspnea, or hoarseness raise the suspicion of malignancy.

Evaluation of a thyroid nodule begins with a diagnostic high-resolution ultrasound. High-risk features on ultrasound include increasing size, hypoechogenicity, irregular margins, size taller than wide, microcalcifications, solid internal structure, extrathyroidal extension, and central vascularity [36].

The decision to subject a nodule for biopsy is based on these radiological criteria, with guidance from Thyroid Imaging Reporting and Data System (TIRADS) or American Thyroid Association (ATA) criteria. The Bethesda System for Reporting Thyroid Cytopathology is a standardized and widely adopted reporting system for thyroid fine needle aspirations [37]. The American Joint Committee (AJCC) TNM staging system is commonly used in clinical practice and is optimized to predict survival in patients with thyroid cancer, based primarily on pathologic findings accumulated preoperatively, intraoperatively, and during the first four months after thyroid surgery [38].

Management

The mainstay of treatment for well differentiated thyroid cancer is surgery. The extent of optimal surgical treatment remains under debate and ranges from thyroid lobectomy in younger patients with small, well encapsulated tumors, to total thyroidectomy with central neck dissection and unilateral or bilateral lymph-node–compartment dissection [24]. Postoperatively, radioactive iodine (131-I) is used for remnant ablation of normal residual thyroid tissue, as adjuvant therapy for subclinical micrometastases, or as treatment of apparent local or distant metastasis [39]. Local recurrences are generally treated surgically.

Patients who develop distant metastatic disease often follow an indolent course, and in many cases, observation is reasonable. Systemic therapy is generally reserved for patients who have metastatic disease that is rapidly progressive, symptomatic, and is not amenable to localized therapies [24]. In patients with indications for systemic therapy, somatic mutation testing and next-generation sequencing can identify oncogenic kinase abnormalities, such as gene rearrangements in *NTRK, RET,* or point mutations in *BRAF* that can guide systemic treatment options.

NTRK gene rearrangements are seen in approximately 6.7% of papillary thyroid cancers and in 20% of poorly differentiated thyroid carcinomas [40]. Larotrectinib, a selective NTRK inhibitor, has a 75% overall response rate. Common side effects included musculoskeletal pain, fatigue, nausea, transaminitis, edema, and gastrointestinal symptoms. The risk of skeletal fractures is increased with larotrectinib therapy [41]. Other NTRK inhibitors include entrectinib and repotrectinib.

Highly selective RET inhibitors such as selpercatinib and pralsetinib are approved for RET fusion-positive progressive radioactive-iodine-refractory differentiated thyroid cancer and for RET-mutant medullary thyroid cancer. Objective response rates are 79–85%. Common adverse events of RET inhibitors include dry mouth, gastrointestinal side effects, elevated transaminases, and QT prolongation [42, 43].

The BRAF inhibitor dabrafenib, used as monotherapy or in combination with the MEK inhibitor trametinib, is used for BRAF V600E-mutated radioactive-iodine-refractory differentiated thyroid cancer with an objective response rate of 42% for dabrafenib monotherapy and 48% for the dabrafenib/trametinib combination [44]. This combination is also approved for the treatment of locally advanced or metastatic BRAF-mutant anaplastic thyroid cancer [45]. Common adverse events include fatigue, pyrexia, chills, hyperglycemia, and skin rash.

For patients with no specific targetable mutation, both sorafenib and lenvatinib are non-specific tyrosine kinase inhibitors approved in the first-line setting. Objective response rates are 12.2% and 64.8% respectively [46, 47]. The most common adverse reactions include hypertension, fatigue, diarrhea, arthralgia/myalgia, decreased appetite, and weight loss. Cabozantinib is approved as second line therapy after progression on an anti-VEGFR therapy and reported an objective response rate of 11%, with a toxicity profile that includes diarrhea, palmar-plantar erythrodysesthesia, and hypertension [48]. These agents cause wound healing impairment through inhibition of vascular endothelial growth factors 1, 2, and 3 and should be withheld for at least 1–3 weeks prior to elective surgery and not resumed for at least 2 weeks after adequate wound healing.

Vandetanib and cabozantinib are approved for locally advanced medullary thyroid carcinoma not amenable to surgery and for patients with metastatic disease. Response rates are 45% and 28% respectively. Common adverse events include diarrhea, palmar-plantar erythrodysesthesia, decreased weight and appetite, nausea, and fatigue [49, 50].

Musculoskeletal Metastasis

Osseous metastases are the second most common site of metastasis in differentiated thyroid cancer after the lung [51]. Osseous metastases arise from a multistep process that involves loss of cell-to-cell adhesion, invasion, and dissemination through the blood and lymphatic system as well as deposition in end organs and tissues. They are observed in all thyroid cancer histologies and convey a worse prognosis. They are more frequently reported in follicular and medullary thyroid cancer patients. Most osseous metastases are initially occult, and the most common presentation is pain from mass effect or periosteal irritation, followed by pathological fractures. Imaging with plain films, CT, and MRI is critical to define the optimal management. Most osseous metastases from thyroid cancer are osteolytic; therefore, bone scans are relatively insensitive [52].

External beam radiation therapy to bone is indicated when pain, risk for fracture, or neurological complications are present or predicted, and is standard of care when surgery is not indicated. Radioactive iodine targets radioactive iodine-avid disease only and is therefore ineffective for radioactive iodine-refractory thyroid cancer, medullary thyroid cancer, and anaplastic thyroid cancer [52, 53].

Kinase inhibitors have proven clinical activity; however, they have not been specifically studied with respect to their effect on osseous metastases and appear to have an attenuated effect in osseous metastases when compared to other metastatic sites [54].

Zoledronic acid and denosumab are approved for the treatment of osseous metastases, and the ATA guidelines recommend that bisphosphonates or denosumab therapy be considered in patients with diffuse or symptomatic osseous metastases from radioactive iodine-refractory differentiated thyroid cancer, alone or in combination with other systemic therapies. Toxicity from bisphosphonates is generally mild and infrequent [55]. Osteonecrosis of the jaw can occur with antiresorptive therapy, and a dental evaluation is recommended before the use of either agent.

Prognosis

Although in general terms, thyroid cancer is a disease with a generally good prognosis, with survival rates of over 90% at 10 years, patients with metastatic osseous involvement have a significantly higher mortality rate when compared to the non-metastatic group, with 5-year and 10-year survival rates of 61% and 27%, respectively. The outcomes drastically decline in patients exhibiting multiorgan metastasis, with a 5-year survival rate of 15.3% [52, 56].

Survivorship and Quality of Life

An increasing incidence of thyroid cancer combined with a high rate of survival results in a high prevalence of thyroid cancer survivors. Findings from the North American Thyroid Cancer Survivorship Study have shown that thyroid cancer survivors experience adverse physical, psychological, social, and spiritual challenges that linger for many years following treatment. It was also found that young age, female sex, decreased education, and participation in survivor groups are associated with decreased quality of life; therefore, quality of life and survivorship issues must continue to be a focus of future research [57].

In summary, thyroid cancers, particularly the papillary and follicular subtypes, are associated with favorable long-term outcomes. Rare and aggressive forms like anaplastic thyroid carcinoma remain a clinical challenge with poor prognosis. Musculoskeletal metastases are associated with significantly worsened survival outcomes and require multidisciplinary management, including radiation therapy, surgery, systemic therapies, and bone-modifying agents. Despite high survival rates, the growing population of thyroid cancer survivors often experiences long-term quality-of-life issues that remain underrecognized. This underscores the importance of survivorship care planning, psychosocial support, and ongoing research into optimizing patient-centered outcomes alongside continued advancements in precision oncology.

References

1. National Cancer Institute Surveillance Epidemiology and End Results Program: Cancer stat facts: Thyroid cancer. https://seer.cancer.gov/statfacts/html/thyro.html
2. Surveillance, Epidemiology, and End Results Program (SEER) Cancer Statistics Review, 1975–2018. http://seer.cancer.gov/csr/1975_2018
3. Siegel RL, Miller KD, Fuchs HE, Jemal A. Cancer statistics, 2022. CA Cancer J Clin. 2022;72:7–33. https://doi.org/10.3322/caac.21708.
4. Lim H, Devesa SS, Sosa JA, et al. Trends in thyroid cancer incidence and mortality in the United States, 1974–2013. JAMA. 2017;317:1338. https://doi.org/10.1001/jama.2017.2719.
5. Powers AE, Marcadis AR, Lee M, et al. Changes in trends in thyroid cancer incidence in the United States, 1992 to 2016. JAMA. 2019;322:2440. https://doi.org/10.1001/jama.2019.18528.
6. Lee M, Powers AE, Morris LGT, Marti JL. Letter to the editor: reversal in thyroid cancer incidence trends in the United States, 2000–2017. Thyroid. 2020;30:1226–7. https://doi.org/10.1089/thy.2020.0321.
7. Noone A-M, Cronin KA, Altekruse SF, et al. Cancer incidence and survival trends by subtype using data from the surveillance epidemiology and End Results Program, 1992–2013. CEBP. 2017;26:632–41. https://doi.org/10.1158/1055-9965.epi-16-0520.
8. Davies L, Morris LGT, Haymart M, et al. American Association of Clinical Endocrinologists and American College of endocrinology disease state clinical review: the increasing incidence of thyroid cancer. Endocr Pract. 2015;21:686–96. https://doi.org/10.4158/ep14466.dscr.
9. Vaccarella S, Franceschi S, Bray F, et al. Worldwide thyroid-cancer epidemic? the increasing impact of overdiagnosis. N Engl J Med. 2016;375:614–7. https://doi.org/10.1056/nejmp1604412.
10. Masone S, Velotti N, Savastano S, et al. Morbid obesity and thyroid cancer rate. A review of literature. J Clin Med. 2021;10:1894. https://doi.org/10.3390/jcm10091894.
11. Mavragani IV, Nikitaki Z, Kalospyros SA, Georgakilas AG. Ionizing radiation and complex DNA damage: from prediction to detection challenges and biological significance. Cancer. 2019;11:1789. https://doi.org/10.3390/cancers11111789.
12. Schneider AB, Ron E, Lubin J, Stovall M, Gierlowski TC. Dose-response relationships for radiation-induced thyroid cancer and thyroid nodules: evidence for the prolonged effects of radiation on the thyroid. J Clin Endocrinol Metab. 1993;77(2):362–9.
13. Ron E, Lubin JH, Shore RE, et al. Thyroid cancer after exposure to external radiation: a pooled analysis of seven studies. Radiat Res. 1995;141:259. https://doi.org/10.2307/3579003.
14. Albright EC. Thyroid carcinoma after radiation therapy for adolescent acne vulgaris. JAMA. 1967;199:280. https://doi.org/10.1001/jama.1967.03120040090024.
15. Ron E, Modan B, Preston D, et al. Thyroid neoplasia following low-dose radiation in childhood. Radiat Res. 1989;120:516. https://doi.org/10.2307/3577801.
16. Hancock SL, Cox RS, McDougall IR. Thyroid diseases after treatment of Hodgkin's disease. N Engl J Med. 1991;325:599–605. https://doi.org/10.1056/nejm199108293250902.
17. Inskip PD. Thyroid cancer after radiotherapy for childhood cancer. Med Pediatr Oncol. 2001;36:568–73. https://doi.org/10.1002/mpo.1132.

18. Furukawa K, Preston D, Funamoto S, et al. Long-term trend of thyroid cancer risk among Japanese atomic-bomb survivors: 60 years after exposure. Int J Cancer. 2012;132:1222–6. https://doi.org/10.1002/ijc.27749.
19. Cardis E, Hatch M. The Chernobyl accident—an epidemiological perspective. Clin Oncol. 2011;23:251–60. https://doi.org/10.1016/j.clon.2011.01.510.
20. Tyer NM, Yu R. Medullary thyroid carcinoma. Endocrine Updates; 2011. p. 171–87. https://doi.org/10.1007/978-1-4614-0875-8_9.
21. Capezzone M, Robenshtok E, Cantara S, Castagna MG. Familial non-medullary thyroid cancer: a critical review. J Endocrinol Invest. 2020;44:943–50. https://doi.org/10.1007/s40618-020-01435-x.
22. Pati S, Irfan W, Jameel A, et al. Obesity and cancer: a current overview of epidemiology, pathogenesis, outcomes, and management. Cancer. 2023;15:485. https://doi.org/10.3390/cancers15020485.
23. Renehan AG, Tyson M, Egger M, et al. Body-mass index and incidence of cancer: a systematic review and meta-analysis of prospective observational studies. Lancet. 2008;371:569–78. https://doi.org/10.1016/s0140-6736(08)60269-x.
24. Fagin JA, Wells SA. Biologic and clinical perspectives on thyroid cancer. N Engl J Med. 2016;375:1054–67. https://doi.org/10.1056/nejmra1501993.
25. Kitahara CM, Schneider AB. Epidemiology of thyroid cancer. CEBP. 2022;31:1284–97. https://doi.org/10.1158/1055-9965.epi-21-1440.
26. Hamidi S, Hofmann M-C, Iyer PC, et al. Review article: new treatments for advanced differentiated thyroid cancers and potential mechanisms of drug resistance. Front Endocrinol. 2023; https://doi.org/10.3389/fendo.2023.1176731.
27. Baloch ZW, LiVolsi VA. Aggressive variants of papillary thyroid carcinoma and poorly differentiated carcinoma. In: Thyroid Cancer; 2016. p. 817–23. https://doi.org/10.1007/978-1-4939-3314-3_80.
28. Toraih EA, Hussein MH, Zerfaoui M, et al. Site-specific metastasis and survival in papillary thyroid cancer: the importance of brain and multi-organ disease. Cancer. 2021;13:1625. https://doi.org/10.3390/cancers13071625.
29. McHenry CR, Phitayakorn R. Follicular adenoma and carcinoma of the thyroid gland. Oncologist. 2011;16:585–93. https://doi.org/10.1634/theoncologist.2010-0405.
30. Nikiforova MN, Lynch RA, Biddinger PW, et al. RAS point mutations and PAX8-PPARΓ rearrangement in thyroid tumors: evidence for distinct molecular pathways in thyroid follicular carcinoma. JCEM. 2003;88:2318–26. https://doi.org/10.1210/jc.2002-021907.
31. Wu M-H, Lee Y-Y, Lu Y-L, Lin S-F. Risk factors and prognosis for metastatic follicular thyroid cancer. Front Endocrinol. 2022; https://doi.org/10.3389/fendo.2022.791826.
32. Leboulleux S, Baudin E, Travagli J, Schlumberger M. Medullary thyroid carcinoma. Clin Endocrinol. 2004;61:299–310. https://doi.org/10.1111/j.1365-2265.2004.02037.x.
33. Dabelić N. Medullary thyroid cancer—feature review and update on systemic treatment. Acta Clin Croat. 2020; https://doi.org/10.20471/acc.2020.59.s1.06.
34. Pavlidis ET, Galanis IN, Pavlidis TE. Update on current diagnosis and management of anaplastic thyroid carcinoma. World J Clinic Oncol. 2023;14:570–83. https://doi.org/10.5306/wjco.v14.i12.570.
35. Smallridge RC, Marlow LA, Copland JA. Anaplastic thyroid cancer: molecular pathogenesis and emerging therapies. Endocr Relat Cancer. 2009;16:17–44. https://doi.org/10.1677/erc-08-0154.
36. Rago T, Vitti P. Risk stratification of thyroid nodules: from ultrasound features to TIRADS. Cancer. 2022;14:717. https://doi.org/10.3390/cancers14030717.
37. Ali SZ, Baloch ZW, Cochand-Priollet B, et al. The 2023 Bethesda system for reporting thyroid cytopathology. Thyroid. 2023; https://doi.org/10.1089/thy.2023.0141.
38. Amin MB, Edge SB, Greene FL, et al., editors. AJCC cancer staging manual. 8th ed. New York: Springer; 2017.

39. Avram AM, Zukotynski K, Nadel HR, Giovanella L. Management of differentiated thyroid cancer: the standard of care. J Nucl Med. 2021;63:189–95. https://doi.org/10.2967/jnumed.121.262402.
40. Pekova B, Sykorova V, Mastnikova K, et al. NTRK fusion genes in thyroid carcinomas: clinicopathological characteristics and their impacts on prognosis. Cancer. 2021;13:1932. https://doi.org/10.3390/cancers13081932.
41. Drilon A, Laetsch TW, Kummar S, et al. Efficacy of larotrectinib in*trk*fusion–positive cancers in adults and children. N Engl J Med. 2018;378:731–9. https://doi.org/10.1056/nejmoa1714448.
42. Demetri GD, De Braud F, Drilon A, et al. Updated integrated analysis of the efficacy and safety of entrectinib in patients with *ntrk* fusion-positive solid tumors. Clin Cancer Res. 2022;28:1302–12. https://doi.org/10.1158/1078-0432.ccr-21-3597.
43. Subbiah V, Hu MI, Wirth LJ, et al. Pralsetinib for patients with advanced or metastatic RET-altered thyroid cancer (arrow): a multi-cohort, open-label, registrational, phase 1/2 study. Lancet Diabet Endocrinol. 2021;9:491–501. https://doi.org/10.1016/s2213-8587(21)00120-0.
44. Busaidy NL, Konda B, Wei L, et al. Dabrafenib versus dabrafenib + trametinib in *braf*-mutated radioactive iodine refractory differentiated thyroid cancer: results of a randomized, phase 2, open-label multicenter trial. Thyroid. 2022; https://doi.org/10.1089/thy.2022.0115.
45. Subbiah V, Kreitman RJ, Wainberg ZA, et al. Dabrafenib and trametinib treatment in patients with locally advanced or metastatic *braf* V600–mutant anaplastic thyroid cancer. J Clin Oncol. 2018;36:7–13. https://doi.org/10.1200/jco.2017.73.6785.
46. Brose MS, Nutting CM, Jarzab B, et al. Sorafenib in radioactive iodine-refractory, locally advanced or metastatic differentiated thyroid cancer: a randomised, double-blind, phase 3 trial. Lancet. 2014;384:319–28. https://doi.org/10.1016/s0140-6736(14)60421-9.
47. Schlumberger M, Tahara M, Wirth LJ, et al. Lenvatinib versus placebo in radioiodine-refractory thyroid cancer. N Engl J Med. 2015;372:621–30. https://doi.org/10.1056/nejmoa1406470.
48. Brose MS, Robinson BG, Sherman SI, et al. Cabozantinib for previously treated radioiodine-refractory differentiated thyroid cancer: updated results from the phase 3 cosmic-311 trial. Cancer. 2022;128:4203–12. https://doi.org/10.1002/cncr.34493.
49. Wells SA, Robinson BG, Gagel RF, et al. Vandetanib in patients with locally advanced or metastatic medullary thyroid cancer: a randomized, double-blind phase III trial. J Clin Oncol. 2012;30:134–41. https://doi.org/10.1200/jco.2011.35.5040.
50. Elisei R, Schlumberger MJ, Müller SP, et al. Cabozantinib in progressive medullary thyroid cancer. J Clin Oncol. 2013;31:3639–46. https://doi.org/10.1200/jco.2012.48.4659.
51. Choksi P, Papaleontiou M, Guo C, et al. Skeletal complications and mortality in thyroid cancer: a population-based study. JCEM. 2017;102:1254–60. https://doi.org/10.1210/jc.2016-3906.
52. Iñiguez-Ariza NM, Bible KC, Clarke BL. Bone metastases in thyroid cancer. J Bone Oncol. 2020;21:100282. https://doi.org/10.1016/j.jbo.2020.100282.
53. Chow E, Zeng L, Salvo N, et al. Update on the systematic review of palliative radiotherapy trials for bone metastases. Clin Oncol. 2012;24:112–24. https://doi.org/10.1016/j.clon.2011.11.004.
54. Cabanillas ME, Hu MI, Jimenez C. Medullary thyroid cancer in the era of tyrosine kinase inhibitors: to treat or not to treat—and with which drug—those are the questions. JCEM. 2014;99:4390–6. https://doi.org/10.1210/jc.2014-2811.
55. Haugen BR, Alexander EK, Bible KC, et al. 2015 American Thyroid Association Management Guidelines for adult patients with thyroid nodules and differentiated thyroid cancer: the American Thyroid Association guidelines task force on thyroid nodules and differentiated thyroid cancer. Thyroid. 2016;26:1–133. https://doi.org/10.1089/thy.2015.0020.
56. Kato S, Demura S, Shinmura K, et al. Current management of bone metastases from differentiated thyroid cancer. Cancer. 2021;13:4429. https://doi.org/10.3390/cancers13174429.
57. Aschebrook-Kilfoy B, James B, Nagar S, et al. Risk factors for decreased quality of life in thyroid cancer survivors: initial findings from the North American Thyroid Cancer Survivorship Study. Thyroid. 2015;25:1313–21. https://doi.org/10.1089/thy.2015.0098.

Medical Management of Multiple Myeloma: An Orthopedic Blood Cancer

30

Yonatan Najman-Licht and James E. Hoffman

Introduction

Multiple myeloma (MM) is a heterogenous malignancy of plasma cells, generally detected in the bone marrow. In myeloma, malignant/clonal plasma cells generally overproduce *M Proteins* (monoclonal proteins)—abnormal immunoglobulins that are unable to fight infection and, in excess, can cause renal dysfunction or other bodily harm directly. Multiple myeloma compromises only 2% of cancer cases in the US, but is the second most common hematologic malignancy. It is slightly more common in men than women (the American Cancer Society estimates that 19,520 men and 16,260 women will be diagnosed with MM in 2024) [1], and is most common in males over 40 in the African American community [2]. Internationally, multiple myeloma accounts for 0.8% of new cancer diagnoses, which is approximately 86,000 new cases per year. It is also responsible for roughly 63,000 annual deaths globally [3]. It remains a highly treatable, but incurable cancer, with patients living more than 8 years on average (with prognosis improving over time, given the rapid pace of drug development and improved diagnostic tools).

Genetic Drivers/Prognosis

The exact cause of multiple myeloma is unknown, and it is not hereditary in most cases. There are certain acquired gene mutations that have prognostic, and even at times, predictive ramifications. Translocations account for many of the MM mutations, and translocations involving chromosome 14 are most common, being found in about 40% of myeloma patients [4]. The promoter for immunoglobulin

Y. Najman-Licht · J. E. Hoffman (✉)
Sylvester Comprehensive Cancer Center, University of Miami, Miami, FL, USA
e-mail: J.hoffman4@med.miami.edu

J. Pretell-Mazzini, S. A. Lozano Calderón (eds.), *Multidisciplinary Approach of Musculoskeletal Metastasis*, https://doi.org/10.1007/978-3-032-09498-8_30

heavy chains rests on this part of the 14th chromosome and is a highly active promoter (as immunoglobulins are robustly produced). When this hyperactive promoter gets translocated in front of an oncogene, this can trigger myeloma to develop or progress. The most common translocations are t(4;14), t(14;16) and t(11;14). The t(11;14) translocation, also found in mantle cell lymphoma, is unique [5]. The oncogene on chromosome 11 is cyclin D1, and myeloma clones driven by Cyclin D1 have unique behaviors and different susceptibilities to some of the typical treatments. It is not a high-risk finding, unlike the other noted translocations [6]. The most common genetic pattern seen in MM is hyperdiploidy (HDMM), and is found in around 50% of diagnoses, connoting a good prognosis [7]. The highest risk genotype involves mutations or deletions of chromosome 17— all or in part (17p), which renders the tumor suppressor gene P53 ineffective or absent [8]. Abnormalities in chromosome 1 are generally high risk as well (e.g., +1q (particularly amplification as opposed to duplication) and -1p). Having more than one high-risk factor, sometimes referred to as "double hit" or "triple hit," connotes a particularly poor prognosis. The standard staging systems incorporate 2 blood tests—B2M and albumin—but are less helpful prognostically than the genetics (see more on staging below).

Risk Factors

Risk factors for multiple myeloma involve older age, African American ethnicity, male sex, family history, previous "precursor" plasma cell conditions, and certain chemical/radiation exposures. Patients under 35 account for only roughly 1% of MM cases; it is most common in individuals over 65. MM is twice as prevalent in African Americans as in white Americans, and is more common in North Africa, the Middle East, and the Mediterranean, but the reason for this is unknown [2]. All cases of myeloma are preceded by a diagnosed or undiagnosed precursor condition called "monoclonal gammopathy of undetermined significance (MGUS). In MGUS, patients have (generally) non-malignant, but abnormal and immortalized plasma cells that, like MM, produce excess M protein. MGUS should be untreated, and patients are, by definition, asymptomatic. Approximately 1% of patients with MGUS will develop MM yearly. Figure 30.1 depicts the progression of a normal cell to MGUS and MM. A more advanced condition, called smoldering myeloma (SMM), involves a higher burden plasma cell neoplasm (with Mspike >3 or plasma cell % in the marrow >10%) without symptoms/"CRAB" criteria (see below) [9]. These patients are usually monitored, occasionally treated (this is controversial), and have an approximately 10% per year risk of developing symptomatic MM within 5 years. After 5 years, the risk drops to approximately 3% annually [10]. As with many cancers, radiation increases the risk of MM as well as being exposed to chemicals, such as benzene, some pesticides and herbicides. Agent Orange is an example of a proven cause, and it is notable that the 9/11 responders have higher rates of MM due to related exposures [11].

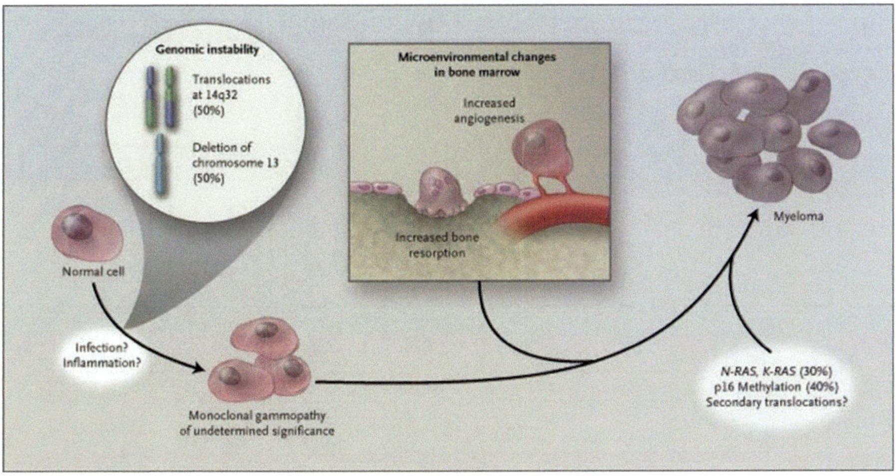

Fig. 30.1 Mechanisms of Disease Progression in the Monoclonal Gammopathies: Percentages are the proportions of patients with the abnormality. Increased bone resorption and the induction of angiogenesis are microenvironmental changes that can cause progression to myeloma. First, affected cells are MGUS, which produce M protein and are not malignant. Then, in only 1% of cases per year, it progresses to myeloma [26]

Classification

Multiple myeloma is classified by the type of heavy chain and light chain that the plasma cells produce. Notably, the antibodies produced are all the same in any one patient (M protein = monoclonal protein, or a singular protein from a singular clone). Heavy chains- from most common to least common- are G, A, M, and D. They are referred to using the prefix "Ig" (immunoglobulin), as IgG, IgA, IgM, and IgD. The most common type of MM is IgG kappa, meaning two IgG heavy chains and two kappa light chains. Figure 30.2 highlights an example of IgG lambda MM, showing a spike in gamma heavy chains in a patient's serum. IgM MM is very rare, so elevated levels of IgM in the blood or urine often indicate Waldenstrom's Macroglobulinemia (or other non-Hodgkin lymphomas), which is a B-cell malignancy, with myeloma being far less likely. Another classification of MM is "light chain myeloma." In these cases, the plasma cells only make light chains- kappa or lambda- which can be directly measured, but do not have a monoclonal spike (indicating lack of a clonal complete antibody). Historically, some patients were thought to have had "non-secretory myeloma;" however, free light chains are usually abnormal in these cases, indicating that the clonal plasma cells do secrete, just not a complete antibody. Free light chains can be detected in the urine (the Bence-Jones proteins) but are most easily detected and followed with the serum "Free Light Assay," which detects free light chains in the blood (distinct from light chains attached to heavy chains in complete immunoglobulins). Nowadays, urine testing is not as important, and in fact, can often be deferred. True "non-secretory" myeloma still exists but is exceedingly rare.

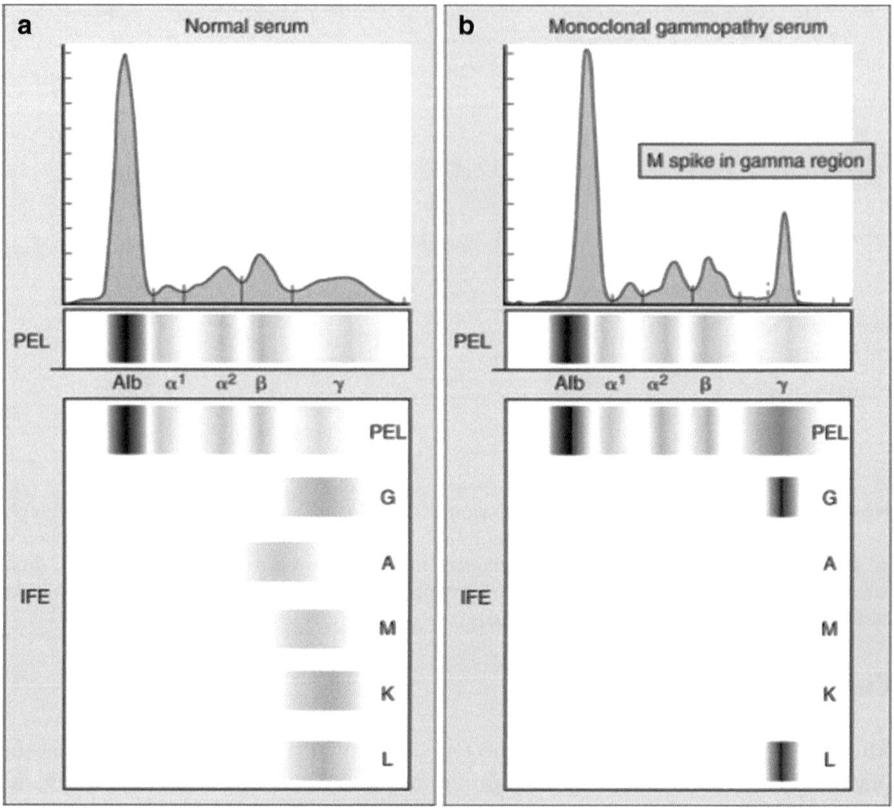

Fig. 30.2 M-spike in serum sample using protein electrophoresis: G, A, M refer to heavy chains gamma, alpha, and mu, respectively, while K, L refer to light chains kappa and lambda, respectively. Normal serum shows little concentration in the gamma region. The myeloma serum sample reveals spikes in the gamma and lambda regions [27]

Extramedullary (EM) myeloma is another, very high-risk, subtype. In EM disease, malignant plasma cells grow outside the marrow, e.g., in organs, spinal fluid, or bloodstream. Plasma cells that have become so undifferentiated that they grow in places healthy plasma cells shouldn't, represent a high-risk scenario.

Solitary plasmacytoma is another unique subtype. In such cases, a singular plasma cell tumor develops without symptomatic systemic disease. The marrow in such cases can be uninvolved or minimally involved, and treatment is generally localized with radiation or, rarely, surgical resection. Most such cases will still relapse over time, but some will not, and recurrences can happen years in the future, delaying toxic systemic therapy.

Diagnosis and Staging

Typically, multiple myeloma is diagnosed and monitored using various lab tests and imaging. The lab test includes standard CBC and CMP, as well as a panel focused on the identification and measurement of the paraprotein (clonal antibody protein).

This panel includes quantitative immunoglobulins, serum protein electrophoresis (SPEP), serum immunofixation (SIF) and serum free light chains (FLC). LDH, B2M, and Vitamin D levels are also routinely checked initially as they have some prognostic and treatment-related relevance.

When diagnosing MM, physicians look for four factors related to symptoms or bodily damage (to distinguish MM from its precursor conditions MGUS and Smoldering myeloma)- making up the acronym CRAB—C: hypercalcemia, R: renal failure, A: anemia, B: bone lesions.

Hypercalcemia

High calcium levels are a common metabolic indicator of MM and are found in about 30% of patients at the time of diagnosis [12]. Myeloma cells increase the rate at which osteoclasts break down bone, and as such, more calcium is released into the blood. Even though hypercalcemia is not always present at the time of diagnosis, patients may develop it later with relapses, and it is associated with a poor prognosis. At times, such hypercalcemia can represent an emergency with patients presenting with lethargy and/or arrhythmias.

Renal Failure

Multiple myeloma frequently leads to renal complications — induced by the pathologic clonal antibodies, with renal failure occurring in about 20–50% of patients at diagnosis [8]. Renal involvement significantly worsens prognosis, as renal failure in multiple myeloma is associated with lower survival rates and limits treatment options. The primary mechanism is linked to excessive production of monoclonal immunoglobulins, particularly free light chains, which accumulate in the kidneys, causing tubular damage. This condition, known as "cast nephropathy," occurs when filtered light chains precipitate within the renal tubules, forming obstructive casts that lead to tubular injury and interstitial inflammation. Additional contributors include hypercalcemia, dehydration, infection, and hyperuricemia, all of which compound renal injury. Pathologically, renal biopsies reveal tubulointerstitial nephritis with casts within the tubules. Treatment involves controlling multiple myeloma itself through fast-acting chemotherapeutic agents like proteasome inhibitors (e.g., bortezomib) and steroids- sometimes with cyclophosphamide-, along with supportive care such as hydration, dialysis, and occasionally plasmapheresis to reduce free light chains. Early and aggressive management of multiple myeloma can improve renal outcomes, though renal recovery is variable and depends on the degree of existing damage and the response to therapy [13–15].

Anemia

One core indicator of symptomatic MM is found in the complete blood count (CBC). If a patient has hemoglobin levels 2 g/dL below normal (14–18 g/dL for men and 12–16 g/dL for women), that is indicative of anemia and is a common sign of MM [16]. MM cells in the bone marrow can interfere with red blood cell (RBC) production, resulting in low hemoglobin levels. Symptoms of anemia can include rapid heartbeat, shortness of breath, dizziness, and fatigue. Anemia is incredibly common in myeloma patients and is found in 60–70% of patients at the time of diagnosis, and many others experience it at some point while having the disease [17]. One additional reason for the drop in hemoglobin in myeloma patients is renal injury. Regulators of erythropoiesis, GATA1 and KLF1, are downregulated in myeloma hematopoietic stem and progenitor cells (HSPCs) [18]. The damage to the kidney done by MM leads to a decrease in production of erythropoietin and thus RBC generation.

Bone Lesions

The standard way to detect and evaluate bone lesions in multiple myeloma is through radiologic scans. X-rays, MRI, CT, and PET scans can all be used to identify bone lesions and can indicate the presence of the disease. X-rays, while still sometimes used (as part of the classic "skeletal survey"), are not as sensitive for bone lesions and require a substantial part of the bone to be missing before the scan can sense it. At least 30% of the trabecular bone (where myeloma grows) must be degraded for the X-ray to pick it up. As such, they should not be relied upon to rule out lytic lesions. MRI can sensitively pick up on focal lesions in the skeletal bone and even the bone marrow, even early in the course of disease [19]. A limitation of MRI is that they don't perfectly distinguish between active lesions and treated ones. CT scans are successful in detecting small bone lesions, including ones that do not appear in X-ray images. Low-dose radiation whole body CT scans (WBLDCT) are often recommended at diagnosis because it is fast, is sensitive to lesions and soft tissue masses, use significantly less radiation than normal CT, and do not require contrast. PET scans can also be used to detect bone lesions through myeloma cell uptake of fluorodeoxyglucose (FDG). The scan can provide real-time images of the myeloma and can help set a baseline at the time of diagnosis and identify the severity of the disease. PET scans are ideal to distinguish active disease from treated disease, by the amount of FDG uptake. As a limitation, due to the high uptake of FDG in the brain, there is concern that skull lesions can be overlooked and missed. Newer types of PET scan, utilizing monoclonal antibody-based tracers, as opposed to glucose, offer promise to better allow these scans to differentiate between active disease and benign areas of high sugar uptake (e.g., infection or inflammation) [20].

After scans indicate a bone lesion, lesional biopsies can be used to conclusively prove myeloma; in the right clinical context, with paraprotein and a positive bone marrow biopsy, lesional biopsies are often not needed. Bone marrow aspiration and core biopsies are not targeted to any lesion and involve a random marrow biopsy, usually in the posterior iliac crest. Testing involves standard immunohistochemistry, slow cytometry, cytogenetics, and FISH testing. If at least 10% of the cells in the bone marrow are myeloma, it can be detected [21].

Table 30.1 Staging for multiple myeloma. Staging is based on serum levels of albumin, beta-2-microglobulin, and LDH, as well as cytogenetic risk [25]

Revised international staging system for myeloma
Stages
Stage I
All of the following:
Serum albumin 3.5 gm/dL
Serum beta-2-microglobulin <3.5 mg/L
No high risk cytogenetics
Normal serum lactate dehydrogenase level
Stage II
Not fitting Stage I or III
Stage III
Both of the following:
Serum beta-2-microglobulin <5.5 mg/L
High risk cytogenetics [t(4;14), t(14;16), or del(17p)] or elevated serum lactate dehydrogenase level

Staging

Staging in solid tumors often factors in tumor size and location, with a focus on metastases that render the cancer not surgically curable. In liquid tumors, like myeloma, they are systemic diseases in general, making localized therapy unhelpful. As such, stage does NOT focus on size/spread of tumors, but rather on characteristics of the disease that might make it easier or harder to systemically control with medications. There are four main metrics used to stage myeloma patients: (1) Levels of Beta 2 microglobulin (2) Levels of Albumin (3) Levels of Lactate dehydrogenase (4) Genetic mutations (acquired) (Table 30.1). There have been many staging systemics and modifications over time, but simply speaking:

Stage 1—Beta 2 microglobulin is under 3.5 mg/L, Albumin is over 3.5 g/dL, LDH is normal, and no high-risk chromosome changes.

Stage 2—Either Beta 2 microglobulin is 3.5–5.5 mg/L and Albumin is at any level, or Beta 2 microglobulin is under 3.5 mg/L, but Albumin is less than 3.5 g/dL, LDH is normal, and absence of high-risk genetic changes.

Stage 3—Beta 2 microglobulin is over 5.5 mg/L, LDH is high, or high-risk genetics.

Management

Systemic Therapy

The cornerstone of treating MM is systemic therapy, as it is generally speaking a systemic, incurable disease. Until approximately the year 2000, the core treatments were steroids and cytotoxic chemotherapy (e.g., melphalan, cyclophosphamide, and doxorubicin). Melphalan is still used, primarily in the form of high-dose therapy

(HDT) with autologous stem cell rescue (also commonly referred to as autologous stem cell transplant- or auto-SCT). Otherwise, the modern treatment of MM is more targeted, effective, and gentle.

Treatment is generally given in phases. Induction is the cyclical administration of medications to drive the MM down. Consolidation is considered, with auto-SCT for those eligible and agreeable. Finally, maintenance therapy is continuous low intensity therapy, often oral, that is often continued until progression.

There are several classes of medication that are used regularly. Steroids, usually dexamethasone, are a part of most treatment regimens. Lower doses have been used over time- 20 mg of dexamethasone weekly is a typical dose and schedule. Cytotoxic chemotherapy (e.g., cyclophosphamide) is used, though outside of melphalan-based autoSCT, they are used sparingly. Auto-SCT still has a role in consolidating responses to induction therapy, but the risk/benefit is debated, as there is no clear evidence of an overall survival benefit in the era of modern therapy (the primary benefit being on progression-free survival). Modern therapy took its form in the early 2000s with the discovery of thalidomide's effectiveness. This drug was originally tried based on a suggestion of a patient's spouse and based on its anti-angiogenic effects. The true mechanism involves the protein cereblon, and it is considered an immunomodulator, but a deeper discussion is beyond the scope of this chapter. Two newer forms of immunomodulators have been developed sequentially, lenalidomide and pomalidomide, with each newer drug having both tolerability and efficacy benefits when compared with its predecessor (in fact, thalidomide is not often used in the United States for this reason). These are oral meds that share risks of causing rash, blood clots, and all notorious teratogens (so many safeguards exist to keep them away from women who may be pregnant) (Table 30.2).

Another critical class of medicine is the proteasome inhibitors (PI). Plasma cells are prolific protein (antibody) producers and therefore do not tolerate the paralysis of the cellular garbage can (the proteasome). Bortezomib was the first in this class and is generally given subcutaneously. Its most difficult side effect is peripheral neuropathy (PN). Carfilzomib is a newer drug within the same class that is more potent (an irreversible PI) and does not carry PN risk (Table 30.2). However, it can cause fluid overload/dyspnea and must be given IV.

The final core class of systemic therapy is the monoclonal antibody (mAb) class. The core drug here is daratumumab, a mAb targeting CD38, a protein found on the surface of plasma cells. This is typically given subcutaneously with premedication to mitigate reactions to the first few doses. It induces hypogammaglobulinemia and dampens antibody generation from vaccines- but is generally very effective and an excellent partner with other drugs. The classes of medicines and surgical considerations when taking them are listed in Table 30.2.

A typical induction regimen for myeloma is a "quad" combination (mAb, PI, immunomodulator, and dexamethasone) for several months. Then consideration of auto-SCT (or possibly stem cell harvest for future auto-SCT)- then maintenance

Table 30.2 Common drug categories used for MM—along with their clinical uses, side effects, and related surgical considerations

Drug class (name)	Clinical use	Side effects	Surgical considerations
Immunomodulator (lenalidomide, pomalidomide)	Used in all lines of therapy	Venous thromboembolism, rash, cytopenias, diarrhea	Require anticoagulant—patients cannot be on them when AC is contraindicated
Proteasome inhibitor (bortezomib, carfilzomib)	Used in all lines of therapy	Peripheral neuropathy (bortezomib), fluid overload/dyspnea (carfilzomib), shingles	none
Monoclonal antibody (daratumumab- to CD38)	Used in all lines of therapy	Infusion reactions, hypogammaglobulinemia	Daratumumab will attach to RBCs and suggest autoimmune hemolytic anemia, making typing for transfusions more complicated (though low clinical risk of transfusion reactions with 'best typed units')
Steroids (dexamethasone)	Used in all lines of therapy	Hyperglycemia, insomnia, PCP, muscle/bone weakness, glaucoma/ cataracts	Wound healing impact
Cytotoxic chemo (cyclophosphamide, melphalan)	Used in relapsed disease or in the form of auto-SCT as consolidation (melphalan)	Cytopenias, immunosuppression	Low blood counts common
T cell based—CAR-T and bispecifics (teclistamab, elranatamab, talquetamab)	Relapsed disease	Cytokine release syndrome (CRS), immunosuppression	None
Bisphosphonates (zoledronic acid), RANKL inhibitors (denosumab)	Bone strengthening, hypercalcemia	Osteonecrosis of the jaw, hypocalcemia, renal injury (zoledronic acid)	Induce delayed bone healing- ideally held as long as possible

immunomodulator. In the US, daratumumab-bortezomib- lenalidomide-dex is a very common induction regimen. Drugs from these classes can be paired and re-utilized for relapse. For multiple relapsed patients, new T cell-directed therapies have made a tremendous impact. Specifically chimeric antigen receptor T cell therapy (CAR-T) and bispecific/T cell engagers have both revolutionized the management of relapsed patients and are being tested in earlier lines of therapy. Both of these approaches provoke T cells to fight the MM. They are associated with immune reactions called cytokine release syndrome (CRS), neurologic reactions (ICANS), and infectious risks.

Stem Cell Transplant

After initial medical therapy (induction), consideration is given to consolidation with high-dose melphalan and autologous stem cell rescue- commonly referred to as autologous stem cell transplant. This is a short-term, intense treatment that is usually associated with a 2–3 week hospitalization, hair loss, and at least 3 months out of work. It prolongs progression-free survival (PFS), and in some settings has shown improvement in overall survival (OS). Patients who are treated with typical myeloma drugs and a stem cell transplant are at higher risk of secondary cancers, including myelodysplastic syndrome (MDS) and acute leukemia. Because of its toxicity and the abundance of less intensive treatment options, the use of SCT has waned somewhat over recent years and is projected to continue to do so. It still has an important role in the treatment of myeloma, and decisions about its use come down to the patient and medical team having discussion about risks and benefits and priorities.

Bone Focused Therapy

In MM, there are also drugs used to specifically target bone damage. One category of drugs used to treat bones is bisphosphonates. The most common ones, Zoledronic acid (Zometa) and pamidronate (Aredia), slow down the activity of osteoclasts, which help prevent fractures, reduce pain, and lower calcium levels in the blood. These drugs are given through IV and take about 15 min to administer. Potential side effects of these drugs, while rare, include jawbone decay (osteonecrosis of the jaw/ONJ) and spontaneous femur fractures. The ONJ risk is associated with cumulative exposure and dental work being done while on drugs. Another common drug used to treat bone damage in MM is denosumab (Xgeva). Denosumab is a biologic drug- a RNKL inhibitor- blocking signals sent to osteoclasts to break down bones. It is more costly than bisphosphonate therapy and has fewer short-term side effects, and tends to be used primarily for bisphosphonate intolerance or advanced renal failure (where bisphosphonate therapy is contraindicated).

Musculoskeletal therapies for MM also include invasive procedures such as kyphoplasty, vertebroplasty, rods/screws in the axial skeleton, and prophylactic fixations with intramedullary nails, plates/screws +/− cement, as well as endoprosthesis in the appendicular skeleton [22]. Of course, effective systemic therapy is the key to the best surgical outcomes, and generally, bone strengthening is critical.

Radiation Therapy

Like many other malignancies, MM can be treated using radiation. Plasmacytomas that are either solitary or cause local symptoms not alleviated with systemic therapy are amenable to radiation-based approaches. Radiation in general is associated with secondary malignancies and needs to be used with discretion- especially in a disease that is systemic most of the time [23].

Response Assessments

Patients undergoing treatment are monitored primarily with blood testing, following the trajectory of the paraprotein. As the paraprotein improves by 50%, patients are deemed to be in “partial remission” or PR, at a 90% drop- a “very good partial remission” or VGPR, and when it is completely absent- a “complete remission” or CR. Long-term outcomes are greatly affected by the depth of response [24]. Modern technology allows an even deeper assessment of the status of myeloma called “minimal residual disease”- or MRD- assessments. These MRD assessments are usually done via follow-up bone marrow biopsies using flow cytometry or genetically based markers. Patients who are MRD negative, especially if they are repeatedly MRD negative, have a particularly good prognosis. For rising paraprotein (or clinical progression- such as new bone tumors), salvage treatment options are employed. Patients historically have had cycles or response/relapse of shorter durations until they develop life threatening complications of disease, but in the current era, with new immunotherapies and targeted therapies, many patients will have more potent options available for a relapse than they did before, giving hope for longer subsequent response times.

References

1. Key statistics for multiple myeloma. Key Statistics for Multiple Myeloma | American Cancer Society. (n.d.). https://www.cancer.org/cancer/types/multiple-myeloma/about/key-statistics.html
2. Kazandjian D. Multiple myeloma epidemiology and survival: a unique malignancy. Seminar Oncol. 2016;43(6):676–81. https://doi.org/10.1053/j.seminoncol.2016.11.004.
3. Fett-Conte, et al. New complex chromosome abnormalities in multiple myeloma associated with a poor prognosis: a case report. Leuk Res. 2012;36(7):e140–2. https://doi.org/10.1016/j.leukres.2012.02.033.
4. Kalff A, Spencer A. The T(4;14) translocation and FGFR3 overexpression in multiple myeloma: Prognostic implications and current clinical strategies. Blood Cancer J. 2012;2(9) https://doi.org/10.1038/bcj.2012.37.
5. Fonseca R, et al. International myeloma working group molecular classification of multiple myeloma: spotlight review. Leukemia. 2009;23(12):2210–21. https://doi.org/10.1038/leu.2009.174.
6. Bolli N, et al. Analysis of the genomic landscape of multiple myeloma highlights novel prognostic markers and disease subgroups. Leukemia. 2018;32(12):2604–16. https://doi.org/10.1038/s41375-018-0037-9.
7. Prideaux SM, Conway O’Brien E, Chevassut TJ. The genetic architecture of multiple myeloma. Adv Hematol. 2014;2014:1–16. https://doi.org/10.1155/2014/864058.
8. Hanamura I. Multiple myeloma with high-risk cytogenetics and its treatment approach. Int J Hematol. 2022;115(6):762–77. https://doi.org/10.1007/s12185-022-03353-5.
9. Mateos M-V, Landgren O. Mgus and smoldering multiple myeloma: Diagnosis and epidemiology. In: Cancer Treatment Res; 2016. p. 3–12. https://doi.org/10.1007/978-3-319-40320-5_1.
10. Kyle RA, Dispenzieri A. Monoclonal gammopathies. In: Clin Immunol; 2013. p. 975–86. https://doi.org/10.1016/b978-0-7234-3691-1.00094-5.
11. Ola Landgren M. Multiple myeloma among firefighters exposed to the World Trade Center disaster. JAMA Oncol. 2018, June 1; https://jamanetwork.com/journals/jamaoncology/fullarticle/2678962

12. Buege MJ, et al. Corrected calcium versus ionized calcium measurements for identifying hypercalcemia in patients with multiple myeloma. Cancer Treat Res Commun. 2019;21:100,159. https://doi.org/10.1016/j.ctarc.2019.100159.
13. Stringer S, Cook M, Cockwell P. Achieving an early myeloma response in patients with kidney impairment. Adv Chronic Kidney Dis. 2012;19(5):303–11. https://doi.org/10.1053/j.ackd.2012.07.005.
14. Dimopoulos MA, et al. International myeloma working group recommendations for the diagnosis and management of myeloma-related renal impairment. J Clin Oncol. 2016;34(13):1544–57. https://doi.org/10.1200/JCO.2015.65.0044.
15. Rajkumar SV, Kumar S. Multiple myeloma: diagnosis and treatment. Mayo Clinic Proc. 2016;91(1):101–19. https://doi.org/10.1016/j.mayocp.2015.11.007.
16. Guideline on haemoglobin cutoffs to define anaemia in individuals and populations. (2024). World Health Organization.
17. Multiple myeloma anemia | international myeloma foundation. https://www.myeloma.org/multiple-myeloma-anemia
18. Liu L, et al. Multiple myeloma hinders erythropoiesis and causes anemia owing to high levels of CCL3 in the bone marrow microenvironment. Sci Rep. 2020;10(1):20,508. https://doi.org/10.1038/s41598-020-77450-y.
19. Derlin T. Imaging of multiple myeloma: current concepts. World J Orthoped. 2014;5(3):272. https://doi.org/10.5312/wjo.v5.i3.272.
20. Bailly C, Chalopin B, Gouard S, Carlier T, Saëc PR, Marionneau-Lambot S, Moreau P, Touzeau C, Kraeber-Bodere F, Bodet-Milin C, Chérel M. ImmunoPET in multiple myeloma—what? so what? now what? Cancers. 2020;12(6):1467. https://doi.org/10.3390/cancers12061467.
21. Albagoush SA (2023, January 30) Multiple myeloma. StatPearls [Internet]. https://www.ncbi.nlm.nih.gov/books/NBK534764/#:~:text=Historically%2C%20we%20have%20diagnosed%20MM,%2Dout%2C%20round%2C%20radiolucent%20lesions
22. Longo UG, et al. The effectiveness and safety of vertebroplasty for osteoporotic vertebral compression fractures. A double blind, prospective, randomized, controlled study. Clinical cases in mineral and bone metabolism. Off J Ital Soc Osteoporosis Miner Metab Skeletal Dis. 2010;7(2):109–13.
23. Talamo G, Dimaio C, Abbi KK, Pandey MK, Malysz J, Creer MH, Zhu J, Mir MA, Varlotto JM. Current role of radiation therapy for multiple myeloma. Front Oncol. 2015;5:40. https://doi.org/10.3389/fonc.2015.00040.
24. Tate JR. The paraprotein—an enduring biomarker. Clin Biochem Rev. 2019;40(1):5–22.
25. Rajkumar SV. Multiple myeloma: 2022 update on diagnosis, risk stratification, and management. Am J Hematol. 2022;97(8):1086–107. https://doi.org/10.1002/ajh.26590.
26. Kyle RA, Rajkumar SV. Multiple myeloma. New England J Med. 2004;351(18):1860–73. https://doi.org/10.1056/nejmra041875.
27. Rajkumar SV, Dispenzieri A. Multiple myeloma and related disorders. Abeloff Clin Oncol. 2020; https://doi.org/10.1016/b978-0-323-47674-4.00101-8.

Radiation Oncology Strategies for the Management of Musculoskeletal Metastasis

31

Matthew D. Hall, Keitrina Mair, and Adam L. Holtzman

Introduction

Bone metastases denote one of the most common sites of metastatic disease and pose a common problem for cancer patients. While some patients may remain asymptomatic from metastatic disease for long periods, bone metastases can progress, cause pain, and lead to skeletal-related events (SRE) such as pathological fracture, hypercalcemia, spinal cord compression, and cauda equina syndrome. Such symptoms can degrade patients' health-related quality of life (HRQOL), functional independence and mobility, and capacity to receive ongoing oncologic care. In addition, both pain and SREs contribute to hospitalization, need for palliative interventions and procedures, increased utilization of healthcare costs, and time away from family at end of life.

Palliative radiotherapy plays a clear and important role in the management of pain and symptoms in patients with metastatic disease. Pain, pathologic fracture, and spinal cord compression are common indications for radiotherapy, but selected patients with metastatic disease may also benefit from prophylactic palliation—treatment for asymptomatic metastases to prevent future complications. The aim of this chapter is to review the common roles and indications for radiotherapy in

M. D. Hall (✉)
Department of Radiation Oncology, Miami Cancer Institute, Baptist Health South Florida, Miami, FL, USA

Department of Oncological Sciences, Florida International University Herbert Wertheim College of Medicine, Miami, FL, USA
e-mail: matthewha@baptisthealth.net

K. Mair
Department of Radiation Oncology, Miami Cancer Institute, Baptist Health South Florida, Miami, FL, USA

A. L. Holtzman
Department of Radiation Oncology, Mayo Clinic Florida, Jacksonville, FL, USA

J. Pretell-Mazzini, S. A. Lozano Calderón (eds.), *Multidisciplinary Approach of Musculoskeletal Metastasis*, https://doi.org/10.1007/978-3-032-09498-8_31

patients with bone metastases and examine emerging indications for its use in patients with advanced cancers.

Radiotherapy Indications for Metastatic Disease

In this chapter, the most common clinical scenarios for metastatic disease are considered, including pain, impending or existing pathologic fracture, radiotherapy following surgical stabilization, spinal cord compression, and cauda equina syndrome as well as emerging indications for prophylaxis to prevent SREs and the role of stereotactic ablative radiotherapy.

Radiotherapy for Pain Relief

Pain relief and reducing the risk of worsening symptoms from local disease progression are the most common radiotherapy indications for metastases. In patients with uncomplicated bone metastases, defined by the absence of cortical destruction, structural degradation, instability, and fracture, randomized clinical trials demonstrated similar response rates for pain relief between fractionated palliative radiotherapy schedules of 20 Gy in 5 fractions and 30 Gy in 10 fractions compared to 8 Gy in one fraction. After radiotherapy, maximal pain relief is realized approximately 4–6 weeks following treatment. Roughly 20–25% of patients with bone metastases from solid tumors experience complete pain relief permitting discontinuation of narcotic analgesics, while another 50% have clinically significant improvements in pain scores after this intervention [1, 2]. Pain relief conveys significant improvements in functional independence and HRQOL measured in both the physical and emotional domains in treated patients [3].

Of note, the mechanism of pain relief following radiotherapy is not exclusively explained by tumor cell kill, but rather by downregulating levels of proinflammatory cytokines and osteoclast signaling pathways in the treatment volume. Specifically, palliative radiotherapy for bone metastases reduces levels of tumor necrosis factor alpha (TNF-α), interleukin 1 (IL-1), and osteoclast stimulating factor, which mediates bone resorption. Declines in the levels of these proinflammatory cytokines after radiotherapy nicely correlate with the timeline for pain response [4, 5]. This biological phenomenon may also explain the activity of bisphosphonates and other bone-modifying agents on pain response in patients with metastatic disease [6, 7]. Further, the knockdown of cytokines appears similar in magnitude between single fraction and short courses of palliative radiotherapy, which helps to explain the similar pain responses and time courses of relief observed between palliative radiotherapy regimens of differing lengths and the similar pain responses observed in radiation sensitive and radioresistant tumor histologies [8].

Approximately 30–40% of patients will develop a pain flare within one week of beginning palliative radiotherapy, where pain levels transiently rise and then return to pretreatment baseline within several days. In a randomized trial, a five-day course of dexamethasone starting on the first day of radiotherapy reduced the incidence of pain flare from 35% to 26% with few grade 3–4 hyperglycemia events (3 cases out of 148 patients randomized to dexamethasone). Importantly, development of pain flare does not predict the ultimate pain response following treatment. This risk reduction strategy should be considered in all patients receiving palliative radiotherapy for bone metastases [9, 10]. It may be particularly useful in patients already in significant pain, where worsening may prompt the need for narcotic pain medications or lead to hospitalization. Once pain flares develop, nonsteroidal anti-inflammatory drugs and acetaminophen are considered first-line, with utilization of more potent medications if clinically indicated. If not already prescribed before starting radiotherapy, patients who develop worsening pain after beginning radiotherapy may benefit from starting a five-day course of dexamethasone after onset of the pain flare.

Patient selection is an important consideration in palliative radiotherapy for patients with bone and musculoskeletal metastases. Of note, older age and radioresistant histologies do not correlate with inferior pain responses [11], although hematologic tumors appear to have superior response rates [12]. Radiographic variables, such as osteoblastic vs. osteolytic lesions, have been examined but did not correlate with outcomes [13]. Clinical decision-making should consider life expectancy juxtaposed against the expected time course for symptomatic relief. Few patients develop symptomatic improvement from palliative radiotherapy within the first two weeks, and prudence demands that physicians consider the duration of interventions delivered at the end of life. Lower pain response rates were observed in patients with shortened life expectancy, which can be explained by the delayed time for patients to experience relief after radiotherapy [4] and the oftentimes multifocal symptoms patients have at the end of life. Clinicians should not refuse palliative radiotherapy to patients with shortened life expectancy, but consideration should be given as to whether patients have a reasonable likelihood of benefiting from the proposed treatment [14]. Palliative radiotherapy can be particularly meaningful in patients who will be discontinuing anticancer therapy, and single fraction treatment should likely be prioritized in patients who have one to several months to live. At the same time, no treatment may be reasonable in patients with <1 month to live.

Figure 31.1 illustrates an example palliative radiotherapy plan treating an uncomplicated bone metastasis at L2.

Impending or Existing Pathologic Fracture of Appendicular Bones and Postoperative Radiotherapy

Bone stability is an important consideration in patients with metastatic disease, and the management of impending or existing pathologic fractures represents a common clinical scenario that requires multidisciplinary care. The first step is often to consider whether patients should be managed with or without surgery. The potential advantages of surgical intervention for bone metastases include fixation and/or

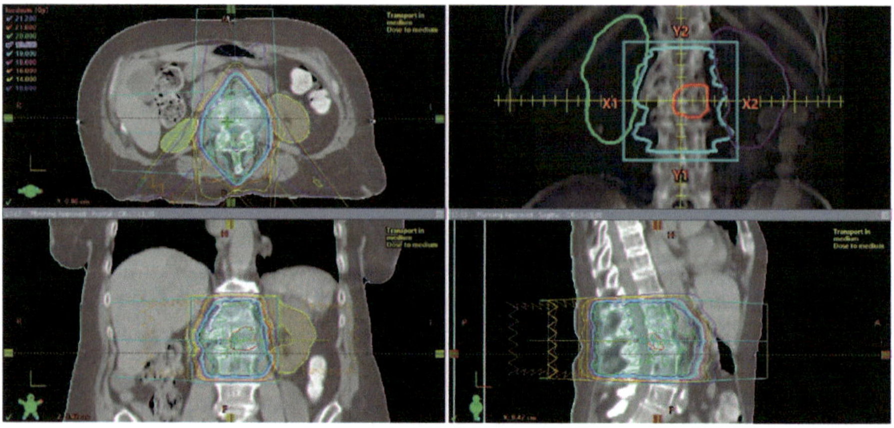

Fig. 31.1 Uncomplicated L2 bone metastasis. Palliative radiotherapy delivered using a 3D conformal plan treating the L1–L3 vertebra to 20 Gy in 5 fractions. Two posterior lateral oblique fields and one posterior field were given to reduce anterior bowel dose compared to an AP/PA plan. Kidney sparing was still achieved, with mean doses of 5.9 and 6.9 Gy, respectively. Note that pain response is comparable between 8 Gy in 1 fraction and fractionated regimens. Due to favorable performance status and expectation of survival >12 months, a 5-fraction regimen was selected in this case

augmentation of an impending or existing fracture, stabilization to permit future weight-bearing, and management of pain. The decision to pursue surgery should consider life expectancy, existing and future function, and capacity for improvement [15]. In addition, the type of intervention, which may take the form of fixation versus replacement, should also be selected to balance form and function over the patient's remaining lifetime relative to comorbidities [16] (see Chap. 12 for further detail).

Radiotherapy is regularly applied after surgery and is associated with improvements in pain and function beyond surgical intervention alone. In a retrospective study of 64 patients who underwent fixation surgery for pathologic fracture or impending fracture of long bones, Townsend and colleagues demonstrated that postoperative radiotherapy was significantly associated with the achievement of normal use following surgery, with full function achieved in 53% of patients compared to 11.5% managed with surgery alone ($p < 0.01$). Further, patients who received postoperative radiotherapy were significantly less likely to require a second orthopedic procedure, likely related to local disease progression at the site and loosening of the implanted hardware if no adjuvant therapy was administered [17]. Postoperative radiotherapy is similarly indicated following surgery for vertebral body compression fractures. While radiotherapy for vertebral compression fractures and spinal cord compression is considered in a later section, prophylaxis against neurological compromise represents a strong indication for the regular utilization of radiotherapy for spine disease.

Controversy remains regarding the need and utility of treating the entire length of the metal hardware in the postoperative setting, with the rationale that tumor seeding may occur during instrumentation. During radiation planning, pretreatment imaging should be reviewed to assess for sites of pretreatment disease involvement, including soft tissue extension. In addition, the operative report should be reviewed for important findings; for example, reaming of the femoral canal during hemiarthroplasty surgery may be employed to widen a narrow femoral canal before insertion, which may seed disease distally even beyond the hardware [18]. While certain operative approaches may minimize this risk, contemporary data continues to suggest a benefit from routine coverage of the entire implanted hardware in long bones. Rosen and colleagues reviewed 138 patients with 145 long bone metastases who received postoperative radiotherapy and identified 36 lesions that developed a local recurrence following median doses of 20–30 Gy. The authors found that coverage of the entire hardware was associated with reduced rates of local recurrence. Using propensity score matching to mitigate the effect of confounding variables, the authors reported that whole hardware coverage decreased the risk of recurrence by a factor of 4 (hazard ratio 0.24, 95% confidence interval [CI], 0.07–0.84%) [19]. Figure 31.2 depicts an example radiotherapy case following hip replacement for a pathologic fracture.

At present, it remains unknown if postoperative radiotherapy can be deferred in the setting of contemporary systemic therapies [20], which may have greater response rates than older agents. While this subject would be optimally addressed in a prospective manner, postoperative radiotherapy remains a common indication following fixation surgery to facilitate healing and maximize functional recovery.

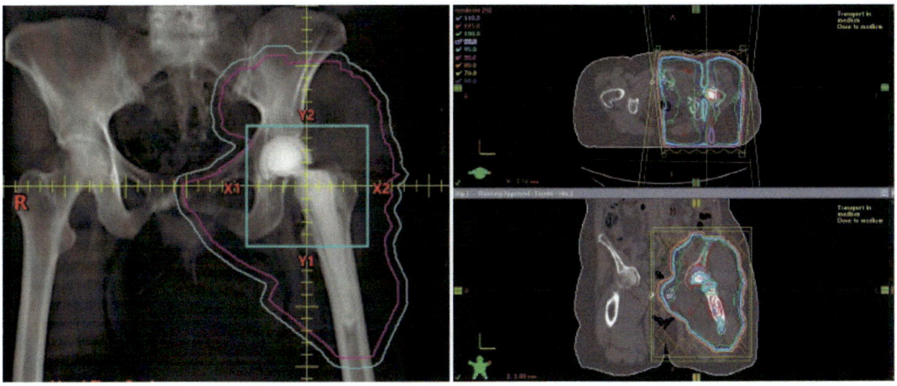

Fig. 31.2 Postoperative radiotherapy following pathologic fracture of left femur (peritrochanteric) and hip hemiarthroplasty. Palliative radiotherapy delivered using a 3D conformal plan to 20 Gy in 5 fractions. The radiotherapy plan covers the residual 4.5 cm soft tissue mass around the left femoral head as well as lytic disease in the left pubic bone. The entire surgical hardware is covered. In the upper right image, note that dose coverage of the radiotherapy plan is somewhat degraded near the surgical implant due to the effects of metal

Spinal Cord Compression and Cauda Equina Syndrome

Spine metastatic disease commonly occurs in solid tumor patients and can lead to vertebral compression fractures and epidural disease causing neurological deficits. In symptomatic patients with spinal cord compression, decompression surgery is recommended and is associated with significant improvements in ambulatory status and functional recovery compared to radiotherapy alone [21]. Surgical approaches include separation surgery, which may be clinically useful to deliver stereotactic ablative radiotherapy (SABR), stabilization, and the more minimally invasive vertebroplasty and kyphoplasty procedures. Surgical evaluation and treatment, if indicated, is essential because pain related to structural instability is not improved with radiotherapy alone [22, 23]. Surgery should be employed in appropriate patients to maintain and maximize function.

Pain response for spine metastases is comparable to nonspine bone metastases and is similar with single and fractionated radiotherapy courses [9]. For soft tissue involvement, fractionated radiotherapy may improve local control for metastatic disease and delay the time interval to subsequent development of spinal cord compression [24]. Therefore, a short course of fractionated palliative radiotherapy should be considered in patients with favorable prognosis [25].

Spinal cord compression and cauda equina syndrome represent two radiotherapy emergencies. After prompt evaluation and a multidisciplinary assessment regarding the potential role of surgery, palliative radiotherapy should be delivered in an expedited manner if no surgery is planned. Short course fractionated radiotherapy is favored in patients with good prognosis, but in patients with limited life expectancy, single fraction treatment is equally efficacious for these conditions. In patients with short, expected survival, single fraction radiotherapy for spinal cord compression provides similar symptomatic benefit in the short run, but may prove insufficient in patients with longer survival [26, 27].

Prevention of Skeletal-Related Events

The above clinical scenarios illustrate the typical indications for palliative radiotherapy: in patients with pre-existing symptoms from metastatic disease. Instead of relieving current symptoms or preventing further clinical deterioration, prophylactic radiotherapy is given to asymptomatic patients to prevent future complications. Treatment of spine disease to reduce the risk of subsequent spinal cord compression is an example of this strategy. Prophylactic treatment to prevent SREs represents an emerging paradigm in patients with bone metastases.

SREs, which include but are not limited to pathologic fractures and spinal cord compression, represent a potentially serious complication from bone metastases that severely affect function and HRQOL and can lead to hospitalization and excess costs [28, 29]. In a retrospective study, Shulman and colleagues reported that patients with breast, prostate, and lung cancer who received prophylactic radiotherapy for asymptomatic metastases had a significantly longer time to the development of symptoms and SREs than patients who were managed without radiotherapy [30]. In a randomized trial of 78 patients, Gillespie and colleagues reported that prophylactic radiotherapy for asymptomatic but high-risk bone metastases yielded a 1.6% risk of developing an SRE at one year compared to 29% with usual care ($p < 0.001$). In this study, high-risk lesions included (a) bulky bone sites (≥2 cm), hip (acetabulum, femoral head, and femoral neck), shoulder (acromion, glenoid, and humeral head), or sacroiliac joint involvement, disease in long bones (humerus, radius, ulna, clavicle, femur, tibia, fibula, metacarpals, and phalanges) with 1/3–2/3 cortical thickness, vertebral body disease in the junctional spine (C7-T1, T12-L1, and L5-S1), and any vertebral metastases with posterior element involvement. Notably, all trial patients had >5 metastatic lesions, which was required to avoid overlap with a competing oligometastatic trial at the institution. In the study, fewer patients were hospitalized for an SRE following prophylactic radiotherapy than in the expectant management group. The authors reported that patients who received prophylactic radiotherapy for high-risk sites survived longer than patients who received usual care [31]. Future clinical trials may better elucidate the optimal role of radiotherapy in asymptomatic patients and if this survival benefit can be reproduced. At present, the current data suggests that prophylactic radiotherapy may reduce the development of SREs in selected patients.

Stereotactic Ablative Radiotherapy

The treatment of bone metastasis using conventional radiotherapy is supported by randomized phase III data and consensus guidelines [32]. Despite its demonstrated efficacy, the optimal dose-fractionation regimen for palliative radiotherapy remains controversial, and retreatment is needed in up to 20% of patients. In addition, because of continued advancements in systemic therapies for metastatic cancer leading to improvements in long-term survival, the clinical rationale for dose escalation and robustness of treatment response is under renewed focus. The individualized goals of therapy for some patients may extend beyond simple short-term palliation to durable symptom management and mitigation of future late toxicities from conventional radiotherapy, which may interact with subsequent lines of systemic therapy. In this setting, some patients may benefit from more focal, targeted, and durable palliative treatments. In this space, stereotactic body radiotherapy (SBRT) or stereotactic ablative radiotherapy (SABR) may be warranted in appropriately selected patients.

SABR is a technique which delivers an ultrahigh dose of radiation to a highly conformal, precisely delineated target over five or fewer treatments. The treatment area is intended to receive an ablative high biologic-equivalent (BED) dose, producing a higher level of tumor cell kill than can be achieved with conventional radiotherapy using lower doses per fraction. Additionally, because of advances in the technical delivery of radiotherapy with intensity modulation and both conformal and dynamic arc approaches, this high-dose, high BED intervention is delivered with a steep dose gradient and rapid dose fall off; hence, it can be delivered safely even to within a few millimeters of adjacent critical organs, such as the spinal cord and esophagus. To achieve this level of precision, SABR is performed with stereotactic image-guidance, robust motion management, and immobilization techniques and can be delivered to any part of the body. When SABR is delivered in the central nervous system, it is often referred to as stereotactic radiosurgery (SRS), particularly when given in one fraction. Extracranial sites, such as the musculoskeletal system, generally receive 2–5 fractions as part of this treatment, which is instead coined SBRT or SABR [33]. SABR can be applied in particular clinical scenarios to achieve more durable pain responses or to achieve metastasis-directed tumor control in oligometastatic and oligoprogressive patients.

Song and colleagues performed a systematic review and meta-analysis comparing SABR to conventional radiotherapy for bone metastases [34]. The primary endpoint was pain relief, and secondary endpoints included change in pain score, local progression-free survival, rates of reirradiation, and toxicities. Six randomized controlled trials were included along with 1 prospective cohort study and 3 retrospective observational studies, totaling 448 patients who received SABR, and 445 patients who received conventional radiotherapy. Pooled results based on all included studies indicated that SABR was associated with a higher overall rate of pain relief ($p < 0.001$ at 3 months; $p = 0.015$ at 6 months) and complete pain relief ($p = 0.029$ at 1 month; $p < 0.001$ at 6 months). Pooled results based on the included randomized controlled trials indicated that at 3 and 6 months, SABR was also associated with both a higher overall pain relief rate ($p < 0.001$ and $p = 0.017$, respectively) and complete relief rate ($p < 0.001$ and $p < 0.001$, respectively). Subgroup analyses indicated that reduction in patient analgesic use was more obvious when spinal lesions were irradiated with SABR.

In general, the complete and partial response rates for pain are approximately 30% and 80%, respectively. General treatment practice patterns for SABR [35] include dose ranges higher than the equivalent of 18 Gy in one fraction ($BED_{10} = 50$ Gy10). For de novo spine metastases, dose-escalated spine SABR regimens include 20 Gy and 24 Gy in 1 fraction, 12 Gy × 2 fractions, 10 Gy × 3 fractions, and 7–8 Gy × 5 fractions. With these doses, the expected local control ranges from 80% to 90% at 1–2 years.

In regard to efficacy comparing SABR to conventional radiotherapy, phase II/III data from Sahgal and colleagues demonstrated a statistically significant improvement in complete response to pain at 3 months: 35% with SABR versus 14% with conventional radiotherapy (risk ratio 1.33; $p = 0.0002$) [36]. This study randomized 229 patients from 18 Canadian and Australian institutions to 12 Gy × 2 fractions versus a conventional palliative regimen of 20 Gy in 5 daily fractions. This significant difference persisted on multivariable analysis (odds ratio 3.47; $p = 0.0003$). The most common severe adverse event was grade 3 pain, which occurred in approximately 5% in each cohort and was not different between SABR and conventional radiotherapy. No treatment-related deaths were observed.

These results were discordant with the NRG Oncology phase III study by Ryu and colleagues, who evaluated 339 patients in a 2:1 randomization to 16–18 Gy in 1 fraction with SABR versus 8 Gy in 1 fraction with conventional radiotherapy. The primary endpoint was patient-reported pain response at 3 months, which favored conventional radiotherapy (41.3% for SABR vs. 60.5% for conventional radiotherapy; 1-sided $p = 0.99$; 2-sided $p = 0.01$). No differences in the proportion of acute or late adverse effects or the incidence of vertebral compression fractures were observed between the two arms [37]. While this study was published after Sahgal and colleagues [36], the trial in actuality was conducted earlier. First, in the NRG trial, a lower BED dose of radiation was employed. Second, the inclusion criteria, baseline performance status, and baseline pain scores, as well as how and when the primary endpoint of pain response at 3 months after treatment were measured may help to explicate the differing results [38]. Importantly, the NRG trial did not distinguish between oncologic and mechanical pain etiologies, which is important because radiotherapy is not expected to provide any meaningful change in mechanical pain at 3 months after treatment. Based on the data from these two studies, SABR may be beneficial for pain in appropriately selected patients. These patients should not have signs of spinal instability, no or minimal epidural disease, $\leq$3 contiguous vertebral segments involved, and good life expectancy. This subpopulation includes potential candidates for SABR. Future data on durability of response, local control, and retreatment may better inform patient selection.

In regard to long bone metastases, Madani and colleagues published data from 111 patients with 114 metastases in the femur (n = 84, 74%), humerus (n = 26, 23%), and tibia (n = 4, 4%) that were treated with SABR at 7 different institutions. While the majority were treated for oligometastatic disease, more than 20% of the cohort was treated with SABR for the indication of palliation alone. The most common primary tumor sites were prostate (41%) and breast cancer (18%); nearly all patients had a Karnofsky performance status $\geq$70%. Treatment schedules ranged from 1 to 10 fractions with a median BED_{10} of 56.7 Gy. With a median follow-up of 21 months, the cumulative incidence of local failure was 7.2% at 2 years and 13.5% at 3 years. The 2- and 3-year cumulative incidence of fracture was 6.1% and 9.8%, respectively. On multivariate analysis, extraosseous disease extension was significantly associated with both fracture ($p = 0.001$) and local failure ($p = 0.02$) [39].

Tinkle and colleagues published a multi-institutional report evaluating SABR in 55 pediatric and young adult patients with a total of 107 extracranial lesions. The median dose/fraction was 7 Gy × 5 fractions. With a median follow-up of 11 months, the radiographic response rates of bone and soft tissue lesions were 90.6% and 76.7%, respectively. Symptom improvement was observed in 62% of symptomatic sites. The 1-year estimated cumulative local failure rate was 25.2% with higher local progression in soft tissue lesions relative to bone on multivariable analysis ($p = 0.04$). Treatment was well tolerated with no grade 4 toxicities; the estimated 1-year cumulative incidence of late toxicity of any grade was 7.5% [40].

Side Effects

Both conventional radiotherapy and SABR for bone metastases are well-tolerated. Side effects are generally demarcated into acute (<90 days) and late (>90 days) from the completion of radiotherapy. Table 31.1 summarizes the potential and expected toxicities during and after palliative radiotherapy along with general management strategies. Acute side effects are limited only to the areas of treatment and may include localized skin reactions, fatigue, typically mild gastrointestinal symptoms, and transient pain flare [41]. Treatment interventions for cutaneous effects include topical creams and ointments, topical steroids, and silver sulfadiazine in more severe cases. To minimize bolus effect from treatment, it is generally advised to have the skin surface free of topical coverings during radiotherapy fractions, unless approved by the manufacturer. Fatigue can be reduced with physical activity and exercise, which mediates endorphin release [42]. If abdominal structures are adjacent to the treatment area, transient and mild nausea or bowel irritability may occur. Management with antiemetics or antidiarrheal medications may be indicated based on the location, treatment dose, and duration. Patients who experience nausea and/or vomiting related to radiotherapy, which typically develops within 1–4 h of the first treatment fraction and all subsequent fractions, should receive prophylactic antiemetics before each later fraction.

In general, delayed long-term side effects are uncommon following both conventional radiotherapy and SABR for patients with metastatic disease, due to the low doses used and highly conformal nature of the treatments, respectively. For spinal metastases, risk of vertebral compression fracture can occur from tumor regression and the direct pretreatment bone destruction. There is no compelling evidence that shows the risk of vertebral compression fracture is higher with the use of SABR compared to conventional radiotherapy, except when using exceptionally high doses of 24 Gy in 1 fraction [34–37]. Radiation myelopathy, an irreversible injury to the spinal cord leading to neurological deficits, can occur but is exceptionally rare (<0.1%) with careful treatment planning [35]. Since such toxicities are particularly hazardous, stringent quality control practices to meet dose constraints for the spinal cord and other critical structures and daily setup verification and reproducibility are of high importance when treating the spine, particularly with SABR [35, 43]. Treatment of long bones with SABR can increase the risk of pathologic fractures, especially in weight-bearing regions. The risk of fracture is generally estimated to be <10% at 3 years. Given that a significant contribution to fracture risk derives from tumor extension and pre-existing bone destruction, patients at risk for instability should be evaluated by orthopedic surgery for stabilization ideally before or soon after radiotherapy. Other long-term side effects can include poor wound healing, soft tissue fibrosis, and persistent pain in the irradiated area. The incidences of these toxicities are fortunately low, especially when treatment volumes are limited and in patients with short life expectancy.

Reirradiation

Selected patients may benefit from a repeat course of radiotherapy for palliation. This scenario may arise more frequently in the future as systemic therapies and survival times with disease improve. While no clear consensus exists for the optimal

Table 31.1 Overview of acute and late toxicities following palliative radiotherapy for bone metastases

Toxicity	Timing	Frequency	Duration	Treatment
Skin reaction	Acute (<90 days)	Common	Short-term (resolves postradiotherapy)	Topical creams (e.g., aloe, petroleum jelly, topical steroids); silver sulfadiazine; skin should be free of topicals during treatment unless approved by the manufacturer
Fatigue	Acute (<90 days)	Common	Short-term (during radiotherapy, may persist for a few weeks postradiotherapy)	Exercise, rest, and maintain hydration
Gastrointestinal: nausea/vomiting and bowel frequency	Acute (<90 days)	Increased risk with stomach/bowel exposure and brainstem (area postrema)	Short-term (resolves postradiotherapy)	Antiemetics for nausea; anti-diarrheal medications for bowel irritability (especially when abdominal structures are near the treatment area); consider prophylactic antiemetics if radiating stomach/bowel
Pain flare	Acute (<90 days)	30–40%	Short-term (during radiotherapy, resolves postradiotherapy)	Nonsteroidal anti-inflammatory drugs (NSAIDs) or acetaminophen first-line; oral corticosteroids; narcotics if needed
Vertebral compression fracture	Late (>90 days)	<20%	Long-term (may occur months to years postradiotherapy)	Neurosurgery evaluation for kyphoplasty/augmentation/stabilization; symptoms can be difficult to distinguish clinically from treatment effect or tumor regression
Radiation myelopathy	Late (>90 days)	Rare (<0.1%)	Long-term (permanent neurological deficits)	Strict adherence to dose constraints and high precision treatment delivery; first-line management includes symptomatic treatment of neurological deficits and oral corticosteroids
Pathologic fractures (long bones)	Late (>90 days)	<10% at 3 years)	Long-term (may occur months to years postradiotherapy)	Orthopedic evaluation for stabilization before or after radiotherapy

(continued)

Table 31.1 (continued)

Toxicity	Timing	Frequency	Duration	Treatment
Impaired wound healing	Late (>90 days)	Rare (<5%)	Long-term (persistent)	Wound care management; may require surgical intervention in severe cases
Soft tissue fibrosis	Late (>90 days)	Rare (<5%)	Long-term (persistent)	Physical therapy, stretching exercises, and symptomatic management
Persistent pain	Late (>90 days)	Rare, variable	Long-term (persistent)	Pain management strategies including NSAIDs, narcotics, or other analgesics; referral to pain management specialist; consider referral for rehabilitation medicine evaluation; consider surgery referral to assess for mechanical instability (spine)

time interval between radiotherapy courses, reirradiation is unlikely to improve symptoms in patients treated <6 months previously for the same indication. Most clinicians prefer to wait 1.5–2.0 years after prior radiotherapy to consider reirradiation, particularly for tumors near spinal cord. Animal models demonstrated lower rates of spinal cord injury than expected following reirradiation after 1.5–2.0 years, suggesting that some repair of spinal cord damage occurs following the first radiotherapy course [44, 45]. It is important to note that subclinical damage repair has not been well modeled in other organs, but the available data still informs clinical decision-making about reirradiation. In addition to potential reduction in toxicities, a longer duration of benefit from the first course of radiotherapy may portend greater likelihood of benefit after a second.

Chow and colleagues randomized 950 patients to a second course of palliative radiotherapy to 8 Gy in 1 fraction vs. 20 Gy in 5 fractions. The authors reported that pain response was similar between 1 and 5 fractions with a lower incidence of radiation-related toxicities following single fraction treatment. Of note, 45–50% of patients experienced an overall pain response [46], which was quantitatively reduced compared to patients receiving a first course of palliative radiotherapy. The risk of pathologic fracture was similar between both regimens, occurring in 50 out of 950 patients (5.3%). Clinicians should follow patients carefully for toxicities after a second course of radiotherapy and remember that risk of complications increases with longer attained survival.

SABR shows promise based on the aforementioned evidence in reirradiation. Its potential advantages, including higher rates of overall and complete pain response and better local control, must be weighed against the potential for increased in-field toxicity associated with higher-dose radiotherapy, heightened demands on staffing and equipment, and the longer time required for treatment delivery compared to conventional radiotherapy. SABR may be particularly useful for radioresistant histologies and could enhance the response rate and the durability of pain relief compared to conventional radiotherapy [33, 34, 47, 48]. Future research focused on patient selection through biomarkers [49, 50], radiomics [51, 52], and the integration of artificial intelligence [53] will be crucial in developing personalized treatment strategies to optimize the therapeutic ratio and cost-effectiveness.

Radionuclides

While external beam radiotherapy is regularly applied in patients with metastatic disease, one common limitation of this modality surfaces in patients with diffuse, symptomatic disease. Treatment of one or two index pain sites may control symptoms locally yet still leave patients equally debilitated from the numerous other sites untreated with radiotherapy. Although most treatment appointments for palliative radiotherapy (room entry, setup and positioning, treatment, and exit) average 30 minutes, this time quickly balloons when treating multiple sites. Further, symptomatic patients will take longer to set up and treat in the radiotherapy vault, particularly if pain remains uncontrolled on medical therapy and if assistance is needed getting on/off the table due to functional limitations caused by their disease burden. Against this reality, radionuclides represent a useful treatment option for selected patients with diffuse bone metastases, with the advantage of treating multiple symptomatic metastases at the same time.

Radionuclide agents combine a radioactive isotope attached to a carrier molecule that localizes to a particular biologic (most frequently bone) or molecular target. The currently available drugs used in patients with metastatic disease are administered intravenously. Older agents, including samarium, strontium, and rhenium, used beta-emitters for the radioactive payload, with low linear energy transfer (LET) that resulted in more limited anticancer activity and a longer pathlength of dose deposition. These drugs preferentially target bone, and their utility was severely limited by bone marrow suppression (especially thrombocytopenia followed by lymphopenia) [54]. Due to the combination of toxicity, lack of prolonged clinical benefit, and their radiation safety obstacles, beta-emitter agents like samarium and strontium never achieved substantive use in clinic.

Radium-223, a newer bone-targeting agent, employs an alpha-emitter, which has a significantly shorter pathlength than beta-emitters, leading to less nontarget bone marrow suppression. Alpha-emitters also have high LET, increasing tumoricidal activity [52]. Even though toxicity is reduced with these new agents, bone marrow suppression remains the major dose-limiting toxicity and treatment duration is limited to 4–6 cycles at most. In the ALSYMPCA trial, 921 castration-resistant prostate cancer patients with ≥2 bone metastases, no visceral disease, and either prior treatment or contraindication to docetaxel were randomized to Radium-223 versus placebo. Both arms also received other standard of care interventions as indicated (including androgen deprivation therapy, antiandrogens, corticosteroids, and external beam radiotherapy). Dosing was every 4 weeks for up to 6 cycles. In this randomized trial, Radium-223 was associated with a 3.6-month improvement in overall survival compared to standard of care therapy with similar incidence of grade 3–4 hematologic toxicity and all nonhematologic toxicities [55]. More importantly, however, treated patients were also more likely to experience meaningful pain relief compared to placebo.

Despite this overall survival advantage, newer systemic therapies, such as Lutetium 177, supplanted the use of Radium-223 in castration-resistant prostate cancer patients. In addition, even more novel agents are under development. Instead of localizing to bone, Lutetium 177 instead targets a protein that is highly expressed in prostate cancer, prostate-specific membrane antigen (PSMA), which may lead to superior activity than prior nonselective agents. Similar targeted therapies are in active development for other solid tumors. Table 31.2 describes the historical and contemporary radionuclides and their properties.

Table 31.2 Radionuclides for metastatic bone disease [50]

Radionuclide	Half-life	Primary emitter	Dose range	Dosing
Strontium-89	50.6 days	Beta	2.4 mm	148 MBq every 12 weeks
Samarium-153	1.9 days	Beta	0.6 mm	37 MBq/kg every 8 weeks
Rhenium-186	3.7 days	Beta	2.4 mm	2405 MBq every 8 weeks
Radium-223	11.4 days	Alpha	<100 μm	55 MBq every 4 weeks

Summary

Radiotherapy is a safe and effective treatment option for patients with metastatic disease and should continue to play a role in minimizing symptoms and maximizing function in this population. Typical dosing schemas for external beam radiotherapy include 30 Gy in 10 fractions, 20 Gy in 5 fractions, and 8 Gy in 1 fraction. For cancer-related pain, 8 Gy in 1 fraction provides similar pain relief compared to longer fractionation schedules. Maximum pain response following treatment of simple bone metastases occurs 4–6 weeks following completion of radiation, with few patients experiencing meaningful pain relief within 2 weeks. In patients with symptomatic bone metastases where survival is estimated to be less than one month, no radiotherapy or single fraction radiotherapy may be appropriate. In patients with short life expectancy (<3 months), 8 Gy in 1 fraction again provides comparable outcomes to longer schedules, including in symptomatic spinal cord compression. A short, fractionated course of radiotherapy (5–10 fractions) may provide more durable tumor and symptom control at the treated sites in patients with longer anticipated survival, but further study is needed to evaluate the optimal dose/fractionation schedule in these patients. In our practice, most patients treated with surgery for spinal cord compression or instability or for impending or pre-existing pathologic fracture receive 20 Gy in 5 fractions.

Emerging data suggests that SABR may improve rates of maximum pain relief compared to conventional radiotherapy for spine metastases. Further, selected patients with limited volume metastatic disease may benefit from ablative radiotherapy to prolong survival or provide a meaningful interval of time before need for subsequent systemic therapy. Prophylactic palliation, with the objective to prevent major SREs and their associated morbidities, represents another developing indication for radiotherapy in this population. Finally, reirradiation will be increasingly relevant with the continued advancement of systemic therapies and longer survival with disease. Reirradiation can benefit selected patients, but great care must be taken to mitigate the risk of injury. Future reirradiation trials should examine optimal dose/fractionation schedules, planning techniques, and toxicity.

References

1. Steenland E, Leer JW, van Houwelingen H, Post WJ, van den Hout WB, Kievit J, de Haes H, Martijn H, Oei B, Vonk E, van der Steen-Banasik E, Wiggenraad RG, Hoogenhout J, Wárlám-Rodenhuis C, van Tienhoven G, Wanders R, Pomp J, van Reijn M, van Mierlo I, Rutten E. The

effect of a single fraction compared to multiple fractions on painful bone metastases: a global analysis of the Dutch bone metastasis study. Radiother Oncol. 1999;52(2):101–9.
2. Hartsell WF, Scott CB, Bruner DW, Scarantino CW, Ivker RA, Roach M, Suh JH, Demas WF, Movsas B, Petersen IA, Konski AA, Cleeland CS, Janjan NA, DeSilvio M. Randomized trial of short- versus long-course radiotherapy for palliation of painful bone metastases. J Natl Cancer Inst. 2005;97(11):798–804. https://doi.org/10.1093/jnci/dji139. PMID: 15928300
3. McDonald R, Ding K, Brundage M, Meyer RM, Nabid A, Chabot P, Coulombe G, Ahmed S, Kuk J, Dar AR, Mahmud A, Fairchild A, Wilson CF, Wu JSY, Dennis K, DeAngelis C, Wong RKS, Zhu L, Chan S, Chow E. Effect of radiotherapy on painful bone Metastases: a secondary analysis of the NCIC clinical trials group symptom control trial SC.23. JAMA Oncol. 2017;3(7):953–9.
4. Hoskin PJ, Stratford MR, Folkes LK, Regan J, Yarnold JR. Effect of local radiotherapy for bone pain on urinary markers of osteoclast activity. Lancet. 2000;355(9213):1428–9.
5. Donaubauer AJ, Deloch L, Becker I, Fietkau R, Frey B, Gaipl US. The influence of radiation on bone and bone cells-differential effects on osteoclasts and osteoblasts. Int J Mol Sci. 2020;21(17):6377.
6. Eisen A, Somerfield MR, Accordino MK, Blanchette PS, Clemons MJ, Dhesy-Thind S, Dillmon MS, D'Oronzo S, Fletcher GG, Frank ES, Hallmeyer S, Makhoul I, Moy B, Thawer A, Wu JY, Van Poznak CH. Use of adjuvant bisphosphonates and other bone-modifying agents in breast cancer: ASCO-OH (CCO) guideline update. J Clin Oncol. 2022;40(7):787–800.
7. Jing D, Zhao Q, Zhao Y, Lu X, Feng Y, Zhao B, Zhao X. Management of pain in patients with bone metastases. Front Oncol. 2023;13:1156618.
8. Price P, Hoskin PJ, Easton D, Austin D, Palmer SG, Yarnold JR. Prospective randomised trial of single and multifraction radiotherapy schedules in the treatment of painful bony metastases. Radiother Oncol. 1986;6(4):247–55.
9. Gomez-Iturriaga A, Cacicedo J, Navarro A, Morillo V, Willisch P, Carvajal C, Hortelano E, Lopez-Guerra JL, Illescas A, Casquero F, Del Hoyo O, Ciervide R, Irasarri A, Pijoan JI, Bilbao P. Incidence of pain flare following palliative radiotherapy for symptomatic bone metastases: multicenter prospective observational study. BMC Palliat Care. 2015;14:48.
10. Chow E, Meyer RM, Ding K, Nabid A, Chabot P, Wong P, Ahmed S, Kuk J, Dar AR, Mahmud A, Fairchild A, Wilson CF, Wu JSY, Dennis K, Brundage M, DeAngelis C, Wong RKS. Dexamethasone in the prophylaxis of radiation-induced pain flare after palliative radiotherapy for bone metastases: a double-blind, randomised placebo-controlled, phase 3 trial. Lancet Oncol. 2015;16(15):1463–72.
11. Cacicedo J, Gómez-Iturriaga A, Navarro A, Morillo V, Willisch P, Lopez-Guerra JL, Illescas A, Casquero F, Del Hoyo O, Ciervide R, Martinez-Indart L, Bilbao P, Rades D. Analysis of predictors of pain response in patients with bone metastasis undergoing palliative radiotherapy: does age matter? J Med Imaging Radiat Oncol. 2018;23
12. Saito T, Toya R, Tomitaka E, Matsuyama T, Ninomura S, Oya N. Predictors of pain palliation after radiation therapy for painful tumors: a prospective observational study. Int J Radiat Oncol Biol Phys. 2018;101(5):1061–8.
13. Bala W, Chiu N, Tao MJ, Lam H, Chow E, Probyn L. Diagnostic imaging modalities to assess treatment response of bone metastasis in patients receiving palliative radiotherapy: a scoping review of the literature. Can Assoc Radiol J. 2020;71(4):495–504.
14. Jones JA, Lutz ST, Chow E, Johnstone PA. Palliative radiotherapy at the end of life: a critical review. CA Cancer J Clin. 2014;64(5):296–310.
15. Metastatic Spinal Cord Compression: Diagnosis and Management of Patients at Risk of or with Metastatic Spinal Cord Compression. Cardiff (UK): National Collaborating Centre for Cancer (UK); 2008 Nov.
16. Mirels H. Metastatic disease in long bones. A proposed scoring system for diagnosing impending pathologic fractures. Clin Orthop Relat Res. 1989;249:256–64.
17. Townsend PW, Rosenthal HG, Smalley SR, Cozad SC, Hassanein RE. Impact of postoperative radiation therapy and other perioperative factors on outcome after orthopedic

stabilization of impending or pathologic fractures due to metastatic disease. J Clin Oncol. 1994;12(11):2345–50.
18. Boden AL, Patel M, Hoyt A, Subhawong T, Conway S, Pretell-Mazzini J. Development of distal femoral metastasis is rare in cases of isolated proximal femoral metastases. J Am Acad Orthop Surg. 2021;29(9):e465–70. https://doi.org/10.5435/JAAOS-D-20-00315.
19. Rosen DB, Haseltine JM, Bartelstein M, Flynn JR, Zhang Z, Kohutek ZA, Yamada Y, Schmitt A, Higginson DS, Vaynrub M, Yang JT, Gillespie EF. Should postoperative radiation for long bone Metastases cover part or all of the orthopedic hardware? Results of a large retrospective analysis. Adv Radiat Oncol. 2021;6(6):100756.
20. Tseng YD. Radiation therapy for painful bone Metastases: fractionation, recalcification, and symptom control. Semin Radiat Oncol. 2023;33(2):139–47.
21. Patchell RA, Tibbs PA, Regine WF, Payne R, Saris S, Kryscio RJ, Mohiuddin M, Young B. Direct decompressive surgical resection in the treatment of spinal cord compression caused by metastatic cancer: a randomised trial. Lancet. 2005;366(9486):643–8.
22. Fisher CG, DiPaola CP, Ryken TC, Bilsky MH, Shaffrey CI, Berven SH, Harrop JS, Fehlings MG, Boriani S, Chou D, Schmidt MH, Polly DW, Biagini R, Burch S, Dekutoski MB, Ganju A, Gerszten PC, Gokaslan ZL, Groff MW, Liebsch NJ, Mendel E, Okuno SH, Patel S, Rhines LD, Rose PS, Sciubba DM, Sundaresan N, Tomita K, Varga PP, Vialle LR, Vrionis FD, Yamada Y, Fourney DR. A novel classification system for spinal instability in neoplastic disease: an evidence-based approach and expert consensus from the spine oncology study group. Spine (Phila Pa 1976). 2010;35(22):E1221–9.
23. Confavreux CB, Follet H, Mitton D, Pialat JB, Clézardin P. Fracture risk evaluation of bone metastases: a burning issue. Cancers (Basel). 2021;13(22):5711.
24. Roos DE, Turner SL, O'Brien PC, Smith JG, Spry NA, Burmeister BH, Hoskin PJ, Ball DL, Trans-Tasman radiation oncology group, TROG 96.05. Randomized trial of 8 Gy in 1 versus 20 Gy in 5 fractions of radiotherapy for neuropathic pain due to bone metastases (trans-Tasman radiation oncology group, TROG 96.05). Radiother Oncol. 2005;75(1):54–63.
25. Spratt DE, Beeler WH, de Moraes FY, Rhines LD, Gemmete JJ, Chaudhary N, Shultz DB, Smith SR, Berlin A, Dahele M, Slotman BJ, Younge KC, Bilsky M, Park P, Szerlip NJ. An integrated multidisciplinary algorithm for the management of spinal metastases: an International Spine Oncology Consortium report. Lancet Oncol. 2017;18(12):e720–30.
26. Rades D, Šegedin B, Conde-Moreno AJ, Garcia R, Perpar A, Metz M, Badakhshi H, Schreiber A, Nitsche M, Hipp P, Schulze W, Adamietz IA, Norkus D, Rudat V, Cacicedo J, Schild SE. Radiotherapy with 4 Gy × 5 versus 3 Gy × 10 for metastatic epidural spinal cord compression: final results of the SCORE-2 trial (ARO 2009/01). J Clin Oncol. 2016;34(6):597–602.
27. Hoskin PJ, Hopkins K, Misra V, Holt T, McMenemin R, Dubois D, McKinna F, Foran B, Madhavan K, MacGregor C, Bates A, O'Rourke N, Lester JF, Sevitt T, Roos D, Dixit S, Brown G, Arnott S, Thomas SS, Forsyth S, Beare S, Reczko K, Hackshaw A, Lopes A. Effect of single-fraction vs multifraction radiotherapy on ambulatory status among patients with spinal canal compression from metastatic cancer: the SCORAD randomized clinical trial. JAMA. 2019;322(21):2084–94.
28. Costa L, Lipton A, Hadji P, Chen YM, Kosmidis P. Treatment of bone metastases before the onset of pain. Int J Clin Oncol. 2013;18(3):531–8.
29. Weinfurt KP, Li Y, Castel LD, Saad F, Timbie JW, Glendenning GA, Schulman KA. The significance of skeletal-related events for the health-related quality of life of patients with metastatic prostate cancer. Ann Oncol. 2005;16(4):579–84.
30. Shulman RM, Meyer JE, Li T, Howell KJ. External beam radiation therapy (EBRT) for asymptomatic bone metastases in patients with solid tumors reduces the risk of skeletal-related events (SREs). Ann Palliat Med. 2019;8(2):159–67.
31. Gillespie EF, Yang JC, Mathis NJ, Marine CB, White C, Zhang Z, Barker CA, Kotecha R, McIntosh A, Vaynrub M, Bartelstein MK, Mitchell A, Guttmann DM, Yerramilli D, Higginson DS, Yamada YJ, Kohutek ZA, Powell SN, Tsai J, Yang JT. Prophylactic radiation therapy versus standard of care for patients with high-risk asymptomatic bone metastases: a multicenter, randomized phase II clinical trial. J Clin Oncol. 2024;42(1):38–46.

32. Alcorn S, Cortés ÁA, Bradfield L, Brennan M, Dennis K, Diaz DA, Doung YC, Elmore S, Hertan L, Johnstone C, Jones J, Larrier N, Lo SS, Nguyen QN, Tseng YD, Yerramilli D, Zaky S, Balboni T. External beam radiation therapy for palliation of symptomatic bone metastases: an ASTRO clinical practice guideline. Pract Radiat Oncol. 2024;14(5):377–97.
33. Simone CB 2nd. Stereotactic body radiation therapy versus multi-fraction radiation therapy for bone metastases. Ann Palliat Med. 2019;8(4):360–3.
34. Song X, Wei J, Sun R, Jiang W, Chen Y, Shao Y, Gu W. Stereotactic body radiation therapy versus conventional radiation therapy in pain relief for bone metastases: a systematic review and meta-analysis. Int J Radiat Oncol Biol Phys. 2023;115(4):909–21.
35. Guckenberger M, Andratschke N, Belka C, Bellut D, Cuccia F, Dahele M, Guninski RS, Josipovic M, Mancosu P, Minniti G, Niyazi M, Ricardi U, Munck Af Rosenschold P, Sahgal A, Tsang Y, Verbakel W, Alongi F. ESTRO clinical practice guideline: stereotactic body radiotherapy for spine metastases. Radiother Oncol. 2024;190:109966.
36. Sahgal A, Myrehaug SD, Siva S, Masucci GL, Maralani PJ, Brundage M, Butler J, Chow E, Fehlings MG, Foote M, Gabos Z, Greenspoon J, Kerba M, Lee Y, Liu M, Liu SK, Thibault I, Wong RK, Hum M, Ding K, Parulekar WR, trial investigators. Stereotactic body radiotherapy versus conventional external beam radiotherapy in patients with painful spinal metastases: an open-label, multicentre, randomised, controlled, phase 2/3 trial. Lancet Oncol. 2021;22(7):1023–33.
37. Ryu S, Deshmukh S, Timmerman RD, Movsas B, Gerszten P, Yin FF, Dicker A, Abraham CD, Zhong J, Shiao SL, Tuli R, Desai A, Mell LK, Iyengar P, Hitchcock YJ, Allen AM, Burton S, Brown D, Sharp HJ, Dunlap NE, Siddiqui MS, Chen TH, Pugh SL, Kachnic LA. Stereotactic radiosurgery vs conventional radiotherapy for localized vertebral metastases of the spine: phase 3 results of NRG oncology/RTOG 0631 randomized clinical trial. JAMA Oncol. 2023;9(6):800–7.
38. Redmond KJ, Hattangadi-Gluth J, Pollum EL, Trifiletti DM, Kim MM, Milano M. Navigating the spinal frontier: recent data on stereotactic body radiation therapy for spine metastases. Int J Radiat Oncol Biol Phys. 2024;118(2):313–7.
39. Madani I, Sahgal A, Erler D, Stish BJ, Olivier KR, Park SS, Eppinga WSC, Seravalli E, Redmond KJ, Cao Y, Siva S, Chang D, Nguyen TK, O'Neil M, Guckenberger M. Stereotactic body radiation therapy for metastases in long bones. Int J Radiat Oncol Biol Phys. 2022;114(4):738–46.
40. Tinkle CL, Singh C, Lloyd S, Guo Y, Li Y, Pappo AS, SG DB, Lucas JT Jr, Haas-Kogan DA, Terezakis SA, Braunstein SE, Krasin MJ. Stereotactic body radiation therapy for metastatic and recurrent solid tumors in children and young adults. Int J Radiat Oncol Biol Phys. 2021;109(5):1396–405.
41. Wang K, Tepper JE. Radiation therapy-associated toxicity: etiology, management, and prevention. CA Cancer J Clin. 2021;71(5):437–54.
42. Hermann AL, Fell GL, Kemény LV, Fung CY, Held KD, Biggs PJ, Rivera PD, Bilbo SD, Igras V, Willers H, Kung J, Gheorghiu L, Hideghéty K, Mao J, Woolf CJ, Fisher DE. β-Endorphin mediates radiation therapy fatigue. Sci Adv. 2022;8(50):eabn6025.
43. Sahgal A, Chang JH, Ma L, Marks LB, Milano MT, Medin P, Niemierko A, Soltys SG, Tomé WA, Wong CS, Yorke E, Grimm J, Jackson A. Spinal cord dose tolerance to stereotactic body radiation therapy. Int J Radiat Oncol Biol Phys. 2021;110(1):124–36.
44. Ang KK, Price RE, Stephens LC, Jiang GL, Feng Y, Schultheiss TE, Peters LJ. The tolerance of primate spinal cord to re-irradiation. Int J Radiat Oncol Biol Phys. 1993;25(3):459–64.
45. Kirkpatrick JP, van der Kogel AJ, Schultheiss TE. Radiation dose-volume effects in the spinal cord. Int J Radiat Oncol Biol Phys. 2010;76(3 Suppl):S42–9.
46. Chow E, van der Linden YM, Roos D, Hartsell WF, Hoskin P, Wu JS, Brundage MD, Nabid A, Tissing-Tan CJ, Oei B, Babington S, Demas WF, Wilson CF, Meyer RM, Chen BE, Wong RK. Single versus multiple fractions of repeat radiation for painful bone metastases: a randomised, controlled, non-inferiority trial. Lancet Oncol. 2014;15(2):164–71.
47. Macià I, Garau M. Radiobiology of stereotactic body radiation therapy (SBRT). Rep Pract Oncol Radiother. 2017;22(2):86–95.

48. Singh R, Valluri A, Didwania P, Lehrer EJ, Baliga S, Hiniker S, Braunstein SE, Murphy ES, Lazarev S, Tinkle C, Terezakis S, Palmer JD. Efficacy and safety of stereotactic body radiation therapy for pediatric malignancies: the LITE-SABR systematic review and meta-analysis. Adv Radiat Oncol. 2023;8(2):101123.
49. Dai YH, Wang YF, Shen PC, Lo CH, Yang JF, Lin CS, Chao HL, Huang WY. Radiosensitivity index emerges as a potential biomarker for combined radiotherapy and immunotherapy. NPJ Genom Med. 2021;6(1):40.
50. Iuliani M, Simonetti S, Ribelli G, Napolitano A, Pantano F, Vincenzi B, Tonini G, Santini D. Current and emerging biomarkers predicting bone metastasis development. Front Oncol. 2020;10:789.
51. Wakabayashi K, Koide Y, Aoyama T, Shimizu H, Miyauchi R, Tanaka H, Tachibana H, Nakamura K, Kodaira T. A predictive model for pain response following radiotherapy for treatment of spinal metastases. Sci Rep. 2021;11(1):12908.
52. Sanli I, Osong B, Dekker A, TerHaag K, van Kuijk SMJ, van Soest J, Wee L, Willems PC. Radiomics biopsy signature for predicting survival in patients with spinal bone metastases (SBMs). Clin Transl Radiat Oncol. 2022;33:57–65.
53. Vu E, Steinmann N, Schröder C, Förster R, Aebersold DM, Eychmüller S, Cihoric N, Hertler C, Windisch P, Zwahlen DR. Applications of machine learning in palliative care: a systematic review. Cancers (Basel). 2023;15(5):1596.
54. Challapalli A, Aziz S, Khoo V, Kumar A, Olson R, Ashford RU, Gabbar OA, Rai B, Bahl A. Spine and non-spine bone metastases—current controversies and future direction. Clin Oncol (R Coll Radiol) 2020;32(11):728–744
55. Zustovich F, Barsanti R. Targeted α therapies for the treatment of bone Metastases. Int J Mol Sci. 2017;19(1):74.

Part V

Rehabilitation and Post-operative Care

Physical Therapy Strategies

32

Kathryn Quaglia, Greer Gilbert, and David Crandell

Introduction

Understanding current trends and future directions of physical therapy (PT) strategies in patients with musculoskeletal metastasis begins with gaining insights from the past. Historically, patients with musculoskeletal metastasis were advised to restrict their activity and consider bed rest. Immobilization, however, is not without complications. These complications include deconditioning, joint contractures, pressure wounds, and an increased risk of thromboembolic disease. Over 40 years ago, clinical researchers at our institution prospectively studied 54 consecutive patients with bony metastasis to determine the risk of producing pathologic fracture during the rehabilitation program [1]. Patients participating in "active" rehabilitation on an inpatient unit performed at least one physical therapy session, five times a week throughout the patient's length of stay. Session duration and length of stay were not specified. Sixteen fractures were observed in 12 patients; however, only one fracture occurred while the patient was receiving PT. This fracture occurred while the patient was attempting a slide board transfer and resulted in further compression of an L4 vertebral fracture. This fracture did not impact the patient's ability to progress toward safe mobility, and in time, the patient was discharged home. Our study demonstrated that new fractures could occur even if the patient is safe in bed

K. Quaglia
Mass General Brigham, Spaulding Rehabilitation, Boston, MA, USA
e-mail: kquaglia1@mgb.org

G. Gilbert
Mass General Brigham, Martha's Vineyard Hospital, Oak Bluffs, MA, USA
e-mail: gmgilbert@mgb.org

D. Crandell (✉)
Mass General Brigham, Spaulding Rehabilitation Hospital, Department of Physical Medicine and Rehabilitation, Harvard Medical School, Boston, MA, USA
e-mail: DCRANDELL@mgh.harvard.edu

J. Pretell-Mazzini, S. A. Lozano Calderón (eds.), *Multidisciplinary Approach of Musculoskeletal Metastasis*, https://doi.org/10.1007/978-3-032-09498-8_32

and the majority of patients with bony metastases who are candidates for active rehabilitation can be treated successfully with a low risk of pathologic fracture.

Bone metastases and the combination of ongoing cancer treatments can often lead to a decrease in a patient's physical activity and functional independence. The design of a rehabilitation program must be personalized and based on the location and level of involvement of their lesions with input from the multidisciplinary team to decrease skeletal-related events (SREs).

More recent randomized controlled trials involving patients with bone metastases from prostate cancer can be used to determine the best physical therapy strategies [2, 3]. Galvao and colleagues in 2017 evaluated the efficacy and safety of a supervised multimodal exercise program (aerobic, resistance, and flexibility training three times per week) compared to usual care in 57 patients with prostate cancer metastatic to bone [2]. Spinal flexion/extension/rotation stretches were excluded in patients with axial or widespread metastases. Sessions were 60 min in length and completed for 3 months. After 3 months, compared to the control group, the supervised multimodal exercise program group had higher self-reported improvements in physical function (via the SF-36 questionnaire) and higher objectively measured muscle strength in the lower extremities (based on 1 repetition (rep) max testing on leg extension exercise). In addition, there were no skeletal fractures or increased bone pain in the exercise group compared to the control group. In 2013, Cormie et al. evaluated the efficacy and safety of a 12-week resistance training program compared to usual-care or nonexercise group in 20 men with established bone metastases secondary to prostate cancer [3]. The resistance training program involved twice-weekly hour-long sessions targeting the major muscle groups of the upper and lower body. Primary findings of the study indicated improved physical function, physical activity levels, and lean mass in the resistance exercise group compared to the usual-care group. These studies have shown that appropriately designed and supervised exercise regimens are well tolerated and do not significantly increase SREs.

PT primarily focuses on strategies that improve bone health in otherwise healthy individuals. Many wonder if this can also be applied to patients with impaired bone health and decreased capacity for bone healing secondary to metastatic disease burden or those undergoing cancer treatment. Strength and cardiorespiratory function have been shown to improve in patients with bone metastases who perform aerobic and strength training [4]. There is inconclusive evidence regarding if exercise is effective in increasing bone density in those with skeletal metastases; however, it does demonstrate a low-risk profile when completing a supervised exercise program [4]. It should be noted that participants selected for clinical trials are often not the most frail or sick cancer patients, and recommendations should be made based on understanding of each patient's specific circumstances and metastatic sites.

Bone metastases and the combination of ongoing cancer treatments can often lead to a decrease in a patient's physical activity and functional independence. Rehabilitation clinicians must consider the location of bone metastases to ensure affected regions do not experience excessive mechanical shear or compressive force [5]. Lesions of the hip and pelvis can be managed with prophylactic surgical fixation or reconstructive arthroplasty and patients are often placed on standard hip precautions with early mobilization and weight bearing as tolerated postoperatively [6, 7]. Spinal fixation is also possible as a prophylactic treatment for axial metastases. Keeping patient's mobile with safe transition to full weight bearing is a crucial step in this population.

Cancer diagnoses and treatment can have a negative effect on many aspects of a patient's life. This includes decreased mobility, increased pain, inability to complete activities of daily living (ADLs), instrumental activities of daily living (IADLs), a reduced quality of life, and a need for assistive devices for mobility including a wheelchair, walker, crutches, a cane, bracing, or a combination of several devices [8]. Based on a clinical practice guideline by Segal and colleagues in 2017 [9], it was determined that people diagnosed with cancer can safely engage in moderate amounts of exercise while on active treatment or after completion of treatment, and evidence suggests that exercise can improve quality of life as well as muscular and aerobic performance. Moderate-intensity exercise refers to physical activity 3–6 times the intensity of baseline, which requires a moderate amount of effort and noticeably accelerates the heart rate (3–6 METs) [9]. Exercise has been recommended by the American Cancer Society for patients with cancer with the following guidelines [10]:

(a) Avoid inactivity and return to daily activities as soon as possible after diagnosis and treatment.
(b) Take part in regular physical activity.
(c) Start slowly and build up the amount of physical activity over time.
(d) Build up to 150–300 min of moderate or 75–150 min of vigorous intensity activity each week.
(e) Exercise several times a week for at least 10 min at a time.
(f) Include resistance training exercise at least 2 days per week.
(g) Complete stretching exercises at least 2 days each week.

PT Strategies

Initial PT Assessment

In order to analyze PT strategies when working with patients with metastatic cancer, it is crucial to first understand the role of PT in the continuum of care. The current focus will be on the inpatient rehabilitation facility setting (IRF setting), also referred to as acute rehabilitation (rehab) and the role of physical therapists involved in the interdisciplinary team. The American Physical Therapy Association (APTA) has defined the vision for physical therapists as "transforming society by optimizing movement to improve the human experience" [15]. This can have a wide variety of meanings depending on the setting in which the physical therapist works; however, it connects all practitioners with a common goal—a patient/client focused approach to rehabilitation. Physical therapists are tasked with identifying impairment level deficits, connecting them to activity limitations, and addressing participation restrictions (Fig. 32.1). Based on the impairments that are identified, a plan of care is developed with goals targeted toward reaching an independent functioning level.

Upon admission to an IRF, patients will undergo a comprehensive assessment to determine impairments, activity limitations, and participation restrictions that will prevent a safe discharge home. PT programs should include consultation with the team of care providers to identify risks and benefits to the prescribed exercise regimen [8, 11]. Information that should be obtained from the patient's oncological team includes specific locations of axial or appendicular metastatic sites, pain management plan, prognostic factors, relevant past medical history and current and future treatment plans. It is also important to complete a thorough interview with the patient to obtain relevant medical and functional information as well as their baseline functional level and personal goals. This will help establish the intervention plan set forth by the physical therapist. Inpatient rehab physical therapists determine realistic functional goals based on patient impairment and utilize various interventions to maximize independent mobility. The identification of functional barriers assists with conversations regarding improving patient movement quality and quantity.

Patients with skeletal metastases require assessment of their preserved lower and upper extremities strength to determine the best method of mobility. For those with appendicular metastases specifically, there may be weight bearing or range of motion restrictions for the involved lower extremity and care should be taken to avoid stress on the location of the metastasis during strength testing and training. For example, in the case of femoral metastases, providing resistance during knee extension or using a leg press machine should be avoided. Manual muscle testing should not be used at all in the affected limb if the bone is at greater risk of pathologic fracture or the patient has already sustained a pathologic fracture in that extremity. If this precaution is indicated, a bodyweight assessment of the patient's lower extremity strength is sufficient.

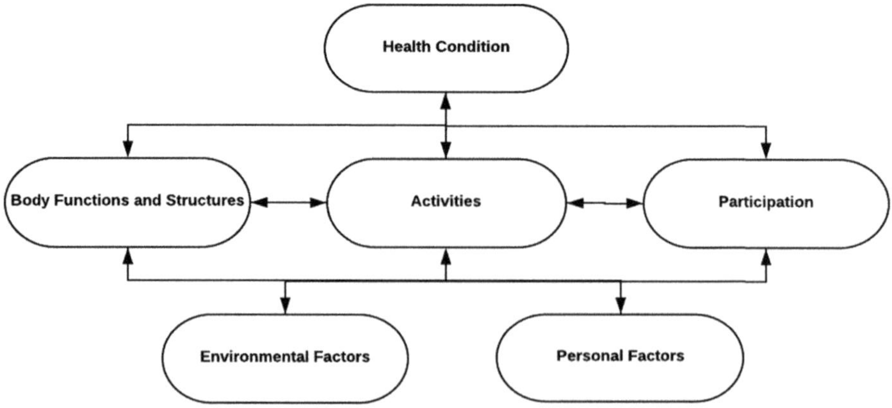

Fig. 32.1 The International Classification of Functioning, Disability, and Health (ICF) model [16]

Mobility Devices

The physical assessment will also help physical therapists determine appropriate assistive devices needed and transfer techniques to ensure patients will maintain their weight-bearing restrictions and promote functional independence. One main question the physical therapist will seek to answer is if a patient will be able to hop safely. If they possess the upper and lower extremity strength and balance to do so, a rolling walker is typically the first assistive device trialed. A rolling walker allows patients to push the walker forward instead of picking the device up with each hop and it provides more external stability compared to axillary or Lofstrand crutches. With higher level mobility patients, crutches may be an acceptable assistive device; however, a rolling walker will routinely be trialed first. Patients that do not possess the strength, balance, or endurance to be ambulatory will benefit from wheelchair propulsion and safe transfer training. Even if they will be able to hop or ambulate safely, due to cancer-related fatigue or pain, patients may prefer to have the option to mobilize via a wheelchair for energy conservation and to increase independence over indoor as well as outdoor terrain. Community wheelchair use for patients that are ambulatory can also prevent overuse injuries in their unaffected lower extremity and upper extremities. The decision regarding assistive devices to be used for mobility is unique to physical therapists and should be done under the direct guidance of a PT (Fig. 32.2).

If it has been determined that a wheelchair will be beneficial as either a primary or secondary means of mobility, it is then appropriate to assess the need for a custom wheelchair versus a standard style wheelchair. A standard manual wheelchair may be sufficient for most patients with adequate upper extremity strength and core control. Modifications can be made to these standard wheelchairs to

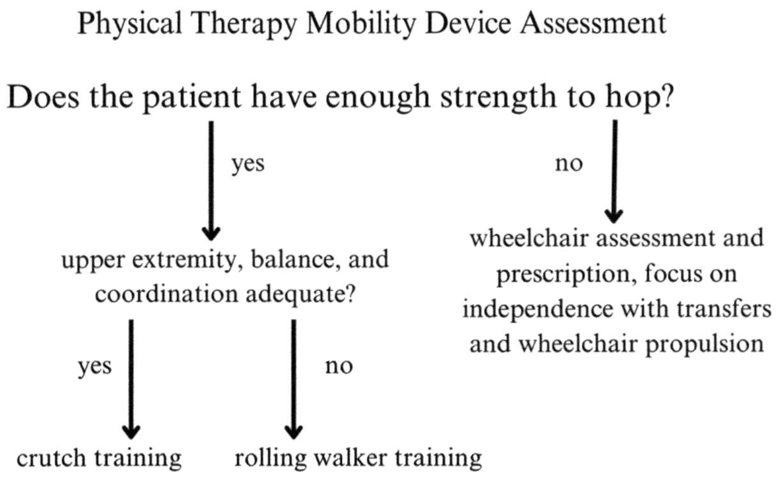

Fig. 32.2 Decision tree for determining appropriate assistive device in patients with appendicular metastases

support limb loss or if additional limb supports are needed; however, these are often not permanent changes and will need to be replaced over time. Patients may need pillows, towels, or bolsters along the lateral trunk to promote additional trunk stability and proper positioning to prevent contractures or skin breakdown. Specialty cushions may be necessary for proper positioning if there is limb loss or limb difference that has occurred because of cancer treatment or if the patient has sacral or pelvic involvement and sitting on a standard foam cushion is uncomfortable. Due to the weight of the standard manual wheelchair, they may not be an ideal solution for those with poor endurance or who will be using a wheelchair as their primary means of mobility. In these instances, a custom manual wheelchair may be recommended, and a referral to a wheelchair specialist is indicated. Patients who have undergone an external hemipelvectomy as a treatment for their primary or secondary skeletal metastases or as a result of complications from skeletal metastases should be specifically considered for a custom specialty wheelchair. In some circumstances, patients may qualify for use of a power wheelchair. Although patients must meet specific insurance criteria, a power wheelchair should be considered if the patient is unable to propel a manual wheelchair safely or independently within their home.

PT Interventions

After initial assessments have been completed, physical therapists will develop a comprehensive plan of care that includes interventions to target the patient's impairments and goals. This will likely include strength training, balance training, endurance training, cancer-related fatigue interventions, pain management, and, as a result, functional mobility training.

Strength Training

When developing a strength training program in the inpatient rehab setting, the focus is most often on functional movements that target large muscle groups such as the quadriceps, hamstrings, and glutes. For patients with skeletal metastases, proper technique and alignment is paramount, starting with bodyweight or low resistance exercises and adding weight or resistance if there is no exacerbation of pain or other adverse effects [11]. If axial metastases are present, it is recommended to avoid excessive spinal flexion, extension, and rotation as well as high impact activities such as hopping. Isometric spinal and core stabilization interventions are the best for spinal involvement [8, 12]. Hook lying (supine positioning with knees flexed) is preferred for core strengthening as opposed to quadruped (hands and knees) due to additional support provided to the spine by the firm surface the patient is lying on. If appendicular metastases are present, physical therapists should avoid prescribing activities with high loads/resistance to joint movements. For femoral metastases, resisted knee extension should be avoided to minimize risk of pathologic fracture.

Balance Training

The impacts of axial metastases on neurological function should be considered. Patients may present with altered sensation and proprioception in their extremities with compression of the spinal cord or spinal nerve roots as a result of the metastases or cancer treatments. Compensatory mechanisms or orthotics may need to be utilized to reduce the risk of falls due to neuropathy and impaired proprioceptive input. To compensate for sensation and proprioception loss, patients can benefit from balance training, especially using visual compensation to offset the decreased sensory input [13]. Both anticipatory and reactive balance training can help mitigate the occurrence of falls in a patient's home or community setting. In order to help the patient return to their baseline activities, physical therapists should attempt to progressively increase the task demands that appropriately challenge a patient's balance.

Due to the potential neurological changes in lower extremity strength, sensation, and proprioception in patients with axial metastases, referral to orthotists whether inpatient or outpatient should be considered to determine if bracing (whether custom or prefabricated) is warranted. Use of an ankle-foot orthosis (AFO) can improve patients' safety and independence with mobility by providing additional stability at the knee and/or ankle if knee extension or ankle dorsiflexion is particularly weak and is resulting in knee buckling or foot drop.

Endurance Training

Chemotherapy and deconditioning can also significantly impact cardiorespiratory fitness in patients diagnosed with cancer. Participation in aerobic exercise during chemotherapy has demonstrated a conservation of, or an improvement in, cardiorespiratory fitness [11]. Physical therapists can educate patients on compensatory strategies or pacing techniques for patients to maximize their endurance throughout the day, especially when treatments are at their most impactful phase. Engaging in endurance and aerobic exercise requires close attention to a patient's specific history to avoid exacerbating cancer-related fatigue.

Standardized Outcome Measures

Standardized outcome measures may be used to track functional strength, balance, and endurance over time [17]. Commonly, these outcome measures have clinical validation in neurologic conditions. To assess functional strength, the 5 Times Sit to Stand test (5×STS) can be used. Common standardized balance assessments include the Timed Up and Go test (TUG), Berg Balance Scale (BBS), and Functional Gait Assessment (FGA). Physical therapists often utilize the 6-min walk test (6MWT) as an endurance outcome measure and the 10-meter walk test to assess overall gait speed and risk of falls. Using these standardized assessments, physical therapists can conduct a baseline measure, perform weekly reassessments which can demonstrate quantitative change for insurance approval, or provide numerical data to show patients their functional improvement over time.

Cancer-Related Fatigue

Cancer-related fatigue has been defined as an "overwhelming and sustained exhaustion and decreased capacity for physical and mental work not relieved by rest" [13]. Fatigue can impact physical, cognitive, and emotional domains and can have a significant impact on a patient's quality of life [14]. Studies have examined the prevalence of cancer-related fatigue, ranging from 19% to 82% depending on the measures used to define and examine fatigue, the type of cancer being studied, and the timing of examination [14]. Typically, increasing the amount of rest is not effective in combating the symptoms of cancer-related fatigue, but reducing medical and functional comorbidities can have an effect. Treatment of depression and chronic pain has also shown to improve general cancer-related symptoms, and exercise has been shown to assist with minimizing fatigue [13]. Determining if a patient has cancer-related fatigue is key for treatment planning by physical therapists.

Pain Management

Pain management often requires an interdisciplinary approach between the medical team and the physical therapists. It is estimated that 30–50% of patients undergoing cancer treatment will experience pain, and up to 70% of those with metastatic disease will have pain [13]. This pain can be as a result of the malignancy itself, medications, or surgical interventions that may have been performed such as mastectomy or amputation [13]. If an amputation has been performed, pain may also manifest in the form of surgical, neuropathic or phantom pain. Physical therapists have many tools to address a patient's pain, but there are some that should be avoided to reduce risk to the malignant sites. Despite possible pain modulation effects, it is recommended to avoid modalities that increase local blood flow such as electrical stimulation, thermotherapy, massage, and ultrasound [12]. Modalities that increase local heating can increase blood flow which can theoretically increase tumorigenesis [19]. Instead, interventions can target stretching activities, gentle strengthening, virtual reality, visualization and meditation, and transcutaneous electrical nerve stimulation (TENS) [18].

Interdisciplinary Rehabilitation

In the acute rehab setting, physical therapists work together with other health care practitioners: physiatry, nursing, occupational therapy, speech language pathology, case management, nutrition, social work, psychology/psychiatry, and therapeutic recreation to create a diverse, interdisciplinary, patient-centered network. An interdisciplinary approach to patient care can help to ensure a thorough and exceptional healthcare plan. Patients with a chronic or life-threatening disease can have difficulty coping with their disease or specific prognosis. Clinical specialists in psychology and psychiatry can be consulted when mental or psychological barriers limit a patient's functional progress.

Following surgical procedures, the mobility and lifting restrictions can limit the patient's ability to participate in ADLs, such as bathing and dressing, and IADLs, such as cooking and cleaning. Occupational therapists are experienced clinicians that educate and train patients on adaptive equipment and strategies to promote functional independence for ADLs and IADLs while also maintaining orthopedic precautions.

Following inpatient rehabilitation, the continuum of care is essential to work toward maintaining independence outside of a hospital setting. Home PT can assist in determining home modifications and adaptations to further promote independence in the home. Outpatient PT will focus on specific extremity strengthening and meeting community reintegration goals.

Prognosis

In addition to physical impairments, physical therapists will assess other prognostic factors that may impact recovery. Prior level of function and whether the patient used an assistive device at baseline may indicate their capacity for recovery. If the patient was a community ambulator at baseline, they may have a better chance of returning to their baseline function level compared to those who were primarily household ambulators only. A patient's home setup and design can be conducive or detrimental to patient success at home, and home modifications may be needed for the patient to safely access their home. Ramps to traverse stairs, widening doorways for wheelchair use, addition of grab bars in bathroom and moving a patient's bed to the first level of the home may be necessary. The financial implications of cancer treatment can also impact recovery. In the current financial landscape, many patients and family members rely heavily on working full time to support their families and taking time off of work to physically care for themselves or their family members can present a significant financial strain. There are various programs, both public and private, that can assist families financially after a cancer diagnosis and involvement of a social worker in the patient's care team is paramount to ensuring they are aware of these opportunities. The presence of comorbid conditions unrelated to cancer treatment may also impact prognosis and should be considered when determining likelihood of achieving independent level of function. For patients that may require hands-on assistance or supervision with mobility at discharge, physical therapists will not only educate and train the patient, but family members and/or private caregivers as well. Teaching effective education regarding transfer set up, transfer techniques and safe body mechanics are important to facilitate a safe home discharge plan to home and minimize the patient's risk of falls. For patients with brain metastases that may affect decision making or judgment, caregiver training would be a high priority prior to discharge.

Conclusion

Physical therapists in the inpatient rehabilitation setting work with patients at a critical point in their recovery. Acute care pain medications have often been reduced, and the reality of managing their surgical outcome or severity of diagnosis commonly sets in. The goal now shifts from medical and pain management to functional outcomes and progression to the next stage of their recovery. Physical therapists focus on the patient's current mobility status and subsequently build a plan of care based on their findings, while also providing the foundation of the equipment, exercise, and self-management education of life going forward. As new medications and surgical procedures advance to help patients survive cancer diagnoses and treatments, the quality of rehabilitation must meet these medical developments. Patients with musculoskeletal metastasis can present with an array of obstacles and challenges that can be successfully addressed through the dedicated efforts of a multidisciplinary team.

References

1. Bunting R, Lamont-Havers W, Schweon D, Kliman A. Pathologic fracture risk in rehabilitation of patients with bony metastases. Clin Orthop Relat Res. 1985;192:222–7.
2. Galvão DA, Taaffe DR, Spry N, et al. Exercise preserves physical function in prostate cancer patients with bone metastases. Med Sci Sports Exerc. 2018;50(3):393–9.
3. Cormie P, Newton R, Spry N, et al. Safety and efficacy of resistance exercise in prostate cancer patients with bone metastases. Prostate Cancer Prostatic Dis. 2013;16:328–35.
4. Avancini A, Benato G, Borsati A, et al. Exercise and bone health in cancer: enemy or ally? Cancers (Basel). 2022;14(24):6078. https://doi.org/10.3390/cancers14246078.
5. Sheill G, Guinan EM, Peat N, et al. Considerations for exercise prescription in patients with bone metastases: a comprehensive narrative review. PM&R. 2018;10(8):843–64.
6. Cheville AL, Murthy NS, Basford JR, et al. Imaging and clinical characteristics predict near-term disablement from bone metastases: implications for rehabilitation. Arch Phys Med Rehabil. 2016;97:53–60.
7. Bunting RW, Shea B. Bone metastases and rehabilitation. Cancer. 2001;92(4 Suppl):1020–8.
8. Locke S, Hanstein A. Physical therapy for patients with bone metastases. Radiat Ther. 2024;33(2):118–24.
9. Segal R, Zwaal C, Green E, et al. Exercise for people with cancer: a clinical practice guideline. Curr Oncol. 2017;24(1):40–6. https://doi.org/10.3747/co.24.3376.
10. American Cancer Society. Physical Therapy and the Person With Cancer. Revised March 16, 2022. https://www.cancer.org/cancer/survivorship/be-healthy-after-treatment/physical-activity-and-the-Cancer-patient.html.
11. Campbell KL, Cormie P, Weller S, et al. Exercise recommendation for people with bone metastases: expert consensus for health care providers and exercise professionals. JCO Oncol Pract. 2022;18(5):e697–709. https://doi.org/10.1200/OP.21.00454.
12. Keilani M, Kainberger F, Pataraia A, et al. Typical aspects in the rehabilitation of cancer patients suffering from metastatic bone disease or multiple myeloma. Wien Klin Wochenschr. 2019;131(21–22):567–75. https://doi.org/10.1007/s00508-019-1524-3.
13. Silver JK, Baima J, Mayer RS. Impairment-driven cancer rehabilitation: an essential component of quality care and survivorship. CA Cancer J Clin. 2013;63(5):295–317. https://doi.org/10.3322/caac.21186.

14. Harrington SE, Fisher MI, Lee JQ, Cohn J, Malone D. Knowledge regarding cancer-related fatigue: a survey of physical therapists and individuals diagnosed with cancer. Physiother Theory Pract. 2023;39(9):1964–73. https://doi.org/10.1080/09593985.2022.2056554.
15. APTA Mission, Vision, and Strategic Plan 2013, 2018. https://www.apta.org/apta-and-you/leadership-and-governance/vision-mission-and-strategic-plan.
16. Sykes C. Health classifications 1 – an introduction to the ICF. WCPT Keynotes. World Confederation for Physical Therapy; 2006.
17. Rehabilitation Measures. Chicago (IL): Shirley Ryan AbilityLab; [cited 2025 May 7]. Available from: https://www.sralab.org/rehabilitation-measures.
18. Yimu Vassantachart A, Yeo E, Chau B. Virtual and augmented reality-based treatments for phantom limb pain: a systematic review. Innov Clin Neurosci. 2022;19(10–12):48–57. https://research.ebsco.com/linkprocessor/plink?id=f3d6a754-15e2-3a63-b73a-44b7baf55bcc. Accessed 30 Oct 2024.
19. Song CW. Effect of local hyperthermia on blood flow and microenvironment: a review. Cancer Res. 1984;44(10 Suppl):4721s–30s.

Pain Management Protocols

33

Rafael Vazquez, Brinda Kamdar, and Ana Gutierrez-Hasbach

Introduction

Orthopedic oncologic surgery (OOS) encompasses a diverse patient population, including adolescents, adults, and the elderly. While some patients present asymptomatically, others experience pain as an initial manifestation of disease or as a chronic consequence of metastatic bone involvement. As disease burden progresses, additional comorbidities—such as escalating opioid requirements—can complicate both surgical management and postoperative recovery.

Pain plays a central and multifaceted role in the OOS patient experience, manifesting as a complex interplay of physical, psychological, and emotional factors, expressed through both verbal and nonverbal means. The biopsychosocial model of pain highlights a wide range of biological, psychological, and social factors that influence the intensity, duration, and overall impact of pain on an individual [1]. Its perception is highly individualized, shaped by prior experiences, psychosocial context, and disease-related factors. Although surgical resection is central for oncologic control, equally important is complete clinical recovery with preservation of quality of life—minimizing pain, alleviating suffering, and reducing reliance on opioid therapies. Suboptimal pain control can result in clinically significant consequences such as increased hospital length of stay (LOS), adverse mental health outcomes, decreased mobility, reduced likelihood of returning to work, and higher likelihood of cardiopulmonary complications [2].

In certain vulnerable patient populations with pre-existing pain phenotypes, the perioperative period poses a heightened risk for inadequate pain control and further

R. Vazquez (✉) · B. Kamdar
Mass General Brigham Anesthesiology, Harvard Medical School, Boston, MA, USA
e-mail: Rvazquez@mgh.harvard.edu; BKamdar@mgh.harvard.edu

A. Gutierrez-Hasbach
Morrissey College of Arts and Sciences, Boston College, Newton, MA, USA
e-mail: gutierbm@bc.edu

J. Pretell-Mazzini, S. A. Lozano Calderón (eds.), *Multidisciplinary Approach of Musculoskeletal Metastasis*, https://doi.org/10.1007/978-3-032-09498-8_33

opioid escalation [1]. Without proactive strategies, acute postoperative pain can evolve into chronic postsurgical pain (CPSP), a major complication after surgery contributing to long-term functional impairment and opioid dependence [3, 4]. Therefore, a multidisciplinary, anticipatory approach to pain management is essential to reduce these risks and support recovery at an individual and personalized level.

This chapter will explore the pathophysiology of pain related to OOS, focusing on the biopsychosocial mechanisms that influence pain transmission and modulation. It will examine patient-specific factors that predispose individuals to increased pain sensitivity and chronicity. Additionally, evidence-based strategies for preventing and managing acute and chronic pain will be presented, including multimodal pharmacologic regimens, regional anesthesia techniques, and interdisciplinary care coordination. By integrating these principles, clinicians can personalize and standardize perioperative pain management to reduce opioid exposure, enhance recovery, and ultimately improve long-term patient outcomes and satisfaction by mitigating the risk of CPSP.

Pain Physiology

Understanding Pain Pathways

Pain is defined as "an unpleasant sensory and emotional experience associated with, or resembling that associated with, actual or potential tissue damage" [5]. This definition by the International Association for the Study of Pain (IASP) aims to encompass the full complexity of pain—both its psychological and physical dimensions—and is referred to as the "neuromatrix" model [6, 7] (Fig. 33.1). Most

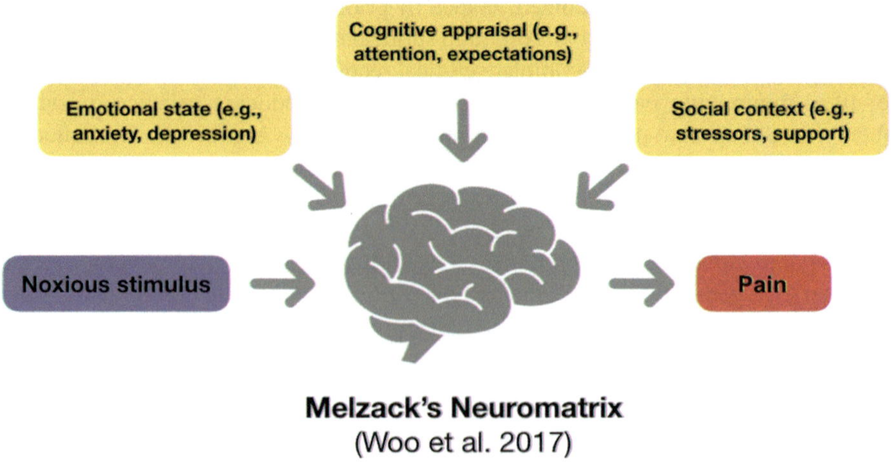

Fig. 33.1 Melzack's neuromatrix model as conceptualized by Woo et al. depicting the physiological mechanisms of pain and neurophysiological process that contribute to pain perception especially as it relates to chronic pain. An optimal pain management pathway should address both mechanisms

importantly, pain is a fundamental biological mechanism that allows the human body to detect and respond to noxious or potentially harmful stimuli. This sensory input prompts protective withdrawal reflexes, initiates physiological processes for wound healing, and, at the cognitive level, facilitates learning to avoid similar dangers in the future.

Appreciation of the neuromatrix pain theory is essential in recognizing that pain is not merely the brain's perception of physical injury but is influenced by a complex interplay of factors including sensory input (nociception), cognitive, emotional, and psychosocial factors. This concept supports the biopsychosocial model of pain which emphasizes that patient factors such as age, sex, pain threshold, depression, anxiety, education, and social support all contribute to the neural, non-physical experience of pain in the brain [1]. Therefore, optimal pain management protocols should incorporate both the physiological mechanisms of pain perception—such as nociception—and the neurophysiological processes that contribute to pain perception and the development of chronic pain. The physiological component can be addressed through standardization of anesthetic and surgical protocols, while the neurophysiological aspect emphasizes identifying patient-specific pain phenotypes to enhance personalized perioperative pain management strategies (Fig. 33.2).

As a starting point, reviewing pain at a fundamental and simplified level is essential in gaining a clear understanding of its management. Somatic pain is a

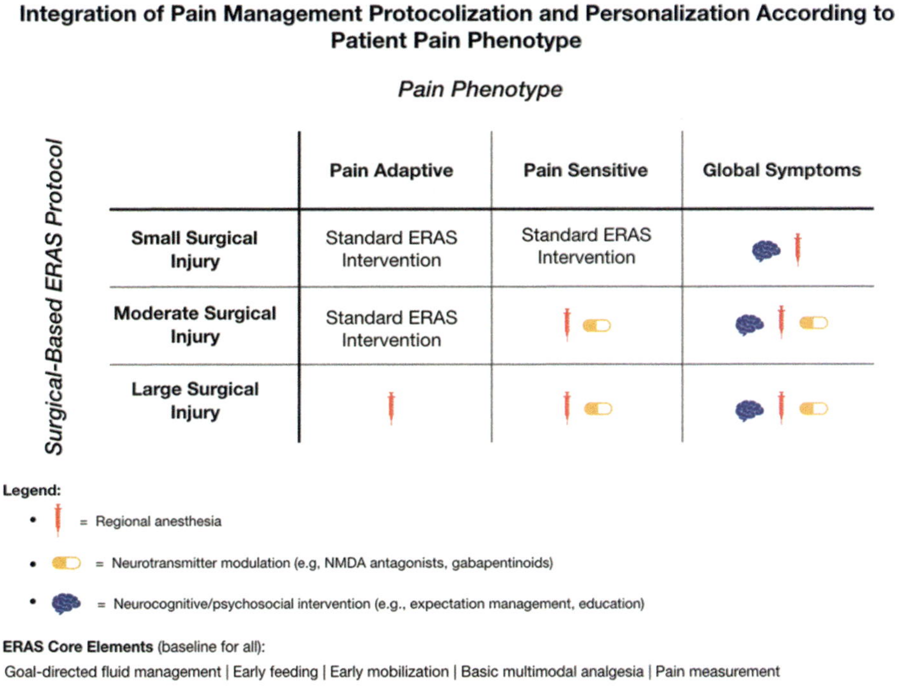

Fig. 33.2 Integration of protocolization and personalization of perioperative orthopedic oncologic surgical patients based on surgical injury extent and patient pain phenotype

type of nociceptive pain that originates from the skin, muscles, joints, bones, and connective tissue. It occurs when nociceptors (pain receptors) in these tissues are activated due to injury, inflammation, or mechanical damage [8]. Noxious stimuli resulting from bone invasion, nerve compression, and cytokine release result in acute pain. With surgical intervention, further pain results from extensive tissue manipulation, including bone resection. This is defined as nociceptive pain. With tissue trauma, there is release of neurotransmitters, cytokines, and activation of an inflammatory cascade, releasing inflammatory mediators from surrounding cells, including immune cells. The mediators include ATP, bradykinin, and prostaglandin E2. Cyclo-oxygenase-2 enzyme is also activated, and immune cells release interleukins. Substance P, calcitonin gene-related peptide (CGRP) neurokinin A, and nitric oxide (NO) are released from C-fiber nociceptors in a process called neurogenic inflammation. This results in an amplification of the process and activation of "silent" C fibers as well as expression of additional receptors and upregulation of nociceptive sodium ion channels. Over time, this leads to a lowered pain threshold and heightened neural excitability, a phenomenon known as hypersensitization [9–12]. This can be thought of as the two phases of surgical pain: phase 1—incisional pain; phase 2—inflammatory pain mediated by neurogenic inflammation (Fig. 33.3).

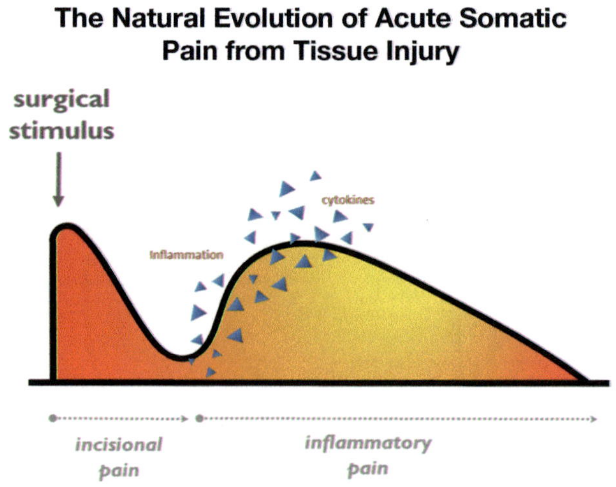

Fig. 33.3 Surgical pain can be thought of as having two phases. The phase is marked by tissue injury, which results in the release of local tissue inflammatory mediators such ATP, bradykinin, prostaglandin E2, histamine, and serotonin. Subsequently, cyclo-oxygenase-2 (COX 2) enzyme is activated and cytokines are released, which further activate nociceptors to release further inflammatory mediators such as substance P. This is termed "neurogenic inflammation." This results in further activation of additional nociceptors, with expression of additional receptors and upregulation of nociceptive sodium ion channels. This results in reduction of pain threshold and increased excitability termed hypersensitization. This process has two components peripheral sensitization and central sensitization which play an important role in the pathophysiology of pain

Nociceptive pain is conducted through specialized peripheral neurons known as nociceptors, which are located in the periphery and form the ascending somatic pain pathway leading to the cerebral cortex. These nociceptors are afferent first-order neurons that synapse to the dorsal horn of the spinal cord with second-order neurons within the substantia gelatinosa, also known as the Rexed laminae I and II. This region is an important site for signal modulation and pain perception. These postsynaptic neurons respond to neurotransmitters—primarily glutamate and substance P—released by the first-order neurons. It is here that glutamate targets 3-hydroxy-5-methyl-4-isoxazolenpropionic acid (AMPA) and N-methyl-D-aspartate (NMDA) receptors on the second-order neurons. Substance P activates receptors within lamia I. Both neurotransmitters also activate microglial cells in the CNS, which have implications for central modulation of pain regulation, especially inhibition [10]. The second-order neurons in turn traverse the spinal cord and travel in the contralateral anterolateral system: spinothalamic tract (STT) where they ascend to the brain stem. At the brain stem level, pain signals are further relayed through the spinoreticular tract, which is responsible for individualized pain responses relating to emotion. After reaching the brain stem, second-order neurons synapse at the thalamus with third-order neurons. These neurons synapse with the somatosensory portion of the brain's pain neuromatrix to be processed as "pain" [13, 14]. It is here that individualized emotional responses to pain are generated (Fig. 33.4).

It is important to recognize that central mechanisms exist to regulate the ascending transmission of nociceptive signals. At the level of the brain, the neuromatrix, rostral ventromedial medulla (RVM), and reticular formation are important for ascending nociceptive pathway regulation. There is a regulatory inhibition of the ascending pathways at the spinal cord level [15]. This process is well known as the gate control theory, postulated by Melzack and Wall [16, 17]. Specifically, gamma-aminobutyric acid (GABA) and glycine interneurons inhibit second-order nociceptors at the spinal cord level to mediate nociceptive transmission. The interplay between second-order neurons transmitting nociceptive peripheral input and these interneurons activated centrally in the brain regions is important in pain homeostasis (Fig. 33.4).

The descending inhibitory pathways travel down to the dorsal horn of the spinal cord and release neurotransmitters norepinephrine and serotonin to further regulate the ascending nociceptive pathway (Fig. 33.3). Endogenous opioids and opioid receptors found in the brain centrally modulate this inhibitory pathway [15].

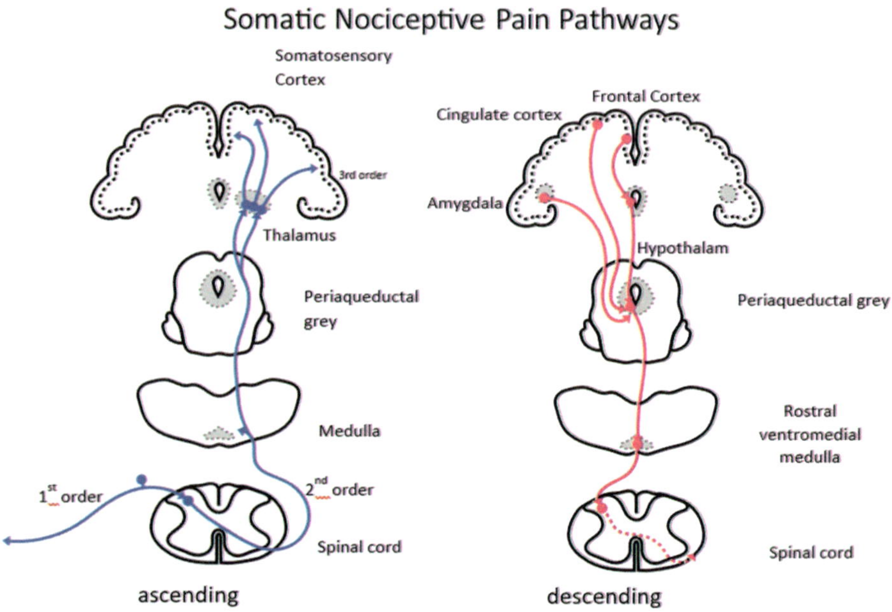

Fig. 33.4 In this simplified figure, the ascending and descending nociceptive pathways are important to understand in the context of somatic pain regulation and homeostasis. Peripheral noxious stimulus is transmitted by first-order nociceptors after tissue injury to the dorsal horn of the spinal cord. They synapse centrally with second-order neurons which then travel up the spinal thalamic pathway centrally to the brainstem where they give off several tracts important in the emotional response of pain (spinoreticular tract). Ultimately the second-order neurons synapse with third-order neurons in the thalamus which in turn project up to the somatosensory cortex and the brain's "neuromatrix," where pain perception is processed. The frontal cortex of the brain including the neuromatrix and brainstem (periaqueductal gray, rostral ventromedial medulla, and reticular formation) forms the descending pathway which is important in regulation of the ascending pathway at the level of the dorsal horn via norepinephrine and serotonin neurotransmitters. The interneurons at the level of the spinal cord also regulate and inhibit the ascending pathway via GABA and glycine at the level of the dorsal horn. This regulation is known as the gate theory of pain

Pain Pathophysiology

Pain becomes pathological when it persists beyond the expected course of healing or contributes to psychological distress and functional impairment. The interaction between physiological nociceptive pathways and the psychological components of pain—often referred to as the neuromatrix—can become dysregulated. This dysregulation may form a self-perpetuating feedback loop: pain induces emotional distress, which in turn amplifies pain perception through central sensitization and maladaptive cortical responses [3].

Pain is typically classified by its duration and underlying mechanisms:

1. *Acute Pain*: Limited in duration, acute pain is typically nociceptive and related to direct tissue injury. It serves a protective function and generally resolves within a predictable timeframe, ranging from minutes to less than 3 months [18].
2. *Chronic Pain*: Defined as pain that persists or recurs for longer than 3 months, chronic pain is often associated with persistent nerve dysfunction and central processing changes. It is commonly neuropathic and/or nociplastic in origin, reflecting the complex interplay between somatosensory, affective, and cognitive pathways [18].
3. *Acute on Chronic Pain*: This describes acute exacerbations of pain episodes in individuals with preexisting chronic pain conditions, complicating both diagnosis and treatment.

Chronic postsurgical pain (CPSP) has been recognized as a new type of chronic pain. It is defined as "chronic pain that develops or increases in intensity after a surgical procedure or a tissue injury and persists at least 3 months after surgery or tissue trauma." [19]. It is the direct result of a controlled incision(s) in the health care setting, with subtypes after specific surgeries, including postamputation and spinal surgery. The pathophysiological underpinnings of the transition from acute pain to (CPSP) are well established, and targeted pharmacologic therapies have been developed to address specific nociceptive pathways. However, severe acute pain and CPSP persist. In recent years, growing attention has been directed toward psychosocial mechanisms—including pain modulation, catastrophizing, and resilience—which help explain the wide variability in pain experiences following similar surgical insults. These insights underscore the need for an integrative approach to pain management that addresses both peripheral and central mechanisms of pain perception [1, 20].

Inflammation and Tissue Damage

Persistent nociceptive stimulation that results from incisional tissue damage leads to distinct pathophysiological changes in the afferent nociceptive pathways due to neuroplasticity—that is, physiologic neurobiological changes in cells that can become transient or permanent. Specifically, inflammation and nerve injury arising from incisions lead to long-term synaptic plasticity that amplifies and maintains pain signaling. This is different from other pain models [19, 21]. Initial peripheral tissue injury triggers the release of inflammatory mediators, which in turn lead to a series of events involving the primary afferent nociceptors and immune cells to release cytokines and neurotransmitters. This triggers a positive feedback loop, which leads to nociceptors releasing inflammatory mediators and recruiting more nociceptors. The mediators also alter nociceptor function, increasing their excitability and lowering their firing threshold, resulting in additional nociceptive input. This process is known as primary sensitization [9, 12].

Centrally, in the process of secondary sensitization, neurotransmitters play a central role. Glutamate has an excitatory effect on NMDA receptors located on second-order neurons in the spinal cord. Continuous release causes an extended depolarization of the second afferent neurons in the dorsal horn, facilitating the removal of magnesium from the NMDA receptors. This allows glutamate to bind to these receptors, which in turn allows any subsequent activation of nociceptors to cause pain in response to nonnoxious stimuli, resulting in allodynia [12]. Glial cells found in the CNS are activated by second-order neurons via substance P, NO, and prostaglandin release. Activated glial cells upregulate COX-2 to produce prostaglandin E2 and release a variety of cytokines. This leads to further activation of COX-2, reduced activation thresholds in second-order neurons, and upregulation of AMPA and NMDA receptors, leading to more allodynia. The resulting sensitization of the CNS plays a key role in the development of neuropathic pain and contributes to increased sensitivity to noxious stimuli in areas surrounding the injury—a phenomenon known as secondary hyperalgesia [10] (Fig. 33.5).

Cancer Versus Noncancer Pain

While CDC opioid-prescribing guidelines designate cancer pain as a distinct category, its underlying physiological mechanisms—nociception and central sensitization—mirror those of non-cancer chronic pain. There is no pathophysiological basis for divergent analgesic strategies based solely on the presence of malignancy. What does differ are the psychosocial burdens faced by cancer patients: emotional distress stemming from prognostic uncertainty, concerns about mortality, and the impact on functional capacity and family dynamics [22, 23].

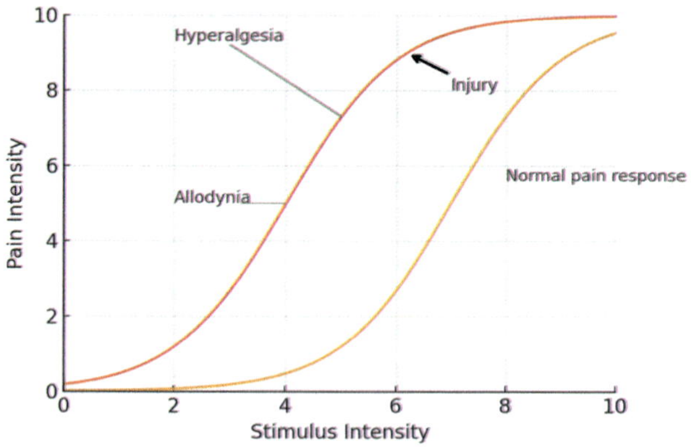

Fig. 33.5 The phenomenon of hyperalgesia is depicted. Hyperalgesia is an augmented response to a noxious stimulus. Surgical trauma results in peripheral and central changes due to the release of inflammatory and pain mediators in a process known as sensitization. This can also occur in response to high opioid administration during surgery. Normally, the hyperalgesic response is self-limiting. However, if it persists more than 3 months due to permanent central nervous system changes (plasticity), it becomes pathologic and transitions to chronic postsurgical pain (CPSP)

Principles of Acute Pain Management in OOS

Acute pain management is founded on the physiological and pathophysiologic principles of nociception. Pharmacologic agents, regional anesthetic techniques, and neurocognitive therapies collectively address the biopsychosocial dimensions of pain. Although the neurobiology of nociception is identical in oncologic and nononcologic patients, the added psychosocial burdens of a cancer diagnosis—prognostic uncertainty, morbidity, and mortality anxieties—heighten vulnerability, making interventions like cognitive-behavioral therapy essential for those with a susceptible pain phenotype.

Pharmacologic and regional strategies aim to curb the inflammatory and neuronal injury induced by surgical insult, thereby attenuating peripheral and central sensitization driven by cytokines, neurotransmitters, and inflammatory mediators. By dampening these processes, postoperative pain intensity decreases, resulting in improved postoperative pain control, reduced opioid requirements, related dependency risks in high-risk populations, and decreased the likelihood of chronic postsurgical pain [19] (Fig. 33.6). Key components of acute pain protocols in orthopedic oncologic surgery include an overarching preventive philosophy that incorporates multimodal analgesia and is tailored to the patient's pain phenotype for the most optimal recovery outcomes.

Presurgical Evaluation and Development of Personalized Pain Protocols

Pain is highly individualized, particularly in the context of an oncologic diagnosis. Surgical procedures already place patients in a vulnerable state, but the psychological impact and stress are significantly amplified when cancer is involved. How each patient processes this stress varies tremendously, and this variability in coping cannot be overstated. All three components of the biopsychosocial model of pain—biological, psychological, and social—are equally important. The brain and mind's influence on nociception is just as significant as the physiological components. Orthopedic surgery is among the most painful types of surgery performed [24], making it essential to tailor pain management protocols to each orthopedic oncologic surgery to optimize periprocedural pain outcomes.

The biophysiological bases of nociception are very well understood. The efficacy and effectiveness of specific analgesics and nerve blocks have been studied and standardized to offer a uniform patient experience. From the surgical perspective, there have been surgery-specific evidence-based protocols devised to optimize patient recovery through the implementation of enhanced recovery after surgery (ERAS) protocols. Yet, interpatient variability of postprocedure acute pain outcomes exists [2, 25, 26]. Some patients benefit, while others develop severe pain and resultant sequelae of postprocedure complications that can transition to chronic pain. This is true for orthopedic oncologic patients. Thus, the focus has now shifted from analgesic interventions and surgical procedures to understanding. Specifically,

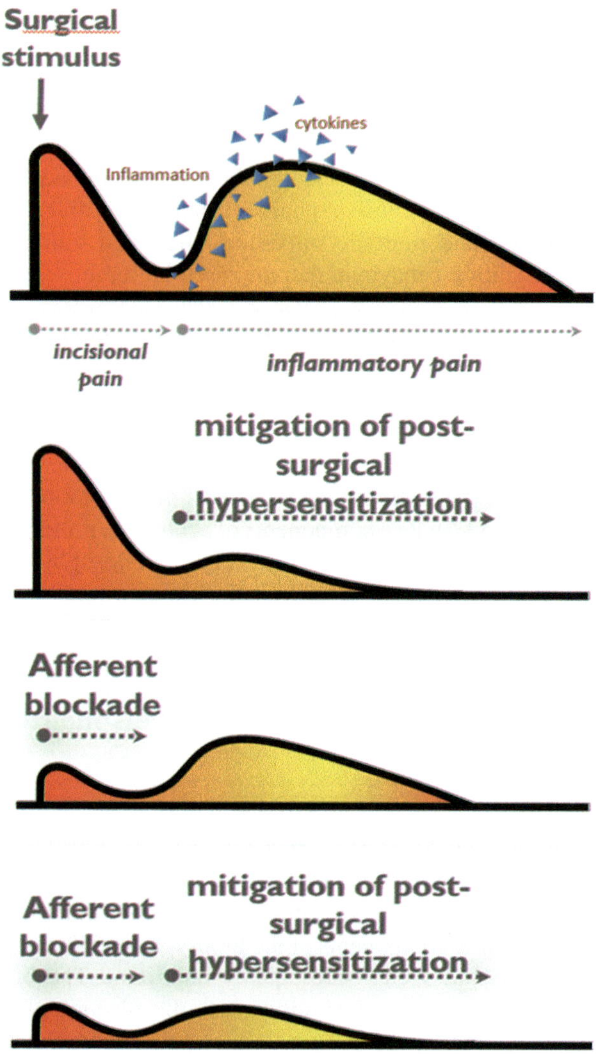

Fig. 33.6 Acute pain management strategies in orthopedic oncology focus on reducing both peripheral and central sensitization following surgical injury. These strategies utilize pharmacologic agents, regional anesthesia, and cognitive interventions to limit the cascade of events—such as inflammation and cytokine release—that drive acute pain and increase the risk of chronic post-surgical pain (CPSP). The top panel illustrates the progression from incisional pain to inflammatory pain, driven by surgical stimulus and ongoing inflammatory mediators. Without adequate intervention, persistent sensitization can lead to heightened and prolonged pain responses. Afferent blockade (using medications or regional anesthesia) can dampen the initial noxious input, but if not sustained, hypersensitization and pain may still develop. The lower panels demonstrate that an optimal multimodal, preventive analgesic protocol extends throughout all perioperative phases, maintaining analgesia well into the postoperative period. This comprehensive approach provides continuous pain control, attenuates hypersensitization, and reduces the risk of transitioning to CPSP

on identifying vulnerable patient populations and tailoring patient specific beneficial protocols to mitigate severe acute and chronic pain [20, 27].

In recent years, psychometric and biopsychosocial measures have been shown to help clinicians pinpoint patient phenotypes who may require more intensive perioperative pain management or who are especially likely to benefit from a particular set of analgesic interventions [28]. Risk factors identified have been depression, anxiety, negative affect, and sleep disturbance [29–31]. The principal challenge is to establish a concise, readily measurable set of parameters that can guide perioperative teams in customizing analgesic regimens to both the specific procedure and the individual patient. Psychometric measures assess the psychological characteristics that the patient brings to the perioperative situation. Biophysical tools are more objective tools that enable the evaluation of an individual's sensitivity to pain perception [20, 32].

Psychometric Measures

There a many psychometric measures, but perhaps the most relevant is the Pain Catastrophizing Scale (PCS) is a 13-question self-assessment questionnaire designed to assess a patient's experience of various negative affective and cognitive responses related to pain [27, 33, 34]. That is an exaggerated negative orientation to noxious stimuli that involves rumination about painful sensations, magnification of the threat value of the pain stimulus, and perceived inability to control pain. There are three dimensions, each with associated items: (1) rumination, (2) magnification, and (3) magnification. Patients are asked to self-rate thoughts or feelings from 0 (not at all) to 4 (all the time), the degree to which each item has occurred when they have experienced pain [35]. Surgical outcomes in patients undergoing mastectomy with high catastrophizing scores are associated with higher acute and chronic pain levels as well as greater prolonged opioid use post [34, 36]. However, these patients demonstrate a benefit from regional anesthesia [37]. Brief Symptom Index (BSI) 18-Somatization Scale can be used to measure somatization or distress about bodily sensations [38].

Biophysical Measures

Preoperative quantitative sensory testing (QST) is defined as quantifiable mechanical, thermal, or electrical stimuli, or pressure. It systematically characterizes responses to sensory and pain perception. Electrical stimuli and pressure pain threshold specifically can be useful in predicting postoperative pain intensity after musculoskeletal surgery [25, 39]. Temporal summation of pain (TSP) is also a useful tool and been shown to be associated with acute pain intensity and chronic pain intensity [36, 39].

Preoperative biopsychometric results can be used to identify patient pain phenotypes as well as clusters which can ultimately aid in guiding perioperative pain protocols. Phenotypes such as high catastrophizing can benefit from perioperative cognitive-behavioral therapy (CBT) and more aggressive interventions such as regional blocks. Patients with high pain amplification (TSP) would benefit from pharmacologic interventions such as ketamine to decrease central sensitization. Low-risk patients would not require additional interventions beyond standard care.

Pain Phenotyping

There are comprehensive validated questionnaires such as the Brief Pain Inventory (BPI) and the Patient Reported Outcome Measurement Information System (PROMIS) that can be administered in the preoperative setting [40, 41]. The PROMIS form can measure baseline depressive symptoms, anxiety, and sleep disturbance (23 questions). These are labor intensive and impractical in day-to-day clinical practice. Shorter and more concise questionnaires are being developed to capture the same breath of clinical three-item questionnaire and a more involved 19-item questionnaire that can give as much comprehensive information as multiple questionnaires (PROMIS and BPI) [17, 42].

Diagnosis and treatment of pain rely principally on self-reported symptoms because biological tests for pathophysiological mechanisms that support individualized treatments are currently not available. Biomarkers can be used to identify patients at susceptible for acute and chronic pain. The NIH set up the Acute to Chronic Pain Signatures (A2CPS) program to study patients perioperatively prior to and after surgery in an effort to examine biological and psychological correlates of pain and resilience. Specific biomarkers include catechol-O-methyltransferase (COMT) and m-opioid receptor 1 (OPRM1). Proteins, extracellular RNA, and metabolites that have been identified as being associated with chronic pain may be detected with neuroimaging. Magnetic resonance imaging (MRI) is a primary tool for neuroimaging studies and can offer a wealth of imaging-based markers for investigating the acute-to-chronic pain transition. There's an effort to integrate all the tools, including biopsychosocial measures, to create a biosignature for an individual patient undergoing surgery [43].

At the present moment, integration and analysis of key biophysical and psychosocial characteristics can generate patient phenotype clusters similar to the biosignatures, except without the biomarkers. These phenotypes can be used to effectively predict pain trajectory. Furthermore, they can be used to risk-stratify patients to care pathways based on the complexity of the surgical procedure [44] (Fig. 33.2). This has been successfully applied to the patients undergoing breast cancer surgery and can be adapted to the orthopedic oncology patient population [45].

Clusters that have been identified based on preoperative scores of anxiety, depression, somatization, and trapezius pressure pain threshold (PPT) are:

1. Pain adaptive (low psychosocial scores, high PTT)
2. Pain sensitive (moderate psychosocial scores, low PPT)
3. Global pain symptoms (high psychosocial scores, moderate PPT)

Orthopedic oncologic patients often experience significant levels of stress, depression, and anxiety [46]. Psychosocial and psychophysical assessments should be performed, utilizing tools such as the PROMIS Short Form to evaluate depression and anxiety over the preceding week, along with the Brief Symptom Index (BSI) 18-Somatization Scale. Based on these assessments, patients can be stratified into appropriate clusters. The chosen cluster, together with the specific surgical procedure, should guide the selection of analgesic interventions. This individualized approach helps optimize the perioperative experience and reduces complications associated with severe postoperative pain.

Preventive Analgesia

Pre-emptive analgesia is the concept of administering analgesic interventions before a noxious stimulus to decrease downstream physiologic changes of inflammation and neuronal damage, including neuroplasticity, that results in hypersensitization of peripheral and central neurons [47]. In theory, optimally timed pharmacologic and regional techniques should lessen acute postoperative pain and lower the risk of chronic postsurgical pain (CPSP). Yet, animal models and human studies have generally failed to show a durable reduction in incision-related pain with purely preemptive strategies, except in very young rats, suggesting an age-dependent effect [48, 49]. There is a beneficial effect of preemptive administration for COX-2 inhibitors and gabapentinoids per the joint guidelines on management of postoperative pain [50]. Because incisional pain is unique in that it's maintained by ongoing nociceptor activation and dorsal-horn transmission, both the efficacy and duration of analgesic regimens are critical. This insight gave rise to "preventive analgesia," a more expansive perioperative approach in which multimodal interventions extend well into the postoperative period—beyond each drug's half-life—to attenuate surgery-induced neuroplasticity and thereby reduce postoperative pain intensity and related complications [21, 51].

Multimodal Analgesia

Orthopedic surgeries are associated with significant nociceptive pain [24, 52]. Initially, there was a heavy reliance on opioid monotherapy for moderate to severe pain; however, it was associated with detrimental side effects, including respiratory depression, nausea, vomiting, constipation, tolerance, physical dependence, and abuse. In vulnerable patient populations, there is more potential for abuse that can evolve into an opioid use disorder, as evident by the opioid epidemic observed in the younger population in the United States [53]. There's also the phenomenon of opioid induced hyperalgesia (OIH) when high-dose short-acting opioids are administered prior to the end of surgery (not observed when opioids are administered in the postoperative period) [54]. There has been an emphasis on minimizing unnecessary opioid use in favor of nonopioid adjunct analgesics. Multimodal analgesia has been defined as the use of 2 or more analgesics or techniques that target different mechanisms or pathways in the nociceptive system [52]. A more comprehensive multimodal strategy incorporates nonpharmacological interventions (i.e., behavioral therapy) that can be of great benefit to oncologic patients. Multimodal analgesia allows for the use of lower doses of drugs yet potentially provides greater pain relief and fewer analgesic adverse effects than can be achieved with monotherapy. Multimodal analgesia optimizes pain relief by treating pain through complementary mechanisms of action along multiple sites of the nociceptive pathway (Fig. 33.7).

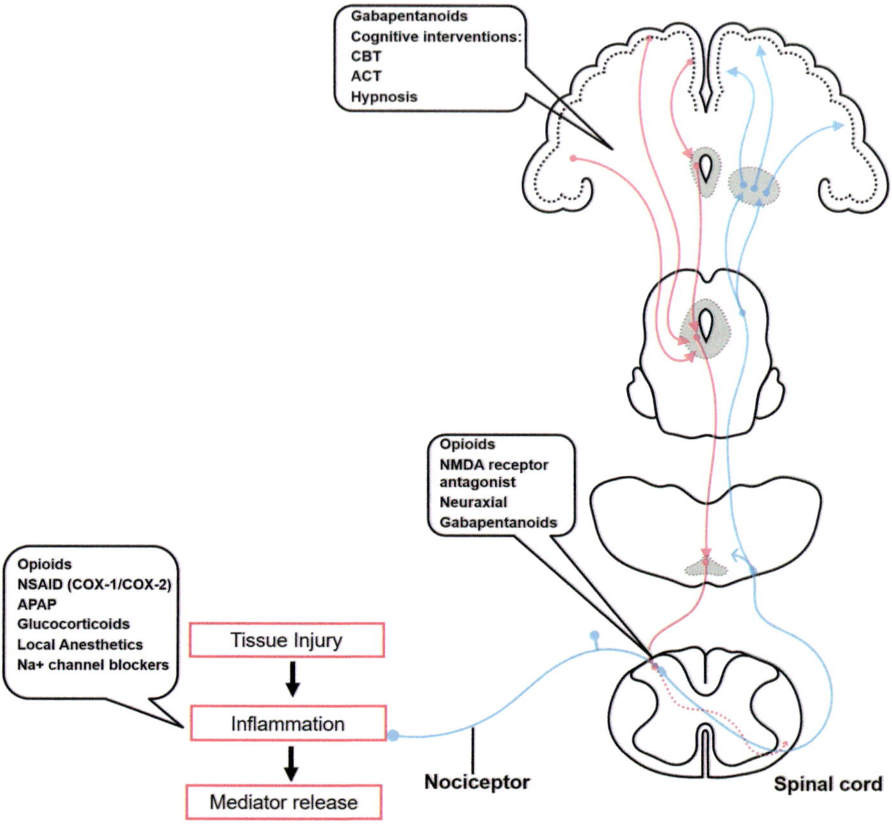

Fig. 33.7 The site of action of the various analgesic interventions is seen on the peripheral and central nervous system. CBT cognitive-behavioral therapy, ACT acceptance and commitment therapy, APAP acetaminophen

Pharmacological agents include traditional agents such as local anesthetics (nonselective sodium channel blockers), acetaminophen, nonsteroidal anti-inflammatory drugs, and opioids, as well as adjuvant agents including anticonvulsants, N-methyl-D-aspartate [NMDA]-receptor antagonists like ketamine and methadone, alpha-2 adrenergic agonists (e.g., dexmedetomidine), and antidepressants. NMDA receptor antagonist work centrally and can inhibit neuroplasticity that occurs from the release of glutamate and activation of the NMDA receptors. It is important to note that the side effects from each of these medications still occur such as sedation and effects on organ systems (renal effects of NSAIDs) [55] (Tables 33.1 and 33.2). Most recently, suzetrigine, a selective sodium channel blocker, was approved by the FDA as a nonopioid analgesic for the treatment of moderate to severe pain. It's available as a tablet. It targets peripheral $Na_v1.8$ channels, and it can be incorporated into a multimodal analgesic regimen.

Table 33.1 Common IV analgesics and adjuvants that can be utilized for intraoperative multimodal analgesia

IV Nonopioid Analgesics				
Drug class	Dose (daily max)	Site of action	Side effects	Contraindications
Acetaminophen	1000 mg every 6–8 hrs (2–4 g)	COX pathway	Increased LFTs	Fulminant liver failure
NSAIDS				
Ketorolac	15–30 mg every 4–6 hrs (120 mg)	COX-1 & COX-2	Antiplatelet myocardial infarction	Renal failure Bleeding diathesis
Opioid				
Methadone	0.2 mg/kg prior to incision	NMDA receptors	Somnolence Constipation Respiratory depression	Prolonged QT Hypersensitivity
Adjuvant analgesics				
Magnesium	50 mg/kg bolus→8–15 mg/kg/hr. infusion	NMDA receptor	Hypotension Respiratory depression Prolonged NMB	Renal failure Heart block
Ketamine	0.5 mg/kg bolus→5–10 mcg/kg/min infusion	NMDA receptor	Hypertension Psychosis Seizure	Active psychiatric history Increased ICP & Ocular pressure
Dexmedetomidine	1.0 mcg/kg bolus over 10 min → 0.3–1.0 mcg/kg/min infusion	Alpha-2 agonist	Bradycardia Hypotension	Impaired cardiac function
Lidocaine	1.5 mg/kg bolus→1–2 mg/min infusion (therapeutic 2–10 μg/ml)	Na channels	Local anesthetic toxicity Lethargy	Severe hepatic impairment

Nonpharmacological interventions, which have varying degrees of evidence regarding efficacy, include transcutaneous electrical nerve stimulation (TENS), and psychological and cognitive interventions such as cognitive-behavioral therapies such as cognitive-behavioral therapy (CBT), acceptance and commitment therapy (ACT), hypnosis, acupuncture, heat therapy, and therapeutic touch [56, 57] (Fig. 33.7).

Multimodal analgesia should be the standard for surgical pain management protocols. By characterizing patient pain phenotypes alongside surgical complexity, analgesic regimens can be tailored effectively. More aggressive regimens, although potentially associated with side effects such as sedation, may offer superior analgesic outcomes and should be used for pain-sensitive patients undergoing complex oncologic procedures. Conversely, less aggressive approaches are suitable for pain-resilient patients regardless of surgical complexity (Fig. 33.2). Opioids should be reserved as a rescue option as needed.

Table 33.2 PO analgesics in acute pain management

PO analgesics in acute pain management				
Drug class	Dose (max)	Site of action	Side effects	Contraindications
Acetaminophen (PO)	975 mg q8h (2 g–4 g)	COX inhibition	Increased LFTs	Fulminant liver failure
NSAIDS				
Celecoxib (PO)	200 mg q12h (400 mg)	COX-2	GI upset Renal impairment	Sulfa allergy Severe CAD Renal impairment Extreme age
Ibuprofen (PO)	200–400 q8h (3200 mg)	COX-1 & COX-2	GI upset Renal impairment	GI pathology Renal impairment
Indomethacin (PO)	100 mg (200 mg)	COX-1 & COX-2	GI upset Renal impairment	GI pathology Renal impairment
Adjuvant analgesics				
Gabapentin (PO)	100–300 mg/day (1200 mg)	Calcium channels	Drowsiness	Hypersensitivity
Opioids				
Tramadol (PO)	50–100 mg q4-6h (300 mg)	Neuropathic pain Nociceptive pain	Drowsiness	MAOI use GI obstruction
Oxycodone (PO)	5–10 mg q4-6h	Opioid mu receptor	Nausea/vomiting Constipation	Intolerance

Neuraxial Anesthesia

When executed properly, epidural anesthesia (EA) can provide effective analgesia for surgeries of the lower appendicular skeleton as well as of the axial skeleton, including hip and pelvis surgeries. It reduces periprocedural pain, impacts the inflammatory response centrally, decreases opioid administration, and the surgical stress response. There is evidence that it can have an impact on cancer recurrence outcomes, but there is no definitive randomized controlled data [58, 59]. EA is also associated with decreased CPSP for oncologic breast and thoracic cases. There may be a benefit for orthopedic oncologic cases. However, EA has the disadvantage of impairing lower extremity motor function and thus early mobilization for patients in an era where mobilization is recognized as a pillar of recovery. EA is also associated with hypotension but is less likely for low thoracic epidurals. Finally, there are risks if anticoagulation is planned immediately postoperatively. EA still has immense benefits; however, it should be utilized for high-risk pain phenotype patients discussed earlier (Table 33.3).

Regional Anesthesia

Regional anesthesia techniques enhance perioperative pain control. In the incisional pain model, there is constant nociceptive input to the CNS. Therefore, efficacy and duration of analgesic regimens are important. There is a body of evidence indicating

Table 33.3 Lower extremity surgeries and analgesic options

Lower extremity amputation surgeries & analgesic interventions	
Amputation	Interventional options
Hemipelvectomy/hip disarticulation	1. Epidural 2. Lumbar plexus + sacral plexus blockade
Above the knee	1. Epidural 2. Lumbar plexus + sacral plexus blockade 3. Femoral (saphenous for below-knee) + sciatic block
Through the knee	1. Epidural 2. Lumbar plexus + sacral plexus blockade 3. Femoral + sciatic block
Below the knee/ankle disarticulation	1. Sciatic nerve block 2. ± femoral/saphenous nerve block
Partial foot amputation	1. Sciatic nerve block 2. ± femoral/saphenous nerve block

a benefit, especially for long-lasting nerve blocks and continuous catheter techniques, in decreasing CPS by attenuating nociceptive and inflammatory inputs. Local anesthetics are also thought to play a role by decreasing neuronal inflammation and central activation of interneurons (glial cells) responsible for neuroplasticity. Additionally, regional anesthesia is associated with a decrease in intraoperative opioid administration, which can decrease opioid induce hyperalgesia (OIH) [55]. There is some evidence that regional may have an impact on cancer recurrence by preserving the cell-mediated immunity [58].

Oncologic patients are at high risk for developing severe acute pain postoperatively, given their cancer diagnosis. Patients with high pain catastrophizing scale (PCS), can benefit from regional anesthesia techniques [37]. Peripheral nerve blocks (PNB) are most applicable for tumors of the appendicular skeleton. For amputations specifically, CPSP in the form of persistent phantom limb pain (PLP) and residual limb pain (RLP) can range from 51% to 85%. Regional anesthesia is beneficial for both amputations of the upper extremity and lower extremity, especially with patients with high pain phenotypes, i.e., high preamputation pain intensity, preoperative opioid use, and anxiety [60]. Single-shot regional injections can be beneficial in the immediate postoperative period as part of a multimodal regimen by reducing opioid analgesic requirements, postoperative nausea and vomiting, and hospital discharge. One of the recognized side effects is the phenomenon of rebound pain, which occurs transiently after the resolution of the regional blockade. It is marked by intense pain perception. It seems it is the unmasking of the expected nociceptive response to the surgical stimulus in the absence of adequate multimodal analgesia [61]. Risk factors include surgery on bone tissue and female sex. Intraoperative dexamethasone administration and an ongoing multimodal pain regimen are key in mitigating this side effect [62] (Tables 33.3 and 33.4).

PNB catheters have an established role in surgical analgesia similar to single-shot PNBs for surgeries of the appendicular skeleton. Other benefits include decreased association with rebound pain. They are appropriate for complex surgeries, including amputations of the extremities, including shoulder disarticulations, and have been shown to decrease PLP [60]. Surgically placed distal perineural catheter is a technique that can be used as an alternative to distal PNBs. It has been

Table 33.4 Upper extremity regional anesthesia options

Upper extremity surgeries and regional anesthesia options	
Surgery type/amputation	Interventional options
Shoulder disarticulation	Interscalene + superficial cervical plexus block (consider cervical paravertebral approach) ± T2 paravertebral block
Above the elbow	Interscalene block Supraclavicular block Infraclavicular block (limited proximal humerus coverage) ± T2 paravertebral block
Through/below the elbow	Infraclavicular block Supraclavicular block Axillary block (poor site for continuous infusion)
Wrist/partial hand/finger amputation	Isolated digital (finger) nerve block Infraclavicular block Supraclavicular block Axillary block (poor site for continuous infusion)

applied to lower extremity amputations with variable results. When compared to neuraxial anesthesia patients, however, they had a significant opioid sparing effect, but no impact on acute pain scores [63].

Continuous wound infiltration of local anesthetic is a technique that has been used as an alternative to PNB. This can be achieved through a catheter placed over the wound. Studied demonstrate positive outcomes with decreased postprocedure hyperalgesia and decreased pain after 3 months for patients undergoing iliac crest grafting [19].

Local infiltration analgesia (LIA) involves the injection of a multimodal "cocktail" of local anesthetic (ropivacaine) combined with adjuvants (ketorolac, epinephrine, and clonidine) in the surgical wound by the surgeon. This is commonplace for joint replacement surgeries. Large studies have found this to be beneficial as a component of multimodal which promotes early mobilization, confers analgesia, and decreases hospital length of stay. It can be supplemented with regional anesthesia blocks, such as an adductor canal block (ACB) for maximum benefit of total knee replacement surgeries [52].

Psychological and Behavioral Interventions

In patients who have been identified as high risk for intense acute postoperative pain and CPSP, cognitive interventions such as cognitive-behavioral therapy (CBT) and acceptance and commitment therapy (ACT) have been associated with improved patient outcomes [25]. CBT is a control-oriented treatment that focuses on reducing pain and distress severity through modifying physical sensations, catastrophic thinking, and maladaptive behaviors. ACT treatments focus on all negative and positive acceptance of experiences, identification of life values, and actions toward achieving those goals. Pain models have demonstrated that acceptance is associated with pain tolerance when compared to distraction or control strategies [64]. In patients with chronic pain, these techniques are associated with reduced postoperative pain, disability, depression, psychological resilience, and pain catastrophizing. These benefits have also been observed in patients without chronic pain [56, 65]. In patients with high sleep disturbance, preoperative sleep hygiene and/or exercise can be recommended to enhance perioperative acute pain outcomes [1]. Hypnosis as a perioperative intervention for acute pain has shown benefit and comparable results to psychological and behavioral therapies, especially in the setting of administering multiple sessions preoperatively [57].

Enhanced Recovery After Surgery (ERAS) in Orthopedic Surgical Oncology

Enhanced Recovery After Surgery (ERAS) pathways comprise standardized, evidence-based perioperative practices designed to optimize patient outcomes by reducing length of stay, infection rates, and improving patient satisfaction. These pathways include three distinct phases—preoperative, intraoperative, and postoperative—each encompassing multiple components of patient care. The objective is the consistent application of these components, ranging from preoperative counseling and fasting guidelines to specific surgical and anesthetic techniques, extending through to postoperative mobilization [66]. The selection of ERAS components is dynamic, continuously adapting to reflect the latest clinical evidence. For oncologic orthopedic surgery, there is a paucity of specific ERAS protocols, except for some arthroplasty protocols that are adapted from nononcologic protocols [63].

Personalizing Pain Protocols

Despite standardizing interventions, variability persists in patient-reported pain due to individual differences in nociceptive responses. As pain remains central to patient care, this chapter emphasizes tailoring pain management protocols within the ERAS framework based on the most current evidence, advocating for personalized approaches to optimize outcomes and reducing complications associated with postoperative pain.

A personalized evidence-based pain protocol for oncology orthopedic patients would encompass the 3 phases of care: preoperative, intraoperative, and postoperative. Each phase would incorporate the elements discussed. This would be stratified based on procedure complexity (simple lipoma excision vs. hemipelvectomy).

In the preoperative phase, patients' pain phenotype and cluster would be characterized by employing psychometric and biophysical measures. They would be assigned to an appropriate cluster by utilizing the PROMIS short form to measure depression and anxiety, the Brief Symptom Index (BSI) 18-Somatization Scale to measure somatization, and the Pain Catastrophizing Scale (PCS) to measure catastrophizing. Biophysical assessment would be made by administering the pressure pain threshold (PPT) test. The patients would be stratified into 3 clusters as outlined by Wilson et al.: Adaptive, Pain Sensitive, and Global Symptoms [45] (Fig. 33.2).

Based on these clusters, the patients would then have standard ERAS protocol procedural and anesthetic care based on the preventive analgesic principles discussed previously. A high-risk patient for post op acute pain would receive an aggressive multimodal regimen of cognitive-behavioral therapy (CBT), ketamine, methadone, and dexmedetomidine. Based on the surgery, they would receive a PNB, or an epidural could also be used. Opioids would be used sparingly as needed (Fig. 33.8). It's important to note that the analgesics must continue throughout the recovery period and as the patient begins to transition to PO medications, appropriate dosages are utilized.

Special Considerations

Amputations

Patients facing amputation present a unique combination of psychological vulnerability and risk for severe pain. As noted earlier, they are particularly susceptible to phantom limb pain (PLP), and higher levels of preoperative pain intensity significantly increase their likelihood of experiencing PLP after surgery. To mitigate both intense acute pain and the progression to chronic postsurgical pain manifesting as PLP, these patients require specialized perioperative care. Preoperatively, comprehensive assessments—including psychometric evaluations and biophysical measurements—should identify those at greatest risk (for example, younger patients, those with longstanding pain, or high catastrophizing scores). High-risk individuals may benefit from proactive interventions such as cognitive-behavioral therapy (CBT) or acceptance and commitment therapy (ACT) before surgery. Intraoperatively, the anesthetic plan should follow enhanced recovery after surgery (ERAS) principles—employing multimodal analgesia and adjunctive agents like ketamine infusions, magnesium, gabapentin, and dexmedetomidine. Regional anesthesia should be optimized with long-acting nerve blocks or continuous perineural catheters for several days. If such blocks are not feasible, local anesthetic infiltration and perineural techniques offer important alternatives [60] (see Fig. 33.9).

Opioid-Tolerant or Chronic Pain Patients

Patients maintained on long-term opioid therapy present unique challenges—including opioid tolerance, opioid-induced hyperalgesia, and an elevated risk of severe postoperative pain—which demand thorough preoperative planning. Early collaboration between acute and chronic pain specialists is essential, with a comprehensive review of the patient's opioid history, comorbidities, and psychosocial factors guided by the American Pain Society and ASA postoperative pain practice guidelines [4, 10]. Wherever feasible, tapering toward the lowest effective dose and, for those on high or multiple opioid regimens, rotating to a single long-acting formulation can help stabilize plasma levels before surgery [67]. Concurrently, screening for anxiety, depression, and catastrophizing should prompt referrals for

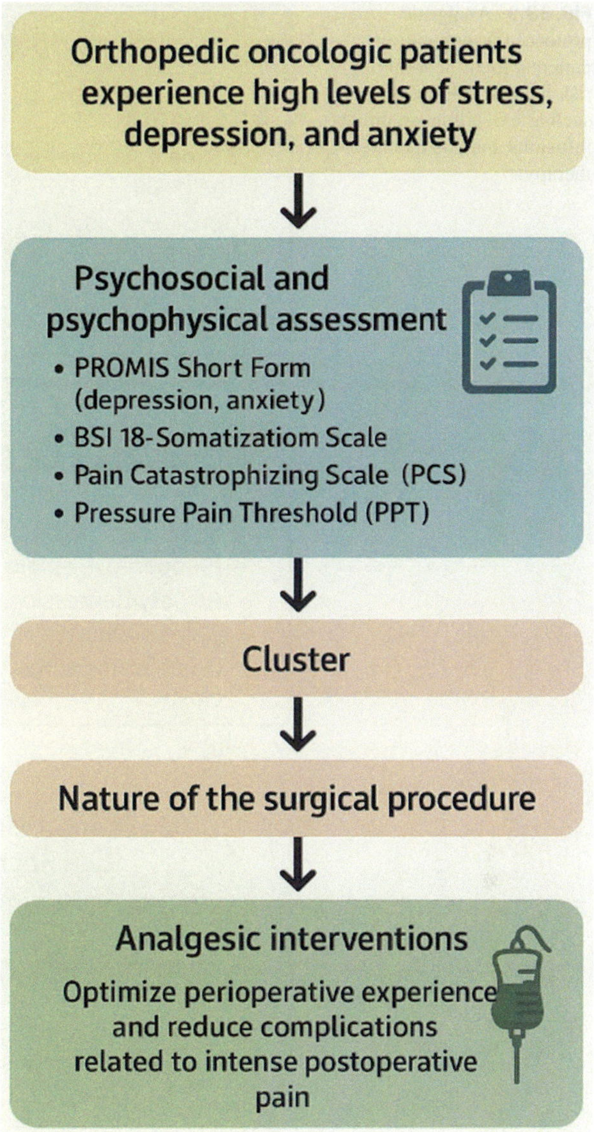

Fig. 33.8 An example of the stratification of pain phenotypes of oncologic orthopedic patients by utilizing psychosocial and psychophysical metrics to cluster patient pain phenotypes. The complexity of the surgical procedure will dictate the analgesic interventions and duration

cognitive-behavioral therapy or acceptance and commitment therapy to bolster coping strategies such CBT and reduce pain catastrophizing [68].

Similar to patients undergoing amputations intraoperatively, an aggressive ERAS-based multimodal analgesic regimen should include scheduled nonopioid agents—acetaminophen and NSAIDs or COX-2 inhibitors—to diminish reliance on opioids and limit peripheral nociception. To further blunt central sensitization, adjuncts such as ketamine infusions, intravenous lidocaine, dexmedetomidine, and

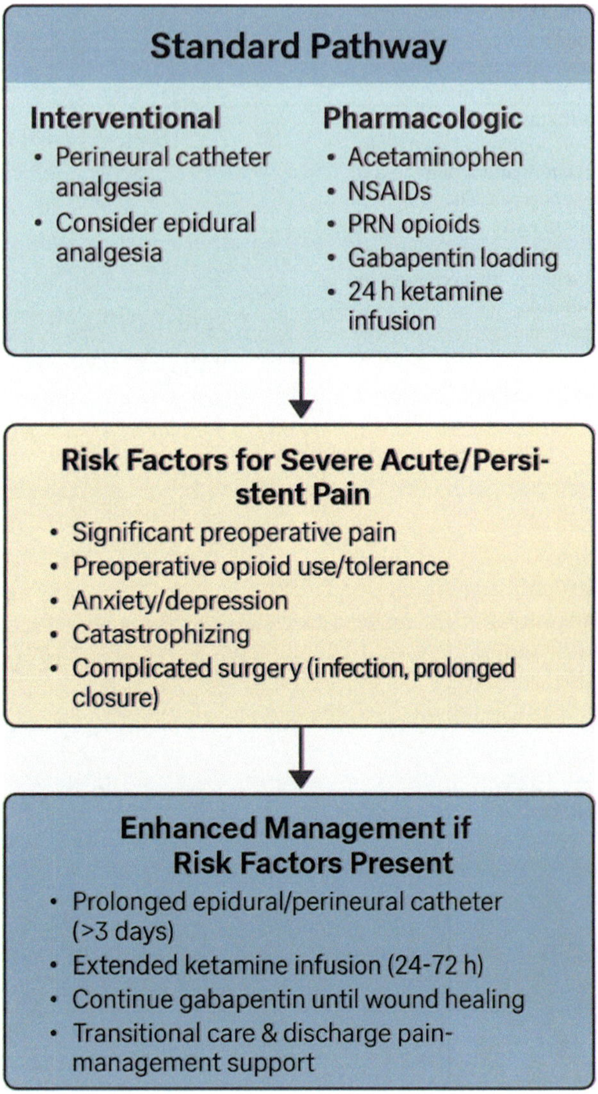

Fig. 33.9 Analgesic protocol for amputation patients. Identification of risk factors and/or options such as extending ketamine infusions and gabapentin therapy

gabapentinoids (gabapentin or pregabalin) are invaluable for minimizing intraoperative and early postoperative opioid requirements [4].

Optimizing regional techniques is another cornerstone of opioid-sparing care. Whenever possible, employ long-acting peripheral nerve blocks or continuous perineural catheters to deliver potent, site-specific analgesia for several days. If these approaches are contraindicated or impractical, judicious local anesthetic infiltration or shorter-acting perineural techniques still offer meaningful opioid-sparing benefits [4].

Postoperatively, maintain a structured, scheduled multimodal plan combining acetaminophen, NSAIDs or COX-2 inhibitors, and adjunctive agents while resuming the patient's baseline opioid regimen in a controlled-release format. Under chronic pain team oversight, initiate a gradual taper—typically 10–20% dose reductions per week—monitoring closely for withdrawal or rebound hyperalgesia. Early multidisciplinary follow-up with pain specialists, physical therapists, and psychosocial support services reinforces functional recovery and promotes a smoother, opioid-sparing trajectory toward long-term wellness [10].

Conclusion

Tailored, multimodal pain protocols are a cornerstone of modern orthopedic oncology care. By combining pharmacologic agents with regional techniques, psychosocial support, and perioperative optimization, we address the diverse mechanisms that underlie both acute and chronic postsurgical pain (CPSP). Customizing analgesic regimens to each patient's risk profile—whether through targeted nerve blocks, judicious use of ketamine or gabapentinoids, or preoperative behavioral therapies—allows us to prevent pain chronification and minimize opioid exposure. When these evidence-based components are incorporated into a structured pathway, we see tangible benefits: faster functional recovery, shorter hospital stays, lower rates of CPSP, and ultimately, enhanced quality of life. In sum, a proactive, individualized, and protocol-driven approach not only mitigates suffering in the perioperative period but also promotes durable improvements in long-term patient outcomes.

References

1. Schreiber KL, Wilson JM, Chen YYK. Recognizing pain phenotypes: biopsychosocial sources of variability in the transition to chronic postsurgical pain. Reg Anesth Pain Med. 2025;50(2):86–92.
2. Gerbershagen H, Aduckathil S, van Wijck A, Peelen L, Kalkman C, Meissner W. Pain intensity on the first day after surgery a prospective cohort study comparing 179 surgical procedures. Anesthesiology. 2013;118:934–44.
3. Chapman CR, Vierck CJ. The transition of acute postoperative pain to chronic pain: an integrative overview of research on mechanisms. J Pain. 2017;18(4):359.e1–359.e38.
4. Fregoso G, Wang A, Tseng K, Wang J. Transition from acute to chronic pain: evaluating risk for chronic postsurgical pain. Pain Physician. 2019;22(5)
5. Raja SN, Carr DB, Cohen M, Finnerup NB, Flor H, Gibson S, et al. The revised International Association for the Study of Pain definition of pain: concepts, challenges, and compromises. Pain. 2020;161:1976.
6. Melzack R. From the gate to the neuromatrix. Pain. 1999;82(SUPPL.1)
7. Woo CW, Schmidt L, Krishnan A, Jepma M, Roy M, Lindquist MA, et al. Quantifying cerebral contributions to pain beyond nociception. Nat Commun. 2017;8
8. Loeser J, Melzack R. Pain: an overview. Lancet. 1999;353:1607.
9. Chapman CR, Tuckett RP, Song CW. Pain and stress in a systems perspective: reciprocal neural, endocrine, and immune interactions. J Pain. 2008;9(2):122–45.
10. Feizerfan A, Sheh G. Transition from acute to chronic pain. Contin Educ Anaesth Crit Care Pain. 2015;15(2):98–102.

11. Breitinger U, Breitinger HG. Excitatory and inhibitory neuronal signaling in inflammatory and diabetic neuropathic pain. Mol Med. 2023;29:53. BioMed Central Ltd
12. McGreevy K, Bottros MM, Raja SN. Preventing chronic pain following acute pain: risk factors, preventive strategies, and their efficacy. Eur J Pain Suppl. 2011;5(2):365–76.
13. Tracey I, Mantyh PW. The cerebral signature for pain perception and its modulation. Neuron. 2007;55:377–91.
14. Fields HL. Pain modulation: expectation, opioid analgesia and virtual pain. Prog Brain Res. 2000;122
15. D'Mello R, Dickenson AH. Spinal cord mechanisms of pain. Vol. 101, Br J Anaesth Oxford University Press; 2008. p. 8–16.
16. Melzack R, Wall PD. Pain mechanisms: a new theory. Science. 1965;150
17. Pan PH, Tonidandel AM, Aschenbrenner CA, Houle TT, Harris LC, Eisenach JC. Predicting acute pain after cesarean delivery using three simple questions. Anesthesiology. 2013;118(5):1170–9.
18. Treede RD, Rief W, Barke A, Aziz Q, Bennett MI, Benoliel R, et al. Chronic pain as a symptom or a disease. In: The IASP classification of chronic pain for the international classification of diseases (ICD-11), vol. 160. Pain; 2019. p. 19–27.
19. Richebé P, Capdevila X, Rivat C. Persistent postsurgical pain: pathophysiology and preventative pharmacologic considerations. Anesthesiology. 2018;129:590–607.
20. Lirk P, Schreiber KL. Lessons learnt in evidence-based perioperative pain medicine: changing the focus from the medication and procedure to the patient. Reg Anesth Pain Med. 2024;49:688.
21. Pogatzki-Zahn EM, Zahn PK, Brennan TJ. Postoperative pain-clinical implications of basic research. Best Pract Res Clin Anaesthesiol. 2007;21:3–13.
22. Dowell D, Haegerich TM, Chou R. CDC guideline for prescribing opioids for chronic pain-United States, 2016. J Am Med Assoc. 2016;315(15):1624–45.
23. Peppin JF, Schatman ME. Terminology of chronic pain: the need to "level the playing field.". J Pain Res. 2016;9:23–4.
24. Pitchon DN, Dayan AC, Schwenk ES, Baratta JL, Viscusi ER. Updates on multimodal analgesia for orthopedic surgery. Anesthesiol Clin. 2018;36:361–73.
25. Warner DS, Werner MU, Mjö HN, Nielsen PR, Rudin Å. Prediction of postoperative pain a systematic review of predictive experimental pain studies [Internet]. Anesthesiology. 2010;112. Available from: www.anesthesiology.org.
26. Correll DJ, Vlassakov KV, Kissin I. No evidence of real progress in treatment of acute pain, 1993–2012: scientometric analysis. J Pain Res. 2014;7:199–210.
27. Strulov L, Zimmer EZ, Granot M, Tamir A, Jakobi P, Lowenstein L. Pain catastrophizing, response to experimental heat stimuli, and post-cesarean section pain. J Pain. 2007;8(3):273–9.
28. Tighe PJ, Le-Wendling LT, Patel A, Zou B, Fillingim RB. Clinically derived early postoperative pain trajectories differ by age, sex, and type of surgery. Pain. 2015;156(4):609–17.
29. Orbach-Zinger S, Fireman S, Ben-Haroush A, Karoush T, Klein Z, Mazarib N, et al. Preoperative sleep quality predicts postoperative pain after planned caesarean delivery. Eur J Pain (UK). 2017;21(5):787–94.
30. Pan PH, Coghill R, Houle TT, Seid MH, Lindel WM, Parker RL, et al. Multifactorial preoperative predictors for postcesarean section pain and analgesic requirement [Internet]. Anesthesiology. 2006;104. Available from: www.anesthesiology.org.
31. Franqueiro AR, Wilson JM, He J, Azizoddin DR, Karamnov S, Rathmell JP, et al. Prospective study of preoperative negative affect and postoperative pain in patients undergoing thoracic surgery: the moderating role of sex. J Clin Med. 2024;13(19)
32. Edwards RR, Dworkin RH, Turk DC, Angst MS, Dionne R, Freeman R, et al. Patient phenotyping in clinical trials of chronic pain treatments: IMMPACT recommendations, vol. 157. Pain; 2016. p. 1851–71.
33. Van Damme S, Crombez G, Bijttebier P, Goubert L, Van Houdenhove B. A confirmatory factor analysis of the pain catastrophizing scale: invariant factor structure across clinical and non-clinical populations [Internet]. 2002. Available from: www.elsevier.com/locate/pain.

34. Sullivan MJ, Thorn B, Haythornthwaite JA, Keefe F, Martin M, Bradley LA, et al. Theoretical perspectives on the relation between catastrophizing and pain. Clin J Pain. 2001:52.
35. Sullivan MJL, Bishop SR, Pivik J. The pain catastrophizing scale: development and validation. Psychol Assess. 1995;7:524.
36. Schreiber KL, Zinboonyahgoon N, Xu X, Spivey T, King T, Dominici L, et al. Preoperative psychosocial and psychophysical phenotypes as predictors of acute pain outcomes after breast surgery. J Pain. 2019;20(5):540–56.
37. Zinboonyahgoon N, Vlassakov K, Lirk P, Spivey T, King T, Dominici L, et al. Benefit of regional anaesthesia on postoperative pain following mastectomy: the influence of catastrophising. Br J Anaesth. 2019;123(2):e293–302.
38. Durá E, Andreu Y, Galdón MJ, Ferrando M, Murgui S, Poveda R, et al. Psychological assessment of patients with temporomandibular disorders: confirmatory analysis of the dimensional structure of the brief symptoms inventory 18. J Psychosom Res. 2006;60(4):365–70.
39. Sangesland A, Storen C, Vaegter HB. Preoperative quantitative sensory testing (QST) predicting postoperative pain: image or mirage? Scand J Pain. 2017;15:44–52.
40. Cella D, Riley W, Stone A, Rothrock N, Reeve B, Yount S, et al. The patient-reported outcomes measurement information system (PROMIS) developed and tested its first wave of adult self-reported health outcome item banks: 2005–2008. J Clin Epidemiol. 2010;63(11):1179–94.
41. Cleeland CS, Ryan KM. Pain assessment: global use of the brief pain inventory. Ann Acad Med Singap. 1994;23(2):129–38.
42. He J, Wilson JM, Fields KG, Flowers Zachos KM, Franqueiro AR, Reale SC, et al. Brief assessment of patient phenotype to explain variability in postsurgical pain and opioid consumption after cesarean delivery: performance of a novel brief questionnaire compared to long questionnaires. Anesthesiology. 2024:701–14.
43. Sluka KA, Wager TD, Sutherland SP, Labosky PA, Balach T, Bayman EO, et al. Predicting chronic postsurgical pain: current evidence and a novel program to develop predictive biomarker signatures. Pain. 2023;164(9):1912–26.
44. Bair E, Gaynor S, Slade GD, Ohrbach R, Fillingim RB, Greenspan JD, et al. Identification of clusters of individuals relevant to temporomandibular disorders and other chronic pain conditions: the OPPERA study. Pain. 2016;157(6):1266–78.
45. Wilson JM, Colebaugh CA, Flowers KM, Overstreet D, Edwards RR, Maixner W, et al. Applying the rapid OPPERA algorithm to predict persistent pain outcomes among a cohort of women undergoing breast cancer surgery. J Pain. 2022;23(12):2003–12.
46. Weng Hong C, Ajit Singh V, Amar Singh AK. Psychological disorders in patients with orthopaedic oncological diseases and their coping strategies. J Orthop Surg. 2022;30(1)
47. Woolf C, Chong M. Preemptive analgesia—treating postoperative pain by preventing the establishment of central sensitization. Anesth Analg. 1993;77(8):362–79.
48. Warltier DC, Møiniche S, Kehlet H, Dahl JB. A qualitative and quantitative systematic review of preemptive analgesia for postoperative pain relief the role of timing of analgesia [Internet]. 2002. Available from: www.cochrane.org.
49. Ong CKS, Lirk P, Seymour RA, Jenkins BJ. The efficacy of preemptive analgesia for acute postoperative pain management: a meta-analysis. Anesth Analg. 2005;100(3):757–73.
50. Chou R, Gordon DB, De Leon-Casasola OA, Rosenberg JM, Bickler S, Brennan T, et al. Management of postoperative pain: a clinical practice guideline from the American pain society, the American society of regional anesthesia and pain medicine, and the American society of anesthesiologists' committee on regional anesthesia, executive committee, and administrative council. J Pain. 2016;17(2):131–57.
51. Rosero EB, Joshi GP. Preemptive, preventive, multimodal analgesia: what do they really mean? Plast Reconstr Surg. 2014;134(4):85S.
52. O'Neill A, Lirk P. Multimodal analgesia. Anesthesiol Clin. 2022;40:455–68.
53. Lirk P, Rathmell JP. Opioid-free anaesthesia: Con: it is too early to adopt opioid-free anaesthesia today. Eur J Anaesthesiol. 2019;36:250–4.
54. Suzan E, Pud D, Eisenberg E. A crucial administration timing separates between beneficial and counterproductive effects of opioids on postoperative pain. Pain. 2018;159(8):1438–40.

55. Kianian S, Bansal J, Lee C, Zhang K, Bergese SD. Perioperative multimodal analgesia: a review of efficacy and safety of the treatment options. Anesthesiol Perioper Sci. 2024;2(1)
56. Wang L, Chang Y, Kennedy SA, Hong PJ, Chow N, Couban RJ, et al. Perioperative psychotherapy for persistent post-surgical pain and physical impairment: a meta-analysis of randomised trials. Br J Anaesth. 2018;120:1304–14.
57. Kendrick C, Sliwinski J, Yu Y, Johnson A, Fisher W, Kekecs Z, et al. Hypnosis for acute procedural pain: a critical review. Int J Clin Exp Hypn. 2016;64:75–115.
58. Kim R. Anesthetic technique and cancer recurrence in oncologic surgery: unraveling the puzzle. Cancer Metastasis Rev. 2017;36(1):159–77.
59. Weng M, Chen W, Hou W, Li L, Ding M, Miao C. The effect of neuraxial anesthesia on cancer recurrence and survival after cancer surgery: an updated meta-analysis [Internet]. Oncotarget. 7. Available from: www.impactjournals.com/oncotarget/
60. Kent ML, Hsia HLJ, van De Ven TJ, Buchheit TE. Perioperative pain management strategies for amputation: a topical review. Pain Med (US). 2017;18:504–19.
61. Muñoz-Leyva F, Cubillos J, Chin KJ KJ. Managing rebound pain after regional anesthesia. Vol. 73, Korean J Anesthesiol; 2020. p. 372–383.
62. Barry GS, Bailey JG, Sarhinha J, Brousseau P, Uppal V. Rebound pain: distinct pain phenomenon or nonentity? Br J Anaesth. 2021;126:761–3.
63. Borghi B, D'Addabbo M, White PF, Gallerani P, Toccaceli L, Raffaeli W, et al. The use of prolonged peripheral neural blockade after lower extremity amputation: the effect on symptoms associated with phantom limb syndrome. Anesth Analg. 2010;111(5):1308–15.
64. Wetherell JL, Afari N, Rutledge T, Sorrell JT, Stoddard JA, Petkus AJ, et al. A randomized, controlled trial of acceptance and commitment therapy and cognitive-behavioral therapy for chronic pain. Pain. 2011;152(9):2098–107.
65. Powell R, Scott NW, Manyande A, Bruce J, Vögele C, Byrne-Davis LMT, et al. Psychological preparation and postoperative outcomes for adults undergoing surgery under general anaesthesia, vol. 2016. Cochrane Database Syst Rev; 2016.
66. Visioni A, Shah R, Gabriel E, Attwood K, Kukar M, Nurkin S. Enhanced recovery after surgery for noncolorectal surgery? Ann Surg. 2018;267(1):57–65.
67. Pozek JPJ, Goldberg SF, Baratta JL, Schwenk ES. Practical management of the opioid-tolerant patient in the perioperative surgical home. Adv Anesth. 2017;35:175–90.
68. Gatchel RJ, Reuben DB, Dagenais S, Turk DC, Chou R, Hershey AD, et al. Research agenda for the prevention of pain and its impact: report of the work group on the prevention of acute and chronic pain of the federal pain research strategy. J Pain. 2018;19(8):837.

34 Multidisciplinary Care Clinics for Patients with Metastatic Bone Disease

Adrian Torres, Carl Quinion, and John Alexander

Introduction

Multidisciplinary care is an approach to patient care that involves a group of professionals of different disciplines working in coordination to optimize patient care through collaboration. This approach involves providers reaching a consensus regarding the treatment of complex patients through either "board room" discussions or in the form of multidisciplinary clinics. In this approach, a patient should receive comprehensive and higher quality care than if they pursued treatment from independent providers for the same disease. Utilizing a multidisciplinary team to approach a single disease has been employed in multiple different pathologies, including Parkinson's disease, chronic kidney disease, amyotrophic lateral sclerosis, and cystic fibrosis, with many of these initiatives leading to increased survival and improved quality of life metrics [1–4].

The history of multidisciplinary care in oncology began in the form of tumor boards, a forum where physicians and other health care professionals met to discuss complex cancer cases and devise individualized treatment plans for patients [5]. The introduction of tumor boards and the idea that multiple specialists should be involved in the development of a patient-specific treatment plan began as early as the 1920s with Dr. Gustave Roussy, a pathologist and founder of one of Europe's first cancer centers. While investigating the radiosensitivity of tumor cells, he proposed that radiosensitivity findings could direct tumor treatment, which at the time consisted of surgery, radiotherapy, or chemotherapy. The idea that multiple different strategies could be utilized to target a patient's cancer spawned meetings where specialists trained in these treatment strategies could discuss their utility. It was under Roussy's direction at the Institut National du Cancer of Villejuif where one of

A. Torres · C. Quinion · J. Alexander (✉)
Division of Musculoskeletal Oncology, Department of Orthopaedics,
The Ohio State University Wexner Medical Center, Columbus, OH, USA
e-mail: Adrian.Torres@osumc.edu; Carl.quinion@osumc.edu; John.alexander@osumc.edu

J. Pretell-Mazzini, S. A. Lozano Calderón (eds.), *Multidisciplinary Approach of Musculoskeletal Metastasis*, https://doi.org/10.1007/978-3-032-09498-8_34

the first interdisciplinary tumor boards was instituted consisting initially of a pathologist, radiation oncologist, and an internist [6, 7]. This multidisciplinary tumor board was then adopted by a few large institutions over the following decades, but did not become widespread until the 1990s with the introduction of quality improvement studies [8, 9]. By the mid-1990s, the Calman-Hine report in the UK and *the Tumor Board Textbook* released by the Commission on Cancer (CoC) in the United States made multidisciplinary care the standard for treating cancer patients [10–12]. The ability to form and hold weekly to monthly tumor boards is now a key part of the accreditation process of CoC institutions [13].

As cancer care continues to evolve with an expanding arsenal of therapeutic strategies, multidisciplinary care has become more critical> Multidisciplinary clinics incorporate real-time clinical assessments of a patient by multiple subspecialty providers during a single-day clinic visit. In its optimal format, a multidisciplinary clinic should consist of a patient seeing multiple different specialists at one time, whether in person or virtually, followed by an immediate discussion of the case and development of a consensus treatment recommendation. These clinics have been utilized in complex cancers, such as hepatocellular carcinoma (HCC) and pancreatic cancer [14, 15], demonstrating improved survival and a more efficient means of facilitating clinical assessments and thorough case discussions resulting in clearly defined therapeutic plans [14, 15].

While these multidisciplinary clinics have been applied to disease-specific cancer treatment, their application to metastatic bone disease (MBD) is a novel initiative. The metastatic bone disease patient population is precariously poised between the primary oncologists, treating orthopedic surgeons, radiation oncologists, palliative medicine teams, and other rehabilitative providers, creating an at-risk environment that demands complex decision making through multidisciplinary collaboration. The number of people living with metastatic disease continues to grow. This is a direct result of the increasing incidence of cancer as well as improving survival among patients with MBD, with many of these patients now living beyond five years [16]. Ensuring our society is equipped to treat this patient population is crucial as MBD has a significant impact on quality of life [17] and healthcare costs [18].

To address this growing need multidisciplinary MBD clinics have emerge as means to deliver comprehensive care to this vulnerable patient population [16]. These patients frequently do not receive what would be considered standard of care including the use of and timing of adjuvant radiation therapy and antiresorptive therapy [19–21]. In light of this suboptimal treatment, medical society initiatives, similar to the American Orthopaedic Association's "Own the Bone" program, have been proposed including the Musculoskeletal Tumor Society (MSTS) who recently establishing clinical practice guidelines for the management of various aspects of MBD [22]. These guidelines were produced by multidisciplinary teams and proposed shared decision-making among the patient and their providers. In order to be equipped to care for the growing number of patients with MBD, it's essential to recognize that, for the most part, cancer is not a disease to be treated by an individual clinician. It is a disease that can be so pervasive throughout a patient's life, beyond just their physical symptoms, that it requires a team that can adequately address care in a holistic approach. This chapter highlights the importance of multidisciplinary care clinics and how to best establish them. Orthopedic oncologists should be the champions of this service line providing a central resource for medical oncologists, radiation oncologists, and orthopedic surgeons within institutions, but more importantly serving their local and regional communities.

Standard of Care

Patient convenience and efficient healthcare delivery as reasons to establish a multidisciplinary metastatic bone disease clinic may create difficulties when justifying the clinic to a health system. Thus, it is critical to first identify standards of care and subsequently, the metrics of care that should be met when caring for this patient population. These include appropriate indications for surgery, initiation of antiresorptive therapy, and appropriate use and timing for radiation therapy. As evidenced by the clinical practice guidelines established by the Musculoskeletal Tumor Society, there is a need for higher levels of evidence to guide the care of musculoskeletal oncology patients, a common theme in the field of orthopedic oncology. However, there is strong evidence to support certain standards of care for metastatic bone disease and documented evidence that these standards are not being met in this patient population.

When considering metastatic bone disease, the strongest evidence supports the use of antiresorptive therapy. These pharmacologic agents are associated with a decrease in the incidence or and time to skeletal-related events as well as increased survival across multiple cancer types [23–25]. Current evidence predominately supports the use of denosumab, a RANK-ligand inhibitor, and zoledronic acid, a bisphosphonate. A recent database study by Agarwal et al. revealed that among patients with osseous metastases, the risk of pathologic fracture was lower in patients on antiresorptive therapy; however, only 7.7% of patients were on antiresorptive therapy [19]. This clearly documents the underutilization of antiresorptive therapy in patients with metastatic bone disease. The underlying reasons for this are unclear. However, Agarwal et al. propose that it is due to a lack of communication and ultimately ownership of prescribing antiresorptive therapy. Unfortunately, due to perceived barriers including the need for dental screening and unfamiliarity with prescribing antiresorptive therapies, orthopedic surgeons do not feel comfortable with owning this prescribing responsibility. However, the orthopedic surgeon can play a critical role of communicating the importance of this intervention to referring medical oncologists and the patient. With respect to the latter, patients are often reluctant to initiate antiresorptive therapy after hearing the risks for adverse events, in particular osteonecrosis of the jaw. Orthopedic surgeons have the opportunity to provide reassurance and educate patients regarding the risks and benefits.

Finally, the role of radiation therapy in the treatment of metastatic bone disease was established in the 1990s when Townsend et al. demonstrated improved functional recovery and a lower risk of second surgeries [26]. This may be outdated as the role of radiation therapy is more nuanced based on considerations of histology and anticipated systemic therapeutic responses. This was recently investigated by Willeumeir et al., who concluded that overall the evidence is poor; the paucity of high-quality literature was also found by Ramadan et al. [27, 28]. Another layer of the discussion regarding adjuvant radiation therapy is with respect to the delivery, dose, and number of fractions required, with evidence supporting various strategies each with their own advantages and disadvantages [29]. Ultimately, there clearly is a role for postoperative radiation therapy in many patients undergoing surgical stabilization of metastatic bone disease; however, the decision is complex and requires an understanding of the unique disease processes and the anticipated patient treatment courses, as well as open communication between medical oncologists, radiation oncologists, and orthopedic surgeons.

Although ideally a multidisciplinary metastatic bone disease clinic can be leveraged for the initial decision making regarding osseous metastatic disease, it may also serve patients who have already undergone surgical treatment of their metastatic lesions. This allows multidisciplinary teams led by musculoskeletal oncologists to ensure patients are receiving evidence-based care for their metastatic bone disease.

Current Evidence of Multidisciplinary Care in Metastatic Disease

Benefits

MBD costs have been estimated to account for 17% of the direct medical expenditures toward cancer care according to the National Institutes of Health (NIH) [30]. While commanding a large percentage of the direct services patients receive for their care, MBD also accounts for notable indirect costs, which including morbidity costs defined as time away from work or expenditures attributed to transportation and childcare to attend appointments. Indirect costs that may be more difficult to calculate are the mortality costs, which are the costs associated with loss of societal productivity due to death [18, 31]. Multidisciplinary clinics can aid in reducing these costs by simplifying the referral process and coordinating visits to take place over a single day to avoid unnecessary transportation or childcare costs accrued during multiple individual visits. These clinics also improve the usage of clinical resources by encouraging collaboration between specialists ensuring appropriate imaging and minimizing unnecessary testing [10]. Another key benefit of centralized collaboration is decreasing time to treatment, which has been associated with higher patient satisfaction scores and increased survival [32–34]. In chronic kidney disease, multidisciplinary clinics have been linked to a decrease in days spent in the hospital, and in esophageal and gastric cancer, early utilization of interdisciplinary support was found to improve overall survival and lead to improved emotional and cognitive function. Many of the benefits of multidisciplinary care stem from the ability to quickly disseminate knowledge across the individual providers and the patient and their family. This improves patient education, provides resources to the family, allows for quick enrollment in clinical trials, and brings the patient to a single venue which allows interaction with specialists such as nutritionists and social work that may otherwise be more difficult to schedule individually [10, 14].

Pitfalls

As a concept, a multidisciplinary clinic provides a well-coordinated, comprehensive, and holistic single-day appointment where a patient may meet with all their providers in succession and have many of their treatments and concerns addressed in a single visit. However, to perform this function, a provider must be readily available to meet with the patient once they have finished their previous appointment of the day, and due to this constraint, this may limit the number of patients able to be seen in a day by a single multidisciplinary clinic due to appointments commonly running over, or patients showing up late. Given the coordination required to have a single patient seen by multiple providers in a single day, a provider may have to extend the appointment time, leaving fewer time slots and decreasing provider volume. Each provider may also require different amounts of time with patients or require additional tests such as imaging, phlebotomy, and medication administration, each of which can cause delays in care if not accounted for prior to the visit. Due to these setbacks and the current compensation models for many healthcare providers, it may not be practical to have these clinics on a frequent basis. In addition to scheduling difficulties among providers, one of the larger limiting factors is the burden placed on patients who are required to attend numerous appointments and/or treatments in a single day. A long, single day of appointments can be taxing on patients and their families, especially those with extensive disease burden or those traveling long distances. However, this multiple-hour affair in a single day likely offsets the multiple days of visits and hours of travel experienced by patients and their families.

Additional limitations to the multidisciplinary clinic include inconsistency of follow-up plans for each specialty. Multiple providers involved in a single patient's care can cause each to feel less individual responsibility for clinical decisions, and ultimately, the availability of these multidisciplinary clinics may be restricted to larger academic institutions creating access issues for patients.

Specialists Involved in Metastatic Bone Disease Care Teams

At initial cancer diagnosis, bone metastases are diagnosed in 5–23% of patients with solid organ primary malignancies, with lung cancer having the highest percentage of MBD at initial diagnosis [35]. Musculoskeletal pain is a common presenting complaint for many patients, and an orthopedic surgeon may be the initial point of contact for many osseous abnormalities, including metastatic bone lesions. These

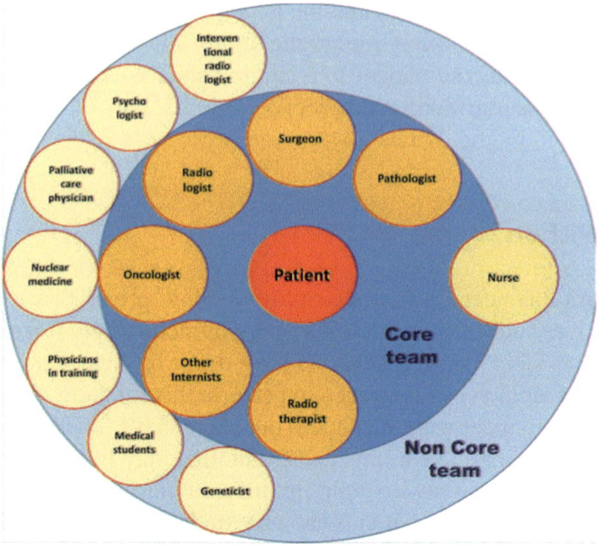

Fig. 34.1 Patient-centered metastatic bone disease care team [37]

interactions can make orthopedists key meditators in coordinating the initial care for these patients. The treatment of MBD, however, involves multiple specialists spanning the medical spectrum. The direct treatment of MBD often includes systemic therapy, radiation therapy, and surgery. However, the effect of MBD is wide ranging, infiltrating every aspect of a patient's life. MBD is associated with severe pain in up to 70–80% of patients, and along with pathological fractures, it can cause significant functional impairment and impact quality of life [35, 36]. Outside of the physical burden, metastatic disease also places a large financial and psychological burden on patients and their families [30]. Given the stresses placed on the patient, the proposed list of providers who should be involved in a multidisciplinary clinic may include oncologists, orthopedic surgeons, pathologists, radiologists, radiation oncologist, nutritionists, palliative care teams, endocrinologists, social workers, and physical and occupational therapy (Fig. 34.1).

Core Specialists

Orthopedic Oncologist Key roles include establishment of bone biopsy proven metastatic disease in newly diagnosed patients. Based on a study of multidisciplinary roles, the oncologic surgeon's role is also believed to involve explaining and evaluating imaging and test results, prognosticating, surgical treatment of oligometastatic disease and pathologic fracture management or prophylactic treatment [38–41]. The decision regarding the potential need for surgery is often necessary prior to initiating other treatments including radiation and chemotherapy.

Medical Oncologist Guides medical management of the primary malignancy utilizing systemic therapy. Medical oncologists are also involved in prognostication, evaluating and explaining imaging and test results, as well as pursing further testing, and aid in palliating symptoms. Medical oncologists can also initiate antiresorptive therapy including bisphosphonates and denosumab to lower pathologic fracture risk, combat hypercalcemia of malignancy, and aid with pain control due to inhibition of bone resorption [17, 19, 42, 43].

Radiation Oncologist Provide consultation to assess the need for palliative and postoperative radiation therapy for painful metastatic lesions [42, 44]. Interestingly, stereotactic ablative radiotherapy has also been shown to improve survival in patients with oligometastatic cancers [44, 45].

Bone Health Providers This can include endocrinologists or advance practice providers with experience in treating metabolic bone disease, including metastatic bone disease. These providers are critical for ensuring patients receive antiresorptive therapy.

Clinic Nurse/Nurse Navigator Key roles include coordinating and facilitating clinical care among specialists, assessing patient support and needs, conducting patient intake and laboratory testing procurement, and aiding in patient education.

Social Work Assess patients social, home, and psychological needs. Lead coordination of home care and aiding patients and their family in accessing available community resources [46].

Dietician Perform initial nutritional assessments to identify needs and signs of malnutrition as well as setting nutritional goals. The dietician can also review patients' home vitamins and supplements [47].

Physical Therapy (PT) Assess strength, endurance, mobility, range of motion, and fall risks. Create exercise programs to help patients reach strength and mobility goals.

Occupational Therapy (OT) Assess and assist with the functional return to activities of daily living such as bathing, toileting, and getting dressed. OT can also help address cognitive impairment, cancer fatigue, fine motor skills, as well as activities that aid in body image and self-identity such as returning to hobbies or sexual activity [48, 49].

Palliative Care Critical in addressing physical symptoms of disease and cancer treatments such as pain, appetite loss, nausea, and vomiting. Palliative care can also address family and caregiver needs as well as recruit pastoral or spiritual care to meet the patients emotional and spiritual needs. Importantly, palliative care specialists can aid the team in providing care that is aligned with the patient's beliefs and their own treatment goals [50].

Psychologist Psychosocial oncology providers meet with patients to assess psychological needs and aid patients in overcoming psychological distress including the anxiety and depression that can accompany a cancer diagnosis or terminal illness [18, 51].

Musculoskeletal Radiologist Radiologists are essential in evaluating oncologic patients and may be used to justify advanced imaging or guide appropriate imaging modalities. Input from radiologists at multidisciplinary conferences has also led to changes in management, which stemmed from reporting unreported findings, discovering missed findings, and correcting incorrect findings [52, 53].

Available as Needed

Financial Counselor Cancer treatment can be prohibitively expensive, especially in patients with MBD and the care requirements they may face. These expenses may also be passed onto the caregivers and lead to increased financial hardship, which can contribute to the psychological distress imposed upon these patients [18]. As such, it is essential to have financial counseling services available to help navigate and estimate these costs, as well as to review patient accounts and insurance benefits regularly and set up a financial plan with patients to help ease the stress imposed by the potential financial burden of treatment [54].

Pathologist Primary role in initial diagnosis and staging of cancer. In MBD, pathologists may be available for discussion of further testing, prognostication, tumor board discussions, and molecular assessments for potential targeted therapies.

Interventional Radiologist To be involved as necessary to obtain core needle biopsies through clear windows or in areas not amendable to open biopsy [55].

Conclusion

The exact constitution of multidisciplinary metastatic bone disease treatment teams can be tailored based on local community needs and institutional resources. The goal of the multidisciplinary MBD clinics is to ultimately increase access for patients and ensure that they are receiving standard of care treatments including compliance with antiresorptive therapy, appropriate use of radiation therapy, and timely surgical interventions with the goal of preventing pathologic fractures and

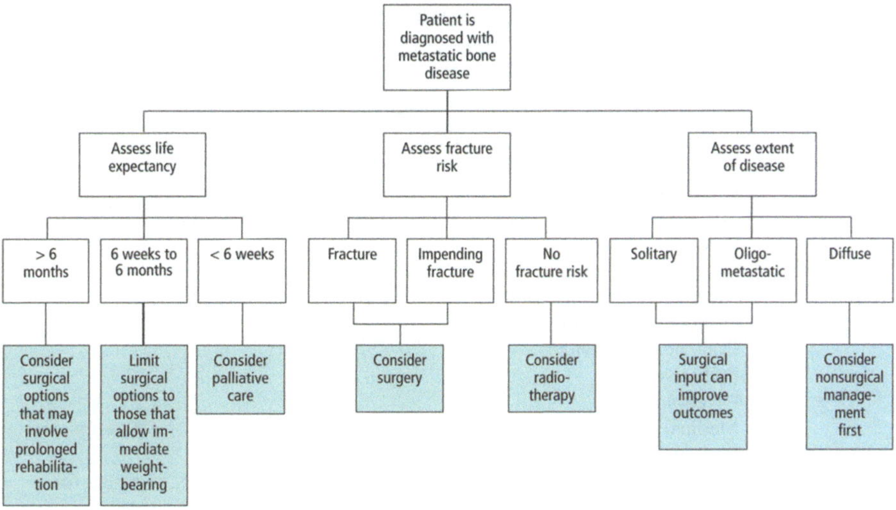

Fig. 34.2 Patient is diagnosed with metastatic bone disease [17]

maintaining functional independence. Regardless of the make-up of the team, the orthopedic oncologist plays an essential role in the care of metastatic bone disease (Fig. 34.2).

Proposed Multidisciplinary Patient Schedule

Below is a proposed patient schedule from a table adapted from Pawlik et al. [14]. Necessary imaging and laboratory studies will be obtained before meeting with physicians or advanced practice providers (APPs). However, meetings between supportive services and specialists with the patient may occur in a different order, pending availability. Time can be provided for patients to take breaks for lunch or throughout the day to allow for rest in between meetings. These breaks can be taken in between obtaining imaging and laboratory studies, meetings with supportive services, and individual specialists. Breaks can also take place in between individual specialist meetings (Table 34.1).

Table 34.1 Metastatic disease multidisciplinary clinic patient schedule

Time period	Objective
0700–0900	Patient arrives, obtains imaging and laboratory studies
0900–1000	Patient provided with overview of support services in 10–15-minute briefings Nurse navigator Dietician Social work Financial counselor[a]
1000–1300	Patients seen by physicians, fellows, residents, APPs of individual specialty services in 30-min encounters Orthopedic oncology Bone health providers Palliative care Psychologist Medical oncologist[a] Radiation oncologist[a]
1300–1400	Patients seen and evaluated by PT and OT in 30-min increments, to assess for outpatient therapy needs
1400–1600	Formal case review by multidisciplinary board to reach consensus recommendation and recommendations discussed with the patient by staff physician

[a]Designates as needed specialists who may not be present at every individual meeting

References

1. van der Marck MA, Munneke M, Mulleners W, Hoogerwaard EM, Borm GF, Overeem S, et al. Integrated multidisciplinary care in Parkinson's disease: a non-randomised, controlled trial (IMPACT). Lancet Neurol. 2013;12(10):947–56.
2. de Almeida FEO, do Carmo Santana AK, de Carvalho FO. Multidisciplinary care in amyotrophic lateral sclerosis: a systematic review and meta-analysis. Neurol Sci. 2021;42(3):911–23.
3. Wang SM, Hsiao LC, Ting IW, Yu TM, Liang CC, Kuo HL, et al. Multidisciplinary care in patients with chronic kidney disease: a systematic review and meta-analysis. Eur J Intern Med. 2015;26(8):640–5.
4. Goetz D, Ren CL. Review of cystic fibrosis. Pediatr Ann. 2019;48(4):e154–61.
5. National Cancer Institute. Tumor board review. In: National Cancer Institute dictionary of cancer terms. 2024.
6. Del Regato JA. Gustave Roussy. Int J Radiat Oncol Biol Phys. 1988;15(5):1223–32.
7. Goker B, Shea M, Zhang R, Wang J, Ferrena A, Chae SS, et al. The evolution of the multidisciplinary tumor board in orthopedic oncology: from its historical roots to its future potential. Holist Integ Oncol. 2024;3(1):36.
8. O'Brien JC. History of tumor site conferences at Baylor University Medical Center. Baylor Univ Med Cent Proc. 2006;19(2):130–1.
9. Wessells ZE. A study of medical social services to tumor patients. Soc Casework. 1949;30(9):375–80.
10. Fleissig A, Jenkins V, Catt S, Fallowfield L. Multidisciplinary teams in cancer care: are they effective in the UK? Lancet Oncol. 2006;7(11):935–43.
11. Greene, Frederick L. American College of Surgeons. 2022. The Commission on Cancer: a 10-decade perspective.
12. Haward RA. The Calman–Hine report: a personal retrospective on the UK's first comprehensive policy on cancer services. Lancet Oncol. 2006;7(4):336–46.

13. Scher KS, Tisnado DM, Rose DE, Adams JL, Ko CY, Malin JL, et al. Physician and practice characteristics influencing tumor board attendance: results from the provider survey of the Los Angeles Women's Health Study. J Oncol Pract. 2011;7(2):103–10.
14. Pawlik TM, Laheru D, Hruban RH, Coleman J, Wolfgang CL, Campbell K, et al. Evaluating the impact of a single-day multidisciplinary clinic on the management of pancreatic cancer. Ann Surg Oncol. 2008;15(8):2081–8.
15. Yopp AC, Mansour JC, Beg MS, Arenas J, Trimmer C, Reddick M, et al. Establishment of a multidisciplinary hepatocellular carcinoma clinic is associated with improved clinical outcome. Ann Surg Oncol. 2014;21(4):1287–95.
16. Gallicchio L, Devasia TP, Tonorezos E, Mollica MA, Mariotto A. Estimation of the number of individuals living with metastatic cancer in the United States. J Natl Cancer Inst. 2022;114(11):1476–83.
17. Ardakani AHG, Faimali M, Nystrom L, Mesko N, Mughal M, Ware H, et al. Metastatic bone disease: early referral for multidisciplinary care. CCJM. 2022;89(7):393–9.
18. DiCaprio MR, Murtaza H, Palmer B, Evangelist M. Narrative review of the epidemiology, economic burden, and societal impact of metastatic bone disease. Ann Jt. 2022;7:28.
19. Agarwal AR, LiBrizzi CL, Wessel L, Thakkar SC, Levin AS. The low and disproportionate utilization of antiresorptive therapy in patients with osseous metastasis. J Bone Oncol. 2023;43:100507.
20. Chen AM, Harris JP, Tjoa T, Haidar Y, Armstrong WB. Racial disparities in the timely receipt of adjuvant radiotherapy for head and neck cancer. Oral Oncol. 2023;147:106611.
21. Seneviratne S, Campbell I, Scott N, Kuper-Hommel M, Round G, Lawrenson R. Ethnic differences in timely adjuvant chemotherapy and radiation therapy for breast cancer in New Zealand: a cohort study. BMC Cancer. 2014;14:839.
22. Musculoskeletal Tumor Society. Guidelines and Evidence Based Medicine [Internet]. 2024. Available from: https://www.msts.org/index.php/education/evidence-based-medicine.
23. Hortobagyi GN, Van Poznak C, Harker WG, Gradishar WJ, Chew H, Dakhil SR, et al. Continued treatment effect of Zoledronic acid dosing every 12 vs 4 weeks in women with breast cancer metastatic to bone: the OPTIMIZE-2 randomized clinical trial. JAMA Oncol. 2017;3(7):906.
24. Henry DH, Costa L, Goldwasser F, Hirsh V, Hungria V, Prausova J, et al. Randomized, double-blind study of Denosumab versus Zoledronic acid in the treatment of bone metastases in patients with advanced cancer (excluding breast and prostate cancer) or multiple myeloma. JCO. 2011;29(9):1125–32.
25. Bozzo A, Deng J, Abbas U, Bhasin R, Deodat M, Wariach S, et al. Which bone-modifying agent is associated with better outcomes in patients with skeletal metastases from lung cancer? a systematic review and network meta-analysis. Clin Orthop Relat Res. 2021;479(9):2047–57.
26. Townsend PW, Rosenthal HG, Smalley SR, Cozad SC, Hassanein RE. Impact of postoperative radiation therapy and other perioperative factors on outcome after orthopedic stabilization of impending or pathologic fractures due to metastatic disease. J Clin Oncol. 1994;12(11):2345–50.
27. Willeumier JJ, van der Linden YM, Dijkstra PDS. Lack of clinical evidence for postoperative radiotherapy after surgical fixation of impending or actual pathologic fractures in the long bones in patients with cancer; a systematic review. Radiother Oncol. 2016;121(1):138–42.
28. Ramadan S, Arifin AJ, Nguyen TK. The role of post-operative radiotherapy for non-spine bone metastases (NSBMs). Cancers (Basel). 2023;15(13):3315.
29. Kraus RD, Weil CR, Wells S, Tward JD, Groundland JS, Jones KB, et al. Radiation therapy in conjunction with surgical stabilization of impending or pathologic fractures secondary to metastasis: is there a difference between single and multifraction regimens? Adv Radiat Oncol. 2022;7(2):100795.
30. Schulman KL, Kohles J. Economic burden of metastatic bone disease in the U.S. Cancer. 2007;109(11):2334–42.
31. Yabroff KR, Lund J, Kepka D, Mariotto A. Economic burden of cancer in the United States: estimates, projections, and future research. Cancer Epidemiol Biomarkers Prev. 2011;20(10):2006–14.

32. Gabel M, Hilton NE, Nathanson SD. Multidisciplinary breast cancer clinics. Do they work? Cancer. 1997;79(12):2380–4.
33. Birchall M, Bailey D, King P, South West Cancer Intelligence Service Head and Neck Tumour Panel. Effect of process standards on survival of patients with head and neck cancer in the south and west of England. Br J Cancer. 2004;91(8):1477–81.
34. Saqlain F, Shalhout SZ, Flaherty KT, Emerick KS, Miller DM. REDCap-based operational tool to guide care coordination in a multidisciplinary cutaneous oncology clinic. JCO Oncol Pract. 2021;17(9):527–33.
35. Colosia A, Njue A, Bajwa Z, Dragon E, Robinson RL, Sheffield KM, et al. The burden of metastatic cancer-induced bone pain: a narrative review. J Pain Res. 2022;15:3399–412.
36. Janssen SJ, Pereira NRP, Thio QCBS, Raskin KA, Bramer JAM, Lozano-Calderon SA, et al. Physical function and pain intensity in patients with metastatic bone disease. J Surg Oncol. 2019;120(3):376–81.
37. Berardi R, Morgese F, Rinaldi S, Torniai M, Mentrasti G, Scortichini L, et al. Benefits and limitations of a multidisciplinary approach in cancer patient management. Cancer Manag Res. 2020;12:9363–74.
38. Kafchinski LA. Metastasectomy for oligometastatic bone disease of the appendicular skeleton: a concise review. J Surg Oncol. 2023;128(3):438–44.
39. Catt S, Fallowfield L, Jenkins V, Langridge C, Cox A. The informational roles and psychological health of members of 10 oncology multidisciplinary teams in the UK. Br J Cancer. 2005;93(10):1092–7.
40. Mosher ZA, Patel H, Ewing MA, Niemeier TE, Hess MC, Wilkinson EB, et al. Early clinical and economic outcomes of prophylactic and acute pathologic fracture treatment. J Oncol Pract. 2019;15(2):e132–40.
41. Hammad A, Ahmed O, Connell PP, Olson D, Balach T. Team approach: management of pathologic fractures. JBJS Rev. 2023;11(1)
42. Molloy AP, O'Toole GC. Orthopaedic perspective on bone metastasis. World J Orthop. 2013;4(3):114–9.
43. Kimura T. Multidisciplinary approach for bone metastasis: a review. Cancers. 2018;10(6):156.
44. Tsai CJ, Gomez DR, Yang TJ. Metastatic disease as a distinct discipline in radiation oncology. JAMA Oncol. 2021;7(1):21–2.
45. Palma DA, Olson R, Harrow S, Gaede S, Louie AV, Haasbeek C, et al. Stereotactic ablative radiotherapy versus standard of care palliative treatment in patients with oligometastatic cancers (SABR-COMET): a randomised, phase 2, open-label trial. Lancet. 2019;393(10185):2051–8.
46. Angelique Caba. The Value of Oncology Social Workers [Internet]. Cancer Care. 2020. Available from: https://www.cancercare.org/publications/262-the_value_of_oncology_social_workers.
47. Bramlet K. Meeting with a dietitian during cancer treatment: what to expect. The University of Texas MD Anderson Cancer Center; 2017.
48. Garcia L. Occupational therapy for cancer patients: what is it, and who needs it? The University of Texas MD Anderson Cancer Center; 2022.
49. Pergolotti M, Williams GR, Campbell C, Munoz LA, Muss HB. Occupational therapy for adults with cancer: why it matters. Oncologist. 2016;21(3):314–9.
50. Ferrell BR, Temel JS, Temin S, Alesi ER, Balboni TA, Basch EM, et al. Integration of palliative care into standard oncology care: american society of clinical oncology clinical practice guideline update. J Clin Oncol. 2017;35(1):96–112.
51. Taberna M, Gil Moncayo F, Jané-Salas E, Antonio M, Arribas L, Vilajosana E, et al. The multidisciplinary team (MDT) approach and quality of care. Front Oncol. 2020;10:85.
52. Komarraju A, Van Rilland EZ, Gebhardt MC, Anderson ME, Heincelman C, Wu JS. What is the value of radiology input during a multidisciplinary orthopaedic oncology conference? Clin Orthop Relat Res. 2023;481(10):2005–13.
53. Miller BJ. Use of imaging prior to referral to a musculoskeletal oncologist. J Am Acad Orthop Surg. 2019;27(22):e1001–8.

54. Margaret Liang. The evolving role of the financial counselor in an oncology care model pilot program [Internet]. Healthcare Delivery Research Program; 2020 May 26; National Cancer Institute. Available from: https://healthcaredelivery.cancer.gov/media/counselor.html.
55. Ogilvie CM, Torbert JT, Finstein JL, Fox EJ, Lackman RD. Clinical utility of percutaneous biopsies of musculoskeletal tumors: clinical orthopaedics and related research. Clin Orthop Relat Res. 2006;450:95–100.

Part VI

Oligometastatic Disease

Metastasectomy for Musculoskeletal Oligometastases—Surgical Point of View

35

Yvette Ho

Historical Background

Historically, presence of metastatic disease has been considered incurable and associated with a dismal prognosis. Oncologic treatment has been based on two major prevailing theories that have persisted through the years regarding the development of metastatic disease. In 1894, in a paper detailing his experience in treating a series of breast cancer patients, W.S. Halsted theorized that cancer spreads in a systematic way from primary site through lymphatics and then to distant sites. This theory was the basis of supporting the radical mastectomy, which includes resecting the breast with the pectoralis musculature as well as the axillary lymph nodes. This principle of en bloc resection has formed the cornerstone of oncologic surgery. Theoretically, with Halsted's contiguous theory of metastatic disease, if tumor is still localized, then it could be curable with en bloc resection. Otherwise, after spread through the lymphatics, it is considered widely metastatic [1–3].

On the other hand, the systemic theory postulates that any clinically apparent cancer is evidence of systemic disease. According to this theory, even if only a small primary tumor is identifiable, the disease is merely early in its presentation, as micrometastastases are presumed to exist. Similarly, nodal involvement is simply another manifestation of widespread disease [4].

With both the systemic and contiguous theories, once the disease has progressed to the metastatic state, it is assumed to be widespread and not curable. Thus traditionally, management of metastatic disease in the past has been restricted mostly to palliative treatment, in the interest of avoiding overly invasive or aggressive treatment when cure is not possible.

Y. Ho (✉)
Department of Orthopedic Oncology & Sarcoma Surgery, Carol G. Simon Cancer Center, Morristown Medical Center, Atlantic Health, Morristown, NJ, USA

J. Pretell-Mazzini, S. A. Lozano Calderón (eds.), *Multidisciplinary Approach of Musculoskeletal Metastasis*, https://doi.org/10.1007/978-3-032-09498-8_35

In 1995, Hellman and Weichselbaum published a paper delineating the concept of oligometastases. They postulated that both the systemic and contiguous theories of metastatic disease were too restrictive, and that there exists an intermediate state between completely localized disease and widespread metastatic disease. Larger tumor size and higher grade were thought to be associated with higher likelihood of widespread disease [3, 5]. The idea of oligometastases also incorporates the "seed and soil" hypothesis, which was proposed by Stephen Paget in 1889. He theorized that the propensity for certain malignancies to metastasize to specific organs is due to specific characteristics of cancer cells ("seeds") to undergo interactions with specific microenvironments in organs ("soil") [3, 6, 7].

Even at the time of publication of Hellman's theory of oligometastases, it had already been observed that more aggressive treatment of limited pulmonary metastases in sarcomas and renal cell carcinoma was associated with improved survival [8–10]. Since then, though, much has been published about oligometastases, as the concept changed the landscape of metastatic disease management from one of mostly palliative treatment only to a setting with possible prolonging of survival or even hope for cure [11–13]. There has been a significant increase in metastasectomies across the United States, as well, for various tumor types. The frequencies have been in keeping with the relative incidences of the primary tumors, as the most common primary diagnoses for which metastasectomies have been performed were colorectal, lung, breast, and melanoma.

Unfortunately, it is still difficult at this time to distinguish between true oligometastatic disease state versus widespread micrometastatic disease that is not yet clinically evident, as no reliable biomarkers exist to determine this disease state [14–20].

Defining the Oligometastatic State

Attempts are being made at standardizing terminology and defining the parameters of oligometastatic disease to develop more effective algorithms for treatment. Oligometastatic disease is generally defined to be fewer than five metastases, ideally isolated to one organ, that are amenable to local therapy [5, 11–13, 21–24].

The European Society for Radiotherapy and Oncology and European Organization for Research and Treatment of Cancer published a consensus recommendation on how oligometastases should be defined. They determined that multiple criteria needed to be addressed to define a patient's oligometastatic state. These characteristics could be classified into four categories: (1) descriptive tumor, (2) quantitative, (3) developmental, and (4) metastases-specific characteristics [25].

The descriptive tumor characteristics included primary tumor diagnosis, site, and stage; history of progression and time to metastasis from first diagnosis; as well as any disease-free or treatment-free intervals. The primary tumor treatment should be noted, including how the treatment and response and whether treatment had been performed with curative or palliative intent. Prior systemic treatment should also be considered. Method for determining oligometastases should also be considered, including imaging modality, and the involved organs should be identified. Quantitative characteristics include the number of metastases, number of organs involved, and total tumor size or volume [25].

Developmental characteristics include multiple aspects. The first element to identify is if the patient had a history of polymetastatic disease. As opposed to true de novo oligometastases, "induced oligometastatic disease" indicates previous polymetastases that responded to treatment, with residual persistent lesions. Additionally, history of oligometastases should be noted. Formation of recurrent oligometastases holds a better prognosis than polymetastatic disease. Metastases that are synchronous (within 6 months) with the primary diagnosis instead of metachronous (>6 months after primary diagnosis) should also be considered. Synchronous metastases have been thought to have a worse prognosis, though there has been discordance in some studies [26–29]. Another distinction to be made is whether the patient was on systemic treatment or not when the oligometastases formed. Oligoprogression indicates development of oligometastases while on systemic treatment, while oligorecurrence is defined as redevelopment of oligometastases while off treatment. Finally, progressive disease on current imaging should be considered versus oligopersistence, which indicates a few persistent metastases after systemic treatment. They also point out that a patient can pass through various stages of different oligometastatic states over time [25].

The reality is that while the true de novo oligometastatic state could theoretically allow for cure with elimination of gross tumor and delivery of effective systemic therapy, improved survival can often be achieved in other disease states besides pure de novo oligometastatic disease [12, 21, 30]. Thus, it is important to make the distinction and determine the appropriate goals of care for each patient.

Musculoskeletal Oligometastases

Many of the studies regarding metastasectomy for oligometastases are focused on pulmonary metastases, as lung is the most common site of metastatic disease across all cancer types [14]. When discussing musculoskeletal oligometastases specifically, the focus should primarily be on osseous metastases, as they are much more common than soft tissue sites. Soft tissue metastases are not only rare compared to bony sites, but they rarely occur as oligometastases; they are more often seen in disseminated disease. Regardless, it should be noted that soft tissue metastases are likely underdetected and underreported due to the relative lack of morbidity compared to bone and visceral involvement. They can still be associated with pain and functional impairment, and even fungation and infection, depending on location.

While bone metastases are quite common, their occurrence as oligometastases with no other organ involvement is relatively unusual. In one published series of 1453 patients with bone metastases, only 27 (1.8%) presented with a solitary bone lesion [31]. Traditionally, surgical treatment for bone metastases has been limited to palliative therapy, and thus, for osseous disease, the indications for surgery were limited to stabilizing impending or complete pathologic fractures. The focus of the surgery was usually less about ablation of the tumor itself, as it was presumed to be widespread already. Now in this era of recognizing oligometastatic disease, wide resection of oligometastatic bone disease is more frequently being considered to extend survival or perhaps help bring about cure. This has long been proposed with some types of relatively radioresistant tumors, such as renal cell carcinoma [8, 32–34] or soft tissue sarcomas [9, 16, 17, 23, 35–39]. Wide resection does also mitigate the bleeding issues related to high vascularity of renal cell carcinomas, as it avoids surgically entering the lesion as with intralesional curettage or even intramedullary fixation through the lesion [32, 34]. The National Comprehensive Cancer Network (NCCN) guidelines for renal cell carcinoma suggest metastasectomy for oligometastases and SBRT (stereotactic body radiation therapy) or ablative techniques such as radiofrequency ablation or cryoablation only if deemed to be poor surgical candidates [40]. Similarly, the NCCN guidelines for soft tissue sarcoma also suggest metastasectomy with or without radiation therapy for metachronous or synchronous oligometastases with "limited tumor bulk that is amenable to local therapy" [41]. Nonetheless, bone metastases rarely present in soft tissue sarcomas without lung metastases; they are synchronous with lung involvement in 87% of cases [37].

In a true de novo oligometastatic state, theoretically wide resection could improve survival or even lead to possible cure. Unfortunately, no established algorithms exist to clearly define how to best obtain the gold standard of local control: a margin-negative, R0 resection [16–20, 33, 39, 42–44]. There is no established universally optimal margin distance. Deciding on reconstruction of a defect requires careful consideration of the proximity of the nearest joint and the amount of residual bone. Generally, nonbiologic reconstruction with a metal endoprosthesis can obviate the need for the extent of healing required with a structural bone allograft reconstruction. Amputation would usually only be considered for a large destructive acrometastasis, or a tumor with too much involvement of the neurovascular bundle and/or skin to reliably salvage the extremity. Neither is likely to occur in a truly oligometastatic state.

Deciding on surgery for an oligometastasis requires consideration of not only the tumor- and metastasis-specific parameters to clarify the oligometastatic state, [24, 25] but also many patient-specific factors as well. Patient input is important, as patients may have strong preferences for or against certain treatment modalities. The patient‘s overall functional status, as can be evaluated using the ECOG performance status score, [45] American Society of Anesthesiologists (ASA) score, [46] and Karnofsky Performance Status, [47] is useful in determining if a patient is even a surgical candidate or not [47, 48]. Especially when considering the type of surgery to perform, consideration must be made for the patient's ability to recover and rehabilitate from the surgery itself. Again, this needs to be decided in conjunction with medical oncology, as some systemic therapies are more detrimental to wound healing. Thus larger, deeper resections/reconstructions need to be considered carefully, due to the longer surgeries with increased morbidity—particularly increased risk of infection, which is particularly calamitous with the metal megaprostheses or large structural allografts required for reconstruction. The length of time required for healing and possibly extended interruption to systemic therapy also needs to be discussed. Obviously, the patient's overall condition could adversely affect their ability to tolerate systemic treatments as well, which may limit the treatability of their disease, even if they have true de novo oligometastases.

The same principles apply in the pelvis as long bones. Wide resection of a pelvic metastasis is possible; reconstruction options have improved over time with the advent of custom endoprostheses that can be designed from the patient's imaging to resect the tumor and preserve as much normal bone as possible, with options for patient-specific cutting guides and intraoperative navigation assistance for screw placement. Nonetheless, as these surgeries are still more extensive and are associated with more morbidity and longer recovery time, the need for multidisciplinary discussion is even more critical, as nonsurgical options such as stereotactic ablative body radiotherapy (SABR) have also shown promise in providing local control [49–51]. Radiation would not provide any structural stability in metastatic bony lesions associated with impending pathologic fractures, but as much progress has been made in percutaneous treatments for periacetabular metastases [52–54], there may be a future in combining these less invasive, less morbid surgeries with ablative procedures.

In summary, defining the patient's oligometastatic state to determine the goal of surgical treatment is quite complex and nuanced, but critical in decision making to avoid subjecting patients to unnecessary invasive procedures that may not provide them with clinical benefit. Additionally, as other ablative strategies and radiation have also been shown to be effective treatments for local control in multiple diseases, multidisciplinary collaboration is critical to determine optimal management for these patients.

References

1. Halsted W. The results of operations for the cure of cancer of the breast performed at the Johns Hopkins Hospital from June, 1889, to January, 1894. AnnSurg. 1894;20:497–555.
2. Hellman S. Natural history of small breast cancers. J Clin Oncol. 1994;12:2229–34.
3. Hellman S, Weichselbaum RR. Oligometastases. J Clin Oncol. 1995;13(1):8–10.
4. Fisher B. Laboratory and clinical research in breast cancer—a personal adventure: the David A. Karnofsky Memorial Lecture. Cancer Res. 1980;40(11):3863.
5. Laurie SA, Banerji S, Blais N, Brule S, Cheema PK, Cheung P, et al. Canadian consensus: oligoprogressive, pseudoprogressive, and oligometastatic non-small-cell lung cancer. Curr Oncol. 2019;26(1):81.
6. Fidler IJ. The pathogenesis of cancer metastsis: the "seed and soil" hypothesis revisited. Nat Rev. 2003 Jun;3:1–6.
7. Paget S. The distribution of secondary growths in cancer of the breast. Cancer Metastasis Rev. 1889;8(2):98–101.
8. Dernevik L, Berggren H, Larsson S, Roberts D. Surgical removal of pulmonary metastases from renal cell carcinoma. Scand J Urol Nephrol. 1985;19(2):133.
9. Martini N, Bains MS, Huvos AG, Beattie EJ. Surgical treatment of metastatic sarcoma to the lung. Surg Clin North Am. 1974;54(4):841–8.
10. O'dea MJ, Zincke H, Utz DC, Bernatz PE. The treatment of renal cell carcinoma with solitary metastasis. J Urol. 1978;120(5):540–2.

11. Kaneda H, Saito Y. Oligometastases: defined by prognosis and evaluated by cure. Cancer Treat Commun. 2015;3:1–6.
12. Niibe Y, Hayakawa K. Oligometastases and oligo-recurrence: the new era of cancer therapy. Jpn J Clin Oncol. 2010;40:107–11.
13. Niibe Y, Jingu K, Onishi H. Oligometastases: History and future vision of breast cancer. Transl Cancer Res. 2020;9:5028–31.
14. Beckers P, Berzenji L, Yogeswaran SK, Lauwers P, Bilotta G, Shkarpa N, et al. Pulmonary metastasectomy in colorectal carcinoma. J Thorac Dis. 2021;13:2628.
15. Colman MW, Kirkwood JM, Schott T, Goodman MA, McGough RL. Does metastasectomy improve survival in skeletal melanoma? Melanoma Res. 2014;24(4):354–9.
16. Abdalla EK, Pisters PWT. Metastasectomy for limited metastases from soft tissue sarcoma. Curr Treat Options in Oncol. 2002;3:497–505.
17. Stamenovic D, Hohenberger P, Roessner E. Pulmonary metastasectomy in soft tissue sarcomas: a systematic review. J Thorac Dis. 2021;13(4):2649.
18. Bartlett EK, Simmons KD, Wachtel H, Roses RE, Fraker DL, Kelz RR, et al. The rise in metastasectomy across cancer types over the past decade. Cancer. 2015;121(5):747.
19. Rajwa P, Yanagisawa T, Gruber M, Heidenreich A, Joniau S, Briganti A, et al. Surgical metastasectomy for visceral and bone prostate cancer metastases: a mini-review. Eur Urol Focus. 2023;9:232.
20. Milosevic M, Edwards J, Tsang D, Dunning J, Shackcloth M, Batchelor T, et al. Pulmonary metastasectomy in colorectal cancer: updated analysis of 93 randomized patients—control survival is much better than previously assumed. Color Dis. 2020;22(10):1314.
21. Niibe Y, Chang JY. Novel insights of oligometastases and oligo-recurrence and review of the literature. Pulm Med. 2012;2012:1.
22. Szturz P, Nevens D, Vermorken JB. Oligometastatic disease management: finding the sweet spot. Front Oncol. 2020;10:617793.
23. Janes LA, Angeles CV. The role of surgery in oligometastatic retroperitoneal sarcoma. Curr Oncol. 2023;30:5240–50.
24. Lievens Y, Guckenberger M, Gomez D, Hoyer M, Iyengar P, Kindts I, et al. Defining oligometastatic disease from a radiation oncology perspective: an ESTRO-ASTRO consensus document. Radiother Oncol. 2020;148:157.
25. Guckenberger M, Lievens Y, Bouma AB, Collette L, Dekker A, deSouza NM, et al. Characterisation and classification of oligometastatic disease: a European Society for Radiotherapy and Oncology and European Organisation for Research and Treatment of Cancer consensus recommendation. Lancet Oncol. 2020;21:e18.
26. Gofrit ON, Gofrit B, Roditi Y, Popovtzer A, Frank S, Sosna J, et al. The different clonal origins of metachronous and synchronous metastases. J Cancer Res Clin Oncol. 2023;149(13):11085.
27. Reboux N, Jooste V, Goungounga J, Robaszkiewicz M, Nousbaum JB, Bouvier AM. Incidence and survival in synchronous and metachronous liver metastases from colorectal cancer. JAMA Netw Open. 2022;5(10):e2236666.
28. Tsai MS, Su YH, Ho MC, Liang JT, Chen TP, Lai HS, et al. Clinicopathological features and prognosis in resectable synchronous and metachronous colorectal liver metastasis. Ann Surg Oncol. 2007;14(2):786.
29. Fleckenstein J, Petroff A, Schäfers HJ, Wehler T, Schöpe J, Rübe C. Long-term outcomes in radically treated synchronous vs. metachronous oligometastatic non-small-cell lung cancer. BMC Cancer. 2016;16(1):348.
30. Reyes DK, Pienta KJ. The biology and treatment of oligometastatic cancer. Oncotarget. 2015;6(11):8491.
31. Hosaka S, Katagiri H, Honda Y, Wasa J, Murata H, Takahashi M. Clinical outcome for patients of solitary bone only metastasis. J Orthop Sci. 2016;21(2):226.
32. Higuchi T, Yamamoto N, Hayashi K, Takeuchi A, Kato S, Miwa S, et al. The efficacy of wide resection for musculoskeletal metastatic lesions of renal cell carcinoma. Anticancer Res. 2018;38(1):577–82.

33. Meagher MF, Mir MC, Autorino R, Minervini A, Kriegmair M, Maurer T, et al. Impact of metastasectomy on cancer specific and overall survival in metastatic renal cell carcinoma: analysis of the REMARCC registry. Clin Genitourin Cancer. 2022;20(4):326–33.
34. Higuchi T, Yamamoto N, Hayashi K, Takeuchi A, Abe K, Taniguchi Y, et al. Long-term patient survival after the surgical treatment of bone and soft-tissue metastases from renal cell carcinoma. Bone Joint J. 2018;100B(9):1241–8.
35. Hayashi K, Hiraoka K, Akiyama T, Iwata S, Tsuchiya H, Kawai A. Benefit of surgical resection of distant metastasis in soft tissue sarcoma: a systematic review. Jpn J Clin Oncol. 2021;51:1088–93.
36. Gonzalez MR, Rizk P, Hodo TW, Bedi A, Karczewski D, Lozano-Calderon SA. Metastatic appendicular soft tissue sarcoma: treatment and survival outcomes of 2,553 patients from the SEER database. J Cancer Metastasis Treat. 2023;9
37. Pretell-Mazzini J, Seldon CS, D'Amato G, Subhawong TK. Musculoskeletal metastasis from soft-tissue sarcomas: a review of the literature. J Am Acad Orthop Surg. 2022;30(11):493–503.
38. Nakamura T, Matsumine A, Yamakado K, Takao M, Uchida A, Sudo A. Clinical significance of radiofrequency ablation and metastasectomy in elderly patients with lung metastases from musculoskeletal sarcomas. J Cancer Res Ther. 2013;9(2):219.
39. Delisle M, Alshamsan B, Nagaratnam K, Smith D, Wang Y, Srikanthan A. Metastasectomy in Leiomyosarcoma: a systematic review and pooled survival analysis. Cancers. 2022;14:3055.
40. Motzer RJ, Memorial Sloan Kettering Cancer Center Þ, Jonasch E, Agarwal N, Alva A, Bagshaw H, et al. NCCN Guidelines Version 2.2025 Kidney Cancer Continue NCCN Guidelines Panel Disclosures [Internet]. 2024. Available from: https://www.nccn.org/home/.
41. Armstrong SA, Balach T, Bishop AJ, Buehler D, Carr-Ascher J, Choy E, et al. NCCN guidelines version 3.2024 soft tissue sarcoma [Internet]. 2024. Available from: https://www.nccn.org/home/member-
42. Berger Y, Spivack JH, Heskel M, Aycart SN, Labow DM, Sarpel U. Extrahepatic metastasectomy for hepatocellular carcinoma: predictors of long-term survival. J Surg Oncol. 2016;114(4):469.
43. Fottner A, Szalantzy M, Wirthmann L, Stähler M, Baur-Melnyk A, Jansson V, et al. Bone metastases from renal cell carcinoma: patient survival after surgical treatment [Internet]. BMC Musculoskelet Disord. 2010;11. Available from: http://www.biomedcentral.com/1471-2474/11/145.
44. Nistor CE, Ciuche A, Cucu AP, Nitipir C, Slavu C, Serban B, et al. Management of lung cancer presenting with solitary bone metastasis. Medicina (B Aires). 2022;58:1463.
45. Azam F, Latif MF, Farooq A, Tirmazy SH, Alshahrani S, Bashir S, et al. Performance status assessment by using ECOG (Eastern Cooperative Oncology Group) score for cancer patients by oncology healthcare professionals. Case Rep Oncol. 2019;12(3):728–36.
46. Sankar A, Johnson SR, Beattie WS, Tait G, Wijeysundera DN. Reliability of the American Society of Anesthesiologists physical status scale in clinical practice. Br J Anaesth. 2014;113(3):424.
47. Mor V, Laliberte L, Morris JN, Wiemann M. The Karnofsky performance status scale. Cancer. 1984;53:2002.
48. Gripp S, Moeller S, Bölke E, Schmitt G, Matuschek C, Asgari S, et al. Survival prediction in terminally ill cancer patients by clinical estimates, laboratory tests, and self-rated anxiety and depression. J Clin Oncol. 2007;25(22):3313.
49. Lindsay AD, Haupt EE, Chan CM, Spiguel AR, Scarborough MT, Zlotecki RA, et al. Treatment of sarcoma lung metastases with stereotactic body radiotherapy. Sarcoma. 2018;2018:9132359.
50. Salim N, Tumanova K, Popodko A, Libson E. Second chance for cure: stereotactic ablative radiotherapy in oligometastatic disease. JCO Glob Oncol. 2024;10:e2300275.
51. David S. Stereotactic ablative body radiotherapy (SABR) for bone only oligometastatic breast cancer: a prospective clinical trial. Breast. 2020;49:55–62.
52. Fourman MS, Ramsey DC, Newman ET, Raskin KA, Tobert DG, Lozano-Calderon S. How I do it: percutaneous stabilization of symptomatic sacral and periacetabular metastatic lesions with photodynamic nails. J Surg Oncol. 2021;124(7):1192.

53. Lee L, Brown A, Lerman DM. Minimally invasive stabilization of pelvic metastatic bone disease: a review of an emerging technique. J Surg Oncol. 2023;128:445.
54. Yang R, Goch A, Murphy D, Wang J, Charubhumi V, Fox J, et al. A novel tripod percutaneous reconstruction technique in periacetabular lesions caused by metastatic cancer. J Bone Joint Surg. 2020;102(7):592–9.

36 Principles of Surgical Management: Ablative Techniques for Skeletal Metastases

Andrew D. Kleven and John C. Neilson

Epidemiology

Bone metastases are a rising concern in orthopedic oncology, with bone being the third most common site for metastatic spread. Approximately 68% of bony metastases are attributed to three primary cancers: breast, prostate, and lung cancer [1, 2]. As multimodal cancer therapies advance, patient survival rates have improved, and many go on to develop bony metastases [3, 4]. Improved survival naturally leads to patient desire to lead a functional life with minimal pain. Skeletal metastases, a common complication of advanced malignancies, significantly impact patient quality of life due to pain, fracture risk, and reduced mobility. These metastases are also often resistant to traditional methods of conservative treatment due to bony anatomy and local bone-tumor physiology [5].

Indications

All attempts at cancer local control are essentially ablative treatments. Since the advent of modern surgery and radiation therapy, metastatic disease to the skeleton has undergone effective ablations to achieve local control, functional improvement and pain relief. Image-guided thermal (energy) ablations are a result of decades of technological advancement in local therapies and are now commonly applied to skeletal metastases to achieve the same goals. The first commonly used thermal ablative therapy to the skeletal system was the use of radiofrequency ablation for osteoid osteoma, a painful benign skeletal tumor in 1990s [6, 7].

Thermal ablation destroys tissue via extreme heat or cold, causing cellular apoptosis or necrosis. The effectiveness depends on energy delivery, tissue thermal conductivity, and proximity to critical structures. Thermal ablation therapies are often

A. D. Kleven · J. C. Neilson (✉)
Department of Orthopaedic Surgery, Medical College of Wisconsin, Milwaukee, WI, USA

J. Pretell-Mazzini, S. A. Lozano Calderón (eds.), *Multidisciplinary Approach of Musculoskeletal Metastasis*, https://doi.org/10.1007/978-3-032-09498-8_36

termed minimally invasive due to the generally percutaneous nature of the treatments but are very powerful and potentially dangerous tools due to the ability to destroy large amounts of tissue using overlapping ablation zones. These powerful tools can completely destroy tumors and have the potential to damage any surrounding structures including skin, nerves, and blood vessels.

Thermal ablative therapies, including radiofrequency ablation (RFA), microwave ablation (MWA), cryoablation, and high-intensity-focused ultrasound (HIFU), leverage localized thermal energy to destroy metastatic tumor tissues while sparing surrounding structures. Image guidance using fluoroscopy, ultrasound, computed tomography (CT), or magnetic resonance imaging (MRI) is integral to ensuring precision, efficacy, and safety. These approaches were initially used to treat patients who may not tolerate or benefit from traditional open treatments but are now commonly used in an integrated fashion or as first line treatments for some disease types. They are advantageous due to the small skin incisions and generally quicker healing times with a rapid return to life saving or life prolonging adjunct therapies, especially when compared to open surgery [8].

Pain from skeletal metastases is present in 75% of patients and significantly impacts their quality of life [9]. Pain generators related to skeletal metastasis are varied and complex but generally can be described as a combination of background, spontaneous and incidental (movement evoked) pain. These could be related to periosteal expansion or irritation, neural origin (compression or inflammatory) and mechanical/weight-bearing pain from periosteal stress, respectively [10]. Thermal ablation therapies can assist in the reduction of all these pain generators. It can be effective at destruction of painful nerve endings at the periosteum. When used around a major nerve, RFA, cryoablation, and HIFU can have well defined thermal zones and can act to decrease the size of the tumor after the initial postablation inflammatory process, thereby decreasing neural compression. Thermal ablation in conjunction with cementation and/or orthopedic fixation can alleviate mechanical/weight-bearing pain. Advancement in thermal ablative therapies has led to the incorporation of these therapies into the National Comprehensive Cancer Network (NCCN) guidelines version 3.2024 for adult cancer pain [11]. In patients with some tumor types such as multiple myeloma, radiation use has decreased due to the desire to preserve any residual marrow function as patients move toward bone marrow transplant. Ablative therapies are particularly effective for radiation resistant tumors as well and can add another layer of treatment modalities for patients whose tumors have not responded to radiation or who are less likely to respond to radiation therapies such as renal cell carcinoma. Complications include thermal injury to adjacent tissues, infection, and fracture risk.

In conjunction with ablative therapies, maintaining the structural integrity of the appendicular and axial skeleton is critical to promoting function and preventing delayed skeletal events. Often, lesions affect the force transfer from the extremities to the axial skeleton. Although ablative therapies can be readily applied, they do not fortify the skeleton thereby, fracture risk and micromotion leading to pain persists. Bony stabilization techniques include the placement of cement, screws, or photodynamic nails and can be used for painful lesions at risk for fracture or large osteolytic lesions in weight-bearing areas of the skeleton. These are easily employed after ablation in the same interventional setting.

The decision to proceed with percutaneous interventions is influenced by various factors, including the primary tumor type, the extent of metastatic spread, the patient's overall health status, and their personal treatment goals. For example, younger patients with localized metastases and a good performance status may be more likely to pursue aggressive surgical options rather than percutaneous ablative techniques. Solitary metastases can be addressed with curative intent with surgical resection and reconstruction depending on the clinical scenario. Conversely, older patients or those with significant comorbidities may prioritize quality of life over aggressive treatment, opting for nonsurgical palliative care or minimally invasive approaches for local disease control and stabilization. Multidisciplinary discussions involving oncologists, surgeons, and palliative care specialists are essential in formulating an individualized treatment plan that aligns with the patient's needs and preferences.

Preoperative Planning

Preoperative planning for interventions in patients with metastatic disease is critical for optimizing outcomes and minimizing complications. This process begins with a comprehensive assessment of the patient's medical history, including the extent of metastatic involvement, previous treatments, and overall functional status. Patient evaluation must identify the location and character of pain, with special attention to pain with weight bearing or mobility. Groin pain may point to hip involvement in the oncologic process. Deep-seated low back or pelvic pain with sitting upright or standing may be indicative of symptomatic sacrospinal metastases. Imaging studies, such as CT or MRI, are essential to evaluate the anatomy of the pelvis, assess the size and location of lesions, plan the surgical margins if indicated, and identify any neurovascular structures at risk or impending pathological fracture. In percutaneous intervention, CT imaging allows surgeons to plan for the location and number of ablative and fixation pathways needed. Three-dimensional reconstructions may help represent complex bony anatomy in the context of bone loss or pathological fracture. MRI or CT with contrast may be considered if vascularity is of interest.

It is critical to establish tissue diagnosis prior to proceeding with surgical intervention as it guides the surgical approach, informs potential complications, and helps tailor postoperative management to the specific characteristics of the disease. Although rare, surgeons should be aware of mimickers or other oncologic processes in patients with metastatic disease prior to proceeding with surgical intervention. Additionally, preoperative laboratory tests should be conducted to assess renal function, coagulation status, and other relevant parameters. Patient education is also vital, as discussing the procedure, expected outcomes, and potential risks can enhance cooperation and reduce anxiety. Overall, thorough preoperative planning ensures that the procedure is safe, targeted, and effective in addressing metastatic lesions.

Table 36.1 Comparison of ablative therapies for treatment of metastatic disease

Parameter	RFA	MWA	Cryoablation	HIFU
Invasiveness	Minimally	Minimally	Minimally	Noninvasive
Tumor size	Small	Medium–large	Medium–large	Small–medium
Heat sink effect	Significant	Minimal	None	None
Precision	Moderate	Moderate	High	High
Imaging required	CT/US	CT/US	CT/MRI	MRI/US

Ablative Options

Ablative interventions for skeletal metastases can be categorized into radiofrequency ablation, microwave ablation, cryoablation, and high-intensity-focused ultrasound. Each of these modalities have their own unique disadvantages and advantages for treating skeletal metastases and are especially convenient for metastatic disease of the pelvis or spine (Table 36.1). Due to the deep anatomy of the pelvis and spine, overlapping symptoms with common pain complaints, and vague presentation, these metastases may be large at the time of diagnosis.

Radiofrequency Ablation (RFA)

Introduced in the 1990s, RFA was initially used for hepatic tumors before its application expanded to bone metastases. RFA uses an electrode at the tip of a needle to create high-frequency (375–500 KHz) waves inducing coagulative necrosis through dielectric hysteresis. This process increases the motion of the polar H_2O molecules surrounding the probes to create extreme heat within the tumor tissue. The heat causes protein denaturation, apoptosis and cellular necrosis, leading to tumor cell destruction. Early devices were monopolar electrodes with a grounding pad attached to the patients. Newer devices are bipolar electrodes with the latest designs incorporating the ability to inject fluid into the area to assist in hysteresis. The treatment is typically guided by CT, fluoroscopy, or ultrasound to ensure accurate needle placement and effective treatment [8]. It is most effective for small to medium-sized tumors and is typically applied to lesions that are accessible through percutaneous approaches. RFA is has well documented use in the spine, pelvis and extremities. RFA probes have a well-defined treatment area dependent on design of the electrode [12].

Studies have demonstrated that RFA can provide significant pain relief, with pain reduction reported in 60–80% of patients. In addition to pain control, RFA can shrink tumors and improve functional status in some patients where local control rates range from 60% to 85%, varying by modality and tumor characteristics. Recent studies, such as the OsteoCool Tumor Ablation Post-Market Study (OPuS One) have shown that RFA plus cement augmentation provides rapid and durable pain relief for lytic bone metastases in the axial and appendicular skeleton, independent of radiation therapy, with significant improvement in quality of life metrics [13]. Secondary analysis of this data demonstrated sustained pain relief at 1-year post procedure with no delayed skeletal events.

As with any treatment modality, RFA offers advantages and disadvantages. RFA is minimally invasive with a relatively low complication rate. It is effective for both pain management and local tumor control, making it a valuable treatment option. Additionally, RFA is suitable for patients with multiple bone metastases. The charring or desiccation of tissue can increase impedance, which reduces the effectiveness of the ablation or be used to an advantage if treating lesions near critical neurovascular structures. No neuromonitoring of critical nerves may be used due to electrical interference from RFA. RFA is less likely to extend past cortical bone, making it ideal when working near joints, such as the acetabulum (Fig. 36.1). Additionally, RFA is less effective for treating osteoblastic lesions, as the dense bone tissue can hinder the proper delivery of heat. It also has limited ability to treat large or deep tumors, which may not be accessible or adequately affected by the procedure. Furthermore, there is a potential for recurrence of the tumor in some cases, limiting the long-term success of the treatment.

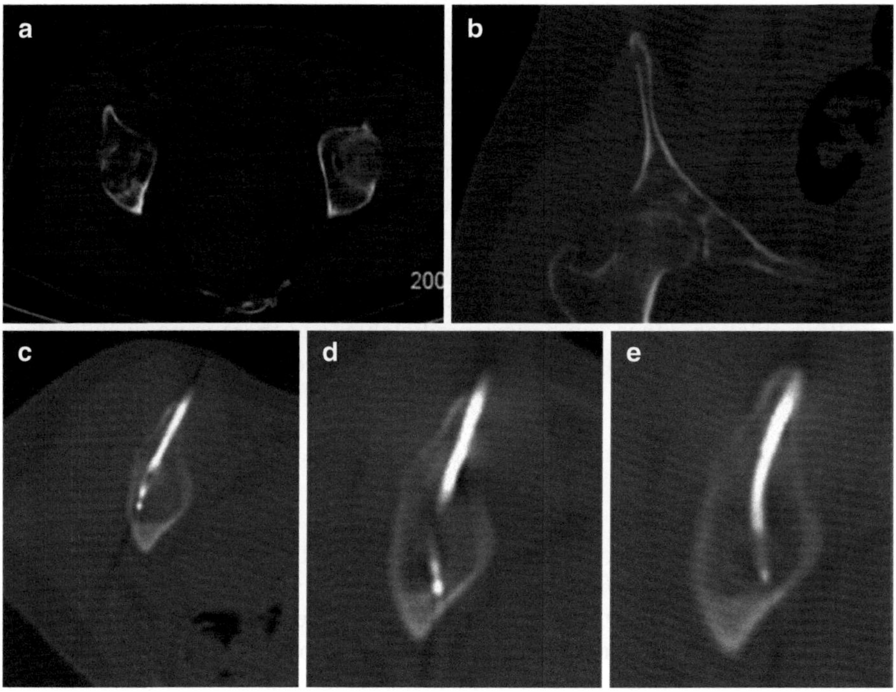

Fig. 36.1 65F Malignant peripheral nerve sheath tumor with metastatic disease to the acetabulum. Patient underwent percutaneous RFA ablation utilizing an articulating probe with overlapping ablation zones. (**a**) Axial CT image. (**b**) Coronal CT image. (**c**–**e**) Lateral to medial overlapping ablation zones seen on axial CT scan performed by using an articulating ablation probe

Microwave Ablation (MWA)

MWA originated in the 1980s as a method for hepatic tumors. Advances in antenna technology and power delivery in the 2000s enhanced its applicability for musculoskeletal tumors. MWA utilizes electromagnetic waves for inducing tumor destruction, commonly utilizing devices with frequencies from 915 MHz or 2.45 GHz. An antenna is inserted into the tumor, and microwave energy is applied to generate temperatures high enough to cause tumor cell death. This coagulative necrosis is induced at higher energy frequencies than RFA which creates primarily dielectric hysteresis [14]. This allows MWA to be used in more tissue types with high impendence, low thermal conductivity, or low electrical conductivity such as dense bone. This allows for generation of higher temperatures and treat larger volumes of tissue. Additionally, MWA can more effectively treat lesions that are located deeper within bone [8]. MWA has been shown to be effective in reducing pain and controlling tumor growth in patients with skeletal metastases. Clinical studies have reported pain relief in up to 85% of patients, with some patients experiencing significant improvements in quality of life.

MWA offers several advantages over other ablation techniques. It generates higher temperatures, making it more effective for treating larger tumors and those located deeper within the body due to the longest available probes. Studies have shown lower recurrence rates with MWA compared to RFA for larger lesions and MWA overcomes the limited efficacy of RFA in water-poor environments by generating high temperatures [15]. MWA also creates larger ablation zones than RFA, and the ability to manipulate the wattage allows for variable size and shape of these zones. Additionally, MWA utilizes larger probes, which are particularly useful for accessing long bones through nail entry sites. However, MWA also has limitations. It is still less effective for osteoblastic lesions and requires specialized equipment and expertise. The resulting ablation zones can be less predictable, and there is no visualization of the ablation on imaging. Finally, MWA is less likely to be affected by tissue desiccation.

Cryoablation (CA)

CA involves the use of extremely cold temperatures to freeze and destroy tumor tissue. CA has roots in the early twentieth century, with liquid nitrogen as the primary cryogen. In 1969, Dr. Marcove published the use of liquid nitrogen circulating probes for primary and metastatic bone tumors [16]. Modern percutaneous systems, developed in the 1990s, use nitrogen- or argon-based cryotherapy for precise ice-ball formation [17, 18]. A cryoprobe is inserted into the tumor, and liquid nitrogen or argon gas is used to cool the probe and surrounding tissue.

The freezing process leads to the formation of intracellular ice crystals, which disrupt cell membranes and cause cellular necrosis. The most common clinically available cryoablation systems utilize the Joule-Thomson effect, which relies on the expansion of a cryogen (argon gas or liquid nitrogen) at the cryoprobe tip to cause internal temperature fluctuation with heat being removed from the surrounding tissue and ice ball formation in a prescribed area around the tip of the cryoprobe [8]. In the region of treatment closest to the probe, intracellular ice formation directly damages cells and cause immediate cell death. Farther from the probe, extracellular ice forms and the resulting osmotic gradient dehydrates the cells. Local effects of CA are important as damage to blood vessels cause platelet activation, thrombosis and ischemia after reperfusion. Mitochondrial damage induces cellular apoptosis. CA is used to treat bone metastases that are located near critical structures often including major vessels and nerves as neuromonitoring is not affected by the ice crystals, whereas local electrical currents with RFA and MWA interfere with monitoring.

CA has been associated with significant pain relief in patients with skeletal metastases, with studies showing a pain reduction rate of 60–90%. Additionally, CA has a lower risk of recurrence for tumors located near critical structures, as the freezing process can effectively destroy both the tumor and its blood supply. A possible additive effect due to a CA-stimulated systemic antitumor immune response has been described [19]. The MOTION trial demonstrated that CA provided durable pain control, improved quality of life, and reduction in opioid use in patients with metastatic bone disease [20]. It can be used for osteoblastic lesions where CA is less affected by cortical bone, and be used for irregular or large lesions with a soft tissue component due to the use of multiple cannulas [21].

CA offers several distinct advantages, including the ability to treat tumors near critical structures and demonstrated lower recurrence rates for certain tumor types [22]. CA may have the advantage of being less painful in the postoperative course, safer in respect of surrounding structures, and unaffected by the lytic or blastic consistency of lesions. This technique can be repeated for recurrent lesions and is considered the best ablation modality for use with image guidance, as the ice zone is visible on CT, ultrasound, and MRI. If performing CA near important nerves, inadvertent injury may be less with CA due to the preserved neuronal architecture [23]. CA is also safe for patients with pacemakers and AICDs, does not interfere with nerve monitoring, and causes less damage to cartilage, ligaments, and tendons. Furthermore, MRI-compatible systems are available. However, CA has limitations. It requires specialized equipment and is less effective for large tumors. There is a potential for injury to surrounding healthy tissue if not performed precisely, and the procedure requires more time due to the necessary freeze-thaw cycles. The solid ablation zone (ice ball) may also create difficulties with hardware placement and cementation, although active warming may assist in mitigating this issue (Table 36.2).

Table 36.2 Technical considerations for application of ablative therapies

Modality	Operating temperature	Key features	Limitations
RFA	50 to 100 °C	Facilitates PMMA for consolidation after ablation; predictable ablation zones; least expensive	Ablation zones limited by tissue charring; lesion size <3 cm
MWA	Can exceed 150 °C	Greatest ability to desiccate tumor; facilitates PMMA for consolidation after ablation; efficient energy deposition; less susceptible to heat sinks	Difficulty predicting ablation zone; ablation zone not visible; caution near joints, connective tissue, and nerves
Cryo	−20 to −40 °C (lethal at −20 °C for most tissues)	Ice ball visible on US, CT, and MRI; most forgiving in event of nerve injury; safest around joints and connective tissue	Challenging to consolidate with PMMA; multiple freeze thaw cycles = longer procedure times

High-Intensity-Focused Ultrasound (HIFU)

HIFU research began in the 1950s, initially for neurosurgical applications [24]. With the advancement of MRI in the 1990s, its use became more practical. This noninvasive thermal ablation uses nonionizing focused acoustic energy as a method to precisely increase the temperature of a treatment site and uses MRI to monitor the intensity of temperature rise. Ultrasound waves converge at a specific focal point to generate temperatures greater than 60 °C causing protein denaturation, coagulative necrosis and acoustic cavitation generating microbubbles which mechanically disrupt cell membranes. HIFU has been particularly effective in pain palliation and tumor control, and can be very effective to the treatment of bone lesion as bone absorbs acoustic energy at 50 times the rate of soft tissue [25]. Given that periosteum is the only innervated portion of the bone it helps explain the pain control associated with HIFU. Clinical studies have demonstrated that HIFU provides significant pain relief for patients with skeletal metastases, with some studies reporting pain reduction in up to 90% of patients [26–29]. Additionally, HIFU has been shown to be effective in reducing tumor size and improving functional status. While the technique is still evolving, it holds great promise as a completely noninvasive alternative to surgery and radiation.

HIFU offers several advantages, including its noninvasive nature, eliminating the need for incisions or needles. It provides real-time imaging for precise targeting and has minimal side effects with quick recovery times. The noninvasive nature of HIFU minimizes the risks associated with percutaneous interventions. However, HIFU also has limitations. It may be less effective for large or deeply located tumors and requires specialized equipment and expertise. Additionally, HIFU is not widely available, and its effectiveness depends on optimal visualization for safety and efficacy. Limited MRI availability and energy attenuation due to bones and air can hinder its use in certain cases, such as if hollow viscera or other bones are in the way. Finally, there is a risk of thermal injury to adjacent nerves or organs when these structures are close.

Ablative therapies are also very promising for the developing world as the access to therapeutic radiation is limited and the upfront investment in radiation is very high. RFA probes are relatively affordable and have a well-established technological methodology that is very portable. While in the developed world, probes are generally one time use, the reuse of one time use equipment is very common in the developing world where resources are limited.

As the field of image-guided thermal ablation continues to evolve, further studies are needed to refine the indications, techniques, and outcomes for these therapies. Comparative studies evaluating long term outcomes of all modalities are important future research goals. Other technological advancement using hybrid imaging systems for improved precision and real-time thermometry for enhanced safety would be valuable. The potential for integration with artificial intelligence for automated planning and monitoring and the integration of thermal ablation with other treatment modalities hold the potential for improving patient outcomes and quality of life.

Bony Stabilization

Skeletal fixation at the time of ablative percutaneous interventions have gained prominence as minimally invasive alternatives to traditional surgery. These techniques include osteoplasty, screw fixation, and photodynamic nail placement. Osteoplasty involves the injection of polymethylmethacrylate (PMMA) into lesions, providing immediate mechanical stabilization. PMMA may also provide local thermal necrosis of the surrounding tumor tissue due to its exothermic reaction during curing [30]. Cement, screw fixation or photodynamic nail placement may be employed to stabilize areas affected by metastasis, often utilized after ablative therapies (Fig. 36.2). These techniques work synergistically due to cavity creation after ablation, but also to prevent hardware failure due to bone loss from local disease progression [23]. English et al. (2021) demonstrated that minimally invasive periacetabular screw stabilization and ablative techniques in treatment of pelvic metastases provided reliable improvements in pain and function with a low surgical risk and rapid initiation of chemotherapy and radiation [31].

Photodynamic nail placement can be a reasonable alternative or adjunct to screw fixation (Figs. 36.3a, 36.3b and 36.3c). The photodynamic bone stabilization system is a minimally invasive light-curable resin that is delivered with an inflatable balloon catheter [32]. The advantages of this technique are that with minimally invasive balloon application and curing of the resin in situ, larger and variable sized bony defects can be filled as compared with screws. In addition, the photodynamic nails can fill nonlinear trajectories and allow customizable starting points and working lengths. This is useful in situations where there is significant bone loss due to metastatic disease or in patients with bony dysplasia. The maximum width and length of screws is limited by the bony corridors

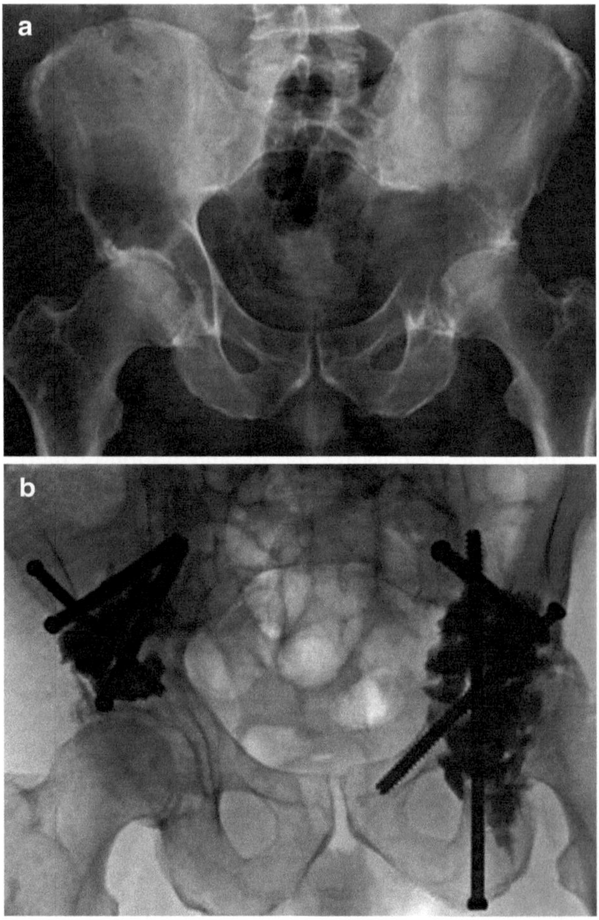

Fig. 36.2 (**a**) 56M with newly diagnosed renal cell carcinoma metastatic to bilateral acetabula. (**b**) Patient underwent embolization and percutaneous ablation with osteoplasty and screw fixation

through which they are applied which vary from patient to patient, but photodynamic nails provide more treatment flexibility. Conveniently, due to their radiolucent nature, application of the nails allows clinical monitoring of metastatic progression using plain radiographs or CT without radiologic scatter as with metal implants. Lozano-Calderon et al. (2024) described that percutaneous

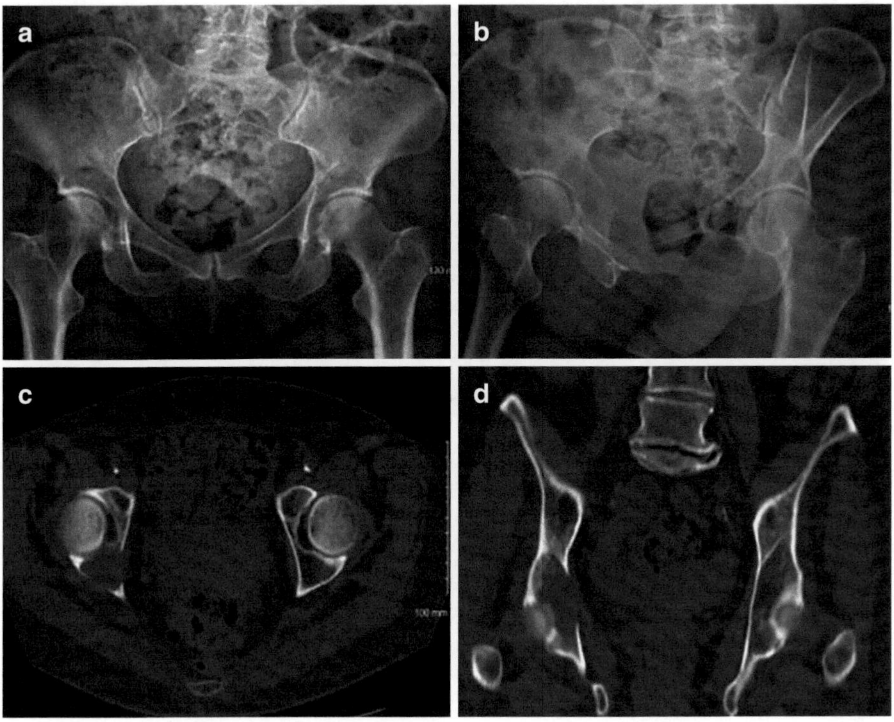

Fig. 36.3a 74F with metastatic lung cancer with painful right periacetabular radiolucent lesion. (**a**) AP pelvis plain radiograph demonstrating right radiolucent periacetabular lesion. (**b**) Iliac oblique plain radiograph demonstrating right radiolucent periacetabular lesion. (**c**) Axial CT scan demonstrating radiolucent lesion of the posterior wall and column of the right acetabulum. (**d**) Coronal CT scan demonstrating radiolucent lesion of the posterior wall and column of the right acetabulum

fixation of pelvic metastatic lesions with photodynamic nails reliably improves pain, function, and weight-bearing status and reduces narcotic pain medication use [33]. These percutaneous stabilization methods not only minimize recovery time and hospital stays, but also reduce surgical risks associated as compared with open procedures while providing local control when used with ablative therapies.

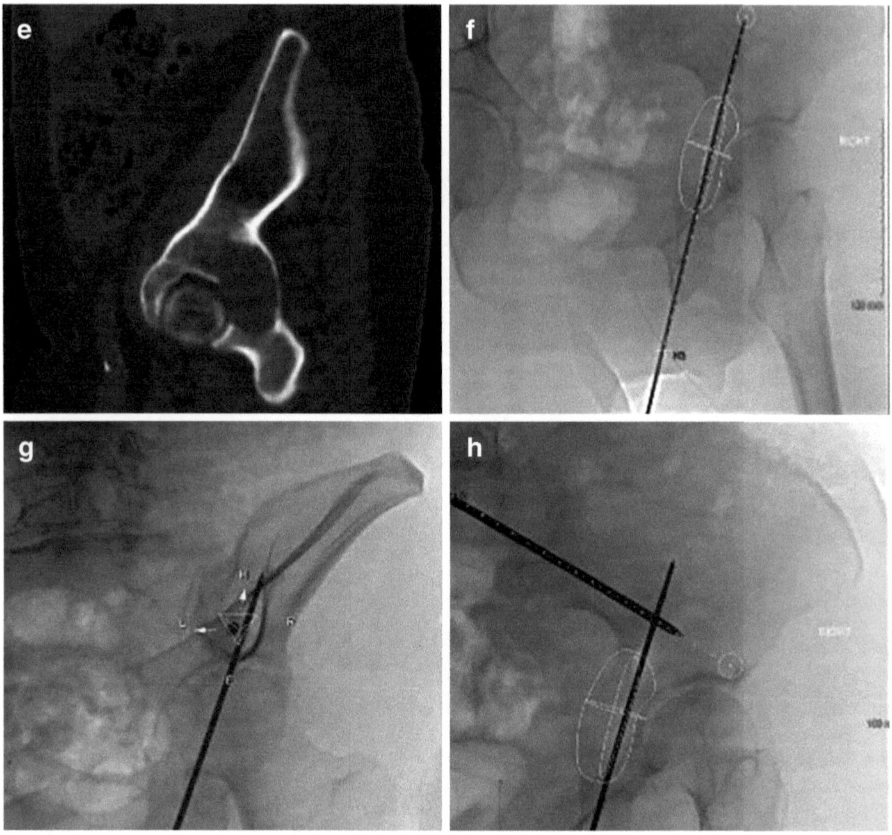

Fig. 36.3b 74F with metastatic lung cancer with painful right periacetabular radiolucent lesion. (**e**) Sagittal CT scan demonstrating radiolucent lesion of the posterior wall and column of the right acetabulum. (**f**) Iliac oblique augmented fluoroscopy demonstrating guidewire placement along the posterior column performed in prone positioning with volumetric object outline of periacetabular lesion (white circle). (**g**) Obturator outlet augmented fluoroscopy demonstrating end point targeting (white triangle) for transcolumnar screw from posterior to anterior, as well as guidewire placement along the posterior column from inferior to superior performed in prone positioning. (**h**) Iliac inlet augmented fluoroscopy demonstrating end point targeting (white circle) for transcolumnar screw from posterior to anterior, as well as guidewire placement along the posterior column from inferior to superior performed in prone positioning

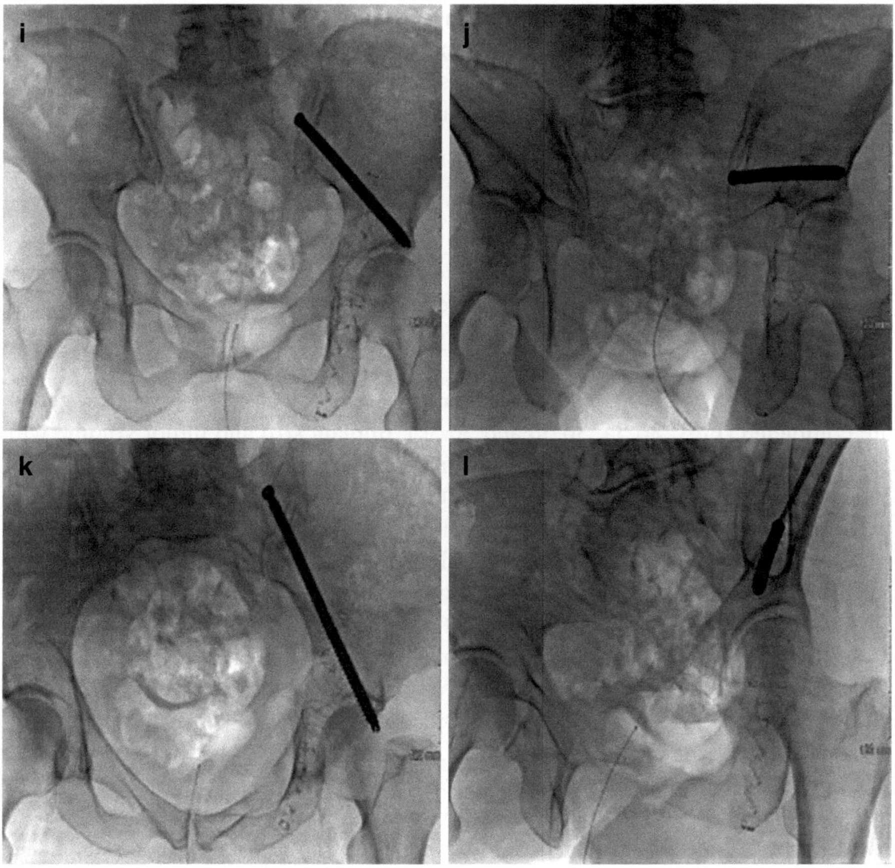

Fig. 36.3c 74F with metastatic lung cancer with painful right periacetabular radiolucent lesion. (**i**) AP pelvis plain radiograph demonstrating final implant positioning of transcolumnar screw and photodynamic nail in the posterior column performed in prone positioning. (**j**) Pelvic outlet plain radiograph demonstrating final implant positioning of transcolumnar screw and photodynamic nail in the posterior column performed in prone positioning. (**k**) Pelvic inlet plain radiograph demonstrating final implant positioning of transcolumnar screw and photodynamic nail in the posterior column performed in prone positioning. (**l**) Obturator oblique plain radiograph demonstrating final implant positioning of transcolumnar screw and photodynamic nail in the posterior column performed in prone positioning

Image Guidance in Percutaneous Therapies

All image guidance at the time of intervention is useful at determining tumor location size and proximity to critical structures. Several imaging modalities are used to guide and monitor ablation procedures. Fluoroscopy is readily available, real-time, and cost-effective, but less precise than CT. Ultrasound offers real-time, cost-effective guidance but is limited by bone interference. CT provides high-resolution imaging for lesion localization, especially in cortical bone, and helps avoid collateral damage due to its axial imaging precision. Finally, MRI offers superior soft tissue contrast and allows for temperature measurement by proton resonance frequency (PRF) MR thermometry. This sequence evaluates the resonance frequency of water protons which decrease as the temperature increases.

Surgical Technique

Various workflows exist, but ablation and skeletal fixation is most easily performed with use of ceiling mounted intraoperative cone beam CT for spine or pelvic metastatic disease due to the complex anatomy (Fig. 36.4). This is most often found in the interventional radiology suite.

After initial CT scan, ablative probe and fixation trajectory mapping, if needed, are completed (Figs. 36.5a, 36.5b and 36.5c). Implant size and tumor volume with any interval disease progression can immediately be assessed using this modality. The number of ablative probes or adjunct temperature probes that are needed can be planned based on overlapping treatment fields and local bony anatomy. This plan is transferred to augmented intraoperative fluoroscopy with needle guidance lines to readily navigate application of cement cannulas, ablative probes, bone trocars, or fixation devices (Figs. 36.6a, 36.6b, 36.6c and 36.6d). This is critical as normal radiograph or palpable bony landmarks may be missing due to metastatic erosion. Our most typical workflow for interventions is ablation, fixation, follow by cementation if needed. Entries into the bone are planned to limit areas of potential cement extravasation. Local anesthetic is injected, and stab incisions (<1 cm) are made over the planned starting points. The cortical bone can be opened with a 10-gauge bone trocar or sharp awl to facilitate the starting point and allow introduction of various sized probes. The bone trocar can be advanced to the lesion of interest and then exchange the stylet for an ablation probe which are more fragile [23]. Care is taken to retract the trocar out of the ablative zone to not affect treatment. Ablative therapy is then applied depending on the modality according to manufacturer guidelines

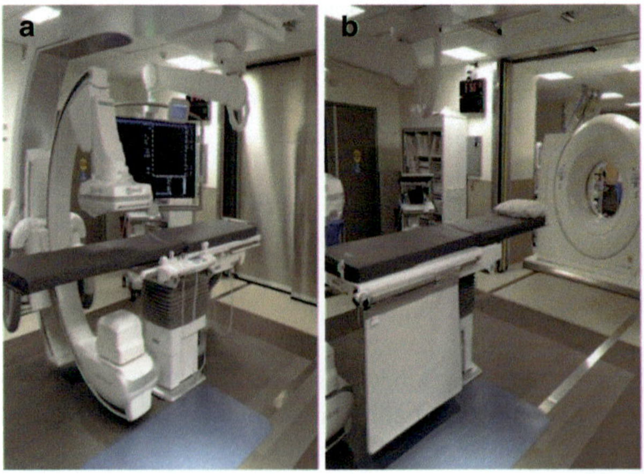

Fig. 36.4 Interventional radiology suite with ceiling-mounted augmented fluoroscopy and cone beam CT

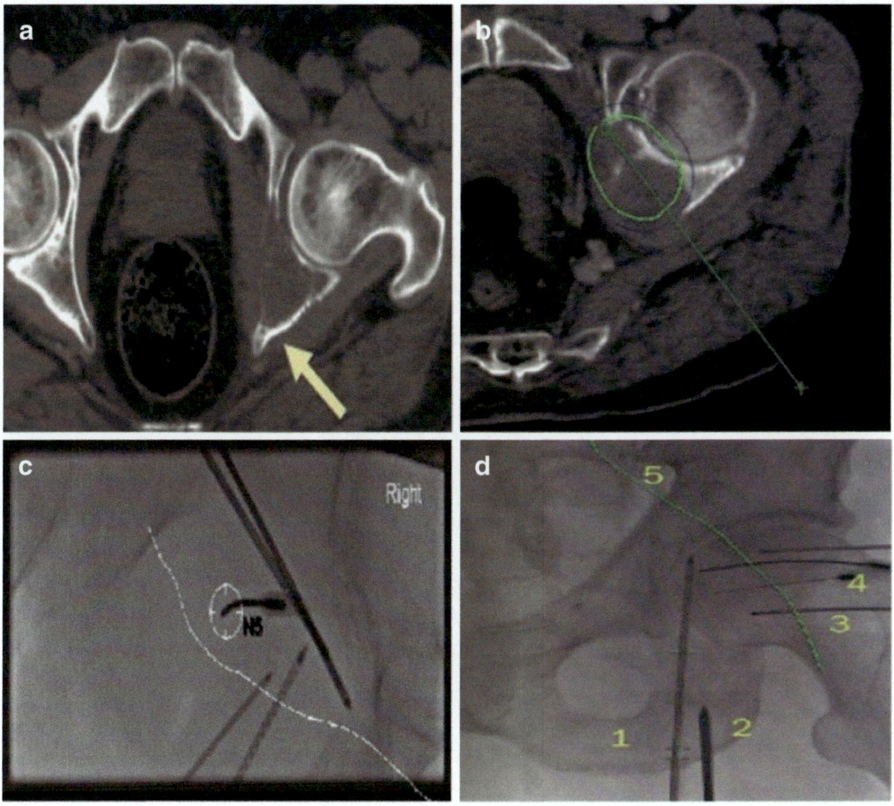

Fig. 36.5a 72M with metastatic renal cell carcinoma with persistent left hip pain after radiation for osteolytic acetabular lesion involving the posterior column. (**a**) Osteolytic acetabular lesion of the posterior column. (**b**) Intraoperative CT scan planning with planned ablative zone (green) and safe margin (blue). (**c**) He underwent cryoablation for tumor control and osteoplasty with screw fixation with sciatic nerve involvement along posterior column (white line overlay). (**d**) Intraoperative augmented fluoroscopy of percutaneous treatment of lesions: (1) cryoablation probe, (2) cement cannula, (3) temperature probes, (4) hydro-dissection probes, (5) sciatic nerve live overlay

(radiofrequency ablation using STAR Tumor Ablation System, DFINE, Inc., San Jose, California; cryoablation using Endocare Cryocare, HealthTronics, Inc., Austin, Texas).

After ablative therapy is complete and skeletal fixation is indicated, a 2.0 mm guide pin or 2.8 mm Steinmann pin is placed manually or with a power drill under fluoroscopic guidance to their planned location through the same ablation entry point. The lengths of the screws should be measured at this time, which can be done manually or with repeat CT. Screws can be advanced along the guide pin until partially or fully seated. For smaller singular lesions, cement application using an 11-gauge bone trocar through the partially advanced cannulated screw can be done and then the screw advanced prior to the cement curing. For larger lesions, we favor

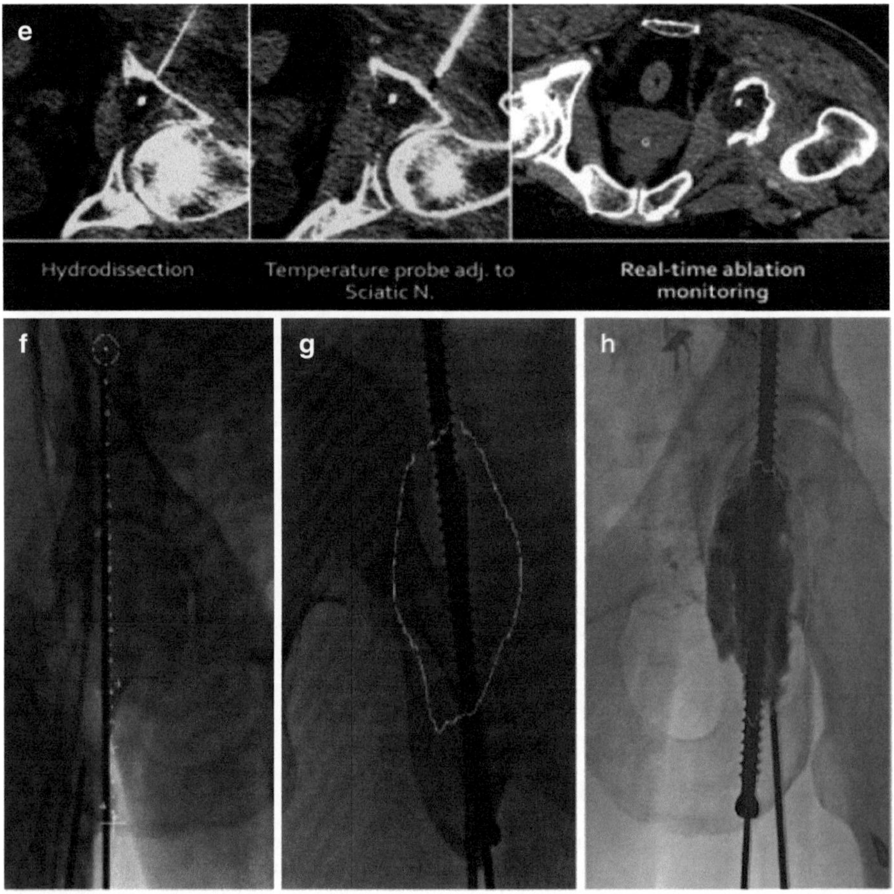

Fig. 36.5b 72M with metastatic renal cell carcinoma with persistent left hip pain after radiation for osteolytic acetabular lesion involving the posterior column. (**e**) Intraoperative CT demonstrating process of hydrodissection to move sciatic nerve away from the lesion area. After hydrodissection, a temperature probe was placed between the lesion and the sciatic nerve to monitor temperatures during treatment. CT was used to monitor cryoablation ice-ball formation. (**f**) Guidewire placed along the posterior column using augmented fluoroscopy along a preplanned pathway. (**g**) After fully seating the screw, a coaxial cement cannula was advanced to the treatment area with 3D volumetric object outline (white lines) of the lytic defect to help assess cementation fill. (**h**) Obturator oblique augmented fluoroscopic view of cementation of lytic defect demonstrating good cement fill of the volumetric object outline

placement of all fixation devices and then using multiple independent coaxial tracks with bone trocars to attain adequate cement fill of larger lesions. The authors typically use AutoPlex with Vertaplex HV cement (Stryker Corporation, Kalamazoo, Michigan). The authors use Stryker Asnis III Cannulated Screws 6.5/8.0 mm or the Synthes Cannulated Screw Set (West Chester, PA). In the case of photodynamic nail placement (IlluminOss, Photodynamic Bone Stabilization System, IlluminOss

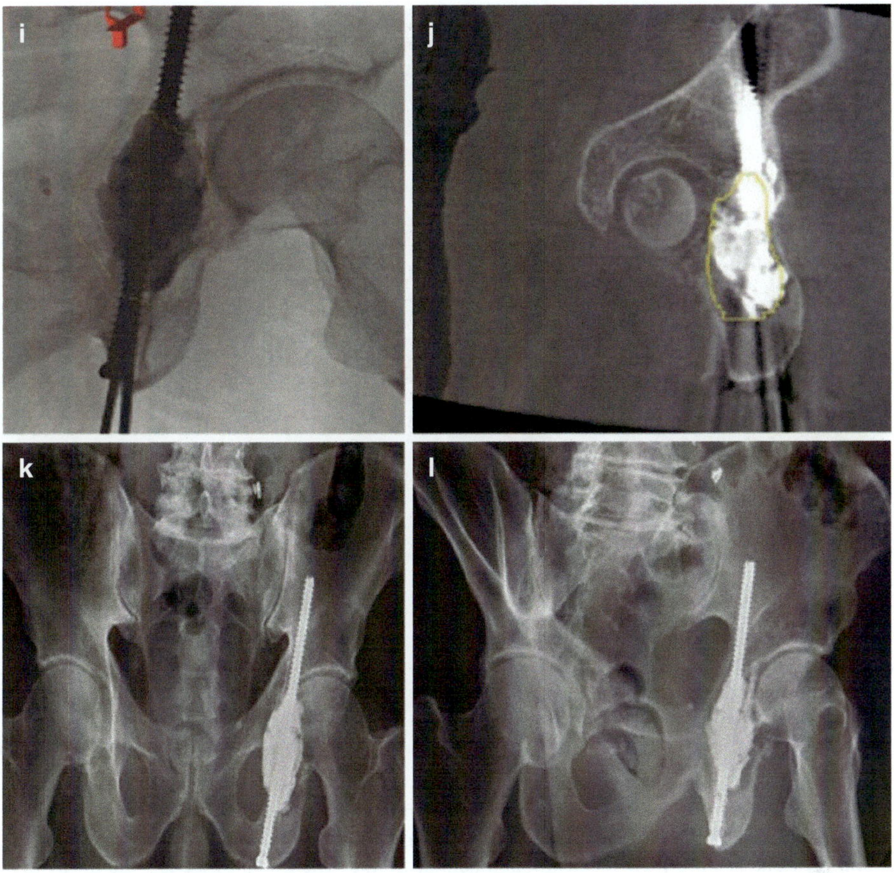

Fig. 36.5c 72M with metastatic renal cell carcinoma with persistent left hip pain after radiation for osteolytic acetabular lesion involving the posterior column. (**i**) Iliac oblique augmented fluoroscopic view of cementation of lytic defect demonstrating good cement fill of the volumetric object outline. (**j**) Postoperative CT scan demonstrating final treatment construct of lytic lesion of the posterior pelvic column. (**k**) AP pelvis plain radiograph for 12-month follow-up after treatment of lytic lesion of posterior column. (**l**) Iliac oblique plain radiograph for 12-month follow-up after treatment of lytic lesion of posterior column

Medical, Inc., East Providence, RI, USA), a 2.5 mm Synthes reaming ball-tipped guidewire is placed along the preplanned path, and then sequential reaming is performed to create space for delivery of the implant. The balloon insertion catheter is placed and inflated with monomer resin to fill the tumor volume which is then cured with light. It is recommended that all guidewires (including guidewires for screws) are placed into the pelvis prior to curing the monomer as it can be difficult to accurately place them through the hardened resin. If using a combination of screws and photodynamic nails, it is recommended that screws be drilled and placed after hardening of the resin as this reduces screw pull out. Final cone beam CT should be performed to confirm appropriate implant positioning and cement fill.

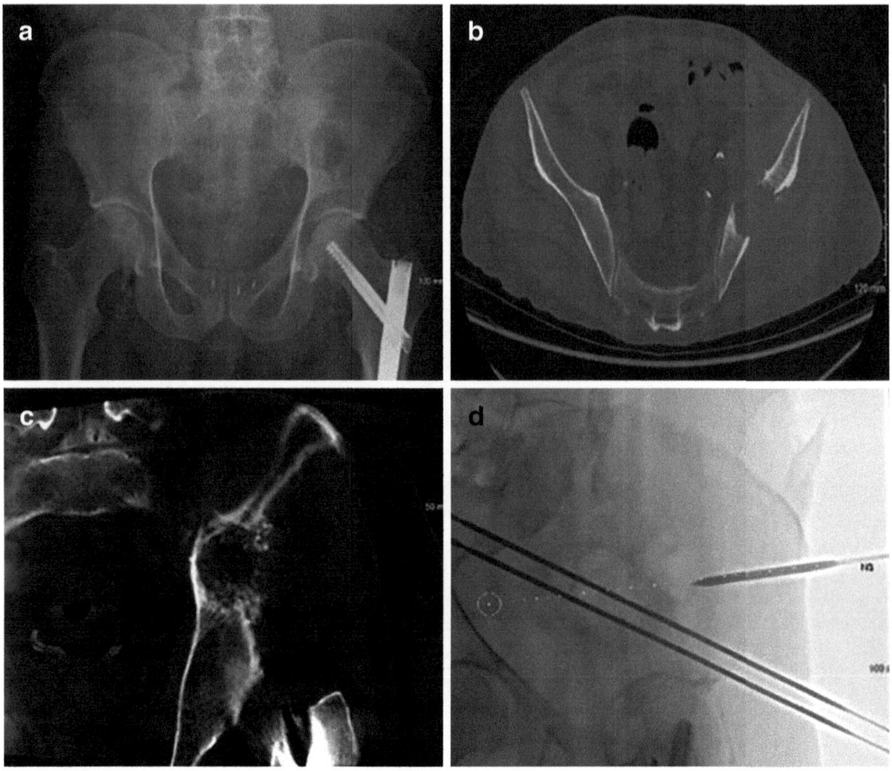

Fig. 36.6a 70M with metastatic lung cancer with left iliac lesion along the sciatic buttress. (**a**) AP pelvis plain radiograph with lytic lesion of left ilium. (**b**) Axial CT scan with lytic lesion of left ilium. (**c**) Cone beam CT 3D reconstruction demonstrating lytic lesion of the left ilium. (**d**) Iliac oblique augmented fluoroscopic demonstrating targeted guidewire placement for screw (white circle) and previously placed guidewires for two photodynamic nails (inferior parallel)

Percutaneous interventions for skeletal metastases can be performed in either the prone or supine position, each offering distinct advantages and disadvantages. Positioning considerations are most indicated for pelvic lesions. Performing the procedure in the prone position with the hips slightly flexed enhances access to posterior column pelvic fixation pathway in the ischium without requiring additional patient movement. In addition, a posterior to anterior screw can be placed in the transcolumnar pathway in the prone position. This can also be done with the patient positioned supine, but intraoperative hip flexion is needed to access to the ischium, or ilium, respectively. However, prone positioning may complicate respiratory function and patient comfort, particularly in those with existing comorbidities or lumbar spine issues, compromised cardiovascular status due to their underlying condition, or pain with prone positioning. Conversely, the supine position generally allows for easier monitoring of vital signs and airway access, but it may limit access to the ischium depending on patient habitus and the need to manipulate the patient's

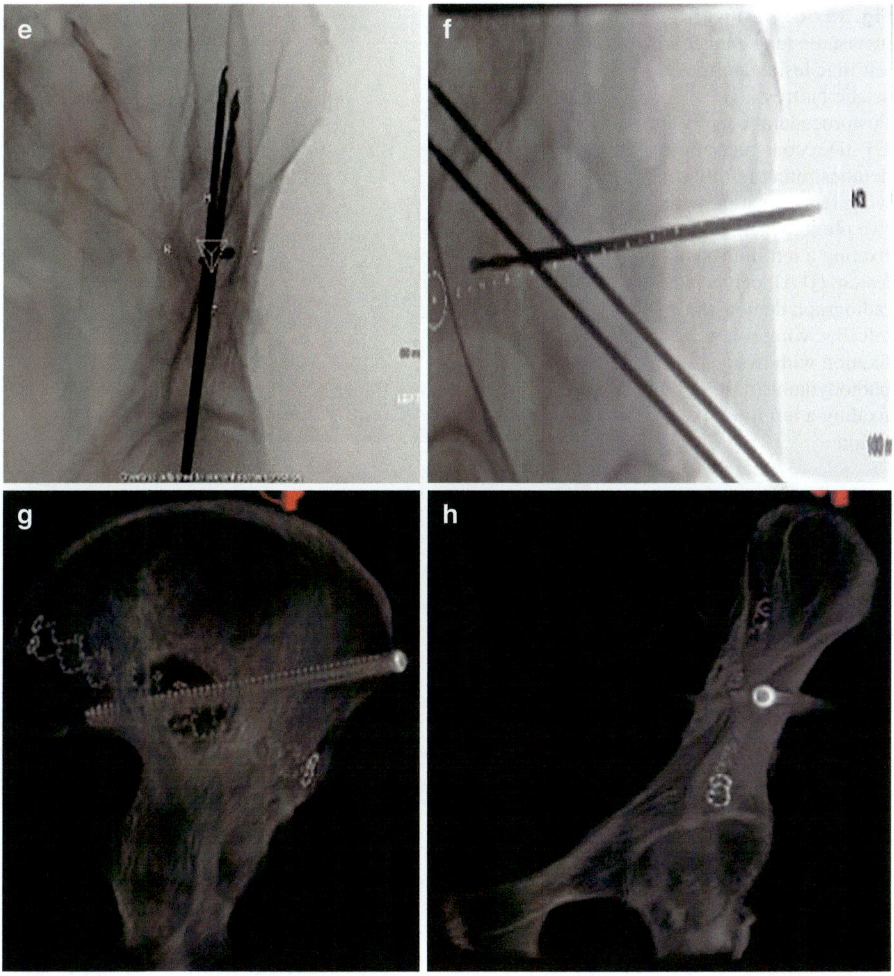

Fig. 36.6b 70M with metastatic lung cancer with left iliac lesion along the sciatic buttress. (**e**) Obturator outlet augmented fluoroscopy demonstrating end point targeting (white triangle) of iliac wing screw fixation from ASIS to posterior medial ilium. (**f**) Iliac oblique augmented fluoroscopic demonstrating targeted guidewire placement for screw (white circle) and previously placed guidewires for two photodynamic nails (inferior parallel). (**g**) Postprocedural cone beam CT 3D reconstruction demonstrating left iliac wing screw fixation with two photodynamic nails fixating a left ilium lytic lesion. (**h**) Postprocedural cone beam CT 3D reconstruction demonstrating left iliac wing screw fixation with two photodynamic nails fixating a left ilium lytic lesion

lower extremity. Ultimately, the choice of position should consider the individual patient's condition, the extent of metastases, and the surgeon's expertise to optimize outcomes and minimize complications.

Similarly, interventions for metastatic disease may be performed under local, general, regional, combined, or monitored anesthesia care depending on the

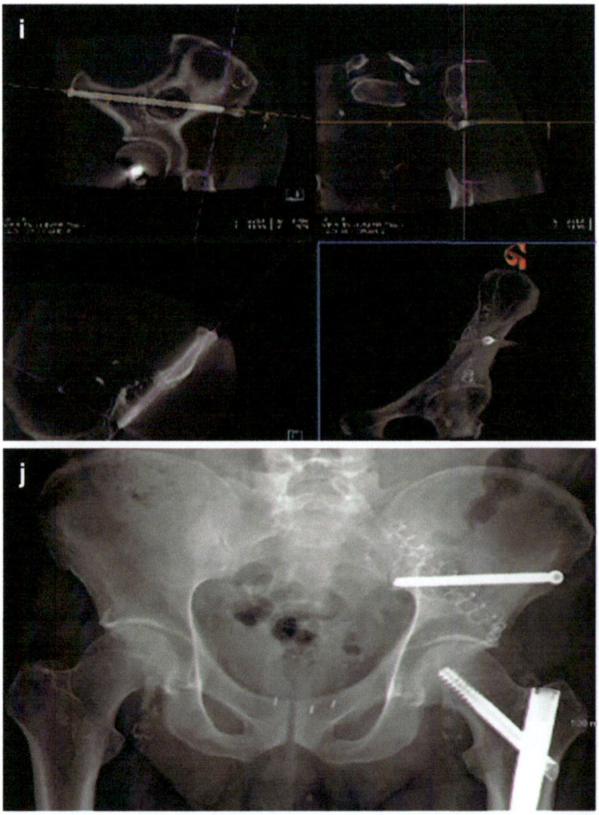

Fig. 36.6c 70M with metastatic lung cancer with left iliac lesion along the sciatic buttress. (**i**) Postprocedural cone beam CT 3D reconstruction demonstrating left iliac wing screw fixation with two photodynamic nails fixating a left ilium lytic lesion. (**j**) AP pelvis plain radiograph demonstrating left iliac wing screw fixation with two photodynamic nails fixating a left ilium lytic lesion

extent of the indicated procedures and medical comorbidities [35, 36]. We favor general anesthesia with a large burden of metastatic disease requiring multiple interventional modalities and if the prone position is indicated to facilitate bony access. When using cone beam CT, patient movement should be limited to prevent misregistration of the navigated pathways which favors general anesthesia. Anesthesia providers should also consider the patient's positioning, particularly if the procedure is performed in the prone position, which can affect respiratory mechanics and venous return. If there is limited burden of metastatic disease not requiring extensive intervention and the patient is relatively healthy, monitored anesthesia care and enhanced recovery protocols with day surgery could be considered.

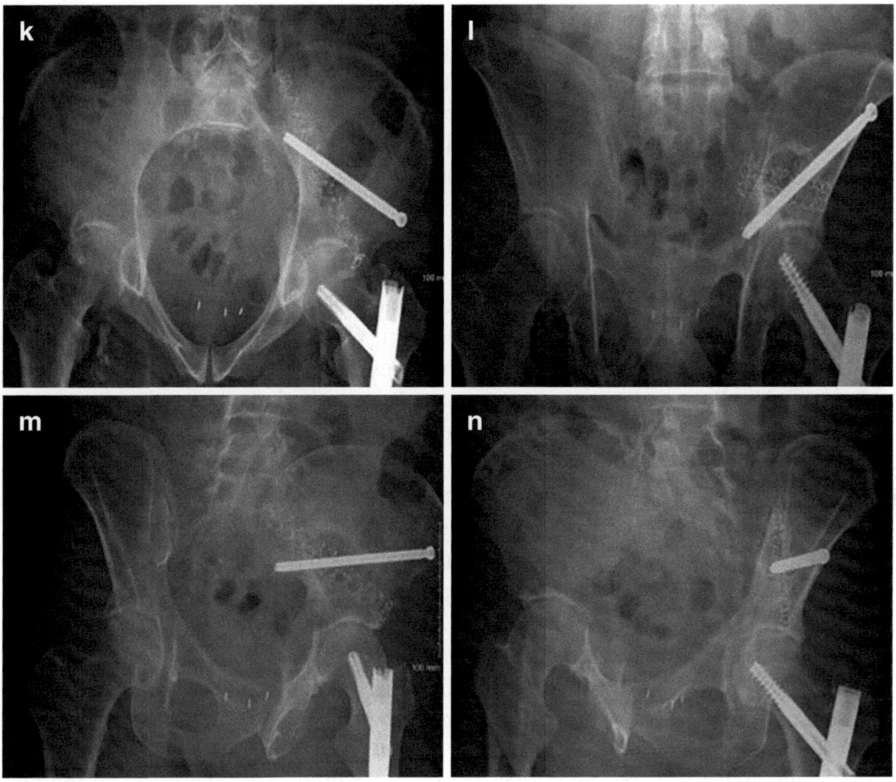

Fig. 36.6d 70M with metastatic lung cancer with left iliac lesion along the sciatic buttress. (**k**) Pelvic inlet plain radiograph demonstrating left iliac wing screw fixation with two photodynamic nails fixating a left ilium lytic lesion. (**l**) Pelvic outlet plain radiograph demonstrating left iliac wing screw fixation with two photodynamic nails fixating a left ilium lytic lesion. (**m**) Iliac oblique plain radiograph demonstrating left iliac wing screw fixation with two photodynamic nails fixating a left ilium lytic lesion. (**n**) Obturator oblique plain radiograph demonstrating left iliac wing screw fixation with two photodynamic nails fixating a left ilium lytic lesion

Technical Pearls

Several important considerations must be considered during ablation procedures. If a neurovascular bundle is adjacent to the treatment area, pretreatment with Decadron (with caution in insulin-dependent diabetics) is advised, along with the use of hydro and CO2 dissection or balloon displacement techniques and nerve monitoring. There is also the potential for significant disruption of articular cartilage, which could lead to arthritis; in these cases, joint irrigation may be an option. Finally, the skin integrity over the lesion should be assessed, and hydro-dissection may be employed.

In the setting of osteoplasty with cement, the authors favor more aggressive use of fixation devices (screw or photodynamic nails) at the index intervention if there is the possibility of metastatic disease progression or if osteoplasty is likely to fail to control symptoms.

This is because attempting screw or photodynamic nail placement through an area previously treated with cement osteoplasty is difficult and inaccurate, thereby increasing risk to the patient if lesions are near critical structures [23]. If there is significant cortical and cancellous bone destruction due to metastatic disease, we recommend using screw fixation or photodynamic nail in isolation after ablative therapy due to the risk of cement leakage into the surrounding soft tissues. Surgeons should attempt to obtain bicortical fixation with screw placement, but when this is not feasible, screws may be potted in cement to limit motion. A Foley catheter for bladder decompression may be needed if there are interventions needed along the superior pubic ramus or sacrum.

Often ablative therapies are used near critical structures. To avoid complications, we recommend application of hydrodissection, CO2, or utilize balloon displacement to move the skin, neurovascular structures or organs away from the planned ablative fields. Percutaneous temperature probes can be independently placed near critical structures to monitor treatment, mainly during CA due to electrical interference with the other modalities.

It is essential to understand the limitations of each respective ablative modality to promote local disease control and palliation of pain. In the setting of CA, targeted tissue must reach a minimum of −20 °C to be lethal. The edge of the ice ball is not necessarily causing tissue death as it is approximately 0 °C. Therefore, it is generally accepted that an ablation margin of 5 mm is used to ensure lethality. This is useful when planning multiple overlapping ablative fields in larger lesions.

In the setting of displaced pathologic fracture, we favor utilizing open reduction with the addition of ablative therapies as percutaneous stabilization may fail to control symptoms without adequate fracture reduction. This is especially important in periarticular pathologic fractures such as the acetabulum or the shoulder, among others. Joint arthroplasty or endoprosthetic reconstruction in conjunction with ablative techniques depending on patient age, function and life expectancy can be considered.

In patients with extensive bone loss of the posterior column due to metastatic disease, there is lack of bony protection for the sciatic nerve. Percutaneous procedures with application of cement to fill bone defects, care must be taken while using imaging to monitor for cement extrusion outside of the defect as this may damage to the sciatic nerve [34]. This is also critical with the anterior column with the femoral or obturator neurovascular bundle.

If there is intra articular extravasation during cement application, it is recommended to perform passive range of motion of the joint to smooth any potential abrasive surfaces. Delayed joint arthroscopy can be considered for symptomatic intra articular cement extravasation. If ablative therapies are used near joint surfaces and cartilage, intra articular irrigation to limit thermal necrosis can be considered.

Postoperative Care

Postoperative care after percutaneous management of metastases should favor rapid progression of the patient's mobility to limit the effects of deconditioning. We favor allowing these patients to weight bear as tolerated with a gait aid. Immediate improvement in pain is not uncommon due to increased stability across the lesion. Sometimes, there is a postablative inflammatory response, most commonly with MWA, that can be treated with oral pain medications and is self-limiting.

Patients are seen at two weeks postoperatively for an incision check, evaluate their mobility progress, and plain radiographs. Regular follow up is indicated to make sure the patient is progressing in their mobility and having reduced pain over time. Patients can usually return to chemotherapy or radiation once the small incisions have healed within two weeks, although it is not uncommon to immediately return to life prolonging and saving treatments based on surgeon preference. Chemical deep venous thrombosis prophylaxis should be considered in pelvic or lower extremity metastatic disease due to the impact on mobility with this high-risk patient population for thrombosis with any intervention.

After ablative therapies, it is critical to identify and manage potential complications. High dose steroids can be considered in patients with neurologic deficits following treatment due to the potential for postoperative edema compressing nearby nerves. If ablation was performed near neurovascular structures and neurologic deficits persist, watchful waiting is recommended as most deficits have some degree of recovery. Persistent neuropathy with motor deficits can be managed with pain medication, orthoses, assistive devices, and physical therapy with nerve conduction studies performed no later than 3 months to characterize the injury. Failure of skeletal fixation with disease recurrence or progression can be treated with major reconstruction or additional osteoplasty with further ablative therapies (Fig. 36.7).

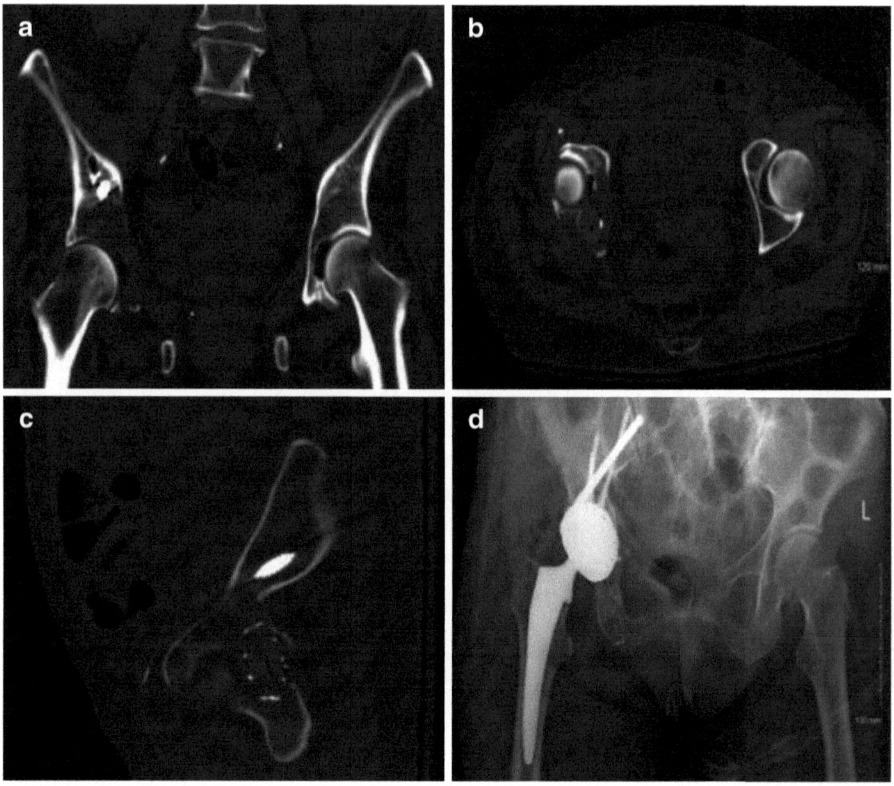

Fig. 36.7 74F with metastatic lung cancer with progressive painful right periacetabular radiolucent lesion status post screw and photodynamic nail fixation. (**a**) Coronal CT scan demonstrating progressive radiolucent lesion of the posterior wall and column of the right acetabulum. (**b**) Axial CT scan demonstrating progressive radiolucent lesion of the posterior wall and column of the right acetabulum. (**c**) Sagittal CT scan demonstrating progressive radiolucent lesion of the posterior wall and column of the right acetabulum. (**d**) AP pelvis plain radiograph status post right anterior total hip arthroplasty with additional periprosthetic screw fixation

References

1. Ashford RU, Benjamin L, Pendlebury S, Stalley PD. The modern surgical and non-surgical management of appendicular skeletal metastases. Orthop Trauma. 2012;26:184–99. https://doi.org/10.1016/j.mporth.2012.02.002.
2. Arneson TJ, Shuling Li, Yi Peng, Weinhandl ED, Blaes A, Cetin K, et al. Estimated number of prevalent cases of metastatic bone disease in the US adult population. CLEP. 2012:87. https://doi.org/10.2147/CLEP.S28339.
3. Papagelopoulos PJ, Mavrogenis AF, Soucacos PN. Evaluation and treatment of pelvic metastases. Injury. 2007;38:509–20. https://doi.org/10.1016/j.injury.2007.01.008.
4. Hage W, Aboulafia A, Aboulafia D. Incidence, location, and diagnostic evaluation of metastatic bone disease. Orthop Clin North Am. 2000;31:515–28.
5. Bagi C. Targeting of therapeutic agents to bone to treat metastatic cancer. Adv Drug Deliv Rev. 2005;57:995–1010. https://doi.org/10.1016/j.addr.2004.12.014.

6. Rosenthal D, Alexander A, Rosenberg A, Springfield D. Ablation of osteoid osteomas with a percutaneously placed electrode: a new procedure. Radiology. 1992;183:29–33. https://doi.org/10.1148/radiology.183.1.1549690.
7. De Filippo M, Russo U, Papapietro VR, Ceccarelli F, Pogliacomi F, Vaienti E, et al. Radiofrequency ablation of osteoid osteoma. Acta Biomedica Atenei Parmensis. 2018;89:175–85. https://doi.org/10.23750/abm.v89i1-S.7021.
8. Ahmed M, Solbiati L, Brace CL, Breen DJ, Callstrom MR, Charboneau JW, et al. Image-guided tumor ablation: standardization of terminology and reporting criteria—a 10-year update. J Vasc Interv Radiol. 2014;25:1691–1705.e4. https://doi.org/10.1016/j.jvir.2014.08.027.
9. Foley KM. Abstract: treatment of cancer-related pain. J Natl Cancer Inst Monogr. 2004;2004:103–4. https://doi.org/10.1093/jncimonographs/lgh034.
10. Falk S, Dickenson AH. Pain and nociception: mechanisms of cancer-induced bone pain. JCO. 2014;32:1647–54. https://doi.org/10.1200/JCO.2013.51.7219.
11. NCCN clinical practice guidelines in oncology: adult cancer pain. National Comprehensive Cancer Network 2024;3.
12. Levy J, Hopkins T, Morris J, Tran ND, David E, Massari F, et al. Radiofrequency ablation for the palliative treatment of bone metastases: outcomes from the multicenter OsteoCool tumor ablation post-market study (OPuS one study) in 100 patients. J Vasc Interv Radiol. 2020;31:1745–52. https://doi.org/10.1016/j.jvir.2020.07.014.
13. Levy J, David E, Hopkins T, Morris J, Tran ND, Farid H, et al. Radiofrequency ablation provides rapid and durable pain relief for the palliative treatment of lytic bone metastases independent of radiation therapy: final results from the OsteoCool tumor ablation post-market study. Cardiovasc Intervent Radiol. 2023;46:600–9. https://doi.org/10.1007/s00270-023-03417-x.
14. Lubner MG, Brace CL, Hinshaw JL, Lee FT. Microwave tumor ablation: mechanism of action, clinical results, and devices. J Vasc Interv Radiol. 2010;21:S192–203. https://doi.org/10.1016/j.jvir.2010.04.007.
15. Hinshaw JL, Lubner MG, Ziemlewicz TJ, Lee FT, Brace CL. Percutaneous tumor ablation tools: microwave, radiofrequency, or cryoablation—what should you use and why? Radio Graphics. 2014;34:1344–62. https://doi.org/10.1148/rg.345140054.
16. Marcove R, Miller T. The treatment of primary and metastatic localized bone tumors by cryosurgery. Surg Clin North Am. 1969;49:421–30. https://doi.org/10.1016/s0039-6109(16)38799-0.
17. Gage A. History of cryosurgery. Semin Surg Oncol. 1998;14:99–109. https://doi.org/10.1002/(sici)1098-2388(199803)14:2<99::aid-ssu2>3.0.co;2-1.
18. Cooper SM, Dawber RPR. The history of cryosurgery. J R Soc Med. 2001;94:196–201. https://doi.org/10.1177/014107680109400416.
19. Mauri G, Orsi F, Sconfienza L. Systemic effects of local tumor ablation: oncogenesis and antitumor induced immunity. Radiology. 2016;279:322–3. https://doi.org/10.1148/radiol.2016151739.
20. Jennings JW, Prologo JD, Garnon J, Gangi A, Buy X, Palussière J, et al. Cryoablation for palliation of painful bone metastases: the MOTION multicenter study. Radiol Imag Cancer. 2021;3:e200101. https://doi.org/10.1148/rycan.2021200101.
21. Çevik HB, Ruggieri P, Giannoudis PV. Management of metastatic bone disease of the pelvis: current concepts. Eur J Trauma Emerg Surg. 2023; https://doi.org/10.1007/s00068-023-02382-x.
22. Callstrom MR, Dupuy DE, Solomon SB, Beres RA, Littrup PJ, Davis KW, et al. Percutaneous image-guided cryoablation of painful metastases involving bone: multicenter trial. Cancer. 2013;119:1033–41. https://doi.org/10.1002/cncr.27793.
23. Lea WB, Neilson JC, King DM, Tutton SM. Minimally invasive stabilization using screws and cement for pelvic metastases: technical considerations for the pelvic "screw and glue" technique. Semin Intervent Radiol. 2019;36:229–40. https://doi.org/10.1055/s-0039-1693982.
24. LELE P. A simple method for production of trackless focal lesions with focused ultrasound: physical factors. J Physiol. 1962;160:494–512. https://doi.org/10.1113/jphysiol.1962.sp006862.

25. Smith N, Temkin J, Shapiro F, Hynynen K. Thermal effects of focused ultrasound energy on bone tissue. Ultrasound Med Biol. 2001;27:1427–33. https://doi.org/10.1016/s0301-5629(01)00454-9.
26. Liberman B, Gianfelice D, Inbar Y, Beck A, Rabin T, Shabshin N, et al. Pain palliation in patients with bone metastases using MR-guided focused ultrasound surgery: a multicenter study. Ann Surg Oncol. 2009;16:140–6. https://doi.org/10.1245/s10434-008-0011-2.
27. Harding D, Giles SL, Brown MRD, Ter Haar GR, Van Den Bosch M, Bartels LW, et al. Evaluation of quality of life outcomes following palliative treatment of bone metastases with magnetic resonance-guided high intensity focused ultrasound: an international multicentre study. Clin Oncol. 2018;30:233–42. https://doi.org/10.1016/j.clon.2017.12.023.
28. Bitton RR, Rosenberg J, LeBlang S, Napoli A, Meyer J, Butts Pauly K, et al. MRI-guided focused ultrasound of osseous metastases: treatment parameters associated with successful pain reduction. Investig Radiol. 2021;56:141–6. https://doi.org/10.1097/RLI.0000000000000721.
29. Hurwitz MD, Ghanouni P, Kanaev SV, Iozeffi D, Gianfelice D, Fennessy FM, et al. Magnetic resonance–guided focused ultrasound for patients with painful bone metastases: Phase III trial results. JNCI J Natl Cancer Inst. 2014:106. https://doi.org/10.1093/jnci/dju082.
30. Lieberman IH, Togawa D, Kayanja MM. Vertebroplasty and kyphoplasty: filler materials. Spine J. 2005;5:S305–16. https://doi.org/10.1016/j.spinee.2005.02.020.
31. English DI, Lea WB, King DM, Tutton SM, Neilson JC. Minimally invasive stabilization with or without ablation for metastatic periacetabular tumors. J Bone Joint Surg. 2021;103:1184–92. https://doi.org/10.2106/jbjs.20.00546.
32. Muir J, Vegt P, Block J. The photodynamic bone stabilization system: a minimally invasive, percutaneous intramedullary polymeric osteosynthesis for simple and complex long bone fractures. MDER. 2014:453. https://doi.org/10.2147/MDER.S71790.
33. Lozano-Calderon SA, Clunk MJ, Gonzalez MR, Sodhi A, Krueger RK, Gruender AC, et al. Assessing pain and functional outcomes of percutaneous stabilization of metastatic pelvic lesions via photodynamic nails: a bi-institutional investigation of Orthopaedic outcomes. JBJS Open Access. 2024;9 https://doi.org/10.2106/JBJS.OA.23.00148.
34. Mastrangelo S, Granruth C, Cipriano C. The Harrington reconstruction for patients with advanced periacetabular metastatic disease. Oncology. 2023;33:71–5.
35. Piccioni F, Poli A, Templeton LC, Templeton TW, Rispoli M, Vetrugno L, et al. Anesthesia for percutaneous radiofrequency tumor ablation (PRFA): a review of current practice and techniques. LRA. 2019;12:127–37. https://doi.org/10.2147/LRA.S185765.
36. Dabu-Bondoc SM. Standard procedures in nonoperating room anesthesia. Curr Opin Anaesthesiol. 2020;33:539–47. https://doi.org/10.1097/ACO.0000000000000898.

Radiation Strategies in Oligometastatic Disease

37

Armando Vera, Lauren Nagel, and Raphael Yechieli

Introduction

The landscape of cancer treatment has evolved significantly, particularly in managing bone metastases. Bone metastases occur when cancer cells from a primary tumor disseminate to the bone, establishing secondary malignant growths. Bone metastases are characterized by an imbalance in the normal bone remodeling process, mediated by interactions between cancer cells, osteoclasts (bone-resorbing cells), and osteoblasts (bone-forming cells). Depending on the nature of these interactions, bone metastases can be osteolytic, where there is increased bone resorption leading to bone destruction, or osteoblastic, where there is increased bone formation resulting in abnormal bone growth. Common signaling pathways involved include the RANK/RANKL/OPG axis, various cytokines (e.g., IL-1, IL-6) and growth factors such as TGF-β and PTHrP [1–6]. The presence of bone metastases indicates advanced disease and can lead to significant clinical complications, including severe pain, pathological fractures, spinal cord compression, and hypercalcemia, all of which necessitate prompt and effective management.

A. Vera
Jackson Health system, University of Miami, Miami, FL, USA
e-mail: axv1117@miami.edu; https://www.radonc.med.miami.edu

L. Nagel
University of Miami, Miami, FL, USA
e-mail: lan109@miami.edu

R. Yechieli (✉)
Sylvester Cancer Center, University of Miami, Miami, FL, USA
e-mail: ryechieli@med.miami.edu

J. Pretell-Mazzini, S. A. Lozano Calderón (eds.), *Multidisciplinary Approach of Musculoskeletal Metastasis*, https://doi.org/10.1007/978-3-032-09498-8_37

Overview of Bone Metastasis

Bone metastases are a common complication of advanced cancers, particularly those originating from the breast, prostate, and lung. These metastases result from a complex interplay of factors that promote cancer cell migration, colonization and survival in the bone. The interactions between cancer cells and the bone microenvironment facilitate tumor growth and bone destruction, leading to significant morbidity.

Oligometastasis

Oligometastasis describes a state of cancer spread where a limited number of metastatic lesions are present, typically defined as up to five metastases in one or a few organs [7]. This concept lies between localized disease and widespread metastatic disease. The hypothesis behind oligometastasis is that the metastatic spread is not yet fully established, and the cancer cells have limited ability to disseminate and colonize multiple sites [8]. Scientifically, oligometastasis is characterized by a lower tumor burden and potentially slower tumor kinetics compared to polymetastatic disease. It suggests a biological heterogeneity within metastatic cancer, where some metastatic cells have not yet acquired the full suite of mutations and adaptations needed for extensive dissemination and colonization [9]. This intermediate state presents a therapeutic window where localized treatments, such as surgical resection, stereotactic body radiation therapy (SBRT), or ablative therapies, can be curative or significantly prolong survival [10]. The biological basis of oligometastasis involves genetic and epigenetic factors that constrain the metastatic potential of cancer cells. Studies have shown that oligometastatic lesions exhibit distinct molecular profiles compared to widely metastatic lesions, with differences in gene expression related to cell adhesion, motility, and immune evasion [11]. Understanding these molecular characteristics is crucial for developing targeted therapies and optimizing treatment strategies for patients with oligometastatic disease [8]. At present, there is no standardized definition. So far, we have defined solid tumors with a certain specific number of lesions, size of lesions, and occasionally histology (specifically prostate, NSCLC, RCC, and breast cancer) [10, 12–15].

Other Definitions

Synchronous oligometastatic disease refers to metastases present at the time of diagnosis. Metachronous oligometastatic disease refers to metastases present after primary treatment, also known as "oligorecurrent disease." Oligoprogressive disease describes metastatic disease that is controlled on systemic therapy except in a limited number of areas.

Management and Prognosis

Management of metastatic disease is rapidly changing. Novel systemic treatments such as immunotherapy and targeted agents have improved patient prognosis and tolerability of therapy. In conjunction with novel systemic therapies, the role of local therapies in metastatic disease is also rapidly evolving. Once thought to play little role beyond palliation of symptoms, local therapies, especially radiation in the form of stereotactic ablative radiation (SABR), are increasingly being used to aggressively manage patients with metastatic disease [16]. Much of this is driven by the spectrum theory of metastasis, a postulation that cancer represents a spectrum of diseases composed of many states, some of which behave more akin to a locoregional process rather than a widely systemic disease [17], meaning that such lesions might benefit from metastasis-directed therapies (MDTs) [18]. In support of such a theory, several prospective trials have now demonstrated that MDT in oligometastases can prolong time to disease progression, initiation of systemic therapy, and even overall survival [19].

Radiation Therapy for Bone Metastases

Conventional radiation therapy remains a cornerstone in the treatment of bone metastases, tailored based on tumor histology and patient factors. It typically involves standard fractionation schemes suitable for patients with a life expectancy of less than three months, poor performance status, and poorly controlled systemic disease. Common fractionation schedules include 30 Gy in 10 fractions, 20 Gy in 5 fractions, and 8 Gy in one fraction. Research has shown no significant difference in effectiveness, overall survival, pain response, or fracture rates among these different fractionation schedules [20].

However, single-fraction treatments tend to have higher retreatment rates (25% vs. 7%) due to a higher incidence of pain recurrence [21, 22]. For patients with diffuse bone pain, large-field or hemibody irradiation, or radiopharmaceuticals like samarium-153 may be considered. Additionally, radiation therapy is often used postfixation in patients with fractures or impending fractures in weight-bearing bones to enhance pain relief and stabilize the skeleton. All these are used for palliative intent.

Lessons from Conventional Radiation Therapy (RT)

Lessons from conventional radiation therapy (RT) include observations that higher baseline pain levels are associated with lower rates of complete pain response, and that increased fractions often result in more durable pain relief. For instance, a regimen of 30 Gy delivered in 10 fractions equates to a BED10 of 39 Gy, whereas 8 Gy delivered in a single fraction yields a BED10 of 14.40 Gy. However, even with these regimens, only 61% of patients typically experience pain relief at one month, highlighting the challenges in achieving consistent dose–response relationships. While there is a recognized dose threshold for pain relief, the exact dose–response remains unclear. Theoretical benefits from dose escalation must be balanced with the need to avoid exceeding spinal cord tolerance [23–25].

Stereotactic Body Radiation Therapy (SBRT) for Oligometastatic Disease

Oligometastatic disease presents both parallel and linear models of metastatic progression, often complicated by intratumor heterogeneity (ITH) observed in preclinical and genomic sequencing data. These models describe how a single metastasis can seed others, suggesting that a few metastases, over time, can lead to a rapid emergence of numerous metastatic sites. A therapeutic opportunity exists for ablative local metastasis-directed therapy (MDT) to potentially control systemic disease. In the linear model, the primary tumor evolves aggressively, generating genetically distinct metastases later in the disease course. Conversely, the parallel model indicates early dissemination of metastases independent of the primary tumor's progression. Analysis from the Cancer Genome Atlas Projects across 12 cancer types highlighted that 86% of primary tumors exhibit at least two progenitor cells, underscoring the role of ITH in disease progression and poor outcomes. For instance, in colorectal cancer (CRC), phylogenetic analysis of a patient with multiple metastatic sites demonstrated that bone and liver metastases initially seeded subsequent metastases in the liver and bones, followed by involvement of the lung and brain [26].

Stereotactic Body Radiation Therapy (SBRT)

Stereotactic body radiation therapy (SBRT) presents an alternative approach. Studies have demonstrated its feasibility, with reported setup and targeting accuracies as precise as 1.5 mm since 2003. Effectiveness is also notable, with median pain control durations of 13.3 months and high rates of local control reaching 95%. For example, delivering 20 Gy over five fractions results in a BED10 of 28 Gy. This approach has been showing improved quality of life [27] and extended survival in oligometastatic patients [28, 29].

Results from a phase II German randomized controlled trial (RCT) published in 2018 [30] further underscored these benefits, with complete pain response rates observed in 52% of SBRT patients at 6 months compared to 10% in conventional EBRT patients ($p = 0.0034$). Overall, 73.7% of SBRT patients showed any pain response compared to 35% in the EBRT group ($p = 0.0154$). SBRT has significantly improved treatment outcomes by delivering cytotoxic doses precisely to tumors [31]. This method allows for higher biologic effective doses and induces lethal double-stranded DNA breaks, endothelial dysfunction, and immunogenic responses [32, 33]. Technological advancements in patient immobilization, imaging modalities, and delivery systems have made SBRT safer and more effective [34, 35].

Principles and Advantages of SBRT

Stereotactic body radiation therapy (SBRT) employs high doses per fraction to target small volumes with remarkable precision, creating steep dose gradients that conform closely to tumor shapes. This precision is achieved through smaller internal target volumes (ITV) and improved dose homogeneity, supported by advanced image guidance and rigid patient immobilization techniques. Key parameters such as the Conformity Index (CI), or R100, measure the ratio of the volume receiving the prescription dose to the planning target volume (PTV), with desirable values generally less than 1.2. Additional metrics like R50, which assesses the ratio of the volume receiving half the prescription dose to the PTV, and D 2 cm, which evaluates the highest dose received within 2 cm from the PTV relative to the prescription dose, ensure rigorous treatment accuracy [36].

SBRT offers numerous advantages, including improved pain response for metastatic lesions and enhanced local control in tumors where local failure is critical. It is particularly effective in treating nonsmall cell lung cancer (NSCLC), prostate cancer, and renal cell carcinoma (RCC), with ongoing Phase III trials further validating its benefits. In prostate cancer, treating the primary tumor with SBRT has been associated with improved survival outcomes. For oligoprogressive disease, SBRT can delay the initiation of systemic therapies by effectively managing progressing lesions, enabling continued tolerance of ongoing systemic treatments. Additionally, SBRT holds potential for inducing an enhanced systemic response, known as the abscopal effect, although current evidence for this phenomenon is largely anecdotal or based on preclinical studies [36].

Clinical Considerations, Patient Selection, and Prognostic Factors

A retrospective study spanning 2004–2010 evaluated outcomes in 104 patients with renal cell carcinoma (RCC), more than half of whom presented with spinal metastases. Treatment consisted of either a single fraction of 24 Gy (yielding a BED10 of 81.6 Gy and BED6 of 120 Gy) or five fractions totaling 30 Gy (BED10 of 48 Gy and BED6 of 60 Gy). The results demonstrated significantly better 3-year local recurrence-free survival (LRFS) rates with the single-dose regimen (88%) compared to fractionated therapy (17%), suggesting that higher single-dose SBRT may provide superior local control in RCC spinal lesions [37].

Eligibility for aggressive treatments such as SBRT in oligometastatic disease depends on high performance status and limited comorbidities. Beyond merely counting metastases, patient selection should integrate clinical factors such as the interval from initial cancer diagnosis to metastasis development, as extended intervals are associated with improved disease-free survival (DFS) and overall survival (OS) following metastasis-directed therapy (MDT) [37].

Emerging biomarkers like microRNAs (miRNAs) may further refine patient selection. These noncoding RNA molecules regulate gene expression at the posttranscriptional level and hold promise for predicting treatment response and outcomes in MDT, offering potential to replicate favorable local control outcomes across broader populations [38].

Prognostic factors such as Distant Metastasis Velocity (DMV) have also proven valuable in assessing outcomes after SBRT. DMV, categorized as <0.5, 0.5–1.5, and >1.5 metastases per month, is a robust independent predictor of OS. Patients with lower DMV exhibit longer median OS, with durations of 37.1, 26.7, and 16.8 months, respectively, after SBRT treatment. Additionally, those with oligometastatic disease at recurrence and longer intervals between SBRT and subsequent recurrence experience significantly improved OS, emphasizing the importance of timing and disease progression kinetics in patient selection and treatment planning [39].

Bone SBRT

Between 2004 and 2007, Erler et al. [40] treated 126 metastases in 103 patients using single-dose stereotactic body radiation therapy (SBRT). The study showed significant improvements in 18-month local recurrence-free survival (LRFS) rates: 82% for doses greater than 22 Gy, 69% for doses between 21 and 22 Gy, and 25% for doses between 18 and 20 Gy ($p < 0.0001$ and $p = 0.04$, respectively). The dose–response was notable for renal cell carcinoma (RCC), with only 37% LRFS for doses below 22 Gy. There were no significant differences observed based on histology, target organ, or target size. Interestingly, colorectal cancer (CRC) patients, primarily with liver metastases, exhibited poorer LRFS rates with 50% at 18 months. The biologically effective dose (BED10) for a single fraction of 22 Gy was calculated as 70.4 Gy, and for 60 Gy in 3 fractions, it was 180 Gy. Nonhepatic metastases showed increased response rates compared to hepatic metastases in this cohort [41]. A planning target volume (PTV) greater than 37.89 cc is associated with worse local control (LC) outcomes. This association is observed independently of factors such as anatomical location, histology, type of bone metastasis, biologically effective dose (BED), and treatment duration [40]. The side effects are related to the treatment site, with overall grade 3 late toxicity less than 4%. For this particular study, there was a grade III esophageal stricture.

For pain relief, SBRT offers significant benefits. A phase 2 randomized controlled trial involving 160 patients compared different radiation regimens: 12 Gy or 16 Gy delivered in a single fraction versus 30 Gy given over 10 fractions. This noninferiority trial aimed for 90% statistical power with a one-sided significance level of 0.2. The trial demonstrated that pain response, defined as complete response (CR) plus partial response (PR), was significantly improved with stereotactic body radiation therapy (SBRT) compared to conventional fractionation at 9 months (77% vs. 46%, $p = 0.04$). Subgroup analysis favored the 16 Gy dose over 12 Gy or 30 Gy in ten fractions for pain relief. Notably, there were no observed differences in quality of life (QoL), treatment toxicities, or overall survival (OS) between the treatment arms. The authors of the phase 2 trial concluded that these findings support SBRT as a potential standard of care for patients with excellent performance status and longer life expectancy, particularly in managing bone metastases where pain control is critical despite competing factors such as patient mortality [42]. Patients with higher life expectancy and better performance status benefit from an optimal medical pain regimen prior to undergoing treatment with stereotactic body radiation therapy (SBRT). MRI imaging is crucial for assessing the intraosseous extent of the tumor; if this extent is unclear, a 3–5 mm bony clinical target volume (CTV) may be considered. For tumors with extraosseous extension, a 1–2 mm CTV extension in the direction of the extension can be applied. Treatment doses vary based on tumor size, location relative to critical organs at risk (OARs), and histology. Dose regimens are often adapted and fractionated for weight-bearing bones and joint spaces, with consideration given to simultaneous integrated boost (SIB) to the gross tumor volume (GTV) if bulky or to adhere to OAR constraints. Prescribing doses may involve targeting D_{min} or ensuring D100% covers over 95% of the target volume. In cases where initial treatment fails, salvage therapy with SBRT is an option, although this approach carries an increased risk of toxicities [43].

Prophylactic Radiation in High-Risk Bone Metastases: Insights from Gillespie et al.

The study by Gillespie et al. [44] highlights the value of prophylactic radiation in managing high-risk bone lesions, offering a proactive approach to reducing morbidity, enhancing quality of life, and potentially improving overall survival (OS). This intervention is particularly important for patients with metastatic bone disease who are at risk of skeletal-related events (SREs) such as fractures, pain, and neurological complications. By addressing these risks early, prophylactic radiation can minimize the need for surgical interventions and hospitalizations, improving outcomes for a broad range of patients.

The Gillespie study provides a clear framework for identifying high-risk lesions that benefit most from prophylactic radiation. High-risk lesions include bulky bone metastases measuring ≥2 cm, lesions involving critical anatomical sites such as the hip, shoulder, or sacroiliac joints, those occupying one-third to two-third of cortical thickness in long bones, and vertebral lesions with posterior element or junctional spine involvement. Depending on the Mirels score, prophylactic fixation may be indicated for weight-bearing extremities to stabilize the bone prior to initiating radiation therapy [44]. Accurate identification of these high-risk lesions is essential, especially for patients with low-volume metastatic disease who have curative potential when treated aggressively.

Prophylactic radiation therapy, as demonstrated in the Gillespie trial [44], is a low-toxicity, cost-effective intervention that offers several benefits. First, it reduces morbidity by preventing fractures, pain, and other complications, which directly translates into fewer hospitalizations and surgical procedures. Second, it enhances patients' quality of life by mitigating SREs and maintaining functional independence. Lastly, while the OS improvement observed in the study is exploratory due to the phase II design, it underscores the potential for early intervention to yield meaningful survival gains. These benefits extend beyond oligometastatic disease to include patients with polymetastatic disease, broadening the applicability of this approach.

The mechanisms underlying the survival benefits of prophylactic radiation are speculative but plausible. Proactively addressing skeletal complications enhances tolerance for systemic therapies, allowing patients to maintain consistent and effective treatment regimens. Additionally, prophylactic radiation may prevent life-threatening complications, such as pulmonary embolisms or sepsis, which are often associated with surgeries or hospitalizations for advanced skeletal disease. Finally, controlling locally aggressive lesions may reduce systemic inflammatory responses and limit further disease progression, contributing to improved outcomes.

Traditionally, radiation therapy for metastatic bone lesions has been reactive, focusing on symptom management. Common regimens include 8 Gy in a single fraction or 20 Gy in five fractions, which are effective for palliative care but not designed for prevention. Gillespie et al. [44] advocate a shift toward proactive management, emphasizing early intervention to prevent complications before they arise. This approach is increasingly relevant as advances in systemic therapies have extended patient survival and created a greater need for durable local control of metastatic lesions.

In terms of fractionation strategies, the Gillespie study [44] identifies 27 Gy in three fractions as a common regimen for prophylactic radiation. This moderately hypofractionated approach balances efficacy and safety, reducing treatment burdens while maintaining high local control rates. Additional evidence supports dose escalation within stereotactic body radiation therapy (SBRT) regimens, which deliver higher biologically effective doses (BED) and may be particularly effective for radioresistant tumors. Ultra-hypofractionated regimens, which deliver higher doses per fraction, also show promise for carefully selected patients, optimizing tumor response while minimizing toxicity.

The broader implications of the Gillespie study challenge traditional paradigms in metastatic bone disease management. By including patients with high-volume metastatic disease, the study expands the eligibility criteria for metastasis-directed therapy (MDT). This inclusion suggests that prophylactic radiation may benefit a wider patient population than previously considered. However, the study also raises important questions. Future research should refine the definitions of high-risk lesions to ensure precise targeting, validate the exploratory OS findings in larger phase III trials, and explore the integration of prophylactic radiation with systemic therapies such as immunotherapy or targeted agents.

Ongoing clinical trials are expected to build on the findings of Gillespie et al. [44], further refining patient selection criteria, fractionation schedules, and combination strategies. These efforts aim to personalize prophylactic radiation, tailoring interventions to tumor biology and individual patient characteristics. The proactive approach advocated by this study may also inspire broader changes in metastatic disease management, such as earlier integration of radiation oncologists and surgeons into multidisciplinary care teams to improve coordination and outcomes.

In conclusion, the Gillespie study provides compelling evidence for the use of prophylactic radiation in managing high-risk bone metastases. By reducing morbidity, enhancing quality of life, and potentially extending survival, this approach represents a significant advancement in metastatic disease treatment. While questions remain regarding optimal patient selection and dose strategies, the findings highlight the importance of early, proactive intervention. As research continues to refine these practices, prophylactic radiation is poised to become a cornerstone of comprehensive metastatic care.

Oligoprogression

Incorporation of Systemic Therapy with Local Therapies

The optimal timing for integrating systemic therapy with local treatments remains unclear. Generally, it is recommended to administer systemic therapy before initiating local therapies (Grade 2C). A typical approach involves a minimum of three months of systemic therapy, although practices may vary. Administering systemic therapy first allows for a better assessment of whether metastatic disease is truly oligometastatic or more widespread, where local therapies may not provide substantial benefit. Alternatively, in cases of oligorecurrence or oligoprogression, deferring systemic therapy until radiographic evidence of widespread disease progression is evident can also be considered. This strategy contrasts with treatment-naïve disease, where systemic therapy may take precedence. The selection of systemic therapy is guided by molecular profiling and aligns with the standard of care for stage IV disease [45].

For patients experiencing oligometastatic recurrence or progression while receiving targeted therapy, systemic therapy is often temporarily held during perioperative periods or radiation therapy (RT) to mitigate adverse effects. Following local treatment, a second-line systemic therapy is typically initiated. However, in cases where the treated site is limited to intracranial metastases, continuation of the original targeted therapy may be appropriate [46].

Reirradiation after SBRT

In a prospective case series involving 40 patients with 56 lesions previously treated with stereotactic body radiation therapy (SBRT), the cohort typically exhibited well-controlled systemic disease and high-performance status. A significant portion, 66.1%, underwent surgical procedures either before the first or second SBRT treatment. Notably, 42.9% had received external beam radiation therapy (EBRT) prior to the initial SBRT session. The median dose for the first SBRT treatment was 24 Gy delivered in 2 fractions, while the median dose for subsequent treatments was 30 Gy over four fractions, with a range of 20–35 Gy across 2–5 fractions. At 24 months, local failure was observed in 28.5% of cases, with a median time to progression of three months. SBRT reirradiation to the spine appears to be safe when delivered at least five months after prior conventional palliative radiotherapy, provided specific dose constraints are met. A reirradiation thecal sac P(max) nBED of 20–25 Gy ($\alpha/\beta = 2$) is acceptable if the cumulative P(max) nBED does not exceed approximately 70 Gy, and the SBRT contribution comprises no more than 50% of that total. Adhering to these thresholds minimizes the risk of radiation myelopathy and supports safe reirradiation in appropriately selected patients [47].

Complications and Limitations of SBRT

High doses of radiation carry a higher risk of toxicity in nearby organs, though this is infrequent [48]. Side effects depend on the treatment area and may include esophagitis, mucositis, dysphagia, diarrhea, paresthesia, transient laryngitis, radiculitis, radiation-induced myelopathy [49], and vertebral compression fractures. Postirradiation fractures of femoral metastases occurred in 7.7% of cases, with a significantly higher incidence in the proximal femur (7.8%) compared to the femoral shaft (1.5%). The majority of fractures occurred within 6 months of radiotherapy, particularly within the first 3 months. While traditional radiologic fracture risk assessments such as Harrington's criteria and Mirels' score showed modest predictive value, their correlation with actual fracture occurrence was limited. Early fractures were often associated with viable tumor cells, whereas those occurring beyond 3 months were more likely related to radiation-induced bone fragility rather than tumor progression. These findings highlight the need for early vigilance and consideration of prophylactic stabilization, especially in the proximal femur, to prevent debilitating complications in patients undergoing radiotherapy for femoral metastases [50, 51]. Established dose constraints help mitigate these risks [52]. The risk of myelopathy associated with stereotactic body radiation therapy (SBRT) is considered low under specific dose constraints. For instance, maintaining a cumulative maximum dose (D_{max}) to the thecal sac below 70 Gy in biologically effective dose (BED) equivalents, calculated based on 2 Gy fractions with an alpha/beta ratio of 2, is recommended. Additionally, to minimize the risk of myelopathy, the maximum SBRT dose to any point on the thecal sac should not exceed 50% of the total BED at that specific location. These guidelines aim to ensure that spinal cord tolerance limits are respected during SBRT treatments [53].

Conclusions

The management of bone metastases has advanced with the use of SBRT, offering durable local control, better pain relief, and survival benefits, particularly in patients with favorable prognostic factors. SBRT's precision enables high-dose delivery with minimal toxicity, and doses ≥22 Gy are linked to improved outcomes. However, risks such as vertebral or long bone fractures highlight the need for careful patient selection and possible prophylactic stabilization. SBRT is especially effective in oligometastatic and oligoprogressive settings, when timed after initial systemic therapy. Prophylactic radiation in high-risk lesions and safe reirradiation practices further enhance its role. Overall, SBRT supports a shift from purely palliative care to long-term control in selected patients.

References

1. Coleman RE. Metastatic bone disease: clinical features, pathophysiology and treatment strategies. Cancer Treat Rev. 2001;27(3):165–76.
2. Sterling JA, Edwards JR, Martin TJ, Mundy GR. Advances in the biology of bone metastasis: how the skeleton affects tumor behavior. Bone. 2011;48(1):6–15.
3. Ikeda T, Utsuyama M, Hirokawa K. Expression profiles of receptor activator of nuclear factor kappaB ligand, receptor activator of nuclear factor kappaB, and osteoprotegerin messenger RNA in aged and ovariectomized rat bones. J Bone Miner Res. 2001;16(8):1416–25.
4. Nakashima T, Hayashi M, Fukunaga T, Kurata K, Oh-Hora M, Feng JQ, et al. Evidence for osteocyte regulation of bone homeostasis through RANKL expression. Nat Med. 2011;17(10):1231–4.
5. Park JH, Lee NK, Lee SY. Current understanding of RANK signaling in osteoclast differentiation and maturation. Mol Cells. 2017;40(10):706–13.
6. Marino S, Roodman GD. Multiple myeloma and bone: the fatal interaction. Cold Spring Harb Perspect Med. 2018;8(8):a031286.
7. Hellman S, Weichselbaum RR. Oligometastases. J Clin Oncol. 1995;13(1):8–10.
8. Niibe Y, Hayakawa K. Oligometastases and oligo-recurrence: the new era of cancer therapy. Jpn J Clin Oncol. 2010;40(2):107–11.
9. Weichselbaum RR, Hellman S. Oligometastases revisited. Nat Rev Clin Oncol. 2011;8(6):378–82.
10. Palma DA, Olson R, Harrow S, Gaede S, Louie AV, Haasbeek C, et al. Stereotactic ablative radiotherapy for the comprehensive treatment of oligometastatic cancers: long-term results of the SABR-COMET phase II randomized trial. J Clin Oncol. 2020;38(25):2830–8.
11. Tree AC, Khoo VS, Eeles RA, Ahmed M, Dearnaley DP, Hawkins MA, et al. Stereotactic body radiotherapy for oligometastases. Lancet Oncol. 2013;14(1):e28–37.
12. Phillips R, Shi WY, Deek M, Radwan N, Lim SJ, Antonarakis ES, et al. Outcomes of observation vs stereotactic ablative radiation for oligometastatic prostate cancer: the ORIOLE phase 2 randomized clinical trial. JAMA Oncol. 2020;6(5):650–9.
13. Deek MP, Van der Eecken K, Sutera P, Deek RA, Fonteyne V, Mendes AA, et al. Long-term outcomes and genetic predictors of response to metastasis-directed therapy versus observation in oligometastatic prostate cancer: analysis of STOMP and ORIOLE trials. J Clin Oncol. 2022;40(29):3377–82.
14. Chmura S, Winter KA, Robinson C, Pisansky TM, Borges V, Al-Hallaq H, et al. Evaluation of safety of stereotactic body radiotherapy for the treatment of patients with multiple metastases: findings from the NRG-BR001 phase 1 trial. JAMA Oncol. 2021;7(6):845–52.
15. Kent CL, McDuff SGR, Salama JK. Oligometastatic breast cancer: where are we now and where are we headed?-a narrative review. Ann Palliat Med. 2021;10(5):5954–68.
16. Mundy GR. Metastasis to bone: causes, consequences and therapeutic opportunities. Nat Rev Cancer. 2002;2(8):584–93.
17. Coleman RE. Clinical features of metastatic bone disease and risk of skeletal morbidity. Clin Cancer Res. 2006;12(20 Pt 2):6243s–9s.
18. Pockett RD, Castellano D, McEwan P, Oglesby A, Barber BL, Chung K. The hospital burden of disease associated with bone metastases and skeletal-related events in patients with breast cancer, lung cancer, or prostate cancer in Spain. Eur J Cancer Care (Engl). 2010;19(6):755–60.

19. Deek MP, Tran PT, Jabbour SK. Metastasis-directed therapy: a moving target advancing Progress forward. J Clin Oncol. 2024;42(1):4–7.
20. Hartsell WF, Scott CB, Bruner DW, Scarantino CW, Ivker RA, Roach M 3rd, et al. Randomized trial of short- versus long-course radiotherapy for palliation of painful bone metastases. J Natl Cancer Inst. 2005;97(11):798–804.
21. Huisman M, van den Bosch MA, Wijlemans JW, van Vulpen M, van der Linden YM, Verkooijen HM. Effectiveness of reirradiation for painful bone metastases: a systematic review and meta-analysis. Int J Radiat Oncol Biol Phys. 2012;84(1):8–14.
22. Howell DD, James JL, Hartsell WF, Suntharalingam M, Machtay M, Suh JH, et al. Single-fraction radiotherapy versus multifraction radiotherapy for palliation of painful vertebral bone metastases-equivalent efficacy, less toxicity, more convenient: a subset analysis of Radiation Therapy Oncology Group trial 97-14. Cancer. 2013;119(4):888–96.
23. van der Linden YM, Steenland E, van Houwelingen HC, Post WJ, Oei B, Marijnen CA, et al. Patients with a favourable prognosis are equally palliated with single and multiple fraction radiotherapy: results on survival in the Dutch bone metastasis study. Radiother Oncol. 2006;78(3):245–53.
24. Meeuse JJ, van der Linden YM, van Tienhoven G, Gans RO, Leer JW, Reyners AK. Efficacy of radiotherapy for painful bone metastases during the last 12 weeks of life: results from the Dutch bone metastasis study. Cancer. 2010;116(11):2716–25.
25. Chow E, Zeng L, Salvo N, Dennis K, Tsao M, Lutz S. Update on the systematic review of palliative radiotherapy trials for bone metastases. Clin Oncol (R Coll Radiol). 2012;24(2):112–24.
26. Correa RJ, Salama JK, Milano MT, Palma DA. Stereotactic body radiotherapy for oligometastasis: opportunities for biology to guide clinical management. Cancer J. 2016;22(4):247–56.
27. Potrebko PS, McCurdy BM, Butler JB, El-Gubtan AS. Improving intensity-modulated radiation therapy using the anatomic beam orientation optimization algorithm. Med Phys. 2008;35(5):2170–9.
28. Ryu S, Rock J, Rosenblum M, Kim JH. Patterns of failure after single-dose radiosurgery for spinal metastasis. J Neurosurg. 2004;101(Suppl 3):402–5.
29. Ryu S, Jin R, Jin JY, Chen Q, Rock J, Anderson J, et al. Pain control by image-guided radiosurgery for solitary spinal metastasis. J Pain Symptom Manag. 2008;35(3):292–8.
30. Sprave T, Verma V, Förster R, Schlampp I, Bruckner T, Bostel T, et al. Randomized phase II trial evaluating pain response in patients with spinal metastases following stereotactic body radiotherapy versus three-dimensional conformal radiotherapy. Radiother Oncol. 2018;128(2):274–82.
31. Chan NK, Abdullah KG, Lubelski D, Steinmetz MP, Benzel EC, Shin JH, et al. Stereotactic radiosurgery for metastatic spine tumors. J Neurosurg Sci. 2014;58(1):37–44.
32. Garcia-Barros M, Paris F, Cordon-Cardo C, Lyden D, Rafii S, Haimovitz-Friedman A, et al. Tumor response to radiotherapy regulated by endothelial cell apoptosis. Science. 2003;300(5622):1155–9.
33. Vignard J, Mirey G, Salles B. Ionizing-radiation induced DNA double-strand breaks: a direct and indirect lighting up. Radiother Oncol. 2013;108(3):362–9.
34. Alongi F, Arcangeli S, Filippi AR, Ricardi U, Scorsetti M. Review and uses of stereotactic body radiation therapy for oligometastases. Oncologist. 2012;17(8):1100–7.
35. Chang BK, Timmerman RD. Stereotactic body radiation therapy: a comprehensive review. Am J Clin Oncol. 2007;30(6):637–44.
36. Videtic GM, Hu C, Singh AK, Chang JY, Parker W, Olivier KR, et al. A randomized phase 2 study comparing 2 stereotactic body radiation therapy schedules for medically inoperable patients with stage I peripheral non-small cell lung cancer: NRG oncology RTOG 0915 (NCCTG N0927). Int J Radiat Oncol Biol Phys. 2015;93(4):757–64.
37. Zelefsky MJ, Greco C, Motzer R, Magsanoc JM, Pei X, Lovelock M, et al. Tumor control outcomes after hypofractionated and single-dose stereotactic image-guided intensity-modulated radiotherapy for extracranial metastases from renal cell carcinoma. Int J Radiat Oncol Biol Phys. 2012;82(5):1744–8.

38. Pitroda SP, Khodarev NN, Huang L, Uppal A, Wightman SC, Ganai S, et al. Integrated molecular subtyping defines a curable oligometastatic state in colorectal liver metastasis. Nat Commun. 2018;9(1):1793.
39. Willmann J, Vlaskou Badra E, Adilovic S, Christ SM, Ahmadsei M, Mayinger M, et al. Distant metastasis velocity as a novel prognostic score for overall survival after disease progression following stereotactic body radiation therapy for oligometastatic disease. Int J Radiat Oncol Biol Phys. 2022;114(5):871–82.
40. Erler D, Brotherston D, Sahgal A, Cheung P, Loblaw A, Chu W, et al. Local control and fracture risk following stereotactic body radiation therapy for non-spine bone metastases. Radiother Oncol. 2018;127(2):304–9.
41. Greco C, Zelefsky MJ, Lovelock M, Fuks Z, Hunt M, Rosenzweig K, et al. Predictors of local control after single-dose stereotactic image-guided intensity-modulated radiotherapy for extracranial metastases. Int J Radiat Oncol Biol Phys. 2011;79(4):1151–7.
42. Nguyen QN, Chun SG, Chow E, Komaki R, Liao Z, Zacharia R, et al. Single-fraction stereotactic vs conventional multifraction radiotherapy for pain relief in patients with predominantly nonspine bone metastases: a randomized phase 2 trial. JAMA Oncol. 2019;5(6):872–8.
43. Nguyen TK, Sahgal A, Dagan R, Eppinga W, Guckenberger M, Kim JH, et al. Stereotactic body radiation therapy for nonspine bone metastases: international practice patterns to guide treatment planning. Pract Radiat Oncol. 2020;10(6):e452–e60.
44. Gillespie EF, Yang JC, Mathis NJ, Marine CB, White C, Zhang Z, et al. Prophylactic radiation therapy versus standard of care for patients with high-risk asymptomatic bone metastases: a multicenter, randomized phase II clinical trial. J Clin Oncol. 2024;42(1):38–46.
45. Tiefenbacher A, Pirker R. Systemic treatment of advanced non-small cell lung cancer: controversies and perspectives. Memo. 2018;11(2):112–5.
46. Myall NJ, Yu H, Soltys SG, Wakelee HA, Pollom E. Management of brain metastases in lung cancer: evolving roles for radiation and systemic treatment in the era of targeted and immune therapies. Neurooncol Adv. 2021;3(Suppl 5):v52–62.
47. Thibault I, Campbell M, Tseng CL, Atenafu EG, Letourneau D, Yu E, et al. Salvage stereotactic body radiotherapy (SBRT) following in-field failure of initial SBRT for spinal metastases. Int J Radiat Oncol Biol Phys. 2015;93(2):353–60.
48. Moussazadeh N, Lis E, Katsoulakis E, Kahn S, Svoboda M, DiStefano NM, et al. Five-year outcomes of high-dose single-fraction spinal stereotactic radiosurgery. Int J Radiat Oncol Biol Phys. 2015;93(2):361–7.
49. Gibbs IC, Patil C, Gerszten PC, Adler JR Jr, Burton SA. Delayed radiation-induced myelopathy after spinal radiosurgery. Neurosurgery. 2009;64(2 Suppl):A67–72.
50. Shimoyama T, Katagiri H, Harada H, Murata H, Wasa J, Hosaka S, et al. Fracture after radiation therapy for femoral metastasis: incidence, timing and clinical features. J Radiat Res. 2017;58(5):661–8.
51. Virk MS, Han JE, Reiner AS, McLaughlin LA, Sciubba DM, Lis E, et al. Frequency of symptomatic vertebral body compression fractures requiring intervention following single-fraction stereotactic radiosurgery for spinal metastases. Neurosurg Focus. 2017;42(1):E8.
52. Schipani S, Wen W, Jin JY, Kim JK, Ryu S. Spine radiosurgery: a dosimetric analysis in 124 patients who received 18 Gy. Int J Radiat Oncol Biol Phys. 2012;84(5):e571–6.
53. Sahgal A, Ma L, Weinberg V, Gibbs IC, Chao S, Chang UK, et al. Reirradiation human spinal cord tolerance for stereotactic body radiotherapy. Int J Radiat Oncol Biol Phys. 2012;82(1):107–16.

Part VII

Especial Scenarios

38 Skeletal Metastasis of Unknown Primary: Now What Should I Do?

Mohamad Cherry and Michael P. Rogers

Introduction

Metastasis, the process by which cancer cells spread from a primary tumor to distant organs, is a defining feature of advanced cancer and plays a major role in cancer-related morbidity and mortality. Every year, over half a million people are diagnosed with metastatic disease and approximately 350,000 people will die with bony metastases [1].

The foremost priority during the initial stages of cancer management is to determine the primary site of origin, as this can affect the therapeutic options available for treatment. Cancer of unknown primary (CUP) is defined as a metastatic condition in which the primary tumor cannot be identified, despite extensive efforts to locate it. CUP accounts for 3–5% of all cancer diagnoses globally, with a median age at diagnosis of 65 years, and it occurs slightly more frequently in men [2, 3].

A high proportion of these patients (10–15%) will also present with skeletal metastasis [4].

A delay in the diagnosis of the primary origin can result in a delay in care for the patient and increase the risk of complications, morbidity from skeletal-related events, or mortality from cancer treatment delay [5].

While metastases can affect almost any tissue or organ, bone metastases are common and often represent a late manifestation of advanced stage solid tumors. When trying to discover the origin of new skeletal metastases it is important to keep a high suspicion of the types of cancers that commonly spread to the bone, including prostate, breast, and lung cancer. Takagi et al. conducted a retrospective

M. Cherry (✉)
Department of Hematology Oncology, Morristown Medical Center, Morristown, NJ, USA
e-mail: Mohamad.Cherry@atlantichealth.org

M. P. Rogers
Department of Surgery, Morristown Medical Center, Morristown, NJ, USA

J. Pretell-Mazzini, S. A. Lozano Calderón (eds.), *Multidisciplinary Approach of Musculoskeletal Metastasis*, https://doi.org/10.1007/978-3-032-09498-8_38

review analyzing 286 cases of skeletal metastases of unknown primary that with further investigation revealed lung cancer as the primary origin in 25.2% of cases [6].

Other common origins include thyroid and kidney cancer. Multiple myeloma is a primary hematologic malignancy with the hallmark feature manifested by diffuse lytic bone lesions [7]. Other hematologic malignancies such as lymphoma and leukemia are rare causes of bone metastasis.

In contrast to bone metastasis, primary metastases to muscles are very unusual, even though the muscle mass makes up approximately 50% of the body weight. The rarity of such occurrence has postulated several theories including the dynamics of muscle contraction/relaxation, muscle metabolism with buildup of lactic acid leading to low pH, the presence of oxygen radicals creating an environment that is not receptive to tumor growth and the presence of barriers such as fascia and other structures [8].

Epidemiology

Metastasis of unknown primary origin is reported in approximately 3–4% of all cancer patients and 10–15% of them represent the bones or muscles [9, 10].

The bone is the third most common site of metastatic cancer of unknown primary origin, after the lymph nodes and the lung [11].

Lung, liver, pancreas, and gastrointestinal tract are common sites of primary occult tumors. Adenocarcinoma is the main histological type, accounting for 70% of all cases, while undifferentiated cancer and squamous cell carcinomas account for 15% and 10%, respectively [4].

Some unknown primary tumors have good prognosis and are treatable, like lymphoma, extragonadal germ cell neoplasms and ovarian cancer, but the majority of cases have a short survival with limited possibilities of employing effective chemotherapy [2].

In one study, lung carcinomas were responsible for 52% of 51 cases of bone lesions from unknown origin, while only 7% of bone metastases were found with a diagnosed primary. Moreover, patients with skeletal metastases from occult carcinomas showed a high incidence of spinal metastases, cord compression, and pathological fractures and a significantly shorter survival compared to bone lesions secondary to known primaries [10].

The detection of bone metastases from occult primaries should raise the suspicion of lung origin and the suspicion should be stronger in relatively young patients (60–65 years) [9].

Subsequently, bone metastases from undiagnosed breast cancer, renal clear cell carcinomas, and prostate and gastrointestinal carcinomas should be considered [12].

The spine is the most common site for bone metastases and is the most common location of bone lesions of CUP, followed by the pelvis and long bones. At these sites (pelvis and long bones), lung and thyroid cancers should be strongly suspected [13, 14].

However, spinal malignancies of unknown origin can arise not only from solid tumors but also from hematological cancers [15]. In one study, myeloma was the most common cause (22%), followed by lung carcinoma, prostate carcinoma, and lymphoma [16].

Serological testing for monoclonal gammopathy was highly effective in diagnosing myeloma in all affected patients [16].

Acrometastases are extremely rare (<0.1%), particularly as the first manifestation of occult cancer, but they are strongly associated with bronchogenic or gastrointestinal cancers [17–19].

In contrast, metastases to the skeletal muscle is rare cancer metastasis including cancer of unknown origin. In one of the large cancer series, out of 2557 patients, only four individuals (0.16%) were confirmed to have muscle metastasis [8].

In another large retrospective series, muscle metastases constitute less than 1% of all malignant metastases of hematogenous origin. It can involve both soft tissues (skeletal muscles) and visceral muscles (such as smooth muscle in the gastrointestinal tract or bladder), with skeletal muscle metastasis being the more commonly observed form [20].

Primary cancers of the lung, breast, blood, gastrointestinal tract and genitourinary tract were the most frequently involved [20, 21].

Pathophysiology

Primary tumors metastasize to the skeletal system through a multistep process including escaping from their origin, entering the vascular system, evading elimination by immune cells, attaching to bone, and finally colonization and uncontrollable proliferation. The vast majority of cancer cells will be eliminated after entering circulation. It is hypothesized that those cancer cells that survive long enough to reach the bone were aggregated within a cluster of platelets to shield the tumor cells from the immune system [22].

Once shielded in the vasculature, there are certain intrinsic properties of the tumor cells that promote their migration to the bone. For example, homing signals such as CXCR4 (CXC motif chemokine 4) on human breast cancer cells have been shown to have an affinity for the CXCL12 signal on bone cells. Kang et al. showed that breast cancer cells with a high expression of CXCR4 have a high propensity to form bone metastases [23].

As the clumped tumor cells approach the narrow sinusoids in the bone metaphysis, blood flow reduces resulting in increased interaction with adhesion proteins on the vessel wall, followed by transmigration [24, 25].

Following movement through the vessel wall, tumor cells enter a period of dormancy for an extended time to become acclimated to the new environment, while evading the immune system. This period can last for years between the time an individual is diagnosed with cancer till bone metastases present clinically, deemed the state of metastatic latency [26].

Although the exact mechanisms of how cancer cells remain dormant in the bone are not well understood, there are multiple hypotheses. The binding of the annexin II receptors on prostate cancer cells to the same receptor on osteoblastic lineage cells results in an increase in transforming growth factor (TGF)-beta-2 signaling, which is associated with dormancy [27].

Ghajar et al. described how breast cancer cells can lay quiescent on the vasculature of metastatic sites due thrombospondin 1, and angiocrine tumor suppressor, secreted from the underlying endothelial cells, and how loss of the thrombospondin 1 is seen closest to outgrowing cancer cells [28].

Initial Evaluation and Diagnostic Approach

A methodical, step-by-step approach to managing a patient with CUP should be utilized to increase the likelihood of uncovering the primary cancer. An accurate diagnosis can guide medical management and improve outcomes in these patients, as individuals treated with the diagnosis of CUP have a significantly poorer survival compared to those with bone metastases of a known primary [10].

The initial workup for CUP, as outlined in the 2024 National Comprehensive Cancer Network (NCCN) guidelines, includes a thorough history and physical examination, basic laboratory tests, computed tomography (CT) scans, clinically directed endoscopy, and testing for microsatellite instability (MSI) and mismatch repair genes. In women, breast imaging is recommended to evaluate breast cancer as a potential primary site, given its commonality and treatability even in advanced stages. Additionally, serum tumor markers may be suggested for certain patients to help identify the primary tumor site. After biopsy, a targeted panel of immunohistochemistry (IHC) tests is advised. Tissue-of-origin testing, often using microarray-based gene expression methods, and next-generation sequencing (NGS) are also used to help diagnose CUP in patients [29].

The diagnosis of bone or muscle metastasis of CUP is particularly challenging because it implies that the metastatic cells have spread to distant tissues without an obvious origin. This clinical scenario is typically characterized by the presence of metastatic disease in either bone or muscle without any identifiable primary tumor, despite exhaustive diagnostic workup. The complexity of managing such cases stems from the diagnostic ambiguity and the need for an integrated, multidisciplinary approach to treatment. The mean survival of these patients has not changed over the past three decades with multiple studies having described an average overall survival between 3 and 12 months in those with SMUP [4, 30, 31], including a recent study by Ren et al. that described an overall one-year survival of 14.5% and cancer-specific survival of 15.9% [32].

Clinical History

In most incidences, a patient with a known history of cancer is a strong rationale to be suspicious of metastatic disease, especially if the previous cancer is one which frequently metastases to bone (lung, prostate, breast, thyroid, and renal) [33].

The age of the patient is also an important predictor, as differentials in patients over the age of 40 with persistent musculoskeletal pain should include metastatic disease, multiple myeloma, and lymphoma [34].

A family history of cancer may help guide the diagnostic process, as certain cancers, such as breast, ovarian, and colorectal cancers, have well-established hereditary components [34].

Physical Examination

The physical examination aims to identify signs of metastasis or an occult primary tumor. Patients with symptomatic skeletal or muscle metastases would most commonly present with progressively worsening pain, swelling, decreased range of motion, or deformity in one or more areas, along with possible soft tissue changes overlying the sites of metastases [35].

An assessment for signs of lymphadenopathy, hepatomegaly, or abdominal masses should be performed to search for other sites of metastasis or to identify possible primary tumor locations. Neurological deficits (e.g., weakness, sensory loss) may suggest spinal cord involvement or direct compression of nerve roots by bone metastases. Although a thorough history and physical exam can aid in the diagnosis, they have not shown to be a good predictor of identifying the origin of cancer, warranting further diagnostic investigative studies [6, 36].

Imaging Studies

Plain Radiographs

X-rays are typically the first imaging modality used to evaluate bone lesions. X-rays of the painful bone or joints should be obtained. The appearance of the lesions, whether osteolytic or osteoblastic, can guide the clinician in the direction of certain carcinomas as the primary origin and help predict results on blood work obtained during the initial visit. An osteolytic appearance of bone metastases is more commonly seen in breast or lung cancer (Fig. 38.1a, b); however, a male patient with an osteoblastic lesion could be cause for alert of prostate or thyroid cancer as the origin [37].

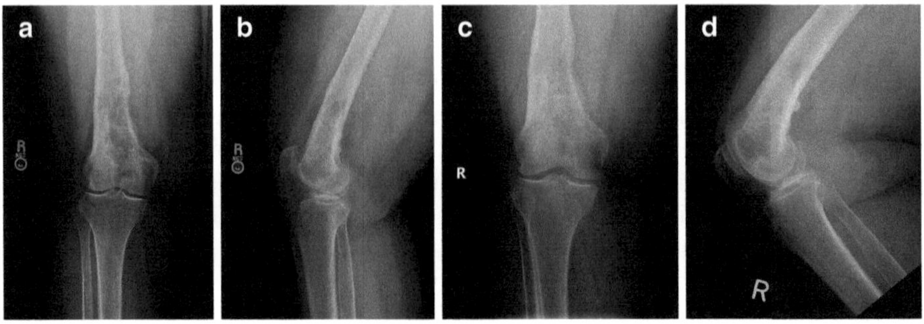

Fig. 38.1 Plain radiographs of a 71-year-old female patient with a past medical history of breast cancer presenting with pain in the right knee. Radiographs demonstrated an osteolytic lesion of the metaphysis of the right distal femur (**a**, **b**). The patient presented one month later with increasing pain in the right leg due to a nondisplaced pathological fracture (**c**, **d**)

Seemingly nonspecific symptoms, such as nausea, constipation, abdominal pain, or mental status changes, could alert the physician to hypercalcemia secondary to extensive bone remodeling. The continued osteoclast activation leaves the bone susceptible to progressive weakening, resulting in an impending fracture, until the bone finally breaks, resulting in a pathological fracture. Multiple reviews have detailed an approximately 8–30% pathological fracture rate in patients with skeletal metastases that most commonly occur at the proximal femur [38, 39].

If the patient presents with a pathological fracture, they are already at a higher risk of increased mortality at one year compared to patients treated for impending fractures (Fig. 38.1c, d). This highlights the importance of taking musculoskeletal pain in individuals with a history of cancer very seriously and have a low threshold to order further diagnostic testing [40].

Some studies have recommended the use of a chest radiograph as a quick and low-cost modality to investigate lung cancer as the primary origin; however, a chest CT may offer more diagnostic value [4].

Computed Tomography (CT) and Magnetic Resonance Imaging (MRI)

These advanced imaging techniques provide complementary roles in evaluating images of bone and soft tissues and are more sensitive than X-rays for detecting bone metastasis. A CT scan can provide a more detailed examination of the cortices at the site of skeletal metastasis to assess for an impending or pathological fracture, if a plain radiograph is not sufficient (Fig. 38.2a, b). It is also a preferred modality in areas such as the ribs or spine to evaluate the lesions without artifacts from surrounding tissue. A CT scan of the chest should be ordered to rule out lung malignancies, as one systematic review revealed this had the second highest likelihood of determining CUP, only behind a biopsy [41].

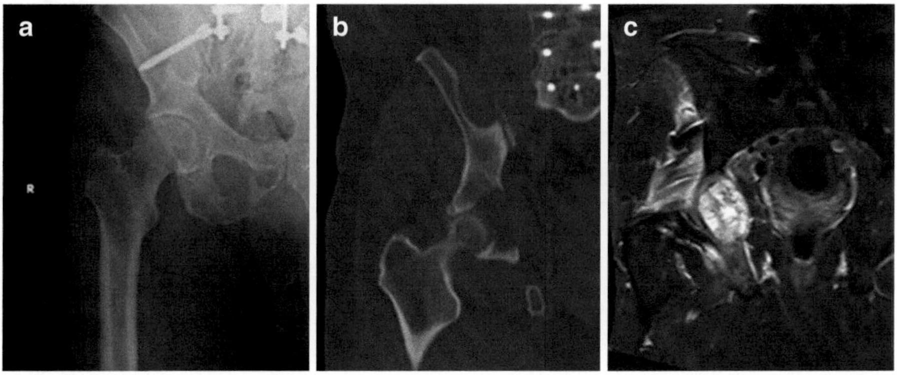

Fig. 38.2 This is a 58-year-old male patient with no pertinent past medical history presenting with right hip pain. Plain radiograph (**a**) of the right hip demonstrated a mass with subtle calcifications in the right acetabulum. Advanced imaging with CT (**b**) and MRI (**c**) better demonstrated the boney destruction and soft tissue components of the tumor. A biopsy of the lesion confirmed metastatic renal cell carcinoma

The decision to include a CT of the abdomen and pelvis should be based on other factors that lead the clinician to believe there is a risk of malignancy in those organs [41].

Muscle metastases typically have five patterns seen on CT including (in order from most to least common) masses with homogeneous contrast enhancement, abscess-like lesions, diffuse infiltration with muscle swelling, intramuscular calcifications, or intramuscular bleeding [20].

MRI is especially useful in assessing the involvement of the spinal cord or soft tissues surrounding the bones. It provides a better image of the surrounding tumor burden, involvement of adjacent structures, such as impinged nerve roots, and offers superior bone marrow evaluation that can identify bone metastases before osseous destruction can be seen (Fig. 38.2c). While investigating the origin of CUP, an MRI can increase the likelihood of establishing breast or head and neck tumors as the origin. In patients presenting with axillary lymphadenopathy, breast MRI has been shown to detect 70–86% of primary breast tumors, with some studies reporting even higher rates of detection [42, 43].

The improved visualization utilizing MRI of the soft tissues of the head and neck have been shown to determine CUP in 16% of cases, with higher detection rates when utilizing diffusion-weighted imaging (DWI; 37%) [44].

Muscle metastases tend to appear hypo- to isointense on T1-weighted images and hyperintense to muscles on a T2-weighted image [20].

Bone Scintigraphy (Bone Scan)

A bone scan using technetium-99 m-labeled diphosphonates is helpful for imaging the entire skeletal system in one scan to detect osteoblastic or osteolytic changes in the bone (Fig. 38.3). It is particularly useful when multiple bone lesions are

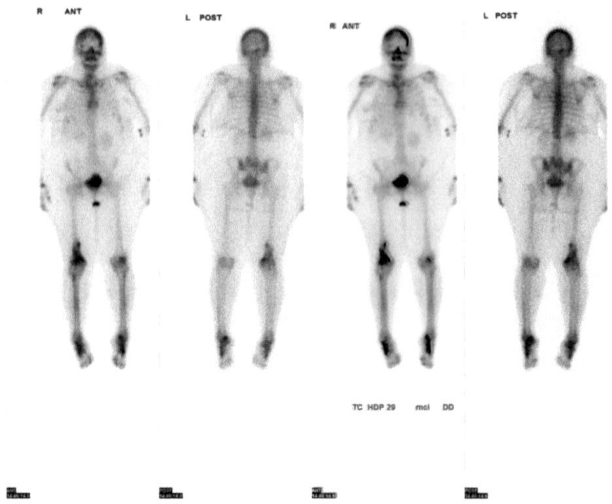

Fig. 38.3 Bone scintigraphy images of a 71-year-old female with metastatic breast cancer to the right distal femur

suspected to determine the full extent of the skeletal metastatic disease, with a sensitivity and specificity of 78% and 48%, respectively [45].

Although bone scans are a commonly used modality, they do have limitations, such as the inability to differentiate multiple metastatic from multiple benign lesions [46], and difficulty revealing primarily osteolytic lesions with minimal reactive osteoblastic processes like renal cell carcinoma [47].

Reduced or absent radiotracer uptake on a follow-up bone scan in areas of known metastases can be a gauge of response to treatment [48].

Positron Emission Tomography Combined with Computed Tomography (PET/CT)

Positron emission tomography combined with computed tomography (PET/CT) is a powerful imaging modality that helps identify areas of increased metabolic activity, which can indicate the presence of hidden primary tumors. PET/CT is increasingly used due to its high sensitivity and specificity to detect bone metastases (100% and 97%) [45] and to evaluate patients with cancer of unknown primary (Fig. 38.4). PET/CT scans have been shown to determine the primary tumor in approximately 40% of patients with CUP [49].

Ultrasound

In cases of muscle metastasis, ultrasound can be used to assess the size, shape, and location of muscle lesions and guide biopsy procedures. Most of these lesions will appear hypoechoic [20].

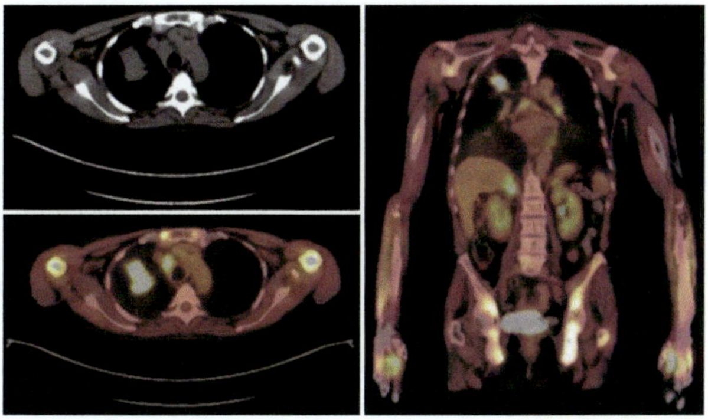

Fig. 38.4 CT/PET imaging of a 67-year-old female patient who initially presented with left hip pain revealed an area of increased metabolic activity in the right upper lung, consistent with lung cancer

Laboratory Studies

Basic Lab Tests

During the initial office visit of a patient diagnosed with CUP, obtaining a blood test should be one of the first steps in the diagnostic workup. A complete blood count (CBC) with differential, basic metabolic panel (CMP), liver function tests (LFTs), kidney function tests, and urinalysis can give the physician insight as to specific organ involvement or other pathologies such as lymphoma, or multiple myeloma. The most common hematological findings for MM include anemia, and elevated creatinine. In the presence of an osteolytic lesion, MM should be high on the differential [50].

Lymphoma can present with a relatively normal hemoglobin with or without lymphocytosis/lymphopenia; however, patients usually are exhibiting classic symptoms, such as fatigue, shortness of breath/cough, night sweats, and weight loss, with palpable lymphadenopathy [51].

In the case of suspected MM, further tests including a serum free light chain level, serum protein electrophoresis, and a 24-hour urine protein electrophoresis should be obtained [52].

Tumor Markers

Elevated serum levels of tumor markers can be helpful in identifying the primary site and monitoring for relapse or metastases. For example, prostate-specific antigen (PSA) may point that prostate cancer, alpha-fetoprotein (AFP), and beta-HCG can suggest testicular cancer, and CA 15–3, CA 27.29, and HER2/neu can help identify breast cancer. A carcinoembryonic antigen (CEA) is not usually produced in an adult patient but can be elevated in the presence of a malignancy. In a male patient with prostate cancer, a PSA >100 ng/mL is very predictive of bone metastases [53].

Biopsy and Histological Examination

If imaging studies confirm the presence of metastatic disease, a biopsy of the lesion (bone or muscle) should be performed to obtain tissue for histopathological analysis. This biopsy can be done percutaneously (e.g., needle biopsy) or through an open surgical procedure if the lesion is accessible. Histopathological examination, including immunohistochemistry (IHC), helps determine the cellular origin of the metastatic lesion. Specific markers can identify the primary tissue type. For instance, CK7, CK20, and TTF-1 can suggest lung or gastrointestinal origin, while ER (estrogen receptor) and PR (progesterone receptor) are helpful for diagnosing breast cancer. PSA and Androgen Receptor can be used to identify prostate cancer.

In some cases, if the biopsy is insufficient or inconclusive, molecular profiling or next-generation sequencing (NGS) may be used to detect gene mutations, copy number variations, and gene expression patterns that could point to the tumor's origin.

Endoscopy and Colonoscopy

In cases where there is suspicion of a primary tumor in the gastrointestinal, genitourinary, or pulmonary systems, endoscopic procedures (e.g., colonoscopy, bronchoscopy, cystoscopy) can be employed to directly visualize potential sources of primary cancer.

Molecular Diagnostics

The advent of molecular technologies, including next-generation sequencing (NGS), has allowed for more precise identification of the primary tumor in cases of CUP. NGS allows for the analysis of genetic alterations, gene expression profiles, and mutation patterns that can provide significant clues about the tissue of origin.

Liquid biopsy, the analysis of circulating tumor DNA (ctDNA) from blood samples, has also gained popularity as a noninvasive diagnostic tool. ctDNA analysis can potentially identify genetic mutations associated with specific cancers, providing another avenue for identifying the primary tumor.

Treatment of Bone or Muscle Metastasis from an Unknown Primary

The management of bone or muscle metastasis of unknown primary is typically palliative, with the main goal being symptom control, prolongation of survival, and improvement in quality of life. Treatment decisions should be individualized, based on the patient's overall health, extent of metastatic disease, and whether the primary site is identified.

Systemic Therapy

Once the diagnosis of metastasis is established, the treatment options primarily consist of systemic therapy, which may include:

Chemotherapy

For many types of metastases, chemotherapy is the mainstay of treatment. The chemotherapy regimen will depend on the histologic features of the tumor and its response to chemotherapy drugs. Chemotherapy can be used in cases of aggressive metastatic cancers that are not amenable to surgical resection.

Targeted Therapy

Targeted therapies are often employed when specific molecular alterations are identified, even in the absence of a known primary tumor. For example, HER2-positive breast cancer can be treated with trastuzumab, and EGFR-mutant lung cancer can be treated with EGFR inhibitors. Immunotherapy with immune checkpoint inhibitors (e.g., nivolumab, pembrolizumab) has shown promise for cancers such as melanoma and non-small cell lung cancer.

Local Therapy

For bone metastasis, local therapy can include radiation therapy or surgical intervention.

Radiation Therapy

Palliative radiation is often used for bone metastasis to reduce pain and prevent fractures. It can be effective in managing spinal cord compression or bone pain associated with metastasis.

Surgical Intervention

Surgery may be required to stabilize bones at risk for fracture or to relieve neurological symptoms caused by spinal metastasis. Classification systems such as the Mirel's Score or Harrington Classification can be used to guide whether or not an impending fracture requires prophylactic fixation. This would most likely be fixated by curettage of the metastatic disease followed by intramedullary fixation.

For muscle metastasis, local excision may be considered in some cases, but this is generally reserved for smaller, isolated lesions where surgical resection is feasible.

Supportive Care

Pain management is a cornerstone of treatment for metastatic bone and muscle cancer. In addition to opioids and NSAIDs, bisphosphonates and denosumab can be used to treat bone pain and reduce skeletal-related events. Physical therapy can improve mobility and function, particularly in cases of muscle metastasis or significant bone involvement.

Prognosis and Follow-Up Care

The prognosis of patients with bone or muscle metastasis of an unknown primary largely depends on the extent of metastasis, the responsiveness of the tumor to therapy, and the identification of the primary tumor. Identifying the primary tumor improves treatment outcomes, while CUP with widespread metastasis generally has a poorer prognosis.

Follow-up care is crucial for monitoring disease progression, managing treatment-related side effects, and ensuring symptom relief. Regular imaging studies, blood tests, and clinical evaluations are necessary to assess treatment response and adjust therapeutic approaches as needed./

Conclusion

Bone and muscle metastasis from an unknown primary is a complex clinical challenge that requires a multidisciplinary approach to diagnosis and treatment. While the primary tumor remains elusive in many cases, advances in molecular diagnostics and imaging have made it possible to identify the origin of metastatic disease in a subset of patients. Treatment is primarily palliative, focusing on symptom control, quality of life, and extending survival. Ongoing research into the molecular characteristics of CUP is paving the way for more personalized therapeutic options and improving outcomes for these patients.

References

1. Huang JF, Shen J, Li X, et al. Incidence of patients with bone metastases at diagnosis of solid tumors in adults: a large population-based study. Ann Transl Med. 2020;8(7):482. https://doi.org/10.21037/atm.2020.03.55.
2. Destombe C, Botton E, Le Gal G, et al. Investigations for bone metastasis from an unknown primary. Joint Bone Spine. 2007;74(1):85–9. https://doi.org/10.1016/j.jbspin.2006.05.007.
3. Pavlidis N, Briasoulis E, Hainsworth J, Greco FA. Diagnostic and therapeutic management of cancer of an unknown primary. Eur J Cancer. 2003;39(14):1990–2005. https://doi.org/10.1016/s0959-8049(03)00547-1.
4. Pavlidis N, Pentheroudakis G. Cancer of unknown primary site. Lancet. 2012;379(9824):1428–35. https://doi.org/10.1016/S0140-6736(11)61178-1.
5. Kato S, Alsafar A, Walavalkar V, Hainsworth J, Kurzrock R. Cancer of unknown primary in the molecular era. Trends Cancer. 2021;7(5):465–77. https://doi.org/10.1016/j.trecan.2020.11.002.
6. Piccioli A, Maccauro G, Spinelli MS, Biagini R, Rossi B. Bone metastases of unknown origin: epidemiology and principles of management. J Orthop Traumatol. 2015;16(2):81–6. https://doi.org/10.1007/s10195-015-0344-0.
7. D'Oronzo S, Coleman R, Brown J, Silvestris F. Metastatic bone disease: pathogenesis and therapeutic options: up-date on bone metastasis management. J Bone Oncol. 2018;15:100224. https://doi.org/10.1016/j.jbo.2018.10.004.

8. Takagi T, Katagiri H, Kim Y, et al. Skeletal metastasis of unknown primary origin at the initial visit: a retrospective analysis of 286 cases. PLoS One. 2015;10(6):e0129428. https://doi.org/10.1371/journal.pone.0129428.
9. Hiraga T. Immune microenvironment of cancer bone metastasis. Bone. 2024;191:117328. https://doi.org/10.1016/j.bone.2024.117328.
10. Tuoheti Y, Okada K, Osanai T, et al. Skeletal muscle metastases of carcinoma: a clinicopathological study of 12 cases. Jpn J Clin Oncol. 2004;34(4):210–4. https://doi.org/10.1093/jjco/hyh036.
11. Hemminki K, Riihimäki M, Sundquist K, Hemminki A. Site-specific survival rates for cancer of unknown primary according to location of metastases. Int J Cancer. 2013;133(1):182–9. https://doi.org/10.1002/ijc.27988.
12. Nottebaert M, Exner GU, von Hochstetter AR, Schreiber A. Metastatic bone disease from occult carcinoma: a profile. Int Orthop. 1989;13(2):119–23. https://doi.org/10.1007/BF00266372.
13. Biermann JS, Holt GE, Lewis VO, Schwartz HS, Yaszemski MJ. Metastatic bone disease: diagnosis, evaluation, and treatment. Instr Course Lect. 2010;59:593–606.
14. Le Chevalier T, Cvitkovic E, Caille P, et al. Early metastatic cancer of unknown primary origin at presentation: a clinical study of 302 consecutive autopsied patients. Arch Intern Med. 1988;148(9):2035–9. https://doi.org/10.1001/archinte.1988.00380090111024.
15. Vandecandelaere M, Flipo RM, Cortet B, Catanzariti L, Duquesnoy B, Delcambre B. Bone metastases revealing primary tumors: comparison of two series separated by 30 years. Joint Bone Spine. 2004;71(3):224–9. https://doi.org/10.1016/S1297-319X(03)00123-4.
16. Ugras N, Yalcinkaya U, Akesen B, Kanat O. Solitary bone metastases of unknown origin. Acta Orthop Belg. 2014;80(1):139–43.
17. Khan MN, Sharfuzzaman A, Mostafa MG. Spinal cord compression as initial presentation of metastatic occult follicular thyroid carcinoma. J Neurosci Rural Pract. 2014;5(2):155–9. https://doi.org/10.4103/0976-3147.131662.
18. Iizuka Y, Iizuka H, Tsutsumi S, et al. Diagnosis of a previously unidentified primary site in patients with spinal metastasis: diagnostic usefulness of laboratory analysis, CT scanning and CT-guided biopsy. Eur Spine J. 2009;18(10):1431–5. https://doi.org/10.1007/s00586-009-1031-2.
19. Cooke KS, Kirpekar M, Abiri MM, Shreefter C. US case of the day: skeletal metastasis from poorly differentiated carcinoma of unknown origin. Radiographics. 1997;17(2):542–4. https://doi.org/10.1148/radiographics.17.2.9084169.
20. Wijayaratna R, Ng JW. Metatarsal metastasis as the presenting feature of occult colorectal carcinoma. BMJ Case Rep. 2013;2013:bcr2013200137. https://doi.org/10.1136/bcr-2013-200137.
21. van Veenendaal LM, de Klerk G, van der Velde D. A painful finger as first sign of a malignancy. Geriatr Orthop Surg Rehabil. 2014;5(1):18–20. https://doi.org/10.1177/2151458514522124.
22. Surov A, Köhler J, Wienke A, et al. Muscle metastases: comparison of features in different primary tumours. Cancer Imaging. 2014;14(1):21. https://doi.org/10.1186/1470-7330-14-21.
23. Ménard O, Parache RM. Muscle metastases of cancers. Ann Med Interne (Paris). 1991;142(6):423–8.
24. Acinas García O, Fernández FA, Satué EG, Buelta L, Val-Bernal JF. Metastasis of malignant neoplasms to skeletal muscle. Rev Esp Oncol. 1984;31(1):57–67.
25. Joyce JA, Pollard JW. Microenvironmental regulation of metastasis. Nat Rev Cancer. 2009;9(4):239–52. https://doi.org/10.1038/nrc2618.
26. Kang Y, Siegel PM, Shu W, et al. A multigenic program mediating breast cancer metastasis to bone. Cancer Cell. 2003;3(6):537–49. https://doi.org/10.1016/s1535-6108(03)00132-6.
27. Glinsky VV. Intravascular cell-to-cell adhesive interactions and bone metastasis. Cancer Metastasis Rev. 2006;25(4):531–40. https://doi.org/10.1007/s10555-006-9029-8.
28. Mazo IB, von Andrian UH. Adhesion and homing of blood-borne cells in bone marrow microvessels. J Leukoc Biol. 1999;66(1):25–32. https://doi.org/10.1002/jlb.66.1.25.
29. Aguirre-Ghiso JA. Models, mechanisms and clinical evidence for cancer dormancy. Nat Rev Cancer. 2007;7(11):834–46. https://doi.org/10.1038/nrc2256.

30. Yu-Lee LY, Yu G, Lee YC, et al. Osteoblast-secreted factors mediate dormancy of metastatic prostate cancer in the bone via activation of the TGFβRIII-p38MAPK-pS249/T252RB pathway. Cancer Res. 2018;78(11):2911–24. https://doi.org/10.1158/0008-5472.CAN-17-1051.
31. Ghajar CM, Peinado H, Mori H, et al. The perivascular niche regulates breast tumour dormancy. Nat Cell Biol. 2013;15(7):807–17. https://doi.org/10.1038/ncb2767.
32. Shih LY, Chen TH, Lo WH. Skeletal metastasis from occult carcinoma. J Surg Oncol. 1992;51(2):109–13. https://doi.org/10.1002/jso.2930510209.
33. Jacobsen S, Stephensen SL, Paaske BP, Lie PG, Lausten GS. Skeletal metastases of unknown origin: a retrospective analysis of 29 cases. Acta Orthop Belg. 1997;63(1):15–22.
34. Ren Y, Qian S, Xu G, Cai Z, Zhang N, Wang Z. Predicting survival of patients with bone metastasis of unknown origin. Front Endocrinol (Lausanne). 2023;14:1193318. https://doi.org/10.3389/fendo.2023.1193318.
35. Vanel D, Ruggieri P, Ferrari S, et al. The incidental skeletal lesion: ignore or explore? Cancer Imaging. 2009;9(Spec No A):S38–43. https://doi.org/10.1102/1470-7330.2009.9011.
36. Lodwick GS. A probabilistic approach to the diagnosis of bone tumors. Radiol Clin North Am. 1965;3(3):487–97.
37. Rougraff BT, Kneisl JS, Simon MA. Skeletal metastases of unknown origin: a prospective study of a diagnostic strategy. J Bone Joint Surg Am. 1993;75(9):1276–81. https://doi.org/10.2106/00004623-199309000-00003.
38. Fujibuchi T, Imai H, Kidani T, Miura H. Effective examination methods for identifying the primary origins of metastatic bone tumors of unknown primary origin during the initial visit: a retrospective chart review study. SAGE Open Med. 2022;10:20503121221097582. https://doi.org/10.1177/20503121221097582.
39. Fang J, Xu Q. Differences of osteoblastic bone metastases and osteolytic bone metastases in clinical features and molecular characteristics. Clin Transl Oncol. 2015;17(3):173–9. https://doi.org/10.1007/s12094-014-1247-x.
40. Higinbotham NL, Marcove RC. The management of pathological fractures. J Trauma. 1965;5(6):792–8. https://doi.org/10.1097/00005373-196511000-00015.
41. Selvaggi G, Scagliotti GV. Management of bone metastases in cancer: a review. Crit Rev Oncol Hematol. 2005;56(3):365–78. https://doi.org/10.1016/j.critrevonc.2005.03.011.
42. Groot OQ, Lans A, Twining PK, et al. Clinical outcome differences in the treatment of impending versus completed pathological long-bone fractures. J Bone Joint Surg Am. 2022;104(4):307–15. https://doi.org/10.2106/JBJS.21.00711.
43. Putro YAP, Aryandono T, Widodo I, et al. Diagnostic approach to bone metastasis of unknown origin: a systematic review. Orthop Rev (Pavia). 2024;16:123441. https://doi.org/10.52965/001c.123441.
44. Orel SG, Weinstein SP, Schnall MD, et al. Breast MR imaging in patients with axillary node metastases and unknown primary malignancy. Radiology. 1999;212(2):543–9. https://doi.org/10.1148/radiology.212.2.r99au40543.
45. Ko EY, Han BK, Shin JH, Kang SS. Breast MRI for evaluating patients with metastatic axillary lymph node and initially negative mammography and sonography. Korean J Radiol. 2007;8(5):382–9. https://doi.org/10.3348/kjr.2007.8.5.382.
46. Gődény M, Lengyel Z, Polony G, et al. Impact of 3T multiparametric MRI and FDG-PET-CT in the evaluation of occult primary cancer with cervical node metastasis. Cancer Imaging. 2016;16(1):38. https://doi.org/10.1186/s40644-016-0097-x.
47. Talbot JN, Paycha F, Balogova S. Diagnosis of bone metastasis: recent comparative studies of imaging modalities. Q J Nucl Med Mol Imaging. 2011;55(4):374–410.
48. Zhang Y, Zhao C, Liu H, Hou H, Zhang H. Multiple metastasis-like bone lesions in scintigraphic imaging. J Biomed Biotechnol. 2012;2012:957364. https://doi.org/10.1155/2012/957364.
49. Rajarubendra N, Bolton D, Lawrentschuk N. Diagnosis of bone metastases in urological malignancies—an update. Urology. 2010;76(4):782–90. https://doi.org/10.1016/j.urology.2009.12.050.

50. Bäuerle T, Semmler W. Imaging response to systemic therapy for bone metastases. Eur Radiol. 2009;19(10):2495–507. https://doi.org/10.1007/s00330-009-1443-1.
51. Bicakci N. Diagnostic and prognostic value of F-18 FDG PET/CT in patients with carcinoma of unknown primary. North Clin Istanb. 2022;9(4):337–46. https://doi.org/10.14744/nci.2021.71598.
52. Cowan AJ, Green DJ, Kwok M, et al. Diagnosis and management of multiple myeloma: a review. JAMA. 2022;327(5):464–77. https://doi.org/10.1001/jama.2022.0003.
53. Paquin AR, Oyogoa E, McMurry HS, Kartika T, West M, Shatzel JJ. The diagnosis and management of suspected lymphoma in general practice. Eur J Haematol. 2023;110(1):3–13. https://doi.org/10.1111/ejh.13863.

Isolated Lytic Lesions: Where Is This Coming from?

39

Alysia K. Kemp and Michelle Ghert

Introduction

Survival rates from cancer have improved significantly [1]. Tumors that frequently metastasize to bone, like breast, prostate, and thyroid, have the highest 5- and 10-year survival rates when compared to other malignancies [1]. As patients live longer, the incidence of metastatic bone disease (MBD) has also increased [2]. Hernandez et al. found that 4.8% of patients with solid tumors will develop bone metastasis within the first year of diagnosis [2]. In their study, the incidence of patients with bone metastasis increased thereafter, reaching 8% at 10 years [2]. MBD can be painful and cause severe functional limitations that lead to a poor quality of life. Depending on the burden of disease, the patient may present with an isolated lytic bone lesion. An in-depth understanding of the pathophysiology, presenting symptoms, diagnosis, work-up, and management of isolated bone lesions is important to achieve the best outcome for patients with metastatic bone disease.

Differential Diagnosis for an Isolated Lytic Bone Lesion

The differential diagnosis of a destructive, lytic, bone lesion identified in an adult (>40 years old) must include metastatic bone disease (MBD). There are five solid tumor types that have a high predilection of metastasis to bone: breast, prostate, lung, renal, and thyroid carcinoma [3], although almost any type of carcinoma can

A. K. Kemp (✉)
The University of Texas MD Anderson Cancer Center, Houston, TX, USA

M. Ghert
University of Maryland School of Medicine, College Park, MD, USA

McMaster University, Hamilton, ON, Canada
e-mail: mghert@som.umaryland.edu

J. Pretell-Mazzini, S. A. Lozano Calderón (eds.), *Multidisciplinary Approach of Musculoskeletal Metastasis*, https://doi.org/10.1007/978-3-032-09498-8_39

metastasize to bone. The tumors spread hematogenously from their site of origin to the bone and attach via integrins, a cell adhesion receptor, to the bone's extracellular matrix (ECM) [4]. Once attached to the ECM, the tumor cells secrete mediators such as parathyroid hormone-related peptide (PTHrP) that stimulate the receptor activator for nuclear factor kB (RANK)-RANK ligand (RANK-L) pathway. RANK is released from osteoblasts and recruit precursor osteoclast cells with RANK-L receptors. Bone resorption is performed by the recruited osteoclasts resulting in the lytic lesions seen on radiographs and pain experienced by patients [5] This process often happens at multiple sites, especially in late-stage disease. However, patients can initially present with an isolated lytic lesion, and it is crucial for physicians to look for additional sites of disease.

Other diagnoses that should be considered for isolated lytic bone lesions include giant cell tumor of bone (GCTB) which is a benign but locally aggressive condition or hematologic malignancies such as multiple myeloma or lymphoma of bone. A patient with a GCTB is often in their third to fourth decade of life with a painful lesion and swelling that involves the knee. The radiographic appearance of GCTB will show an eccentric lytic lesion within the metaphysis of the bone, most frequently the distal femur or proximal tibia. GCTB are often large and can demonstrate erosion of the cortex and pathologic fracture, because of this aggressive appearance they are often mistaken for malignancy. Multiple myeloma is a hematologic malignancy caused by the abnormal proliferation of plasma cells in the bone marrow [6]. Classically, in addition to the destructive bone lesions, patients present with elevated calcium, kidney damage, and anemia [6]. There is a higher incidence of multiple myeloma among patients of the male sex, African American race, and age 65 years or older [6].

Primary lymphoma of bone is a rare malignancy that can present as an isolated lytic bone lesion [7]. The median age of presentation is 30 s to 50 s; however, some studies have demonstrated a peak incidence later in life in the sixth to seventh decade [7]. The most common radiographic appearance of lymphoma of bone is a large lytic bone lesion adjacent to a large soft tissue mass. However, there is often a "mismatch" between the plain radiographs and the MRI appearance. It is not uncommon to see a large soft tissue mass or a large infiltrative, marrow replacing process on MRI, but very subtle changes on the plain x-ray. In a retrospective series of 237 patients, Mulligan et al. found that 70% of patients presented with an ill-defined, permeative appearance of the bone lesion on x-ray [8].

Primary bone sarcomas are a small group of malignant tumors that can radiographically present as an isolated lytic lesion within the bone. Undifferentiated pleomorphic sarcoma (UPS) of bone is a rare diagnosis, the reported incidence is less than 2% of all primary bone sarcomas. UPS of bone is commonly found in the lower extremity with a higher predilection for the long bones such as the femur and tibia. Although UPS of bone is rare, UPS and other primary bone sarcomas should be included in your differential diagnosis for isolated lytic bone lesions.

Finally, osteomyelitis is another diagnosis that must be considered when evaluating a patient with an isolated lytic bone lesion. Osteomyelitis, a deep bone infection, can occur after open fracture, surgical procedures, or progression of soft tissue infections especially in the setting of poorly controlled diabetes. Radiographic changes can take time to develop, patients may have an infection for 10 to 21 days before sufficient bone resorption allows the lesion to be visible on x-ray [9]. Patients with suspected osteomyelitis should be worked up with complete blood count (CBC), erythrocyte sedimentation rate (ESR), C-reactive protein (CRP) labs, and advanced imaging such as imaging with computed tomography (CT) and magnetic resonance imaging (MRI). Often, the treatment of osteomyelitis requires a referral to an infectious disease specialist for recommendations regarding the appropriate antibiotic regimen and orthopedic surgery for an evaluation for surgical debridement of the bone.

Clinical Evaluation

Evaluation of a patient with an isolated lytic bone lesion begins with obtaining a detailed history and performing a thorough physical exam. Patients often present with persistent pain and the duration of the pain can give an indication of how long the lesion has been present. Pain is often insidious and will gradually worsen over time. It is important to elucidate if the patient has night pain, and if the patient has pain with or without weight bearing as the level of pain often correlates with the integrity of the bone. It is important to ask about a patient's personal or family history of cancer, previous radiation or toxin exposure, and surgeries or injuries to the area. On physical exam, inspection of the skin overlying the area of concern is important [3]. Direct palpation can detect an underlying soft tissue mass or fracture. Physical exam should include a neurologic examination to identify motor weakness, sensory deficits, or areas of vascular compromise.

Diagnosis and Work-Up

To identify the etiology of isolated bone lesions, it is important to obtain imaging and laboratory studies. Radiographs are an important first step and should be obtained during the initial consultation with the patient. A full-length x-ray of the bone should include the joint above and below to evaluate the lesion size and extent of tumor involvement. On the plain radiographs, approximately 50% of the trabecular bone must be involved before it becomes evident on the x-ray [10]; therefore, large lytic lesions should be concerning for risk of a pathologic fracture and prompt consultation with an orthopedic surgeon for possible stabilization (Fig. 39.1). Classically, renal, thyroid, and lung metastasis will appear lytic. While prostate and breast cancer can appear blastic or have mixed (lytic and blastic) appearance. However, it should be noted that the sclerotic bone is abnormal, and it is still possible to sustain an atraumatic (pathologic) fracture through these lesions.

Advanced imaging with computed tomography (CT) and magnetic resonance imaging (MRI) are often used during the work-up of lytic bone lesions to identify areas of involvement that may be at risk for fracture (Fig. 39.2). CT can be very helpful for identifying sclerotic lesions and to determine the extent of cortical involvement [3]. CT scans of the chest, abdomen, and pelvis (CT CAP) are often used to determine the extent (stage) of disease and can be used to identify the

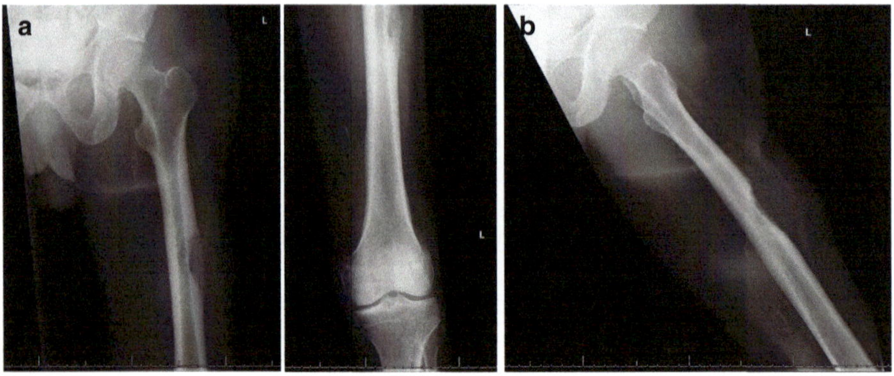

Fig. 39.1 (**a**) Proximal anteroposterior (AP) radiographs of a 65-year-old man with an isolated lytic bone lesion secondary to metastatic lung carcinoma. (**b**) Proximal lateral radiographs of a 65-year-old man with an isolated lytic bone lesion secondary to metastatic lung carcinoma

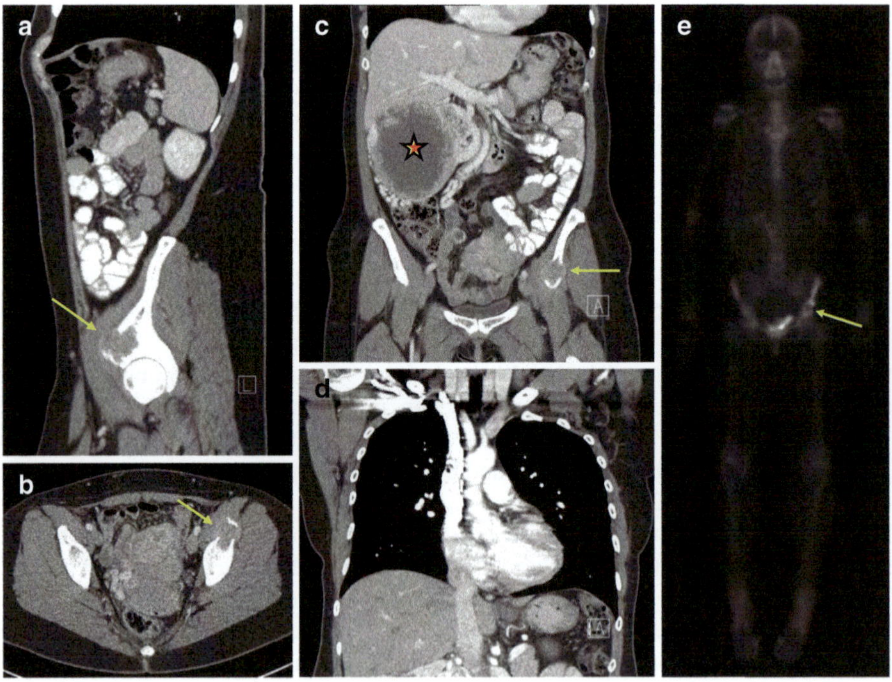

Fig. 39.2 Fifty-three-year-old female with oligometastatic renal cell carcinoma of the left acetabulum. CT of the pelvis demonstrated a lytic lesion in the sagittal (**a**), axial (**b**) and coronal (**c**) planes, which is also appreciated on full-body bone scan (**e**). Staging studies (**c**, **d**, and **e**) revealed the renal primary tumor (red star) and pelvic metastasis (yellow arrows) as the only sites of disease. (These images are courtesy of Dr. Santiago A. Lozano Calderón, MD, PhD; his group owns the copyrights)

primary tumor site in patients that present with metastatic disease. The abnormal lymphadenopathy of lymphoma as well as primary breast, lung, thyroid, renal, and prostate cancers are often detected on initial CTAP studies and should be a part of the patient's diagnostic work-up.

MRI is also used to work-up isolated lytic bone lesions. A recent study by Azad et al. evaluated 92 patients with suspected bone lesions and found that both x-ray and MRI have high diagnostic accuracy (86.95% for x-ray and 87.0% for MRI) but that MRI is most useful for determining the extent of tumor growth and giving precise analysis of tissue involvement [11]. These studies are helpful to identify discontinuous tumor deposits (skip lesions) and soft tissue involvement, helpful for surgical planning, and can assist with implant selection during the preoperative visit.

Bone scintigraphy (bone scan) is also used to detect osseous metastatic disease. Technetium-99 m (Tc99m) is a radioactive isotope that binds to the skeletal bone via a phosphate analog that forms a complex with the crystalline hydroxyapatite of bone [12]. Bone scan is sensitive at detecting bone metastatic disease and may be used for initial staging especially in patients without clinical symptoms. Some notable exceptions are with patients with multiple myeloma or some subtypes of renal cell carcinoma. In these patients, the bone turnover is so rapid that the radioisotope and phosphate analog are not taken up by the host bone, and therefore, the lesions appear inactive or "cold."

Laboratory studies should be obtained on the patients with isolated lytic bone lesions. These tests can help with the diagnosis and work-up of the lesion and give an indication of a patient's fitness for surgery. Anemia and thrombocytopenia are common lab abnormalities among patients with metastatic disease and can be evaluated with a complete blood count (CBC). Often, tumor cells replace the hematopoietic cells of the bone marrow (myelophthisis) leading to suppressed hemoglobin and platelet production and function [10, 13]. Hypercalcemia is also common and is identified on a basic metabolic panel. The increased bone resorption by osteoclasts can lead to toxic levels of calcium that can cause bone and abdominal pain, fatigue, constipation, nausea, vomiting, and confusion. If left unresolved patients can die from hypercalcemia, it is important to screen for it and if diagnosed patients should be immediately admitted to the hospital for intravenous (IV) hydration. Given its relevance, it is important to correct or adjust serum calcium concentrations in relation to albumin. Elevated serum tumor markers like prostate specific antigen (PSA), carcinoembryonic antigen (CEA), or cancer antigen 125 (CA 125 antigen) can be elevated in prostate, gastrointestinal, colorectal, or gynecologic malignancies but may also be associated with benign conditions such as pregnancy, endometriosis, or menstruation. These serum markers may be elevated in patients with isolated lytic bone lesions due to metastatic carcinoma, and when present, can be used to evaluate a patient's response to therapy after treatment has begun [14]. Serum protein electrophoresis (SPEP) including light kappa chains and urine protein electrophoresis (UPEP) are used to detect abnormal monoclonal protein (M-protein) that are produced in patients with multiple myeloma. This test is diagnostic for multiple myeloma and patients with lytic bone lesions and positive SPEP and/or UPEP do not require further work-up with a biopsy.

Bone Biopsy

While most primary tumor sites can be identified by history, physical exam, bloodwork, and CT CAP, a bone biopsy may be needed to confirm the diagnosis and identify the primary tumor site in patients without a known primary tumor diagnosis or previous history of bone metastasis. The biopsy should be performed at a high-volume cancer center where patients will get their definitive treatment. Coordination between the treatment team and the proceduralist performing the biopsy is crucial to avoid soft tissue contamination and to minimize sampling errors. Previous studies have demonstrated that percutaneous biopsies performed at high-volume centers have a diagnostic accuracy between 90% and 95% especially for osteolytic lesions [15]. However, the patients should have a full history, physical exam, staging studies, and advanced local imaging of the area of concern completed prior to their biopsy so that this information can assist the pathologist and improve the accuracy of the results. Open biopsy, if needed, should be performed using a longitudinal incision in line with the incision that will be used for the definitive treatment. During open biopsy, blood loss should be minimized, and when possible, surgeons should use a direct approach through one compartment, to access the tumor. Frozen section can be used, and it is recommended to confirm that sufficient tissue was retrieved to make a diagnosis. Frozen biopsy confirmation is also recommended in percutaneous biopsies [16]. Lastly, at the end of the open biopsy meticulous hemostasis is needed to minimize bleeding and prevent hematoma formation.

Treatment

The treatment of isolated bone lesions in patients with metastatic disease includes, first, identifying the primary site of disease through the modalities discussed previously. Next, the goals are to identify patients at risk for pathologic fracture, when possible, to prevent fracture, restore function, and minimize pain [17]. Recent studies have underscored the importance of early identification of patients at risk for pathologic fracture. A recent study evaluated the American College of Surgeons National Surgical Quality Improvement Program database for patients with metastatic bone disease who had surgical stabilization before and after pathologic long bone fracture [18]. That study found that the risk of major complications, including but not limited to death, was lower among patients who underwent prophylactic fixation [18]. Similarly, Nooh et al. found that patients experienced improvement in their pain, quality of life, and functional outcomes as early as two weeks post op [19]. They advocate for surgical intervention if the life expectancy is anticipated to be longer than 2 weeks [19]. New prognostic tools with machine learning models such PathFX and the SORG calculator can assist in estatimating patient survival. This information can aid in determining the type of surgical intervention recommended.

However, identifying patients at risk for pathologic fracture remains a challenge. The Mirels criteria is a popular and simple tool that many treating teams use to categorize a patient's risk for fracture. The Mirels score is a weighted score based on four criteria: size of the lesion, radiographic appearance, location, and the patient's reported pain. Each criteria is given a weighted score, 1–3, based on the severity of the symptoms, with a higher score (>9) indicating a higher risk of fracture, whereas a lower score (<7) would suggest the patient is not at immediate risk and might not benefit from surgical stabilization. While the sensitivity for Mirels is noted to be high (91%), the specificity is quite low (35%), and thus, more sensitive tests are needed [20]. [GM5] Newer clinical decision-support tools have been developed to predict post-operative mortality for patients with skeletal metastases and can assist surgeons in evaluating the risks associated with surgery. PATHFx version 3.0 is a validated machine-learning decision aide that uses patient age, sex, diagnosis, Eastern Cooperative Oncology Group (ECOG), pre-operative hemoglobin, presence or absence of a fracture, and disease burden to predict the post-operative survival rate at one, three, six, twelve, eighteen, and twenty-four months [21]. PATHFx has been validated using diverse patient populations and recently was validated to predict survival for patients undergoing nonoperative treatment such as external beam radiation [21]. The Skeletal Oncology Research Group machine-learning algorithms (SORG-MLAs) also developedmachine-learning models to predict survival in patients with skeletal metastases [22, 23]. This model uses patient sex, age, histologic subtype, tumor location, disease burden, 12 preoperative lab values, and prior systemic therapy to estimate survival at 90 days and one year [22]. Both the PATHFx and SORG-MLA use objective clinical data to guide surgeon decision making and are available at no cost to the patient or physician.

Selecting the appropriate surgical procedure and construct is also a challenge. Lin et all demonstrated that patients with bone only involvement and limited burden of disease have the most favorable prognosis and long term survival [24]. Therefore, the goal for surgical stabilization of an isolated lytic bone lesion from metastasis is palliative in nature. The construct must be durable and allow for immediate weight bearing as well as alleviate pain and restore the preoperative function, when possible. Many authors have investigated the role for wide resection in metastatic bone disease, especially in the setting of historically radiation and chemotherapy resistant disease like renal cell carcinoma [25]. Langerhuizen et al found reduced local recurrence and improved survival among patients with bone metastasis due to renal cell carcinoma that were treated with negative margin metastasectomy instead of intralesional curettage or stabilization [25]. Conversely, Fuchs et al. analyzed the survival rates of 60 patients with solitary metastatic lesions from renal cell carcinoma and found no survival advantage to wide resection compared to intralesional stabilization [26]. They did, however, find a survival advantage to surgical stabilization over nonoperative management [26]. Additionally, wide resection and endoprosthesis reconstruction has also been associated with higher rates of postoperative complications such as implant dislocation and is most common in the first postoperative year (time to complication 3 to 11.6 months) [27]. Intralesional resection should be performed through an open approach to the lesion, with removal of the lesion by curettage, extended high speed bur, and the use of adjuvant treatments like hydrogen peroxide, argon beam laser, and polymethyl methacrylate (PMMA) [28, 29]. The

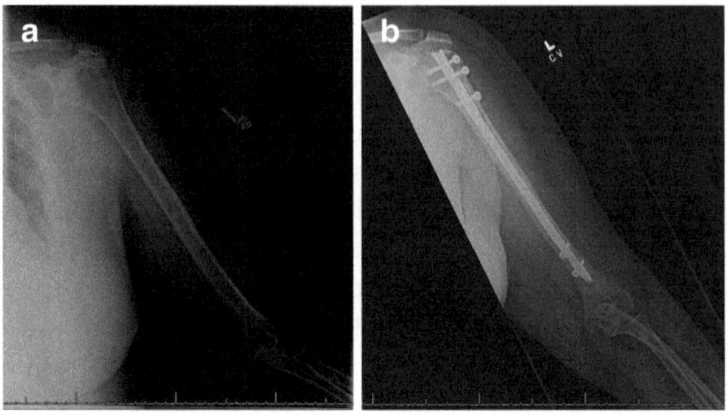

Fig. 39.3 (**a**) Preoperative AP radiographs of a left humerus in a 48-year-old woman treated with a locked intramedullary nail for an isolated lytic bone lesion due to metastatic breast cancer. (**b**) Postoperative AP radiographs of a left humerus in a 48-year-old woman treated with a locked intramedullary nail for an isolated lytic bone lesion due to metastatic breast cancer

goal of intralesional resection is to induce local tumor necrosis and implant survivorship is increased with the use of adjuvants like PMMA. Preoperative cryoablation of painful skeletal metastatic disease has also been used to improve pain control after surgery. A prospective multicenter clinical trial evaluated the use of ablation on painful skeletal metastases and found a 4 to 6 point (on a 10-point scale) decrease in the mean pain score within 3- 6 months post procedure [30].

If the lesion is in the diaphysis of a long bone, stabilization of the lesion can be achieved with a locked intramedullary nail, a plate, or a combination of the two. The use of nails is preferred by many given its load sharing characteristics (Fig. 39.3). If the lesion is in the metaphysis, or rarely the epiphysis, stabilization may be achieved with periarticular plating or endoprosthesis reconstruction. Selecting an implant that will allow for immediate weight bearing is important, as prolonged nonweight bearing and bed rest can lead to premature mortality.

Summary

As survival rates from cancer continue to improve, the incidence of metastatic bone disease is also increasing. Isolated lytic bone lesions can be painful and cause severe functional limitations that lead to a poor quality of life. Prompt diagnosis and work-up will lead to early identification and definitive treatment of patients at high risk for pathologic fracture. Early treatment with durable implant constructs has been associated with improved survival and increased quality of life measures as early as the two weeks post op.

References

1. Sakr RA, Nasr AA, Zineldin EI, Gouda MA. Long-term survival in patients with cancers: surveillance, epidemiology and end results-based analysis. Sultan Qaboos Univ Med J. 2023;23(3):344–50. https://doi.org/10.18295/squmj.1.2023.002. Epub 2023 Aug 28. PMID: 37655083; PMCID: PMC10467541.
2. Hernandez RK, Wade SW, Reich A, Pirolli M, Liede A, Lyman GH. Incidence of bone metastases in patients with solid tumors: analysis of oncology electronic medical records in the United States. BMC Cancer 2018;18(1):44. https://doi.org/10.1186/s12885-017-3922-0. PMID: 29306325; PMCID: PMC5756362.
3. Jayarangaiah A, Kemp AK, Theetha Kariyanna P. Bone Metastasis. 2023 Jul 31. In: StatPearls [Internet]. Treasure Island: StatPearls Publishing; 2024. PMID: 29939688.
4. Hamidi H, Ivaska J. Every step of the way: integrins in cancer progression and metastasis. Nat Rev Cancer. 2018;18(9):533–48. https://doi.org/10.1038/s41568-018-0038-z. Erratum in: Nat Rev Cancer. 2019 Mar;19(3):179. doi: 10.1038/s41568-019-0112-1. PMID: 30002479; PMCID: PMC6629548.
5. Kearns AE, Khosla S, Kostenuik PJ. Receptor activator of nuclear factor kappaB ligand and osteoprotegerin regulation of bone remodeling in health and disease. Endocr Rev. 2008;29(2):155–92. https://doi.org/10.1210/er.2007-0014. Epub 2007 Dec 5. PMID: 18057140; PMCID: PMC2528846.
6. Cowan AJ, Green DJ, Kwok M, Lee S, Coffey DG, Holmberg LA, Tuazon S, Gopal AK, Libby EN. Diagnosis and management of multiple myeloma: a review. JAMA. 2022;327(5):464–77. https://doi.org/10.1001/jama.2022.0003. PMID: 35103762.
7. Krishnan A, Shirkhoda A, Tehranzadeh J, Armin AR, Irwin R, Les K. Primary bone lymphoma: radiographic-MR imaging correlation. Radiographics. 2003;23(6):1371–83; discussion 1384–7. https://doi.org/10.1148/rg.236025056. PMID: 14615550.
8. Mulligan ME, McRae GA, Murphey MD. Imaging features of primary lymphoma of bone. AJR Am J Roentgenol. 1999;173(6):1691–7. https://doi.org/10.2214/ajr.173.6.10584821. PMID: 10584821.
9. Pineda C, Espinosa R, Pena A. Radiographic imaging in osteomyelitis: the role of plain radiography, computed tomography, ultrasonography, magnetic resonance imaging, and scintigraphy. Semin Plast Surg. 2009;23(2):80–9. https://doi.org/10.1055/s-0029-1214160. PMID: 20567730; PMCID: PMC2884903.
10. Riccio AI, Wodajo FM, Malawer M. Metastatic carcinoma of the long bones. Am Fam Physician. 2007;76(10):1489–94. PMID: 18052014.
11. Azad H, Ahmed A, Zafar I, Bhutta MR, Rabbani MA, Kc HR. X-ray and MRI correlation of bone tumors using histopathology as gold standard. Cureus. 2022;14(7):e27262. https://doi.org/10.7759/cureus.27262. PMID: 36039258; PMCID: PMC9403219.
12. Adams C, Banks KP. Bone scan. 2023 Aug 14. In: StatPearls [Internet]. Treasure Island: StatPearls Publishing; 2024. PMID: 30285381.
13. Gangat N, Wolanskyj AP. Anemia of chronic disease. Semin Hematol. 2013;50(3):232–8. https://doi.org/10.1053/j.seminhematol.2013.06.006. Erratum in: Semin Hematol. 2014 Apr;51(2):e1. PMID: 23953340.
14. Jacobs I, Bast RC Jr. The CA 125 tumour-associated antigen: a review of the literature. Hum Reprod. 1989;4(1):1–12. https://doi.org/10.1093/oxfordjournals.humrep.a136832. PMID: 2651469.
15. Filippiadis D, Mazioti A, Kelekis A. Percutaneous, imaging-guided biopsy of bone metastases. Diagnostics (Basel). 2018;8(2):25. https://doi.org/10.3390/diagnostics8020025. PMID: 29670012; PMCID: PMC6023375.
16. Chang CY, Garner HW, Ahlawat S, Amini B, Bucknor MD, Flug JA, Khodarahmi I, Mulligan ME, Peterson JJ, Riley GM, Samim M, Lozano-Calderon SA, Wu JS. Society of Skeletal Radiology- white paper. Guidelines for the diagnostic management of incidental solitary bone lesions on CT and MRI in adults: bone reporting and data system (Bone-RADS). Skeletal

Radiol. 2022;51(9):1743–64. https://doi.org/10.1007/s00256-022-04022-8. Epub 2022 Mar 28. PMID: 35344076; PMCID: PMC9283187.
17. Groot OQ, Lans A, Twining PK, Bongers MER, Kapoor ND, Verlaan JJ, Newman ET, Raskin KA, Lozano-Calderon SA, Janssen SJ, Schwab JH. Clinical outcome differences in the treatment of impending versus completed pathological long-bone fractures. J Bone Joint Surg Am. 2022;104(4):307–15. https://doi.org/10.2106/JBJS.21.00711. PMID: 34851323.
18. El Abiad JM, Raad M, Puvanesarajah V, Rao SS, Morris CD, Levin AS. Prophylactic versus postfracture stabilization for metastatic lesions of the long bones: a comparison of 30-day postoperative outcomes. J Am Acad Orthop Surg. 2019;27(15):e709–16. https://doi.org/10.5435/JAAOS-D-18-00345. PMID: 31344005.
19. Nooh A, Goulding K, Isler MH, Mottard S, Arteau A, Dion N, Turcotte R. Early improvement in pain and functional outcome but not quality of life after surgery for metastatic long bone disease. Clin Orthop Relat Res. 2018;476(3):535–45. https://doi.org/10.1007/s11999.0000000000000065. PMID: 29529637; PMCID: PMC6260046.
20. Damron TA, Morgan H, Prakash D, Grant W, Aronowitz J, Heiner J. Critical evaluation of Mirels' rating system for impending pathologic fractures. Clin Orthop Relat Res. 2003;415 Suppl:S201–7. https://doi.org/10.1097/01.blo.0000093842.72468.73. PMID: 14600611.
21. Anderson AB, Wedin R, Fabbri N, Boland P, Healey J, Forsberg JA. External Validation of PATHFx Version 3.0 in Patients Treated Surgically and Nonsurgically for Symptomatic Skeletal Metastases. Clin Orthop Relat Res. 2020 Apr;478(4):808-818. https://doi.org/10.1097/CORR.0000000000001081. PMID: 32195761; PMCID: PMC7282571.
22. Thio QCBS, Karhade AV, Bindels BJJ, Ogink PT, Bramer JAM, Ferrone ML, Calderón SL, Raskin KA, Schwab JH. Development and Internal Validation of Machine Learning Algorithms for Preoperative Survival Prediction of Extremity Metastatic Disease. Clin Orthop Relat Res. 2020 Feb;478(2):322-333. https://doi.org/10.1097/CORR.0000000000000997. Erratum in: Clin Orthop Relat Res. 2021 Apr 1;479(4):862. doi: 10.1097/CORR.0000000000001678. PMID: 31651589; PMCID: PMC7438151.
23. Bongers MER, Thio QCBS, Karhade AV, Stor ML, Raskin KA, Lozano Calderon SA, DeLaney TF, Ferrone ML, Schwab JH. Does the SORG Algorithm Predict 5-year Survival in Patients with Chondrosarcoma? An External Validation. Clin Orthop Relat Res. 2019 Oct;477(10):2296-2303. https://doi.org/10.1097/CORR.0000000000000748. PMID: 31107338; PMCID: PMC6999936.
24. Lin PP, Mirza AN, Lewis VO, Cannon CP, Tu SM, Tannir NM, Yasko AW. Patient survival after surgery for osseous metastases from renal cell carcinoma. J Bone Joint Surg Am. 2007 Aug;89(8):1794-801. https://doi.org/10.2106/JBJS.F.00603. PMID: 17671020.
25. Langerhuizen DW, Janssen SJ, van der Vliet QM, Raskin KA, Ferrone ML, Hornicek FJ, Schwab JH, Lozano-Calderón SA. Metastasectomy, intralesional resection, or stabilization only in the treatment of bone metastases from renal cell carcinoma. J Surg Oncol. 2016 Aug;114(2):237-45. https://doi.org/10.1002/jso.24284. Epub 2016 May 9. PMID: 27156495.
26. Fuchs B, Trousdale RT, Rock MG. Solitary bony metastasis from renal cell carcinoma: significance of surgical treatment. Clin Orthop Relat Res. 2005;431:187–92. https://doi.org/10.1097/01.blo.0000149820.65137.b4. PMID: 15685074.
27. Di Martino A, Martinelli N, Loppini M, Piccioli A, Denaro V. Is endoprosthesis safer than internal fixation for metastatic disease of the proximal femur? A systematic review. Injury. 2017;48(Suppl 3):S48–54. https://doi.org/10.1016/S0020-1383(17)30658-7. PMID: 29025610.
28. Sim FH, Daugherty TW, Ivins JC. The adjunctive use of methylmethacrylate in fixation of pathological fractures. J Bone Joint Surg Am. 1974;56(1):40–8. PMID: 4812166.
29. Harrington KD, Sim FH, Enis JE, Johnston JO, Diok HM, Gristina AG. Methylmethacrylate as an adjunct in internal fixation of pathological fractures. Experience with three hundred and seventy-five cases. J Bone Joint Surg Am. 1976;58(8):1047–55. PMID: 1002744.
30. Kurup AN, Morris JM, Callstrom MR. Ablation of Musculoskeletal Metastases. AJR Am J Roentgenol. 2017 Oct;209(4):713-721. https://doi.org/10.2214/AJR.17.18527. Epub 2017 Aug 4. PMID: 28777648.

Soft Tissue Metastases from Carcinomas and Sarcomas: What Is This About?

40

Javier Delgado Obando

Introduction

Soft tissue metastases from carcinomas or sarcomas are rare, accounting for approximately 0.2% to 1% of all soft tissue tumors [1–3]. Despite this, some studies have found a higher percentage of soft tissue metastases in autopsies, around 16% to 17.5% [4]. In North America, approximately 13,590 new cases of soft tissue sarcomas are expected to be diagnosed, with an estimated 5200 deaths attributed to this disease [5]. Although soft tissue sarcomas constitute less than 1% of all human cancers, they can have considerable clinical implications, especially when they metastasize to the musculoskeletal system. Generally, studies are standardized to identify the primary tumor and diagnose soft tissue metastases from sarcomas or carcinomas. The metastatic spread of sarcomas or carcinomas to the musculoskeletal system can occur through hematogenous or lymphatic pathways or by direct extension. Initially, distant metastasis usually occurs through lymphatic drainage from the primary tumor to adjacent regional lymphnodes.

For this reason, these patients must be staged, have performance status scores applied, and be treated based on their survival, which is around 55% at one year and a maximum of 33% at two years. Early diagnosis is crucial; the PET scan currently facilitates the timely identification of soft tissue metastases with high sensitivity and specificity. This early detection can significantly impact the quality of life of these patients.

J. D. Obando (✉)
Universidad Austral de Chile, Clínica Alemana of Valdivia, Valdivia Chile Base Hospital, Valdivia, Chile
e-mail: jdelgado@alemana.cl

J. Pretell-Mazzini, S. A. Lozano Calderón (eds.), *Multidisciplinary Approach of Musculoskeletal Metastasis*, https://doi.org/10.1007/978-3-032-09498-8_40

Epidemiology

Morinaga et al. found that the most common carcinomas that metastasize to soft tissues are the lung (38.5%), kidney (27%), digestive tract (15.4%), breast (3.8%), liver (3.8%), and the rest of unknown origin. Glockner et al. [6] reported 1421 patients who presented soft tissue metastases in eleven patients [7]; 58% were males, and 75% had solitary tumors [1, 9].

Another less frequent primary tumor that metastasizes to soft tissues that are described in the literature is a tumor of the skin, as well as of the oropharyngeal tract [8].

The lungs represent approximately 41% of the documented soft tissue metastases [15], followed by metastases from renal cell carcinoma, prostate cancer, and breast cancer, respectively [10].

Skeletal muscle metastases originating from carcinomas are frequently detected in the axial region of the body, encompassing the paravertebral, gluteal, and psoas muscles. Identifying a mass in these anatomical locations should prompt healthcare providers to contemplate metastatic disease in conjunction with primary soft tissue neoplasms. When examining muscle groups individually, the thigh muscles emerge as the most affected, a characteristic they share with soft tissue sarcomas [1].

Pathogenesis and Histopathological Findings

Metastatic behavior can be anticipated based on their propensity to metastasize to various organs, contingent upon their physiological conditions. Carcinomas metastasize frequently to local and regional lymph nodes, subsequently affecting the liver, lungs, and bones. However, metastasis to skeletal muscle is relatively rare due to its specific homeostatic conditions that obstruct tumor cell proliferation; even though skeletal muscle represents 50% of the total body mass and receives a large portion of cardiac output, there are factors that may prevent the arrival of metastases of sarcomas or carcinomas, such as repetitive trauma from cellular muscle contraction, high lactate concentration, high blood flow, low pH, and variability in cellular oxygen supply [12]. In instances where skeletal muscle metastases do occur, hematogenous dissemination is prevalent, along with spread via venous vessels and perineural pathways [11]. The paucity of soft tissue and skeletal muscle metastases can be explained by various theories. One theory suggests that skeletal muscle vessels are conditioned to lactic acid, which inhibits the formation of new blood vessels needed for metastatic growth. Additionally, skeletal muscle produces tumor-inhibiting substances that can prevent the development of metastases. Conversely, changes in the normal physiology of skeletal muscle, such as trauma, may make these sites more susceptible to metastases. Trauma can lead to increased blood flow and the release of factors that promote the growth and trapping of circulating tumor cells, thereby increasing the likelihood of metastatic seeding [13].

All soft tissue tumors must be biopsied to determine whether they are primary tumors or secondary locations of carcinoma or sarcoma, and the basic principles of sampling must be duly considered, whether by core biopsy, incisional biopsy, or excisional biopsy [14–16].

Immunohistochemistry techniques and markers are important for diagnosis: cytokeratin (CK), CK7 and CK20, carcinogenic embryonic antigen (CEA), EMA, or epithelium membrane antigen. MOC31, S-100 protein, HMB-45, Melan-A, CD31, CD34, bcl-2, CD117, and GCDFP-15 are beneficial information in identifying the primary tumor (Table 40.1). Some specific expression patterns, such as TTF-1 (thyroid transcription factor) and PE-10 for lung carcinoma and CDX2 for colorectal carcinoma, confirm specific primary origins. MUC5AC and H1K1083 help confirm gastrointestinal origin and Hep-Par 1 for hepatocarcinoma [17].

These techniques can accurately identify metastatic tumors from carcinomas and sarcomas. However, histopathological studies must be correlated with immunohistochemistry and, in some cases, with genetic translocation tests by FISH to confirm complex diagnoses.

Although a primary tumor is commonly identified, some authors describe 4% to 24% of cases with unknown origin [10, 12].

Table 40.1 Description of different immunohistochemistry stains and tumor markers used to diagnose tumor origin

Immunohistochemistry/ Tumor marker	Use
Cytokeratin	A general epithelial marker; used to identify carcinomas (epithelial tumors). It stains the cytoplasm of epithelial cells and helps in the diagnosis of epithelial origin tumors.
CK7	Used to identify adenocarcinomas of lung, breast, ovarian, and endometrial origin. Positive in many carcinomas, particularly in tumors from the lungs, breast, and gastrointestinal tract.
CK20	Typically used to identify carcinomas from the gastrointestinal tract, including colorectal cancers. Positive in many gastrointestinal, urothelial, and Merkel cell carcinomas.
CEA	A marker used for colorectal cancers and other gastrointestinal carcinomas, and sometimes in lung, pancreatic, and breast cancer. It is also used as a tumor marker for monitoring treatment response and recurrence.
EMA	Primarily used to identify carcinomas, particularly those originating from glandular or ductal epithelium (e.g., breast, gastrointestinal, or salivary gland cancers). It can be a marker for epithelial differentiation.
Epithelium membrane antigen	Often used in identifying epithelial tumors and differentiating them from mesenchymal tumors.
MOC31	A monoclonal antibody used for identifying carcinomas, particularly those of the lung, breast, and colon. It is useful for distinguishing epithelial tumors from others.
S-100 protein	A marker of neuroectodermal differentiation, positive in melanocytes, Schwann cells, neural tumors (e.g., melanoma, Schwannoma, neurofibroma), and some other tumors of neural origin.
HMB-45	A marker specific for melanocytic tumors, especially melanoma, and is used to distinguish melanomas from other tumors.

(continued)

Table 40.1 (continued)

Immunohistochemistry/ Tumor marker	Use
Melan-A	Another melanocytic marker, used to confirm the diagnosis of melanoma and other melanocytic lesions.
CD31	A marker for endothelial cells, commonly used to identify vascular and endothelial differentiation in tumors, especially angiosarcomas, hemangiomas, and other vascular tumors.
CD34	A marker of hematopoietic progenitor cells and endothelial cells; used to identify tumors of endothelial or hematopoietic origin, such as vascular tumors (hemangiomas, angiosarcomas) and dermatofibrosarcoma.
Bcl-2	A marker of apoptosis inhibition, used to differentiate follicular lymphoma from other non-Hodgkin lymphomas. It is also expressed in various types of malignancies, including lymphomas and some carcinomas.
CD117	A marker of stem cells, hematopoietic progenitor cells, and gastrointestinal stromal tumors (GISTs). It is positive in GISTs, mast cell tumors, and some leukemias.
GCDFP-15	A marker primarily used for breast carcinomas, especially to differentiate breast cancer from other metastatic carcinomas. It is positive in most breast cancer cases and may also be seen in apocrine carcinoma.

Clinical Findings

Location Approximately 50% will appear in an axial location, 15% in the upper extremities, and 35% in the lower extremities [3]. The most frequent sites are the abdominal wall (21%), chest wall (10%), upper extremities (13%) and lower extremities (14%), lower back (17%), and less percentage in head and neck, shoulder, and dorsum, respectively [22].

They can appear as single or multiple lesions associated with metastasis to other organs, such as the brain and bones [1, 3, 8, 10]. There is evidence that when masses in the abdominal wall appear, there is a very advanced disease (for example, bladder papillary carcinoma metastasis) [20]. Over 60% to 81% are in skeletal muscle, and only 10 to 20% of cases are in the subcutaneous tissue and deeper areas [13].

Pain and mass or swelling Most will be painless initially, but the distribution was 50% for patients with and without pain. [3] Most painless tumors may become painful as they grow larger. Twenty percent of cases exhibit pain at diagnosis, and 22% present pain in conjunction with a palpable mass or increased volume. In a lesser proportion of cases, additional symptoms, including alterations in visual acuity, redness, decreased joint mobility, weight loss, anorexia, and fever, have been documented [10, 19].

Diagnosis

In orthopedic oncology, diagnosing tumor pathology requires considering the "*Diagnostic Triad,*" which correlates clinical, imaging, and histopathological findings. This triad is essential for achieving an accurate diagnosis and facilitating appropriate treatment.

Assessing soft tissue masses necessitates a cautious approach, beginning with a comprehensive history and physical examination. In instances of persistent uncertainty, imaging studies and biopsies should be contemplated. Soft tissue carcinoma metastasizing to the muscle is uncommon as a primary consideration in the absence of a previously identified primary carcinoma at another site. In a retrospective analysis conducted by Surov et al., 61 cases of muscle metastases were identified among a cohort of 5170 patients presenting with established metastatic disease [21].

There are general recommendations for whole-body staging in sarcoma. Soft tissue sarcomas have a reported incidence of metastatic spread of 11.9% based on data from 2000 to 2018, with distant metastases most common in the lungs, followed by bone, lymph nodes, liver, brain, and subcutaneous tissue. Metastases are more likely in sarcomas located beneath the deep fascia and in moderate or high-grade tumors. The incidence of metastases is influenced by the histological tumor type, with metastases worsening prognosis and resulting in upstaging of patients. Improved outcomes have been reported after metastasectomy, including surgery, radiotherapy, and systemic treatment. Therefore, diagnosing metastases is crucial for effectively managing soft tissue sarcomas [22].

Conventional X-rays permit the evaluation of soft tissue extensions and their correlation with bones in the extremities. It has been described as the presence of calcifications from metastasis from gastric, pancreatic, or colonic cancers [23], which can be useful for diagnosis.

Ultrasound can be used as a first tool in assessing a soft tissue mass in these patients. High-resolution ultrasound is recommended for diagnosing superficial lesions, and ultrasound-guided biopsy can provide a definite diagnosis. Patients with a history of malignant tumors should have newly discovered soft tissue lesions closely followed up by ultrasound and consider timely ultrasound-guided core needle biopsy for evaluation [24]; however, the gold standard for soft tissue tumors evaluation is magnetic resonance imaging (MRI) of the affected area. Total body MRI is effective in the diagnosis of multiple soft tissue metastasis as a protocol for detection.

Staging to look for a primary tumor is imperative. CT of the chest, abdomen, and pelvis is necessary in the search for primary tumors. When a tumor has been detected, it is necessary to complement it with PET-CT to identify soft tissue metastases early.

CT scans have the advantage of a shorter examination time, are cost-effective, and are quite useful in diagnosing pulmonary metastases from soft tissue tumors [25]. Although MRI is superior to CT in representing soft tissue, the intrinsic spatial resolution of CT is higher than that of MRI. In addition, MRI is more expensive and has longer exam times than CT scans. However, after radiotherapy, it is imperative to examine the extent of tumor necrosis in certain tumors, as this may result in inaccuracies. It is crucial to complement this assessment and reevaluate it through MRI. In these cases, magnetic resonance imaging helps determine the nano/micro-anatomical and pathological determinants between irreversible tumor cells and postradiotherapy iatrogenesis.

CT scans facilitate the surgical planning process by comprehensively assessing compartments in the extremities, chest wall, and abdominal wall. Furthermore, using contrast agents can enhance the precision of the surgical planning. Surov et al. identified various patterns of presentation on CT imaging for muscular metastasis. The most common manifestations included a discrete intramuscular mass in 53% of cases, a cystic abscess-like lesion in 33% of cases, and diffuse muscle infiltration in 9% [21] (Fig. 40.1).

MRI is the preferred imaging method for assessing soft tissue masses [26] due to its high tissue contrast and ability to provide detailed anatomical information. It helps determine the extent of tissue involvement and the tumor's relationship to nearby structures like nerves, blood vessels, tendons, or bones, which is crucial for treatment planning. Muscle metastases from carcinoma typically show indistinct boundaries, low signal on T1-weighted images, and high signal on T2-weighted images. In most cases, they often present with significant peritumoral edema, enhancement, and central necrosis [27]. MRI is the preferred imaging modality for assessing soft tissue masses due to its exceptional tissue contrast and ability to show intricate anatomical features. It can determine the extent of tissue involvement and the proximity of tumors to surrounding structures, effectively helping the medical team with a surgical plan or radiation treatments [7].

PET-CT is highly sensitive in detecting skeletal muscle metastasis, often uncovering subclinical lesions that other imaging techniques like MRI and contrast-enhanced CT may miss [28].

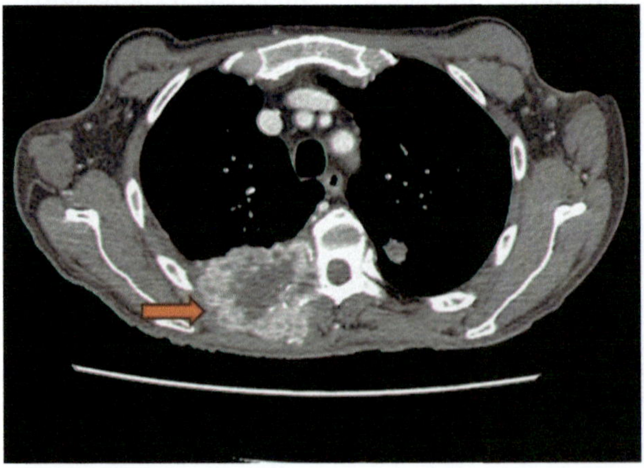

Fig. 40.1 Sixty-five-year-old male presents with thoracic wall mass. Biopsy confirms metastasis from cutaneous myofibroblastic sarcoma diagnosed one year ago (see orange arrow)

There are established guidelines for evaluating soft tissue tumors, including soft tissue metastasis. The updated ESSR agreement for imaging of soft tissue tumors provides expert consensus guidelines for standardized imaging algorithms, techniques, and reporting in soft tissue tumors in adults. It aims to establish best practices for imaging these tumors [29] (Fig. 40.2).

The standardized use of SUVs values in assessing response to treatment has gained importance. In addition, it is beneficial to integrate FDG-PET [30] into current intensity-modulated radiotherapy techniques and their clinical treatment. [18]FDG-PET is useful in the detection of soft tissue metastasis from lung cancer. In a study of a total of 219 lesions, 215 cases were identified with [18]FDG PET/CT, and 139 with CT [31]. Muscle metastases were mainly localized in the hip and upper limb muscles, while subcutaneous soft tissue metastasis was primarily in the chest, abdomen, and back. In the same study, 68 patients had soft tissue metastasis incidentally found during routine PET/CT staging, with isolated soft tissue metastasis identified in 6 patients affecting their staging and treatment. Survival rates did not significantly differ between patients with and without soft tissue metastasis at baseline PET/CT. Brain and adrenal metastases, rather than soft tissue metastasis, were linked to a poorer prognosis in stage 4 lung cancer patients [31].

PET/CT can detect unsuspected soft tissue metastasis, influencing staging and treatment decisions, although soft tissue metastasis is not a useful prognostic marker for advanced lung cancer patients [31].

A biopsy is needed to obtain a final diagnosis. Several techniques can be used, but the most used is percutaneous ultrasound-guided core needle biopsy. In cases where there are nonconclusive results, an incisional biopsy can be attempted [32].

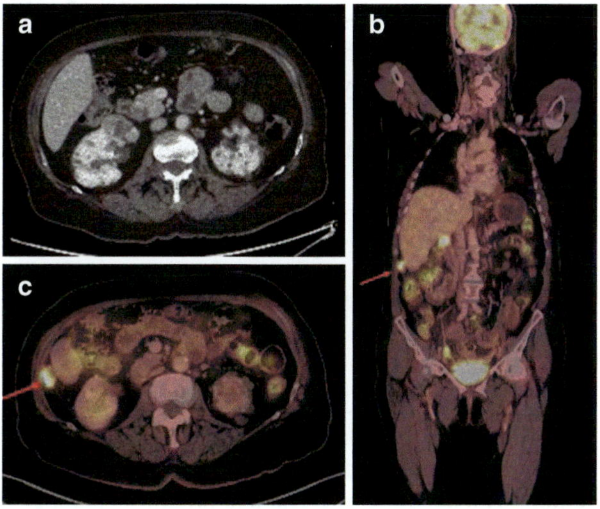

Fig. 40.2 Sixty-eight-year-old female with right renal cell carcinoma, presenting with right subphrenic soft tissue metastasis. (**a**) Right renal cell tumor. (**b**, **c**) Hypermetabolic lesion in right subphrenic wall (red arrows)

Treatment

Several factors should be considered to choose the best treatment options for these patients. These factors can be divided into different categories including but not limited to:

A. Patient characteristics: The patient's comorbidities, expectations, overall degree of functionality, and life expectancy should be discussed with a multidisciplinary team.
B. Tumor characteristics: Type of primary tumor, presence of solitary or multiple metastasis, potential response to systemic and local treatments, and relationship with anatomical structures (bone and neurovascular bundle) in case of surgical intent.

The main goal of treatment in these cases is to improve the patient's quality of life. The patient's general condition and self-sufficiency should be assessed to make therapeutic decisions. The Karnofsky Scores and the Eastern Cooperative Oncology Group ECOG [33] are used for this purpose [37].

Systemic Therapy

Systemic treatment can improve overall survival and further improve patients' socioeconomic situations and/or quality of life. Treatment options are diverse and must be chosen and tailored to the tumor type. In many cases, subsequent systemic treatments must be used due to a lack of response and disease progression [38, 40].

Treatment strategies may vary depending on the type of primary tumor, the patient's age, comorbidities, treatment goals, and tumor genetic profile and include infusion chemotherapy, hormone therapy, targeted therapy, and immunotherapies [41].

Multimodal management and the involvement of a multidisciplinary team lead to personalized treatment for each patient. Interdisciplinary counseling is necessary to help ensure that the treatment strategies chosen are the most appropriate for the patient, encouraging compliance [38].

Surgery

Local control of musculoskeletal metastasis is variable. When considering surgery, it is important to have adequate control of the systemic disease [39]. Surgery excision may be indicated in cases of solitary lesions, after a disease-free interval, tumors with adequate survival and prognosis parallel with adequate systemic treatment of the primary tumor. In the case of multiple metastases, palliative treatment with reduction of tumor mass is indicated if pain and neurovascular damage are becoming clinically significant [41, 42].

Surgery is the main treatment for tumors, aiming to achieve negative margins. Additional treatments like radiation therapy and isolated limb perfusion can be used with surgery to reduce local recurrence and allow for limb-sparing surgery, especially when positive margins are anticipated [43].

Many times, with metastasectomy in extremities, the quality of life is improved, regardless of the poor prognosis that this condition has.

Radiotherapy

The primary focus of radiation therapy is on pain relief. The doses vary according to the planning. They can be fractionated or hypo-fractionated depending on experience, scientific evidence, and the decisions of each oncology committee in their institution [34, 35]. Radiotherapy plays an important role in inaccessible lesions.

When metastasectomy is planned, it is important to define with the team whether pre- or postoperative radiotherapy can be performed, depending on the desired margins, the type of tumor, and the tumor aggressiveness [34]. Another strategy includes local control with ablative therapy or embolization in certain types of tumors (hipervascular renal cell metastasis, hipervascularized soft tissue metastasis) [36].

Conclusion

Soft tissue metastases are rare, accounting for less than 0.2% to 1% of all soft tissue tumors. They are usually a sign of advanced disease and are located mainly in the axial region of the body. Up to 75% of cases present as solitary lesions and are considered in the differential diagnosis with soft tissue sarcomas. The work-up should include a thorough medical history, physical exam, adequate imaging studies, and biopsy. Staging studies for the assessment of extension of disease are crucial to plan the appropriate treatment. Management primarily involves medical treatment, including systemic therapy tailored to the primary tumor type and radiation for local control. In cases where local control cannot be achieved with the previously described strategies, surgery is indicated. Overall, the prognosis is poor. Future priorities include improving trial design, quality of care, and setting guidelines for future treatment strategies to address the condition and its side effects, as well as deciding not just about surgery but also functionality and optimizing patient care.

Bibliography

1. Pretell-Mazzini J, Younis MH, Subhawong T. Skeletal muscle metastases from carcinomas: a review of the literature. JBJS Rev. 2020;8(7):e1900114–8. https://doi.org/10.2106/JBJS.RVW.19.00114.
2. Tuoheti Y, Okada K, Osanai T, Nishida J, Ehara S, Hashimoto M, et al. Skeletal muscle metastases of carcinoma: a clinicopathological study of 12 cases. Jpn J Clin Oncol. 2004;34(4):210–4. https://doi.org/10.1093/jjco/hyh036.

3. Morinaga S, Yamamoto N, Hayashi K, Takeuchi A, Miwa S, Igarashi K, et al. Clinical features of patients with carcinoma soft tissue metastases as surgical indications: a retrospective cohort study. BMC Cancer. 2024;24(1):577. https://doi.org/10.1186/s12885-024-12350-2.
4. García A, Fernández O, Satué FA, Buelta EG, Val-Bernal L. Metastasis of malignant neoplasms to skeletal muscle. Rev Esp Oncol. 1984;31(1):57–67.
5. Key statistics for soft tissue sarcomas. Cancer.org. [citado el 7 de octubre de 2024]. Disponible en: https://www.cancer.org/cancer/types/soft-tissue-sarcoma/about/key-statistics.html
6. Glockner JF, White LM, Sundaram M, McDonald DJ. Unsuspected metastases presenting as solitary soft tissue lesions: a fourteen-year review. Skeletal Radiol. 2000;29(5):270–4. https://doi.org/10.1007/s002560050606.
7. Ye Q, Hu T, Sun Z, Xu P, Wang C, Sun Y, et al. Surgical resection of soft tissue metastasis in cancers: a single-center study of 77 cases over a 7-year period. Cancer Med. 2023;12(24):22081–90. https://doi.org/10.1002/cam4.6808.
8. Plaza JA, Perez-Montiel D, Mayerson J, Morrison C, Suster S. Metastases to soft tissue: a review of 118 cases over a 30-year period: a review of 118 cases over a 30-year period. Cancer. 2008;112(1):193–203. https://doi.org/10.1002/cncr.23151.
9. Hopkins DT, Waters D, Manecksha RP, Lynch TH. Isolated soft tissue mass of the finger as the first presentation of oligometastatic renal cell carcinoma. BMJ Case Rep. 2022;15(5):e248718. https://doi.org/10.1136/bcr-2021-248718.
10. Pretell-Mazzini J, de Neyra JZS, Luengo-Alonso G, Shemesh S. Skeletal muscle metastasis from the most common carcinomas orthopedic surgeons deal with. A systematic review of the literature. Arch Orthop Trauma Surg. 2017;137(11):1477–89. https://doi.org/10.1007/s00402-017-2782-z.
11. Disibio G, French SW. Metastatic patterns of cancers: results from a large autopsy study. Arch Pathol Lab Med. 2008;132(6):931–9. https://doi.org/10.5858/2008-132-931-MPOCRF.
12. Tsuchie H, Emori M, Miyakoshi N, Okada K, Nagasawa H, Murahashi Y, et al. Prognostic factors in patients with distant soft tissue metastasis of carcinoma: a clinicopathological study of 16 cases. Med Princ Pract. 2020;29(6):538–43. https://doi.org/10.1159/000508463.
13. Magee T, Rosenthal H. Skeletal muscle metastases at sites of documented trauma. AJR Am J Roentgenol. 2002;178(4):985–8. https://doi.org/10.2214/ajr.178.4.1780985.
14. Bickels J, Jelinek JS, Shmookler BM, Neff RS, Malawer MM. Biopsy of musculoskeletal tumors. Curr Concepts Clin Orthop Relat Res. 1999;368:212–9.
15. Shives TC. Biopsy of soft-tissue tumors. Clin Orthop Relat Res. 1993:32–5.
16. Mankin HJ, Lange TA, Spanier SS. The hazards of biopsy in patients with malignant primary bone and soft-tissue tumors. J Bone Joint Surg Am. 1982;64(8):1121–7. https://doi.org/10.2106/00004623-198264080-00002.
17. Yoshimura Y, Isobe K-I, Koike T, Arai H, Aoki K, Kato H. Metastatic carcinoma to subcutaneous tissue and skeletal muscle: Clinicopathological features in 11 cases. Jpn J Clin Oncol. 2011;41(3):358–64. https://doi.org/10.1093/jjco/hyq199.
18. Sridhar KS, Rao RK, Kunhardt B. Skeletal muscle metastases from lung cancer. Cancer. 1987;59(8):1530–4. https://doi.org/10.1002/1097-0142(19870415)59:8<1530::aid-cncr2820590824>3.0.co;2-h.
19. Haygood TM, Wong J, Lin JC, Li S, Matamoros A, Costelloe CM, et al. Skeletal muscle metastases: a three-part study of a not-so-rare entity. Skeletal Radiol. 2012;41(8):899–909. https://doi.org/10.1007/s00256-011-1319-8.
20. San Miguel Fraile P, Antón Badiola I, Ortiz Rey JA, Iglesias Rodríguez B, Fernández Fernández G, Zungri Telo E. Metástasis en partes blandas por un carcinoma micropapilar de vejiga. Actas Urol Esp. 2007;31(7):788–91. https://doi.org/10.1016/s0210-4806(07)73725-7.
21. Surov A, Hainz M, Holzhausen H-J, Arnold D, Katzer M, Schmidt J, et al. Skeletal muscle metastases: primary tumours, prevalence, and radiological features. Eur Radiol. 2010;20(3):649–58. https://doi.org/10.1007/s00330-009-1577-1.
22. Trovik C, Bauer H, Styring E. The scandinavian sarcoma group central register: 6000 patients after 25 years of monitoring of referral and treatment of extremity and trunk wall soft-tissue sarcoma. Acta Orthop. 2017;88:341–7.

23. Molina-Garrido MJ, Guillén-Ponce C. Muscle metastasis of carcinoma. Clin Transl Oncol. 2011;13(2):98–101. https://doi.org/10.1007/s12094-011-0625-x.
24. Bai M, Li Z, He P. Ultrasound findings of subcutaneous soft tissue metastasis of renal cell carcinoma: a case report. Radiol Case Rep. 2023;18(1):192–5. https://doi.org/10.1016/j.radcr.2022.09.055.
25. Saifuddin A. The diagnosis of lung metastases on chest CT in primary bone Sarcoma and soft tissue musculoskeletal Sarcoma. Br J Radiol. 2021;94.
26. Soffia SP. Difusión Por resonancia magnética: Bases y aplicaciones oncológicas en órganos extracraneanos. Rev Chil Radiol. 2009;15 https://doi.org/10.4067/s0717-93082009000400004.
27. Sammon J, Jain A, Bleakney R, Mohankumar R. Magnetic resonance imaging appearance of soft-tissue metastases: our experience at an orthopedic oncology center. Skeletal Radiol. 2017;46(4):513–21. https://doi.org/10.1007/s00256-017-2582-0.
28. Nguyen NC, Chaar BT, Osman MM. Prevalence and patterns of soft tissue metastasis: detection with true whole-body F-18 FDG PET/CT. BMC Med Imaging. 2007;7(1):8. https://doi.org/10.1186/1471-2342-7-8.
29. Adult soft tissue tumor imaging: Whole-body staging in Sarcoma, Nonmalignant Entities Requiring Special Algorithms, Traps, and Special Imaging Aspects. 2024 Guidelines of the Sociedad Europea de Radiología Musculoesquelética (ESSR). Radiol Eur. 2024:1–9.
30. Vaarwerk B, Breunis WB, Haveman LM, de Keizer B, Jehanno N, Borgwardt L, et al. Fluorine-18-fluorodeoxyglucose (FDG) positron emission tomography (PET) computed tomography (CT) for the detection of bone, lung, and lymph node metastases in rhabdomyosarcoma. Cochrane Libr. 2021;2021(11) https://doi.org/10.1002/14651858.cd012325.pub2.
31. Xu T, Zhang X, Zhang S, Liu C, Fu W, Zeng C, et al. Imaging features and prognostic value of 18F-FDG PET/CT detection of soft-tissue metastasis from lung cancer: a retrospective study. BMC Cancer. 2020;20(1) https://doi.org/10.1186/s12885-020-07080-0.
32. Tuttle R, Kane JM III. Biopsy techniques for soft tissue and bowel sarcomas. J Surg Oncol. 2015;111(5):504–12. https://doi.org/10.1002/jso.23870.
33. Shvarts O, Lam JS, Kim HL, Han K-R, Figlin R, Belldegrun A. Eastern Cooperative Oncology Group performance status predicts bone metastasis in patients presenting with renal cell carcinoma: implication for preoperative bone scans. J Urol. 2004;172(3):867–70. https://doi.org/10.1097/01.ju.0000135803.91207.b0.
34. Haas RLM, Delaney TF, O'Sullivan B, Keus RB, Le Pechoux C, Olmi P, et al. Radiotherapy for management of extremity soft tissue sarcomas: why, when, and where? Int J Radiat Oncol Biol Phys. 2012;84(3):572–80. https://doi.org/10.1016/j.ijrobp.2012.01.062.
35. Guadagnolo BA, Bassett RL, Mitra D, Farooqi A, Hempel C, Dorber C, et al. Hypofractionated, 3-week, preoperative radiotherapy for patients with soft tissue sarcomas (HYPORT-STS): a single-centre, open-label, single-arm, phase 2 trial. Lancet Oncol. 2022;23(12):1547–57. https://doi.org/10.1016/s1470-2045(22)00638-6.
36. Gronchi A, Guadagnolo BA, Erinjeri JP. Local ablative therapies to metastatic soft tissue sarcoma. Am Soc Clin Oncol Educ Book. 2016;36:e566–75. https://doi.org/10.1200/edbk_157450.
37. İlhan A, Eraslan E, Yildiz F, Arslan ÜY, Alkiş N. Factors affecting prognosis and treatment strategies in metastatic soft tissue sarcomas: twenty years of experience. Eur Rev Med Pharmacol Sci. 2021;25(21):6465–72. https://doi.org/10.26355/eurrev_202111_27090.
38. Damron TA, Heiner J. Management of metastatic disease to soft tissue. Orthop Clin North Am. 2000;31(4):661–73, ix. https://doi.org/10.1016/s0030-5898(05)70183-8.
39. Koike Y, Hatori M, Kokubun S. Skeletal muscle metastasis secondary to cancer--a report of seven cases. Ups J Med Sci. 2005;110(1):75–83. https://doi.org/10.3109/2000-1967-183.
40. Skafida E, Kokkali S, Nikolaou M, Digklia A. Metastatic soft tissue sarcoma: current treatment landscape and future perspectives. Expert Rev Anticancer Ther. 2017;17(6):537–43. https://doi.org/10.1080/14737140.2017.1321989.
41. Perisano C, Spinelli MS, Graci C, Scaramuzzo L, Marzetti E, Barone C, et al. Soft tissue metastases in lung cancer: a review of the literature. Eur Rev Med Pharmacol Sci. 2012;16(14):1908–14.

42. Ramakrishna G, Harms GF, Erlichman C. Metastatic non-small cell lung carcinoma with involvement of extremity musculature: case report and review of the literature. Mayo Clin Proc. 2000;75(12):1333–7. https://doi.org/10.4065/75.12.1333.
43. Borghi A, Gronchi A. Extremity and truncal soft tissue sarcoma: risk assessment and multidisciplinary management. Semin Radiat Oncol. 2024;34(2):147–63. https://doi.org/10.1016/j.semradonc.2023.12.001.

41 Post-Radiation Pathological Fractures - this Bone is not Healing !!!

Nicole Butler and Sean P. Kelly

Introduction/Radiation Effect on Bone

Radiation therapy (RT) is frequently used in the treatment of metastatic bone disease and soft tissue sarcomas; however, its use is not without several adverse effects. Radiation therapy causes damage to the structural integrity of bone and decreases its strength which can have varying levels of regeneration [2]. Not only does RT decrease the strength of bone, but also the ability of bone to heal.

Bone can generally repair structural damage and respond to stressors, but this can be impaired when bone is subjected to ionizing radiation. The reported minimum threshold for radiation to affect bone is variable, but overall, there is decreased bone healing with increasing radiation dosages [5]. Additionally, radiation can cause radiation osteitis, osteopenia, and progress to osteoporosis. These effects are due to both the structural damage to bone and its surrounding vascularity [13]. The overall net effect encountered by the orthopedic surgeon is a tendency for radiated bone to be more likely to fracture and less likely to heal [5].

N. Butler
Tripler Army Medical Center, Honolulu, HI, USA

S. P. Kelly (✉)
Tripler Army Medical Center, University of Hawaii Cancer Center, Honolulu, HI, USA

J. Pretell-Mazzini, S. A. Lozano Calderón (eds.), *Multidisciplinary Approach of Musculoskeletal Metastasis*, https://doi.org/10.1007/978-3-032-09498-8_41

Overall Treatment Management Strategy

Mitigation of Radiation-Associated Fractures (RAF)

Risk for future RAF can be mitigated during the initial treatment of the tumor by limiting the dose of radiation, as higher doses have been associated with increased likelihood of sustaining a RAF [18].

Radiation therapy is commonly given as external beam radiation therapy (EBRT), administered in various techniques. Targeted radiation, with techniques such as intensity modulated radiation therapy (IMRT), has been seen to have potentially lower rates of radiation-associated fractures when compared to older dosing techniques [15]. IMRT, for example, is thought to have a relative decrease in radiation-associated fractures due to the reduced volume and/or circumference of bone subjected to high-dose radiation [3]. Even without the use of IMRT, it has been observed that there is a lower rate of radiation-associated fractures when there is not circumferential radiation of the affected bone [6], suggesting a benefit to focused radiation doses.

Radiation particles with rapid dose drop off, such as proton therapy, may also decrease the risk of fracture in select cases as the far cortex of bone can be spared radiation dose; however, this therapy is limited to geographic availability and primarily limited to tumors of the spine and sacrum [3].

Brachytherapy has also been observed to have a lower incidence of radiation-associated fractures when compared to EBRT due to the potential to directly apply radiation to the desired location [23].

Lastly, in patients that have sustained a radiation-associated fracture that has healed, it is suggested to retain the hardware as the risk for refracture is high [10].

Risk Stratification (Table 41.1)

Patients can be risk stratified for prophylactic treatment based on patient and treatment factors. The incidence of radiation-associated fractures ranges from 2% to 6.3% [6, 18]. With respect to patient characteristics, female gender and increased age have been associated with increased fracture risk [18], but this relationship has not been seen in all studies [6, 17, 31].

As previously mentioned, increased doses of radiation increase the incidence of radiation-associated fractures [18], but there is a disagreement as to what constitutes a high vs. low dose of radiation or how strong a risk factor dosing may actually be [3, 16].

Periosteal stripping has been associated with radiation-associated fractures in soft tissue sarcomas [3, 7, 16, 24] and may be due to impaired vascularity or simply represents a radiation field that involves bone [9].

Regarding tumor characteristics, the presence of the tumor in the anterior compartment of the thigh has an increased risk of fracture compared to medial or posterior [7, 16, 24]. The presence of the tumor in the anterior compartment as a risk

Table 41.1 Factors associated with radiation-associated fracture by author

Author	Age	Gender	Radiation dose	Anterior thigh compartment	Periosteal stripping	Radiation timing	Chemotherapy	Tumor size	Circumferential radiation
Lin [22]	y	y	n	y	y	n	y	n	–
Helmstedter	n	n	–	y	y	–	–	–	–
Holt	y	y	y	n	n	y	–	–	–
Gortzak	y	n	n	y	y	–	–	y	–
Kim	–	n	–	–	n	–	–	–	–
Bishop	n	n	n	–	y	n	y	n	y
Folkert	y	n	n	y	y	–	n	n	–
Pak	y	n	y	–	n	n	n	n	–
Cannon [9]	n	n	–	–	n	–	n	n	y
Sambri	n	y	–	–	–	n	–	n	–
Blaes	n	n	–	y	y	n	n	–	y

factor is thought to be due to interruption of the femoral blood supply to the femur and associated periosteal stripping [7]. Additionally, alteration of the gait after quadriceps impairment may change the rate of loading during the cycle of the gait and may predispose the femur to fracture [23].

Different criteria and scores exist to attempt to stratify patients into either a higher-risk or lower-risk category of pathologic fracture for metastatic disease to bone (such as the Harrington criteria and Mirel score) may predict which patients will fail radiation therapy without prophylactic fixation with fair accuracy [32]. The ability to stratify patients as high risk can avoid the difficulty of treating a radiation-associated fracture.

Prophylactic Treatment

Due to the difficult nature of treating radiation-associated fractures, prophylactic fixation of high-risk segments of bone can be undertaken. Risks inherent to metastatic bone disease include possible seeding or spread of metastatic bone disease, possibly further increasing radiation fields in low-risk lesions [30].

Due to the low rate of fracture (~2%) in soft tissue sarcomas, prophylactic nailing is generally not advised for all patients and brings inherent risk of overtreatment [6]. Additionally, follow-up surveillance MRI can be more difficult and may be more likely to miss recurrence or metastasis [30]. The use of carbon fiber nails has been proposed in order to mitigate this imaging issue. With carbon fiber, there is reduced metal artifact and interference seen with traditional implants and better allow detection of local recurrences or progression of disease [35]. The use of carbon fiber implants in lieu of titanium nails is promising with promising initial results [26].

Even more vexing is that prophylactic nailing is not always successful in preventing radiation-associated fractures as fractures can still occur even with fixation in place [30].

Diagnosis of RAF

The first step in proper management and treatment of RAFs is diagnosis. Often, fracture in a radiation field can occur due to recurrence of tumor, radiation-associated sarcoma, isolated RAFs, or RAFs occurring with a neoplasm, all requiring different treatments [29]. For RAF of long bones, radiographs can often identify the presence of a fracture, typically minimally displaced with a transverse or oblique fracture pattern as a result of either low energy or nontraumatic [17, 23]. CT can be used to better characterize the fracture and to evaluate further for osteolytic lesions or soft tissue mass. Characteristic findings on CT include focal cortical disruption, a fracture line, and endosteal callus formation. However, not all fractures can be easily identified on x-ray or CT. MRI, can be

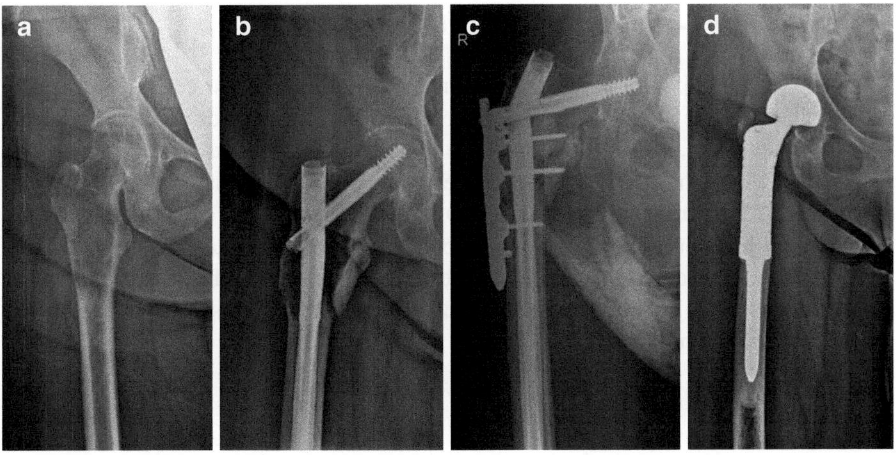

Fig. 41.1 (**a**–**d**) 45F with oligometastatic metastatic thyroid carcinoma. (**a**) Destructive lytic lesion in the right proximal femur treated with 30 Gy and nail placement. (**b**) Radiation-associated fracture around the nail after additional 32 Gy given for recurrence. (**c**) Implant failure after bone grafting and additional fixation for radiation-associated fracture. (**d**) Conversion to proximal femoral replacement

used to diagnose occult RAF, often seen in the pelvis and associated with low signal intensity on T1 images and high signal intensity on STIR images [8]. Overall, imaging can be a helpful diagnostic tool in ruling out fracture from a pure recurrence or radiation-associated sarcoma [8]. Lastly, time from radiation dose is unreliable in the diagnosis as the median time to fracture has been reported to be as short as 4.4 months [32] to as long as 41 months [18].

Overview of Treatment

Treatment of radiation-associated fractures varies depending on the location of the fracture as well as the quality of bone and soft tissue. There are high rates of nonunion and complications, including infection, associated with these fractures, and therefore, careful follow-up is needed [10, 17].

Frequently, the management of these fractures occurs after patients have left specialized cancer centers and are encountered by orthopedic surgeons that may not be as familiar with the effects of radiation on bone and unique complications that may arise with radiation-associated fractures [33].

Surgical treatment exists on a continuum of conventional fracture fixation, bone grafting, vascularization procedures, and lastly megaprosthesis (Fig. 41.1a–d). Treatment should factor in patient survival and favor the principle of immediate weight bearing to maximize quality of life and independence in these patients.

Treatment for Upper Extremity Fractures

Tumors of the upper extremity are more commonly discovered when they are smaller in size. The diagnosis when tumor burden is smaller leads to a smaller radiation treatment area, leading to a decrease in soft tissue damage and to a lower rate of fracture when compared to the lower extremity [23]. Especially in the humerus, vascularized fibular grafts can be used to augment nonunions or replace avascular segments of bone improving the rate of patients achieving union at the expense of time required to heal [12]. Nonoperative management can always be attempted; however, there is a higher chance of nonunion, and the patients may require subsequent surgical treatment [17].

Humerus

For fractures in the proximal humerus near the surgical neck, when treated with ORIF and plate fixation, there are high rates of failure and complications and may ultimately require subsequent resection and reconstruction with a mega-prosthesis [28]. An alternative approach would be to favor humeral head replacement [10] with hemiarthroplasty, anatomic, or reverse total shoulder arthroplasty with long stemmed, cemented components.

In treating fractures of the proximal humeral shaft, intramedullary nails can stabilize fractures with intact bone or lower functional demand while proximal humerus endoprosthesis is reserved for high dose radiation fields and cavitary bone defects. The disadvantage of losing native muscular attachments leads to worse shoulder function, but avoids the risk of radiation-associated nonunion and fixation failure [10, 27].

The humeral shaft favors intramedullary nail fixation [27] in most cases to maintain what vascularity is remaining. However, nonunions or those with significant bone loss may require intercalary megaprosthesis or cement augmentation (Fig. 41.2).

In fractures of the distal humerus, resection of the fracture fragments and reconstruction of the distal humerus with total elbow arthroplasty with or without a distal humerus replacement can be required [10] once progressed to nonunion.

Forearm/Wrist Fractures

In fractures involving the proximal ulna, alternatives include resection of the fracture fragment with those involving the proximal ulna potentially needing long stemmed total elbow arthroplasty [10]. In treatment of RAFs of the forearm, compression plate fixation can be attempted, with higher risk of nonunion [28].

In the distal radius, treatment options exist including primary wrist arthrodesis with possible shortening and arthroplasty with vascularized fibular head autograft have been described in the setting of fixation failure [25].

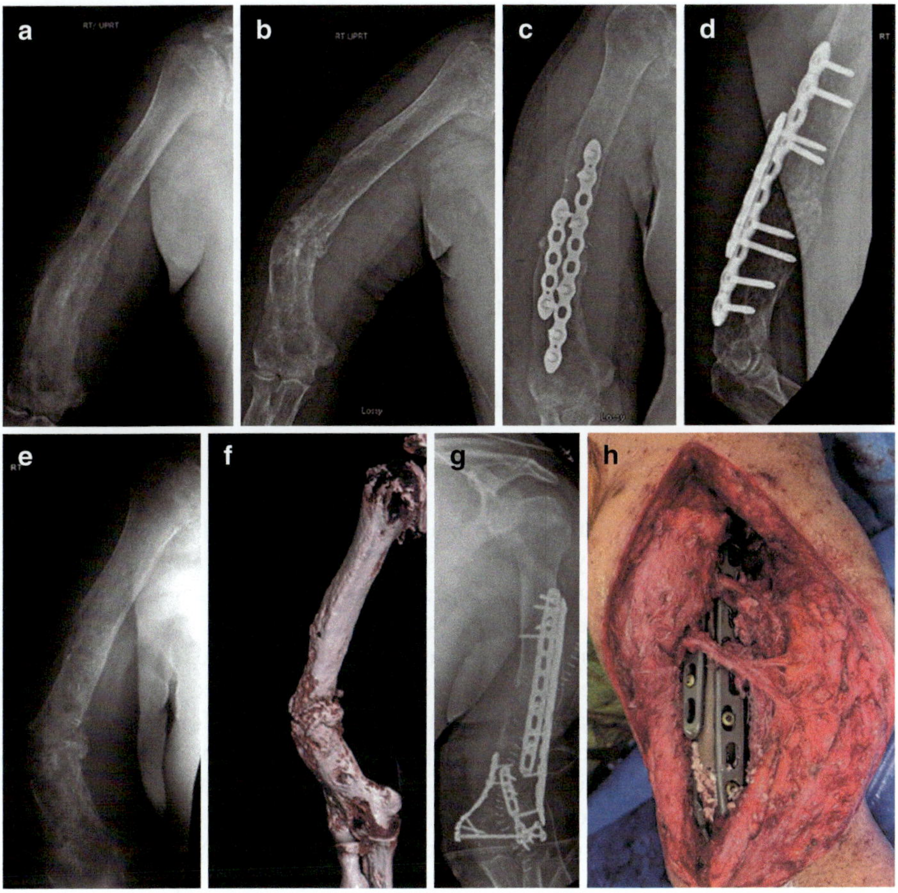

Fig. 41.2 (**a–d**) 55F with metastatic prostate cancer to the right humerus. (**a**) Radiographs of painful, blastic lesion of the humerus. (**b**) Radiographs showing radiation-associated fracture after 35 Gy. (**c, d**) Three-month post-op radiographs after in situ fixation and bone grafting of atrophic malunion. (**e, f**) X-ray and CT after hardware removal for fixation failure showing continued non-union. (**g, h**) Post-op x-ray and intra-op image of intercalary custom 3D-printed poly-ethyl-ketone-ketone (PEKK) implant with resection of the atrophic bone

Finally in patients with extensive soft tissue compromise due to radiation, primary amputation may be required to reduce the need for multiple surgeries in conditions when union is unlikely to occur [28].

Clavicle

RAFs in the clavicle are often asymptomatic and nonoperative management is appropriate even in nonunions. For painful lesions, surgical management of the clavicle has limited soft tissue leading to open reduction and fixation (ORIF)

having a high complication profile. An alternative to ORIF is resection of the clavicle as patients experience relief in pain and still retain adequate functional outcome [10].

Ribs

RAFs in the ribs are common after chest wall radiation. Similar to clavicle fractures, RAF in the ribs generally remain asymptomatic and are frequently incidental. Upon diagnosis of RAF of the rib, observation is the mainstay of treatment [1].

Treatment for Lower Extremity Fractures

Femur

Fractures with an intertrochanteric location and that are nondisplaced may still heal with ORIF and are less likely to need revision than displaced fractures but require close follow-up [20]. For femoral neck fractures, a calcar-replacing prosthesis is often required [10], and cemented implants are preferred.

In diaphyseal locations, intramedullary nail provides coverage of the entire bone and larger diameter nails are recommended due to the increased time to union of radiated bone [10]. Failure rates of nails still support the need for close follow-up and breakage of the nail is not uncommon [22, 23, 31]. Endoprosthesis reconstruction with resection of the radiated bone has been shown to have a favorable complication profile when compared to ORIF [20, 31].

When looking into rates of failure of internal fixation (IMN, DHS, plates) in femur fractures, the rates vary across studies. Sternheim et al. reported nonunion failure rate of 82% (18/22), and 8/18 required conversion to an endoprosthesis [33]. Similarly, Muratori et al. found that of 14 patients undergoing femoral IMN for a RAF, 4 conversions to prosthetic replacement were performed [28]. Of interest, all four of these conversions sustained additional complications compared to only 1 of 3 treated initially with megaprosthesis in their cohort [28]. Building on the idea of a potential lower complication rate with resection is a study by Kim et al. that found a complication rate of 63% for initial stabilization vs. 33% for resection [20]. Additional considerations for fixation compared to resection include functional deficits of resection, length of recovery, and the overall health and survival of each patient.

Tibia

Radiation-associated fractures of the tibia are encountered less often than the femur and is potentially due to fewer tumors in the lower leg overall [23].

Generally, RAFs of the tibia are to be treated with IMN [28]. Nonoperative management with casting frequently leads to nonunion and subsequent surgeries to achieve union [17, 28]. Metaphyseal fractures may require proximal tibial replacement but can lead to extensor lag. In the treatment of distal tibial fractures, prolonged cast immobilization may be attempted, or the fracture may be managed with ORIF and augmentation with bone grafts [4]. As with all RAFs, there is a higher incidence of nonunion, and amputation may ultimately be the definitive management if the soft tissue envelope does not allow for management of the bone nonunion [4].

Bone grafting is a common augment or salvage technique used for RAF. While iliac crest bone graft is commonly seen as an augment during initial fixation of an RAF, there have been multiple studies exploring the use of free vascularized fibular flaps in combination with autogenous iliac crest bone graft in salvage of RAF that have gone on to nonunion. Subsequent union rates of 71–89% have been reported [12, 14] . Generally bulk allograft would be dissuaded from use unless the resection margins are outside the field of radiation due to the risk of nonunion and subsequent failure. Tenets of allograft fixation in these instances would include those of standard tumor reconstructions including compression, importance of graft-host interface, and protection of the graft wither with an intramedullary nail, spanning plates, intramedullary cement, or a long-stemmed arthroplasty type implant.

Acetabulum

Overall, there is a higher rate of failure of total hip arthroplasty in patients that have previous irradiation to the hip. In select patients with lower risk factors including male and with lower radiation exposure, THA may be a viable option [11]. When looking at the use of uncemented, conventional implants, short-term failure rates have been reported as high as 44% and as low as 3% [19, 21]. However, when evaluating the use of revision-style tantalum cups with screw fixation, 10-year survivorship approached 90% and radiographic loosening was not observed in a study by Wellings et al. [34].

Patella

Patellar fractures after radiation are not commonly seen and little literature currently reports the incidence or treatment for patellar radiation-associated fractures. However, as with patellar fractures not associated with radiation, nonoperative management may be appropriate. Even in the case of patellar nonunions, patients may still be relatively asymptomatic and able to ambulate [4, 17].

References

1. Aoki M, et al. Radiation-induced rib fracture after stereotactic body radiotherapy with a total dose of 54-56 Gy given in 9-7 fractions for patients with peripheral lung tumor: impact of maximum dose and fraction size. Radiat Oncol. 2015;10:99.
2. Bakar AAA, et al. Systematic review on multilevel analysis of radiation effects on bone microarchitecture. Biomed Res Int. 2022;2022:9890633.
3. Bartelstein MK, et al. Postradiation fractures after combined modality treatment in extremity soft tissue sarcomas. Sarcoma. 2021;2021:8877567.
4. Bell RS, et al. Fractures following limb-salvage surgery and adjuvant irradiation for soft-tissue sarcoma. Clin Orthop Relat Res. 1991;271:265–71.
5. Berk L. The effects of high-dose radiation therapy on bone: a scoping review. Radiat Oncol J. 2024;42(2):95–103.
6. Bishop AJ, et al. Treatment-related fractures after combined modality therapy for soft tissue sarcomas of the proximal lower extremity: can the risk be mitigated? Pract Radiat Oncol. 2016;6(3):194–200.
7. Blaes AH, et al. Pathologic femur fractures after limb-sparing treatment of soft-tissue sarcomas. J Cancer Surviv. 2010;4(4):399–404.
8. Blomlie V, et al. Incidence of radiation-induced insufficiency fractures of the female pelvis: evaluation with MR imaging. AJR Am J Roentgenol. 1996;167(5):1205–10.
9. Cannon CP, et al. Complications of combined modality treatment of primary lower extremity soft-tissue sarcomas. Cancer. 2006;107(10):2455–61.
10. Cannon CP, et al. Management of radiation-associated fractures. J Am Acad Orthop Surg. 2008;16(9):541–9.
11. Cho MR, Kim SB, Kim JD. Outcome of primary total hip arthroplasty in patients with osteonecrosis of the acetabulum after previous pelvic irradiation. J Korean Orthop Assoc. 2004;39(5):469–75.
12. Claxton MR, et al. Utility of free vascularized fibular flaps to treat radiation-associated nonunions in the upper extremity. J Plast Reconstr Aesthet Surg. 2020;73(4):633–7.
13. Donaubauer AJ, et al. The influence of radiation on bone and bone cells-differential effects on osteoclasts and osteoblasts. Int J Mol Sci. 2020;21(17)
14. Duffy GP, et al. Vascularized free fibular transfer combined with autografting for the management of fracture nonunions associated with radiation therapy. J Bone Joint Surg Am. 2000;82(4):544–54.
15. Folkert MR, et al. Femoral fracture in primary soft-tissue sarcoma of the thigh and groin treated with intensity-modulated radiation therapy: observed versus expected risk. Ann Surg Oncol. 2019;26(5):1326–31.
16. Gortzak Y, et al. Prediction of pathologic fracture risk of the femur after combined modality treatment of soft tissue sarcoma of the thigh. Cancer. 2010;116(6):1553–9.
17. Helmstedter CS, et al. Pathologic fractures after surgery and radiation for soft tissue tumors. Clin Orthop Relat Res. 2001;389:165–72.
18. Holt GE, et al. Fractures following radiotherapy and limb-salvage surgery for lower extremity soft-tissue sarcomas. A comparison of high-dose and low-dose radiotherapy. J Bone Joint Surg Am. 2005;87(2):315–9.
19. Jacobs JJ, et al. Early failure of acetabular components inserted without cement after previous pelvic irradiation. J Bone Joint Surg Am. 1995;77(12):1829–35.
20. Kim HJ, et al. Site-dependent replacement or internal fixation for postradiation femur fractures after soft tissue sarcoma resection. Clin Orthop Relat Res. 2010;468(11):3035–40.
21. Kim KI, et al. Uncemented total hip arthroplasty in patients with a history of pelvic irradiation for prostate cancer. J Bone Joint Surg Am. 2007;89(4):798–805.
22. Lee J, Kim JJ, Kim W. Pathological fractures of the femur after radiation therapy for soft tissue tumor: a case series of seven patients treated with repeated internal fixation. Arch Orthop Trauma Surg. 2021;141(4):629–35.

23. Lin PP, Boland PJ, Healey JH. Treatment of femoral fractures after irradiation. Clin Orthop Relat Res. 1998;352:168–78.
24. Lin PP, et al. Pathologic femoral fracture after periosteal excision and radiation for the treatment of soft tissue sarcoma. Cancer. 1998;82(12):2356–65.
25. Liu W, et al. Wrist reconstruction after En bloc resection of bone tumors of the distal radius. Orthop Surg. 2021;13(2):376–83.
26. Lozano-Calderon SA, et al. Outcomes of long bones treated with carbon-fiber nails for oncologic indications: international multi-institutional study. J Am Acad Orthop Surg. 2024;32(3):e134–45.
27. Mavrogenis AF, et al. Side effects of radiation in musculoskeletal oncology: clinical evaluation of radiation-induced fractures. Int J Immunopathol Pharmacol. 2011;24(1 Suppl 2):29–37.
28. Muratori F, et al. Treatment options in femoral radiation fractures following soft tissue sarcoma: incidence, risk factors, failures and flowchart of treatment. Injury. 2021;52(6):1597–605.
29. Raj R, et al. Clinical characteristics, patterns of care, and treatment outcomes of radiation-associated sarcomas. Cancers (Basel). 2024;16(10)
30. Sabharwal S, Alcorn SR, Morris CD. Treatment of radiation-associated fractures: a critical analysis review. JBJS Rev. 2021;9(8)
31. Sambri A, et al. Femoral fracture in primary soft-tissue sarcoma of the thigh treated with radiation therapy: indications for prophylactic intramedullary nail. Arch Orthop Trauma Surg. 2021;141(8):1277–82.
32. Shimoyama T, et al. Fracture after radiation therapy for femoral metastasis: incidence, timing and clinical features. J Radiat Res. 2017;58(5):661–8.
33. Sternheim A, et al. Internal fixation of radiation-induced pathological fractures of the femur has a high rate of failure. Bone Joint J. 2013;95-b(8):1144–8.
34. Wellings EP, et al. Contemporary porous titanium acetabular components for total hip arthroplasty after pelvic radiation. J Arthroplast. 2021;36(5):1714–8.
35. Zoccali C, et al. The Carbofix™ "Piccolo Proximal femur nail": a new perspective for treating proximal femur lesion. A technique report. J Orthop. 2016;13(4):343–6.

Ethical Considerations in Musculoskeletal Metastasis Management

42

Kyle E. Walker, J. B. Smith, and R. Lor Randall

Introduction

In managing musculoskeletal metastatic disease, a physician must navigate intricate and often challenging ethical decisions. Patients afflicted with metastatic disease can present with varying levels of comorbidities, systemic disease burden, and/or expected overall survival. Furthermore, each patient may have different goals of care influenced by patient-specific factors such as age, social connections, and cultural/religious beliefs [1]. Given these and the other nuances of treating a patient with advanced metastatic disease, clinicians often find the need to take an individualized approach to shared decision-making (SDM) [2]. Moreover, clinicians must use their training and experience to set reasonable patient expectations relative to realistic treatment goals. At times, a physician may have to balance these treatment considerations with appropriate stewardship and allocation of resources. Lastly, when the time comes, a patient with metastatic disease will likely need guidance on numerous aspects of end-of-life decisions. In this chapter, we will evaluate principles and concepts that can drive a clinician's approach with respect to SDM, balancing treatment goals with patient expectations, and end-of-life considerations. We will also introduce tools which may aid a clinician in taking a methodical approach to these areas.

K. E. Walker (✉) · J. B. Smith · R. L. Randall
Department of Orthopaedic Surgery, University of California, Davis Medical Center, Sacramento, CA, USA
e-mail: kywalker@health.ucdavis.edu; jbsm@health.ucdavis.edu; rlrandall@health.ucdavis.edu

J. Pretell-Mazzini, S. A. Lozano Calderón (eds.), *Multidisciplinary Approach of Musculoskeletal Metastasis*, https://doi.org/10.1007/978-3-032-09498-8_42

Shared Decision-Making with Patients

In patients with advanced metastatic disease, the concept of SDM presents itself as uniquely challenging. Historically, medicine was rooted in a more paternalistic approach where physicians made decisions on patient care with little patient input. In the 1970s, toward the end of the twentieth century, the pendulum swung in many regions (i.e., North America, Europe, etc.) to revere patient autonomy above almost all other factors [3]. The informed consent, a term dating back to the early 1900s, gained traction through the 1970s and 1980s due to multiple cultural shifts and in the wake of studies with questionable or completely unethical standing such as the infamous Tuskegee Syphilis Study [4].

While some believe an informed consent is a suitable surrogate for ensuring patient autonomy, there are those who argue "informed consent requirements are neither necessary nor sufficient for safeguarding individual autonomy in the biomedical sphere." [5] Others expand on this concept in theorizing that, in some cases a truly "informed" consent is not possible. In the setting of "emotional overwhelm" or "information overload" [6] states often encountered by patients deciding advanced cancer treatment decisions, even the most well-educated patients may struggle to find the capacity to process the volume and complexity of information they receive. In some cases, a surrogate decision-maker may be employed as an alternative. However, a surrogate decision-maker is not immune to the challenges a patient would face in processing such information but may be useful and necessary in some situations [7].

Due to the challenges and limitations of individualistic autonomy and a truly informed consent, concepts around relational autonomy and SDM have grown in popularity. While term "Shared-Decision-Making" was first introduced in the 1980s, the reactionary shift from the paternalistic approach to medical decision-making resulted in an implementation of largely individualistic or "pure" autonomy [8]. As we entered the twenty-first century, catalyzed by endorsements from the American Medical Association and other medical societies, the pendulum shifted yet again toward a "meeting in the middle" between paternalistic decision-making and individualistic autonomy [9]. Of note, the word "shared" in SDM is not meant to reinstate physician authority over the decision itself but, rather, refers to the sharing of information and the methods by which a patient arrives at a decision which aligns with one's own values [10–12].

A person rarely makes a decision in the vacuum of their own desires and, even so, the term "relational autonomy" was coined to describe how an individual's values and principles are shaped through the relationships we *share* with others [13]. In this context, SDM is meant to support a patients' relational autonomy particularly in navigating complex, emotional decisions [12]. Both concepts are continuously evolving with respect to legal and ethical implications in numerous areas of medicine [13]. In treating patients with advance cancer, however, SDM requires an individualized approach [2] which factors in social/familial roles in the decisions these patients face [14–16].

A healthy clinician–patient relationship is paramount to effectively implement SDM in a clinical setting. A physician's role in SDM is best framed by thinking relationally to listen to a patient's desires, engaging in open dialogue around treatment options, involving friends/family members as needed, and allowing questions. Ultimately, a patient should feel safe to select any treatment option, within reason, despite the physician's recommendation [17]. A clinician must also take care to avoid undue influence, intentional or unintentional, which may compromise patient autonomy. Kehl et al. demonstrated that cancer patients gave lower quality of care and communication scores to physicians who employed physician-controlled decision-making even if doing so aligned with the patient's stated preferences [18].

The importance both family and multidisciplinary care team support in mitigating emotional overwhelm and information overload in patients afflicted with metastatic disease cannot be understated [6]. However, a clinician must similarly be cognizant of sign of undue influence from friends, family, and/or other members of the multidisciplinary team. Signs of undue influence can be subtle and include abrupt changes in a patient's goals of care or newly observing signs of anxiety when discussing treatment plans—especially in the presence of certain family members. More obvious signs include patient visits wherein family members speak on behalf of a patient who otherwise demonstrates capacity or blatantly attempt to change a patient's preferences [19]. Risk factors for undue influence include, but are not limited to, dependency, social isolation, presence of physical disabilities, personality disorders, mental health disorders, and/or the existence of a will which contains "unnatural provisions" [19]. A clinician must address undue influence when it is suspect to appropriately empower the patient to make a decision that is consistent with their own values. Doing so may require conversations with other clinicians more familiar with the patient, speaking with the patient privately, and/or taking time to understand any extenuating circumstances. In order to avoid undue influence from the care team, members of the team should communicate regularly, work to deliver a consistent message, and ensure goals of care are aligned across all specialties. Also, when possible, clinicians should defer nuanced treatment discussions to the most appropriate member of the multidisciplinary team [19].

Appropriately implementing SDM in a clinical practice requires intentional effort. Elwyn et al. [20] developed a three-step model to aid a clinician in establishing SDM to ensure patients are adequately informed and provided an opportunity to "deliberate" on their options. Though clinical interactions are generally fluid, a structured framework can guide the discussion. First, a patient needs awareness that reasonable treatment options exist ("Choice Talk"). A patient may, at times, complete this step prior to an initial in-person visit (e.g., discussion with referring physician, online review, etc.). From there, more in-depth discussions around the available options can occur. The clinician should check the patient's knowledge, list all relevant options, and then highlight risks and benefits. It is important to pause regularly to ensure the patient has an appropriate level of understanding regarding each option ("Option Talk"). Finally, the dialogue can shift to focus on a patient's preferences before arriving at an informed decision which should ultimately align with the patient's values ("Decision Talk").

In the care of a patient with metastatic disease, these steps (Fig. 42.1) may occur over multiple visits and, at times, must be revisited as new information emerges.

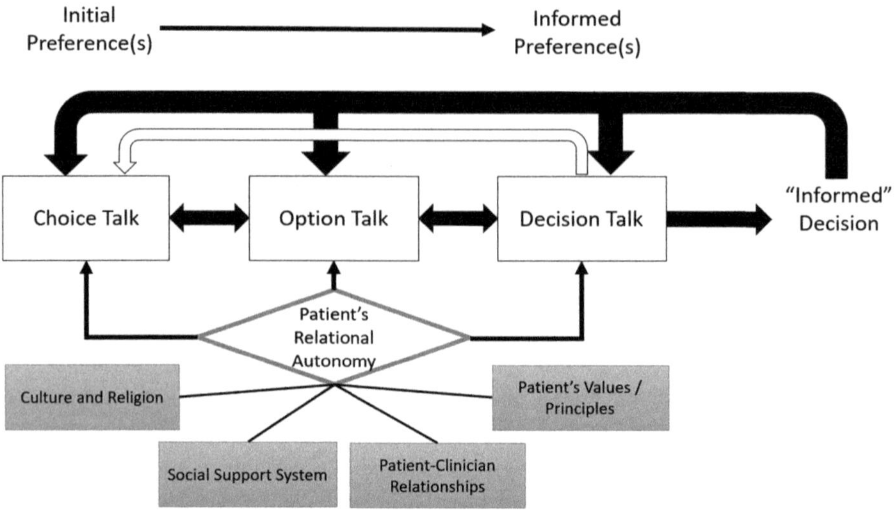

Fig. 42.1 Adaption of a figure from the SDM framework by Elwyn et al. [20] with emphasis on potential to return to earlier steps in the framework as new information presents itself—a factor particularly important in comprehensive cancer care. Additionally, the role of a patient's relational autonomy within the SDM deliberation as well as the factors which provide the foundation for relational autonomy is also demonstrated

However, appropriate application of SDM in a clinical practice can provide the clinician (and patient) confidence that the patient's values are upheld and respected even in the face of complex decisions.

Balancing Treatment Goals and Managing Patient Expectations

Addressing and managing expectations with realistic goals of treatment in patients with an advanced cancer diagnosis is, undoubtedly, a challenge oncology clinicians face regularly. This section will focus primarily on symptomatic and functional treatment modalities as disease prognosis and life-extending treatments will be covered in the "End-of-Life Considerations" section. Factors complicating patient expectations include wide variability in patients' treatment priorities or outlook on meaningful improvement in symptomology [21, 22]. Furthermore, patients frequently underestimate side effects of an intervention [23]. Physician may, in some instances, focus on different priorities than the patients they are treating [22, 24]. Another obstacle in patient expectation management centers around the fact that, not infrequently, a patient's goals of care or treatment expectations will differ from that of their family/caregivers [25, 26]. With these challenges in mind, a patient-focused and personalized approach in managing musculoskeletal metastatic disease is still paramount. Patients may base treatment goals on which symptom(s) they find most debilitating [24, 25]. Tometich et al. investigated patients with metastatic breast cancer and found different "clusters" of patients when assessing how important patients rated improvement across 10 symptoms [21]. Unsurprisingly, the cluster which favored improvement in subjective cognitive and sleeping problems also reported more severe "thinking problems" than patient's clustered by pain symptoms or those that rated improvement in all symptoms equally important. Similarly, patients may hold different views on what treatment effect is clinically meaningful. In the same study, the average symptom reduction needed for patients who deemed treatment successful was almost 50% for "fatigue" and "thinking problems" but only 30% for "pain" and 23% for "hot flashes." With these challenges in mind, a patient-focused and personalized approach in managing musculoskeletal metastatic disease is still paramount.

Physicians' treatment preferences may sometimes differ from those of the patient's preferences. A 2023 meta-analysis of 37 studies demonstrated discrepancies between patient and physician preferences in cancer treatment care. Both groups placed value on "health benefits" when considering treatments, but patients ranked this metric higher than the physicians. Patients secondarily ranked the category "treatment process" or how a selected treatment might affect their daily lives. Physicians, on the other hand, placed higher weight on adverse effects of treatment in their preferences. Altogether, these findings are suggestive that patients had less grasp of potential severity of the adverse effects of treatment or were overall willing to assume more risks to improve their prognosis [24]. Willingness to accept more risks from treatment has been described, particularly in patients with advanced cancer, as "hopeful gambles" and may be driven by a patient's assumption their own response may be above average [27, 28].

A recent study from Memorial Sloan Kettering Cancer Center tied these concepts together in comparison between surgical priorities among patients and musculoskeletal oncology surgeons. Though not exclusive to musculoskeletal metastatic disease, the study analyzed 23 years of patient priority survey results for patients undergoing treatment of extremity tumors. The results were compared to multiple musculoskeletal oncology surgeon responses of the same survey. The two groups aligned in many categories with both groups prioritizing "Physical Activity" and "Pain" above all else in treatment goals. However, the patients exhibited more variability in their prioritization. Also of note, 22–26% of patients ranked "recreational activity," "sexual life," and "financial stability" as top five priorities. Conversely, none of the surgeons surveyed ranked these categories in the top five of importance [22].

Regardless of a negative or positive influence, family and caregivers play an important role in balancing treatment goals and expectations. Clinician's need to be cognizant of familial influence, whether positive or negative, when implementing SDM tactics [16, 29]. The interplay of familial influence in patient treatment considerations is infinitely complex, dynamic, and predicated on a combination of factors: the individual patient, involvement of key family members, and overall family support system.

With some understanding of these complicating factors, clinicians can enter SDM interactions prepared to address any discordant goals or expectations of care. Leveraging a consistent message from members from the multidisciplinary team can help set appropriate expectations. Engaging a team member with a well-established clinician–patient relationship, via tumor board discussions or collaborative clinic visits, may lead to more rapid alignment of treatment goals. Clinicians should also make attempts to individualize treatment while emphasizing the pros and cons of different modalities. In treating symptoms of fatigue in patients with metastatic breast cancer, for instance, some individuals may prefer stimulant medications for more acute improvement even if side effects (i.e., insomnia, anxiousness) are expected [30]. Other patients may opt for nonpharmacologic treatments, like cognitive-behavioral therapy, which has the potential to simultaneously improve insomnia, fatigue, and cognitive impairment [21, 31].

In other cases, a clinician may need to reference evidence-based literature to support recommendations or, when sparse, spend additional time communicating their reasoning. For example, when discussing surgical treatment options for metastatic bone disease, a surgeon might utilize a validated tool like PATHFx v 3.0 [32] which provides estimated survival trajectory based on multiple patient-specific inputs. Equipped with such information, a surgeon can better discuss treatment options and tailor recommendations relative to expected survival. Furthermore, a discussion of this nature can serve as a natural transition toward a conversation around end-of-life care.

End-of-Life Care Considerations

Considerations around end-of-life care require candid, sometimes difficult, conversations between clinicians and patients. As a patient's condition or prognosis changes, discussions may need to be revisited multiple times to ensure alignment in goals of care and appropriate application of medical resources.

Research demonstrates that ~70–85% of patients with advanced cancer demonstrate an inaccurate awareness of disease status (ADS) [33, 34]. Similarly, in patients diagnosed with incurable cancer, 66–87% of patients did not adequately understand that the goal of treatment (e.g., chemotherapy) was not curative intent [23, 26, 33, 35].

In addition to previously discussed factors, such as emotional overwhelm and information overload, which can affect a patient's understanding, research has demonstrated other influences which may lead to a lack of ADS in patients diagnosed with advanced cancer. A clinician may implement SDM strategies, such as "Option Talk" and "Decision Talk," to maximize patient understanding for goal(s) of care. However, a clinician should also be aware of patient risk factors associated with lack of ADS including male gender, less education, and advancing age [36, 37]. A clinician must also be aware that, at times, a prognostic disconnect may represent an avoidant or disengaging coping strategy by patients [33, 38].

Clinician factors to consider are centered namely around poor communication. Gramling et al. [39], for instance, demonstrated race as an independent risk factor for discordance from their clinician's 2-year survival prognosis. In this study, 95% of nonwhite patients (versus 65% of white patients) had differing opinions from treating physicians with regards to their survival prognosis. The authors reached the conclusion that poor clinician communication was most likely the causative factor given the prior literature suggesting racial disparities in end-of-life communication

[40] and the fact nearly 90% of patients reported with a discordant view stated being entirely unaware their prognosis differed from their treating oncologist's prognosis. One outcome of this discrepancy in clinician communication is that black patients are up to 3 times less likely than white patients to receive end-of-life care consistent with their initially indicated preferences [41, 42].

More generally, clinicians may enter end-of-life discussions with consciousness or unconscious biases which result in poor communication, patient ADS, and discordant goals of care between patient and physician. In some instances, a physician may fail to recognize a prognostic disconnect [43]. Other physicians may actively choose to defer the conversation due to insufficient clinical time, overall prognostic uncertainty, or until no other treatment options are available [39, 44]. In a qualitative study investigating clinicians who treat metastatic spinal disease with palliative intent, most reported expectation management was performed by providing general treatment information and frequently failing to even inquire about a patient's individual expectations. Despite acknowledging the importance of these individualized conversations, every surgeon and most radiation oncologists interviewed felt outcomes are difficult to predict in this population and frequently elect to defer discussions on overall prognosis and palliative care to the medical oncologist. Spine surgeons additionally cited time constraints in the outpatient clinic or due to urgency of intervention as additional limitations to setting individualized patient expectations in their practice [45].

Patient optimism may factor into a physician's willingness to clarify a terminal prognosis early in a patient's care and/or a patient's ADS. Patients who are optimistic tend to exhibit better quality of life scores related to spiritual well-being, anxiety, and depression [46, 47]. As such, some physicians may avoid candid end-of-life care conversation in attempts to prevent quashing a patient's optimistic outlook. Others may be attuned to the fact that patients with advanced cancer prefer physicians who deliver a more positive message and view these clinicians as more compassionate [48]. Similarly, perhaps a sequela from receiving unwanted "bad news," there is an association between patients who carry a more accurate interpretation of their poor prognosis giving physician lower ratings in communication scores [35].

The importance of ensuring a patient's accurate perception of their prognosis, despite potential negative consequences, cannot be understated. Patients with an overly optimistic outlook of life expectancy related to their advanced cancer diagnosis may opt for more aggressive life-extending therapy even when such treatments do not portend longer survival [49]. In a randomized, controlled trial, Temel et al. demonstrated patients with metastatic nonsmall cell lung cancer who were given an early introduction to a palliative care team were more likely to reach accurate perception of their prognosis. Less than 10% of this patient cohort received end-of-life intravenous chemotherapy. Conversely, approximately one-third of who did not receive early referral to palliative care and half of the patient's with an inaccurate understanding of their prognosis elected to receive chemotherapy during the last 60 days of life [33].

With knowledge of how a patient's prognostic understanding can alter treatment selection, clinicians can employ tools to better counsel toward decisions that align with a patient's own values. First, utilizing objective measures may aid a clinician in navigating uncertainty around prognosis. For patients with osseous metastatic disease, the Scandinavian Sarcoma Group developed a scoring system where patients receive 1 point for any of the four categories: (1) single skeletal metastatic site, (2) absence of visceral metastatic disease, (3) Karnofsky score ≥ 70, and (4) primary cancer categorized as breast, renal, thyroid, multiple myeloma, or lymphoma. Patients with a score ≥ 3 are expected to survive at least six months with two-third surviving more than 12 months. Those with a score of 1–2 have a 50% chance of surviving more than six months but are expected to survive at least three months. Patients who score 0 are unlikely to survive 3 months [50].

While this scoring system provides a quick assessment that can be memorized, a more precise estimation of survival in the setting of osseous metastatic disease comes from the PATHFx v 3.0 calculator (https://www.pathfx.org). This validated, free, online tool provides a survival trajectory (Fig. 42.2) of patients with metastatic disease based on individual characteristics including patient age, gender, primary tumor type, number of metastatic sites, and functional status [32].

The 2024 version of the website (www.PathFx.org) also provides illustrations of different surgical treatment options for osseous metastatic disease categorized by anatomic location which can aid a clinician in facilitating a comprehensive discussion of potential treatment options. Of note, when multiple treatment options exist, the authors recommend employing a pragmatic approach in the SDM deliberation which includes discussions around postoperative recovery time particularly as related to percentage of overall life expectancy.

Armed with the above resources, clinicians can enter the end-of-life dialogue with an objective foundation, poised to cultivate better patient ADS, and prepared to facilitate SDM treatment decisions that align with a patient's individual values. Physicians can employ additional techniques to further a patient's prognostic understanding. The importance of early engagement of the palliative care team was previously discussed. Similarly, patients newly diagnosed with metastatic cancer were

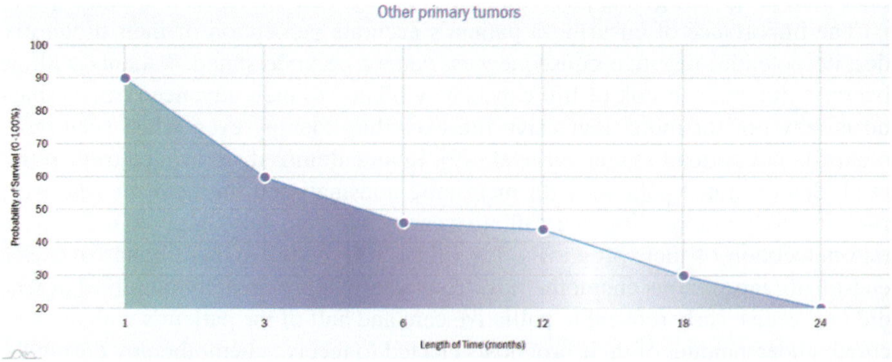

Fig. 42.2 Estimated survival trajectory graph created by PATHFx 3.0 for a 65-year-old female presenting with a realized pathologic fracture due to breast cancer, a solitary osseous metastatic site, positive visceral metastatic disease, and an Eastern Cooperative Oncology Group Score of 1

over three times as likely to demonstrate accurate prognostic awareness when being treated by physicians who favored early discussions on prognosis [33]. Physicians, with good reason, tend to use more optimistic speech when discussing prognosis related to advanced cancer. However, the patient and physician's prognostic estimations more frequently align when patients are presented with at least one pessimistic statement during their encounter [35].

A clinician treating patients with musculoskeletal metastatic disease must also make considerations beyond the specialty-specific treatment they provide. For instance, any member of the multidisciplinary team can ensure the patient and family have had discussions to ease end-of-life care decision-making. Clinicians should seek opportunities to engage patients and their caregivers in dialogue around surrogate decision-making, living wills, and financial burdens associated with end-of-life care [51]. A physician, based on their past experiences, might prompt consideration of scenarios the patient and caregivers failed to contemplate. At other times, a clinician serves as a connection to appropriate supportive care services to ensure their patients' physical, mental, and social needs are met. In serving these many complex roles, clinicians empower patients to focus on their values and truly obtain comprehensive cancer care.

Conclusion

Treating patients with musculoskeletal metastatic disease is challenging. Clinicians must navigate the intricate interplay of patient autonomy and realistic goals of care without avoiding difficult discussions related to end-of-life care. Physicians are meant to support a patient in making educated decisions which concur with their own values. Clear and intentional clinician communication, however, is paramount to facilitating such decisions. An understanding of relational autonomy, established SDM frameworks, and implementation of validated prognostic tools can improve efficiency in deliberation. A patient's ADS, accurate or not, remains a fundamental factor driving treatment decisions. As one source eloquently concluded, "In the end, it is the patient's ADS that will determine treatment and life decisions, as well as the quality of their life and, potentially, the quality of their death" [37]. By promoting patient-centered care that respects individual values, clinicians can guide patients through difficult decisions with empathy and precision.

References

1. Comer A, Fettig L, Torke AM. Identifying goals of care. Med Clin North Am. 2020;104(5):767–75. https://doi.org/10.1016/j.mcna.2020.06.002.
2. Hawley ST, Jagsi R. Shared decision making in cancer care: does one size fit all? JAMA Oncol. 2015;1(1):58–9. https://doi.org/10.1001/jamaoncol.2014.186.
3. Igel LH, Lerner BH. Moving past individual and "pure" autonomy: the rise of family-centered patient care. AMA J Ethics. 2016;18(1):56–62. https://doi.org/10.1001/journalofethics.2017.18.1.msoc1-1601.

4. Kumar NK. Informed consent: past and present. Perspect Clin Res. 2013;4(1):21–5. https://doi.org/10.4103/2229-3485.106372.
5. Pugh J. Informed consent, autonomy, and beliefs. In: Autonomy, rationality, and contemporary bioethics [Internet]. Oxford University Press; 2020. https://www.ncbi.nlm.nih.gov/books/NBK556864/. Accessed 1 Oct 2024.
6. Bester J, Cole CM, Kodish E. The limits of informed consent for an overwhelmed patient: clinicians' role in protecting patients and preventing overwhelm. AMA J Ethics. 2016;18(9):869–86. https://doi.org/10.1001/journalofethics.2016.18.9.peer2-1609.
7. Appelbaum PS. Assessment of patients' competence to consent to treatment. N Engl J Med. 2007;357(18):1834–40. https://doi.org/10.1056/NEJMcp074045.
8. Waldron T, Carr T, McMullen L, et al. Development of a program theory for shared decision-making: a realist synthesis. BMC Health Serv Res. 2020;20(1):59. https://doi.org/10.1186/s12913-019-4649-1.
9. Kon AA. The shared decision-making continuum. JAMA. 2010;304(8):903–4. https://doi.org/10.1001/jama.2010.1208.
10. Childress JF. Needed: a more rigorous analysis of models of decision making and a richer account of respect for autonomy. Am J Bioeth AJOB. 2017;17(11):52–4. https://doi.org/10.1080/15265161.2017.1382168.
11. Elwyn G, Laitner S, Coulter A, Walker E, Watson P, Thomson R. Implementing shared decision making in the NHS. BMJ. 2010;341:c5146. https://doi.org/10.1136/bmj.c5146.
12. Ubel PA, Scherr KA, Fagerlin A. Autonomy: what's shared decision making have to do with it? Am J Bioeth AJOB. 2018;18(2):W11–2. https://doi.org/10.1080/15265161.2017.1409844.
13. Dove ES, Kelly SE, Lucivero F, Machirori M, Dheensa S, Prainsack B. Beyond individualism: is there a place for relational autonomy in clinical practice and research? Clin Ethics. 2017;12(3):150–65. https://doi.org/10.1177/1477750917704156.
14. Zhang AY, Siminoff LA. The role of the family in treatment decision making by patients with cancer. Oncol Nurs Forum. 2003;30(6):1022–8. https://doi.org/10.1188/03.ONF.1022-1028.
15. Walter JK, Ross LF. Relational autonomy: moving beyond the limits of isolated individualism. Pediatrics. 2014;133(Suppl 1):S16–23. https://doi.org/10.1542/peds.2013-3608D.
16. Dijkman BL, Luttik ML, Van der Wal-Huisman H, Paans W, van Leeuwen BL. Factors influencing family involvement in treatment decision-making for older patients with cancer: a scoping review. J Geriatr Oncol. 2022;13(4):391–7. https://doi.org/10.1016/j.jgo.2021.11.003.
17. Entwistle VA, Carter SM, Cribb A, McCaffery K. Supporting patient autonomy: the importance of clinician-patient relationships. J Gen Intern Med. 2010;25(7):741–5. https://doi.org/10.1007/s11606-010-1292-2.
18. Kehl KL, Landrum MB, Arora NK, et al. Association of actual and preferred decision roles with patient-reported quality of care: shared decision making in cancer care. JAMA Oncol. 2015;1(1):50–8. https://doi.org/10.1001/jamaoncol.2014.112.
19. Fuller T, Hendrick J, Beck AC. Undue influence versus relational autonomy in clinical decision making #439. J Palliat Med. 2022;25(6):996–7. https://doi.org/10.1089/jpm.2022.0089.
20. Elwyn G, Frosch D, Thomson R, et al. Shared decision making: a model for clinical practice. J Gen Intern Med. 2012;27(10):1361–7. https://doi.org/10.1007/s11606-012-2077-6.
21. Tometich DB, Mosher CE, Hirsh AT, et al. Metastatic breast cancer patients' expectations and priorities for symptom improvement. Support Care Cancer. 2018;26(11):3781–8. https://doi.org/10.1007/s00520-018-4244-8.
22. Bartelstein M, Lavery J, Chernetskii N. Comparing patient and surgeon priorities in musculoskeletal oncology surgery. Podium Presentation presented at: Musculoskeletal Tumor Society Annual Meeting; September 11, 2024; Chicago, Illinois.
23. Minichsdorfer C, Zeller O, Kirschbaum M, Berghoff AS, Bartsch R. Expectations and perception of cancer treatment goals in previously untreated patients. The EXPECT trial. Support Care Cancer. 2021;29(7):3585–92. https://doi.org/10.1007/s00520-020-05826-x.

24. Zhang M, He X, Wu J, Xie F. Differences between physician and patient preferences for cancer treatments: a systematic review. BMC Cancer. 2023;23:1126. https://doi.org/10.1186/s12885-023-11598-4.
25. Ozdemir S, Chaudhry I, Malhotra C, Teo I, Finkelstein EA, Cost of Medical Care of Patients with Advanced Serious Illness in Singapore (COMPASS) Study Group. Goals of care among patients with advanced cancer and their family caregivers in the last years of life. JAMA Netw Open. 2024;7(4):e245866. https://doi.org/10.1001/jamanetworkopen.2024.5866.
26. Burns CM, Broom DH, Smith WT, Dear K, Craft PS. Fluctuating awareness of treatment goals among patients and their caregivers: a longitudinal study of a dynamic process. Support Care Cancer. 2007;15(2):187–96. https://doi.org/10.1007/s00520-006-0116-8.
27. Chu DT, Kim SW, Hsu HK, et al. Patient attitudes towards chemotherapy and survival: a prospective observational study in advanced non-small cell lung cancer. Lung Cancer. 2009;66(2):250–6. https://doi.org/10.1016/j.lungcan.2009.01.022.
28. Lakdawalla DN, Romley JA, Sanchez Y, Maclean JR, Penrod JR, Philipson T. How cancer patients value hope and the implications for cost-effectiveness assessments of high-cost cancer therapies. Health Aff (Millwood). 2012;31(4):676–82. https://doi.org/10.1377/hlthaff.2011.1300.
29. Laryionava K, Hauke D, Heußner P, Hiddemann W, Winkler EC. "Often relatives are the key […]" –family involvement in treatment decision making in patients with advanced cancer near the end of life. Oncologist. 2021;26(5):e831–7. https://doi.org/10.1002/onco.13557.
30. Carroll JK, Kohli S, Mustian KM, Roscoe JA, Morrow GR. Pharmacologic treatment of cancer-related fatigue. Oncologist. 2007;12(S1):43–51. https://doi.org/10.1634/theoncologist.12-S1-43.
31. Campbell T, Garland S, Johnson J, et al. Sleeping well with cancer: a systematic review of cognitive behavioral therapy for insomnia in cancer patients. Neuropsychiatr Dis Treat. Published online June 2014:1113. https://doi.org/10.2147/NDT.S47790.
32. Anderson AB, Wedin R, Fabbri N, Boland P, Healey J, Forsberg JA. External validation of PATHFx version 3.0 in patients treated surgically and nonsurgically for symptomatic skeletal metastases. Clin Orthop Relat Res. 2020;478(4):808. https://doi.org/10.1097/CORR.0000000000001081.
33. Temel JS, Greer JA, Admane S, et al. Longitudinal perceptions of prognosis and goals of therapy in patients with metastatic non–small-cell lung cancer: results of a randomized study of early palliative care. J Clin Oncol. 2011;29(17):2319–26. https://doi.org/10.1200/JCO.2010.32.4459.
34. Liu PH, Landrum MB, Weeks JC, et al. Physicians' propensity to discuss prognosis is associated with patients' awareness of prognosis for metastatic cancers. J Palliat Med. 2014;17(6):673–82. https://doi.org/10.1089/jpm.2013.0460.
35. Weeks JC, Catalano PJ, Cronin A, et al. Patients' expectations about effects of chemotherapy for advanced cancer. N Engl J Med. 2012;367(17):1616–25. https://doi.org/10.1056/NEJMoa1204410.
36. Pronzato P, Bertelli G, Losardo P, Landucci M. What do advanced cancer patients know of their disease? A report from Italy. Support Care Cancer. 1994;2(4):242–4. https://doi.org/10.1007/BF00365729.
37. Finlayson CS, Rosa WE, Mathew S, Applebaum A, Squires A, Fu MR. Awareness of disease status among patients with cancer. Cancer Nurs. Published online January 16, 2023. https://doi.org/10.1097/NCC.0000000000001170.
38. Fasano J, Shao T, Huang H h, Kessler AJ, Kolodka OP, Shapiro CL. Optimism and coping: do they influence health outcomes in women with breast cancer? A systemic review and meta-analysis. Breast Cancer Res Treat. 2020;183(3):495–501. https://doi.org/10.1007/s10549-020-05800-5.
39. Gramling R, Fiscella K, Xing G, et al. Determinants of patient-oncologist prognostic discordance in advanced cancer. JAMA Oncol. 2016;2(11):1421–6. https://doi.org/10.1001/jamaoncol.2016.1861.

40. Welch LC, Teno JM, Mor V. End-of-life care in black and white: race matters for medical care of dying patients and their families. J Am Geriatr Soc. 2005;53(7):1145–53. https://doi.org/10.1111/j.1532-5415.2005.53357.x.
41. Loggers ET, Maciejewski PK, Paulk E, et al. Racial differences in predictors of intensive end-of-life care in patients with advanced cancer. J Clin Oncol. 2009;27(33):5559. https://doi.org/10.1200/JCO.2009.22.4733.
42. Trice ED, Prigerson HG. Communication in end-stage cancer: review of the literature and future research. J Health Commun. 2009;14 Suppl 1(1):95–108. https://doi.org/10.1080/10810730902806786.
43. Mackillop WJ, Stewart WE, Ginsburg AD, Stewart SS. Cancer patients' perceptions of their disease and its treatment. Br J Cancer. 1988;58(3):355–8. https://doi.org/10.1038/bjc.1988.218.
44. Keating NL, Landrum MB, Rogers SO Jr, et al. Physician factors associated with discussions about end-of-life care. Cancer. 2010;116(4):998–1006. https://doi.org/10.1002/cncr.24761.
45. Versteeg AL, Gal R, Charest-Morin R, et al. Expectations of treatment outcomes in patients with spinal metastases; what do we tell our patients? A qualitative study. BMC Cancer. 2021;21:1263. https://doi.org/10.1186/s12885-021-08993-0.
46. Mazanec SR, Daly BJ, Douglas SL, Lipson AR. The relationship between optimism and quality of life in newly diagnosed cancer patients. Cancer Nurs. 2010;33(3):235. https://doi.org/10.1097/NCC.0b013e3181c7fa80.
47. Applebaum AJ, Stein EM, Lord-Bessen J, Pessin H, Rosenfeld B, Breitbart W. Optimism, social support, and mental health outcomes in patients with advanced cancer. Psychooncology. 2014;23(3):299–306. https://doi.org/10.1002/pon.3418.
48. Tanco K, Rhondali W, Perez-Cruz P, et al. Patient perception of physician compassion after a more optimistic vs a less optimistic message: a randomized clinical trial. JAMA Oncol. 2015;1(2):176–83. https://doi.org/10.1001/jamaoncol.2014.297.
49. Weeks JC, Cook EF, O'Day SJ, et al. Relationship between cancer patients' predictions of prognosis and their treatment preferences. JAMA. 1998;279(21):1709–14. https://doi.org/10.1001/jama.279.21.1709.
50. Ratasvuori M, Wedin R, Keller J, et al. Insight opinion to surgically treated metastatic bone disease: Scandinavian Sarcoma Group Skeletal Metastasis Registry report of 1195 operated skeletal metastasis. Surg Oncol. 2013;22(2):132–8. https://doi.org/10.1016/j.suronc.2013.02.008.
51. Horyna A. Psychosocial considerations for patients with metastatic bone disease. In: Randall RL, editor. Metastatic bone disease: an integrated approach to patient care. Springer; 2016. p. 145–54. https://doi.org/10.1007/978-1-4614-5662-9_14.

Comprehensive Percutaneous Intervention for Skeletal Metastases: Ablation-Osteoplasty-Reinforcement-Internal Fixation (AORIF)

43

Francis Y. Lee

Rationale for AORIF

Orthopedic care for skeletal metastases is an oncological procedure and is not the same as surgeries for traumas, arthritis, and sarcomas. The ideal procedure for skeletal metastases involves decreasing cancer burden, to allow patients to ambulate for continued outpatient oncologic care, and to promote bone reconstitution in a minimally invasive way with very low postoperative complications. To this end, a minimally invasive AORIF procedure is a feasible option in the era of targeted molecular cancer therapies.

Assessment of Pain on Ambulation and Indications

Indications for AORIF are for both palliative salvage and functional cure for any massive defects in the periarticular area including the periacetabular region. The loss of subchondral bone on x-rays or CT does not contraindicate AORIF because there are radiolucent tissues or cartilage that prevent massive cement leaking. When there is a comminuted fracture with a cavitary or linear gap in the subchondral bone, cement leakage is a possible theoretical concern, but restoration of ambulatory function with less pain is more beneficial than radiographic presence of bone cement outside the bone boundary, if it is asymptomatic. Cement leakage is minimized by creating the first cement deposition space with balloon inflation or balloon osteoplasty. Patients with skeletal metastases often have multiple visceral and skeletal metastases for which patients are on chronic pain management plan. Resting pain or

F. Y. Lee (✉)
Department of Orthopaedic Surgery, Yale School of Medicine, New Haven, CT, USA
e-mail: francis.lee@yale.edu

J. Pretell-Mazzini, S. A. Lozano Calderón (eds.), *Multidisciplinary Approach of Musculoskeletal Metastasis*, https://doi.org/10.1007/978-3-032-09498-8_43

night pain without weight-bearing or limb movement could indicate cancer pain, while worsening pain on limb movement or with weight-bearing indicates pain from insufficient bone or pathological fractures. It would be ideal to have an easy scoring system for pain and ambulatory function so that various treatment effects can be assessed before and after the procedures. The function and pain status of each patient can be assessed by using the "Pain and Ambulatory Function Scoring System" that was developed for the skeletal metastases in the lower extremities [1–3] (Fig. 43.1).

Oncologic Considerations

Prevention of Mechanical Spreading of Cancer Cells

Minimizing the cancer spread during the procedure is the basic principle of cancer surgeries. When the presence of a single cancer cell is a major concern for sarcoma surgeries, reaming through the cancer has been traditionally granted for metastatic cancers, which is a paradoxical oncological bias. Open surgical approach and inserting metallic hardware may mechanically spread treatment-resistant cancer cells in the adjacent soft tissues and bones. Ablative therapy may be done first and then

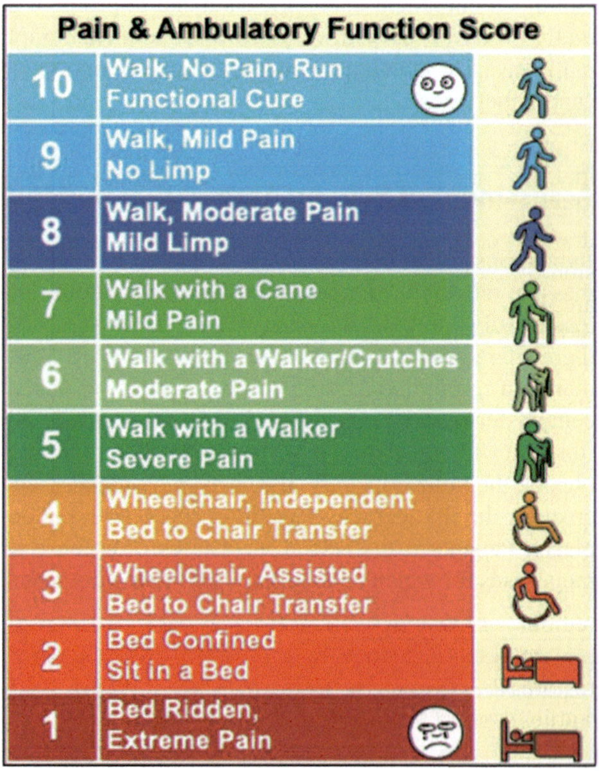

Fig. 43.1 Combined "Pain on Ambulation Functional Scoring System" for the skeletal metastases in the pelvis and lower extremities. From [1], Wolters Kluwer Health, Inc.

passage of surgical instruments or implants may be considered for localized metastatic cancers. AORIF is intended not involve reaming or in-and-out of surgical instruments.

Local Killing of Metastatic Cancer Cells

In many patients with advanced cancers, local progression of metastatic lesions, bone destruction, pain, and fractures cannot be eradicated even after radiation, chemotherapy, or antiresorptive therapies. Therefore, it would be ideal to include therapeutic modalities that would kill metastatic cancer cells locally at the site of painful bone destruction before any stabilization procedures. Cancer killing may be accomplished by various modalities such as radiofrequency ablation, cryoablation, laser ablation, microwave ablation, or electroporation. Local cancer drug delivery may be considered but is not a common practice yet but heat stable antiresorptive drugs such as bisphosphonate may be mixed with PMMA bone cement for local cancer control [1, 2].

Biomechanical Considerations

Immediate Restoration of Bone Strength and Ambulation with Percutaneous PMMA Bone Cement and Screws

"Where to reinforce?" is an important scientific consideration. For example, patients may present with disseminated osteolytic lesions in the bilateral pelvis. The most important consideration is to identify the most critical region that transmits load from the femoral head. This weight-bearing portion of the acetabulum is easily identified a curved sclerotic region in contact with the femoral head. This area is often called as "Acetabular Sourcil." This area typically shows thickened subchondral cortical bone and dense trabecular bone that is in line with the proximal extension of the calcar femorale in the femoral head. Therefore, the author named this area of weight-bearing dense subchondral bone as "calcar acetabuli." Reinforcement of the "calcar acetabuli" region is the first priority for the restoration of the ambulatory function. A similar therapeutic concept is seen in an "ice-cream cone prosthesis" that restores ambulatory function in the absence of the pubic ramus and ischial bones. Restoration and preservation of ambulatory status is critical for medical oncologic care and radiation therapy for patients with advanced cancers. Percutaneous reinforcement allows patients to resume ambulation and anticancer drug therapy faster when compared to open surgical reconstructions using arthroplasty implants or plate-and-screw constructs. Immediate reinforcement of bone with PMMA bone cement is an easy option in the pelvis, sacrum, subchondral regions, or metaphyseal locations of long bones. While screws or any internal fixation implants are good in connecting large pieces of bone, bone cement easily restores the connectivity of the trabecular bones by filling in the gap. Biomechanical studies suggest that addition of metallic implants such as screws provide higher load to failure and resistance to cyclic loading [4]. Therefore, percutaneous hybrid reinforcement with PMMA bone cement and cannulated screws provide an easy way of reinforcing bone. High viscosity cement is used to minimize cement leakage. Cement can be best retained in the target area when the kyphoplasty balloon inflation is used for opening up the space. PMMA bone cement generates heat during polymerization and has secondary effects of thermal necrosis of cancer cells, thermal denervation for pain control, and local immune response.

Biological Considerations for Bone Reconstitution

Osteolytic metastatic cancer cells increase osteoclastic bone resorption and inhibit osteoblastic bone formation. Very frequently, cancer cells in the bone are resistant to radiation and anticancer drug therapies. Ablation of cancer cells can induce bone reconstitution [5–9].

Pain Control Perspectives

Unstable fractures of acetabulum cause enormous pain and limit patients to bed rest. Theoretically, radiofrequency ablation, killing of cancer cells, and generation of heat from polymerization of PMMA bone cement reduce pain by thermal denervation or reducing release of pain mediators from cancer cells or fractures. Therefore, AORIF provides pain control by reducing mechanical pain and cancer-induced pain (Table 43.1).

Table 43.1 Overall survival intervals according to the different types of carcinomas that mostly metastasized to the skeleton

	5-Year survival	5-Year survival	5-Year survival	5-Year survival	5-Year survival	10-Year survival	10-Year survival	10-Year survival	10-Year survival	10-Year survival	20-Year survival	20-Year survival	20-Year survival	20-Year survival	20-Year survival
	1975–1984	1985–1994	1995–2004	2005–2014	All years	1975–1984	1985–1994	1995–2004	2005–2014	All years	1975–1984	1985–1994	1995–2004	2005–2014	All years
Breast	75.50%	84.00%	89.00%	91.10%	86.10%	63.60%	76.10%	83.50%	86.30%	78.80%	53.20%	67.50%	75.70%	N/A	69.80%
Prostate	70.50%	89.10%	98.60%	99.10%	93.40%	55.70%	81.90%	97.90%	99.10%	89.70%	39.60%	72.40%	94.40%	N/A	81.70%
Lung and bronchus	12.70%	13.40%	15.30%	19.60%	15.40%	8.70%	9.00%	10.20%	13.10%	10.40%	4.80%	4.80%	5.50%	N/A	5.60%
Colon and rectum	52.10%	60.00%	64.00%	66.40%	60.80%	46.50%	53.90%	58.40%	60.10%	54.90%	42.60%	49.50%	52.60%	N/A	50.00%
Kidney and renal pelvis	51.50%	58.40%	65.50%	75.10%	65.80%	44.50%	50.80%	57.70%	68.80%	58.40%	36.80%	40.90%	47.00%	N/A	47.60%
Thyroid	92.70%	94.40%	96.50%	98.40%	96.60%	91.30%	93.30%	95.80%	98.40%	96.00%	89.90%	92.10%	94.90%	N/A	95.00%

Technical Details of AORIF for Pelvic Defects and Fractures

Anatomical Considerations

AORIF uses the iliac crest as screw entry sites because the iliac crest is strongest intact superficial entry point and allows multiple screw insertion (Figs. 43.2 and 43.3). By entering the area with cancer involvement from the iliac crest, we can minimize bleeding, cancer contamination, and cement leakage. If the lesion is entered directly from the lateral aspect of the pelvis, leakage and spreading of cancer cells from the lesion contradict the principle of cancer surgery. In addition, hemorrhagic lesions such as renal cell cancer metastases may bleed through bone defects and bleeding control could be problematic. From a biomechanical perspective, screws are best supported by a long bone tunnel surrounded by the strongest intact bone. Direct entry of the periacetabular region from the lateral defective pelvic wall may end up with loose screws or guide wires. Therefore, the anterior iliac crest is most commonly used to enter the lesions with cannulated screws for superior-medial and posterior defects. Additional screws are optionally used only if defects in the superior pubic ramus cause major

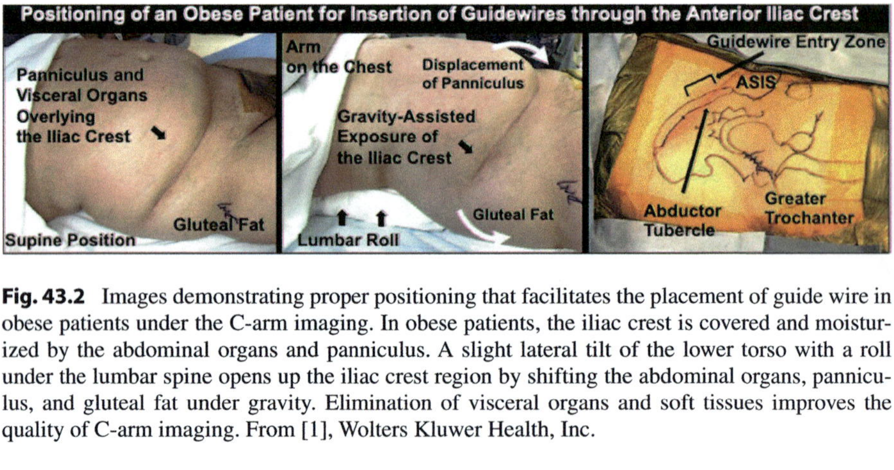

Fig. 43.2 Images demonstrating proper positioning that facilitates the placement of guide wire in obese patients under the C-arm imaging. In obese patients, the iliac crest is covered and moisturized by the abdominal organs and panniculus. A slight lateral tilt of the lower torso with a roll under the lumbar spine opens up the iliac crest region by shifting the abdominal organs, panniculus, and gluteal fat under gravity. Elimination of visceral organs and soft tissues improves the quality of C-arm imaging. From [1], Wolters Kluwer Health, Inc.

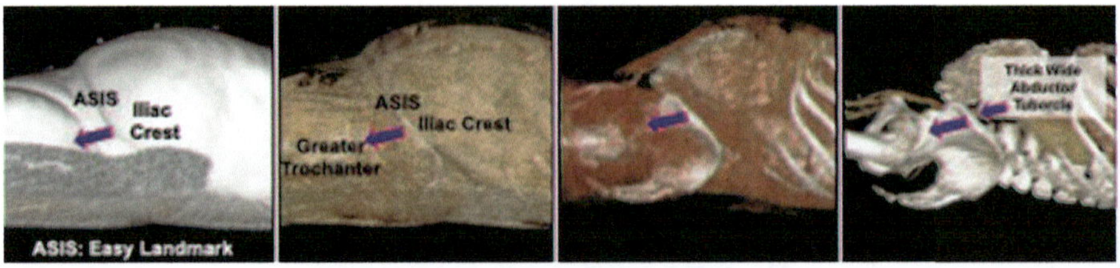

Fig. 43.3 CI images of a patient showing surface anatomy, soft tissues, and bone anatomy for the AORIF procedure. The iliac crest allows the easiest access to the periacetabular region without going through muscle layers

disconnection of the pelvis. As emphasized previously, osteolytic metastases are not traumatic acetabular fractures and routine screw fixation of the superior pubic ramus or ischium is not necessary because two to four screws from the anterior iliac crest can provide screw and cement fixation in the anterior, middle, and posterior part of the acetabulum. When there are defects in the sacroiliac joint and acetabulum, AORIF and sacro-iliac joint fixation can be done in the prone position.

Positioning and Draping

Patients are placed in the supine or prone position for optimal access to the osteolytic defects in the weight-bearing portion of the acetabulum. By placing a bump under the low back and an ipsilateral arm on the chest, visceral organs, abdominal panniculus, and gluteal fat are displaced away from the iliac crest, making iliac crest more palpable and visible (Figs. 43.2 and 43.3). When defects involve posterior iliac wing and sacro-iliac joint, prone position is preferred. In the prone position, posterior acetabular wall and superomedial weight-bearing acetabular dome is also accessible. When there is an acute acetabular fracture or protrusio acetabuli, skin traction may be applied to pull the leg longitudinally to stabilize the acetabulum in a more anatomic position.

Insertion of Guide Wires Under C-Arm Imaging

3D imaging is preferred, but it is not always available. 3D imaging is offered by O-arm with a navigation system, C-arm with 3D image formatting (Ziehm C-arm), cone beam CT imaging technique (Siemens Artis Zeego), or others. Authors prefer to use Artis Zeego CT because of available robotic arm that controls various imaging projection angles instantly by memory. By understanding the regional anatomy, conventional C-arm imaging can be used for the safe placement of screws and cement (Fig. 43.4) [10–12]. Guide wires are inserted through the most intact and healthy bone so that the guide wires stay in place during the procedure. If guide wires are placed directly through the osteolytic lesion, unstable guide wires, cancer leakage, and bleeding can cause unwanted delay and technical difficulties. Furthermore, screws may be even lost inside the osteolytic defects and retrieval of loose migrated screw is not a worthwhile effort. Therefore, AORIF uses the most intact and strongest bone for insertion of guide wires and screws. Outer and inner iliac wall cortical bones are visualized with an oblique view. Posterior acetabular bone is visualized with an iliac wing view. A true lateral view is used to verify the direction of guide wires for anterior, superior-medial, or posterior acetabular defects. A true lateral view is obtained by overlapping sciatic notches and femoral heads. The superior pubic screw is visualized by external oblique, inlet, and outlet views. 3D imaging confirms the location of guide wires best after all.

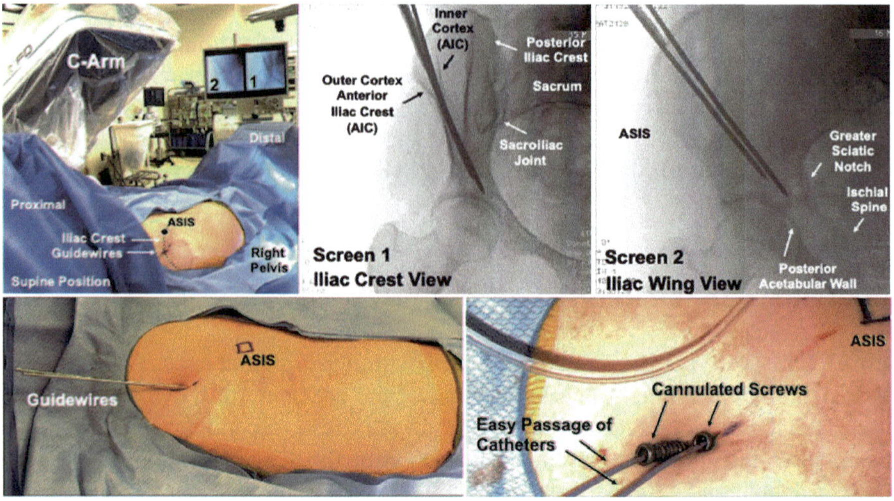

Fig. 43.4 Images demonstrating antegrade iliac crest guide wire placement techniques and screw placement under C-arm imaging guidance. Two or more guide wires are inserted from the anterior iliac crest to provide access from the anterior to the posterior acetabulum. Under the "iliac crest view" showing both the anterior and posterior iliac crest, the guide wires proceed between the outer and inner cortexes to the osteolytic metastasis in the superior aspect of the weight-bearing portion of the acetabulum. Under the iliac wing (internal oblique) view, the posterior acetabular wall, sciatic notch, and ischial spine are visualized. Guide wires are placed between the acetabular wall and the sciatic notch. Cannulated screw heads are visible outside the skin for easier passage of catheters for RFA, balloon osteoplasty, and cement delivery. ASIS = anterior superior iliac spine. From [1], Wolters Kluwer Health, Inc.

Insertion of Fully Threaded Cannulated Screw as a Universal Portal for Instruments

Once placement of the distal tips of guide wires are confirmed by C-arm, cone-beam CT, or 3D imaging, fully threaded cannulated screws are inserted over the guide wire. Fully threaded screws are preferred to partially threaded screws because of better anchoring of screws at the thick bones in the iliac crest and better interdigitation with PMMA bone cements. Screws are fully inserted making sure that screw tips do not directly touch subchondral bones of the acetabulum or far cortical bones. Optimal screw length is usually 1 cm from the subchondral bone so that the distal tip of the screw does not mechanically block the displaced femoral head when cancer progresses locally. Once appropriate screw length is measured, screws are backed out of the skin so that the central hollow of the cannulated screw head is visible and accessible for biopsy, ablation, balloon osteoplasty, and PMMA cement injection (Fig. 43.5).

Biopsy Through Cannulated Screws

Biopsy is an easy extra-step to confirm the diagnosis especially when the lesion is a solitary skeletal metastasis or unsuspected sarcoma is rarely seen in patients with double or triple cancers. Aspiration through a large diameter bore provides ample

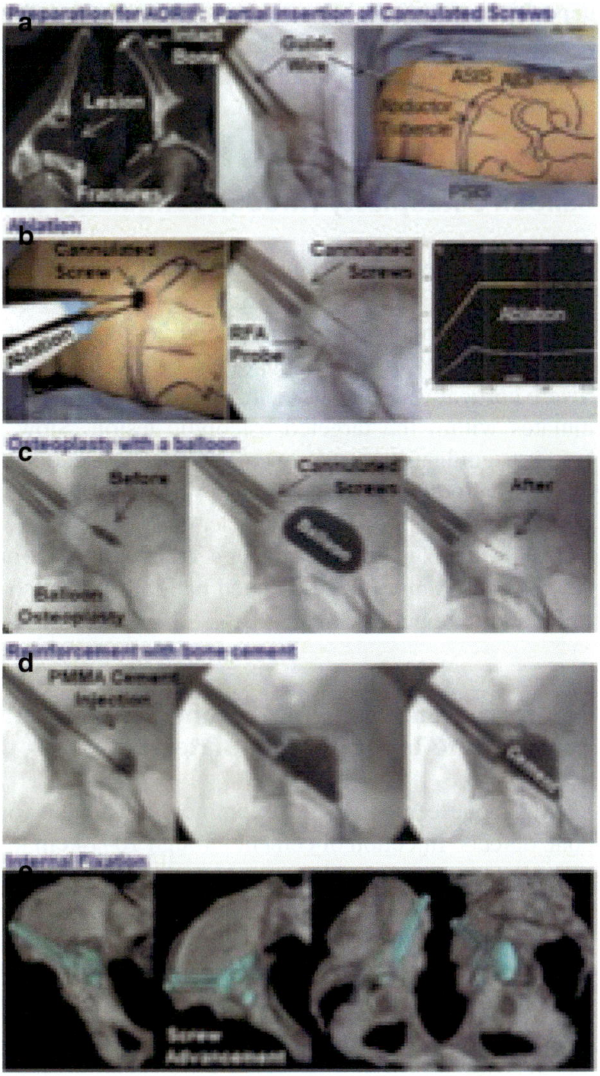

Fig. 43.5 Illustrated details of the AORIF procedure. (**a**) The lesion is entered through intact bone to avoid cancer contamination and bleeding. The most common entry portal is through an abductor tubercle in the anterior iliac crest. A guide wire for cannulated screws is inserted toward the superomedial aspect of the weight-bearing portion of the acetabulum. Cannulated screws are inserted partially so the screw heads stay outside the skin for easy access to the osteolytic lesion. Two screws are usually placed so one screw hole can serve as a vent during balloon osteoplasty and cement injection. (**b–d**) Ablation probes, balloon osteoplasty, and bone cement injection channels are easily inserted through the central channels of the cannulated screws. (**e**) Screws are advanced before the cement is cured. From [3], Elsevier Inc.

amount of specimens for pathology diagnosis. Biopsy specimens are invaluable in the era of targeted therapy because a new subset of cancer cells with different receptors, kinases, or immune checkpoint regulators are identified for changing cancer targeting drugs.

Ablation Through Cannulated Screws

As part of the oncologic care, reducing caner burden by all means is meaningful. Ablation can be done as per institutional choices or availability of various ablation systems. Authors prefer radiofrequency ablation (RFA) to cryoablation because RFA causes liquefaction thermal necrosis that allows balloon inflation and deposition of PMMA bone cement. Instant heat generation by microwave ablation is another alternative option. Freezing theoretically make bone more brittle and may cause thrombosis of the small arteries that supply the femoral head, leading to avascular necrosis (Fig. 43.5).

Balloon Osteoplasty Through Cannulated Screws

Radiolucent metastatic lesions are not empty unlike benign unicameral bone cyst. Apparently radiolucent osteolytic lesions contain cancer tissues, reactive host tissues, residual bone debris, and blood vessels. Percutaneous injection of PMMA cement may not fill inside the target osteolytic defects in a predictable way. For severe cortical wall defects, cement may leak outside the bone before cement accumulates within the targeted osteolytic lesion. Slow inflation of balloons under imaging guidance creates open space where cement can be deposited first. An inflated balloon can occlude tumor vessels as an internal tamponade for bleeding control or prevention of cement leaking into tumor vessels. An inflated balloon forms a wall-like structure so that cement extravasation outside massive osteolytic defects can be minimized. Dilated balloons may also reduce displaced depressed subchondral bone.

Reinforcement with PMMA Bone Cement with Optional Zoledronate Through Cannulated Screws

PMMA bone cement is most commonly available for immediate stabilization of bone defects. Injection of cement through 10–14 cm long, 3.5 mm hollow tunnel of cannulated screws within a few minutes can be time-taking and challenging in terms of efficiency of cement delivery and volume. Therefore, we use a cement injector that is commonly used for vertebroplasty. By using a well-controlled hydraulic pressurized cement injector with a long delivery cannula, cement injection is more accurate and effective in the area where balloon osteoplasty was done. Cement filling could be geographic in the osteolytic area where dilated balloon created an initial deposition cavity (Fig. 43.5). After main cavitary defect lesion is filled out, additional injection of cement inside the trabecular space shows infiltration of cement within Swiss cheese-like osteopenic trabecular space. This additional cement injection in the screw tract supports the proximal part of the screw that is inserted from the iliac crest.

One theoretical concern is extravasation of cements into the hip joint, pelvis through a medial wall defect, lateral wall defect, sciatic notch, and sacral neural structures. Intermittent fluoroscopic guidance is essential to minimize such unwanted cement deposition. For contained cavitary osteolytic lesions intact acetabular subchondral bone, intra-articular extravasation of cement does not occur during the procedure. Small extravasation of cement into the hip joint may occur in patients with cracks or holes in the acetabular subchondral bone and articular cartilage is not detrimental compared to downsides of reconstructive surgeries replacing entire hip joints using mega-prostheses. Patients are asymptomatic when small amount of cement entered the hip joint through comminuted acetabular fractures and remain ambulatory. The author observed that the cement in the synovial cavity is not symptomatic as the cement is collected in the synovial sac or cotyloid fossa. Once life-saving chemotherapy is completed with cure of cancers, formal hip replacement arthroplasty may be considered.

Postoperative Courses

Patients are instructed to go home on the day of the AORIF and to initiate any activities of daily living as tolerated (Fig. 43.6). In-house consultation patients returned to medical oncology floor after AORIF. Anticancer drug therapies resume next day without waiting for would healing. Radiation can start any time after AORIF.

Complications of AORIF

General Complications Pertaining to Any Open Surgeries For acetabular metastatic defects, open trauma-style repair or open arthroplasty reconstructive surgeries using femoral head and acetabular metal prostheses are deferred. Therefore, AORIF precludes any well-known complications pertaining to open surgical approaches such as infection, bleeding, severe postoperative pain, migration of metal implants,

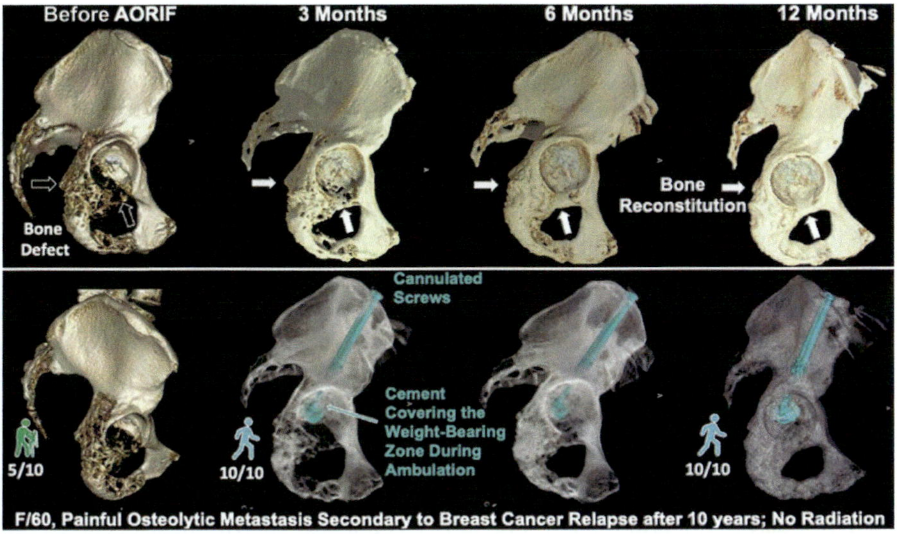

Fig. 43.6 A case demonstrating cancer-induced bone loss that is reversible. The massive bone defect secondary to breast cancer relapse and metastases was reconstituted dramatically after AORIF. From [1], Wolters Kluwer Health, Inc.

implant loosening, thrombosis, prolonged rehabilitation, and others. Most AORIF patients report improved pain and functional ambulatory status.

Cement Leakage When there is a comminuted fracture with a cavitary or linear gap in the subchondral bone, cement leakage into the joint is possible but minimized by creating the first cement deposition space with balloon inflation. The presence of cement in the joint does not cause significant pain that reduces functional ambulatory status. Small particles are collected in the nonweight-bearing synovial sac. Linear congruent cement supports a thin layer of articular cartilage. Orthopedic surgeons are well aware of long-term functional benefits of subchondral bone cement through follow-up cases of giant cell tumor that invariably destroy epiphyseal and subchondral bone. Cement leakage into surrounding soft tissues do not cause problems. When PMMA bone cement is injected into the major defects in the anterior acetabular dome and superior pubic ramus regions, the involved hip joint is slightly flexed to displace femoral vessels and nerve from the cement curing site.

Protrusio Acetabuli or Subluxation (Proximal, Medial, or Lateral Migration of Femoral Head) When there is a notching of the superolateral corner of the femoral head in the presence of acetabular subchondral bone loss or comminuted fractures, this indicates that the femoral head was impacted by the acetabular rim from proximal migration of the femoral head during weight-bearing. This femoral head notch sign indicates development of incongruent femoral head and migration of femoral head toward acetabulum. The presence of this sign does not preclude ambulation. Protrusion acetabuli is one of indications for AORIF. Notching of femoral head by a screw or cement may occur as vertically oriented AORIF screw-cement construct block proximal migration of the femoral head into the acetabulum. This means disease progression or continuation of cancer-induced bone loss without bone healing. When AORIF construct itself is intact, AORIF serves its purpose of preventing catastrophic protrusion acetabuli. Maintenance of ambulatory status in the presence of notching of femoral head or partial loss of the femoral head from concurrent cancer in the femoral head or osteopenia is similar to patients who had Girdlestone operation for infected total hip arthroplasty, femoral head resection for painful neuromuscular hip subluxation secondary to cerebral palsy, or sequalae of avascular necrosis of femoral head. Decreased pain perception could be due to thermal denervation by heat from RFA or polymerization of cement or decreased pain mediators after local cancer ablation and cementation.

Conversion Hip Arthroplasty When the cancer is well-controlled and the patient desires more activities in the presence of noncongruent hip joint or protrusion, a formal hip arthroplasty can be done on top of the AORIF construct like a staged AORIF-delayed hip arthroplasty. Previously conducted AORIF makes delayed hip reconstruction much easier by allowing hemi-arthroplasty or total hip arthroplasty with augmented acetabular cups. The rate of conversion to hip arthroplasty is currently known as less than 5% [1–3].

Dislodgement of AORIF Constructs from the Iliac Crest Theoretically, the entire AORIF screw and cement construct may be loose just like loosening of mega-hip prosthesis in osteolytic bones. There have been no reported cases of dislodgement of AORIF screws from the iliac crest that provides the most possible, longest, and strongest bone support for iliac screws. Local cancer ablation and local delivery of zoledronate reduce regional cancer-induced bone loss.

AORIF in Other Anatomic Locations
AORIF can be done in other periarticular bones such as proximal femur, distal femur, proximal tibia, and proximal humerus [13–16]. Sacral metastases and sacral fractures are other indications for AORIF. AORIF can be done for many different anatomic locations under one anesthesia. AORIF can be done for ipsilateral periacetabular metastases and proximal femur metastases as a joint-preserving and muscle-sparing procedure. AORIF could be a first-line procedure before long intramedullary nailing and prosthetic replacement arthroplasty are considered.

References

1. Imanish J, Yang R, Kawano H, Lee FY. Recent advances in minimally invasive local cancer control and skeletal stabilization of periacetabular osteolytic metastases under C-arm imaging guidance. J Am Academy of Orthop Surg. 2024;33(3):e136–50. https://doi.org/10.5435/JAAOS-D-24-00077. PMID: 39448055
2. Dussik CM, et al. AORIF improves pain, ambulation, and bone in periacetabular osteolytic malignancies. Radiology. 2023;307:221401. https://doi.org/10.1148/radiol.221401. Online ahead of print. PMID: 36916888 (Impact Factor 29.5). This manuscript presents the most state-of-the-art and innovative minimally invasive reconstruction technique and long-term outcomes of periacetabular bone defects secondary to metastatic bone cancers.
3. Lee FY, Latich I, Toombs C, Mungur A, Conway D, Alder K, Ibe I, Lindskog D, Friedlaender G. Minimally invasive image-guided Ablation-Osteoplasty-Reinforcement-Internal fixation (AORIF) for metastatic osteolytic lesions in the pelvis and periarticular regions of weight-bearing bones. J Vasc Interv Radiol. 2020;31(4):649–658.e1. https://doi.org/10.1016/j.jvir.2019.11.029. PMID: 32139256. This manuscript represents the most state-of-the-art and innovative minimally invasive reconstruction of pelvis bone defects secondary to metastatic bone cancers.
4. Morris MT, Alder KD, Alex Moushey BS, Munge AM, Milligan K, Toombs C, Conway D, Lee I, Chen F, Tommasini SM, Lee FY. Biomechanical restoration of metastatic cancer-induced peri-acetabular bone defects by AORIF (Ablation-Osteoplasty-Reinforcement-Internal Fixation) technique: to screw or not to screw? Clin Biomech. 2022;92:105565. PMID: 34999389
5. Back J, Nguyen MN, Li L, Lee S, Lee I, Chen F, Gillinov L, Chung Y-H, Alder KD, Kwon H-K, Yu KE, Dussik CM, Zichen Hao, Michael J Flores, Yoseph Kim, Izuchukwu K Ibe, Alana M Munger, Sung Wook Seo, Francis Y. Lee. Inflammatory conversion of the skeletal compartment by aggressive osteolytic breast cancer cells with high inflammatory kinase activity aggravates bone destruction. Bone Res. 2021;9(1):1–14. PMID: 34588427. PMCID: PMC8481290
6. Jiang W, Caruana DL, Back J, Lee FY. Unique spatial transcriptomic profiling of the murine femoral fracture callus: a preliminary report. Cells. 2024;13(6):522. https://doi.org/10.3390/cells13060522. PMID: 38534368; PMCID: PMC10969736

7. Ibe IK, Sahlstrom A, White A, Henderson SE, Lee FY. AAOS instructional course Lect. Metastatic cancers to bone: an overview and cancer-induced bone loss. 2019;68:547–56. PMID: 32032068
8. Henderson SE, Dussik C, Brand J, Ibe IK, Lee FY. Pharmacologic and radiation therapies for cancer-induced bone loss. AAOS Instr Course Lect. 2019;68:557–66. PMID: 32032067
9. Jiang W, Caruana DL, Dussik CM, Conway D, Latich I, Chapiro J, Lindskog DM, Friedlaender GE, Lee FY. Bone mass changes following percutaneous radiofrequency ablation, osteoplasty, reinforcement, and internal fixation of periacetabular osteolytic metastases. J Clin Med. 2023;12(14):4613. https://doi.org/10.3390/jcm12144613. PMID: 37510728; PMCID: PMC10380351
10. Ibe I, Dussik C, Callan A, Barr J, Lee FY. JBJS current concept review. Emerging technologies of minimally invasive procedures for the treatment of periacetabular osteolytic metastases. J Bone Joint Surg Am. 2023;105(6):479–89. https://doi.org/10.2106/JBJS.22.00694. Online ahead of print. PMID: 36701566
11. Lee FY. Ablation, osteoplasty, reinforcement, and internal fixation as a new alternative first-line management for osteolytic pelvic metastases. AAOS Instructional Course Lecture. 2022;71:213–20. PMID: 35254784
12. Toombs C, Conway D, Munger AM, Alder KD, Latich I, Lee FY. ablation, osteoplasty, reinforcement, and internal fixation for percutaneous endoskeletal reconstruction of periacetabular and other periarticular osteolytic metastases. AAOS Instr Course Lect. 2021;70:503–14. PMID: 33438931
13. Jiang W, Latich I, Lindskog D, Friedlaender G, Lee FY. Comparison of Percutaneous Interventional Ablation-Osteoplasty-Reinforcement-Internal Fixation (AORIF), long intramedullary nailing, and hemiarthroplasty for the treatment of focal metastatic osteolytic lesions in the femoral head and neck. Cardio Vascular Interventional Radiology (CVIR). 2023; Apr 13. PMID: 37052716
14. Jiang W, Latich I, Lee FY. Advancements in minimally invasive interventional oncology procedures for painful sacral metastases under imaging guidance. Eng Med. 2025; (In Press)
15. Jiang W, Latich I, Lindskog D, Friedlaender G, Lee FY. Progression of femoral Osteolytic metastases after intramedullary nailing and subsequent salvage techniques. Cancer. 2024;16:2812. https://doi.org/10.3390/cancers16162812.
16. Henderson SE, Yu K, Alder KD, Ibe IK, Lee FY, Lindskog DM. Femoral and Tibial reconstruction for skeletal metastases and pathologic fractures. AAOS Instr Course Lect. 2019;68:607–12. PMID: 32032089

Part VIII

What Comes After Treatment and Future Directions?

Survivorship and Quality of Life

44

Tanios Dagher, Matthew Hess, and Tessa Balach

Introduction

In 2024, it is projected that more than two million individuals in the United States will be diagnosed with cancer, with each individual holding a lifetime probability of nearly 40% for developing invasive cancer [1]. Although mortality rates are declining, the incidence of 6 of the 10 most common malignancies—including breast, prostate, and kidney cancers—continues to rise. This increase in new diagnoses, alongside increased survivorship, is contributing toward a growing prevalence of metastatic disease [2–4]. Improvements in cancer detection and treatment have also rendered many malignancies to be more often chronic and indolent, further increasing metastatic disease burden. For example, by 2030, it is expected that almost 200,000 men in the United States will be living with metastatic prostate cancer, with approximately 89% of cases involving metastases to the bone [4].

Bone is the third most common site for cancer metastasis, following the lungs and liver [5]. The prevalence of bone metastases is small but significant, affecting approximately 5% of all cancer patients in the United States [6]. The bone's rich arterial supply makes it prone to metastatic spread, and many of the most common malignancies—such as those of the breast, lung, thyroid, kidney, and prostate—have a predilection to metastasize to bone. Specifically, up to 73% of patients with advanced breast cancer, 68% of those with advanced prostate cancer, and 36% of patients with advanced lung cancer develop bone lesions [7].

T. Dagher · T. Balach
Department of Orthopaedic Surgery, The University of Chicago, Chicago, IL, USA
e-mail: tbalach@bsd.uchicago.edu

M. Hess (✉)
Department of Orthopaedic Surgery, University of Alabama-Birmingham, Birmingham, AL, USA
e-mail: mhess@uabmc.edu

J. Pretell-Mazzini, S. A. Lozano Calderón (eds.), *Multidisciplinary Approach of Musculoskeletal Metastasis*, https://doi.org/10.1007/978-3-032-09498-8_44

Although bone metastases are most often associated with these malignancies, they can occur in virtually any cancer, including hematologic malignancies such as multiple myeloma and lymphoma. Prostate cancer is the most likely malignancy to metastasize to bone in men, and breast cancer is the most common source of bone metastases in women [8].

Bone metastases can be classified into three categories based on their action at bone tissue: osteolytic, osteoblastic, or mixed. Osteolytic lesions, which are characterized by bone destruction, are common in multiple myeloma, renal cell cancer, melanoma, and nonsmall cell lung cancer [8]. In contrast, osteoblastic lesions form new bone and are typical in prostate cancer, carcinoid tumors, and small cell lung cancer. Mixed metastases have both osteolytic and osteoblastic components. Breast cancer, while predominantly osteolytic, can also produce osteoblastic or mixed lesions. Osteolytic lesions are more likely to cause rapid bone destruction and result in complications, including pathologic fracture and hypercalcemia. Osteoblastic lesions are more likely to cause malignant spinal cord compression, and still predispose the patient to pathologic fractures, as new bone quality is poor [9].

The most frequent sites for bone metastases include the spine, pelvis, femur, and ribs [10]. Bone metastases generally result from hematogenous spread, though local invasion from adjacent soft tissue tumors has also been implicated. Spinal metastases are often results of hematogenous spread via the venous system. Lung and breast cancers tend to metastasize to the thoracic spine due to venous drainage through the azygous vein, while prostate cancer most often spreads to the lumbosacral spine and pelvis via the pelvic plexus [11] (Fig. 44.1).

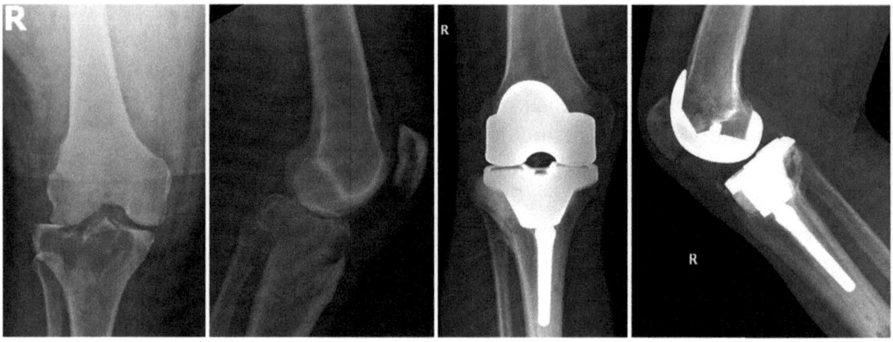

Fig. 44.1 Eighty-three-year-old man with renal cell carcinoma who underwent kidney resection 14 years prior who presented with acute right knee pain and pathologic fracture. Radiolucent destruct bone lesion in the right proximal tibia. He underwent radiation therapy without pain relief. After preoperative embolization procedure he underwent curettage and cemented total knee arthroplasty with a cone for metaphyseal fixation. One year after surgery with no right knee pain and good mobility

Tools for Estimating Survival

Accurately estimating survival in patients with musculoskeletal metastases is useful not only for setting realistic patient expectations, but for guiding clinical decision-making. A clear understanding of prognosis is a component of the decision as to whether a patient should undergo operative intervention and, if so, which intervention may be most beneficial [12, 13]. Generally, patients with an estimated survival <6 weeks should be managed nonoperatively. For patients with projected survival of 6 weeks to 6 months, minimally invasive options, such as intramedullary nailing, are recommended. In cases where survival exceeds 6 months, more durable solutions, including endoprostheses, may be most appropriate [14].

Despite its importance, accurately estimating survival remains a challenge. Historically, prognostication relied heavily on the clinical judgment of physicians, which is unreliable [15, 16]. Over time, prognostic factors have been identified to aid in survival estimates. Some factors are well-established, including age, race, chemotherapy, and the presence of brain and visceral metastases [17–19]. Others, including laboratory values such as hemoglobin and absolute lymphocyte count, demonstrate potential but are still under examination [20, 21].

Several tools have been developed that leverage prognostic factors in making data-driven survival estimates. These models typically focus on either appendicular or spinal metastases. Commonly used tools for estimating prognosis in appendicular metastases include PATH Fx 3.0 [20], the SPRING 2013 nomogram [22], and the OptiModel [17]. Data remain inconclusive as to which tool is most effective [23]. For spinal metastases, tools such as the Global Spine Tumour Study Group Risk Calculator [24], the revised Tokuhashi score [25, 26], and the New England Spinal Metastasis Score [27] are commonly used.

Machine learning algorithms, which are able to integrate and weigh a wide variety of potential prognostic factors, have also emerged as valuable tools for survival estimation. The SORG-MLA model is one of the leading technologies in this domain and has demonstrated excellent results in retrospective cohort studies [28, 29]. Another notable model, METSSS, has also shown promise; however, data suggest inferiority to the SORG model [30]. Despite the advancements in survival estimation models, their full applicability across patient populations is still under investigation. Initial data support the use of the SORG-MLA model in some non-western populations, with few caveats [31]. As the therapeutic landscape continues to evolve, these models may also need to be updated to reflect improved survival [32–34].

Trends in Survivorship

Cancer survival rates have improved for many malignancies over the past three decades, particularly for patients diagnosed early in the disease process. Regardless, metastatic spread has remained a primary cause of mortality, accounting for approximately 90% of cancer-related deaths [35]. Although recent developments in targeted therapies and immunotherapies have begun to extend survival for patients with metastatic disease, the presence of bone metastases continues to confer a poor prognosis [36–38]. In the following paragraphs, we review trends in survival for malignancies that commonly metastasize to bone for patients with and without bone metastasis (Table 44.1).

Breast cancer survival has seen a significant rise over recent decades. In the late 1970s, the 5-year relative survival rate was 74% [39]. Current data indicate that 5-year survival is now 91%, a milestone attributed largely to earlier detection through screening and advancements in multimodal treatment strategies [40, 41]. Despite improved survivorship, disparities persist, with Black women exhibiting lower survival rates at every stage of breast cancer [40]. For metastatic breast cancer, the 5-year survival rate has improved, growing from 46% in 2005–2009 to 57% in 2010–2015 [39]. However, bone metastasis continues to dramatically affect prognosis, with a 5-year survival rate of 8% as of 2011 [36]. Patients with skeletal-related events (SREs) face a particularly grim outlook, with a 5-year survival rates of just 3% [36].

Lung cancer incidence has declined markedly in the United States over the last three decades, especially among men, due to reductions in smoking rates [42]. As a result, mortality has been cut in half. In addition, the 5-year survival rate for lung cancer has increased to 26.7%, representing a significant increase in survivorship across the past decade. The use of targeted therapies and immunotherapy, particularly for those with EGFR and ALK mutations, is partially responsible for this trend [42]. Globally, survival trends remain variable; however, it is clear that females and younger patients tend to have better outcomes [43]. The presence of bone metastases in lung cancer patients is a poor prognostic factor, with median survival of approximately 148 days after bone metastases diagnosis [37]. In these patients, age under 60 years, chemotherapy for skeletal metastases, and surgical intervention for pathological fractures have been associated with longer survival [44].

The 5-year relative survival rate for thyroid cancer has been over 95% each year for the past three decades [45]. However, bone metastases significantly reduce survival, with 5-year survival rates in the low 40% range [48]. Approximately 4% of thyroid cancer patients develop bone metastases, with follicular or medullary thyroid cancer conferring the highest risk [46]. Prognosis is further worsened

Table 44.1 Five-year relative survival rates for cancers that commonly metastasize to bone according to the National Cancer Institute's Surveillance, Epidemiology, and End Results (SEER) database

Year	Breast cancer	Kidney + renal pelvis cancer	Lung cancer	Prostate cancer	Thyroid cancer
1995	87.3%	61.1%	14.4%	95.7%	95.8%
2000	90.4%	66.5%	15.5%	99.1%	96.6%
2005	91.4%	74.9%	17.4%	99.6%	97.5%
2010	91.7%	74.9%	19.8%	99.5%	98.3%
2015	92.9%	78.3%	25.4%	97.2%	98.6%

by other factors, including older age at diagnosis, lymph node involvement, the presence of extraskeletal metastases, and the lack of radioactive iodine (RAI) treatment [46, 47].

Prostate cancer incidence has remained relatively constant over the past two decades, and affected individuals are being diagnosed sooner due to increased screening [48]. Currently, the 5-year relative survival rate for prostate cancer stands at 97.5% and has been stable since the mid-1990s [49]. However, racial disparities persist, with African American men experiencing higher mortality rates [49]. In a disease with such high survival, bone metastases significantly worsen the prognosis, decreasing 5-year survival to only 33% [50]. Autopsy studies reveal that 90% of men with metastatic prostate cancer had bone involvement at the time of death [51]. In prostate cancer patients, data demonstrate that the extent of skeletal metastasis directly correlates with survival [38].

Survival rates for kidney cancer have steadily increased over the past thirty years, from around 60% in 1990 to 78% most recently [52]. This improvement is largely due to the introduction of targeted therapies, particularly antiangiogenic tyrosine kinase inhibitors (TKIs) and mTOR inhibitors, which became available in 2005 and have shown benefit in about 80% of patients [53]. Earlier detection through incidental findings on imaging has also contributed to survivorship [53]. Bone metastases are present in 17% of patients with renal cell carcinoma (RCC), and median overall survival for these patients is 23.2 months. Data show that those with solitary bone lesions survive longer on average compared to those with multiple metastases [54]. Additionally, surgical removal of bone lesions is associated with increased survival in patients with RCC bone metastases [54] (Table 44.3).

Implications of Increased Survival

As patients with metastatic bone disease live longer, orthopedic surgeons are learning to adapt treatment strategies to accommodate not only the longer overall survival and resultant need for increased durability of reconstruction methods but also the risk of local progression of metastatic sites of disease. A generational shift is occurring in which the way many practicing orthopedic surgeons were taught to manage metastatic bone disease is becoming outdated, or at least more nuanced, thus requiring additional insight into the patient's overall oncologic condition. For example, traditional teaching for a patient with metastatic carcinoma and a large destructive lesion in the proximal femur with functional pain would be to consider a prophylactic intramedullary nail. As patient survival increases and available systemic therapies become increasingly complex, the risk of local progression must be considered. Patients need to be individually counseled on a unique case-by-case basis about the risks of fixation failure due to local progression and possible need for secondary surgeries (Fig. 44.2). While the implications of increased survival on surgical management are detailed throughout the remaining chapters of this book, the role of wide resection and reconstruction in the setting of oligometastatic bone disease to improve overall survival in unique scenarios continues to gain support.

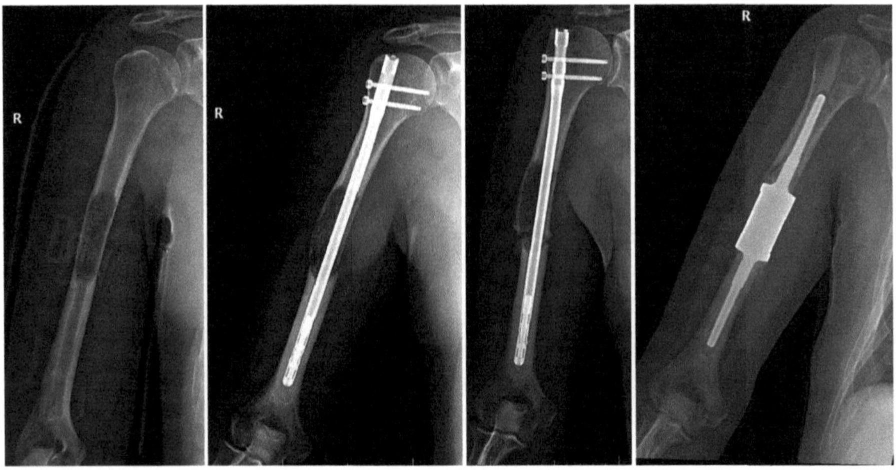

Fig. 44.2 Fifty-three-old-male with metastatic renal cell carcinoma and a right humerus lesion with ongoing pain. He underwent intramedullary nail fixation and radiation therapy. One year after surgery he had local progression and ongoing pain. He underwent preoperative embolization followed by wide resection and intercalary cemented endoprosthetic reconstruction with relief of pain at one year after surgery

The increasingly nuanced care of patients with metastatic bone disease has prompted some to suggest that specialist orthopedic oncologists should assume the majority of care of metastatic bone disease. However, due to the high burden of cancer and metastatic bone disease, a shift in ownership of these patients would necessitate a significant increase in the number of fellowship-trained orthopedic oncologists. Orthopedic oncologic surgeons must continue to innovate and direct education of general orthopedic surgeons regarding the implications of increased survival of patients with metastatic bone disease. Clinical practice guidelines for managing metastatic bone disease are beginning to be developed, such as those recently published regarding the management of metastatic humerus disease [55]. Objective evidence and recommendations currently remain limited by low numbers, but additional high-quality studies are expected as patients survive longer with metastatic bone disease and national groups make management a priority.

Quality of Life

Cancer patients with musculoskeletal metastases are at an increased risk of skeletal-related events (SREs), which may affect their quality of life and/or clinical outcomes [56]. Among these, bone pain is the most common symptom and is reported by approximately 80% of patients [57, 58]. The pain is often intermittent and mild but may become more severe as the cancer progresses. Pathologic fractures, another serious consequence of bone metastases, occur in 9–29% of patients [59]. These fractures not only reduce survival rates [59] but can cause pain and decrease physical functioning [60]. Up to 20% of patients with bone metastases experience metastatic epidural spinal cord compression [61]. Initially presenting with back pain, numbness, and weakness [62], this can progress to loss of bowel or bladder control and eventually paralysis. Time to treatment is critical for these patients, as early intervention is associated with preserved function and quality of life [62, 63]. Hypercalcemia, most commonly associated with breast cancer, lung cancer, and multiple myeloma, is another concern affecting approximately 20% of cancer patients during their treatment course [64–66]. It often manifests as nausea, vomiting, and abdominal pain [67], but in more severe cases, can impair cognition, worsen cardiac function, even lead to coma or kidney failure [68, 69]. Several therapies have been demonstrated promising results for the prevention of SREs, including bone-targeting agents such as bisphosphonates and denosumab as well as external beam radiotherapy [70] (Table 44.2).

Loss of mobility is also a frequent concern related to quality of life in patients with metastatic musculoskeletal disease. Physical impairment can be a direct consequence of the tumor itself or a side effect of cancer therapy [71]. Loss of mobility in patients with bone metastases is often related to pain and neurologic deficits rather than bone tumor burden or location [72]. Many cancer patients also experience poor bone health, and in these patients, progressive immobility further acts to worsen bone and muscle quality [71]. Surgical intervention, most commonly intramedullary nailing or internal fixation, may be indicated for restoration of mechanical stability to enable weight-bearing and preserve quality of life [73–75]. Physical therapy must be cautiously prescribed, especially in patients without surgical stabilization, to optimize mobility while minimizing the risk of adverse events or pain exacerbation [76].

Side effects of chemotherapy and radiation also affect quality of life. Many chemotherapeutic agents are known causes of nausea, vomiting, and fatigue, and some have been shown to impair physical and cognitive functioning [77, 78]. While radiation has demonstrated an ability to reduce pain and even extend survival in some patients with musculoskeletal metastases [79, 80], it may also contribute to fatigue, nausea, and vomiting, often in a dose-dependent manner [81, 82]. While these effects are usually temporary, they may have a considerable impact on a patient's quality of life, and can affect physical, emotional, and cognitive function [82].

Table 44.2 Summary of common skeletally-related events (SREs)

	Bone pain	Pathologic fracture	Spinal cord compression	Hypercalcemia
Incidence	60–84%	9–29%	15–20%	10–30%
Symptoms	Often dull, persistent pain but may become more severe	Sudden and severe vs. dull and persistent pain with minimal or no trauma	Most commonly back pain, patients may also have limb weakness, sensory loss, and bowel/bladder dysfunction	Nausea, vomiting, anorexia, abdominal pain, cognitive impairments, arrhythmia
Management	Most often opioid and/or non-opioid (e.g. NSAID) pain medication, radiotherapy and denosumab may provide benefits in refractory cases	Prophylactic vs. therapeutic surgical fixation	Immediate corticosteroid administration with consideration for radiation therapy and/or surgical decompression	Hydration, loop diuretics, bisphosphonates or denosumab

Table 44.3 Predicted survival for the most common cancers with metastasis to bone

	5-Year survival	5-Year survival	5-Year survival	5-Year survival	5-Year survival	10-Year survival	10-Year survival	10-Year survival	10-Year survival	10-Year survival	20-Year survival	20-Year survival	20-Year survival	20-Year survival	20-Year survival
	1975–1984	1985–1994	1995–2004	2005–2014	All years	1975–1984	1985–1994	1995–2004	2005–2014	All years	1975–1984	1985–1994	1995–2004	2005–2014	All years
Breast	75.50%	84.00%	89.00%	91.10%	86.10%	63.60%	76.10%	83.50%	86.30%	78.80%	53.20%	67.50%	75.70%	N/A	69.80%
Prostate	70.50%	89.10%	98.60%	99.10%	93.40%	55.70%	81.90%	97.90%	99.10%	89.70%	39.60%	72.40%	94.40%	N/A	81.70%
Lung and bronchus	12.70%	13.40%	15.30%	19.60%	15.40%	8.70%	9.00%	10.20%	13.10%	10.40%	4.80%	4.80%	5.50%	N/A	5.60%
Colon and rectum	52.10%	60.00%	64.00%	66.40%	60.80%	46.50%	53.90%	58.40%	60.10%	54.90%	42.60%	49.50%	52.60%	N/A	50.00%
Kidney and renal pelvis	51.50%	58.40%	65.50%	75.10%	65.80%	44.50%	50.80%	57.70%	68.80%	58.40%	36.80%	40.90%	47.00%	N/A	47.60%
Thyroid	92.70%	94.40%	96.50%	98.40%	96.60%	91.30%	93.30%	95.80%	98.40%	96.00%	89.90%	92.10%	94.90%	N/A	95.00%

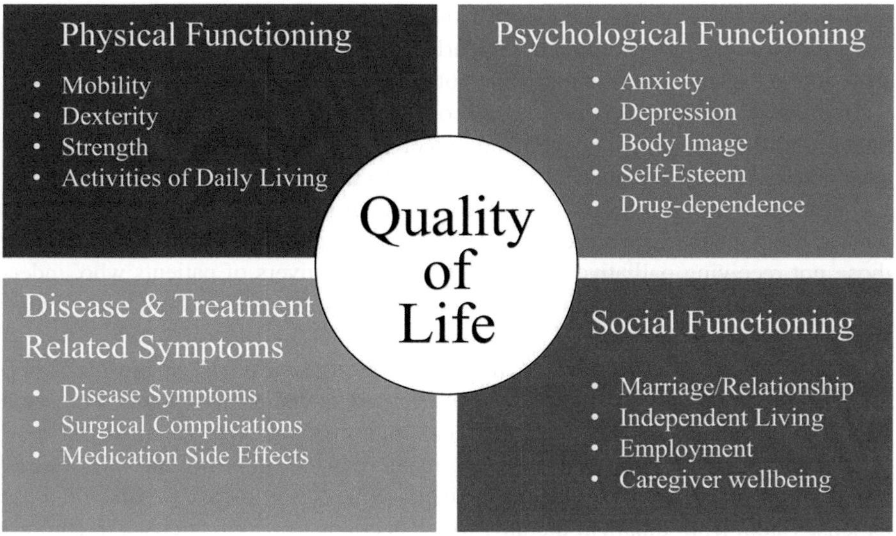

Fig. 44.3 Quality of life

Nonhealth-related stressors may also affect the quality of life of patients receiving cancer therapy. Extended hospital stays for medical and perioperative management are a notable burden for both patients and families [83, 84]. Data demonstrate that patients with metastatic disease in particular face exorbitant medical costs, higher than those without metastases [85]. These additional expenses are partially related to the cost of managing SREs [86] and place considerable economic strain on patients and their families. Caregiver fatigue is another important consideration for family of cancer patients, with one study reporting that 23% of caregivers experience severe fatigue [87]. This fatigue is correlated to patient health-related quality of life, which is often lower in those with musculoskeletal metastases [88]. Caregiver mental health has been found to be interrelated to patient well-being, underscoring the need for interventions aimed at reducing caregiver burden [89] (Fig. 44.3).

Role of Palliative Care

Research suggests that early referral to palliative care may provide meaningful improvements in patient symptom severity, quality of life, and mental health. In one study, patients randomized to an early palliative care intervention group experienced notable reductions in symptom intensity by the four-month mark [90]. Moreover, those receiving integrated palliative and oncologic care reported higher quality of life and an enhanced ability to cope with their illness [91]. Psychological benefits were also evident, as palliative care intervention was associated with reduced anxiety and depressive symptoms [91, 92]. Based on these data, the American Society of Clinical Oncology (ASCO) has formally recommended referral to palliative care for patients with advanced solid tumors and hematologic malignancies [93]. Guidelines particularly emphasize that referral should occur early in the disease course and include both outpatient and inpatient care options.

In addition to improving clinical outcomes, palliative care enhances the overall patient and family experience. Patients receiving palliative services report higher satisfaction with their care and more positive clinician-patient interactions [94]. Palliative care has also been shown to foster better prognostic understanding, with patients demonstrating a more accurate perception of their prognosis compared to those not receiving palliative care [95]. Finally, caregivers of patients who underwent palliative care interventions reported lower levels of stress burden following their loved one's death [96].

Palliative care utilization has been shown to offer modest cost savings. It is well known that the financial burden of end-of-life care is disproportionately large. In a cohort of older adults with cancer, the mean total medical expenses in the last year of life was $33,712, with 63% and 34% of these costs occurring in the last three months and one month of life, respectively [97]. A significant portion of these expenses stem from inpatient therapy, which accounts for 55% of costs, in contrast to hospice care, which represents only 4% [98]. Studies suggest that early palliative care can reduce healthcare expenditures. In a retrospective cohort study, patients who received palliative interventions were less likely to utilize inpatient hospitalization and emergency department visits and were found to incur significantly lower inpatient hospital costs in the final month of life [99].

Conclusion

As therapeutic advances continue to prolong life for patients with metastatic bone disease, the prevalence and complexity of bone metastases have grown. Patients with metastatic bone disease benefit from increasingly accurate prognostication tools that inform individualized decision-making. Surgical interventions are centered around maximizing function and minimizing pain and fracture risk, generally involving durable reconstructive options for patients with longer life expectancy and minimally invasive stabilization for patients with shorter prognoses and/or imminent fracture risk. Multidisciplinary management—including medical oncology, surgical specialists, social work, and early palliative care—is key to providing holistic support for patients and addressing the spectrum of physical and psychosocial needs associated with bone metastases and skeletal-related events.

References

1. Siegel RL, Giaquinto AN, Jemal A. Cancer statistics, 2024. CA Cancer J Clin. 2024;74:12–49. https://doi.org/10.3322/caac.21820.
2. Huang J-F, Shen J, Li X, Rengan R, Silvestris N, Wang M, Derosa L, Zheng X, Belli A, Zhang X-L, Li YM, Wu A. Incidence of patients with bone metastases at diagnosis of solid tumors in

adults: a large population-based study. Ann Transl Med. 2020;8:482. https://doi.org/10.21037/atm.2020.03.55.
3. Gallicchio L, Devasia TP, Tonorezos E, Mollica MA, Mariotto A. Estimation of the number of individuals living with metastatic cancer in the United States. J Natl Cancer Inst. 2022;114:1476–83. https://doi.org/10.1093/jnci/djac158.
4. Devasia TP, Mariotto AB, Nyame YA, Etzioni R. Estimating the number of men living with metastatic prostate cancer in the United States. Cancer Epidemiol Biomarkers Prev. 2023;32:659–65. https://doi.org/10.1158/1055-9965.EPI-22-1038.
5. Coleman RE. Metastatic bone disease: clinical features, pathophysiology and treatment strategies. Cancer Treat Rev. 2001;27:165–76. https://doi.org/10.1053/ctrv.2000.0210.
6. Ryan C, Stoltzfus KC, Horn S, Chen H, Louie AV, Lehrer EJ, Trifiletti DM, Fox EJ, Abraham JA, Zaorsky NG. Epidemiology of bone metastases. Bone. 2022;158:115783. https://doi.org/10.1016/j.bone.2020.115783.
7. Fang J, Xu Q. Differences of osteoblastic bone metastases and osteolytic bone metastases in clinical features and molecular characteristics. Clin Transl Oncol. 2015;17:173–9. https://doi.org/10.1007/s12094-014-1247-x.
8. Macedo F, Ladeira K, Pinho F, Saraiva N, Bonito N, Pinto L, Goncalves F. Bone Metastases: An Overview. Oncol Rev. 2017;11:321. https://doi.org/10.4081/oncol.2017.321.
9. Yin JJ, Pollock CB, Kelly K. Mechanisms of cancer metastasis to the bone. Cell Res. 2005;15:57–62. https://doi.org/10.1038/sj.cr.7290266.
10. Svensson E, Christiansen CF, Ulrichsen SP, Rørth MR, Sørensen HT. Survival after bone metastasis by primary cancer type: a Danish population-based cohort study. BMJ Open. 2017;7:e016022. https://doi.org/10.1136/bmjopen-2017-016022.
11. Jayarangaiah A, Kemp AK, Theetha Kariyanna P. Bone metastasis. In: StatPearls. Treasure Island (FL): StatPearls publishing; 2024.
12. Canavan ME, Wang X, Ascha MS, Miksad RA, Showalter TN, Calip GS, Gross CP, Adelson KB. Systemic anticancer therapy and overall survival in patients with very advanced solid tumors. JAMA Oncol. 2024;10:887–95. https://doi.org/10.1001/jamaoncol.2024.1129.
13. Errani C, Mavrogenis AF, Cevolani L, Spinelli S, Piccioli A, Maccauro G, Baldini N, Donati D. Treatment for long bone metastases based on a systematic literature review. Eur J Orthop Surg Traumatol. 2017;27:205–11. https://doi.org/10.1007/s00590-016-1857-9.
14. Wodajo F, Colman M, Getty P. AAOS clinical practice guideline summary: treatment of metastatic carcinoma and myeloma of the femur. J Am Acad Orthop Surg. 2023;31(3):e118–e129. https://doi.org/10.5435/JAAOS-D-21-00888.
15. Hartsell WF, Desilvio M, Bruner DW, Scarantino C, Ivker R, Roach M, Suh J, Demas WF, Movsas B, Petersen IA, Konski AA. Can physicians accurately predict survival time in patients with metastatic cancer? Analysis of RTOG 97-14. J Palliat Med. 2008;11:723–8. https://doi.org/10.1089/jpm.2007.0259.
16. Nathan SS, Healey JH, Mellano D, Hoang B, Lewis I, Morris CD, Athanasian EA, Boland PJ. Survival in patients operated on for pathologic fracture: implications for end-of-life orthopedic care. J Clin Oncol. 2005;23:6072–82. https://doi.org/10.1200/JCO.2005.08.104.
17. Willeumier JJ, van der Linden YM, van der Wal CWPG, Jutte PC, van der Velden JM, Smolle MA, van der Zwaal P, Koper P, Bakri L, de Pree I, Leithner A, Fiocco M, Dijkstra PDS. An easy-to-use prognostic model for survival estimation for patients with symptomatic long bone metastases. J Bone Joint Surg Am. 2018;100:196–204. https://doi.org/10.2106/JBJS.16.01514.
18. Ren Y, Qian S, Xu G, Cai Z, Zhang N, Wang Z. Predicting survival of patients with bone metastasis of unknown origin. Front Endocrinol (Lausanne). 2023;14:1193318. https://doi.org/10.3389/fendo.2023.1193318.
19. Zhong X, Lin Y, Zhang W, Bi Q. Predicting diagnosis and survival of bone metastasis in breast cancer using machine learning. Sci Rep. 2023;13:18301. https://doi.org/10.1038/s41598-023-45438-z.
20. Anderson AB, Wedin R, Fabbri N, Boland P, Healey J, Forsberg JA. External validation of PATHFx version 3.0 in patients treated surgically and nonsurgically for symptomatic

skeletal metastases. Clin Orthop Relat Res. 2020;478:808–18. https://doi.org/10.1097/CORR.0000000000001081.

21. Thio QCBS, Karhade AV, Ogink PT, Bramer JAM, Ferrone ML, Calderón SL, Raskin KA, Schwab JH. Development and internal validation of machine learning algorithms for preoperative survival prediction of extremity metastatic disease. Clin Orthop Relat Res. 2020;478:322–33. https://doi.org/10.1097/CORR.0000000000000997.
22. Sørensen MS, Gerds TA, Hindsø K, Petersen MM. External validation and optimization of the SPRING model for prediction of survival after surgical treatment of bone metastases of the extremities. Clin Orthop Relat Res. 2018;476:1591–9. https://doi.org/10.1097/01.blo.0000534678.44152.ee.
23. Ben Gal O, Soh TCF, Vaughan S, Jayasanker V, Mahendra A, Gupta S. The prediction of survival after surgical management of bone metastases of the extremities—a comparison of prognostic models. Curr Oncol. 2022;29:4703–16. https://doi.org/10.3390/curroncol29070373.
24. Choi D, Pavlou M, Omar R, Arts M, Balabaud L, Buchowski JM, Bunger C, Chung CK, Coppes MH, Depreitere B, Fehlings MG, Kawahara N, Lee C-S, Leung Y, Martin-Benlloch JA, Massicotte EM, Mazel C, Meyer B, Oner FC, Peul W, Quraishi N, Tokuhashi Y, Tomita K, Ulbricht C, Verlaan J-J, Wang M, Crockard HA. A novel risk calculator to predict outcome after surgery for symptomatic spinal metastases; use of a large prospective patient database to personalise surgical management. Eur J Cancer. 2019;107:28–36. https://doi.org/10.1016/j.ejca.2018.11.011.
25. Tokuhashi Y, Matsuzaki H, Oda H, Oshima M, Ryu J. A revised scoring system for preoperative evaluation of metastatic spine tumor prognosis. Spine (Phila Pa 1976). 2005;30:2186–91. https://doi.org/10.1097/01.brs.0000180401.06919.a5.
26. Quraishi NA, Manoharan SR, Arealis G, Khurana A, Elsayed S, Edwards KL, Boszczyk BM. Accuracy of the revised Tokuhashi score in predicting survival in patients with metastatic spinal cord compression (MSCC). Eur Spine J. 2013;22(Suppl 1):S21–6. https://doi.org/10.1007/s00586-012-2649-5.
27. Schoenfeld AJ, Ferrone ML, Schwab JH, Blucher JA, Barton LB, Tobert DG, Chi JH, Shin JH, Kang JD, Harris MB. Prospective validation of a clinical prediction score for survival in patients with spinal metastases: the New England spinal metastasis score. Spine J. 2021;21:28–36. https://doi.org/10.1016/j.spinee.2020.02.009.
28. Park JH, Won J, Kim H-S, Kim Y, Kim S, Han I. Comparison of survival prediction models for bone metastases of the extremities following surgery. Bone Joint J. 2024;106-B:203–11. https://doi.org/10.1302/0301-620X.106B2.BJJ-2023-0751.
29. de Groot TM, Ramsey D, Groot OQ, Fourman M, Karhade AV, Twining PK, Berner EA, Fenn BP, Collins AK, Raskin K, Lozano S, Newman E, Ferrone M, Doornberg JN, Schwab JH. Does the SORG machine-learning algorithm for extremity metastases generalize to a contemporary cohort of patients? Temporal validation from 2016 to 2020. Clin Orthop Relat Res. 2023;481:2419–30. https://doi.org/10.1097/CORR.0000000000002698.
30. Lee T-Y, Chen Y-A, Groot OQ, Yen H-K, Bindels BJJ, Pierik R-J, Hsieh H-C, Karhade AV, Tseng T-E, Lai Y-H, Yang J-J, Lee C-C, Hu M-H, Verlaan J-J, Schwab JH, Yang R-S, Lin W-H. Comparison of eight modern preoperative scoring systems for survival prediction in patients with extremity metastasis. Cancer Med. 2023;12:14264–81. https://doi.org/10.1002/cam4.6097.
31. Tseng T-E, Lee C-C, Yen H-K, Groot OQ, Hou C-H, Lin S-Y, Bongers MER, Hu M-H, Karhade AV, Ko J-C, Lai Y-H, Yang J-J, Verlaan J-J, Yang R-S, Schwab JH, Lin W-H. International validation of the SORG machine-learning algorithm for predicting the survival of patients with extremity metastases undergoing surgical treatment. Clin Orthop Relat Res. 2022;480:367–78. https://doi.org/10.1097/CORR.0000000000001969.
32. Ragel BT, Mendez GA, Reddington J, Ferachi D, Kubicky CD, Philipp TC, Zusman NL, Klimo P, Hart R, Yoo J, Ching AC. Life expectancy and metastatic spine scoring systems: An academic institutional experience. Clin Spine Surg. 2017;30:335–42. https://doi.org/10.1097/BSD.0000000000000154.

33. Chen JH, Alagappan M, Goldstein MK, Asch SM, Altman RB. Decaying relevance of clinical data towards future decisions in data-driven inpatient clinical order sets. Int J Med Inform. 2017;102:71–9. https://doi.org/10.1016/j.ijmedinf.2017.03.006.
34. Katagiri H, Okada R, Takagi T, Takahashi M, Murata H, Harada H, Nishimura T, Asakura H, Ogawa H. New prognostic factors and scoring system for patients with skeletal metastasis. Cancer Med. 2014;3:1359–67. https://doi.org/10.1002/cam4.292.
35. Chaffer C, Weinberg R. A perspective on cancer cell metastasis. Science (New York). 2011;331 https://doi.org/10.1126/science.1203543.
36. Yong M, Jensen AÖ, Jacobsen JB, Nørgaard M, Fryzek JP, Sørensen HT. Survival in breast cancer patients with bone metastases and skeletal-related events: a population-based cohort study in Denmark (1999–2007). Breast Cancer Res Treat. 2011;129:495–503. https://doi.org/10.1007/s10549-011-1475-5.
37. Pruksakorn D, Phanphaisarn A, Settakorn J, Arpornchayanon U, Tantraworasin A, Chaiyawat P, klangjorhor J, Teeyakasem P. Prognostic score for life expectancy evaluation of lung cancer patients after bone metastasis. J Bone Oncol. 2017;10:1–5. https://doi.org/10.1016/j.jbo.2017.10.001.
38. Liu D, Kuai Y, Zhu R, Zhou C, Tao Y, Han W, Chen Q. Prognosis of prostate cancer and bone metastasis pattern of patients: a SEER-based study and a local hospital based study from China. Sci Rep. 2020;10:9104. https://doi.org/10.1038/s41598-020-64073-6.
39. Guo F, Kuo Y, Shih YCT, Giordano SH, Berenson AB. Trends in breast cancer mortality by stage at diagnosis among US young women. Cancer. 2018;124:3500–9. https://doi.org/10.1002/cncr.31638.
40. Cancer of the Breast (Female)—Cancer Stat Facts. In: SEER. https://seer.cancer.gov/statfacts/html/breast.html. Accessed 11 Oct 2024.
41. Soldato D, Arecco L, Agostinetto E, Franzoi MA, Mariamidze E, Begijanashvili S, Brunetti N, Spinaci S, Solinas C, Vaz-Luis I, Di Meglio A, Lambertini M. The future of breast cancer research in the survivorship field. Oncol Ther. 2023;11:199–229. https://doi.org/10.1007/s40487-023-00225-8.
42. Howlader N, Forjaz G, Mooradian MJ, Meza R, Kong CY, Cronin KA, Mariotto AB, Lowy DR, Feuer EJ. The effect of advances in lung-cancer treatment on population mortality. N Engl J Med. 2020;383:640–9. https://doi.org/10.1056/NEJMoa1916623.
43. Bi J, Tuo J, Xiao Y, Tang D, Zhou X, Jiang Y, Ji X, Tan Y, Yuan H, Xiang Y. Observed and relative survival trends of lung cancer: a systematic review of population-based cancer registration data. Thorac Cancer. 2023;15:142–51. https://doi.org/10.1111/1759-7714.15170.
44. Cho YJ, Cho YM, Kim SH, Shin K-H, Jung S-T, Kim HS. Clinical analysis of patients with skeletal metastasis of lung cancer. BMC Cancer. 2019;19:303. https://doi.org/10.1186/s12885-019-5534-3.
45. Cancer of the Thyroid—Cancer Stat Facts. In: SEER. https://seer.cancer.gov/statfacts/html/thyro.html. Accessed 9 Oct 2024.
46. Iñiguez-Ariza NM, Bible KC, Clarke BL. Bone metastases in thyroid cancer. J Bone Oncol. 2020;21:100282. https://doi.org/10.1016/j.jbo.2020.100282.
47. Piñar-Gutiérrez A, Romero-Lluch AR, Dueñas-Disotuar S, de Lara-Rodríguez I, Gálvez-Moreno MÁ, Martín-Hernández T, García-Alemán J, Pinillos GM, Navarro-González E. Bone metastases from differentiated thyroid cancer: characteristics and prognostic factors in a multicenter series. 2023; https://doi.org/10.1530/ETJ-23-0086.
48. Abudoubari S, Bu K, Mei Y, Maimaitiyiming A, An H, Tao N. Prostate cancer epidemiology and prognostic factors in the United States. Front Oncol. 2023;13:1142976. https://doi.org/10.3389/fonc.2023.1142976.
49. Cancer of the Prostate—Cancer Stat Facts. In: SEER. https://seer.cancer.gov/statfacts/html/prost.html. Accessed 9 Oct 2024.
50. Kingsley LA, Fournier PGJ, Chirgwin JM, Guise TA. Molecular biology of bone metastasis. Mol Cancer Ther. 2007;6:2609–17. https://doi.org/10.1158/1535-7163.MCT-07-0234.

51. Bubendorf L, Schöpfer A, Wagner U, Sauter G, Moch H, Willi N, Gasser TC, Mihatsch MJ. Metastatic patterns of prostate cancer: an autopsy study of 1,589 patients. Hum Pathol. 2000;31:578–83. https://doi.org/10.1053/hp.2000.6698.
52. Cancer of the Kidney and Renal Pelvis—Cancer Stat Facts. In: SEER. https://seer.cancer.gov/statfacts/html/kidrp.html. Accessed 9 Oct 2024.
53. Huang JJ, Hsieh JJ. The therapeutic landscape of renal cell carcinoma: from the dark age to the Golden age. Semin Nephrol. 2020;40:28–41. https://doi.org/10.1016/j.semnephrol.2019.12.004
54. Ruatta F, Derosa L, Escudier B, Colomba E, Guida A, Baciarello G, Loriot Y, Fizazi K, Albiges L. Prognosis of renal cell carcinoma with bone metastases: experience from a large cancer Centre. Eur J Cancer. 2019;107:79–85. https://doi.org/10.1016/j.ejca.2018.10.023
55. Tedesco NS, Mesko N, Wodajo F, Management of Metastatic Humeral Disease Work Group, Staff of the American Academy of Orthopaedic Surgeons and the Musculoskeletal Tumor Society, Management of Metastatic Humeral Disease Work Group and Staff of the American Academy of Orthopaedic Surgeons and the Musculoskeletal Tumor Society. The Musculoskeletal Tumor Society Clinical Practice Guideline on the Management of Metastatic Humeral Disease. J Am Acad Orthop Surg. 2024;32:e482–8. https://doi.org/10.5435/JAAOS-D-24-00053.
56. Tsuzuki S, Park SH, Eber MR, Peters CM, Shiozawa Y. Skeletal complications in cancer patients with bone metastases. Int J Urol. 2016;23:825–32. https://doi.org/10.1111/iju.13170
57. Janjan NA, Payne R, Gillis T, Podoloff D, Libshitz HI, Lenzi R, Theriault R, Martin C, Yasko A. Presenting symptoms in patients referred to a multidisciplinary clinic for bone metastases. J Pain Symptom Manag. 1998;16:171–8. https://doi.org/10.1016/s0885-3924(98)00069-4
58. Coleman RE. Skeletal complications of malignancy. Cancer. 1997;80:1588–94. https://doi.org/10.1002/(sici)1097-0142(19971015)80:8+<1588::aid-cncr9>3.3.co;2-z
59. Böhm P, Huber J. The surgical treatment of bony metastases of the spine and limbs. J Bone Joint Surg Br. 2002;84:521–9. https://doi.org/10.1302/0301-620x.84b4.12495
60. Kirkinis MN, Lyne CJ, Wilson MD, Choong PFM. Metastatic bone disease: a review of survival, prognostic factors and outcomes following surgical treatment of the appendicular skeleton. Eur J Surg Oncol. 2016;42:1787–97. https://doi.org/10.1016/j.ejso.2016.03.036
61. Cole JS, Patchell RA. Metastatic epidural spinal cord compression. Lancet Neurol. 2008;7:459–66. https://doi.org/10.1016/S1474-4422(08)70089-9
62. Prasad D, Schiff D. Malignant spinal-cord compression. Lancet Oncol. 2005;6:15–24. https://doi.org/10.1016/S1470-2045(04)01709-7
63. Sørensen S, Helweg-Larsen S, Mouridsen H, Hansen HH. Effect of high-dose dexamethasone in carcinomatous metastatic spinal cord compression treated with radiotherapy: a randomised trial. Eur J Cancer. 1994;30A:22–7. https://doi.org/10.1016/s0959-8049(05)80011-5
64. Stewart AF. Clinical practice. Hypercalcemia associated with cancer. N Engl J Med. 2005;352:373–9. https://doi.org/10.1056/NEJMcp042806.
65. Horwitz MJ, Hodak SP, Stewart AF. Non-parathyroid hypercalcemia. In: Primer on the metabolic bone diseases and disorders of mineral metabolism. John Wiley & Sons, Ltd. p. 562–71
66. Rosner MH, Dalkin AC. Onco-nephrology: the pathophysiology and treatment of malignancy-associated hypercalcemia. Clin J Am Soc Nephrol. 2012;7:1722–9. https://doi.org/10.2215/CJN.02470312
67. Clines GA. Mechanisms and treatment of hypercalcemia of malignancy. Curr Opin Endocrinol Diabetes Obes. 2011;18:339–46. https://doi.org/10.1097/MED.0b013e32834b4401
68. Wolf ME, Ranade V, Molnar J, Somberg J, Mosnaim AD. Hypercalcemia, arrhythmia, and mood stabilizers. J Clin Psychopharmacol. 2000;20:260–4. https://doi.org/10.1097/00004714-200004000-00022
69. Edelson GW, Kleerekoper M. Hypercalcemic crisis. Med Clin North Am. 1995;79:79–92. https://doi.org/10.1016/s0025-7125(16)30085-2
70. Coleman R, Body JJ, Aapro M, Hadji P, Herrstedt J, ESMO Guidelines Working Group. Bone health in cancer patients: ESMO Clinical Practice Guidelines. Ann Oncol. 2014;25(Suppl 3):iii124–37. https://doi.org/10.1093/annonc/mdu103

71. Sturgeon KM, Mathis KM, Rogers CJ, Schmitz KH, Waning DL. Cancer- and chemotherapy-induced musculoskeletal degradation. JBMR Plus. 2019;3:e10187. https://doi.org/10.1002/jbm4.10187
72. Cheville AL, Murthy NS, Basford JR, Rose PS, Tran K, Pittelkow TP, Ringler MD. Imaging and clinical characteristics predict near-term disablement from bone metastases: implications for rehabilitation. Arch Phys Med Rehabil. 2016;97:53–60. https://doi.org/10.1016/j.apmr.2015.09.011.
73. Cirstoiu C, Cretu B, Iordache S, Popa M, Serban B, Cursaru A. Surgical management options for long-bone metastasis. EFORT Open Rev. 2022;7:206–13. https://doi.org/10.1530/EOR-21-0119.
74. Ehne J, Tsagozis P. Current concepts in the surgical treatment of skeletal metastases. World J Orthop. 2020;11:319–27. https://doi.org/10.5312/wjo.v11.i7.319.
75. Zhang C, Wang J, Wu H, Lin Y, Chekhonin VP, Peltzer K, Bukharov AV, Kaprin AD, Guo X, Liu Z. Ten-year retrospect of the investigation of proximal limbs metastasis in cancer: a multicenter study on survival outcome, limb function status and surgical procedures analysis. BMC Cancer. 2023;23:795. https://doi.org/10.1186/s12885-023-11292-5.
76. Sheill G, Guinan EM, Peat N, Hussey J. Considerations for exercise prescription in patients with bone metastases: a comprehensive narrative review. PM R. 2018;10:843–64. https://doi.org/10.1016/j.pmrj.2018.02.006.
77. Binotto M, Reinert T, Werutsky G, Zaffaroni F, Schwartsmann G (2020) Health-related quality of life before and during chemotherapy in patients with early-stage breast cancer. http://ecancer.org/en/journal/article/1007-health-related-quality-of-life-before-and-during-chemotherapy-in-patients-with-early-stage-breast-cancer. Accessed 3 Oct 2024.
78. Fernandes SS, Saini JK. The impact of chemotherapy on quality of life in advanced-stage lung cancer patients. Indian J Cancer. 2023;60:310. https://doi.org/10.4103/ijc.IJC_1246_20.
79. Wu JS-Y, Wong RK, Lloyd NS, Johnston M, Bezjak A, Whelan T. Radiotherapy fractionation for the palliation of uncomplicated painful bone metastases—an evidence-based practice guideline. BMC Cancer. 2004;4:71. https://doi.org/10.1186/1471-2407-4-71.
80. Gillespie EF, Yang JC, Mathis NJ, Marine CB, White C, Zhang Z, Barker CA, Kotecha R, McIntosh A, Vaynrub M, Bartelstein MK, Mitchell A, Guttmann DM, Yerramilli D, Higginson DS, Yamada YJ, Kohutek ZA, Powell SN, Tsai J, Yang JT. Prophylactic radiation therapy versus standard of care for patients with high-risk asymptomatic bone metastases: a multicenter, randomized phase II clinical trial. J Clin Oncol. 2024;42:38–46. https://doi.org/10.1200/JCO.23.00753.
81. Ahmadsei M, Christ SM, Seiler A, Vlaskou Badra E, Willmann J, Hertler C, Guckenberger M. Quality-of-life and toxicity in cancer patients treated with multiple courses of radiation therapy. Clin Transl Radiat Oncol. 2022;34:23–9. https://doi.org/10.1016/j.ctro.2022.03.006.
82. Yucel B, Akkaş EA, Okur Y, Eren AA, Eren MF, Karapınar H, Babacan NA, Kılıçkap S. The impact of radiotherapy on quality of life for cancer patients: a longitudinal study. Support Care Cancer. 2014;22:2479–87. https://doi.org/10.1007/s00520-014-2235-y.
83. Ristevski B, Jenkinson RJ, Stephen DJG, Finkelstein J, Schemitsch EH, McKee MD, Kreder HJ. Mortality and complications following stabilization of femoral metastatic lesions: a population-based study of regional variation and outcome. Can J Surg. 2009;52:302–8.
84. Mosher ZA, Patel H, Ewing MA, Niemeier TE, Hess MC, Wilkinson EB, McGwin G, Ponce BA, Patt JC. Early clinical and economic outcomes of prophylactic and acute pathologic fracture treatment. J Oncol Pract. 2019;15:e132–40. https://doi.org/10.1200/JOP.18.00431.
85. Schulman KL, Kohles J. Economic burden of metastatic bone disease in the U.S. Cancer. 2007;109:2334–42. https://doi.org/10.1002/cncr.22678.
86. Delea T, McKiernan J, Brandman J, Edelsberg J, Sung J, Raut M, Oster G. Retrospective study of the effect of skeletal complications on total medical care costs in patients with bone metastases of breast cancer seen in typical clinical practice. J Support Oncol. 2006;4:341–7.
87. Peters MEWJ, Goedendorp MM, Verhagen SAHHVM, Smilde TJ, Bleijenberg G, van der Graaf WTA. A prospective analysis on fatigue and experienced burden in informal caregivers

of cancer patients during cancer treatment in the palliative phase. Acta Oncol. 2015;54:500–6. https://doi.org/10.3109/0284186X.2014.953254.

88. Langenberg SMCH, van Herpen CML, van Opstal CCM, Wymenga ANM, van der Graaf WTA, Prins JB. Caregivers' burden and fatigue during and after patients' treatment with concomitant chemoradiotherapy for locally advanced head and neck cancer: a prospective, observational pilot study. Support Care Cancer. 2019;27:4145–54. https://doi.org/10.1007/s00520-019-04700-9
89. Roberson PNE, Tasman J, Cortez G, Lenger KA, Lloyd J. I am feeling stressed, are you?: breast cancer patients and caregivers' inter-psychophysiological symptoms during the first year of breast cancer treatment. J Psychosom Res. 2023;168:111213. https://doi.org/10.1016/j.jpsychores.2023.111213
90. Zimmermann C, Swami N, Krzyzanowska M, Hannon B, Leighl N, Oza A, Moore M, Rydall A, Rodin G, Tannock I, Donner A, Lo C. Early palliative care for patients with advanced cancer: a cluster-randomised controlled trial. Lancet (London). 2014;383 https://doi.org/10.1016/S0140-6736(13)62416-2
91. Greer JA, El-Jawahri A, Pirl WF, Jackson VA, Park ER, Back A, Muzikansky A, Kamdar M, Rinaldi S, Jacobsen J, Gallagher ER, Fishbein J, Eusebio J, VanDusen H, Temel JS. Randomized trial of early integrated palliative and oncology care. JCO. 2016;34:104–14. https://doi.org/10.1200/jco.2016.34.26_suppl.104
92. Bakitas M, Lyons KD, Hegel MT, Balan S, Brokaw FC, Seville J, Hull JG, Li Z, Tosteson TD, Byock IR, Ahles TA. Effects of a palliative care intervention on clinical outcomes in patients with advanced cancer: the project ENABLE II randomized controlled trial. JAMA. 2009;302:741–9. https://doi.org/10.1001/jama.2009.1198
93. Sanders J, Temin S, Ghoshal A, Alesi E, Ali ZV, Chauhan C, Cleary JF, Epstein AS, Janice F I, Jones JA, Litzow MR, Lundquist D, Mardones MA, Nipp RD, Rabow MW, Rosa WE, Zimmermann C, Ferrell BR. Palliative care for patients with cancer: ASCO guideline update. J Clin Oncol. 2024;42 https://doi.org/10.1200/JCO.24.00542
94. Rodin R, Swami N, Pope A, Hui D, Hannon B, Le LW, Zimmermann C. Impact of early palliative care according to baseline symptom severity: secondary analysis of a cluster-randomized controlled trial in patients with advanced cancer. Cancer Med. 2022;11:1869–78. https://doi.org/10.1002/cam4.4565
95. Temel JS, Greer JA, Admane S, Gallagher ER, Jackson VA, Lynch TJ, Lennes IT, Dahlin CM, Pirl WF. Longitudinal perceptions of prognosis and goals of therapy in patients with metastatic non-small-cell lung cancer: results of a randomized study of early palliative care. J Clin Oncol. 2011;29:2319–26. https://doi.org/10.1200/JCO.2010.32.4459
96. Dionne-Odom JN, Azuero A, Lyons KD, Hull JG, Tosteson T, Li Z, Li Z, Frost J, Dragnev KH, Akyar I, Hegel MT, Bakitas MA. Benefits of early versus delayed palliative care to informal family caregivers of patients with advanced cancer: outcomes from the ENABLE III randomized controlled trial. J Clin Oncol. 2015;33:1446–52. https://doi.org/10.1200/JCO.2014.58.7824
97. Jo M, Lee Y, Kim T. Medical care costs at the end of life among older adults with cancer: a national health insurance data-based cohort study. BMC Palliat Care. 2023;22 https://doi.org/10.1186/s12904-023-01197-2
98. Chastek B, Harley C, Kallich J, L N, P C, Teitelbaum A. Health care costs for patients with cancer at the end of life. J Oncol Pract. 2012;8 https://doi.org/10.1200/JOP.2011.000469
99. Wodajo F, Colman M, Getty P. AAOS Clinical Practice Guideline Summary: Treatment of Metastatic Carcinoma and Myeloma of the Femur. J Am Acad Orthop Surg. 2023;31(3):e118-e129. https://doi.org/10.5435/JAAOS-D-21-00888

Future Directions in Multidisciplinary Musculoskeletal Oncology

45

Timothy A. Damron

Introduction

To discuss the future of multidisciplinary care of metastatic bone disease (MBD), we need to take stock of where we are now. Where we are now also reflects the trends that led us here. The trends that led us here provide an opportunity to peek into our potential future as the primary care givers of patients with MBD. In this chapter, we will explore the current state of multidisciplinary care, the forces behind the need for change, the opportunities for change, and potential steps to take to get there.

Where Are We Now in Multidisciplinary Management of Metastases?

Care of MBD patients requires multiple disciplines. The multidisciplinary team is an extension of the members who care for sarcoma patients but with a few differences. The care of both sarcoma patients and those with MBD requires medical oncologists, orthopedic surgeons, pathologists, radiation oncologists, and radiologists. Pediatric sarcoma patients also require pediatric oncologists, social workers, and child life specialists.

Metastatic bone disease care has the need for additional team members not as routinely involved in sarcoma care. Palliative care specialists are crucial to end-of-life decision-making so integral to care of all patients with metastatic cancer [1]. Mental health professionals may be called upon to address the varied responses patients have to their diagnoses, treatments, and outcomes. Other specialists, such as those in pain management, rehabilitation medicine, social work, and hospice,

T. A. Damron (✉)
Department of Orthopedic Surgery, Upstate Medical University, East Syracuse, NY, USA
e-mail: Damront@upstate.edu

J. Pretell-Mazzini, S. A. Lozano Calderón (eds.), *Multidisciplinary Approach of Musculoskeletal Metastasis*, https://doi.org/10.1007/978-3-032-09498-8_45

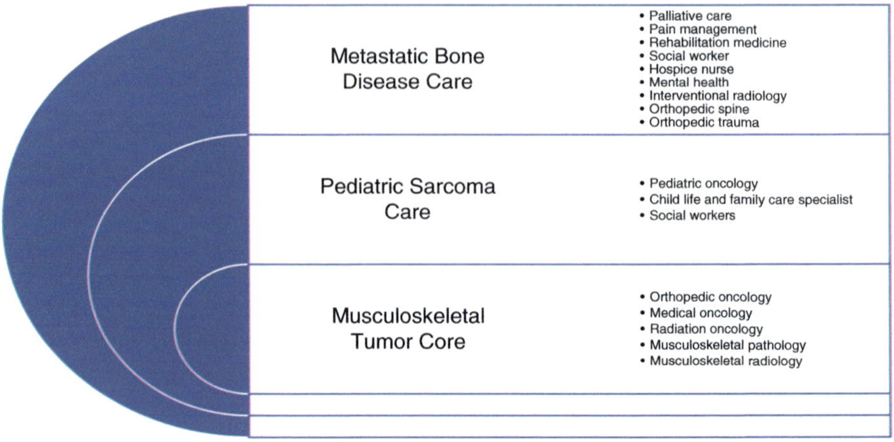

Fig. 45.1 Key players in musculoskeletal oncology patient care

have a role to play. Although interventional radiologists may play an important role in diagnosis for both sarcoma and MBD patients, their role in MBD care is arguably more integral due to their ability to provide ablative therapies and assistance with osteoplasty/cementoplasty (Fig. 45.1).

Even within orthopedics, we may find ourselves increasingly interacting with other subspecialists beyond orthopedic oncology. Spine surgeons, for instance, are sometimes involved in sarcoma care, where their involvement is dependent on the site of the primary and primary spine sarcomas are unusual. In MBD, however, due to the frequency of spinal involvement, their expertise is more frequently needed to address spinal instability, cord compression and fractures. With the current trend toward dealing less invasively with pelvic metastatic disease, our orthopedic trauma colleagues who routinely place percutaneous fixation for traumatic pelvic ring injuries may be called upon at some centers to perform these procedures.

What Is the Need for Change?

The importance of optimizing future multidisciplinary care of MBD patients is widely recognized but bears an up-to-date examination. Bone is the number three site for metastases after lung and liver [2]. However, morbidity to patients in terms of pain and skeletally related events (SRE), including pathologic fractures, spinal cord compression, orthopedic surgical intervention, and palliative radiation to bone, is possibly the single most impactful among sites of metastatic involvement on patients' daily lives in terms of pain, function, and quality of life [4]. The prevalence of MBD increases over time in cancer patients. In an analysis of 382,733 patients, cumulative incidence of bone metastasis was approximately 3% at 30 days, 5% at one year, 6% at two years, 7% at five years, and 8% at ten years [5]. The most prevalent metastatic bone sources are lung. Most at risk for development of MBD over time are patients with prostate cancer (18–29%), followed by lung, renal, and breast cancer, and those with Stage IV disease (any site) at diagnosis, 11% of whom were diagnosed with MBD by 30 days [5].

We are all aware that our cancer patients are living longer. Based on SEER data examining 4,412,024 patients between 1975 and 2014, median overall survival of all cancer patients improved from 35 months between 1975 and 1984 to 101 months between 2005 and 2014 [6]. Pertinent to the most common cancers affecting bone, notably improved survival has been observed for breast, prostate, lung, colon, kidney, and thyroid [6, 7]. Given that the majority of patients with cancers leading to MBD are in older age groups, the shift toward an older American population compounds the prevalence of patients with MBD. Americans at least 65 years of age are projected to increase by 41% from the 2022 population of 58 million to an estimated 82 million by 2050. Patients of this age group represented 17% of the total population in 2022 but are estimated to represent 23% of the population by 2050 [8, 9]. The aging population is not unique to the United States. Across the globe, persons 65 or older are expected to increase by 110%, from 761 million in 2021 to 1.6 billion by 2050 [10]. Globally, cancer cases are projected to increase from 20 million in 2022 to 35 million by 2050 (Table 45.1) [11].

In summary, MBD morbidity is high, prevalence of MBD increases over time in cancer patients, our cancer patients are surviving longer, and our population is becoming older. All of these factors add up to an increased disease burden not only in America but across the globe that must be dealt with in an ever more efficient and effective multidisciplinary fashion.

Table 45.1 Survival data for commonly diagnosed tumors [6, 7]

	5-year survival					10-year survival					20-year survival				
	1975–1984	1985–1994	1995–2004	2005–2014	All years	1975–1984	1985–1994	1995–2004	2005–2014	All years	1975–1984	1985–1994	1995–2004	2005–2014	All years
Breast	75.50%	84.00%	89.00%	91.10%	86.10%	63.60%	76.10%	83.50%	86.30%	78.80%	53.20%	67.50%	75.70%	N/A	69.80%
Prostate	70.50%	89.10%	98.60%	99.10%	93.40%	55.70%	81.90%	97.90%	99.10%	89.70%	39.60%	72.40%	94.40%	N/A	81.70%
Lung and bronchus	12.70%	13.40%	15.30%	19.60%	15.40%	8.70%	9.00%	10.20%	13.10%	10.40%	4.80%	4.80%	5.50%	N/A	5.60%
Colon and rectum	52.10%	60.00%	64.00%	66.40%	60.80%	46.50%	53.90%	58.40%	60.10%	54.90%	42.60%	49.50%	52.60%	N/A	50.00%
Kidney and renal pelvis	51.50%	58.40%	65.50%	75.10%	65.80%	44.50%	50.80%	57.70%	68.80%	58.40%	36.80%	40.90%	47.00%	N/A	47.60%
Thyroid	92.70%	94.40%	96.50%	98.40%	96.60%	91.30%	93.30%	95.80%	98.40%	96.00%	89.90%	92.10%	94.90%	N/A	95.00%

What Are the Opportunities for Change?

Developing Focused MBD Tumor Boards One of the foundational principles of multidisciplinary management (MDM) of all cancer care is the tumor board or multidisciplinary cancer conference (MCC). The importance of the tumor board has been underscored by their inclusion requirement for accreditation by the American College of Surgeons Commission on Cancer Program [12]. The impact of tumor boards remains controversial, though. Advocates point to the influence of tumor boards in changing treatment (43% of breast cancer patients reviewed at the University of Pennsylvania), diagnoses (45% of cancer cases at the University of Michigan), and surgical management (11% at University of Michigan) [13–15]. The Expert Panel on Multidisciplinary Cancer Conference Standards in Ontario, Canada, performed a systematic review to evaluate the effects of MCCs and concluded that only limited evidence suggested a benefit for MCCs, but concluded that they are impactful on changing patient management and are the standard of care internationally, thereby supporting their implementation throughout Ontario [16]. The importance of collaboration with our radiology and pathology colleagues is widely recognized as crucial to cancer care. Diagnosis was changed in 29% of cases by pathology review at the University of Michigan and in 9% of gynecologic tumors after both radiology and pathology review [14, 15]. At most institutions, MCC are cancer type and/or subspecialty based, and in orthopedic oncology, sarcoma MCC are the norm at major institutions, and guidelines exist to direct their development and execution [17].

A natural extension of cancer type and subspecialty-based MCC is the bone metastases cancer board (BMCB) as described by Miyazaki et al. [18]. Their retrospective review demonstrates the quantifiable potential impact of such a board comprised of orthopedic oncologists, rehabilitation medicine specialists, radiologists, radiation oncologists, oncologists, hematologists, pathology, and palliative care. Using posterior spinal decompression and/or instrumentation overall versus emergency surgeries as a measuring stick comparing pre- and postimplementation of the BMCB, the number of annual surgeries increased while the percentage of emergency surgeries decreased after the BMCB was formulated. Mean spinal procedure operative time was 34 min shorter and hospital stay almost 35 days shorter after implementation. Moreover, deterioration rates in Eastern Cooperative Oncology Group Global Performance Status (ECOG-GPS) and Barthel index (BI) for activities of daily living from 1 to 6 months decreased less postoperatively after implementation of the BMCB [18]. The success of the Japanese BMCB in Kobe serves both as a template and as an inspiration for development of BMCB in all of our cancer centers in America as well.

Utilizing Online Resources to Reach Underserved MBD Patients in Need of Orthopedic Care Due to the relative rarity of musculoskeletal oncology problems, orthopedic oncologists are limited in number. Given that most orthopedic oncologists are affiliated with major cancer centers, access elsewhere is limited. This is true both in the United States and abroad. It is generally accepted that sarcoma patients should be treated at cancer centers, but the prevalence of MBD is too great to allow transfer to a major center of every patient in need of treatment. This would quickly overburden the system. With improved training domestically and abroad, general orthopedic, adult reconstructive, and orthopedic trauma surgeons are now more likely to be comfortable handling pathologic and impending fractures. However, for difficult cases or in regions where there is no orthopedic surgeon comfortable taking on these patients' care or no orthopedic surgeon at all, local caregivers may benefit from remote access to the expertise of an orthopedic oncologist or BMCB.

Although everyone knows Internet-based resources are accessible, providing access to orthopedic MBD expertise in a more organized, case-based, patient-specific, and time-sensitive fashion may be an important step to improve care for MBD patients. Each of the Internet-based options currently available has advantages and disadvantages (Table 45.2). Telehealth encounters provide patient-specific data-gathering and real-time analysis by an orthopedic oncologist with immediate recommendations provided in most cases, but setting up patient–physician encounters requires scheduling, connections are sometimes difficult, and just navigating the platform may be difficult or impossible for some. Simultaneous availability of online translators is improving but may be another roadblock to these encounters, particularly for global access. Although there is a plethora of MBD information available on the Internet at any time to anyone with Internet access, independent interpretation of the information without focused training to allow educated clarification may be dangerous. AI-generated information in particular requires professional review. At this time, neither of these tools allows for case-specific review and firm recommendations, either. Videoconferencing has the advantage of allowing multiple participants, but it requires an even higher degree of coordination than Telehealth encounters directly proportionate to the number of physicians involved. Online forums can be valuable to surgeons in getting their colleagues' opinions, but their colleagues may not be able to provide feedback in a timely fashion. Live chat platforms require sponsorship, typically by a health entity, or payment. Regardless of the platform, the relative sparsity of orthopedic oncologists remains at the root of difficulty in providing access to the underrepresented and underserved with MBD in the United States and abroad.

Although provision of healthcare via telemedicine has blossomed since the COVID-19 pandemic, there are still numerous barriers to providing global access that affect potential delivery of orthopedic oncology care to the same extent as other specialties. Some of these barriers are universal, although the extent to which individual countries and regions have developed varies widely. Common threads include infrastructure, availability of EMR, government support, and legislation regarding consent, privacy, and utilization [19].

Employing Big Data and Machine Learning to Improve Care of MBD Patients Collaboration with our research, data analysis and statistics colleagues are needed to improve care of MBD patients by improving diagnostic capabilities,

Table 45.2 Internet resources available to improve orthopedic care of MBD in underserved populations

Internet resources	Provides useful and accurate info	Allows for case-specific review	Availability of access	Timeliness of access
Telehealth encounters	Yes	Yes	Requires patient and physician be present	Scheduling difficulties
Internet-based academic instructional resources	Requires training to recognize appropriate information	No	Widely available with internet access	Available 24–7-365
Internet-based patient educational and AI resources	Unreliable	No	Widely available with internet access	Available 24–7-365
Videoconferencing platforms (zoom, teams)	Yes	Yes	Requires both physicians (or BMCB members) be present along with treating physicians	High level of coordination required
Online forums, message boards, discussion boards (Orthogate, Reddit-Orthopaedics, LinkedIn)	Provides multiple opinions	To the degree allowed by the platform and moderator and provided by the presenter	Limited by physician knowledge of and participation in specific forum or board	Variable response times dependent on physician participation and effort
Live chat platforms [telehealth on-demand] (just answer, amazon one medical)	Based on the training and experience of the doctor available	To the degree allowed by the platform and moderator and provided by the presenter	Available for referrals but unlikely to have access to a qualified physician	Available 24–7-365

treatment planning, drug development, and patient monitoring. Although to some extent machine learning has already been used to predict survival in patients with MBD, there is room not only for improvement in survival prediction but also for expansion of machine learning to other diagnostic questions. There are numerous survival prediction tools available for MBD patients, including those for extremity and spine tumors [20–27]. Further continued refinement to those tools is important.

There are numerous other orthopedic MBD diagnostic and treatment questions to which machine learning can be applied. Some of these are fracture risk prediction in the setting of impending fracture of the femur and humerus, failure prediction of internal fixation devices in the proximal femur to determine when endoprosthetic reconstruction would be advisable as a primary procedure, procedure selection between ablation, cementoplasty, minimally invasive internal fixation, and arthroplasty reconstruction for pelvic periacetabular MBD. There are numerous others.

Increasing International Registry Collaboration to Generate and Interrogate MBD "Big Data" Multidisciplinary collaboration between international data registries would incrementally improve access to robust data and improve power of analysis for numerous musculoskeletal conditions. This sort of multidisciplinary, multi-institutional international registry cooperation will need to overcome substantial organizational, economical, governmental, legal, technological, and logistical hurdles in order to provide much needed collaborative information. In the United States, the AAOS Musculoskeletal Tumor Registry (MsTR) [https://www.aaos.org/registries/registry-program/musculoskeletal-tumor-registry/] of the Musculoskeletal Tumor Society (MSTS) is the primary registry. The International Society of Limb Salvage (ISOLS) [isols.com/societies/] serves as the umbrella organization for the MSTS, European Musculoskeletal Oncology Society (EMSOS) [28], Sociedad Latino Americana de Tumores Musculo Esqueleticos (SLATME) [www.slatme.org], and the Asia Pacific Musculoskeletal Tumour Society, each of which has the potential to provide collaborative database information. Other independent tumor registries such as the Japanese Bone Tumor Registry [29, 30] could contribute important data, too. To contribute meaningful understanding and basis for making impactful changes in the treatment of MBD, tissue banking would be vital, introducing yet another layer of complexity to this process. Ultimately, personalized treatment plans incorporating patient demographics, comorbidities, tumor genetic profiles, radiographic analysis, and treatment outcomes would be the Holy Grail. The first step in engendering this complex cooperation entails the development of a "minimal dataset" across all international registries.

Exploiting Artificial Intelligence to the Benefit of MBD Patients Artificial intelligence can play a role in MBD drug development, drug discovery, and matching patient and tumor biology to currently available drugs applicable to MBD. Wearable AI can allow monitoring of our clinical patients and research subjects to better assess and track treatment results and functional improvements. Changes in activity noted through these devices may identify the need for modification of musculoskeletal care, such as the detection of worsening gait parameters as an indication of progressive bone destruction warranting operative intervention [27]. AI decision-making tools can be incorporated into electronic medical record (EMR) systems to provide information for decision-making in our MBD patients [31]. This could include morbidity and mortality data, individualize survival estimates, and detailed risk–benefit analysis of surgical vs. nonsurgical options. There are already templates for these advances being developed in more common cancers such as lung and breast [32–34]. Advances via AI are being developed currently for lung cancer diagnosis based on imaging and biomarkers, treatment decisions, response assessment and salvage management [32]. Analogous applications exist in the multidisciplinary management of MBD with the ultimate goal of individualized personalized precision medicine [35]. To date, there are barriers to implementation of AI for MBD parallels that of its use in all other aspects of medical care. Steps to overcoming these barriers include improving poor data quality, protecting patient health information privacy, improving compatibility between AI data storage systems and the electronic medical records (EMR), and winning over untrusting patients and physicians [27, 36]. Despite those barriers, as of 2021 there were dozens of FDA-approved AI-based devices already on the market for cancer diagnosis and treatment (Fig. 45.2a, b). However, very few were surgeon based, and very few if any were developed for MBD. This represents an incredible opportunity for growth and improvement in patient care.

Developing Global Cancer Hospitals and Programs as a Platform to Provide MBD Orthopedic Care Global collaboration to address MBD orthopedic care cannot begin in isolation. It requires an organizational structure for outreach. Beyond utilization of Telemedicine to improve medical care abroad, increasing the number of National Cancer Control Programs (NCCP) in low- and middle-income countries (LMC) provides the necessary structure for this. Establishment of a cancer center in a LMC provides leadership to establish cancer control programs and provides a point of contact with the global cancer community for education and outreach. Although orthopedic oncology would not be a major component of a NCCP, it would provide a similar point of contact for the development of more specific educational and outreach efforts. Eventually, it would be hoped that by way of trickle down, an opportunity for improved care of MBD would arise.

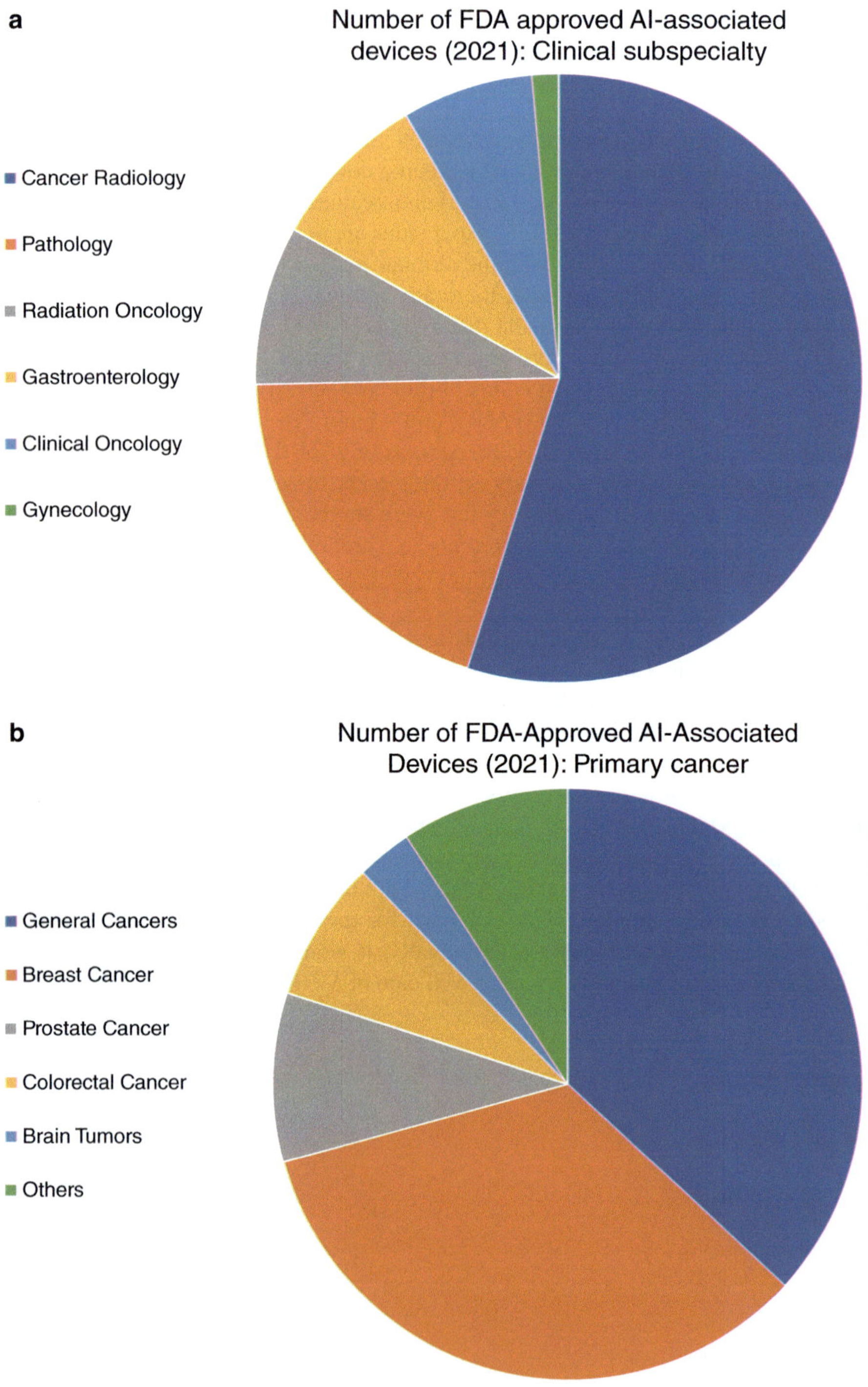

Fig. 45.2 Number of FDA-approved artificial intelligence associated devices in 2021 according to (**a**) clinical subspecialty and (**b**) primary cancer for which the device was developed. (Data from Luchini et al. [37])

What Are the Future Directions?

Future directions in improved orthopedic care of MBD should follow the outline of opportunities previously highlighted. (1) Develop MBD tumor boards. (2) Utilize online resources to reach underserved MBD patients in need of orthopedic care. (3) Employ big data and machine learning to answer clinical MBD orthopedic care questions. (4) Increase international registry collaboration to interrogate MBD data. (5) Exploit AI to the benefit of MBD patients. (6) Develop global cancer hospitals and programs as a platform to provide MBD orthopedic care. Each of these steps requires multidisciplinary cooperation, whether it be at the local, regional, national, or international level, among all the key players. For MBD patients, the key players are hospice nurses, interventional radiologists, medical oncologists, mental health professionals, orthopedic oncologists and other surgeons, pain management teams, palliative care specialists, pathologists, radiation oncologists, radiologists, rehabilitation medicine physicians, and social workers. Exploiting the benefits of AI, machine learning, and big data will require collaboration with our research and statistics colleagues and epidemiologists. Achieving improvements in global MBD orthopedic care will require additional collaboration with our law colleagues to navigate the complex regulatory issues limiting international efforts. All steps forward in improving the multidisciplinary care of MBD patients will benefit from funding efforts, whether organizational, where Sarcoma Strong shines as a glowing example, philanthropic through grateful patients and donors, or governmental grants.

References

Where Are We Now?

1. Roth C, Weiss K. Palliative care needs of patients with musculoskeletal malignancies. Curr Oncol Rep. 2024;26(7):784–90. https://doi.org/10.1007/s11912-024-01543-4. Epub 2024 May 24
2. Jayarangaiah A, Kemp AK, Kariyanna PT. StatPearls [internet]. Treasure Island: StatPearls Publishing; 2024. https://www.ncbi.nlm.nih.gov/books/NBK507911/

MBD Statistics

3. Ryan C, Stoltzfus KC, Horn S, Chen H, Louie AV, Lehrer EJ, Trifiletti DM, Fox EJ, Abraham JA, Zaorsky NG. Epidemiology of bone metastases. Bone. 2022;158:115783. https://doi.org/10.1016/j.bone.2020.115783. Epub 2020 Dec 1. PMID: 33276151
4. Hernandez RK, Wade SW, Reich A, Pirolli M, Liede A, Lyman GH. Incidence of bone metastases in patients with solid tumors: analysis of oncology electronic medical records in the United States. BMC Cancer. 2018;18(1):44. https://doi.org/10.1186/s12885-017-3922-0. PMID: 29306325; PMCID: PMC5756362
5. Sakr RA, Nasr AA, Zineldin EI, Gouda MA. Long-term survival in patients with cancers: surveillance, epidemiology and end results-based analysis. Sultan Qaboos Univ Med

J. 2023;23(3):344–50. https://doi.org/10.18295/squmj.1.2023.002. Epub 2023 Aug 28. PMID: 37655083; PMCID: PMC10467541
6. Siegel RL, Miller KD, Fuchs HE, Jemal A. Cancer statistics, 2022. CA Cancer J Clin. 2022;72(1):7–33. https://doi.org/10.3322/caac.21708. Epub 2022 Jan 12. PMID: 35020204
7. Mather M, Scommegna P. Fact sheet: Aging in the United States. 01–24-Aging-Fact-Sheet-j Population Reference Bureau https://www.prb.org/resources/fact-sheet-aging-in-the-united-states/#_ednref1
8. U.S. Census Bureau., 2023 National Population Projections Tables: Main Series. [cited by Mather 2024].
9. World Social Report: Leaving No One Behind in an Ageing World. United Nations Department of Economic and Social Affairs. https://www.un.org/development/desa/dspd/wp-content/uploads/sites/22/2023/01/WSR_2023_Chapter_Key_Messages.pdf
10. American Cancer Society. Global Cancer Facts & Figures 5th Edition. Atlanta: American Cancer Society; 2024.

Metastatic Cancer Tumor Boards

11. El Saghir NS, Keating NL, Carlson RW, Khoury KE, Fallowfield L. Tumor boards: optimizing the structure and improving efficiency of multidisciplinary management of patients with cancer worldwide. Am Soc Clin Oncol Educ Book. 2014:e461–6. https://doi.org/10.14694/EdBook_AM.2014.34.e461. PMID: 24857140
12. Chang JH, Vines E, Bertsch H, et al. The impact of a multidisciplinary breast cancer center on recommendations for patient management: the University of Pennsylvania experience. Cancer. 2001;91:1231–7.
13. Newman EA, Guest AB, Helvie MA, et al. Changes in surgical management resulting from case review at a breast cancer multidisciplinary tumor board. Cancer. 2006;107:2346–51.
14. Cohen P, Tan AL, Penman A. The multidisciplinary tumor conference in gynecologic oncology-does it alter management? Int J Gynecol Cancer. 2009;19:1470–2.
15. Wright FC, De Vito C, Langer B, Hunter A. Expert panel on multidisciplinary cancer conference standards. Multidisciplinary cancer conferences: a systematic review and development of practice standards. Eur J Cancer. 2007;43(6):1002–10. https://doi.org/10.1016/j.ejca.2007.01.025. Epub 2007 Feb 27. PMID: 17329094
16. Optimizing Multidisciplinary Tumor Board Efficiency Worldwide. asco.org/edbook I 2014 ASCO Educational Book e465 Downloaded from ascopubs.org by 174.231.48.96 on August 17, 2024 from 174.231.048.096 Copyright © 2024 American Society of Clinical Oncology. All rights reserved.
17. Miyazaki K, Kanda Y, Sakai Y, Yoshikawa R, Yurube T, Takeoka Y, Hara H, Akisue T, Kuroda R, Kakutani K. Effect of bone metastasis cancer board on spinal surgery outcomes: a retrospective study. Medicina (Kaunas). 2023;59(12):2087. https://doi.org/10.3390/medicina59122087. PMID: 38138190; PMCID: PMC10744340
18. Bhaskar S, Bradley S, Chattu VK, Adisesh A, Nurtazina A, Kyrykbayeva S, Sakhamuri S, Yaya S, Sunil T, Thomas P, Mucci V, Moguilner S, Israel-Korn S, Alacapa J, Mishra A, Pandya S, Schroeder S, Atreja A, Banach M, Ray D. Telemedicine across the globe-position paper from the COVID-19 Pandemic Health System Resilience PROGRAM (REPROGRAM) International Consortium (Part 1). Front Public Health. 2020;8(16):556720. https://doi.org/10.3389/fpubh.2020.556720. PMID: 33178656; PMCID: PMC7596287

Big Data and Machine Learning

19. Li C, Liu M, Li J, Wang W, Feng C, Cai Y, Wu F, Zhao X, Du C, Zhang Y, Wang Y, Zhang S, Qu J. Machine learning predicts the prognosis of breast cancer patients with initial bone metastases. Front Public Health. 2022;10:1003976. https://doi.org/10.3389/fpubh.2022.1003976. PMID: 36225783; PMCID: PMC9549149
20. Skalitzky MK, Gulbrandsen TR, Groot OQ, Karhade AV, Verlaan JJ, Schwab JH, Miller BJ. The preoperative machine learning algorithm for extremity metastatic disease can predict 90-day and 1-year survival: an external validation study. J Surg Oncol. 2022;125(2):282–9. https://doi.org/10.1002/jso.26708. Epub 2021 Oct 5. PMID: 34608991
21. Nieder C, Mannsåker B, Yobuta R. Independent validation of a comprehensive machine learning approach predicting survival after radiotherapy for bone metastases. Anticancer Res. 2021;41(3):1471–4. https://doi.org/10.21873/anticanres.14905. PMID: 33788739
22. Thio QCBS, Karhade AV, Bindels BJJ, Ogink PT, Bramer JAM, Ferrone ML, Calderón SL, Raskin KA, Schwab JH. Development and internal validation of machine learning algorithms for preoperative survival prediction of extremity metastatic disease. Clin Orthop Relat Res. 2020;478(2):322–33. https://doi.org/10.1097/CORR.0000000000000997. Erratum in: Clin Orthop Relat Res 2021 Apr 1;479(4):862. https://doi.org/10.1097/CORR.0000000000001678. PMID: 31651589; PMCID: PMC7438151
23. Cui Y, Shi X, Wang S, Qin Y, Wang B, Che X, Lei M. Machine learning approaches for prediction of early death among lung cancer patients with bone metastases using routine clinical characteristics: an analysis of 19,887 patients. Front Public Health. 2022;10:1019168. https://doi.org/10.3389/fpubh.2022.1019168. PMID: 36276398; PMCID: PMC9583680
24. Cilla S, Rossi R, Habberstad R, Klepstad P, Dall'Agata M, Kaasa S, Valenti V, Donati CM, Maltoni M, Morganti AG. Explainable machine learning model to predict overall survival in patients treated with palliative radiotherapy for bone metastases. JCO Clin Cancer Inform. 2024;8:e2400027. https://doi.org/10.1200/CCI.24.00027. PMID: 38917384
25. Tseng TE, Lee CC, Yen HK, Groot OQ, Hou CH, Lin SY, Bongers MER, Hu MH, Karhade AV, Ko JC, Lai YH, Yang JJ, Verlaan JJ, Yang RS, Schwab JH, Lin WH. International validation of the SORG machine-learning algorithm for predicting the survival of patients with extremity metastases undergoing surgical treatment. Clin Orthop Relat Res. 2022;480(2):367–78. https://doi.org/10.1097/CORR.0000000000001969. PMID: 34491920; PMCID: PMC8747677
26. Rizk PA, Gonzalez MR, Galoaa BM, Girgis AG, Van Der Linden L, Chang CY, Lozano-Calderon SA. Machine learning-assisted decision making in orthopaedic oncology. JBJS Rev. 2024;12(7) https://doi.org/10.2106/JBJS.RVW.24.00057. PMID: 38991098
27. Longhi A, Bielack SS, Grimer R, et al. Extraskeletal osteosarcoma: a European musculoskeletal oncology society study on 266 patients. Eur J Cancer. 2017;74:9–16. https://doi.org/10.1016/j.ejca.2016.12.016.

Registries

28. Takenaka S, Tamiya H, Wakamatsu T, Nakai S, Imura Y, Outani H, Yagi T, Kawai A. Impact of surgical resection and reasons for poor prognosis of pelvic osteosarcoma based on the bone tumor registry in Japan. Cancer. 2021;13(13):3320. https://doi.org/10.3390/cancers13133320.
29. Ogura K, Higashi T, Kawai A. Statistics of bone sarcoma in Japan: report from the bone and soft tissue tumor registry in Japan. J Orthop Sci. 2017;22(1):133–43. https://doi.org/10.1016/j.jos.2016.10.006. Epub 2016 Nov 12. PMID: 27847134

AI

30. Xie W, Butcher R. Artificial intelligence decision support tools for end-of-life care planning conversations: CADTH horizon scan [Internet]. Ottawa: Canadian Agency for Drugs and Technologies in Health; 2023. Available from: https://www.ncbi.nlm.nih.gov/books/NBK599854/
31. Christie JR, Lang P, Zelko LM, Palma DA, Abdelrazek M, Mattonen SA. Artificial intelligence in lung cancer: bridging the gap between computational power and clinical decision-making. Can Assoc Radiol J. 2021;72(1):86–97. https://doi.org/10.1177/0846537120941434.
32. Hunter B, Hindocha S, Lee RW. The role of artificial intelligence in early cancer diagnosis. Cancer. 2022;14:1524. https://doi.org/10.3390/cancers14061524.
33. Xu F, Sepúlveda M-J, Jiang Z, Wang H, Li J, Liu Z, Yin Y, Roebuck MC, Shortliffe EH, Yan M, Song Y, Geng C, Tang J, Jackson GP, Preininger AM, Rhee K. Effect of an artificial intelligence clinical decision support system on treatment decisions for complex breast cancer. JCO Clin Cancer Inform. 4:824–38. https://doi.org/10.1200/CCI.20.00018.
34. Hamamoto R, Suvarna K, Yamada M, Kobayashi K, Shinkai N, Miyake M, Takahashi M, Jinnai S, Shimoyama R, Sakai A, et al. Application of artificial intelligence technology in oncology: towards the establishment of precision medicine. Cancer. 2020;12:3532. https://doi.org/10.3390/cancers12123532.
35. Heudel P-E, et al. Impact of artificial intelligence in transforming the doctor–cancer patient relationship. ESMO Real World Data and Digital Oncology. 3:100026.
36. Luchini C, Pea A, Scarpa A. Artificial intelligence in oncology: current applications and future perspectives. Br J Cancer. 2022;126:4–9. https://doi.org/10.1038/s41416-021-01633-1.

Index

A
Ablation, 205, 206, 208, 212
Ablation-Osteoplasty-Reinforcement-Internal Fixation (AORIF), 208
  biomechanical considerations, 669, 670
  bone reconstitution, biological considerations for, 670
  complications, 677–679
  pain assessment, 667–669
  pain control perspectives, 670
  pelvic defects and fractures, techniques, 672–677
  rationale for, 667
Ablative techniques, 205–206
Ablative therapy, 577, 668
Abscopal effect, 593
Academy of Nutrition and Dietetics (AND), 73
Acceptance and commitment therapy (ACT), 532
Acetabular Sourcil, 669
Acetabulum, 651
Acetaminophen, 437
Acral metastases, in foot and ankle, 369, 370
  diagnosis, 372, 373
  literature review for, 371
  management, 376
    calcanectomy, 381
    distal fibula, 382, 383
    distal tibia, 381, 382
    forefoot, 376
    hindfoot, 379
    Midfoot, 378, 379
    ray amputations, 377
    Syme amputation, 379–381
    tarsometarsal amputations, 377
    transmetatarsal amputations, 377
  non-operative treatment, 374
  strategies, 376, 382
  surgical procedures, 374–376
  treatment, 373
Acrometastases, 607
Activities of daily living (ADL), 50, 53–55, 503
Acute leukemia, 476
Acute pain, 516, 519, 521, 531, 532
Acute pain management, 522
  OOS, 521, 522
    amputations, 532, 534
    ERAS pathways, 531
    multimodal analgesia, 525–527
    neuraxial analgesia, 528, 529
    opioid-tolerant or chronic pain patients, 532, 533, 535
    personalizing pain protocols, 531–533
    PO analgesics, 528
    pre-surgical evaluation and development of, 521, 523–525
    preventive analgesia, 525
    psychological and behavioral interventions, 530, 531
    regional analgesia, 528–530
Acute to Chronic Pain Signatures (A2CPS) program, 524
ADC therapy, 394
Adenocarcinoma, 428
Advanced disease, 634
Advanced practice providers (APPs), 547

J. Pretell-Mazzini, S. A. Lozano Calderón (eds.), *Multidisciplinary Approach of Musculoskeletal Metastasis*, https://doi.org/10.1007/978-3-032-09498-8

Advanced renal cell carcinoma (RCC), management, 447
FDA-approved agents for treatment, 447
immunotherapy, 447
monotherapy, 447
next-generation sequencing, 447
radionuclide therapy, 451
radiotherapy, 450
supportive therapy, 451, 452
surgery, 450
systemic therapy, 449
TKI, 450, 451
treatment selection, 447
Alkaline phosphatase (ALP), 405
ALK-positive non-small cell lung cancer (ALK+ NSCLC), 433
American Joint Committee (AJCC) TNM staging system, 460
American Physical Therapy Association (APTA), 503
American Society for Parenteral and Enteral Nutrition (ASPEN), 73
American Society of Clinical Oncology (ASCO), 691
American Thyroid Association (ATA) criteria, 460
Amputations, 285–290, 374, 532, 534, 558
Analgesics, 414–416
Anaplastic thyroid cancer, 459
Anaplastic thyroid carcinoma, 458
Androgen deprivation therapy (ADT), 403
Anemia, 471, 472
Angiosarcoma, 132
Ankle-foot orthosis (AFO), 507
Anterior cervical corpectomy, 193
Anterior inferior iliac spine (AIIS), 211, 212
Anterior lumbar corpectomy, 194
Anterior tibial tuberosity (ATT), 340, 341
Antibiotic prophylactic, 346
Antiresorptive therapy, 541
Appropriate Use Criteria (AUC), 405
Aquamantys system, 334
Arthrodesis, 375
Artificial intelligence (AI), 706–708
Artificial neural network (ANN), 152
Atypical femoral fractures (AFFs), 37, 398
Australia-modified KPS (AKPS), 56
Autologous stem cell transplant (auto-SCT), 474
Axial metastases, 507
Axial skeleton, 104

B
Balance training, 507
Balloon-assisted osteoplasty, 213
Barthel index (BI), 54, 55, 703
Bayesian belief network (BBN), 152
Big data, 704–706
Biologically effective dose (BED), 594
Biophysical assessment, 531
Biophysical measures, 523, 531, 532
Biopsy, 105, 406, 431, 626, 636, 637, 674
bone biopsy, 105
coaxial technique, 107
components, 109
diagnostic accuracy, 109
genetic testing, 108
image-guided biopsy, 106
medial approach, 107
MRI, 106
MRI/PET-CT, 106
percutaneous needle biopsy, 107
pre-biopsy planning, 105
procedures, 107
soft-tissue structures, 107
spring-loaded cutting mechanism, 108
surgical biopsy, 111
ultrasound, 106, 108
Biopsy technique, 390
Biopsychosocial model of pain, 515
Bipolar hemiarthroplasty, 203
Bisphosphonates, 282, 412, 451
BOMET-QoL-10, 52, 54
Bone allograft reconstruction, 558
Bone biopsy, 104, 431, 544
Bone destruction, 202, 203
Bone focused therapy, 476
Bone Health Providers, 545
Bone metastases, 18, 33, 34, 117, 119, 383, 389, 403, 429, 502, 557, 558, 590, 594, 684
ablative interventions for, 566
biopsy, 105
bony stabilization, 571–575
conventional radiation therapy, 591
cryoablation, 568–570
diagnostic and interventional, 104
diagnostic workup for, 430
epidemiology, 563
HIFU, 570, 571
image guidance, 575, 576
imaging modalities, characteristics of, 430
incidence of, 429
indications, 104, 563–565

management and prognosis, 591
in mCRPC, 415
MWA, 568
National Comprehensive Cancer Network (NCCN) guidelines, 564
palliative radiotherapy for, 489–490
pathologic fracture, 104
pathophysiology, 429
percutaneous interventions for, 580–582
postoperative care, 585, 586
preoperative planning, 565
prophylactic radiation in, 595–597
radiation therapy, 591
in RCC, 446
RFA, 566, 567
surgical technique, 576–582
techniques, 583–585
Bone metastases cancer board (BMCB), 703
Bone microenvironment, 404
Bone mineral density (BMD), 391, 393
Bone morphogenetic proteins (BMPs), 7, 20
Bone/muscle metastases of unknown primary
biopsy and histological examination, 614
CT and MRI, 611
endoscopy and colonoscopy, 614
epidemiology, 606, 607
follow-up care, 616
imaging studies
bone scan, 611, 612
CT and MRI, 610, 611
PET/CT, 612
plain radiographs, 609, 610
ultrasound, 612
initial evaluation and diagnostic approach, 608
history, 609
physical examination, 609
laboratory tests, 613
molecular technologies, 614
pathophysiology, 607, 608
prognosis, 616
treatment
local therapy, 615
supportive care, 615
systemic therapy, 615
tumor markers, 613
Bone reconstruction, 226
Bone resorption, 622
Bone sarcoma, 134
CT scan, 135
Ewing sarcoma, 135
MRI, 135
PET/CT, 135
Bone SBRT, 594
Bone scan (technetium-99m scintigraphy), 430
Bone scintigraphy, 49, 95–97, 391, 611, 612, 625
Bone stability, 481
Bone-modifying agents (BMA), 36, 390
indications, 36
SREs, 36
Bone-only metastases (BOM), 389
Bone-targeted agents (BTA), 173
BPI (Brief Pain Inventory), 417
Brachytherapy, 644
BRAF inhibitor dabrafenib, 461
Brain metastases, 509
Breast cancer, 323, 396, 683, 684, 686, 689, 690
Breast Cancer Bone Metastasis (BC BM), 397
Breast cancer cells, 608
Brief Pain Inventory (BPI) scores, 412

C

Cabazitaxel, 409
Cabozantinib, 461
Cachexia, 74
Calcanectomy, 381
Calcar acetabuli, 669
Cancer, 540, 668
diagnoses and treatment, 503
treatment, 509
Cancer care, 540
Cancer of unknown primary (CUP), 605
Cancer related fatigue, 508
Cancer survival rates, 685
Cancer therapy, 689
Cancer versus non-Cancer Pain, 520
Cancer-induced bone pain, 416, 437–439
Cannulated screws, 206–207
Carbon fiber (CF), 260, 356
Carbon fiber intramedullary implants (CFI), 311, 312
Carbon fiber plates, 260
Carcinoma, 18, 120, 632
adenocarcinoma, 120
biliary tract carcinomas, 122
cytokeratins, 120, 121
mucoepidermoid carcinoma, 123
lytic lesion, 135
PET/CT, 136
SCC, 120–122
Caregiver mental health, 691

Carfilzomib, 474
C-arm imaging, 672
Catastrophizing personality, 519, 523
Cauda equina syndrome, 479, 480, 484
CDK4/6 inhibitors, 392, 396
Cement extravasation, 189
Cement osteoplasty, 584
Cementation, 203, 205, 208, 212
Cemented reverse endoprosthesis, 242
Cementoplasty, 174, 205, 213, 564, 569, 576, 584
Central nervous system, 486
Central sensitization, 516, 518, 520, 521, 523, 533
Cephalic vein, 263
CHAARTED criteria, 407
Chemotherapeutic agents, 142–143
Chemotherapy, 392, 396, 432, 507, 615
Chest radiograph, 610
Chimeric antigen receptor T cell therapy (CAR-T), 475
Chronic pain, 519
Chronic pain management plan, 667
Chronic postsurgical pain (CPSP), 514, 519, 522, 525, 529, 530, 535
Circulating tumor DNA (ctDNA), 392, 614
Clavicle, 649
Clear cell renal cell carcinoma (ccRCC), 443
Clear cell sarcoma, 132
Clinical prediction of survival (CPS), 58
Cognitive behavioral therapy (CBT), 143, 532
Colonoscopy, 614
Commission on Cancer (CoC), 540
Common Terminology Criteria for Adverse Events (CTCAE), 76
Complete blood count (CBC), 47
Compression fractures, 189
Computed tomography (CT), 48, 87, 90, 91, 131, 391, 430, 444, 610, 611, 635
Conformality Index (CI), 593
Continuum of care, 509
Contralateral anterolateral system, 517
Conventional C-arm imaging, 673
Conventional chemotherapy, 394
Conventional external beam radiotherapy (cEBRT), 182
Conventional imaging techniques, 405
Conventional radiation therapy (RT), 591
Conventional radiotherapy, 444, 485
Conventional X-rays, 635
Conversion hip arthroplasty, 678
Cooled systems, 206
Core needle biopsies (CNB), 390
Coronal PET-CT, 6
Corpectomy, 189, 193, 194
COVID pandemic, 139
CRAB' criteria, 468
C-reactive protein, 75
Cryoablation (CA), 205, 206, 212, 213, 568–570
CT CAP, 623
CT-based biomechanical methods, 170
CT-based rigidity analysis (CTRA), 352
CT-finite-element analysis (CT-FEA), 171
CT-rigidity analysis (CT-RA), 170, 171
Cyclooxygenase-2 (COX-2) inhibitors, 154, 516
Cyclophosphamide, 471
Cytokeratins, 120
Cytokine release syndrome (CRS), 475
Cytotoxic chemotherapy, 173, 474

D

Data-driven survival model, 685
Decision-making, 685
Decompression, 189, 190, 193
Deep vein thrombosis (DVT), 174, 186, 362
Denosumab, 412, 413, 452, 462, 476
Destructive acrometastasis, 558
Destructive bone lesion, 687
Dietician, 545
Diffusion-weighted imaging (DWI), 92
Dirty necrosis, 122
Distal exposure, 264
Distal fibula, 382, 383
Distal radial ulnar joint (DRUJ), 288
Distal tibia, 381, 382
Distant metastasis velocity (DMV), 594
Distant metastatic disease, 460
Docosahexaenoic acid (DHA), 81

E

Eastern Clinical Oncology Group (ECOG), 32
Eastern Cooperative Group Performance Status (ECOG PS), 56, 57
Eastern Cooperative Oncology Group (ECOG), 11, 31, 56, 185
Eastern cooperative oncology group global performance status (ECOG-GPS), 703
EGFR tyrosine kinase inhibitors (EGFR TKIs), 432
EGFR-positive non-small cell lung cancer (EGFR+ NSCLC), 432
Eicosapentaenoic acid (EPA), 81
Elbow

epidemiology, 269
evaluation, 270
implant selection
humerus and ulnar components, 277
patient positioning, 276
setting of, 275
surgical approach, 276, 277
internal fixation
intra-lesional curettage, 275
patient positioning, 274
surgical approach, 274, 275
post-operative care, 277, 278
pre-operative planning, 271, 272
surgical indications/non-surgical indications, 270, 271
surgical treatment options
internal fixation +/- intralesional curettage, 272, 273
modular endoprostheses, 273
Emotional Distress Thermometer (EDT), 140
Endocrine therapy (ET), 392, 396
End-of-life care considerations, 658, 660–663
Endoprosthesis, 241–244
Endoprosthetic reconstruction, 258, 375, 627
Endoscopy, 614
Endurance training, 507
Enhanced Recovery After Surgery (ERAS) pathways, 521, 531
Enteral nutrition (EN), 79, 80
Enzalutamide, 407, 409
Epidermal growth factor receptor (EGFR) gene, 428, 432
Epidural anesthesia (EA), 528
Epidural spinal cord compression (ESCC), 181, 182, 184, 187, 189, 190
Epithelial-to-mesenchymal transition (EMT), 14, 21
Epithelioid sarcoma, 132
Esophageal and gastric cancer, 542
Estrogen receptor (ER), 50, 150
Estrogen receptor protein (ER), 126
Ethical considerations
end-of-life care, 660–663
patient expectations, 655, 658–660
prognostic uncertainty, 661
SDM, 656
clinician-patient relationship, 656
family and multidisciplinary care, 657
relational autonomy and, 656
European Musculoskeletal Oncology Society (EMSOS), 706
European Society for Clinical Nutrition and Metabolism (ESPEN), 79
European Society of Parenteral and Enteral Nutrition (ESPEN), 70
Ewing sarcoma, 124
External beam radiation therapy (EBRT), 154, 172, 462, 644
External beam radiotherapy (EBRT), 37, 407, 491
Extracellular matrix (ECM), 14, 17
Extramedullary (EM) myeloma, 470

F

FACT-BP PROM, 52
FACT-P (Functional Assessment of Cancer Therapy – Prostate), 417
Familial non-medullary thyroid carcinoma (FNMTC), 458
Femoral diaphysis
epidemiology, 307, 308
indications, 308, 309
pre-operative planning and surgical options
intramedullary rod fixation, 310–312
metastatectomy, 309, 310
resection and intercalary reconstruction, 313–315
surgical techniques
intercalary reconstruction, 317, 318
metastatectomy with intercalary prosthetic reconstruction, 315, 316
positioning, surgical approach and resection, 316, 317
Femur, 650
Fine needle aspirates (FNA), 105
First line therapy, 392
Fixation techniques, 174
Fluorodeoxyglucose (FDG), 472
18-F-Fluoroestradiol (FES) PET, 391
18F-fluoromethylcholine PET/CT (18F-FCH), 405
18flourinine-flourodeoxyglucose (F18-FDG), 96
18-Fluorodeoxyglucose positron emission tomography with computed tomography (FDG PET/CT), 372–373, 391, 445, 637
Follicular cells, 458
Follicular thyroid cancer, 459
Forearm, *see* Hand
Forearm/wrist fractures, 648
Fracture, 391, 623
"Free Light Assay", 469
Free light chains, 469
Functional pain, 169

G
Ga68-PSMA, 96
Gadolinium-containing contrast material (GCM), 92, 94
Gamma-aminobutyric acid (GABA), 517
Gene mutations, 467
Generalized Anxiety Disorder (GAD-7) Anxiety Scale, 140
Genetic mutations, 473
Genome-wide association studies (GWAS), 403
Genomic profiling technology, 127
Giant cell tumor of bone (GCTB), 622
Gleason-based grading, 406

H
Hand
  epidemiology, 281, 282
  post-operative care, 293
  pre-operative planning, 282–284
  surgical indications/non-surgical indications, 282
  surgical options
    amputation, 284
    complications, 286
    contraindications, 285
    reconstruction, 285
    risks, 285
  surgical technique, 290
    advantage and disadvantage, 291
    bone reconstruction, 287–289
    patient positioning and surgical incision, 286
    soft tissue reconstruction, 289, 290
    tumor resection, 286, 287
    wound closure, 290
Health-related quality of life (HRQoL), 52, 186, 479
Hematopoietic stem and progenitor cells (HSPCs), 472
Hemiarthroplasty endoprosthesis, 242–244, 248
Hepatocellular carcinoma (HCC), 540
HER2 monoclonal antibodies, 394
Hereditary leiomyomatosis and RCC (HLRCC) syndrome, 445
High dose therapy (HDT), 474
High-Intensity Focused Ultrasound (HIFU), 570, 571
Hodgkin lymphoma, 458
Hormone receptor (HR), 389
Hormone therapy, 173
Hounsfield Units (HU), 91
Human epidermal growth factor receptor 2 (HER2), 389
Humeral diaphysis
  epidemiology, 253
  nonoperative treatment, 254
  operative treatment, 254, 255
    facture risk, 255, 256
    function, 255
    pain, 255
    prognosis, 255, 256
    tumor histology, 256
  postoperative care, 265
  pre-operative planning
    diagnosis, 257
    fixation, 257, 258
    metastasectomy, 257
    radiolucent implants, 259–262
    reconstruction, 257–259
  surgical technique
    fixation, 264
    metastatectomy and reconstruction, 264, 265
    patient positioning, 262–264
    soft tissue closure, 265
Humerus, 648
Hurtle cell carcinoma, 459
β-Hydroxy β-methylbutyrate (HMB), 81
Hyperalgesia, 520
Hypercalcemia, 397, 471, 689
Hypercalcemia of malignancy, 439
Hyperdiploidy (HDMM), 468
Hypogammaglobulinemia, 474

I
Ice-cream cone prosthesis, 669
IlluminOss System, 260
IMDC (International Metastatic Renal Cell Carcinoma Database Consortium), 446
Immobilization, 345
Immunohistochemistry (IHC), 608, 614, 633
Immunotherapy, 392, 409, 447, 449
Impending pathological fractures, of long bones
  diagnosis, 165
    history and physical examination, 165
    imaging, 166, 167
    laboratory testing & biopsy, 167–168
  epidemiology and pathophysiology, 164–165
  non-operative treatment
    radiation therapy, 172–173
    systemic therapy, 173

operative treatment, 173, 174
post-treatment considerations, 174–175
risk assessment, 168
current risk stratification systems, 168–170
imaging-based strategies, 170–171
machine learning models, 171
treatment, 171, 172
Incisional pain, 516, 525
Incisional pain model, 528
Indirect calorimetry (IC), 70
Informed consent, 656
Instrumental activities of daily living (IADL), 53, 503
Insulin-dependent diabetics, 583
Integrated approach, 406
Intensity modulated radiation therapy (IMRT), 644
Intercalary endoprosthesis, 361
Intercalary reconstruction, 258, 358
Interleukin 1 (IL-1), 480
International Association for the Study of Pain (IASP), 514
International Classification of Functioning, Disability and Health (ICF) Model, 504
International Myeloma Working Group (IMWG), 95
Intramedullary nail (IMN), 38, 239–240, 258, 263, 350
Intramedullary rod fixation, 310–312
Intraoperative image guided percutaneous instrumentation, 188
Intra-tumor heterogeneity (ITH), 592
Iodinated contrast media (ICM), 90
Isolated lytic bone lesions
bone biopsy, 626
challenges, 627
clinical evaluation, 623
diagnosis and work-up, 623–625
proximal AP radiographs, 624
proximal lateral radiographs, 624
differential diagnosis for, 621–623
post-operative complications, 627
treatment, 626–628
post-operative AP radiographs, 628
pre-operative AP radiographs, 628
ISUP Grade Group system, 406

K

kappa chains, 625
Karnofsky performance score (KPS), 31, 32, 56, 57, 185, 558
Karnofsky score, 57
Katz ADL, 55, 56
Kidney cancer, 683, 687, 690
Knee
clinical manifestations, 324
guidelines, 323
local control, 324
pain, 324
post-operative care, 345, 346
pre-operative planning
computed tomography (CT) scan, 327
intraoperative guidance, 328
magnetic resonance imaging (MRI), 327
MDT, 326
plain radiographs, 326, 327
preoperative evaluation, 328, 329
surgical decision-making, 328
site for, 324
surgical indications/non-surgical indications, 325, 326
surgical options
distal femur and proximal tibia, 330, 331
patella, 329, 330
surgical technique
arthroplasty revision techniques, 334–336
intratumoral resection and cement reconstruction, 333, 334
patella, 331, 332
proximal tibia, 340–344
resection and prothesis reconstruction, 336–340
retrograde IM nailing, 332, 333
thyroid and renal metastases, 324
KRAS-positive non-small cell lung cancer (KRAS+ NSCLC), 433
Kyphoplasty, 189, 195

L

Larotrectinib, 460
Leptomeningeal carcinomatosis (LC), 396
Limb salvage, 374
Linear energy transfer (LET), 492
Local anesthetic, 530
Local cancer drug delivery, 669
Local Infiltration Analgesia (LIA), 530
Local therapy, 615
Lower extremity fractures, treatment for
acetabulum, 651
femur, 650
patella, 651
tibia, 650, 651

Lung biopsy, 104
Lung cancer, 149, 150, 157, 158, 369, 683, 684, 686, 689
  biomarker-driven NSCLC, treatment of, 432–434
  bone biopsy, indications for, 431
  bone metastases, 429
    assessing response to treatment, 434
    diagnostic workup for, 430
    imaging modalities, characteristics of, 430
    pathophysiology, 429
    systemic therapies, 431, 433
  epidemiology, 427
  histology and staging, 428
  oligometastatic NSCLC, management of, 434
  prognosis, 439, 440
  risk factors, 427
  skeletal complications
    cancer-induced bone pain, 437–439
    hypercalcemia of malignancy, 439
  SREs, management of, 435
    pathologic fractures, 435
    SCC, 436
    surgical prediction models, 435–437
Luteinizing hormone-releasing hormone (LHRH), 406
Lymphoma, 613
Lytic metastases, 88

M

M Proteins (monoclonal proteins), 467–469
Machine learning (ML) algorithms, 33, 171, 685, 704, 705
Macrophage colony-stimulating factor (M-CSF), 48
Magnetic resonance imaging (MRI), 87, 92, 94, 131, 352, 373, 391, 430, 610, 611, 625, 635, 636
Major depressive disorder (MDD), 139
Maladaptive cortical responses, 518
Malignant hematolymphoid tumors, 126
Malnutrition, 74, 297
Malnutrition Screening Tool (MST), 71, 72
Mantle cell lymphoma, 468
Matrix metalloproteinases (MMPs), 14
Mechanical pain, 165
Median survival time, 158
Medical nutrition therapy (MNT), 76
Medullary thyroid cancer, 459
Medullary thyroid carcinoma, 458
Megaprosthesis, 297
Melanoma
  genotypic markers, 124
  immunohistochemical markers, 124, 125
Melanomas, 124
Melzack's neuromatrix model, 514
Memorial Sloan Kettering Cancer Center (MSKCC), 156, 446
Merkel cell carcinoma, 124
Metachronous oligometastatic disease, 590
Metastasectomy, 557, 558, 639
Metastasis, 605
Metastasis-directed therapy (MDT), 592, 596
Metastatic acetabular classification (MAC), 202–204, 219
Metastatic adenocarcinoma, 121
Metastatic bone disease (MBD), 18, 29, 45, 149, 150, 370, 540, 621, 688, 699
  aspects of, 540, 541
  clinical assessment
    biochemical profile, 47, 48
    bone scintigraphy, 49
    CBC, 47
    CT imaging, 48, 49
    medical history, 47
    MRI, 49
    osteoblastic pattern, 46
    pathophysiology, 48
    PET scans, 49, 50
    PET/CT scans, 50
    plain radiograph, 48
    SREs, 47
  diagnosis, 150
  functional assessment, 50
    ADL assessment, 53–55
    oncologic diseases, 56
    palliative care, 58
    PaP, 58
    PPI, 58, 60
    prognostic tool, 56
    PROMIS, 51
    QoL, 51, 52
    treatment, 60, 61
  isolated lytic bone lesion, differential diagnosis, 621
  life expectancy, 157, 158
  multidisciplinary care
    benefits, 542
    patient schedule, 547, 548
    risks of, 543
    treatment, 543–547
  outcomes, 157
  radionuclides for, 493
  skeletal metastases, 157
  skeleton, 45

SREs, 150
standard of care, 541, 542
treatment indications, 151
ANN models, 152
less invasive surgical interventions, 155
more invasive surgical interventions, 155, 156
non-surgical palliative care, 154
OPTIModel, 151
orthopedic surgeons, 153
PATHFx, 152
pathologic fracture, 154
preoperative planning, 151
prognostic information, 152
PTIModel, 151
surgical indications, 155
survival time, 151
Metastatic bone pain, 352
Metastatic breast cancer (MBC)
BOM, 389
diagnosis and staging, 390–392
HER2+, 393, 394
HR+/HER2- and HR+/HER2 low, 392, 393
management, 392
metastatic TNBC, 394
NCCN guidelines, 391, 392
SREs, 395–398
systemic therapy for, 392
Metastatic breast carcinoma, 122
Metastatic cancer, 3
CBT, 143
exercise, 144
mindfulness, 143
pharmachologic interventions, 141–143
process of monitoring, 131
psycho-oncology, 139
role of biopsy, 137
screening, 140
staging of
bone sarcoma, 135
bone sarcomas, 134
carcinoma, 135, 136
Myxoid liposarcoma, 133
soft tissue sarcoma, 131, 132
symptoms, 140
treatment planning, 131
Metastatic carcinoma, 122, 687
Metastatic castrate-resistant prostate cancer (mCRPC), 403, 406–411
Metastatic castration-resistant disease, 413
Metastatic disease, 172–174, 404
appendicular bones and postoperative radiotherapy, pathologic fracture of, 481–483
interventions for, 581
radiotherapy indications for, 480
pain relief, 480, 481
radionuclides, 491, 492
reirradiation, 488, 491
SABR, 485–487
side effects, 487, 488
skeletal-related events, prevention of, 484, 485
spinal cord compression and cauda equina syndrome, 484
Metastatic epidural spinal cord compression (MESCC), 181
Metastatic lung carcinoma, 624
Metastatic prostate cancer, 649
Metastatic renal cell carcinoma, 448–449
Metastatic TNBC (mTNBC), 394
Meticulous hemostasis, 111
METSSS model, 308
Micrometastases, 555
MicroRNAs (miRNAs), 593
Microsatellite instability (MSI), 608
Microwave ablation (MWA), 568
Midfoot disease, 378
Mimic metastatic disease
autoimmune and congenital conditions, 98
benign bone lesions, 99
degenerative conditions, 98
infection, 97
sagittal CT, 98
sarcoidosis, 98
trauma, 97
Minimal residual disease (MRD), 477
Minimally invasive, 564
Minimally invasive AORIF procedure, 667
Minimally invasive approaches, 375, 565
Minimally invasive pelvic surgery
ablative techniques, 205–206
anatomy of, 202–203
cementoplasty, 205
complications, 213
pearls and pitfalls, 213–214
percutaneous fixation
advantages, 207
AORIF, 208
cannulated screws, 206–207
disadvantage, 207
photodynamic nails, 207–208
postoperative care, 212–213
pre-operative planning, 204
prevalence, 201
surgical indications, 203–204
surgical technique, 208–212

Minimally Invasive Surgery (MIS), 173, 187–191
Mirel's Score or Harrington Classification, 615
Mirels scoring system, 169, 237, 255, 308, 309
Mismatch repair-deficient (dMMR) tumors, 409
Mitogen-activated protein kinase (MAPK) signaling pathway, 459
Moderate-intensity exercise, 503
Modular endoprostheses, 273
Molecular profiling/next-generation sequencing (NGS), 614
Monitored anesthesia care (MAC), 105
Monoclonal gammopathy, 469, 607
Monoclonal gammopathy of undetermined significance (MGUS), 468
Monotherapy, 447
M-protein, 625
MSI-high (MSI-H), 409
Multidisciplinary approach, 351, 404, 445
Multidisciplinary cancer conference (MCC), 703
Multidisciplinary care, 539, 540, 657
- MBD
  - benefits, 542
  - patient schedule, 547, 548
  - risks of, 543
  - treatment, 543–547
- with musculoskeletal metastases
  - case example, 35, 36
  - clinical decision-making, 34–36
  - ECOG, 32
  - impending fractures, 33
  - KPS, 31, 32
  - life expectancy and prognosis, 30
  - lung carcinoma, 35
  - metastatic bone disease, 30
  - pathologic fractures, 33
  - preventing SREs, 33
  - radiation therapy, 37–38
  - surgery, 38
  - surgical procedure, 33
  - systemic treatment, 36

Multi-disciplinary collaborative approach, 396
Multi-disciplinary management (MDM), 699, 700, 703
Multidisciplinary musculoskeletal oncology, 700–702, 709
- opportunities for change
  - artificial intelligence, 706–708
  - big data, 704, 705
  - developing global cancer hospitals and programs, 707
  - international registry collaboration to generate and interrogate MBD big data, 706
  - machine learning, 704, 705
  - MBD tumor boards, developing, 703
  - online resources to reach underserved MBD patients in orthopedic care, 703–705

Multidisciplinary teams (MDT), 153, 502, 659
Multimodal analgesia, 521, 525, 527, 529, 532
Multiparametric MRI (mpMRI), 92
Multiple myeloma (MM), 46, 95, 355, 467, 606, 622, 689
- anemia, 471, 472
- bone lesions in, 472
- classification, 469, 470
- diagnosis and staging, 470, 471
- drug categories, 475
- epidemiology, 467
- genetic drivers/prognosis, 467, 468
- hypercalcemia, 471
- management
  - bone focused therapy, 476
  - radiation therapy, 476
  - response assessments, 477
  - stem cell transplant, 476
  - systemic therapy, 473–475
- renal failure, 471
- risk factors, 468
- staging, 473

Multiple screening tools, 140
Musculoskeletal lesion, 105, 106
Musculoskeletal metastases, 103
- bone destruction, 6
- bone metastasis, 4, 16, 18
- carcinomas, 18
- carcinomas *vs.*sarcomas, 20, 21
- clinical impacts, 22, 25
- colonization, 14
- dissemination, 14
- epidemiology, 4
- key genes and mutations, 14
- invasion, 14
- lung carcinoma, 5
- management of, 10, 11
- mechanisms of, 19
- metastatic bone disease, 5
- musculoskeletal tumors, 4, 5
- osteoblastic lesions, 6
- pathogenesis, 5, 6
- pattern of, 17, 18
- phenotypic plasticity, 21
- primary tumor diagnosis, 5
- process of remodeling, 4

route of dissemination, 22
signaling pathways, 15
skeletal muscle, 20
soft tissue metastasis, 17
soft tissue sarcomas, 4, 5
symptoms and skeletal related events, 7, 8
TME, 15
treatment of, 9, 10
tumor microenvironment, 22
vicious cycle, 19–20
Musculoskeletal oncology, 49
Musculoskeletal oncology patient care, 700
Musculoskeletal pain, 543
Musculoskeletal sarcoma, 106
Musculoskeletal system
artificial intelligence, 118
bone metastases, 117, 119
carcinomas, 120
cytology, 118
digital pathology, 118
genomic profiling technology, 127
image-guided biopsy, 118
malignant hematolymphoid tumors, 126
melanoma, 124
neuroendocrine tumors, 124
sarcomas, 125
skeleton, 126
tumor type, 118
Musculoskeletal Tumor Registry (MsTR), 706
Musculoskeletal Tumor Society (MSTS), 540, 541, 706
Musculoskeletal tumors, 105, 111
Mutational signatures (MSs), 127
Myelodysplastic syndrome (MDS), 476
Myxoid liposarcoma, 133

N

Nail fixation, 296
National Comprehensive Cancer Network (NCCN) guidelines, 390, 558, 608
National Institute for Health and Clinical Excellence (NICE), 144
National Institutes of Health (NIH), 542
Neoplastic pain, 165
Neuraxial anesthesia, 528
Neuroblastoma/Wilms tumor, 458
Neuroendocrine carcinomas, 124
Neuroendocrine tumors, 124
Neurogenic inflammation, 516
Neurologic, Oncologic, Mechanical, and Systemic (NOMS) decision framework, 182, 184
Neuromatrix, 514, 517, 518
Neuromatrix pain theory, 515
Neurophysiological processes, 515
Neurotransmitters, 520
New England Spinal Metastasis Score (NESMS), 186
Next-generation sequencing (NGS), 127, 447, 614
Nociception, 515, 520, 521, 533
Non-clear cell renal cell carcinoma (nccRCC), 443
Non-health-related stressors, 691
Non-metastatic PC, 412
Non-opioid analgesics, 527
Nonpharmacological interventions, 527
Non-small cell lung cancer (NSCLC), 428, 429, 593
biomarker-driven, treatment of, 432–434
oligometastatic, management of, 434
treatment algorithm for first line treatment selection in, 432
Non-squamous NSCLC, 428
Non-steroidal anti-inflammatory drugs (NSAIDs), 415, 437
Nonsurgical palliative care, 565
Normal bone matrix, 203
NTRK gene rearrangements, 460
Nuclear medicine, 95
bone scan, 96
oncotropic radioisotope, 96
osteotropic radiotracer, 95
Nutrition Focused Physical Exam (NFPE), 72
Nutrition impact symptoms (NIS)
anorexia, 77
constipation, 77
diarrhea, 77
dysgeusia, 78
fatigue, 78
nausea, 78
nutrition recommendations, 76
oncology treatment, 76
palliative treatment, 76
parosmia, 78
systematic assessment, 75
xerostomia, 78
Nutrition support
enteral and parenteral nutrition, 80
oral nutrition supplements, 79
patients with advanced cancer, 80
Nutritional challenges
cachexia, 74
malnutrition, 74
precachexia, 74
refractory cachexia, 74
sarcopenia, 74, 75

Nutritional interventions, 416
Nutritional requirements
energy needs, 70
ESPEN guidelines, 70, 71
fluid management, 71
Nutritional Risk Screening (NRS), 71
Nutritional screening
AND/ASPEN diagnosis, 73
MST, 72
NFPE, 72
nutrient deficiency, 72
PG-SGA, 71

O

Obesity, 458
Occupational therapists, 509
Occupational therapy (OT), 545
Olaparib, 409
Oligmetastases
American Society of Anesthesiologists (ASA) score, 558
characteristics, 556, 557
definition, 556
ECOG performance status score, 558
goal of surgical treatment, 559
Hellman's theory of, 556
historical background, 555, 556
metastasectomy for, 557, 558
reconstruction, 558
Oligo progression, 597–598
Oligometastasis, 590
Oligometastatic disease, 173, 174, 434, 544
SBRT, 592
Oligometastatic metastatic thyroid carcinoma, 647
Oligoprogressive disease, 590
Oligorecurrence/oligoprogression, 557, 597
Oligorecurrent disease, 590
Omega-3 fatty acids, 81
Oncocytic carcinoma, 459
Open decompression, 189, 193–195
Open reduction and internal fixation (ORIF), 38
Open surgical approach, 668
Opioid induced hyperalgesia (OIH), 525
Opioids, 415, 532
OPTIModel, 151, 685
Oral nutrition supplements (ONSs), 79
Orthopedic oncologic surgery (OOS), 513, 688
acute pain management, 521, 522
amputations, 532, 534
ERAS pathways, 531
multimodal analgesia, 525–527
neuraxial analgesia, 528, 529
opioid-tolerant or chronic pain patients, 532, 533, 535
personalizing pain protocols, 531–533
PO analgesics, 528
pre-surgical evaluation and development of, 521, 523–525
preventive analgesia, 525
psychological and behavioral interventions, 530, 531
regional analgesia, 528–530
Orthopedic oncologists, 46
os coxae, 202
Osseous metastases, 461, 557
Osseous metastatic disease, 662
Osseous microenvironment, 404
Osteoblastic lesions, 684
Osteoblastic metastases, 404
Osteoclast signaling pathways, 480
Osteoclast stimulating factor, 480
Osteoclasts resorb bone, 451
Osteolytic lesions, 684
Osteomyelitis, 622
Osteonecrosis of jaw, 462
Osteonecrosis of the jaw (ONJ), 37, 397
Osteoplasty, 571, 583

P

Pain
classification, 518, 519
definition, 514
pathophysiology, 518, 519
cancer versus non-cancer pain, 520
inflammation and tissue damage, 519, 520
physiology, pathways, 515–518
Pain and Ambulatory Function Scoring System, 668
Pain assessment, 667, 668
Pain catastrophizing scale (PCS), 523, 529, 531
Pain management, 508, 615
Pain on Ambulation Functional Scoring System, 668
Pain phenotypes, 513, 515, 521, 523–525, 527–529, 531, 533
Pain relief, radiotherapy, 480, 481
Palliative care, 545, 661, 662, 691, 692
Palliative Performance Scale (PPS), 56
Palliative Prognostic Index (PIP), 56
Palliative Prognostic Score (PaP), 56, 59
Palliative radiotherapy, 479, 481, 482
for bone metastases, 489–490

Pamidronate (Aredia), 476
Pancreatic cancer, 540
Papillary thyroid cancers, 459
Parafollicular or C-cells, 458
Parathyroid hormone-related peptide (PTHrP), 6, 16, 48, 81, 397
Parenteral nutrition (PN), 80
Patellar fractures, 651
PATHFx tool, 171, 308
Patient expectations, 658–661
Patient Health Questionnaire (PHQ)-9, 140
Patient Reported Outcome Measurement Information System (PROMIS), 51, 524
Patient-focused and personalized approach, 658
Patient-Generated Subjective Global Assessment (PG-SGA), 71
Pedicle flap technique, 342
Pelvic defects and fractures, 672–677
Pelvic metastasis
- advantages, 230
- anatomic considerations, 217–218
- disadvantages, 230–231
- epidemiology, 217
- postoperative care, 231
- preoperative planning, 219–220
- surgical classifications, 219
- surgical indications/non-surgical indications, 218–219
- surgical options
  - acetabular reconstruction, 222
  - cement augmentation, 221
  - curettage and bone filling, 221
  - en bloc resection, 224
  - girdlestone (resection arthroplasty), 224
  - prophylactic stabilization with internal fixation, 221, 222
  - sacroiliac fixation and sacral tumor resection, 223
- surgical technique, 220
  - bone reconstruction, 226–227
  - patient positioning and surgical incision, 225–226
  - soft tissue reconstruction, 227
  - tumor resection, 226
  - wound closure, 227

Pembrolizumab monotherapy, 432
Percutaneous fixation
- advantages, 207
- AORIF, 208
- cannulated screws, 206–207
- disadvantage, 207
- photodynamic nails, 207–208

Percutaneous instrumentation- fluoroscopic guided, 191
Percutaneous interventions, 565
Percutaneous reinforcement, 669
Peripheral nerve blocks (PNB), 529
Peripheral neuropathy (PN), 474
Peripheral noxious stimulus, 518
Peripheral sensitization, 516
Personalized pain protocols, 521, 523
Pharmacologic agents, 142–143
Photodynamic balloon stabilization, 174
Photodynamic bone stabilization, 240–241
Photodynamic nail stabilization, 212
Photodynamic nails (PDNs), 207–208, 213, 260, 263
Physical therapy (PT) strategies, 174, 501, 502, 545, 689
- exercise guidelines for cancer patients, 503
- initial PT assessment, 503, 504
- interdisciplinary rehabilitation, 508, 509
- interventions, 506
  - balance training, 507
  - cancer related fatigue, 508
  - endurance training, 507
  - pain management, 508
  - standardized outcome measures, 507
  - strength training, 506
- mobility devices, 505, 506
- prognosis, 509
- strength and cardiorespiratory function, 502

Plain radiographs, 610
Plasma cell neoplasms, 126
Platelet count, 396
PMMA bone cement, 669, 670, 676
Polymethyl methacrylate (PMMA), 155, 205, 208, 212, 213, 335, 375, 571
Positron emission tomography (PET), 9, 87, 95, 96, 370, 372
Positron emission tomography combined with computed tomography (PET/CT), 34, 430, 612, 636, 637
Posterior inferior iliac spine (PIIS), 211, 212
Postoperative medical management, 249
Postoperative radiation therapy (PORT), 363
Postoperative rehabilitation, 249
Pre-emptive analgesia, 525
Preoperative embolization, 174
Preoperative quantitative sensory testing (QST), 523
Pressure pain threshold (PPT), 524
Primary cancer, 607, 608, 614
Primary lymphoma of bone (PLB), 46, 622
Primary tumor, 633, 638

Proactive approach, 596
Progesterone receptor protein (PR), 126
Progression free survival (PFS), 392
Pro-inflammatory cytokines, 480
Prophylactic fixation, versus after-fracture care, 174
Prophylactic radiation therapy, 595
Prostaglandin E2 (PGE2), 48
Prostate and breast cancer, 623
Prostate cancer (PC), 502, 593, 683, 684, 687
  biological and molecular mechanisms, 404
  cancer-related mortality, 403
  classification, 403
  diagnosis and staging, 404–406
  pathophysiology, 404
  prognosis, 417
    quality of life assessment and supportive interventions, 417, 418
    recommendation, 417
    SREs and symptom management, 417
  skeletal related events, management of, 409
    analgesics, 414–416
    osteoclast inhibitors, 412, 413
    radiopharmaceuticals, 414, 415
    supportive measures, 416
  systemic treatment, 406
    mCRPC, 407, 409–411
    mCSPC, 406–408
Prostate Cancer Clinical Trials Working Group, 417
Prostate cancer prognosis, 406
Prostate-specific antigen (PSA), 7, 136, 404, 613
Prostate-specific membrane antigen (PSMA), 50, 121, 405, 492
Proteasome inhibitors (PI), 474
Protein electrophoresis, 470
Proton density images (PD), 92
Proton resonance frequency (PRF), 576
Proximal femur
  advantages and disadvantages, 296
  complications, 297
  contraindications, 297
  epidemiology, 295
  post-operative care, 304, 305
  preoperative planning, 296
  surgical indications/non-surgical indications, 295, 296
  surgical options, 296
  surgical technique
    advantages and disadvantages, 303, 304
    megaprosthesis, 301, 302
    patient positioning and surgical incision, 297, 298
    soft tissue reconstruction, 302
    tumor resection, 298–301
    wound closure, 303
Proximal humerus reconstruction, 248–250
Proximal tibia, 340–344
"Pseudohypoxic" state, 443
PSMA PET-CT, 405
Psychometric measures, 523
PTIModel, 151
Pulmonary embolism (PE), 174, 362

Q

Quality of life (QoL), 29, 50–52, 389, 403, 416, 417, 440, 451, 462, 463, 503, 594, 631, 638, 639, 688, 689, 691
Quantitative immunoglobulins, 471

R

Radiation associated fractures (RAF), 38, 647
  diagnosis, 646, 647
  factors, 645
  lower extremity fractures, 650, 651
  mitigation of, 644
  prophylactic treatment, 646
  quality of bone and soft tissue, 647
  risk stratification, 644, 646
  upper extremity fractures, 648–650
Radiation osteitis, 643
Radiation therapy (RT), 37–38, 154, 168, 171–173, 476, 541, 558, 615, 639, 644
  effect on bone, 643
Radioactive iodine (RAI), 460, 687
Radiofrequency ablation (RFA), 205, 206, 212, 213, 566, 567
Radiographic surveys, 94
Radiographs, 48
Radioligand therapy, 409
Radiological assessment
  blastic metastases, 89
  CT, 87, 90, 91
  lytic metastases, 88
  mixed lytic and blastic lesions, 89
  MRI, 87, 92, 94
  osseous metastases, 87
  PET, 87
  PET-MRI, 87
  sclerotic lesion, 89
  wb-CT, 94, 95
  whole-body imaging, 94
  X-ray, 87, 88, 90
Radionuclide therapy, 451
Radionuclides, 491, 492

for metastatic bone disease, 493
Radiopharmaceuticals, 414, 415
Radiotherapy, 379, 450
appendicular bones and postoperative radiotherapy, pathologic fracture of, 481–483
for pain relief, 480, 481
radionuclides, 491, 492
reirradiation, 488, 491
SABR, 485–487
side effects, 487, 488
skeletal-related events, prevention of, 484, 485
spinal cord compression and cauda equina syndrome, 484
Radium-223 (Ra-223), 409, 414, 492
Randomized controlled trials (RCTs), 36
RANKL-mediated osteoclast activation, 412
Ray amputations, 377
Receptor activator nuclear factor-κ B ligand (RANKL) pathway, 6, 7, 19, 33, 48, 374, 452
Receptor tyrosine kinase inhibitors (RTKIs), 447
RECIST (Response Evaluation Criteria in Solid Tumors), 434
Recommended daily allowance (RDA), 71
Red blood cell (RBC), 472
Regenerative peripheral nerve interface (RPNI), 285
Regional anesthesia, 528–530
Registered dietitian nutritionists (RDNs), 73, 76
Reirradiation, 493
Relational autonomy, 656, 658, 663
Renal cell and thyroid carcinoma, 186
Renal cell carcinoma (RCC), 123, 374, 593, 594
diagnosis and staging, 444, 445
clinical factors, 446
metastatic sites, 445, 446
epidemiology, 443
screening in general population, 445
somatic or germline genomic alterations, 443, 444
Renal failure, 471
Resistance training program, 502
Resting energy expenditure (REE), 70
Reverse shoulder arthroplasty endoprosthesis, 242
Rexed laminae I and II, 517
Rhabdomyosarcoma, 132
Ribs, 650
Risk stratification, 644, 646
Robotic assisted instrumentation, 191–193
Role of biopsy, 137
Rostral ventromedial medulla (RVM), 517
Roswell Park Memorial Institute (RPMI) solution, 108
Rougraff criteria, 235, 236

S

Samarium-153, 414
Sarcomas, 125
metastatic osteosarcoma, 125, 126
pathological examination, 125
round cell sarcomas, 126
Scandinavian Sarcoma Group (SSG) score, 151, 157
SDF-1–CXCR-4 pathway, 404
"Seed and soil" hypothesis, 404, 556
Sensitization, 520
Separation surgery, 182, 187, 190–191
Sepsis, 297
Serotonin reuptake inhibitors (SSRIs), 141
Serum free light chains (FLC), 471
Serum immunofixation (SIF), 471
Serum protein electrophoresis (SPEP), 471, 625
Shared decision making (SDM), 655
ethical considerations, 656
clinician-patient relationship, 656
family and multidisciplinary care, 657
relational autonomy and, 656
pragmatic approach, 662
Shoulder gridle metastatic disease
operative management, 237
preoperative planning, 237–238
prevalence, 235
surgical techniques
anesthesia considerations, 245
incision/exposure, 246–247
patient positioning, 244–245
postoperative medical management, 249
postoperative rehabilitation, 249
proximal humerus reconstruction, 248–250
tumor resection, 247–248
wound closure, 248
surgical treatment options, 239–244
endoprosthesis, 241–244
intramedullary nail, 239–240
ORIF with locking plate and cementation, 239
photodynamic bone stabilization, 240–241
work-up and exam, 235–237

Single-photon emission-CT (SPECT-CT)-modified Mirel's score, 170
Skeletal metastases, 45, 104, 119, 157, 164, 165, 171–175
Skeletal muscle metastases, 632
Skeletal Oncology Research Group (SORG) Nomogram, 185, 186
Skeletal related events (SREs), 29, 30, 47, 150, 155, 363, 390, 404, 479, 502, 686, 701
  management of, 409, 435
    analgesics, 414–416
    BMAs, 397, 398
    hypercalcemia of malignancy, 397
    osteoclast inhibitors, 412, 413
    pain from BC BM, 397
    pathologic fractures, 395, 396, 435
    prognosis, 398, 399
    radiopharmaceuticals, 414, 415
    SCC, 396, 436
    supportive measures, 416
    surgical prediction models, 435–437
Skeleton, 126
Skin incision, 263
Small cell lung cancer, 428
Smoldering myeloma (SMM), 468
Social work, 545
Soft tissue metastases
  adjacent regional lymph nodes, 631
  clinical findings, 634
  diagnosis, 635–637
  epidemiology, 632
  pathogenesis and histopathological findings, 632–634
  treatment, 638
    radiation therapy, 639
    surgery, 638, 639
    systemic therapy, 638
Soft tissue reconstruction, 225, 227
Soft tissue sarcoma (STS), 25, 631, 635, 639, 646
Soft tissues, 105–109
Solid tumor malignancies, 409
Solitary plasmacytoma, 470
SORG-MLA model, 685
Spinal cord compression (SCC), 396, 409, 436, 450, 479, 480, 482, 484, 493
Spinal Instability Neoplastic Score (SINS), 169, 170, 436
Spinal malignancies, 607
Spinal metastases, 165, 169
Spinal metastatic disease
  epidemiology, 181–182
  post-operative care, 195
  pre-operative planning, 186–187
  surgical indications, 182, 184–186
  surgical options
    kyphoplasty/vertebroplasty, 189
    minimally invasive surgery, 187–189
    open decompression, 189
    separation surgery, 187
  surgical technique
    kyphoplasty/vertebroplasty, 195
    open decompression, 193–194
    percutaneous instrumentation-fluoroscopic guided, 191–193
    robotic assisted instrumentation, 191–193
    separation surgery, 190–191
Spinal stabilization, 187
Spine metastatic disease, 484
Spine-specific scoring systems, 436
SPRING 2013 nomogram, 685
Squamous cell carcinoma (SCC), 120, 432
Stabilization, 623, 628
Standard of care, 541, 542
Standard staging systems, 468
Standardized outcome measures, 507
Stem cell transplant, 476
Stereotactic ablative radiotherapy (SABR), 484–487, 493, 559, 591
Stereotactic body radiation therapy (SBRT), 51, 153, 154, 182, 184, 187, 434, 485, 558, 590, 592, 594
  benefits, 594
  clinical considerations, patient selection, and prognostic factors, 593, 594
  complications and limitations, 598
  in oligometastatic disease, 593
  principles and advantages, 593
  re-irradiation after, 597, 598
  treatment, 595
  treatment failure, 595
Stereotactic radiosurgery (SRS), 486
Strength training program, 506
Strontium-89, 414
Surgery, 38, 435, 450, 638
Surgical approach–curative-intent metastasectomy, 173
Surgical biopsy, 111
Surgical intervention, 615
Syme amputation, 379–381
Symptomatic skeletal events (SSEs), 414
Symptomatic spinal metastases, 181, 187
Symptoms and skeletal related events (SREs), 7

bone marrow, 8
bone scans, 8
CT scans, 8
hypercalcemia, 8
imaging, 8
MRI, 9
pathologic fractures, 8
PET scans, 9
spinal cord compression, 8
Synchronous metastases, 557
Synchronous oligometastatic disease, 590
Synovial sarcoma, 132
Systemic anti-cancer therapies, 173
Systemic theory, 555
Systemic therapy, 392, 407, 449, 460, 473–475, 597, 615, 638

T
T1-weighted sequences (T1w), 92
Tailored analysis, 170
Targeted muscle reinnervation (TMR), 285
Targeted therapy, 392, 615
Tarsal navicular and calcaneocuboid joints amputation (Chopart), 378–379
Tarsometarsal amputations, 377
Tc99-MDP, 95, 96
Technetium-99 m (^{99m}Tc)-diphosphonate bone scintigraphy, 444
Technetium-99m (Tc99m), 95, 625
Telehealth, 704
Therapeutic strategies, multidisciplinary care, 540
Thermal ablation, 563, 564
Three-dimensional advanced imaging, 170
3D computer models, 171
3D evaluation, 91
3D imaging, 673
3-D metal reconstruction, 381
Thromboembolic events, 174
Thromboembolic prophylaxis, 345
Thyroid cancer, 686, 690
Thyroid carcinoma, 457
classification, 458–460
diagnosis and staging, 460
diagnostic techniques, 457
incidence of, 457
prognosis, 462
quality of life, 462, 463
risk factors, 458
survivors experience, 462
treatment, 460–462
Thyroid Imaging Reporting and Data System (TIRADS), 460
Thyroid parenchymal cells, 457
Thyroid transcription factor-1 (TTF-1), 122
Tibia, 650, 651
Tibial metastatic lesions
epidemiology, 349, 350
post-operative care, 362–364
surgical management techniques, 362
techniques
intercalary endoprosthesis, 361
intramedullary stabilization of, 360
open curettage and cementation of, 359, 360
treatment
non-surgical & surgical indications, 351, 352
pre-operative planning, 352, 353
surgical treatment, 353–359
TNM classification system, 404
Tokuhashi Scoring System, 437
Total energy expenditure (TEE), 70
Total hip arthroplasty (THA), 651
Transcutaneous electrical nerve stimulation (TENS), 508
Transforming growth factor (TGF)-beta-2 signaling, 16, 608
Translocations, 467, 468
Transmetatarsal amputations, 377
Transpedicular decompression, 189
Triceps fascia, 264
Triplet therapy, 407, 408
Tripod percutaneous screw fixation, 212
“Tripod” technique, 206
Tumor cells approach, 607
Tumor markers, 613
Tumor microenvironment (TME), 13, 15, 22, 444
blood supply and lymphatic drainage, 16
tumor-stroma interaction, 15
Tumor necrosis factor (TNF), 48
Tumor necrosis factor alpha (TNF-α), 480
Tumor-associated macrophages (TAMs), 15
Tyrosine kinase inhibitors (TKI), 449–451

U
Ulnar nerve, 264
Ultrasound, 575, 612, 635
Undifferentiated pleomorphic sarcoma (UPS), 622
Undifferentiated thyroid carcinoma, 459
Upper extremity, 269, 271, 274, 277

Upper extremity fractures, 648
clavicle, 649
forearm/wrist fractures, 648
humerus, 648
ribs, 650
Urine protein electrophoresis (UPEP), 625

V
Vandetanib, 461
Vascular endothelial-growth factor (VEGF) monoclonal antibodies, 435
Vastus lateralis muscle, 299
Vertebral compression fractures, 189
Vertebroplasty, 189, 195
Very good partial remission (VGPR), 477
VHL silencing, 443
"Vicious cycle" hypothesis, 445
Vitamin D toxicity, 81

W
Waldenstrom's Macroglobulinemia, 469
Weight bearing, 345
Whole body CT scans (WBLDCT), 94, 472
Whole-body MRI (wb-MRI), 95, 405
Wnt pathways, 16
Wound closure, 290
Wrist, *see* Hand

X
X-rays, 430, 609

Z
Zoledronic acid (Zometa), 462, 476

9783032094971